AF719808

Paratuberculosis

Organism, Disease, Control

2nd Edition

Paratuberculosis

Organism, Disease, Control

2nd Edition

Edited by

Marcel A. Behr
McGill University, Montreal, Canada

Karen Stevenson
Moredun Research Institute, Edinburgh, UK

and

Vivek Kapur
Pennsylvania State University, Pennsylvania, USA

CABI

CABI is a trading name of CAB International

CABI Head Office
Nosworthy Way
Wallingford
Oxfordshire OX10 8DE
UK
Tel: +44 (0)1491 832111
Fax: +44 (0)1491 833508
E-mail: cabi@cabi.org
Website: www.cabi.org

CABI
WeWork, One Lincoln St
24th Floor
Boston, MA 02111
USA
Tel: +1 617 682 9015
E-mail: cabi-nao@cabi.org

© CAB International 2020. All rights reserved. No part of this publication may be reproduced in any form or by any means, electronically, mechanically, by photocopying, recording or otherwise, without the prior permission of the copyright owners.

A catalogue record for this book is available from the British Library, London, UK.

Library of Congress Cataloging-in-Publication Data

Names: Behr, Marcel A., editor.
Title: Paratuberculosis : organism, disease, control / edited by Marcel A. Behr, McGill University, Montreal, Canada, Karen Stevenson, Moredun Research Institute, Edinburgh, UK, and Vivek Kapur, Pennsylvania State University, Pennsylvania.
Description: Second edition. | Wallingford, Oxfordshire ; Boston, MA : CAB International, [2020] | Revised edition of: Paratuberculosis / edited by Marcel A. Behr and Desmond M. Collins. c2010. | Includes bibliographical references and index. | Summary: "This new edition is the only comprehensive text on Paratuberculosis, providing historical context and state-of-the-art knowledge. It examines epidemiology, the organism that causes the disease, and practical aspects of its diagnosis and control, as well as the link between paratuberculosis in the food chain and human health implications"-- Provided by publisher.
Identifiers: LCCN 2020024158 (print) | LCCN 2020024159 (ebook) | ISBN 9781789243413 (hardback) | ISBN 9781789243420 (ebook) | ISBN 9781789243437 (epub)
Subjects: LCSH: Paratuberculosis.
Classification: LCC SF809.P375 P37 2020 (print) | LCC SF809.P375 (ebook) | DDC 636.2/089634--dc23
LC record available at https://lccn.loc.gov/2020024158
LC ebook record available at https://lccn.loc.gov/2020024159

ISBN-13: 9781789243413 (hardback)
9781789243420 (ePDF)
9781789243437 (ePub)

Commissioning Editor: Alex Lainsbury
Editorial Assistant: Lauren Davies
Production Editor: Tim Kapp

Typeset by Exeter Premedia Services Pvt Ltd, Chennai, India

Contents

Contributors

David C. Alexander, University of Manitoba, Winnipeg, Manitoba, Canada. E-mail: david.alexander@gov.mb.ca

Angelika Agdestein, Norwegian Veterinary Institute, Oslo, Norway. E-mail: angelika.agdestein@vetinst.no

Christina Ahlstrom, USGS Alaska Science Center, Anchorage, Alaska. E-mail: cahlstrom@usgs.gov

John P. Bannantine, National Animal Disease Center, USDA-ARS, Ames, Iowa, USA. E-mail: john.bannantine@usda.gov

Herman W. Barkema, University of Calgary, Alberta, Canada. E-mail: barkema@ucalgary.ca

Raúl G. Barletta, University of Nebraska, Lincoln, Nebraska, USA. E-mail: rbarletta1@unl.edu

Douglas Begg, University of Sydney, Sydney, Australia. E-mail: doug.begg@outlook.com

Marcel A. Behr, McGill University, Montreal, Canada. E-mail: marcel.behr@mcgill.ca

Evan P. Brenner, Michigan State University, East Lansing, Michigan, USA. E-mail: brenne72@msu.edu

Tim Bull, St George's University, London, UK. E-mail: tim.bull@sgul.ac.uk

Ofelia Chacon, Medical Science Magnet, School District of Palm Beach, Palm Beach Gardens, Florida, USA.

Paul Coussens, Michigan State University, East Lansing, Michigan, USA. E-mail: coussens@msu.edu

Ross S. Davidson, SRUC, Edinburgh, UK. E-mail: ross.davidson@sruc.ac.uk

Justin L. DeKuiper, Michigan State University, East Lansing, Michigan, USA. E-mail: dekuipe5@msu.edu

Kumudika de Silva, University of Sydney, Sydney, Australia. E-mail: kumi.desilva@sydney.edu.au

Berit Djønne, Norwegian Veterinary Institute, Oslo, Norway. E-mail: berit.djonne@vetinst.no

Karsten Donat, Thuringian Animal Disease Fund, Animal Health Service, Jena, Germany and Justus-Liebig-University, Gießen, Hesse, Germany. E-mail: kdonat@thtsk.de

Shannon C. Duffy, McGill University, Montreal, Canada. E-mail: shannon.duffy@mail.mcgill.ca

Natalia Elguezabal, NEIKER, Derio, Bizkaia, Spain. E-mail: nelguezabal@neiker.eus

Susanne Eisenberg, Animal Disease Fund of Lower Saxony, Hannover, Germany. E-mail: susanne.eisenberg@ndstsk.de

Marie-Eve Fecteau, University of Pennsylvania, Kennett Square, Pennsylvania, USA. E-mail: mfecteau@vet.upenn.edu

Naomi J. Fox, SRUC, Edinburgh, UK. E-mail: Naomi.Fox@sruc.ac.uk
Joseba M. Garrido, NEIKER, Derio, Bizkaia, Spain. E-mail: jgarrido@neiker.eus
Irene R. Grant, Queen's University, Belfast, Northern Ireland. E-mail: i.grant@qub.ac.uk
J. Frank Griffin, University of Otago, Dunedin, New Zealand. E-mail: frank.griffin@otago.ac.nz
Murray E. Hines II, Department of Pathology, University of Georgia, Athens, Georgia, USA. E-mail: mhinesii@uga.edu
Michael R. Hutchings, SRUC, Edinburgh, UK. E-mail: mike.hutchings@sruc.ac.uk
Jamie Imada, University of Guelph, Ontario, Canada. E-mail: imadaj@uoguelph.ca
Ramón A. Juste, NEIKER, Derio, Spain. E-mail: rjuste@neiker.eus
Edward Kabara, Michigan State University, East Lansing, Michigan, USA.
Vivek Kapur, Pennsylvania State University, State College, Pennsylvania, USA. E-mail: vxk1@psu.edu
Annette H. Kampen, Norwegian Veterinary Institute, Oslo, Norway. E-mail: annette.kampen@vetinst.no
David F. Kelton, University of Guelph, Ontario, Canada. E-mail: dkelton@uoguelph.ca
Jennifer N. Kiser, Washington State University, Washington, USA. E-mail: jennifer.kiser@wsu.edu
Ad Koets, Wageningen Bioveterinary Research, Lelystad, The Netherlands and Faculty of Veterinary Medicine, Utrecht University, The Netherlands. E-mail: ad.koets@wur.nl
Elise A. Lamont, University of Minnesota, St Paul, Minnesota, USA. E-mail: lamo0062@umn.edu
Kari R. Lybeck, Norwegian Veterinary Institute, Oslo, Norway. E-mail: kari.lybeck@vetinst.no
Colin G. Mackintosh, AgResearch, Invermay, New Zealand. E-mail: colin.mackintosh@agresearch.co.nz
Glenn Marion, Biomathematics and Statistics Scotland, Edinburgh, UK. E-mail: glenn.marion@bioss.ac.uk
Ian Marsh, Elizabeth Macarthur Agricultural Institute, Menangle, NSW Department of Primary Industries Menangle, Australia. E-mail: ian.marsh@dpi.nsw.gov.au
Holly L. Neibergs, Washington State University, Washington, USA. E-mail: neibergs@wsu.edu
Søren Saxmose Nielsen, University of Copenhagen, Copenhagen, Denmark. E-mail: saxmose@sund.ku.dk
Rory O'Brien, DRL, Invermay Agricultural Centre, Mosgiel, New Zealand. E-mail: rory@drl.net.nz
Karren M. Plain, Sydney School of Veterinary Science, Faculty of Science, University of Sydney, Sydney, Australia. E-mail: karren.plain@sydney.edu.au
Auriol C. Purdie, Sydney School of Veterinary Science, Faculty of Science, University of Sydney, Sydney, Australia. E-mail: auriol.purdie@sydney.edu.au
Govardhan Rathnaiah, University of Nebraska, Lincoln, Nebraska, USA. E-mail: gopichowvet@gmail.com
Iker A. Sevilla, NEIKER, Derio, Bizkaia, Spain. E-mail: isevilla@neiker.eus
Fernanda M. Shoyama, Michigan State University, East Lansing, Michigan, USA. E-mail: miyagaki@msu.edu
Lesley A. Smith, SRUC, Edinburgh, UK. E-mail: lesley.smith@sruc.ac.uk
Srinand Sreevatsan, Michigan State University, East Lansing, Michigan, USA. E-mail: sreevats@msu.edu
Judith R. Stabel, USDA-ARS, Ames, Iowa, USA. E-mail: judy.stabel@ars.usda.gov
Karen Stevenson, Moredun Research Institute, Penicuik, UK. E-mail: karen.stevenson@moredun.ac.uk
Jaryd R. Sullivan, McGill University, Montreal, Canada. E-mail: jaryd.sullivan@mail.mcgill.ca
Adel M. Talaat, University of Wisconsin, Madison, Wisconsin, USA. E-mail: atalaat@wisc.edu
Girum T. Tessema, Norwegian Veterinary Institute, Oslo, Norway. E-mail: girum.tessema@vetinst.no
Christine Y. Turenne, University of Manitoba, Winnipeg, Manitoba, Canada. E-mail: cturenne@sharedhealthmb.ca
Richard Whittington, University of Sydney, Sydney, Australia. E-mail: richard.whittington@sydney.edu.au
Chia-wei Wu, University of Wisconsin-Madison, Wisconsin, USA. E-mail: chiaweiwu@wisc.edu
Denise K. Zinniel, University of Nebraska, Lincoln, Nebraska, USA. E-mail: dzinniel2@unl.edu

Preface

In the preface to the first edition, it was noted, with perhaps only a hint of irony, that paratuberculosis is 'not a disease on which fast progress has been made'. This remains a truism, and during the past decade and a half since the first edition was written and published, much has changed, yet much remains the same. To capture the state of knowledge, thought leaders in the field of paratuberculosis detail in the 23 chapters (condensed from 29 chapters) of this second edition, all the many improvements in the understanding of the epidemiology of the pathogen, the organism and its association with various hosts, as well as how to better diagnose infection and control disease. In so many ways, the progress is remarkable – yet much remains the same. Paratuberculosis is still stubbornly endemic across most of the world, and progress in control of the spread is dependent almost entirely on avoiding infection in the first instance through what might be considered largely as good husbandry practices that limit opportunities of transmission. Vaccines are still rarely used, not because they are entirely ineffective, but rather because they confound diagnosis of tuberculosis and the business case for their use remains unclear. Importantly as well, the long-hypothesized association of *Mycobacterium avium* subsp. *paratuberculosis* with Crohn's disease remains just that. However, there are several good reasons that this is just the right time to collectively take stock of what we have learnt and how far we have come; to begin to harmonize use of common terminology to facilitate better understanding; as well as to share perspectives on what might be achieved in the coming decades and how we might get there.

In this new edition of the book we have reorganized some of the chapters to incorporate and place in context additional information that has come to light since the last edition. With the focus being on new information, the chapter on the history of paratuberculosis has been omitted since the advances of the past two decades are implicit in the chapters in the new edition. With more researchers using whole genome sequencing, there is a plethora of new data on phylogenomics, epidemiology and strain diversity, and this information has been captured in Chapter 6 (Stevenson) rather than being distributed across chapters on the genome, on strain characterization and on comparative differences between strains. We have combined the first-edition chapters on experimental ruminant and small animal models. Likewise, the three chapters on control of paratuberculosis in Europe, the USA and Australia have been combined, together with new information from other countries, to provide a more balanced overview.

Readers of the second edition will notice that to enable better harmonization and standardization, throughout the book, we have edited *Mycobacterium avium* subsp. *paratuberculosis* to *MAP*. This was done in the first edition, and we continue this approach, to avoid the unnecessary use of 50 characters in the place of three. The editors recognize that this is not a standardized abbreviation in microbial nomenclature, so we have not extended this to other organisms. Where the authors refer specifically to *M. avium hominissuis*, we have left this name as is; when referring to *M. avium* other than *MAP*, we have edited this to *M. avium avium*, in keeping with the latest recommended nomenclature (see Chapter 5, Turenne). Likewise, *M. tuberculosis* is not *Mtb*. Within *MAP*, there are different lineages that have been called strains by others. In some papers, these are called type I/II/III; in others, *MAP*-C and *MAP*-S. To harmonize across chapters, we have edited the usage to indicate that there is a C-lineage and an S-lineage, and these can be shortened to *MAP*-C and *MAP*-S (see Chapter 6, Stevenson, for more details on this). Similarly, when *MAP* causes infection and disease, different descriptive terms are used by different groups. The word 'latent' appeared in several draft chapters, and it often was not clear whether this referred to an undiagnosed infection, immunological reactivity in the absence of clinical signs, a dormant bacteriological process or the interval between being infected and becoming infectious. The problems associated with the use of this term have been recently highlighted in the case of 'latent TB', and so we have tried to remove this word when the meaning was potentially ambiguous. We also encountered terms such as 'silent' and 'subclinical' infection, but the distinction between these entities was not always clear. We considered adopting throughout the book the recently proposed three-stage terminology (Whittington *et al.*, *BMC Veterinary Research*, 2017, 13(1), 328): (i) subclinical infection, which is defined as *MAP* infection without demonstrable pathology in tissues; (ii) subclinical disease, which is defined as the presence of pathology in tissues without weight loss or diarrhoea; and (iii) clinical disease, which is defined as the presence of pathology + weight loss and/or diarrhoea. However, several authors have chosen to refer to four stages of disease, as has been classically done in the prior literature: silent, subclinical, clinical and advanced clinical. We decided to leave the four stages as written but encourage the field to consider ways to best harmonize the terminology regarding the progress of infection to disease. Related to this, some use Johne's disease (JD) to describe all of these stages while others have used the term Johne's disease to describe severe disease only. We have tried to use the term paratuberculosis throughout the book, and avoid the use of Johne's disease, as it is arguably a subset of the overall process.

The novel information on paratuberculosis, together with new technologies developed in the past decade, offer new opportunities and perspectives for the future. Areas of potential interest include using genomic comparisons to determine the age of *MAP*, the age of *MAP-C* divergence from *MAP-S* and the timing of introduction of *MAP* into regions and countries, in order to date the problem. This has been performed for *Mycobacterium tuberculosis* and *Mycobacterium leprae*, and while this often corroborates pre-existing wisdom, there are occasional surprises, such as a pinnepid origin of human TB in pre-Columbian Peru. Other areas of potential interest are how one might differentiate infected but recovered or self-cured animals from those that are infectious or diseased, and how one might predict which infected animals are likely to shed and progress to clinical disease. For reasons not entirely clear, advances in understanding the immunology of host responses to *MAP* have somewhat lagged behind some of the more rapid advances in understanding the organism itself including diagnostics. Perhaps this is because of the lower-hanging fruit associated with technical advances in genomics and other *omics* fields. Another understudied area is the environment, particularly in terms of environmental transmission and survival of the organism. Plugging knowledge gaps in this area would better inform management and control.

What is not here, and might present in the future, should the science so develop, includes new fundamental science (microbiome, epigenomics, microRNAs, organoid culture), and emergent translational tools in diagnostics (phage, point-of-care tests, sensor technologies, digital polymerase chain reaction (PCR)) and potentially treatment (probiotics, phage therapy). What is also not here and is much needed is fit-for-purpose interventions for better control of paratuberculosis in low and middle income countries where *MAP* is endemic, but traditional approaches to control (test and remove) are unfeasible due to socioeconomic considerations.

Finally, a word about standardization and why we wrote this book. As stated in the American Academy of Microbiology report on *MAP* (https://www.asmscience.org/content/report/colloquia/colloquia.33), there is a lot we do *not* know, in part because the research is not sufficiently reproducible. There are numerous reasons for this conundrum. For instance, *MAP* is a slow-growing organism that is difficult to enumerate. Borders block the flow of pathogens, constraining research opportunities. *MAP* infects different host species, with variable outcomes of infection, etc. Despite these differences, there is a lot that we *do* know, as articulated in the 23 chapters written by content experts, compiled here. For those coming into the *MAP* domain, for research or veterinary practice, we hope you can make use of this information, these methods, these current concepts, to fill the knowledge gaps. Work on *MAP* is slow and challenging enough; let's not have everyone start from zero, when there is a *MAP* research community that can offer its lessons, its reagents and its insights to those who wish to join. These were the intentions of our colleague Des Collins, who guided the first edition to completion and who provided generous and wise counsel to the three of us for this second edition. We hope this book serves its stated purpose.

Marcel A. Behr
Karen Stevenson
Vivek Kapur
January 2020

1 Epidemiology, Global Prevalence and Economics of Infection

Jamie Imada[1]*, David F. Kelton[1] and Herman W. Barkema[2]
[1]University of Guelph, Guelph, Ontario, Canada; [2]University of Calgary, Calgary, Alberta, Canada

1.1 Introduction

First described in the late 19th century (Johne and Frothingham, 1895), Johne's disease (JD) or paratuberculosis is a chronic, progressive and untreatable disease, mainly affecting ruminants and caused by *Mycobacterium avium* subspecies *paratuberculosis* (*MAP*). This bacterium enters through the gastrointestinal tract and can persist within the host for extended intervals (years) before clinical signs arise. Infected animals may excrete *MAP* prior to detection (diagnostics or clinical manifestations). *MAP* has demonstrated the ability to survive for extended periods of time in the environment, 152–246 days in pastures, and in water for up to 6–18 months (Lovell *et al.*, 1944; Whittington *et al.*, 2004; Singh *et al.*, 2013). This persistence poses concerns for the application of contaminated manure on feed crops (Obasanjo *et al.*, 1997). Disease characteristics of paratuberculosis as well as prolonged environmental survivability of *MAP* have made this a challenging pathogen to study and control.

1.2 Transmission

The main mode of transmission for *MAP* is faecal-oral, although cows in late stages of infection may excrete *MAP* in colostrum and milk (Pithua *et al.*, 2011; Stabel *et al.*, 2014). Furthermore, late-stage cases can transmit infection *in utero*, with 20–60% of calves born to clinical dams being infected (Donat *et al.*, 2016). Although direct excretion of *MAP* from the mammary gland does occur, it is believed that poor udder hygiene and inadequate milking routines result in significant contamination of milk and colostrum with faecal material, and are thus the mechanism for potential transmission (McAloon *et al.*, 2016a). Calf-to-calf transmission can occur; experimentally inoculated calves infected exposed pen mates, causing faecal shedding, highlighting the utility of individual housing of calves as a means of decreasing the risk of infection (Mortier *et al.*, 2014; Corbett *et al.*, 2017). It is believed that young animals are at greater infection risk due to their 'open gut', specifically the presence of specialized lymphoid tissue (Peyer's patches) that accept maternal antibodies early in life. Although very young animals are still believed to be most susceptible to *MAP* infection, there is an apparent dose dependence, with older animals being susceptible to infection with higher doses (Windsor and Whittington, 2010; Mortier *et al.*, 2013). There is also evidence that mature cows can become infected by exposure to clinically affected animals, with newly infected

*Corresponding author: imadaj@uoguelph.ca

© CAB International 2020. *Paratuberculosis: Organism, Disease, Control, 2nd Edition* (eds M.A. Behr *et al.*)

cows diagnosed by tissue culture at slaughter (Schukken *et al.*, 2015). The infective dose also seems to affect onset of clinical paratuberculosis; high-prevalence herds with high infective doses in the environment usually have a faster onset of clinical disease than low-prevalence herds (Lombard *et al.*, 2005; Weber *et al.*, 2010).

Recent evidence that *MAP* may be able to sporulate under certain conditions (Lamont *et al.*, 2012), be carried in dust and infect animals through the respiratory tract (Rowe and Grant, 2006), suggests alternate transmission routes. Infected animals may not manifest clinical signs until 2–10 years after infection, often at times of physiological stress, e.g. calving (Tiwari *et al.*, 2006). This long period between infection and clinical signs allows subclinical animals that were infected early in life to excrete the organism intermittently in their faeces; thereby acting as carriers, exposing their herd mates and avoiding detection until their immune system reacts and they either produce detectable antibody or display clinical signs. As infection progresses, the rate of faecal excretion increases, with advanced clinical cows representing a large source of infectious material (Mitchell *et al.*, 2015).

With clinical cows representing a major source of contamination for an infected herd, test and cull can be useful for paratuberculosis control programmes. In addition, based on susceptibility of young stock to infection, management practices, biosecurity and biocontainment also have important roles. However, the more recent evidence that mature cows are also susceptible to infection may reduce effectiveness of traditional control practices.

1.3 Stages of Paratuberculosis

Paratuberculosis infection has been described in four stages: silent, subclinical, clinical and advanced clinical (Tiwari *et al.*, 2006). Of note, a recently proposed reclassification suggests that the silent and the subclinical stages can be binned together, as these first two stages both represent *MAP* infection without any noticeable clinical signs. However, the classical proposal of four stages is detailed below.

Animals in the silent stage have been infected with *MAP* but do not shed the bacteria, nor do they have a detectable immune response. Therefore, diagnosis is based on tissue culture/histology/polymerase chain reaction (PCR) (Whittington *et al.*, 2017). Animals in the subclinical stage may be detected through diagnostic tests that detect presence of the organism, but these methods often yield a high proportion of false-negative test results, due to intermittent shedding of low amounts of *MAP*. Enzyme-linked immunosorbent assays (ELISAs) can also be used at the subclinical stage but have lower sensitivity; however, sensitivity improves as the animal approaches the clinical stage (Tiwari *et al.*, 2006). Subclinically affected animals have decreased milk production that is positively associated with ELISA status and test optical density (OD) (Sorge *et al.*, 2011). The clinical stage is characterized by decrease in body condition despite a normal or increased appetite, intermittent to persistent diarrhoea with increased thirst, decreased milk production and reduced fertility. Diagnostic tests evaluating antibody responses are most reliable for animals in the clinical stage of infection (Tiwari *et al.*, 2006). Reduced productivity plus other clinical signs (diarrhoea, weight loss) often prompt the removal of these animals from the herd before they enter the final stage. Animals entering the advanced clinical stage of infection are often weak, lethargic and emaciated with 'pipe stream' diarrhoea. Eventually they may develop mandibular oedema (bottle jaw) from enteric protein loss and the subsequent loss of oncotic pressure (Tiwari *et al.*, 2006).

Often paratuberculosis is referred to as a disease that exhibits the iceberg phenomenon: for every advanced clinical case, there are one or two clinical cases, four to eight subclinical cases and 10–14 in the silent stage of *MAP* infection. However, based on models by Magombedze *et al.* (2013), the proportion of silent infections may be greatly overestimated. They reported that the calculated ratio was highly dependent on how long the disease had been present on the farm, but in all instances, the number of animals in the silent stage was much less than that in the subclinical stage. This could possibly be due to past studies misclassifying subclinical animals as silent infections (Magombedze *et al.*, 2013). This has implications for disease modelling and control, in that the iceberg ratio has been used to provide a rough estimate of disease burden and disease risk. With misclassification of subclinical shedding animals into a silent stage of disease, we underestimate the risk for disease spread. More recently, two distinct types of shedding patterns of paratuberculosis have been described, progressive and non-progressive (Schukken *et al.*, 2015; Beaver *et al.*, 2017). These non-progressive shedders shed low numbers of *MAP* in their faeces intermittently and do not mount any appreciable humoral immune response, in contrast to progressive shedders, which slowly increase the amount of *MAP* shed in their faeces and have seroconversion of immunoglobulins. Proportions of these progressive and non-progressive shedders within an infected herd may be important in future paratuberculosis research and control programmes.

1.4 Tests Used in Prevalence Studies

An essential step in understanding and quantifying the magnitude of the problem with paratuberculosis requires identification of infected animals. Diagnostic tests mainly approach identification of *MAP* infection with two distinct methods. The first method is detection of the organism itself (bacterial culture or PCR), whereas the other is detection of immunological response to the organism using ELISA, agar gel immunodiffusion (AGID) or complement fixation tests (CFTs). The latter two tests are less commonly used compared with ELISA (Tiwari *et al.*, 2006; Whittington *et al.*, 2019). Further details on these specific tests are described in Chapters 18, 19 and 20 of this book. These tests can either be performed as individual tests on single animals, or as tests on pooled samples (including bulk milk). The high cost and long diagnostic waiting time are disadvantages of faecal culture; to offset costs, a pooled sampling approach with five cows/sample is used (Collins *et al.*, 2006). There is also the approach to using ELISA tests in pooled milk samples such as the bulk tank milk (BTM) ELISA. The major limitation in the use of these diagnostic tools for identifying *MAP* infection is low sensitivity. That diagnostic test evaluations are often done in comparison to faecal culture, which has variable sensitivity (19–70%) (Whitlock *et al.*, 2000; Tiwari *et al.*, 2006; Nielsen and Toft, 2008), represents a challenge. Lack of a standardized test with 100% sensitivity makes it difficult to assess true prevalence of disease.

The use of the ELISA on its own results in only 10–25% of infected animals being detected; therefore, most estimates of animal and herd level prevalence based on these numbers are likely underestimates (Whitlock *et al.*, 2000; Barkema *et al.*, 2010). Collins *et al.* (2006) reported higher sensitivities for milk and serum ELISA between 25 and 35%. The sensitivity of faecal culture is slightly higher at 55–65% and faecal PCR tests have reported sensitivities in the same range as ELISA (25–35%) (Collins *et al.*, 2006). To maximize sensitivity, the recommended time to conduct milk ELISA is either within the first 2 weeks or after the 45th week of lactation (Lombard *et al.*, 2006). This is an attempt to capture increased immunoglobulin production in colostrum or to avoid the dilution effects of peak milk production. Sensitivity of ELISA does increase as disease progresses, with a positive correlation between ELISA OD and *MAP* faecal shedding (Lombard *et al.*, 2006). The probability of detecting infected animals also increased with age (Nielsen and Toft, 2008; Zare *et al.*, 2013; Sun *et al.*, 2015). Due to the overall low sensitivity of ELISA, it is often recommended that positive ELISA animals are confirmed with faecal culture. Apart from imperfect sensitivity, a limitation of faecal culture for identifying paratuberculosis is intermittent shedding (Donat *et al.*, 2015). It is important to note that the performance of these tests is dependent on within-herd prevalence; therefore, as prevalence decreases during a control programme, diagnostic test performance also decreases.

The use of faecal culture and PCR on environmental samples has been proposed to help identify a herd's paratuberculosis status, but there are false-negatives when within-herd prevalence is low (Donat *et al.*, 2015). To maximize sensitivity, site selection of samples is important; therefore, it is recommended to collect from manure storage and high-traffic areas (Raizman *et al.*, 2004; Smith *et al.*, 2011; Lavers *et al.*, 2013; Wolf *et al.*, 2015). Klawonn *et al.* (2016) reported that environmental cultures had a sensitivity of ~64% for beef cow-calf herds. Whereas specificity is often cited to be 98–100%, there is some concern over cross-reactivity with other mycobacteria and 'passive shedding' of *MAP* (Nielsen and Toft, 2008; Nielsen, 2014).

BTM ELISA tests have also been evaluated for paratuberculosis monitoring/screening. However, results will be influenced by the number of shedders relative to the overall herd size, the proportion of BTM that infected cows contribute and the stage of lactation of the infected animals. The reported range in sensitivity of bulk tank ELISA is similar to that of individual ELISA (25%) (Stabel *et al.*, 2002). To address this low sensitivity, there have been modified techniques of current bulk tank ELISA tests proposed (Nielsen *et al.*, 2000; Slana *et al.*, 2008). It is recommended that these BTM tests be used for monitoring regional/national prevalence, but their interpretation on an individual-farm basis should be used with caution.

Interferon-γ assays are a newer diagnostic approach for detection of cell-mediated responses to paratuberculosis. Estimates for the sensitivity ranged from 13–85% (Nielsen and Toft, 2008; Nielsen, 2014).

1.5 Global Prevalence

Paratuberculosis is on the World Organisation for Animal Health (OIE) list of notifiable animal diseases, despite a lack of surveillance programmes in many countries. Most countries that have identified paratuberculosis as an endemic disease have a herd prevalence ranging from 10–70% (Figs 1.1–1.3); however, a few countries have been declared free (Whittington *et al.*, 2019). Control of this endemic disease has been particularly difficult due to poor test performance and subclinical animals intermittently shedding the pathogen. Pooled sampling techniques, e.g. BTM ELISA and environmental cultures, have promise for estimating nation-wide prevalence (Raizman *et al.*, 2004; Wilson *et al.*, 2010).

1.6 Herd-Level Prevalence

1.6.1 Dairy cattle herd-level prevalence

Due to the intensive nature of dairy farming and the long incubation period of paratuberculosis, *MAP* infection is prevalent and increasing in the dairy industry. Very few countries have established low prevalence (<1%) in their dairy herds (Thailand, Norway, Sweden), with many having herd level prevalences >20% (Belgium, Canada, Chile, Denmark, France, Germany, India, Israel,

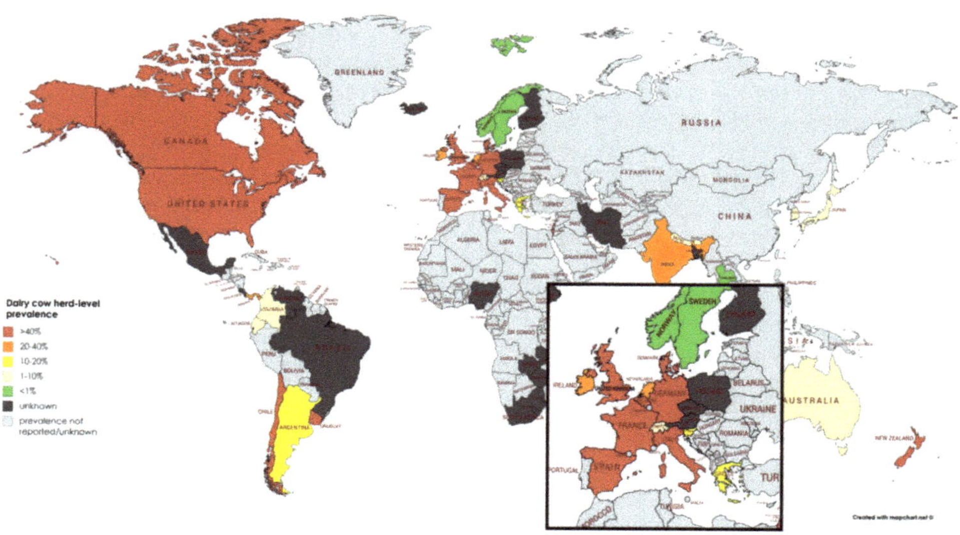

Fig. 1.1. Global dairy cattle herd-level prevalence of MAP infection. (Adapted from Whittington *et al.*, 2019.)

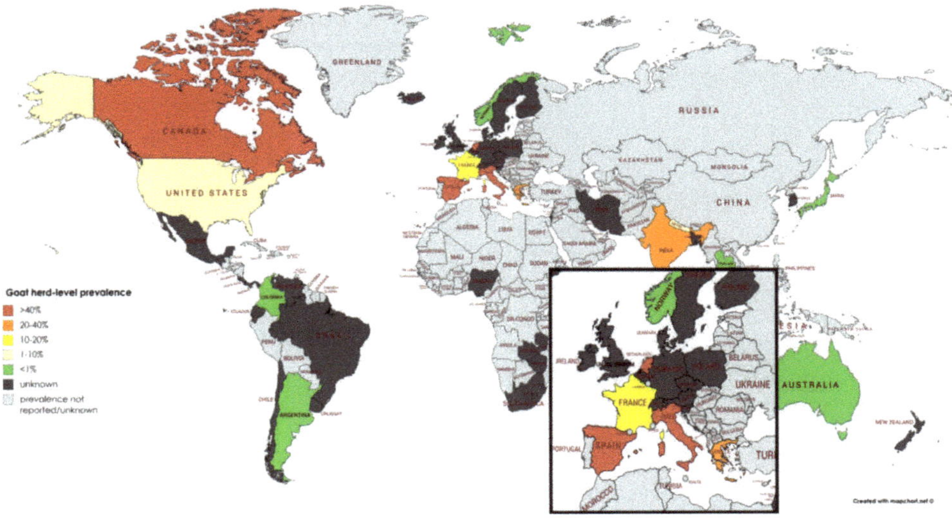

Fig. 1.2. Global goat herd-level prevalence of MAP infection. (Adapted from Whittington *et al.*, 2019.)

Italy, the Netherlands, New Zealand, Panama, Republic of Ireland, Spain, UK, USA, Uruguay) (Whittington *et al.*, 2019). Lombard *et al.* (2013) reported an apparent herd-level prevalence of 70% among American dairy farms in 2007, with Bayesian estimates being closer to 91%. A true herd-level prevalence was determined to be 46% for Canadian dairy farms using environmental cultures (Corbett *et al.*, 2018). Elsewhere, true herd-level prevalence of *MAP*-infected herds was estimated to be 85% of dairy herds in Denmark, 20–71% of herds in the Netherlands and approximately 28–43% of UK herds (Geraghty *et al.*, 2014). Some regions in Germany have an apparent herd-level seroprevalence of up to 85% (Khol and Baumgartner, 2011). Channel Island breeds are often considered at an increased risk for *MAP* infection;

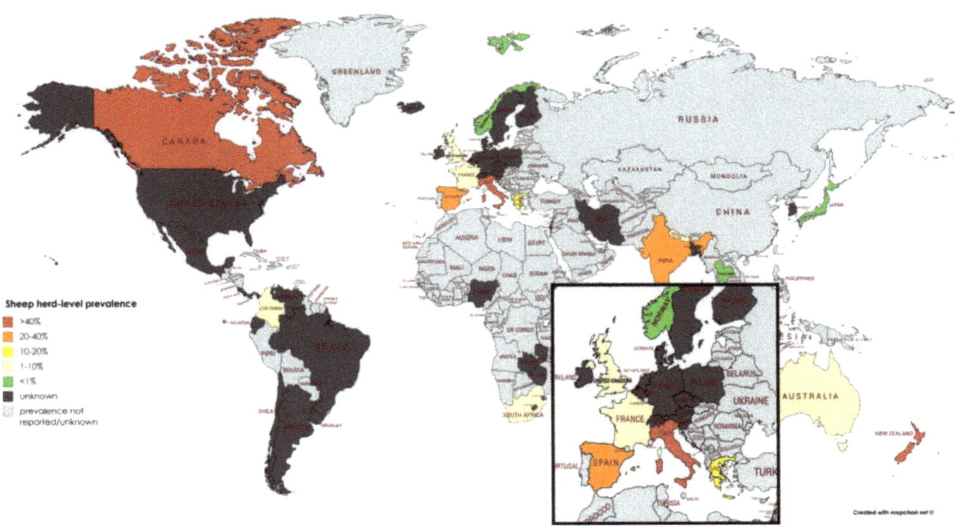

Fig. 1.3. Global sheep herd-level prevalence of MAP infection. (Adapted from Whittington *et al.*, 2019.)

however, it is unclear whether this is due to increased susceptibility or due to increased exposure due to other farm-level factors (Sorge *et al.*, 2012).

1.6.2 Beef cattle herd-level prevalence

Research investigating the prevalence of *MAP* infection in beef cows is limited. Prevalence is believed to be lower than that of dairy herds, possibly due to more extensive management, reducing transmission risk. Furthermore, lifespan of feedlot cattle (15–28 months) (Drouillard, 2018) is often much shorter than that of dairy cattle (5–7 years) (Hare *et al.*, 2006) providing less opportunity for the infection to progress to clinical manifestations. In contrast to this, cow-calf herds may keep brood cows much longer (8–12 years) (Rae *et al.*, 1993; Selk, 2018), potentially allowing for the manifestation, persistence and spread of paratuberculosis. The use of cull dairy cattle as either nurse cows or embryo transfer recipients may be a potential risk factor for introducing *MAP* into beef herds (Roussel *et al.*, 2005). Herd-level prevalence is often estimated near 7% (7.9% Canadian and US beef herds) (Geraghty *et al.*, 2014); however, these are likely underestimates (Dargatz *et al.*, 2001). Current estimates are even higher for purebred cattle, with ~44% of American and British beef herds being positive (Roussel *et al.*, 2005; Lawson and Caldow, 2014). This breed effect may be due to genetic susceptibility or the trade in purebred genetic stock.

1.6.3 Within-herd prevalence

A lower proportion of beef cattle is thought to be infected with *MAP* than dairy cattle; in a Western Canadian study, there was an apparent cow-level prevalence of 0.8% (Pruvot *et al.*, 2014a). By comparison, apparent cow-level seroprevalence in Canadian dairy cows ranges from 1.3–7% (Tiwari *et al.*, 2006). This lower prevalence in beef cattle could be due to the extensive management style of beef herds (Dargatz *et al.*, 2001). Infected herds with older animals can be expected to have more visible clinical signs than herds who remove animals earlier than the typical *MAP* incubation period. Herds with more intensive management may expect more transmission between animals and contamination of the environment. Cow-level prevalence is underestimated due to imperfect diagnostics. Where 2.2% of cows were detected as faecal shedders, 17% of them were tissue culture-positive at slaughter (Beaver *et al.*, 2017). A regional study in the USA estimated a true prevalence of 8.8% of all beef cattle in a brucellosis certification programme (Hill *et al.*, 2003). Similar results were reported from Missouri (USA) with an estimated true prevalence of 8% in beef cattle and 16% in dairy cattle, based on serum ELISA (Thorne and Hardin, 1997). In another regional US study, 9.6% of cull dairy cows and 4.0% of beef cows were serum ELISA-positive (Pence *et al.*, 2003).

1.6.4 Sheep, goats and other ruminants

Mortality due to paratuberculosis in Australian sheep flocks is considerable, ranging from 6–8% (Bush *et al.*, 2006). Regional herd-level prevalence in Australia ranges from 0–60%, with >5% of their national herd believed to be infected (2117 known *MAP* infected flocks in 2011). Australia is successfully using vaccination in paratuberculosis control (Windsor, 2015). There has been a relative lack of data on risk factors for paratuberculosis in goats (Gautam *et al.*, 2018). However, there has been a successful eradication of *MAP* infection from a goat herd using extensive testing and removal of positive animals (Gavin *et al.*, 2018). Similarly, through an intensive national campaign, Norway was able to eradicate *MAP* infections in their goat herds (Whittington *et al.*, 2019). Infection with *MAP* in small ruminants often does not result in the same progressive diarrhoea as in cattle. These animals usually have progressive weight loss and, in sheep, impairments in their wool production (Salem *et al.*, 2012). In New Zealand, 76% of sheep flocks and 46% of deer flocks are considered infected. In Ontario, Canada, true herd-level prevalence was 67% for dairy sheep and 83% for dairy goats, whereas within-herd prevalence was 35% for *MAP*-infected goat flocks and 48% for infected sheep (Bauman *et al.*, 2016). Without many large-scale studies on regional prevalence of *MAP* infection, comparisons are difficult; however, with its lack of pathognomonic clinical signs it may be reasonable to expect similar findings of relatively high prevalence within small ruminant populations without appropriate biosecurity or paratuberculosis control programmes in place (Windsor, 2015; Bauman *et al.*, 2016). Consequences of paratuberculosis could be more troubling in developing countries that are heavily reliant on small ruminants for nutrition and trade but lack adequate infrastructure for detection and control.

1.6.5 Wildlife

Mycobacterium avium subspecies *paratuberculosis* has been identified in >100 vertebrate species, of these, five (white tailed deer, roe deer, red deer, fallow deer, wild hare) are considered potential reservoirs for domestic cattle (Carta *et al.*, 2013). *MAP* was also isolated from rabbit faeces and urine; contaminated pastures could act as a potential reservoir for infection of cattle grazing (Daniels *et al.*, 2003; Shaughnessy *et al.*, 2013). There has also been a proposed risk for cattle and elk comingling (Pruvot *et al.*, 2014b). Further, mature deer seem to be equally at risk for infection as younger counterparts.

1.7 Economic Consequences

Monetary losses due to paratuberculosis have been difficult to quantify. Due to the unknown magnitude of subclinical effects, as well as the inability to determine true prevalence, economic studies have been at best rough estimations and likely underestimate true costs. Economic losses because of the disease are usually attributed to either milk loss, early culling, decreased carcass weight, increased susceptibility to other diseases or reduced fertility (Johnson *et al.*, 2001; Mckenna *et al.*, 2006; Raizman *et al.*, 2009; Vázquez *et al.*, 2012). For the US dairy industry, losses due to paratuberculosis have been estimated to be ~US\$200 ± 160 million per year (Losinger, 2005), whereas for the Canadian dairy industry, costs would be close to CA\$15 million per year (Mckenna *et al.*, 2006). Milk loss due to paratuberculosis are estimated to be ~2 kg/cow/day for *MAP*-infected animals (Sorge *et al.*, 2011; McAloon *et al.*, 2016b). Faecal culture-positive cows are removed from the herd 124 days earlier and produce 11% less milk than faecal culture-negative herd mates (Raizman *et al.*, 2007). Interestingly, there was increased milk production in cows prior to testing positive, suggesting a genetic susceptibility component to high producers (Sorge *et al.*, 2011; Garcia and Shalloo, 2015; McAloon *et al.*, 2016b).

Cost estimates of paratuberculosis are mainly based on the reduced milk production, reduced carcass weights and increased disease susceptibility; there is, however, also a potential public health concern. There have been numerous studies on the viability of *MAP* in processed food. *MAP* has been detected in both milk and meat products and can survive both high temperature short time (HTST) pasteurization and desiccation (Savi *et al.*, 2015; McAloon *et al.*, 2019) as well as processing of some meat products and, although it has been demonstrated that cooking meat to well done will render it free of the bacterium (Mutharia *et al.*, 2010), its presence in the raw product is a concern. *MAP* has long been associated with Crohn's disease in humans, and there have also been studies that have demonstrated links to type 1 diabetes (Barkema *et al.*, 2010). The current theory is that there is a genetic susceptibility to *MAP* and the development of autoimmune disorders (Sechi and Dow, 2015). If this theory is supported, this could harm future success and growth of the dairy industry. With a perceived public health risk, consumer confidence and demand for dairy will decline, resulting in potentially enormous economic losses (Groenendaal and Galligan, 2003). There are similar concerns in beef herds; however, the main economic consequence is reduced weaning weights, due to decreased milk production and direct effects of *MAP* infection (Roy *et al.*, 2017). Estimated losses from these adjusted weaning weights amount to US\$57/calf from cows with elevated ELISA titres to US\$157/calf from cows that are heavy shedders (Bhattarai *et al.*, 2013).

Costs of implementing a paratuberculosis control programme must be balanced with disease cost (Pillars *et al.*, 2009). The most effective control programmes at reducing disease prevalence are sometimes not the most economical (Webb Ware *et al.*, 2012; Smith *et al.*, 2017). Cost–benefit analysis often shows poor economic benefit to utilizing control programmes especially in low-prevalence herds (Collins and Morgan, 1991; Wolf *et al.*, 2014; Kirkeby *et al.*,

2016). Some economic models predicted only minor benefit in instituting control measures, even with subsidies (Groenendaal and Wolf, 2008; Wolf *et al.*, 2014). Chiu *et al.* (2018) reported that utilizing a test and cull strategy to control paratuberculosis was less economically beneficial compared with culling based off low production in herds known to have *MAP*; however, these herds did not achieve eradication. Costs of culling cows as well as the low apparent prevalence of disease within the herd and the long lag until benefits of required management changes are evident are barriers to the adoption of paratuberculosis control programmes (Arrigoni *et al.*, 2014; Cardwell *et al.*, 2016; McAloon *et al.*, 2017; Ritter *et al.*, 2017).

1.8 Knowledge Gaps

There are still many questions left unanswered with regards to paratuberculosis and its control or eradication (Barkema *et al.*, 2018). With such an endemic disease and decades of attempted control programmes, is eradication a reasonable goal? There remains a need for more accurate diagnostic tests, especially to detect early stages of infection. Performance of current diagnostic tests hinders accurate estimates in regional and global prevalence. A better understanding of the significance of the global prevalence of paratuberculosis is required to better estimate economic losses. Further research is also required in development of control programmes with good economic returns. What role can vaccines play in the control of paratuberculosis? How much risk does wildlife have in *MAP* transmission to livestock? More answers are needed as to the role of genetics: the importance of genetic susceptibility and genotypes involved in the two observed shedding patterns. Further, more complete understanding of psychosocial determinates of producer behaviour and producer participation in control programmes must be acquired.

References

Arrigoni, N., Ruocco, L., Paternoster, G. and Tamba, M. (2014) 4th ParaTB forum. *Bovine Paratuberculosis in Italy: Building Infrastructures and Defining the Guidelines for Control and for Ranking of Herds* (June), 53–59.

Barkema, H.W., Hendrick, S., De Buck, J.M., Ghosh, S. and Kaplan, G.G. (2010) Crohn's disease in humans and Johne's disease in cattle-linked diseases? In: Krause, D.O. and Hendrick, S. (eds) *Zoonotic Pathogens in the Food Chain*. CAB International, Wallingford, UK, pp. 197–213.

Barkema, H.W., Orsel, K., Nielsen, S.S., Koets, A.P., Rutten, V.P.M.G. *et al.* (2018) Knowledge gaps that hamper prevention and control of *Mycobacterium avium* subspecies *paratuberculosis* infection. *Transboundary and Emerging Diseases* 65, 125–148.

Bauman, C.A., Jones-Bitton, A., Menzies, P., Toft, N., Jansen, J. *et al.* (2016) Prevalence of paratuberculosis in the dairy goat and dairy sheep industries in Ontario, Canada. *Canadian Veterinary Journal* 57(2), 169–175.

Beaver, A., Sweeney, R.W., Hovingh, E., Wolfgang, D.R., Gröhn, Y.T. *et al.* (2017) Longitudinal relationship between fecal culture, fecal quantitative PCR, and milk ELISA in *Mycobacterium avium* ssp. *paratuberculosis*-infected cows from low-prevalence dairy herds. *Journal of Dairy Science* 100(9), 7507–7521.

Bhattarai, B., Fosgate, G.T., Osterstock, J.B., Fossler, C.P., Park, S.C. *et al.* (2013) Comparison of calf weaning weight and associated economic variables between beef cows with and without serum antibodies against or isolation from feces of *Mycobacterium avium* subsp *paratuberculosis*. *Journal of the American Veterinary Medical Association* 243(11), 1609–1615.

Bush, R.D., Windsor, P.A. and Toribio, J.A. (2006) Losses of adult sheep due to ovine Johne's disease in 12 infected flocks over a 3-year period. *Australian Veterinary Journal* 84(7), 246–253.

Cardwell, J.M., Van Winden, S., Beauvais, W., Mastin, A., De Glanville, W.A. *et al.* (2016) Assessing the impact of tailored biosecurity advice on farmer behaviour and pathogen presence in beef herds in England and Wales. *Preventive Veterinary Medicine* 135, 9–16.

Carta, T., Álvarez, J., Pérez de la Lastra, J.M. and Gortázar, C. (2013) Wildlife and paratuberculosis: a review. *Research in Veterinary Science* 94(2), 191–197.

Chiu, L.J. V., Tauer, L.W., Al-Mamun, M.A., Kaniyamattam, K., Smith, R.L. *et al.* (2018) An agent-based model evaluation of economic control strategies for paratuberculosis in a dairy herd. *Journal of Dairy Science* 101(7), 6443–6454.

Collins, M.T. and Morgan, I.R. (1991) Economic decision analysis model of a paratuberculosis test and cull program. *Journal of the American Veterinary Medical Association* 199(12), 1724–1729.

Collins, M.T., Gardner, I.A., Garry, F.B., Roussel, A.J. and Wells, S.J. (2006) Consensus recommendations on diagnostic testing for the detection of paratuberculosis in cattle in the United States. *Journal of the American Veterinary Medical Association* 229(12), 1912–1919.

Corbett, C.S., De Buck, J., Orsel, K. and Barkema, H.W. (2017) Fecal shedding and tissue infections demonstrate transmission of *Mycobacterium avium* subsp. *paratuberculosis* in group-housed dairy calves. *Veterinary Research* 48(1), 27.

Corbett, C.S., Naqvi, S.A., Bauman, C.A., De Buck, J., Orsel, K. *et al.* (2018) Prevalence of *Mycobacterium avium* ssp. *paratuberculosis* infections in Canadian dairy herds. *Journal of Dairy Science* 101(12), 11218–11228.

Daniels, M.J., Henderson, D., Greig, A., Stevenson, K., Sharp, J.M. *et al.* (2003) The potential role of wild rabbits *Oryctolagus cuniculus* in the epidemiology of paratuberculosis in domestic ruminants. *Epidemiology and Infection* 130(3), 553–559.

Dargatz, D.A., Byrum, B.A., Hennager, S.G., Barber, L.K., Kopral, C.A. *et al.* (2001) Prevalence of antibodies against *Mycobacterium avium* subsp *paratuberculosis* among beef cow-calf herds. *Journal of the American Veterinary Medical Association* 219(4), 497–501.

Donat, K., Kube, J., Dressel, J., Einax, E., Pfeffer, M. *et al.* (2015) Detection of *Mycobacterium avium* subspecies *paratuberculosis* in environmental samples by faecal culture and real-time PCR in relation to apparent within-herd prevalence as determined by individual faecal culture. *Epidemiology and Infection* 143(5), 975–985.

Donat, K., Schmidt, M., Köhler, H. and Sauter-Louis, C. (2016) Management of the calving pen is a crucial factor for paratuberculosis control in large dairy herds. *Journal of Dairy Science* 99(5), 3744–3752.

Drouillard, J.S. (2018) Current situation and future trends for beef production in the United States of America – a review. *Asian-Australasian Journal of Animal Sciences* 31(7), 1007–1016.

Garcia, A.B. and Shalloo, L. (2015) Invited review: the economic impact and control of paratuberculosis in cattle. *Journal of Dairy Science* 98(8), 5019–5039.

Gautam, M., Ridler, A., Wilson, P.R. and Heuer, C. (2018) Control of clinical paratuberculosis in New Zealand pastoral livestock. *New Zealand Veterinary Journal* 66(1), 1–8.

Gavin, W.G., Porter, C.A., Hawkins, N., Schofield, M.J. and Pollock, J.M. (2018) Johne's disease: a successful eradication programme in a dairy goat herd. *Veterinary Record* 182(17), 483.

Geraghty, T., Graham, D.A., Mullowney, P. and More, S.J. (2014) A review of bovine Johne's disease control activities in 6 endemically infected countries. *Preventive Veterinary Medicine* 116(1–2), 1–11.

Groenendaal, H. and Galligan, D.T. (2003) Economic consequences of control programs for paratuberculosis in midsize dairy farms in the United States. *Journal of the American Veterinary Medical Association* 223(12), 1757–1763.

Groenendaal, H. and Wolf, C.A. (2008) Farm-level economic analysis of the US National Johne's Disease Demonstration Herd Project. *Journal of the American Veterinary Medical Association* 233(12), 1852–1858.

Hare, E., Norman, H.D. and Wright, J.R. (2006) Survival rates and productive herd life of dairy cattle in the United States. *Journal of Dairy Science* 89(9), 3713–3720.

Hill, B.B., West, M. and Brock, K. V. (2003) An estimated prevalence of Johne's disease in a subpopulation of Alabama beef cattle. *Journal of Veterinary Diagnostic Investigation* 15(1), 21–25.

Johne, H.A. and Frothingham, L. (1895) Ein eigenthuemlicher fall von tuberculose beim Rind. *Deutsche Zeitschrift für Tiermedizin und Pathologie* 21, 438–454.

Johnson, Y.J., Kaneene, J.B., Gardiner, J.C., Lloyd, J.W., Sprecher, D.J. *et al.* (2001) The effect of subclinical *Mycobacterium paratuberculosis* infection on milk production in Michigan dairy cows. *Journal of Dairy Science* 84(10), 2188–2194.

Khol, J.L. and Baumgartner, W. (2011) Examples and suggestions for the control of paratuberculosis in European cattle. *Japanese Journal of Veterinary Research* 60(Suppl.), 1–7.

Kirkeby, C., Græsbøll, K., Nielsen, S.S., Christiansen, L.E., Toft, N. *et al.* (2016) Simulating the epidemiological and economic impact of paratuberculosis control actions in dairy cattle. *Frontiers in Veterinary Science* 3, 90.

Klawonn, W., Einax, E., Pützschel, R., Schmidt, M. and Donat, K. (2016) Johne's disease: Reliability of environmental sampling to characterize *Mycobacterium avium* subspecies *paratuberculosis* (MAP) infection in beef cow-calf herds. *Epidemiology and Infection* 144(11), 2392–2400.

Lamont, E.A., Bannantine, J.P., Armién, A., Ariyakumar, D.S. and Sreevatsan, S. (2012) Identification and characterization of a spore-like morphotype in chronically starved *Mycobacterium avium* subsp. *paratuberculosis* cultures. *PloS one* 7(1), 1–10.

Lavers, C.J., McKenna, S.L.B., Dohoo, I.R., Barkema, H.W. and Keefe, G.P. (2013) Evaluation of environmental fecal culture for *Mycobacterium avium* subspecies *paratuberculosis* detection in dairy herds and association with apparent within-herd prevalence. *Canadian Veterinary Journal* 54(11), 1053–1060.

Lawson, J.B. and Caldow, G.L. (2014) A study of thirteen years of a beef breed society's Johne's disease control initiative. *Cattle Practice* 22(1), 35–41.

Lombard, J.E., Garry, F.B., McCluskey, B.J. and Wagner, B.A. (2005) Risk of removal and effects on milk production associated with paratuberculosis status in dairy cows. *Journal of the American Veterinary Medical Association* 227(12), 1975–1981.

Lombard, J.E., Byrem, T.M., Wagner, B.A. and McCluskey, B.J. (2006) Comparison of milk and serum enzyme-linked immunosorbent assays for diagnosis of *Mycobacterium avium* subspecies *paratuberculosis* infection in dairy cattle. *Journal of Veterinary Diagnostic Investigation* 18(5), 448–458.

Lombard, J.E., Gardner, I.A., Jafarzadeh, S.R., Fossler, C.P., Harris, B. *et al.* (2013) Herd-level prevalence of *Mycobacterium avium* subsp. *paratuberculosis* infection in United States dairy herds in 2007. *Preventive Veterinary Medicine* 108(2–3), 234–238.

Losinger, W.C. (2005) Economic impact of reduced milk production associated with Johne's disease on dairy operations in the USA. *Journal of Dairy Research* 72(4), 425–432.

Lovell, R., Levi, M. and Francis, J. (1944) Studies on the survival of Johne's bacilli. *Journal of Comparative Pathology and Therapeutics* 54, 120–129.

Magombedze, G., Ngonghala, C.N. and Lanzas, C. (2013) Evalution of the 'Iceberg Phenomenon' in Johne's Disease through Mathematical Modelling. *PLoS ONE* 8(10).

McAloon, C.G., Doherty, M.L., Donlon, J., Lorenz, I., Meade, J. *et al.* (2016a) Microbiological contamination of colostrum on Irish dairy farms. *Veterinary Record* 178(19), 474.

McAloon, C.G., Whyte, P., More, S.J., Green, M.J., O'Grady, L. *et al.* (2016b) The effect of paratuberculosis on milk yield--A systematic review and meta-analysis. *Journal of Dairy Science* 99(2), 1449–1460. DOI: 10.3168/jds.2015-10156.

McAloon, C.G., Macken-Walsh, Á., Moran, L., Whyte, P., More, S.J. *et al.* (2017) Johne's disease in the eyes of Irish cattle farmers: A qualitative narrative research approach to understanding implications for disease management. *Preventive Veterinary Medicine* 141, 7–13.

McAloon, C.G., Roche, S., Ritter, C., Barkema, H.W., Whyte, P. *et al.* (2019) A review of paratuberculosis in dairy herds - Part 1: Epidemiology. *The Veterinary Journal* 246, 59–65. DOI: 10.1016/j.tvjl.2019.01.010.

Mckenna, S.L.B., Keefe, G.P., Tiwari, A., Vanleeuwen, J. and Barkema, H.W. (2006) Johne's disease in Canada Part 2: Disease impacts, risk factors, and control programs for dairy producers. *Canadian Veterinary Journal* 47(11), 1089–1099.

Mitchell, R.M., Schukken, Y., Koets, A., Weber, M., Bakker, D. *et al.* (2015) Differences in intermittent and continuous fecal shedding patterns between natural and experimental *Mycobacterium avium* subspecies *paratuberculosis* infections in cattle. *Veterinary Research* 46(1), 66.

Mortier, R.A.R., Barkema, H.W., Orsel, K., Wolf, R., Atkins, G.A. *et al.* (2013) Evaluation of age-dependent susceptibility in calves infected with two doses of *Mycobacterium avium* subspecies *paratuberculosis* using pathology and tissue culture. *Veterinary Research* 44(1), 94.

Mortier, R.A.R., Barkema, H.W., Orsel, K., Wolf, R. and De Buck, J. (2014) Shedding patterns of dairy calves experimentally infected with *Mycobacterium avium* subspecies *paratuberculosis*. *Veterinary Research* 45(1), 71.

Mutharia, L.M., Klassen, M.D., Fairles, J., Barbut, S. and Gill, C.O. (2010) *Mycobacterium avium* subsp. *paratuberculosis* in muscle, lymphatic and organ tissues from cows with advanced Johne's disease. *International Journal of Food Microbiology* 136(3), 340–344.

Nielsen, S.S. (2014) Developments in diagnosis and control of bovine paratuberculosis. *CAB Reviews* 9(012), 1–12.

Nielsen, S.S. and Toft, N. (2008) Ante mortem diagnosis of paratuberculosis: a review of accuracies of ELISA, interferon-γ assay and faecal culture techniques. *Veterinary Microbiology* 129(3–4), 217–235.

Nielsen, S.S., Thamsborg, S.M., Houe, H. and Bitsch, V. (2000) Bulk-tank milk ELISA antibodies for estimating the prevalence of paratuberculosis in Danish dairy herds. *Preventive Veterinary Medicine* 44(1–2), 1–7.

Obasanjo, I.O., Gröhn, Y.T. and Mohammed, H.O. (1997) Farm factors associated with the presence of *Mycobacterium paratuberculosis* infection in dairy herds on the New York State Paratuberculosis Control Program. *Preventive Veterinary Medicine* 32(3–4), 243–251.

Pence, M., Baldwin, C. and Black, C.C. (2003) The seroprevalence of Johne's disease in Georgia beef and dairy cull cattle. *Journal of Veterinary Diagnostic Investigation* 15(5), 475–477.

Pillars, R.B., Grooms, D.L., Wolf, C.A. and Kaneene, J.B. (2009) Economic evaluation of Johne's disease control programs implemented on six Michigan dairy farms. *Preventive Veterinary Medicine* 90(3–4), 223–232.

Pithua, P., Wells, S.J. and Godden, S.M. (2011) Evaluation of the association between fecal excretion of *Mycobacterium avium* subsp *paratuberculosis* and detection in colostrum and on teat skin surfaces of dairy cows. *Journal of the American Veterinary Medical Association* 238(1), 94–100.

Pruvot, M., Kutz, S., Barkema, H.W., De Buck, J. and Orsel, K. (2014a) Occurrence of *Mycobacterium avium* subspecies *paratuberculosis* and *Neospora caninum* in Alberta cow-calf operations. *Preventive Veterinary Medicine* 117(1), 95–102.

Pruvot, M., Kutz, S., Van Der Meer, F., Musiani, M., Barkema, H.W. *et al.* (2014b) Pathogens at the livestock-wildlife interface in Western Alberta: does transmission route matter? *Veterinary Research* 45(1), 18.

Rae, D.O., Kunkle, W.E., Chenoweth, P.J., Sand, R.S. and Tran, T. (1993) Relationship of parity and body condition score to pregnancy rates in Florida beef cattle. *Theriogenology* 39(5), 1143–1152.

Raizman, E.A., Wells, S.J., Godden, S.M., Bey, R.F., Oakes, M.J. *et al.* (2004) The distribution of *Mycobacterium avium* ssp. *paratuberculosis* in the environment surrounding Minnesota dairy farms. *Journal of Dairy Science* 87(9), 2959–2966.

Raizman, E.A., Fetrow, J., Wells, S.J., Godden, S.M., Oakes, M.J. *et al.* (2007) The association between *Mycobacterium avium* subsp. *paratuberculosis* fecal shedding or clinical Johne's disease and lactation performance on two Minnesota, USA dairy farms. *Preventive Veterinary Medicine* 78(3–4), 179–195.

Raizman, E.A., Fetrow, J.P. and Wells, S.J. (2009) Loss of income from cows shedding Mycobacterium avium subspecies paratuberculosis prior to calving compared with cows not shedding the organism on two Minnesota dairy farms. *Journal of Dairy Science* 92(10), 4929–4936.

Ritter, C., Jansen, J., Roche, S., Kelton, D.F., Adams, C.L. *et al.* (2017) Invited review: Determinants of farmers' adoption of management-based strategies for infectious disease prevention and control. *Journal of Dairy Science* 100(5), 3329–3347.

Roussel, A.J., Libal, M.C., Whitlock, R.L., Hairgrove, T.B., Barling, K.S. *et al.* (2005) Prevalence of and risk factors for paratuberculosis in purebred beef cattle. *Journal of the American Veterinary Medical Association* 226(5), 773–778.

Rowe, M.T. and Grant, I.R. (2006) *Mycobacterium avium* ssp. *paratuberculosis* and its potential survival tactics. *Letters in Applied Microbiology* 42(4), 305–311.

Roy, G.L., De Buck, J., Wolf, R., Mortier, R.A.R., Orsel, K. *et al.* (2017) Experimental infection with *Mycobacterium avium* subspecies *paratuberculosis* resulting in decreased body weight in Holstein-Friesian calves. *Canadian Veterinary Journal* 58(3), 296–298.

Salem, M., Heydel, C., El-Sayed, A., Ahmed, S.A., Zschöck, M. *et al.* (2012) *Mycobacterium avium* subspecies *paratuberculosis*: an insidious problem for the ruminant industry. *Tropical Animal Health and Production* 45(2), 351–366.

Savi, R., Ricchi, M., Cammi, G., Garbarino, C., Leo, S. *et al.* (2015) Survey on the presence of *Mycobacterium avium* subsp. *Paratuberculosis* in ground beef from an industrial meat plant. *Veterinary Microbiology* 177(3–4), 403–408.

Schukken, Y.H., Whitlock, R.H., Wolfgang, D., Grohn, Y., Beaver, A. *et al.* (2015) Longitudinal data collection of *Mycobacterium avium* subspecies *Paratuberculosis* infections in dairy herds: the value of precise field data. *Veterinary Research* 46(1), 65.

Sechi, L.A. and Dow, C.T. (2015) *Mycobacterium avium* ss. *paratuberculosis* zoonosis – The hundred year war – Beyond Crohn's disease. *Frontiers in Immunology* 6, 96.

Selk, G. (2018) Cow age and cow productivity (When is she too old?). Cow/calf corner – The newsletter, (Oct). Available at: http://agecon.okstate.edu/livestock/files/COW%2010-1-18.pdf

Shaughnessy, L.J., Smith, L.A., Evans, J., Anderson, D., Caldow, G. *et al.* (2013) High prevalence of paratuberculosis in rabbits is associated with difficulties in controlling the disease in cattle. *Veterinary Journal* 198(1), 267–270.

Singh, S. V., Singh, A. V., Kumar, A., Singh, P.K., Deb, R. *et al.* (2013) Survival mechanisms of *Mycobacterium avium* subspecies *paratuberculosis* within host species and in the environment - A review. *Natural Science* 5(6), 710–723.

Slana, I., Paolicchi, F., Janstova, B., Navratilova, P. and Pavlik, I. (2008) Detection methods for *Mycobacterium avium* subsp. *paratuberculosis* in milk and milk products: a review. *Veterinarni Medicina* 53(6), 283.

Smith, R.L., Schukken, Y.H., Pradhan, A.K., Smith, J.M., Whitlock, R.H. *et al.* (2011) Environmental contamination with *Mycobacterium avium* subsp. *paratuberculosis* in endemically infected dairy herds. *Preventive Veterinary Medicine* 102(1), 1–9.

Smith, R.L., Al-Mamun, M.A. and Gröhn, Y.T. (2017) Economic consequences of paratuberculosis control in dairy cattle: a stochastic modeling study. *Preventive Veterinary Medicine* 138, 17–27.

Sorge, U.S., Lissemore, K., Godkin, A., Hendrick, S., Wells, S. *et al.* (2011) Associations between paratuberculosis milk ELISA result, milk production, and breed in Canadian dairy cows. *Journal of Dairy Science* 94(2), 754–761.

Sorge, U.S., Lissemore, K., Godkin, A., Jansen, J., Hendrick, S. *et al.* (2012) Risk factors for herds to test positive for *Mycobacterium avium* ssp. *paratuberculosis*-antibodies with a commercial milk enzyme-linked immunosorbent assay (ELISA) in Ontario and Western Canada. *Canadian Veterinary Journal* 53(9), 963–970.

Stabel, J.R., Bradner, L., Robbe-Austerman, S. and Beitz, D.C. (2014) Clinical disease and stage of lactation influence shedding of *Mycobacterium avium* subspecies *paratuberculosis* into milk and colostrum of naturally infected dairy cows. *Journal of Dairy Science* 97(10), 6296–6304.

Stabel, J.R., Wells, S.J. and Wagner, B.A. (2002) Relationships between fecal culture, ELISA, and bulk tank milk test results for Johne's disease in US dairy herds. *Journal of Dairy Science* 85(3), 525–531.

Sun, W.W., Lv, W.F., Cong, W., Meng, Q.F., Wang, C.F. *et al.* (2015) *Mycobacterium avium* subspecies *paratuberculosis* and Bovine Leukemia Virus seroprevalence and associated risk factors in commercial dairy and beef cattle in Northern and Northeastern China. *BioMed Research International* 2015, 315173.

Thorne, J.G. and Hardin, L.E. (1997) Estimated prevalence of paratuberculosis in Missouri, USA cattle. *Preventive Veterinary Medicine* 31(1–2), 51–57.

Tiwari, A., Vanleeuwen, J.A., Mckenna, S.L., Keefe, G.P. and Barkema, H.W. (2006) Johne's disease in Canada Part I: clinical symptoms, pathophysiology, diagnosis, and prevalence in dairy herds. *Canadian Veterinary Journal* 47(9), 874–882.

Vázquez, P., Garrido, J.M. and Juste, R.A. (2012) Effects of paratuberculosis on friesian cattle carcass weight and age at culling. *Spanish Journal of Agricultural Research* 10(3), 662–670.

Webb Ware, J.K., Larsen, J.W.A. and Kluver, P. (2012) Financial effect of bovine Johne's disease in beef cattle herds in Australia. *Australian Veterinary Journal* 90(4), 116–121.

Weber, M.F., Kogut, J., de Bree, J., van Schaik, G. and Nielen, M. (2010) Age at which dairy cattle become *Mycobacterium avium* subsp. *paratuberculosis* faecal culture positive. *Preventive Veterinary Medicine* 97(1), 29–36.

Whitlock, R.H., Wells, S.J., Sweeney, R.W. and Van Tiem, J. (2000) ELISA and fecal culture for paratuberculosis (Johne's disease): sensitivity and specificity of each method. *Veterinary Microbiology* 77(3–4), 387–398.

Whittington, R.J., Marshall, D.J., Nicholls, P.J., Marsh, I.B. and Reddacliff, L.A. (2004) Survival and dormancy of *Mycobacterium avium* subsp. *paratuberculosis* in the environment. *Applied and Environmental Microbiology* 70(5), 2989–3004.

Whittington, R.J., Begg, D.J., de Silva, K., Purdie, A.C., Dhand, N.K. *et al.* (2017) Case definition terminology for paratuberculosis (Johne's disease). *BMC Veterinary Research* 13(1), 328.

Whittington, R.J., Donat, K., Weber, M.F., Kelton, D., Nielsen, S.S. *et al.* (2019) Control of paratuberculosis: who, why and how. A review of 48 countries. *BMC Veterinary Research* 15(1), 198.

Wilson, D.J., Rood, K., Biswas, P. and Byrem, T.M. (2010) Herd-level prevalence of Johne's disease in Utah and adjacent areas of the intermountain west as detected by a bulk-tank milk surveillance project. *Journal of Dairy Science* 93(12), 5792–5797.

Windsor, P.A. (2015) Paratuberculosis in sheep and goats. *Veterinary Microbiology* 181(1–2), 161–169.
Windsor, P.A. and Whittington, R.J. (2010) Evidence for age susceptibility of cattle to Johne's disease. *Veterinary Journal* 184(1), 37–44.
Wolf, R., Clement, F., Barkema, H.W. and Orsel, K. (2014) Economic evaluation of participation in a voluntary Johne's disease prevention and control program from a farmer's perspective – the Alberta Johne's Disease Initiative. *Journal of Dairy Science* 97(5), 2822–2834.
Wolf, R., Barkema, H.W., De Buck, J. and Orsel, K. (2015) Sampling location, herd size, and season influence *Mycobacterium avium* ssp. *paratuberculosis* environmental culture results. *Journal of Dairy Science* 98(1), 275–287.
Zare, Y., Shook, G.E., Collins, M.T. and Kirkpatrick, B.W. (2013) Evidence of birth seasonality and clustering of *Mycobacterium avium* subspecies *paratuberculosis* infection in US dairy herds. *Preventive Veterinary Medicine* 112(3–4), 276–284.

2 *Mycobacterium avium* subsp. *paratuberculosis* in Animal-Derived Foods and the Environment

Irene R. Grant*
Queen's University, Belfast, Northern Ireland

2.1 Introduction

Animals infected by *Mycobacterium avium* subsp. *paratuberculosis* (*MAP*), whether clinically or subclinically affected, can shed live *MAP* in both their faeces and milk. If these animals are farmed for food production, the safety of foods derived from them becomes an important consideration, because *MAP* has been hypothesized to be linked with Crohn's disease in humans (see Chapter 3, this volume). Infected animals also contaminate their surrounding environment, increasing the risk of spread of paratuberculosis at the farm level and potentially contaminating watercourses used for abstraction of drinking water. This chapter summarizes current evidence for the presence of *MAP* in animal-derived foods, describes the effect of various dairy processes on *MAP* survival, and reviews the reservoirs of *MAP* infection in the environment and the various mechanisms potentially aiding its survival for long periods. Shedding of *MAP* by infected animals has implications for food and water safety, as illustrated in Fig. 2.1.

2.2 Evidence of *MAP* in Animal-Derived Foods

The current evidence for *MAP* contamination of animal-derived foods, both raw and processed, is summarized in Table 2.1. The results presented are based on cultural, polymerase chain reaction (PCR or quantitative PCR) and phage-based detection of *MAP*. The starting points for this summary table were the outcomes of a comprehensive review and meta-analysis of the evidence for *MAP* in animal-derived foods by Waddell *et al.* (2016); the additional information is from other published studies between 2011 and 2018. To the best of the author's knowledge, all published food surveillance studies have been included.

2.2.1 Milk and dairy products

2.2.1.1 Raw milk

Raw cow's milk has been the focus of most *MAP* surveillance because it is recognized as a major factor in the transmission of paratuberculosis from cow to calf (Nielsen *et al.*, 2008).

*i.grant@qub.ac.uk

© CAB International 2020. *Paratuberculosis: Organism, Disease, Control, 2nd Edition* (eds M.A. Behr *et al.*)

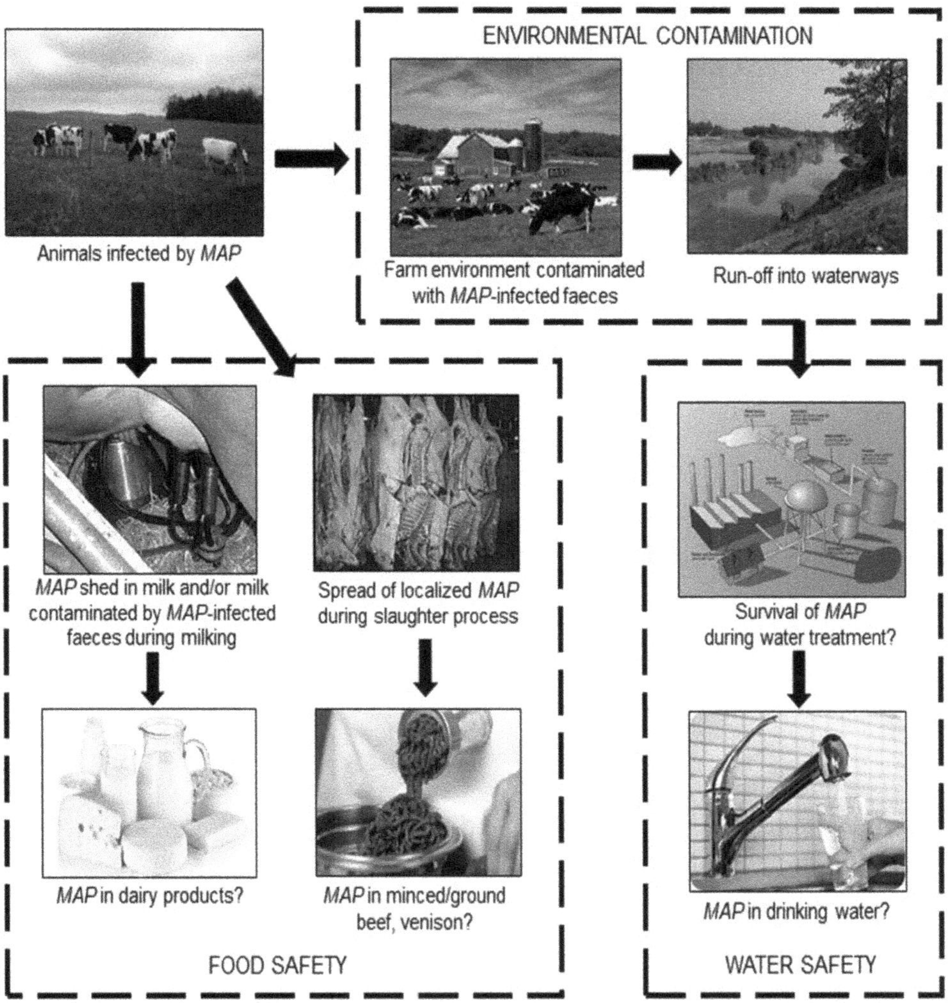

Fig. 2.1. Spread of *Mycobacterium avium* subsp. *paratuberculosis* (*MAP*) shed by infected animals in faeces and milk, and potential routes of human exposure to *MAP* via animal-derived foods and water.

Published studies have tested raw milk from individual animals, bulk tank at farm level or bulk silo milk prior to processing. Until the advent of quantitative real-time PCR and phage amplification assays for *MAP*, information on the levels of *MAP* in raw milk at these different points was limited. Early estimates of *MAP* levels in milk from sub-clinically affected cows derived by culture (e.g. Sweeney *et al.* (1992) reported two to eight colony-forming units (CFU) *MAP*/50 ml milk) are likely to have been underestimates for two reasons. First, chemical decontamination is generally applied to milk samples and this, depending on the method employed, negatively impacts the estimated counts of *MAP* (Dundee *et al.*, 2001; Bradner *et al.*, 2013). Second, in studies concerning milk from individual animals, the milk tested was obtained after thorough cleaning and disinfection of the exterior of the udder, removing the possibility of faecal contamination. Faecal contamination can and does occur during the milking process (Vissers *et al.*, 2007), and different cleaning regimes applied to the udder affect the degree of faecal contamination

Table 2.1. A summary of the evidence for *Mycobacterium avium* subsp. *paratuberculosis* (*MAP*) contamination of raw and processed animal-derived foods obtained via surveillance studies. Findings of the meta-analysis by Waddell *et al.* (2016) are listed first for each food category – milk, various dairy products and meat – followed by additional studies published since 2013.

Food	Size of survey	% PCR- or qPCR-positive	% Culture-positive	% Phage-based assay positive	No. of *MAP* reported	Reference
Milk and dairy products						
Raw cow's milk (individual animals)	Multiple studies	19.3 (14.8, 23.8)	10.9 (3.4, 21.4)	–[a]	–[a]	Waddell *et al.* (2016)[b]
	63	20.6	–[a]	–[a]	–[a]	Khol *et al.* (2013)
	160	17.5	3.7	–[a]	–[a]	Hanifian and Khani (2016)
	88	23.9	–[a]	–[a]	–[a]	Saad *et al.* (2018)
	146	9.1	11.6	21.2	6–948 PFU/50 ml[c]	Foddai and Grant (2017)
	40	–[a]	25.0	32.5	8–94 PFU/50 ml[c]	O'Brien *et al.* (2018)
	121	16.5 PCR, 28.1 qPCR	–[a]	–[a]	36–368 cells/ml[d]	de Albuquerque *et al.* (2017)
	147	54.4	1.4	–[a]	–[a]	Gilardoni et al. (2016)
Raw cow's milk (bulk tank, farm level)	Multiple studies	37.1 (26.3, 47.9)	3.5 (0.2, 9.1)	–[a]	–[a]	Waddell *et al.* (2016)[b]
	225	–[a]	0.9	22.0	2–320 PFU/50 ml[c]	Botsaris *et al.* (2013)
	2834	26.9 qPCR	–[a]	–[a]	<15–32.4cells/ml[d]	Ricchi *et al.* (2016)
	22	45.4	50.0	59.1	18–685 PFU/50 ml[c]	Foddai and Grant (2017)
	80	63.8	21.3	–[a]	–[a]	Hanifian and Khani (2016)
Raw cow's milk (processing level)	Multiple studies	14.1 (2.5, 24.8)	10.2 (0, 32.8)	–[a]	–[a]	Waddell *et al.* (2016)[b]
	83	4.8	4.8	–[a]	–[a]	Boulais *et al.* (2011)

Continued

Table 2.1. Continued

Food	Size of survey	% PCR- or qPCR-positive	% Culture-positive	% Phage-based assay positive	No. of *MAP* reported	Reference
	75	82.7	18.7	–[a]	–[a]	Hanifian and Khani (2016)
Pasteurized cow's milk	Multiple studies	13.1 (8.7, 17.5)	5.3 (1.9, 10.0)	–[a]	–[a]	Waddell *et al.* (2016)[b]
	70	–[a]	2.9	–[a]	–[a]	Paolicchi *et al.* (2012)
	300	10.7	–[a]	–[a]	–[a]	Anzabi and Hanifian (2012)
	37	–[a]	2.7	–[a]	–[a]	Carvalho *et al.* (2012)
	160	–[a]	0	–[a]	–[a]	Serraino *et al.* (2017)
	368	–[a]	–[a]	10.3	1–32 PFU/50 ml[c]	Gerrard *et al.* (2018)
	120	0.83	0	–[a]	–[a]	Liandris *et al.* (2014)
Raw goat's milk (individual animals)	Multiple studies	35.7 (19.7, 51.6)	0.7 (0.0, 12.5)	–[a]	–[a]	Waddell *et al.* (2016)[b]
Raw goat's milk (bulk tank, farm level)	Multiple studies	14.4 (0.0, 54.9)	0	–[a]	–[a]	Waddell *et al.* (2016)[b]
Cheese						
Raw milk cheeses	Multiple studies	7.7 (4.3, 11.9)	1.7 (0, 12.0)	–[a]	–[a]	Waddell *et al.* (2016)[b]
	30	10.0	3.3	–[a]	–[a]	Faria *et al.* (2014)
	7 goats, 25 sheep	57.1 50.0	14.2 0	• –[a] • –[a]	• –[a] • –[a]	Galiero *et al.* (2016)
Pasteurized milk cheeses	Multiple studies	17.2 (1.8, 32.7)	1.1 (0.0, 10.9)	–[a]	–[a]	Waddell *et al.* (2016)[b]
Yoghurt or yoghurt-based products	130	1.54	0	–[a]	–[a]	Liandris *et al.* (2014)

Continued

Table 2.1. Continued

Food	Size of survey	% PCR- or qPCR-positive	% Culture-positive	% Phage-based assay positive	No. of *MAP* reported	Reference
Powdered milk products						
Infant formula	32	21.9	9.4	12.5	9–95 PFU/g[c]	Botsaris *et al.* (2016)
	122	4.9	0	–[a]	>5 *MAP*/1.5 g[d]	Acharya *et al.* (2017)
	51	35.2	0	–[a]	–[a]	Hruska *et al.* (2011)
Calf milk replacer	83	8.4	14.5	20.5	6–1212 PFU/50 ml[c]	Grant *et al.* (2017)
	15	0	0	–[a]	–[a]	Khol *et al.* (2017)
Meat (including beef, lamb and pork)						
Raw beef and sheep meat	Multiple studies	25.5 (5.7, 50.8)	3.3 (0.5, 6.2)	–[a]	–[a]	Waddell *et al.* (2016)[b]
Raw, semi-finished meat products	45	26.7	0	–[a]	$2.9 \times 10^2 – 1.4 \times 10^3$ cells/g[d]	Lorencova *et al.* (2014)
Ready-to-eat meat products	44	4.5	0	–[a]		

[a]not tested or not reported.
[b]For data from Waddell *et al.* (2016), the figures are prevalence from a meta-analysis of multiple studies with 95% credible interval in brackets.
[c]counts determined by plaque assay; PFU is plaque-forming units.
[d]counts determined by quantitative real-time polymerase chain reaction (qPCR).

(Gibson *et al.*, 2008). It is clear from Table 2.1 that qPCR and phage-based detection of *MAP* in milk and dairy products consistently provide higher estimates of the proportion of samples positive for *MAP* than cultural methods; with percentage figures generally highest for phage assay results when all three methods are applied (Foddai and Grant, 2017; Grant *et al.*, 2017). While acknowledging that PCR does not differentiate between viable and dead bacteria, *MAP* cells detected in raw milk by qPCR are likely to be viable, and these numbers will not have been adversely affected by chemical decontamination (as CFU counts are). Phage amplification assays indicate numbers of only viable *MAP* cells present (by counting of plaque-forming units (PFU) produced), and, for this test also, the counts will not have been adversely affected by prior decontamination of the sample. The numbers of viable *MAP* in raw milk from individual animals and in bulk tank milk reported when qPCR and phage-based assays were employed (summarized in Table 2.1) indicate considerably higher levels in raw milk than previously thought.

The level of *MAP* contamination of bulk tank milk at farm level is influenced by both infection status of the animals in a herd and hygiene practices employed during milking. Okura *et al.* (2013) used a modelling approach to estimate the concentration of *MAP* in bulk tank milk in dairy herds with within-herd Johne's disease prevalences of 7.5–60%; the estimated median loads of *MAP* were 0.54–7.53 CFU/ml (or 27–377 CFU/50 ml). However, their model indicated that maximum concentration at a within-herd prevalence of 60% could be 1186 CFU/ml (or 59,300 CFU/50 ml) bulk tank milk, caused by shedding of high numbers of *MAP* in faeces of infected cows. Recently, Rani *et al.* (2019) used a Monte Carlo simulation model to predict the effect of hygiene practices (udder washing prior to milking, use of in-line milk filters, cleaning of milking equipment) on the levels of *MAP* contamination in bulk tank milk. Results indicated that 93% of *MAP* load in bulk tank milk is from faecal contamination, with the remainder contaminating via internal (direct shedding into milk) and environmental routes. Both Okura *et al.* (2013) and Rani *et al.* (2019) commented that a very few 'super-shedding' animals in a herd may play a major role in raw milk *MAP* contamination levels.

The expectation has always been that there are two dilution steps for *MAP* in milk, towards very low or non-detectable levels, before processing: first, when milk from individual infected animals is mixed with milk of non-infected herdmates, and, second, when milk from one farm is mixed with milk from other farms before dairy processing. This may or may not be the case. Khol *et al.* (2013) reported the expected reduction in *MAP* levels from individual cow milks to bulk tank milk, whereas Hanifian and Khani (2016) reported a significant increase in *MAP* levels ($p < 0.01$) from 'hygienically obtained' quarter milks to bulk tank. The latter authors concluded that *MAP* levels in raw milk are very dependent on the care and attention of farmers during the milking process (i.e. hygiene practices), in order to avoid faecal contamination.

Raw milk from sheep and goats has not been as extensively tested as cow's milk. In the developed world at least, these types of milk would tend to be consumed as dairy products rather than as liquid milk. The situation in developing countries is probably the reverse (Chaubey *et al.*, 2017). There have been no recent surveillance studies of sheep and goat's milk. The review by Waddell *et al.* (2016) reported *MAP* mean PCR positivity of 35.7% in individual sheep and goat's milk and 14.4% in bulk tanks at farm level, but much lower culture positivity – 0.7% for individual milks only (Table 2.1). Greater difficulties in culturing isolates of *MAP*-S, as opposed to *MAP*-C (detailed in Chapters 12 and 18), may partly explain the low culture positivity reported for raw sheep or goat's milk surveys.

2.2.1.2 *Pasteurized milk*

The question of whether *MAP* can survive commercial pasteurization processes remains a contentious topic for the dairy industry (Robertson *et al.*, 2017; Mullan, 2019). To date, 13 studies have reported the presence of viable *MAP* in retail pasteurized milk in different countries of the world. Waddell *et al.* (2016) included seven pasteurized milk surveys within their meta-analysis and reported mean prevalence of *MAP* of 13.1% by PCR (95% CI: 8.7, 17.5) and 5.3% by culture (95% CI: 1.9, 10.0). Since then a further six pasteurized milk surveys have been reported in different parts of the world (Table 2.1), two of which reported isolation of viable *MAP* in 2.9%

(Paolicchi *et al.*, 2012) and 2.7% (Carvalho *et al.*, 2012) of samples tested. A third study reported detection of viable *MAP* by the phage-PCR assay in 10.3% of retail semi-skimmed pasteurized milk samples in the UK (Gerrard *et al.*, 2018), at levels of 1–32 PFU/50 ml milk in test positive samples. On the basis of the above surveillance data, it is difficult to conclude that humans are not being periodically exposed to some level of viable *MAP* via consumption of pasteurized milk in countries where paratuberculosis exists in dairy herds. However, it is still proving difficult for dairy processors to understand what might be the explanation for *MAP* survival when technical and engineering aspects of the milk pasteurization processes applied have been scrutinized and not found to be defective or deficient (Robertson *et al.*, 2017; Mullan, 2019). The infective dose for humans is currently unknown, as is the potential risk that comes from repeated exposure to viable *MAP*.

2.2.1.3 Various dairy products

Cheese (both raw and pasteurized milk varieties), powdered milk products (infant milk formula and calf milk replacer, specifically), and yoghurt or yoghurt-based products are the dairy products that have received attention as possible vehicles of human (or animal) exposure to *MAP*. Apart from a study of raw milk cheeses made from sheep's milk carried out in Italy (Galiero *et al.*, 2016), *MAP* has consistently been isolated by culture (despite its recognized shortcomings) from raw and pasteurized milk cheeses, but PCR positivity of the cheeses was always much higher (Table 2.1). A single Greek study of 130 yoghurt or yoghurt-based products marketed for consumption by children (Liandris *et al.*, 2014) found only PCR evidence of *MAP* presence (1.54%), but no *MAP* was cultured. There have been two reports of isolation of viable *MAP* in surveys of powdered milk products to date. A survey of European Union (EU) infant formula products by Botsaris *et al.* (2016) reported 9.4% of samples *MAP* positive by culture and 12.5% by phage-PCR assay, and a survey of US calf milk replacer products by Grant *et al.* (2017) reported 14.5% of samples *MAP* positive by culture and 20.5% by PMS-phage assay. Two other surveys of powdered infant formula (Hruska *et al.*, 2011; Acharya *et al.*, 2017) and one of calf milk replacer (Khol *et al.*, 2017) failed to culture viable *MAP*, although, with the exception of the latter study, there was PCR evidence of the presence of *MAP* cells in the powdered milk samples tested (Table 2.1).

2.2.2 Meat and meat products

MAP contamination of animal carcasses can arise from three sources: (i) faecal contamination on hides being transferred to carcasses during slaughter; (ii) disseminated infection via the bloodstream; and (iii) localized foci of infection, such as lymph nodes, being disrupted, such that the contents are spread to other parts of the carcass or mixed into minced/ground meat. The risk of *MAP* contaminating beef is possibly higher when old dairy cows are slaughtered rather than beef cattle, as the latter are generally slaughtered before lesions are evident (Adam and Brülisauer, 2010). However, currently there appear to be no restrictions on paratuberculosis positive cows being sent to slaughter for human consumption, and these animals will pose the greatest risk of *MAP* transmission via beef, regardless of whether they are from dairy or beef herds. The Waddell *et al.* (2016) review and meta-analysis considered seven raw beef and sheep studies published up to 2011 and reported a prevalence of 25.5% by PCR methods and 3.3% by culture (Table 2.1). The only *MAP* and meat survey since that review appears to be Lorencova *et al.* (2014), who reported the findings of a survey of retail raw, semi-finished and ready-to-eat meat products, comprising minced beef and pork and/or offal in the Czech Republic. They reported a similar percentage of *MAP* positivity (26.7%) by qPCR to previous raw meat studies (summarized by Waddell *et al.*, 2016) for raw, semi-finished meat products, but lower qPCR positivity for ready-to-eat meat products (4.5%). Numbers of *MAP* estimated by qPCR to be present in the retail meat products ranged from $2.9 \times 10^2 - 1.4 \times 10^3$ cells/g. Viable *MAP* was not isolated from either type of meat product by culture, but the decontamination method employed (HCl and neutralization with NaOH) may not have been optimal for *MAP* recovery from meat samples.

Paratuberculosis is also a problem in farmed deer (Power *et al.*, 1993; Fawcett *et al.*, 1995;

de Lisle *et al.*, 2003; Kopecna *et al.*, 2008; de Albuquerque *et al.*, 2017) and so there is a theoretical risk of venison being contaminated by *MAP* in the same way as meat from other infected species. However, no venison surveys have been reported to date. The incidence of *MAP* in farmed rabbits has been investigated (Arrazuria *et al.*, 2015) because paratuberculosis can also be a problem in rabbits. *MAP* was not detected in gut-associated lymphoid tissue of 66 rabbits in commercial rabbit farms in Spain by either culture or PCR, but *M. avium* subsp. *avium* (10/66, 15.2%) and *M. avium* subsp. *hominissuis* (1/66, 1.51%) were, by PCR only.

Generally meat is cooked before consumption and, thus, the effect of cooking on survival of *MAP* is of interest. A single study, Hammer *et al.* (2013), has investigated the effect of cooking of hamburger patties made with *MAP*-spiked beef (at three levels – 10^2, 10^4 and 10^6 CFU/g) for 2–6 min on a hotplate operating at 177°C, with flipping once during cooking. Results in terms of log kill of *MAP* achieved were 'highly variable' according to the authors. Cooking to well done, 6 min for 70 g patties achieved a four-$\log_{10}$ reduction, and 5 or 6 min for 50 g patties achieved a more than five-$\log_{10}$ reduction. Internal temperature of the patties was data-logged during cooking, which revealed that not all points in a hamburger achieve the same temperature during cooking. Undercooked, or unevenly cooked, hamburgers may, therefore, represent a food safety risk if high numbers of *MAP* were present in the raw meat.

2.3 Survival of *MAP* During Dairy Processing

Numerous pasteurization studies involving *MAP* have been reported in the scientific literature. Space does not permit the listing of all these studies here, so readers are directed to a review and critique of these studies by Robertson *et al.* (2017). The studies involved different types of heating apparatus, different *MAP* strains prepared in different ways and different culture methodologies after heating. Consequently, it is very difficult to compare the studies or to reach a consensus opinion on the effect of commercial high-temperature, short-time (HTST) pasteurization conditions (72°C for 15 s) on *MAP* viability. The findings can best be described as conflicting: some researchers report inactivation of more than seven $\log_{10}$ *MAP*, whereas others consistently report a more modest four-$\log_{10}$ reduction (Cerf *et al.*, 2007). Hammer *et al.* (2014) reported that homogenization applied before, during or after HTST pasteurization did not impact *MAP* inactivation significantly, which is contrary to a previous report by Grant *et al.* (2005), which found that homogenization before or during HTST pasteurization led to greater *MAP* inactivation.

A recently published review by Mullan (2019) concluded that there are at least eight possible reasons why viable *MAP* is being reported to be present in retail pasteurized milk. These are as follows: (i) the remarkable heat resistance of *MAP*; (ii) the potential existence of a heat-resistant subpopulation; (iii) the location of *MAP* cells in somatic cells in milk affording protection during heating; (iv) the non-homogeneous distribution of *MAP* in milk; (v) leakage within, or incorrect operation of, pasteurizers; (vi) poorly designed pasteurizing systems without over-pressure or differential pressure systems; (vii) high concentrations of *MAP* in raw milk and clumping of *MAP* cells; and (viii) possible false positive results with the newer phage-based assays. Two pasteurization process modelling studies, which specifically considered survival of *MAP*, should also make interesting reading for dairy processors. Salgado *et al.* (2011, p. E500) suggested that there is '68.4% probability that the 15 s HTST pasteurization process would not achieve at least five decimal reductions in *MAP* counts'. Chandrakash and Davey (2017, p. 11) concluded that 'any pasteurization process is a mix of failure and success, with around 5.7% failure events suggested, equating to ~21 pasteurization failures, with unwanted *MAP* survival, each year averaged over an extended period of daily batch-continuous operations'.

In view of the fact that studies have suggested that HTST pasteurization may not completely eliminate viable *MAP*, the effect of novel milk processing techniques has been studied. The use of pulsed electric fields to destroy pathogenic bacteria, due to electrical breakdown of the cell membrane and electroporation, has been investigated for *MAP* inactivation. Rowan *et al.* (2001) observed a 5.9-$\log_{10}$ reduction in viable *MAP* when spiked cow's milk was subjected to 2500 pulses at 30 kV/cm in a 25-min period, which represented a greater kill than was achieved by laboratory pasteurization (2.4 $\log_{10}$). Stabel *et al.* (2001) reported that application of 5, 10 or 15 kGy of gamma radiation achieved a six-$\log_{10}$ reduction in *MAP* in raw milk. In contrast, ultraviolet light treatment of *MAP* in milk had minimal effect on viability (0.5–1.0-$\log_{10}$ reduction per 1000 mJ/ml; Altic *et al.*, 2007). Treatment of *MAP*-spiked milk with high hydrostatic pressure (500 MPa for 10 min) achieved a four- to six-$\log_{10}$ kill (López-Pedemonte *et al.*, 2006; Donaghy *et al.*, 2007), a similar reduction to HTST pasteurization. Peterz *et al.* (2016) studied the effect of direct steam injection at 105°C for 3 s using pilot-scale equipment and were unable to recover viable *MAP* from spiked milk (10^{5-6} CFU/ml) after treatment. None of the alternative processes investigated to date appears to offer a viable alternative to HTST pasteurization, with the possible exception of direct steam injection. However, more research on the latter dairy processing technology is required.

Limited research on *MAP* survival during production of other dairy products is available. During cheesemaking there is an approximately ten-fold concentration of any *MAP* present in the milk upon curd formation (Donaghy *et al.*, 2004). During the subsequent ripening period, some inactivation of *MAP* occurs, principally dictated by pH, salt concentration, length of ripening period and presence of lactic acid cultures (Sung and Collins, 2000; Spahr and Schafroth, 2001; Donaghy *et al.*, 2004; Hanifian, 2014). During yoghurt production, numbers of viable *MAP* remained unchanged in two studies (Van Brandt *et al.*, 2011; Klanicova *et al.*, 2012). However, during storage of fermented or probiotic-containing products (e.g. kefir, acidophilus milk and yoghurt), longer exposure to pH <4.0 led to non-culturability of *MAP* (Klanicova *et al.*, 2012) and 1.2–≥3.8-$\log_{10}$ reductions in viable *MAP* (Van Brandt *et al.*, 2011).

2.4 Contamination of the Farm Environment

Infected animals shed *MAP* in their faeces, so they will contaminate their immediate environment whether housed indoors or grazed outside. Cattle are commonly classified as light, moderate or heavy shedders on the basis of the numbers of *MAP* in their faeces (<300 CFU/g, 300–3000 CFU/g and >3000 CFU/g, respectively; van Schaik *et al.*, 2003). Hovingh *et al.* (2006) identified a further category of infected animal – the *MAP* supershedder – which can shed >1,000,000 CFU/g faeces. Hovingh *et al.* (2006) found that 10–15% of animals in four infected herds were supershedders. They calculated that a single supershedder would shed more *MAP* than 2000 moderate or 20,000 light shedders. Okura *et al.* (2013) and Rani *et al.* (2019) concluded the same about the relative contribution of faecal shedding by supershedders to levels of *MAP* contamination of raw milk at farm level. This situation has major implications for levels of *MAP* contamination in bulk tank milk of infected farms, and potential for environmental transmission of paratuberculosis within either dairy or beef herds.

2.4.1 Spread and survival of *MAP* in the environment

Cattle are generally not housed all the time, and movement of animals around the farm results in contamination of outdoor areas. Raizman *et al.* (2004) and Lombard *et al.* (2006) tested environmental samples from various locations around dairy operations in the USA. Farm locations commonly contaminated by *MAP* were parlour exits, floors of holding pens, common alleyways, lagoons, manure spreaders and manure pits. When animals are grazing on pastures, their faeces contaminate soil and grass. Whittington *et al.* (2004) studied survival in faeces in the Australian environment, and *MAP* was cultured for up to 55 weeks from dry, fully shaded locations and for much shorter time periods in unshaded conditions. They postulated that diurnal temperature flux due to infrared radiation, rather than ultraviolet inactivation, influenced *MAP* survival. In a subsequent study of survival of *MAP* in dam water in shaded or exposed water troughs, Whittington *et al.* (2005) recorded survival times of up to 48 weeks and 36 weeks, respectively, and for 12–26 weeks longer in the dam sediment. In both studies, Whittington and colleagues obtained results suggestive of dormancy, i.e. *MAP* detection followed by disappearance and then detection again after a period of time. Numerous invertebrate and protozoal species were observed to be present in the dam water (Whittington *et al.*, 2005). It has been suggested that interaction with nematodes, insects or protozoa (Whan *et al.*, 2006) may enable *MAP*, an intracellular pathogen, to survive and/or multiply in the environment. Other potential survival mechanisms of *MAP* in the environment (dormancy, aerosolization and biofilm formation) were reviewed by Rowe and Grant (2006).

MAP on contaminated pasture can run off into watercourses when it rains. Studies by Pickup *et al.* (2005, 2006) presented evidence of runoff from hills grazed by *MAP*-infected sheep into the Taff and Tywi rivers in South Wales, UK, especially after periods of high rainfall. More recent studies by the same UK research group (Rhodes *et al.*, 2014; Richardson *et al.*, 2019) reported that aerosols above the Rivers Taff and Tywi contain *MAP*, and the break of foam bubbles in the water may cause *MAP* to be aerosolized leading to potential human exposure along these rivers. Richardson *et al.* (2019) studied these geographic locations over a 10-year period and demonstrated persistence of *MAP* correlating with levels of paratuberculosis in British herds over the same period. They also reported *MAP* levels in the river water estimated by qPCR of up to 10^5 cell equivalents/l (which is equal to 10^8*MAP* per m^3 of river water). Singh *et al.* (2012) reported detection of *MAP* DNA by PCR in 10% of river waters tested in parts of India. The above findings with regard to *MAP* in raw (untreated) water raise questions about the ability of water treatment processes to remove or inactivate *MAP* before it reaches the consumer. In laboratory simulations, chlorination of *MAP*-spiked water at 2 μg/ml for 30 min resulted in a maximum 2.8-$\log_{10}$ reduction in numbers (Whan *et al.*, 2001), which means that, in common with other mycobacteria, *MAP* is chlorine resistant. Another water treatment process (COCODAFF) physically removes *MAP* along with suspended solids (Pickup *et al.*, 2006). However, the contaminated slurry removed may be disposed of back on to the land, creating a cycle of environmental persistence. Waddell *et al.* (2016) reviewed five untreated water studies and the prevalence of *MAP* from their meta-analysis was 8.7% by culture (95% CI: 2.5, 7.0) and 42.5% by PCR detection (95% CI: 25.5, 60.4). Similar to milk pasteurization, the efficacy of water treatment processes will depend on the numbers of *MAP* present in raw waters being treated, which will be influenced by prevalence of *MAP* infection in animals grazing surrounding land and rainfall patterns leading to runoff into rivers.

Waddell *et al.* (2016) reported a meta-analysis of six drinking water (i.e. treated water) surveys and reported prevalences of 2.3% by culture (95% CI: 0.0, 66.8) and 35.7% by PCR (95% CI: 21.5, 49.8). Beumer *et al.* (2008) reported that *MAP* can frequently be detected in biofilm samples from taps. A study by Sarmento *et al.* (2018) of drinking water and an untreated domestic water supply in the Porto area of Portugal showed higher levels of *MAP* contamination in samples of drinking water than in domestic water, and detected *MAP* DNA at a higher frequency in tap biofilms than in the corresponding collected water, which is in agreement with the US study. A study of water supply systems in the Czech Republic by Klanicova *et al.* (2013) found no evidence of *MAP* at any point (watershed, reservoir, drinking water treatment plant, drinking water or household supply) by culture or qPCR, although several other *Mycobacterium* spp. were cultured and non-tuberculous mycobacteria were detected in 76.7% of water samples by the PCR methods employed.

2.5 Conclusions

Viable *MAP* have been successfully cultured from, or *MAP* DNA detected in, a range of animal-derived food products and drinking water. It is difficult to conclude that humans are not being exposed to *MAP*-contaminated milk and dairy products on the basis of this evidence. However, whether the level of MAP exposure exceeds a safe level remains unknown. Results of culture-based studies do not paint an accurate picture of the risk of consuming viable *MAP*, since these methods underestimate the true number of organisms present in any particular sample. Factors such as the adverse effect of chemical decontamination on the viability of some of the *MAP* present and overgrowth of cultures by other bacteria potentially masking the presence of *MAP* colonies may also lead to underestimates of *MAP* presence. In recent years, more sensitive and rapid phage-based and real-time qPCR methods for detection and quantitation of *MAP* have been developed. Despite the fact that further validation of the phage-based assays is still needed, as these improved detection methods have been adopted for food and drinking water surveillance, a clearer picture regarding levels of *MAP* present in, and hence risk of human exposure to this pathogen via, animal-derived foods and the environment is emerging. Milk and dairy products, and meat and meat products, remain the foods considered to be at the greatest risk of contamination by *MAP*. Other potential food or environmental sources of *MAP* may exist, but they simply have not been identified or studied yet. For a safer food supply from a *MAP* perspective, the focus needs to be on controlling paratuberculosis in food animals coupled with strict hygiene practices during milking and slaughter to reduce levels of contamination of raw milk and meat going forward.

References

Acharya, K.R., Dhand, N.K., Whittington, R.J. and Plain, K.M. (2017) Detection of *Mycobacterium avium* subspecies *paratuberculosis* in powdered infant formula using IS*900* quantitative PCR and liquid culture media. *International Journal of Food Microbiology* 257, 1–9. DOI: 10.1016/j.ijfoodmicro.2017.06.005.

Adam, K. and Brülisauer, F. (2010) The application of food safety interventions in primary production of beef and lamb: a review. *International Journal of Food Microbiology* 141(Suppl.), S43–S52. DOI: 10.1016/j.ijfoodmicro.2009.12.020.

Altic, L.C., Rowe, M.T. and Grant, I.R. (2007) UV light inactivation of *Mycobacterium avium* subsp. *paratuberculosis* in milk as assessed by FASTPlaqueTB phage assay and culture. *Applied and Environmental Microbiology* 73(11), 3728–3733. DOI: 10.1128/AEM.00057-07.

Anzabi, Y. and Hanifian, S. (2012) Detection of *Mycobacterium avium* subspecies *paratuberculosis* in pasteurized milk by IS*900* PCR and culture method. *African Journal of Microbiology Research* 6(7), 1453–1456.

Arrazuria, R., Sevilla, I.A., Molina, E., Pérez, V., Garrido, J.M. *et al.* (2015) Detection of *Mycobacterium avium* subspecies in the gut associated lymphoid tissue of slaughtered rabbits. *BMC Veterinary Research* 11(1), 130. DOI: 10.1186/s12917-015-0445-2.

Beumer, A., King, D. and Pfaller, S.L. (2008) Detection of *Mycobacterium avium* subsp. *paratuberculosis* in drinking water and biofilms using quantitative PCR. *Abstracts of the 108th General Meeting of the American Society for Microbiology, abstract Q-487*, American Society for Microbiology, Washington, DC.

Botsaris, G., Liapi, M., Kakogiannis, C., Dodd, C.E.R. and Rees, C.E.D. (2013) Detection of *Mycobacterium avium* subsp. *paratuberculosis* in bulk tank milk by combined phage-PCR assay: evidence that plaque number is a good predictor of *MAP*. *International Journal of Food Microbiology* 164(1), 76–80. DOI: 10.1016/j.ijfoodmicro.2013.03.023.

Botsaris, G., Swift, B.M.C., Slana, I., Liapi, M., Christodoulou, M. *et al.* (2016) Detection of viable *Mycobacterium avium* subspecies *paratuberculosis* in powdered infant formula by phage-PCR and confirmed by culture. *International Journal of Food Microbiology* 216, 91–94. DOI: 10.1016/j.ijfoodmicro.2015.09.011.

Boulais, C., Wacker, R.O.N., Augustin, J.C., Ben Cheikh, M.H. and Peladan, F. (2011) Modeling the occurrence of *Mycobacterium avium* subsp. *paratuberculosis* in bulk raw milk and the impact of management options for exposure mitigation. *Journal of Food Protection* 74(7), 1126–1136. DOI: 10.4315/0362-028X.JFP-11-005.

Bradner, L., Robbe-Austerman, S., Beitz, D.C. and Stabel, J.R. (2013) Chemical decontamination with N-acetyl-L-cysteine-sodium hydroxide improves recovery of viable *Mycobacterium avium* subsp. *paratuberculosis* organisms from cultured milk. *Journal of Clinical Microbiology* 51(7), 2139–2146. DOI: 10.1128/JCM.00508-13.

Carvalho, I.A., Pietralonga, P.A.G., Schwarz, D.G.G., Faria, A.C.S. and Moreira, M.A.S. (2012) Short communication: recovery of viable *Mycobacterium avium* subspecies *paratuberculosis* from retail pasteurized whole milk in Brazil. *Journal of Dairy Science* 95(12), 6946–6948. DOI: 10.3168/jds.2012-5657.

Cerf, O., Griffiths, M. and Aziza, F. (2007) Assessment of the prevalence of *Mycobacterium avium* subsp. *paratuberculosis* in commercially pasteurized milk. *Foodborne Pathogens and Disease* 4(4), 433–447. DOI: 10.1089/fpd.2007.0028.

Chandrakash, S. and Davey, K.R. (2017) Advancing the Fr 13 risk framework to an integrated three-step microbiological failure synthesis of pasteurization of raw milk containing *Mycobacterium avium* subsp. *paratuberculosis* (MAP). *Chemical Engineering Science* 171, 1–18. DOI: 10.1016/j.ces.2017.05.020.

Chaubey, K.K., Singh, S.V., Gupta, S., Singh, M., Sohal, J.S. *et al.* (2017) *Mycobacterium avium* subspecies *paratuberculosis* – an important food borne pathogen of high public health significance with special reference to India: an update. *Veterinary Quarterly* 37(1), 282–299. DOI: 10.1080/01652176.2017.1397301.

de Albuquerque, P.P., Santos, A.S., Souza Neto, O.L., Kim, PdeC.P., Cavalcanti, E.F. *et al.* (2017) Detection of *Mycobacterium avium* subsp. *paratuberculosis* in bovine milk from the state of Pernambuco, Brazil. *Brazilian Journal of Microbiology* 48(1), 113–117. DOI: 10.1016/j.bjm.2016.10.010.

de Lisle, G.W., Yates, G.F. and Montgomery, H. (2003) The emergence of *Mycobacterium paratuberculosis* in farmed deer in New Zealand - a review of 619 cases. *New Zealand Veterinary Journal* 51(2), 58–62.

Donaghy, J.A., Totton, N.L. and Rowe, M.T. (2004) Persistence of *Mycobacterium paratuberculosis* during manufacture and ripening of cheddar cheese. *Applied and Environmental Microbiology* 70(8), 4899–4905. DOI: 10.1128/AEM.70.8.4899-4905.2004.

Donaghy, J.A., Linton, M., Patterson, M.F. and Rowe, M.T. (2007) Effect of high pressure and pasteurization on *Mycobacterium avium* ssp. *paratuberculosis* in milk. *Letters in Applied Microbiology* 45(2), 154–159. DOI: 10.1111/j.1472-765X.2007.02163.x.

Dundee, L., Grant, I.R., Ball, H.J. and Rowe, M.T. (2001) Comparative evaluation of four decontamination protocols for the isolation of *Mycobacterium avium* subsp. *paratuberculosis* from milk. *Letters in Applied Microbiology* 33(3), 173–177. DOI: 10.1046/j.1472-765x.2001.00979.x.

Faria, A.C.S., Schwarz, D.G.G., Carvalho, I.A., Rocha, B.B., De Carvalho Castro, K.N. *et al.* (2014) Short communication: viable *Mycobacterium avium* subspecies *paratuberculosis* in retail artisanal Coalho cheese from northeastern Brazil. *Journal of Dairy Science* 97(7), 4111–4114. DOI: 10.3168/jds.2013-7835.

Fawcett, A., Goddard, P., McKelvey, W., Buxton, D., Reid, H. *et al.* (1995) Johne's disease in a herd of farmed red deer. *Veterinary Record* 136(7), 165–169. DOI: 10.1136/vr.136.7.165.

Foddai, A.C.G. and Grant, I.R. (2017) Sensitive and specific detection of viable *Mycobacterium avium* subsp. *paratuberculosis* in raw milk by the peptide-mediated magnetic separation-phage assay. *Journal of Applied Microbiology* 122(5), 1357–1367. DOI: 10.1111/jam.13425.

Galiero, A., Fratini, F., Mataragka, A., Turchi, B., Nuvoloni, R. *et al.* (2016) Detection of *Mycobacterium avium* subsp. *paratuberculosis* in cheeses from small ruminants in Tuscany. *International Journal of Food Microbiology* 217, 195–199. DOI: 10.1016/j.ijfoodmicro.2015.10.029.

Gerrard, Z.E., Swift, B.M.C., Botsaris, G., Davidson, R.S., Hutchings, M.R. *et al.* (2018) Survival of *Mycobacterium avium* subspecies *paratuberculosis* in retail pasteurised milk. *Food Microbiology* 74, 57–63. DOI: 10.1016/j.fm.2018.03.004.

Gibson, H., Sinclair, L.A., Brizuela, C.M., Worton, H.L. and Protheroe, R.G. (2008) Effectiveness of selected premilking teat-cleaning regimes in reducing teat microbial load on commercial dairy farms. *Letters in Applied Microbiology* 46(3), 295–300. DOI: 10.1111/j.1472-765X.2007.02308.x.

Gilardoni, L.R., Fernández, B., Morsella, C., Mendez, L., Jar, A.M. *et al.* (2016) *Mycobacterium paratuberculosis* detection in cow's milk in Argentina by immunomagnetic separation-PCR. *Brazilian Journal of Microbiology* 47(2), 506–512. DOI: 10.1016/j.bjm.2016.01.013.

Grant, I.R., Williams, A.G., Rowe, M.T. and Muir, D.D. (2005) Efficacy of various pasteurization time-temperature conditions in combination with homogenization on inactivation of *Mycobacterium avium* subsp. *paratuberculosis* in milk. *Applied and Environmental Microbiology* 71(6), 2853–2861. DOI: 10.1128/AEM.71.6.2853-2861.2005.

Grant, I.R., Foddai, A.C.G., Tarrant, J.C., Kunkel, B., Hartmann, F.A. *et al.* (2017) Viable *Mycobacterium avium* ssp. *paratuberculosis* isolated from calf milk replacer. *Journal of Dairy Science* 100(12), 9723–9735. DOI: 10.3168/jds.2017-13154.

Hammer, P., Walte, H.G.C., Matzen, S.N., Hensel, J. and Kiesner, C. (2013) Inactivation of *Mycobacterium avium* subsp. *paratuberculosis* during cooking of hamburger patties. *Journal of Food Protection* 76(7), 1194–1201. DOI: 10.4315/0362-028X.JFP-12-474.

Hammer, P., Kiesner, C. and Walte, H.-G.C. (2014) Short communication: effect of homogenization on heat inactivation of *Mycobacterium avium* subspecies *paratuberculosis* in milk. *Journal of Dairy Science* 97(4), 2045–2048. DOI: 10.3168/jds.2013-7361.

Hanifian, S. (2014) Survival of *Mycobacterium avium* subsp. *paratuberculosis* in ultra-filtered white cheese. *Letters in Applied Microbiology* 58(5), 466–471. DOI: 10.1111/lam.12215.

Hanifian, S. and Khani, S. (2016) Tracking of *Mycobacterium avium paratuberculosis* load in milk production chain: a real-time qPCR and culture assay. *Journal of Food Safety* 36(1), 136–141. DOI: 10.1111/jfs.12243.

Hovingh, E., Whitlock, R.H., Sweeney, R.W., Fyock, T., Wolfgang, D.R. *et al.* (2006) Identification and implications of *MAP* supershedders. *Journal of Animal Science* 84(Suppl. 1), 134.

Hruska, K., Slana, I., Kralik, P. and Pavlik, I. (2011) *Mycobacterium avium* subsp. *paratuberculosis* in powdered infant milk: F57 competitive real time PCR. *Veterinární Medicína* 56(5), 226–230. DOI: 10.17221/1563-VETMED.

Khol, J.L., Wassertheurer, M., Sodoma, E., Revilla-Fernández, S., Damoser, J. *et al.* (2013) Long-term detection of *Mycobacterium avium* subspecies *paratuberculosis* in individual and bulk tank milk from a dairy herd with a low prevalence of Johne's disease. *Journal of Dairy Science* 96(6), 3517–3524. DOI: 10.3168/jds.2012-6466.

Khol, J.L., Braun, A.L., Slana, I., Kralik, P. and Wittek, T. (2017) Testing of milk replacers for *Mycobacterium avium* subsp. *paratuberculosis* by PCR and bacterial culture as a possible source for Johne's disease (paratuberculosis) in calves. *Preventive Veterinary Medicine* 144, 53–56. DOI: 10.1016/j.prevetmed.2017.05.013.

Klanicova, B., Slana, I., Roubal, P., Pavlik, I. and Kralik, P. (2012) *Mycobacterium avium* subsp. *paratuberculosis* survival during fermentation of soured milk products detected by culture and quantitative real time PCR methods. *International Journal of Food Microbiology* 157(2), 150–155. DOI: 10.1016/j.ijfoodmicro.2012.04.021.

Klanicova, B., Seda, J., Slana, I., Slany, M. and Pavlik, I. (2013) The tracing of mycobacteria in drinking water supply systems by culture, conventional, and real time PCRs. *Current Microbiology* 67(6), 725–731. DOI: 10.1007/s00284-013-0427-1.

Kopecna, M., Parmova, I., Dvorska-Bartosova, L., Moravkova, M., Babak, V. *et al.* (2008) Distribution and transmission of *Mycobacterium avium* subspecies *paratuberculosis* in farmed red deer (*Cervus elaphus*) studied by faecal culture, serology and IS*900* RFLP examinations. *Veterinarni Medicina* 53, 510–523.

Liandris, E., Gazouli, M., Taka, S., Andreadou, M., Vaiopoulou, A. *et al.* (2014) Evaluation of the microbial safety of child food of animal origin in Greece. *Journal of Food Science* 79(3), M362–M368. DOI: 10.1111/1750-3841.12366.

Lombard, J.E., Wagner, B.A., Smith, R.L., McCluskey, B.J., Harris, B.N. *et al.* (2006) Evaluation of environmental sampling and culture to determine *Mycobacterium avium* subspecies *paratuberculosis* distribution and herd infection status on US dairy operations. *Journal of Dairy Science* 89(11), 4163–4171. DOI: 10.3168/jds.S0022-0302(06)72461-4.

López-Pedemonte, T., Sevilla, I., Garrido, J.M., Aduriz, G., Guamis, B. *et al.* (2006) Inactivation of *Mycobacterium avium* subsp. *paratuberculosis* in cow's milk by means of high hydrostatic pressure at mild temperatures. *Applied and Environmental Microbiology* 72(6), 4446–4449. DOI: 10.1128/AEM.01924-05.

Lorencova, A., Vasickova, P., Makovcova, J. and Slana, I.V.A. (2014) Presence of *Mycobacterium avium* subspecies and hepatitis E virus in raw meat products. *Journal of Food Protection* 77(2), 335–338. DOI: 10.4315/0362-028X.JFP-13-252.

Mullan, W.M.A. (2019) Are we closer to understanding why viable cells of *Mycobacterium avium* subsp. *paratuberculosis* are still being reported in pasteurised milk? *International Journal of Dairy Technology* 72(3), 332–344. DOI: 10.1111/1471-0307.12617.

Nielsen, S.S., Bjerre, H. and Toft, N. (2008) Colostrum and milk as risk factors for infection with *Mycobacterium avium* subspecies *paratuberculosis* in dairy cattle. *Journal of Dairy Science* 91(12), 4610–4615. DOI: 10.3168/jds.2008-1272.

Okura, H., Nielsen, S.S. and Toft, N. (2013) Modeling the effect of direct and indirect contamination of on-farm bulk tank milk with *Mycobacterium avium* subsp. *paratuberculosis*. *Foodborne Pathogens and Disease* 10(3), 270–277. DOI: 10.1089/fpd.2012.1280.

O'Brien, L.M., McAloon, C.G., Stewart, L.D., Strain, S.A.J. and Grant, I.R. (2018) Diagnostic potential of the peptide-mediated magnetic separation (PMS)-phage assay and PMS-culture to detect *Mycobacterium avium* subsp. *paratuberculosis* in bovine milk samples. *Transboundary and Emerging Diseases* 65(3), 719–726. DOI: 10.1111/tbed.12794.

Paolicchi, F., Cirone, K., Morsella, C. and Gioffré, A. (2012) First isolation of *Mycobacterium avium* subsp *paratuberculosis* from commercial pasteurized milk in Argentina. *Brazilian Journal of Microbiology* 43(3), 1034–1037. DOI: 10.1590/S1517-83822012000300028.

Peterz, M., Butot, S., Jagadeesan, B., Bakker, D. and Donaghy, J. (2016) Thermal inactivation of *Mycobacterium avium* subsp. *paratuberculosis* in artificially contaminated milk by direct steam injection. *Applied and Environmental Microbiology* 82(9), 2800–2808. DOI: 10.1128/AEM.04042-15.

Pickup, R.W., Rhodes, G., Arnott, S., Sidi-Boumedine, K., Bull, T.J. *et al.* (2005) *Mycobacterium avium* subsp. *paratuberculosis* in the catchment area and water of the river Taff in South Wales, United Kingdom, and its potential relationship to clustering of Crohn's disease cases in the city of Cardiff. *Applied and Environmental Microbiology* 71(4), 2130–2139. DOI: 10.1128/AEM.71.4.2130-2139.2005.

Pickup, R.W., Rhodes, G., Bull, T.J., Arnott, S., Sidi-Boumedine, K. *et al.* (2006) *Mycobacterium avium* subsp. *paratuberculosis* in lake catchments, in river water abstracted for domestic use, and in effluent from domestic sewage treatment works: diverse opportunities for environmental cycling and human exposure. *Applied and Environmental Microbiology* 72(6), 4067–4077. DOI: 10.1128/AEM.02490-05.

Power, S., Haagsma, J. and Smyth, D. (1993) Paratuberculosis in farmed red deer (*Cervus elaphus*) in Ireland. *Veterinary Record* 132(9), 213–216. DOI: 10.1136/vr.132.9.213.

Raizman, E.A., Wells, S.J., Godden, S.M., Bey, R.F., Oakes, M.J. *et al.* (2004) The distribution of *Mycobacterium avium* ssp. *paratuberculosis* in the environment surrounding Minnesota dairy farms. *Journal of Dairy Science* 87(9), 2959–2966. DOI: 10.3168/jds.S0022-0302(04)73427-X.

Rani, S., Beaver, A., Schukken, Y.H. and Pradhan, A.K. (2019) Modeling the effects of infection status and hygiene practices on *Mycobacterium avium* subspecies *paratuberculosis* contamination in bulk tank milk. *Food Control* 104, 367–376. DOI: 10.1016/j.foodcont.2019.04.031.

Rhodes, G., Richardson, H., Hermon-Taylor, J., Weightman, A., Higham, A. *et al.* (2014) *Mycobacterium avium* subspecies *paratuberculosis*: Human exposure through environmental and domestic aerosols. *Pathogens* 3(3), 577–595. DOI: 10.3390/pathogens3030577.

Ricchi, M., Savi, R., Bolzoni, L., Pongolini, S., Grant, I.R. *et al.* (2016) Estimation of *Mycobacterium avium* subsp. *paratuberculosis* load in raw bulk tank milk in Emilia-Romagna Region (Italy) by qPCR. *MicrobiologyOpen* 5(4), 551–559. DOI: 10.1002/mbo3.350.

Richardson, H., Rhodes, G., Henrys, P., Sedda, L., Weightman, A.J. *et al.* (2019) Presence of *Mycobacterium avium* subspecies *paratuberculosis* monitored over varying temporal and spatial scales in river catchments: persistent routes for human exposure. *Microorganisms* 7(5), 136. DOI: 10.3390/microorganisms7050136.

Robertson, R.E., Cerf, O., Condron, R.J., Donaghy, J.A., Heggum, C. *et al.* (2017) Review of the controversy over whether or not *Mycobacterium avium* subsp. *paratuberculosis* poses a food safety risk with pasteurised dairy products. *International Dairy Journal* 73, 10–18. DOI: 10.1016/j.idairyj.2017.04.009.

Rowan, N.J., MacGregor, S.J., Anderson, J.G., Cameron, D. and Farish, O. (2001) Inactivation of *Mycobacterium paratuberculosis* by pulsed electric fields. *Applied and Environmental Microbiology* 67(6), 2833–2836. DOI: 10.1128/AEM.67.6.2833-2836.2001.

Rowe, M.T. and Grant, I.R. (2006) *Mycobacterium avium* ssp. *paratuberculosis* and its potential survival tactics. *Letters in Applied Microbiology* 42(4), 305–311. DOI: 10.1111/j.1472-765X.2006.01873.x.

Saad, N.M., El-Prince, E., Nasr, E.A. Abd El-Kader, M.G.E.-D. and El-Din, M.G. (2018) Detection of *Mycobacterium avium* subsp. *paratuberculosis* in cow's milk. *Journal of Microbiology, Biotechnology and Food Sciences* 7(6), 562–565. DOI: 10.15414/jmbfs.2018.7.6.562-565.

Salgado, D., Torres, J.A., Welti-Chanes, J. and Velazquez, G. (2011) Effect of input data variability on estimations of the equivalent constant temperature time for microbial inactivation by HTST and retort thermal processing. *Journal of Food Science* 76(6), E495–E502. DOI: 10.1111/j.1750-3841.2011.02265.x.

Sarmento, A., Sousa, T., Cabeda, J., Abreu, C. and Manso, C. (2018) *Mycobacterium avium paratuberculosis* persistence in drinking and untreated water of Porto area of Portugal – a public health issue? *Journal of Community Medicine Health Education* 8, 22.

Serraino, A., Bonilauri, P., Giacometti, F., Ricchi, M., Cammi, G. *et al.* (2017) Short communication: investigation into *Mycobacterium avium* ssp. *paratuberculosis* in pasteurized milk in Italy. *Journal of Dairy Science* 100(1), 118–123. DOI: 10.3168/jds.2016-11627.

Singh, S.V., Tiwari, A., Singh, A.V., Singh, P.K., Singh, B. *et al.* (2012) Contamination of natural resources (soil and river water) with *Mycobacterium avium* subspecies *paratuberculosis* in three districts of Uttar Pradesh: a pilot study. *Haryana Veterinarian* 51, 1–5.

Spahr, U. and Schafroth, K. (2001) Fate of *Mycobacterium avium* subsp. *paratuberculosis* in Swiss hard and semihard cheese manufactured from raw milk. *Applied and Environmental Microbiology* 67(9), 4199–4205. DOI: 10.1128/AEM.67.9.4199-4205.2001.

Stabel, J.R., Waldren, C.A. and Garry, F. (2001) Gamma-radiation effectively destroys *Mycobacterium paratuberculosis* in milk. *Journal of Dairy Science* 84(Suppl.1), 27.

Sung, N. and Collins, M.T. (2000) Effect of three factors in cheese production (pH, salt, and heat) on *Mycobacterium avium* subsp. *paratuberculosis* viability. *Applied and Environmental Microbiology* 66(4), 1334–1339. DOI: 10.1128/AEM.66.4.1334-1339.2000.

Sweeney, R.W., Whitlock, R.H. and Rosenberger, A.E. (1992) *Mycobacterium paratuberculosis* cultured from milk and supramammary lymph nodes of infected asymptomatic cows. *Journal of Clinical Microbiology* 30(1), 166–171. DOI: 10.1128/JCM.30.1.166-171.1992.

Van Brandt, L., Coudijzer, K., Herman, L., Michiels, C., Hendrickx, M. *et al.* (2011) Survival of *Mycobacterium avium* ssp. *paratuberculosis* in yoghurt and in commercial fermented milk products containing probiotic cultures. *Journal of Applied Microbiology* 110(5), 1252–1261. DOI: 10.1111/j.1365-2672.2011.04979.x.

van Schaik, G., Stehman, S.M., Schukken, Y.H., Rossiter, C.R. and Shin, S.J. (2003) Pooled fecal culture sampling for *Mycobacterium avium* subsp. *paratuberculosis* at different herd sizes and prevalence. *Journal of Veterinary Diagnostic Investigation* 15(3), 233–241. DOI: 10.1177/104063870301500304.

Vissers, M.M.M., Driehuis, F., Te Giffel, M.C., De Jong, P. and Lankveld, J.M.G. (2007) Short communication: quantification of the transmission of microorganisms to milk via dirt attached to the exterior of teats. *Journal of Dairy Science* 90(8), 3579–3582. DOI: 10.3168/jds.2006-633.

Waddell, L., Rajić, A., Stärk, K. and McEwen, S.A. (2016) *Mycobacterium avium* ssp. *paratuberculosis* detection in animals, food, water and other sources or vehicles of human exposure: a scoping review of the existing evidence. *Preventive Veterinary Medicine* 132, 32–48. DOI: 10.1016/j.prevetmed.2016.08.003.

Whan, L.B., Grant, I.R., Ball, H.J., Scott, R. and Rowe, M.T. (2001) Bactericidal effect of chlorine on *Mycobacterium paratuberculosis* in drinking water. *Letters in Applied Microbiology* 33(3), 227–231. DOI: 10.1046/j.1472-765x.2001.00987.x.

Whan, L., Grant, I.R. and Rowe, M.T. (2006) Interaction between *Mycobacterium avium* subsp. *paratuberculosis* and environmental protozoa. *BMC Microbiology* 6(1), 63. DOI: 10.1186/1471-2180-6-63.

Whittington, R.J., Marshall, D.J., Nicholls, P.J., Marsh, I.B. and Reddacliff, L.A. (2004) Survival and dormancy of *Mycobacterium avium* subsp. *paratuberculosis* in the environment. *Applied and Environmental Microbiology* 70(5), 2989–3004. DOI: 10.1128/AEM.70.5.2989-3004.2004.

Whittington, R.J., Marsh, I.B. and Reddacliff, L.A. (2005) Survival of *Mycobacterium avium* subsp. *paratuberculosis* in dam water and sediment. *Applied and Environmental Microbiology* 71(9), 5304–5308. DOI: 10.1128/AEM.71.9.5304-5308.2005.

3 Paratuberculosis and Crohn's Disease

Shannon C. Duffy* and Marcel A. Behr
McGill University, Montreal, Quebec, Canada

3.1 Introduction

The potential aetiological relationship between *Mycobacterium avium* subsp. *paratuberculosis* (*MAP*) and Crohn's disease was first proposed over 100 years ago (Dalziel, 1913). This hypothesis was stimulated by clinical and pathological similarities observed between paratuberculosis and Crohn's disease. Clinically, both conditions are characterized by intermittent diarrhoea and weight loss; pathologically, the conditions often present with disease within the ileocecal area, and the formation of ulcerations and intramural granulomas (Savarino *et al.*, 2019). Yet, several differences between the two diseases have also been noted (Sartor, 2005).

The cause of Crohn's disease remains unknown, but it is generally considered to involve a genetically determined immune dysregulation to an intestinal microbiological exposure (Abraham and Cho, 2009; Stappenbeck *et al.*, 2011; Kostic *et al.*, 2014). Of note, while a number of *MAP*-centred reviews discuss the possibility of a *MAP*–Crohn's link (Gitlin *et al.*, 2012; Liverani *et al.*, 2014; Bach, 2015), with varying degrees of conviction, most Crohn's aetiology reviews do not mention *MAP*, as this is not presently a leading hypothesis among inflammatory bowel disease (IBD) researchers. The reasons for this general lack of interest in the *MAP* hypothesis are variable but appear to include intellectual fatigue (the hypothesis has not advanced in a century), uncertainty (inconsistent results from different studies) and the limited effect of anti-mycobacterial drugs (16% absolute efficacy of anti-*MAP* treatment in a clinical trial published by Selby *et al.*, 2007). Each of these will be addressed below, after a review of the data that have accumulated since the previous edition of this book. Our objective is to provide an up-to-date summary of the latest research, which provides evidence for and against the potential relationship between *MAP* and Crohn's disease (Tables 3.1 and 3.2). In doing so, we intend to identify what remains unresolved and what further studies could ultimately answer this century-old question, one way or the other.

3.2 Clinical and Pathological Comparison of Paratuberculosis and Crohn's Disease

Crohn's disease is a chronic systemic inflammatory condition presenting primarily with gastrointestinal pathology. It is a debilitating disease, which typically leads to weight loss, abdominal pain, diarrhoea and obstruction, among others (Chiodini, 1989). The diseased tissue in both Crohn's and paratuberculosis is characterized by transmural ulceration, non-caseating granulomas, infiltration of lymphocytes and macrophages, mucosal cobblestoning and creeping mesenteric fat (Chacon *et al.*, 2004). More recent

*Corresponding author: shannon.duffy@mail.mcgill.ca

© CAB International 2020. *Paratuberculosis: Organism, Disease, Control, 2nd Edition* (eds M.A. Behr *et al.*)

Table 3.1. Pros. Arguments in favour of the potential aetiological role of *Mycobacterium avium* subsp. *paratuberculosis* (*MAP*) in Crohn's disease. Based on Sartor, 2005.

Arguments	Responses from emerging data	References
1. Clinical and pathological similarities of paratuberculosis and Crohn's disease	Confirmed	Zarei-Kordshouli *et al.*, 2019
2. Presence in food chain (milk, meat) and water supplies	Confirmed	Beumer *et al.*, 2010; Carvalho *et al.*, 2012; Rhodes *et al.*, 2014; Espeschit *et al.*, 2018; Gerrard *et al.*, 2018; Lorencova *et al.*, 2019
3. Increased detection of *MAP* in Crohn's disease tissues by culture, PCR, FISH	Confirmed	Feller *et al.*, 2007; Abubakar *et al.*, 2008; Waddell *et al.*, 2015
4. Positive blood cultures of *MAP* in Crohn's disease patients	Unresolved	Parrish *et al.*, 2009; Mendoza *et al.*, 2010
5. Increased serological responses to *MAP* in Crohn's disease patients	Confirmed	Verdier *et al.*, 2013; Xia *et al.*, 2014; Zamani *et al.*, 2017
6. Therapeutic responses to combination anti-tuberculosis therapy that include macrolide antibiotics	Confirmed	Feller *et al.*, 2010; Khan *et al.*, 2011

FISH, fluorescence in situ hybridization; MAP, *Mycobacterium avium* subsp. *paratuberculosis*; PCR, polymerase chain reaction.

comparative pathology studies have emphasized that both disease tissues display relatively the same frequency of granulomatous enteritis, lymphoplasmacytic enteritis, oedema and lymphangiesctasia (Zarei-Kordshouli *et al.*, 2019).

However, differences have been observed between these two conditions. A comparative pathological study between 29 Crohn's disease, 17 ulcerative colitis and 35 bovine paratuberculosis samples described Crohn's disease lesions as more destructive than paratuberculosis, as defined by the degree of damage to the epithelium, ulcer and fissure formation, among others (Momotani *et al.*, 2012). Conversely, a recent study found that vasculitis was observed in nearly all paratuberculosis tissues and only a small percentage of Crohn's disease tissues. The most prominent pathological difference between classical paratuberculosis and Crohn's disease is the ability to readily visualize acid-fast bacteria in the diseased tissue from cattle (Zarei-Kordshouli *et al.*, 2019). The detection of *MAP* in tissues of humans with Crohn's disease has proven inconsistent or unsuccessful, a fact that will be discussed in more detail in a future section.

Together, these observations demonstrate that the rationale implicating *MAP* in Crohn's disease is backed by the many similarities with paratuberculosis. However, it is noted that granulomatous pathology is not pathognomonic for a single aetiology; other granulomatous conditions range from a wood splinter to berylliosis and include bacterial infections (e.g. leprosy) and fungal infections (e.g. histoplasmosis). Furthermore, while differences in pathological presentation between paratuberculosis and Crohn's disease might argue against a common aetiological agent, Crohn's disease is generally accepted to develop due to a combination of genetic, immune and environmental factors. Given the differences in genetics, immunology and environmental exposures between humans and cattle, the final clinical and pathological outcomes of a similar microbial exposure need not be identical.

3.3 Epidemiological Evidence Investigating Transmissible Infection

An extensive amount of recent research has supported the presence of *MAP* in several potential routes of human exposure. Both viable *MAP* and *MAP* DNA have been detected in milk (Boulais *et al.*, 2011; Okura *et al.*, 2012)

Table 3.2. Cons. Arguments against the potential aetiological role of *Mycobacterium avium* subsp. *paratuberculosis* (*MAP*) in Crohn's disease. Based on Sartor, 2005.

Arguments	Responses from emerging data	References
1. Difference in clinical and pathological responses in paratuberculosis and Crohn's diseases	Confirmed	Momotani *et al.*, 2012; Zarei-Kordshouli *et al.*, 2019
2. Lack of epidemiological support of transmissible infection	Unresolved	Pierce, 2009; Pierce *et al.*, 2011; Pistone *et al.*, 2012; Opstelten *et al.*, 2016
3. No evidence of transmission to humans in contact with animals infected with *MAP*	Unresolved	Qual *et al.*, 2010; Singh *et al.*, 2011a
4. Genotypes of Crohn's disease and bovine *MAP* isolates not similar	Refuted	Paustian *et al.*, 2008; Singh *et al.*, 2009; Bannantine *et al.*, 2014; Wynne *et al.*, 2011; Timms *et al.*, 2015
5. Variability in detection of *MAP* by PCR (0–100% in Crohn's disease and ulcerative colitis tissues) and serological testing	Confirmed	Waddell *et al.*, 2015
6. No evidence of mycobacterial cell wall by histochemical staining	Unresolved	Jeyanathan *et al.*, 2007; Banche *et al.*, 2015; Zarei-Kordshouli *et al.*, 2019
7. No worsening of Crohn's disease with immunosuppressive agents or HIV infection	Unresolved	Allen *et al.*, 2011
8. No documented cell mediated immune responses to *MAP* in patients with Crohn's disease	Refuted	Sibartie *et al.*, 2010; Olsen *et al.*, 2009, 2013
9. No therapeutic response to traditional antimycobacterial antibiotics	Unresolved	Selby *et al.*, 2007; Behr and Hanley, 2008; RedHill Biopharma, 2018a, b

MAP, *Mycobacterium avium* subsp. *paratuberculosis*; PCR, polymerase chain reaction.

including in commercially available milk (Carvalho *et al.*, 2012; Gerrard *et al.*, 2018) revealing the ability of this pathogen to withstand pasteurization. *MAP* has also been detected in meat (Lorencova *et al.*, 2019), cheese (Botsaris *et al.*, 2010; Faria *et al.*, 2014), infant formula (Botsaris *et al.*, 2016; Acharya *et al.*, 2017), water sources (Beumer *et al.*, 2010; Rhodes *et al.*, 2013; Espeschit *et al.*, 2018) and aerosols (Rhodes *et al.*, 2014). From these data, it can be agreed upon that there are numerous opportunities for humans to be exposed to *MAP*. Yet the ability to detect *MAP* in the environment does not provide direct evidence of transmission of bacteria into humans, nor does it demonstrate that the burden in these sources exceeds the unknown infectious dose for humans.

A 2009 report described three individuals who developed Crohn's disease that lived in close proximity to one another but otherwise had no recorded contact. The author hypothesized that exposure may have occurred due to a shared water system (Pierce, 2009). Another study published by the same author described seven unrelated children who developed Crohn's disease after moving to Forest, Virginia, where the number of cases of IBD was 47 times more than expected. Of these seven individuals, five (71.4%) were positive for *MAP*-specific antibodies (anti-p35 and anti-p36) (Pierce *et al.*, 2011). *MAP* infection was suggested to have occurred due to contaminated water from nearby dairy farms. Although *MAP* has been reported in drinking water, neither of these studies investigated whether *MAP* was present in the water system of the patients' homes. In contrast to these results, a questionnaire-based assessment of the risk of IBD development and milk consumption in over 400,000 participants found that those consuming more milk were at a significantly reduced risk for development of Crohn's disease (Opstelten *et al.*, 2016). Again however, the presence of *MAP* was not evaluated in this study. One study, which did measure *MAP* presence in both patients and the potential source of exposure, was conducted in a region of Italy where *MAP* is endemic in cattle. The authors found *MAP* DNA in 23/35 (65.7%) biopsies from patients with Crohn's disease and in the circulating tap water (Pistone *et al.*, 2012). It is important to note that *MAP* was detected in 7/35 (20%) controls in this study. Although the detection of *MAP* was significantly greater in patients with Crohn's disease, these findings reinforce that exposure is not sufficient to cause disease.

An argument raised in regard to the epidemiology of the disease is that there is no evidence of an increased risk to developing Crohn's disease associated with persons in contact with *MAP*-infected animals such as farmers or veterinarians (Jones *et al.*, 2006). Since the previous edition, there have been few published studies that have addressed this argument. A 2010 cross-sectional survey of 1476 cattle farmers and veterinarians found no relationship between Crohn's disease and *MAP* exposure (Qual *et al.*, 2010), although there were only seven cases of Crohn's disease recorded in total. In contrast, a study conducted on 58 animal attendants with reported gastrointestinal problems cultured more *MAP* compared with controls (Singh *et al.*, 2011a). These data suggest a role of additional confounding factors such as location or the type of livestock.

Additional research is required in order to gain a clearer picture of the epidemiology surrounding the link between *MAP* and Crohn's disease. But there are several circumstances concerning these types of studies that should be considered. First, research aimed at associating exposure to *MAP* (today) with presentation of disease (later) is impeded by the unknown incubation period; in cattle bacteria typically persist undetected in a subclinical state for 2–5 years before disease manifests (United States Department of Agriculture, 2008). Therefore, studies starting with ascertained *MAP* exposure may fail to link cases, which present years later; conversely, studies starting with Crohn's disease cases are unable to do microbial analysis of the environment to which one was exposed in the preceding years. Second, the idea that contact with infected animals must lead to an increased risk in development of the disease is not found with other bacteria whose aetiological role is accepted. For example, cattle are the known reservoirs for *E. coli* O157:H7, however no increased risk was associated with disease development in those in direct contact with infected cattle (Rangel *et al.*, 2005). Third, Crohn's disease development is thought to involve a combination of factors including genetic susceptibility and immune dysregulation. If *MAP* only plays a pathogenic role in such susceptible hosts, the vast majority of those exposed would not develop disease. Future research aimed at rigorously evaluating this argument must be long-term and evaluate both bacteriological exposure and host predisposition.

3.4 Detection of *MAP* in Crohn's Disease Patients

Earlier systematic reviews have concluded that *MAP* detection is significantly enriched in Crohn's disease patients (Feller *et al.*, 2007; Abubakar *et al.*, 2008; Waddell *et al.*, 2008). An updated systematic review conducted by Waddell *et al.* (2015) also found a significant association across several methods, but they noted heterogeneity between studies detecting *MAP* by PCR vs hybridization. It is important to note that the primary data compiled in such systematic reviews may be subject to publication bias. As it is more difficult to publish a negative result, investigators in the field need to be reminded that negative data remain critical in the future to updating the collective literature on *MAP* and Crohn's disease.

Isolation in pure culture from diseased tissue is classically the gold standard for demonstrating the presence of a bacterium and serves as the first step in determining if the bacterium is associated with a particular disease (if found to be more common in diseased than in control subjects). However, this method comes with a set of challenges, some inherent to *MAP* microbiology and some inherent to research involving human subjects. *MAP* is among the slowest-growing culturable mycobacteria, and proper decontamination is difficult to achieve in tissue samples containing many other competing bacteria. Yet, other chapters of this book provide details on how to isolate this same slow-growing bacterium from manure, demonstrating technical feasibility. Complicating factors in human studies include the nature of the sample and the status of the patient. In Crohn's patients who are clinically stable, one can only obtain mucosal biopsies or stool samples. If the organisms reside beneath the mucosa (e.g. the muscularis, the mesenteric lymph nodes), then the chance of detecting them is greatly reduced. In most Crohn's patients who have progressed to the point of requiring a surgical excision, there has been a trial of pre-operative antibiotics to avoid surgery (e.g. metronidazole and ciprofloxacin) and there are usually peri-operative antibiotics given to reduce the risk of a post-operative infection (Weiser *et al.*, 2019). The effect of these *in vivo* antibiotic treatments on the ability to isolate *MAP* in culture from the sample is unknown.

Overall, recent reports have detected *MAP* by culture in a small percentage of Crohn's disease patients. One study cultured *MAP* from 2/18 (11.1%) Crohn's patients, although none could be cultured from ulcerative colitis or non-IBD controls (Zamani *et al.*, 2017). Banche *et al.* (2015) cultured *MAP* from 13/76 (17.1%) samples from patients with Crohn's disease compared with 2/44 (4.5%) controls. Another study did not culture *MAP* from Crohn's disease patients, however only five patients were included in the study (Carvalho *et al.*, 2016). In a larger study including 75 individuals recently diagnosed with Crohn's disease, two (2.7%) were *MAP* positive by culture compared with 2/135 (1.5%) controls (Ricanek *et al.*, 2010).

Due to the difficulties with *MAP* culture, most studies combine culture with molecular methods such as IS*900* PCR to detect *MAP*. However, the presence of bacterial DNA does not equate with actively growing bacteria, as exemplified by the use of ancient DNA in anthropology. The majority of molecular method-based studies have reported a higher frequency of *MAP* in Crohn's disease patients, but they are also subject to considerable heterogeneity. Indeed, many recent studies have reported a significant association with Crohn's disease (Singh *et al.*, 2011b; Banche *et al.*, 2015; Khan *et al.*, 2016; Zamani *et al.*, 2017); however, PCR detection rates continued to range from 0–100% across studies (Parrish *et al.*, 2009; Sasikala *et al.*, 2009; Mendoza *et al.*, 2010). A 2015 systematic review indicated that a significant amount of heterogeneity could be explained by choice of primers and the sampling frame (Waddell *et al.*, 2015). Even within the rubric of PCR-based detection of IS*900*, there are technical variants, such as the nested IS*900* PCR assay. Several recent studies have applied this method, nearly all of which reported a positive association (Lee *et al.*, 2011; Molicotti *et al.*, 2013; Nazareth *et al.*, 2015a; Timms *et al.*, 2016; Sharp *et al.*, 2018; Zarei-Kordshouli *et al.*, 2019). However, one study did not detect *MAP* in 75 recently diagnosed Crohn's disease patients using nested IS*900* PCR (Ricanek *et al.*, 2010).

Another method used to detect *MAP* in tissue is staining. Staining for acid-fast bacilli and looking under oil-immersion microscopy was described by Jeyanathan *et al.* (2007), who applied methods previously validated on tissue from mice experimentally infected with *MAP* (positive control) and *M. tuberculosis* (negative control) (Jeyanathan *et al.*, 2006). Banche *et al.* (2015) reported non-significant differences in detection where 3/76 (3.95%) Crohn's disease samples were positive by acid-fast staining compared with 0/44 (0%) non-IBD controls. Zarei-Kordshouli *et al.* (2019) detected acid-fast bacteria in 2/30 (7%) Crohn's disease subjects compared with 0/30 (0%) non-IBD controls and 27/30 (90%) caprine paratuberculosis samples. Studies using immunohistochemistry also failed to detect differences between samples from Crohn's disease and controls (Magin *et al.*, 2013; Khan *et al.*, 2016). However, *MAP* infection in humans is often observed to be paucibacillary, which can be more difficult to visualize using acid-fast staining. In addition, some studies have suggested that this lack of visualization is due to *MAP* entering a cell wall-deficient spheroplast form in human tissue, which will not be detected by acid-fast staining (Chiodini *et al.*, 1986). Given that cell wall-deficient forms of *MAP* have not to our knowledge been described in the natural bovine and ovine hosts, this would be a unique situation in clinical microbiology where a zoonotic agent undergoes a morphologic transition during the human spillover infection. An alternative possibility is that *MAP* does undergo a transformation into a cell wall-deficient form in its natural hosts, but this has not been reported to date.

There is some evidence to suggest intra-study variability in results due to changes in *MAP* status in individuals. In 2009, Kirkwood *et al.* reported significant *MAP* association in biopsies and peripheral blood mononuclear cells (PBMCs) of children with Crohn's disease by IS*900* PCR and culture compared with non-IBD controls (Kirkwood *et al.*, 2009). The children participating in this study were then followed up for up to 4 years. *MAP* was found to persist only within a subset of patients and the authors suggested that infection status at disease onset may be an important factor in predicting infection persistence (Wagner *et al.*, 2011). Overall, the recent data suggest that, although a positive association between *MAP* and Crohn's disease has been consistently reported by systematic reviews, heterogeneity between and within studies continues to be reported. Future studies applying multiple externally validated and standardized methods may allow for more robust cross-study comparison.

3.5 Comparative Genotyping of Human and Bovine-Derived *MAP* Isolates

An argument previously raised against the potential link between *MAP* and Crohn's disease was the perception that the genotypes of human-derived *MAP* isolates are significantly different from those of bovine origin (Sartor, 2005). This argument originated from research using multiplex PCR of IS*900* integration loci (MPIL) fingerprinting, which found that bovine isolates clustered together, whereas human and ovine isolates were more diverse (Motiwala *et al.*, 2003). Many research studies followed reporting similarity between *MAP* cattle and human isolates, using several different comparative genomics methods: DNA microarrays, IS*1311* PCR restriction enzyme analysis, short-sequence repeat genotyping and, most recently, whole genome sequencing (Paustian *et al.*, 2008; Singh *et al.*, 2009; Wynne *et al.*, 2011; Bannantine *et al.*, 2014; Timms *et al.*, 2015). The latter, most definitive method, has repeatedly shown that human *MAP* isolates are similar to bovine isolates, in different countries. Bannantine *et al.* (2014) reported the whole genome sequence of *MAP* isolated from the breast milk of a person with Crohn's disease and determined the isolate was highly similar to those of bovine origin. Another study that sequenced the genome of a human *MAP* isolate 43,545 also found that it was most closely related to bovine type (Timms *et al.*, 2015). Wynne *et al.* (2011) performed comparative genomics on seven human *MAP* isolates (four from Crohn's disease patients), two bovine isolates and one ovine isolate. Sequencing again revealed that all human-derived isolates were similar to one another and closely related to a bovine isolate.

When considering the presented evidence, it can be concluded that most human *MAP* isolates studied have the genotype of bovine *MAP*. While some variation has been described between human isolates (Wynne *et al.*, 2014), it is noted that single nucleotide polymorphism (SNP) genotyping of *MAP* isolates from 26 dairy herds has revealed heterogeneity, both within and between herds (Amin *et al.*, 2015). Unless *MAP* is proposed to have 'jumped' once into humans, and that clone subsequently spread uniquely in humans, there is little reason to anticipate that there should be a human-specific genotype. Additionally, if a *MAP* lineage associated with sheep was isolated from a human sample, this would not invalidate the zoonotic potential of *MAP*; rather it would point to a different potential source.

3.6 *MAP* Immune Responses in Crohn's Disease Patients

Immune responses that might be detectable in Crohn's disease include humoral and cell-mediated responses. In serological studies, Verdier *et al.* (2013) found significantly increased IgG against four *MAP* antigens (glycosyl-transferase-d, Johnin-purified protein derivative (PPD), heparin-binding haemagglutinin, L5P) in gut lavage fluids. Another study also found a positive association in antibody titres against two *MAP* antigens ($MAP3865C_{133\text{-}141}$ and $MAP3865_{125\text{-}133}$) (Zamani *et al.*, 2017). Lefrançois *et al.* (2011) described a *MAP* laminin-binding/histone-like protein (Lbp/Hlp) that is bound by serum antibodies of patients with Crohn's disease. In addition, Xia *et al.* (2014) found that anti-PtpA antibodies were significantly increased in Crohn's disease patients compared with healthy controls. However, many of these antigens are also found in other mycobacterial species. Therefore, seropositivity may not indicate reactivity to *MAP* but rather cross-reactivity due to previous exposure with another mycobacteria. These studies are also limited by the fact that detection of *MAP*-reactive antibodies indicates prior exposure to the bacteria but does not specify ongoing infection. In animals with paratuberculosis, detection of *MAP* antibodies is often a lagging indicator, where culture positivity precedes seropositivity (Li *et al.*, 2017). When considered with the above culture data, it does raise the possibility of some Crohn's disease occurring as a post-infectious pathology, given that the prevalence of antibodies to *MAP* is generally much greater than the prevalence of a live *MAP* culture. To our knowledge, only one study has looked at serology and culture in the same matched subjects, so this hypothesis is difficult to explore from the published literature. Zamani *et al.* (2017) found that 13/28 (46%) and 11/28 (39%) of Crohn's disease patients were seropositive for MAP3865c$_{133\text{-}141}$ and MAP3865c$_{125\text{-}133}$ peptides, respectively. *MAP* was cultured from two of these patients.

Crohn's disease lesions typically have an abundance of CD4+ T cells that will produce inflammatory cytokines including IL-17 and IFN-γ, which are characteristic of a Th1/Th17 immune response. In contrast to the expanding literature on a humoral immune response to *MAP* in Crohn's disease, there is a relative paucity of evidence for *MAP*-reactive T cells present in patients with Crohn's disease. Sibartie *et al.* (2010) found that PBMCs co-incubated with *MAP* produced significantly more T cells in patients with Crohn's disease compared with ulcerative colitis or controls. Similarly, T-cell lines were shown to proliferate more in response to *MAP* when isolated from Crohn's disease patients compared with ulcerative colitis. These T cells produced both IL-17 and IFN-γ, supporting the presence of a Th1/Th17 immune response (Olsen *et al.*, 2009). In intestinal biopsies of Crohn's disease patients, the frequency of T cells reactive to mycobacteria was found to be significantly higher than those reactive against *Escherichia coli*. These T-cell lines were also shown to produce both IL-17 and IFN-γ (Olsen *et al.*, 2013). In contrast however, Nakase *et al.* (2011) compared the cytokine production of monocytes from Crohn's disease patients in response to *MAP* and *M. avium*. Although significantly more tumour necrosis factor α (TNF-α) was produced by monocytes infected with *MAP*, both IL-12p40 and IL-6 were not produced, arguing against the induction of a Th1/Th17 response.

A key component of the rationale linking *MAP* to Crohn's disease concerns the gene mutations associated with an increased risk of disease. Many of these genes are involved in innate immune recognition of intracellular bacteria, suggesting that those who develop Crohn's disease are at an increased risk of certain types of infection. Loss of function mutations in *NOD2*, an intracellular pathogen recognition receptor, are significantly associated with Crohn's disease (Ogura *et al.*, 2001; Hugot *et al.*, 2001). The NOD2 receptor will bind to bacterial muramyl dipeptide (MDP). Mycobacteria are known to produce a glycolylated form of MDP, which is a more potent and efficacious activator of NOD2 (Coulombe *et al.*, 2009; Vinh and Behr, 2013). Other genes include *IRGM1*, which encodes a GTPase involved in resistance to intracellular pathogens (Singh *et al.*, 2006; Parkes *et al.*, 2007) including mycobacteria (MacMicking *et al.*, 2003; Feng *et al.*, 2004), and *ATG16L1*, which is involved in induction of autophagy and the formation of the autophagosome (Cooney *et al.*, 2010; Travassos *et al.*, 2010). Taken together, one interpretation of the genomic data suggests that Crohn's disease occurs in individuals with a genetically determined susceptibility to infection with an intracellular bacterium that produces glycolyl-MDP, such as *MAP*. However, several studies aimed at determining the relationship between *MAP* infection status and Crohn's disease susceptibility genes including *NOD2* have not found a positive association between them (Bernstein *et al.*, 2007; Mendoza *et al.*, 2010). An alternative interpretation of the data is that the genetic predisposition to Crohn's disease overlaps not with the risk of developing a persistent infection, but rather with the risk of developing an inflammatory reaction to that intracellular pathogen. Research conducted by Fava *et al.* (2016) has shown that the predisposition to leprae reactions, which are inflammatory reactions during or after treatment of leprosy, have a remarkably similar genetic profile to Crohn's disease. Supporting this paradigm, Wagner *et al.* (2013) found that susceptibility gene mutations in *TLR4*, which promotes proinflammatory responses in the gut (Lu *et al.*, 2018), and *IL-10Ra*, the receptor for an immunoregulatory cytokine (Nunberg *et al.*, 2018), were significantly associated with *MAP*-positive Crohn's disease patients.

Overall, recent research has provided more evidence of antibodies to *MAP* antigens in Crohn's disease along with more limited evidence of a potential induction of a Th1/Th17 immune response against *MAP* infection in Crohn's disease patients. More data are required concerning host immune responses to *MAP* in order to gain a clearer picture of how *MAP* could contribute to the development of Crohn's disease.

3.7 The Effect of Immunosuppressive Therapy on *MAP* Infection

Anti-inflammatory agents such as anti-TNF-α and corticosteroids are common effective treatments for Crohn's disease. These treatments are also recognized to have immunosuppressive effects. In the case of tuberculosis, it is known that these treatments are associated with risk of disseminated disease (Gitlin *et al.*, 2012). Indeed, Crohn's disease patients treated with TNF-α inhibitors are at an increased risk of tuberculosis when compared with untreated individuals (Cao *et al.*, 2018). The lack of reported disseminated *MAP* infection following immunosuppressive therapy has served as a strong argument against its potential link with Crohn's disease. However, there is evidence to suggest that *MAP* responses to immunosuppressive therapy may not follow the established paradigm of other mycobacteria. Research using calf infection models have suggested that prior depletion of CD4 cells has no effect on *MAP* disease progression (Allen *et al.*, 2011); this stands in stark contrast to the expectation from the human literature on AIDS-associated CD4 depletion and *M. avium* disease. An additional observation is that *M. leprae*, the cause of leprosy, has not been reported to cause disseminated disease following anti-TNF-α and steroid treatments (Athreya, 2007); indeed, anti-TNF-α has been used as an adjunctive therapy to manage hyperinflammatory phenotypes of leprosy, albeit while also receiving antimicrobial therapy. In this section we will review the recent research which has aimed to understand anti-TNF-α in *MAP* infection.

Several studies have suggested that Crohn's disease immunosuppressive therapies lead to a measurable decrease in *MAP* infection status. Bach *et al.* (2011) demonstrated that subjects with Crohn's disease had significantly higher titres of antibodies against the *MAP* proteins PtpA and PknG when compared with controls and that these antibodies were decreased in those who were treated with infliximab (Bach *et al.*, 2012). Similarly, Xia *et al.* (2014) found that anti-PtpA antibodies were significantly decreased in Crohn's disease patients treated with the immunosuppressant azathioprine, but not 5-aminosalycilic acid, steroids or their combination with azathioprine. In a study involving experimental infection of macrophages with *MAP*, macrophages from IBD patients treated with infliximab retained significantly lower *MAP* colony-forming units when compared with those without treatment or healthy controls. The authors suggested that this finding could be a result of an increased induction of *MAP* dormant forms, which may not proliferate well despite immunosuppression (Nazareth *et al.*, 2015b), an explanation that warrants further investigation.

In order to understand the effect of TNF-α inhibitors, more research is required to delineate the role of TNF-α in host responses to *MAP*. Recent studies have suggested that *MAP* infection modulates TNF-α production to support its survival in the host, potentially explaining how anti-TNF-α treatments may negatively impact *MAP*. The gut culture supernatant TNF-α levels are increased in Crohn's disease patients positive for *MAP* when compared with *MAP*-negative patients (Clancy *et al.*, 2007). Human macrophages have been shown to secrete increased levels of TNF-α when infected with *MAP*, but not when infected with other mycobacteria such as *M. avium* (Nakase *et al.*, 2011). Additionally, two SNPs in the TNF-α receptor, which were previously associated with undesired outcomes of anti-TNF-α therapy, have been associated with increased susceptibility to infection with *MAP* (Qasem *et al.*, 2019). In opposition to this argument however, Campos *et al.* (2011) found that macrophages from Crohn's disease patients had a defective TNF-α response to *MAP* compared with healthy controls, although this effect was not specific to bacterial challenge with *MAP*.

Taken together, recent data suggest that a lack of observed *MAP* dissemination following immunosuppressive treatments neither proves nor disproves a potential role of *MAP* in Crohn's disease. More research is required to elucidate the role of TNF-α in *MAP* pathogenesis and how this may further our understanding of patient responses to anti-TNF-α therapy.

3.8 The Use of Anti-Mycobacterial Therapy for Treatment of Crohn's Disease

Another argument against the role of *MAP* in Crohn's disease is that anti-mycobacterial therapies are ineffective at treating patients with Crohn's disease. This argument primarily stems from the results of a double-blind randomized control trial conducted in 2007, where 213 Crohn's disease patients were treated with anti-mycobacterial drugs or placebo (Selby *et al.*, 2007). The authors reported an absolute benefit of 16% at week 16, but no evidence of a sustained benefit through the remainder of the 2-year study. However, an intention-to-treat reanalysis of the data later revealed that the benefit observed at 16 weeks was maintained at weeks 52 and 104 (Behr and Hanley, 2008). Notwithstanding the duration of the observed benefit, a major constraint of this trial was the lack of evaluation of *MAP* status prior to inclusion in the study. The study evaluated whether patients got better, which is a clinically valid outcome, but could not provide insights into a pathogenic role of *MAP*.

The varying interpretations of these data has left the potential efficacy of antimycobacterial therapy unresolved. A systematic review conducted by Feller *et al.* (2010) evaluated the outcomes of Crohn's disease remission and relapse for patients treated with antibiotics vs placebo. This study found a significant benefit in studies treating patients with clofazimine, an antibiotic known to be active against mycobacteria, but no benefit was observed in studies involving classic anti-tuberculosis drugs. A second systematic review conducted in 2011 evaluated the effect of antibiotic therapy in IBD treatment using only randomized control trials. The authors found a statistically significant effect of antibiotics over placebo in treating both active and quiescent Crohn's disease (Khan *et al.*, 2011). It is important to note however, that these systematic reviews included a variety of antibiotics, including those active and those inactive against mycobacteria.

In 2013, RedHill Biopharma began a phase III double-blind placebo-controlled randomized control trial to assess the effect of RHB-104 vs placebo for treatment of moderate to severe active Crohn's disease. An important difference from the 2007 study is that the changes in *MAP* blood status would be evaluated by PCR, although *MAP* positivity was not a requirement for inclusion in the trial (clinicaltrials.gov NCT01951326). RHB-104 is a formulation of rifabutin, clofazimine, and clarithromycin shown to be effective against *MAP* strains isolated from Crohn's disease patients (Alcedo *et al.*, 2016; Qasem *et al.*, 2016). RedHill Biopharma has announced that the treatment group was superior to placebo in achieving remission at week 26 (37% vs 23%) with statistically greater responses. They also reported those receiving RHB-104 experienced a significant benefit in attaining an earlier remission and in maintaining remission to week 52 (RedHill Biopharma, 2018a). Additionally, the company stated that RHB-104 was shown to have a greater calculated maintenance of remission over other standard-of-care therapies for Crohn's disease such as Infliximab (RedHill Biopharma, 2018b). At the time of writing, the RedHill study remains unpublished.

Overall, these results indicate a statistically significant, although clinically modest benefit of antimycobacterial therapy in the treatment of Crohn's disease. More data are required to determine whether this clinical benefit supports a direct involvement of *MAP* in development of Crohn's disease.

3.9 Conclusions and Directions Forward

Although there are many recognizable similarities between paratuberculosis and Crohn's disease, several inconsistencies exist. There have been ~200 papers published on the potential aetiological role of *MAP* in the development of Crohn's disease since the first edition of this chapter. Yet, despite this new information, the hypothesized relationship has yet to be proven or invalidated. At the time of the previous edition, there were several outstanding arguments that argued against a link. Although one argument has been largely refuted by recent data – the differences in the genotypes of human and bovine *MAP* isolates – the remaining arguments have yet to be resolved. The intent of this chapter is to emphasize these pending areas of uncertainty and highlight what types of future research studies would be most beneficial to ultimately resolve this debate.

Epidemiological research is limited by the many challenges involved in associating *MAP* exposure to disease outcomes, including its slow growth rate and the multifactorial nature of Crohn's disease. Although clusters of Crohn's disease have been described in the population, how these relate to *MAP* is uncertain. Future studies would benefit from implementing bacterial genotyping in order to establish a transmissible infection from the suggested source of exposure. Additionally, the majority of research investigating a relationship between occupational exposure and disease development have not found an association with direct contact with livestock. However, the notion that occupational risk should be considered a defining point for establishing an aetiological relationship is not consistently found in other types of intestinal infection. Therefore, whether this argument should play a significant role in evaluating this theory going forward should be considered. Studies seeking to detect *MAP* in humans have demonstrated a positive association between *MAP* and Crohn's disease as described by multiple systematic reviews. However, the interpretation of the results across multiple studies is largely impeded by small sample sizes and the use of different, non-standardized detection methods. To rigorously evaluate an association between *MAP* and Crohn's disease, future studies must follow a recognized standard method in order to ease cross-comparison analysis. In addition, the discrepancy observed between acid-fast staining detection of *MAP* in paratuberculosis and Crohn's disease could theoretically be explained by the formation of a cell wall-deficient form in humans, however what role (if any) this form plays in paratuberculosis must be investigated if the theory of a spheroplast infection in humans is to be supported.

Key knowledge gaps exist in our understanding of the immunology of *MAP* infection in humans, which limits our ability to determine its involvement in the development of Crohn's disease. Recent research has broadly established the potential for a *MAP*-induced Th1/Th17 immune response in humans. Future studies should aim to delineate the role of TNF-α as a host defence mechanism or as a response initiated by *MAP* to modulate the host immunity to promote its own infection. This may provide evidence to explain the counterintuitive host responses to Crohn's disease immunosuppressive therapies if *MAP* is implicated in disease aetiology. Additionally, many of the genes positively associated with disease are involved in innate immune responses to intracellular bacterial infection such as with *MAP*. However, no relationship with *MAP*-positive status has been established with a number of susceptibility genes. As research in this field has been primarily focused on searching for *MAP*, we may be overlooking other potential candidates, which involve similar host immune responses.

In addition to the need for further bench-level research, incoming data from the recently concluded RHB-104 phase III trial will be a key piece of information in determining how the elimination of *MAP* infection affects outcomes of Crohn's disease. Both this study and a previous report have shown a 15–20% absolute benefit of antibiotic therapy in Crohn's disease treatment. However, the direct measurement of *MAP* and its relationship with disease severity has not yet been evaluated. If reversal of *MAP* infection status is shown to have a significant impact on the benefit observed from RHB-104, this would provide strong support of a pathogenic role of *MAP* in development of Crohn's disease.

References

Abraham, C. and Cho, J.H. (2009) Inflammatory bowel disease. *New England Journal of Medicine* 361(21), 2066–2078. DOI: 10.1056/NEJMra0804647.

Abubakar, I., Myhill, D., Aliyu, S.H. and Hunter, P.R. (2008) Detection of *Mycobacterium avium* subspecies *paratuberculosis* from patients with Crohn's disease using nucleic acid-based techniques: a systematic review and meta-analysis. *Inflammatory Bowel Diseases* 14(3), 401–410. DOI: 10.1002/ibd.20276.

Acharya, K.R., Dhand, N.K., Whittington, R.J. and Plain, K.M. (2017) Detection of *Mycobacterium avium* subspecies *paratuberculosis* in powdered infant formula using IS900 quantitative PCR and liquid culture media. *International Journal of Food Microbiology* 257, 1–9. DOI: 10.1016/j.ijfoodmicro.2017.06.005.

Alcedo, K.P., Thanigachalam, S. and Naser, S.A. (2016) RHB-104 triple antibiotics combination in culture is bactericidal and should be effective for treatment of Crohn's disease associated with *Mycobacterium paratuberculosis*. *Gut Pathogens* 8(1), 32. DOI: 10.1186/s13099-016-0115-3.

Allen, A.J., Park, K.-T., Barrington, G.M., Lahmers, K.K., Abdellrazeq, G.S. *et al.* (2011) Experimental infection of a bovine model with human isolates of *Mycobacterium avium* subsp. *paratuberculosis*. *Veterinary Immunology and Immunopathology* 141(3-4), 258–266. DOI: 10.1016/j.vetimm.2011.03.014.

Amin, A.S., Hsu, C.-Y., Darwish, S.F., Ghosh, P., AbdEl-Fatah, E.M. *et al.* (2015) Ecology and genomic features of infection with *Mycobacterium avium* subspecies *paratuberculosis* in Egypt. *Microbiology* 161(Pt 4), 807–818. DOI: 10.1099/mic.0.000051.

Athreya, S.P.K. (2007) Azathioprine in controlling type 2 reactions in leprosy: a case report. *Leprosy Review* 78(3), 290–292.

Bach, H. (2015) What role does *Mycobacterium avium* subsp. *paratuberculosis* play in Crohn's disease? *Current Infectious Disease Reports* 17(2), 463. DOI: 10.1007/s11908-015-0463-z.

Bach, H., Ko, H.H., Raizman, E.A., Attarian, R., Cho, B. *et al.* (2011) Immunogenicity of *Mycobacterium avium* subsp. *paratuberculosis* proteins in Crohn's disease patients. *Scandinavian Journal of Gastroenterology* 46(1), 30–39. DOI: 10.3109/00365521.2010.513061.

Bach, H., Rosenfeld, G. and Bressler, B. (2012) Treatment of Crohn's disease patients with infliximab is detrimental for the survival of *Mycobacterium avium* ssp. *paratuberculosis* within macrophages and shows a remarkable decrease in the immunogenicity of mycobacterial proteins. *Journal of Crohn's and Colitis* 6(5), 628–629. DOI: 10.1016/j.crohns.2012.01.011.

Banche, G., Allizond, V., Sostegni, R., Lavagna, A., Bergallo, M. *et al.* (2015) Application of multiple laboratory tests for *Mycobacterium avium* ssp. *paratuberculosis* detection in Crohn's disease patient specimens. *New Microbiologica* 38(3), 357–367.

Bannantine, J.P., Li, L., Mwangi, M., Cote, R., Raygoza Garay, J.A. *et al.* (2014) Complete genome sequence of *Mycobacterium avium* subsp. *paratuberculosis*, isolated from human breast milk. *Genome Announcements* 2(1), e01252-13 DOI: 10.1128/genomeA.01252-13.

Behr, M.A. and Hanley, J. (2008) Antimycobacterial therapy for Crohn's disease: a reanalysis. *The Lancet Infectious Diseases* 8(6), 344. DOI: 10.1016/S1473-3099(08)70104-X.

Bernstein, C.N., Wang, M.-H., Sargent, M., Brant, S.R. and Collins, M.T. (2007) Testing the interaction between NOD-2 status and serological response to *Mycobacterium paratuberculosis* in cases of inflammatory bowel disease. *Journal of Clinical Microbiology* 45(3), 968–971. DOI: 10.1128/JCM.02062-06.

Beumer, A., King, D., Donohue, M., Mistry, J., Covert, T. *et al.* (2010) Detection of *Mycobacterium avium* subsp. *paratuberculosis* in drinking water and biofilms by quantitative PCR. *Applied and Environmental Microbiology* 76(21), 7367–7370. DOI: 10.1128/AEM.00730-10.

Botsaris, G., Slana, I., Liapi, M., Dodd, C., Economides, C. *et al.* (2010) Rapid detection methods for viable *Mycobacterium avium* subspecies *paratuberculosis* in milk and cheese. *International Journal of Food Microbiology* 141, S87–S90. DOI: 10.1016/j.ijfoodmicro.2010.03.016.

Botsaris, G., Swift, B.M.C., Slana, I., Liapi, M., Christodoulou, M. *et al.* (2016) Detection of viable *Mycobacterium avium* subspecies *paratuberculosis* in powdered infant formula by phage-PCR and confirmed by culture. *International Journal of Food Microbiology* 216, 91–94. DOI: 10.1016/j.ijfoodmicro.2015.09.011.

Boulais, C., Wacker, R.O.N., Augustin, J.C., Cheikh, M.H. and Peladan, F. (2011) Modeling the occurrence of *Mycobacterium avium* subsp. *paratuberculosis* in bulk raw milk and the impact of management options for exposure mitigation. *Journal of Food Protection* 74(7), 1126–1136. DOI: 10.4315/0362-028X.JFP-11-005.

Campos, N., Magro, F., Castro, A.R., Cabral, J., Rodrigues, P. *et al.* (2011) Macrophages from IBD patients exhibit defective tumour necrosis factor-α secretion but otherwise normal or augmented pro-inflammatory responses to infection. *Immunobiology* 216(8), 961–970. DOI: 10.1016/j.imbio.2011.01.002.

Cao, B.L., Qasem, A., Sharp, R.C., Abdelli, L.S. and Naser, S.A. (2018) Systematic review and meta-analysis on the association of tuberculosis in Crohn's disease patients treated with tumor necrosis factor-α inhibitors (Anti-TNFα). *World Journal of Gastroenterology* 24(25), 2764–2775. DOI: 10.3748/wjg.v24.i25.2764.

Carvalho, I.A., Pietralonga, P.A.G., Schwarz, D.G.G., Faria, A.C.S. and Moreira, M.A.S. (2012) Short communication: recovery of viable *Mycobacterium avium* subspecies *paratuberculosis* from retail pasteurized whole milk in Brazil. *Journal of Dairy Science* 95(12), 6946–6948. DOI: 10.3168/jds.2012-5657.

Carvalho, I.A., Schwarz, D.G.G., Pietralonga, P.A.G., Faria, A.C.S., Braga, I.F.E. *et al.* (2016) Presence of *Mycobacterium avium* subsp. *paratuberculosis* (MAP) in Brazilian patients with inflammatory bowel diseases and in controls. *Sao Paulo Medical Journal* 134(1), 13–19. DOI: 10.1590/1516-3180.2014.8251809.

Chacon, O., Bermudez, L.E. and Barletta, R.G. (2004) Johne's disease, inflammatory bowel disease, and *Mycobacterium paratuberculosis*. *Annual Review of Microbiology* 58(1), 329–363. DOI: 10.1146/annurev.micro.58.030603.123726.

Chiodini, R.J. (1989) Crohn's disease and the mycobacterioses: a review and comparison of two disease entities. *Clinical Microbiology Reviews* 2(1), 90–117. DOI: 10.1128/CMR.2.1.90.

Chiodini, R.J., Van Kruiningen, H.J., Thayer, W.R. and Coutu, J.A. (1986) Spheroplastic phase of mycobacteria isolated from patients with Crohn's disease. *Journal of Clinical Microbiology* 24(3), 357–363. DOI: 10.1128/JCM.24.3.357-363.1986.

Clancy, R., Ren, Z., Turton, J., Pang, G. and Wettstein, A. (2007) Molecular evidence for *Mycobacterium avium* subspecies *paratuberculosis* (MAP) in Crohn's disease correlates with enhanced TNF-α secretion. *Digestive and Liver Disease* 39(5), 445–451. DOI: 10.1016/j.dld.2006.12.006.

Cooney, R., Baker, J., Brain, O., Danis, B., Pichulik, T. *et al.* (2010) NOD2 stimulation induces autophagy in dendritic cells influencing bacterial handling and antigen presentation. *Nature Medicine* 16(1), 90–97. DOI: 10.1038/nm.2069.

Coulombe, F., Divangahi, M., Veyrier, F., de Léséleuc, L., Gleason, J.L. *et al.* (2009) Increased NOD2-mediated recognition of N-glycolyl muramyl dipeptide. *The Journal of Experimental Medicine* 206(8), 1709–1716. DOI: 10.1084/jem.20081779.

Dalziel, T.K. (1913) Chronic interstitial enteritis. *British Medical Journal* 2, 1068–1070.

Espeschit, I.F., Souza, M.C.C., Lima, M.C. and Moreira, M.A.S. (2018) First molecular typing of *Mycobacterium avium* subspecies *paratuberculosis* identified in animal and human drinking water from dairy goat farms in Brazil. *Brazilian Journal of Microbiology* 49(2), 358–361. DOI: 10.1016/j.bjm.2017.06.005.

Faria, A.C.S., Schwarz, D.G.G., Carvalho, I.A., Rocha, B.B., De Carvalho Castro, K.N. *et al.* (2014) Short communication: viable *Mycobacterium avium* subspecies *paratuberculosis* in retail artisanal Coalho cheese from northeastern Brazil. *Journal of Dairy Science* 97(7), 4111–4114. DOI: 10.3168/jds.2013-7835.

Fava, V.M., Manry, J., Cobat, A., Orlova, M., Van Thuc, N. *et al.* (2016) A missense LRRK2 variant is a risk factor for excessive inflammatory responses in leprosy. *PLOS Neglected Tropical Diseases* 10(2), e0004412. DOI: 10.1371/journal.pntd.0004412.

Feller, M., Huwiler, K., Stephan, R., Altpeter, E., Shang, A. *et al.* (2007) *Mycobacterium avium* subspecies *paratuberculosis* and Crohn's disease: a systematic review and meta-analysis. *The Lancet Infectious Diseases* 7(9), 607–613. DOI: 10.1016/S1473-3099(07)70211-6.

Feller, M., Huwiler, K., Schoepfer, A., Shang, A., Furrer, H. *et al.* (2010) Long-term antibiotic treatment for Crohn's disease: systematic review and meta-analysis of placebo-controlled trials. *Clinical Infectious Diseases* 50(4), 473–480. DOI: 10.1086/649923.

Feng, C.G., Collazo-Custodio, C.M., Eckhaus, M., Hieny, S., Belkaid, Y. *et al.* (2004) Mice deficient in LRG-47 display increased susceptibility to mycobacterial infection associated with the induction of lymphopenia. *The Journal of Immunology* 172(2), 1163–1168. DOI: 10.4049/jimmunol.172.2.1163.

Gerrard, Z.E., Swift, B.M.C., Botsaris, G., Davidson, R.S., Hutchings, M.R. *et al.* (2018) Survival of *Mycobacterium avium* subspecies *paratuberculosis* in retail pasteurised milk. *Food Microbiology* 74, 57–63. DOI: 10.1016/j.fm.2018.03.004.

Gitlin, L., Borody, T.J., Chamberlin, W. and Campbell, J. (2012) *Mycobacterium avium* SS *paratuberculosis*-associated diseases. *Journal of Clinical Gastroenterology* 46(8), 649–655. DOI: 10.1097/MCG.0b013e31825f2bce.

Hugot, J.-P., Chamaillard, M., Zouali, H., Lesage, S., Cézard, J.-P. *et al.* (2001) Association of NOD2 leucine-rich repeat variants with susceptibility to Crohn's disease. *Nature* 411(6837), 599–603. DOI: 10.1038/35079107.

Jeyanathan, M., Alexander, D.C., Turenne, C.Y., Girard, C. and Behr, M.A. (2006) Evaluation of in situ methods used to detect *Mycobacterium avium* subsp. *paratuberculosis* in samples from patients with Crohn's disease. *Journal of Clinical Microbiology* 44(8), 2942–2950. DOI: 10.1128/JCM.00585-06.

Jeyanathan, M., Boutros-Tadros, O., Radhi, J., Semret, M., Bitton, A. *et al.* (2007) Visualization of *Mycobacterium avium* in Crohn's tissue by oil-immersion microscopy. *Microbes and Infection* 9(14–15), 1567–1573. DOI: 10.1016/j.micinf.2007.09.001.

Jones, P.H., Farver, T.B., Beaman, B., Cetinkaya, B. and Morgan, K.L. (2006) Crohn's disease in people exposed to clinical cases of bovine paratuberculosis. *Epidemiology and Infection* 134(1), 49–56. DOI: 10.1017/S0950268805004681.

Khan, K.J., Ullman, T.A., Ford, A.C., Abreu, M.T., Abadir, A. *et al.* (2011) Antibiotic therapy in inflammatory bowel disease: a systematic review and meta-analysis. *American Journal of Gastroenterology* 106(4), 661–673. DOI: 10.1038/ajg.2011.72.

Khan, I.A., Pilli, S., A, S., Rampal, R., Chauhan, S.K. *et al.* (2016) Prevalence and association of *Mycobacterium avium* subspecies *paratuberculosis* with disease course in patients with ulcero-constrictive ileocolonic disease. *PloS one* 11(3), e0152063. DOI: 10.1371/journal.pone.0152063.

Kirkwood, C.D., Wagner, J., Boniface, K., Vaughan, J., Michalski, W.P. *et al.* (2009) *Mycobacterium avium* subspecies *paratuberculosis* in children with early-onset Crohn's disease. *Inflammatory Bowel Diseases* 15(11), 1643–1655. DOI: 10.1002/ibd.20967.

Kostic, A.D., Xavier, R.J. and Gevers, D. (2014) The microbiome in inflammatory bowel disease: current status and the future ahead. *Gastroenterology* 146(6), 1489–1499. DOI: 10.1053/j.gastro.2014.02.009.

Lee, A., Griffiths, T.A., Parab, R.S., King, R.K., Dubinsky, M.C. *et al.* (2011) Association of *Mycobacterium avium* subspecies *paratuberculosis* with Crohn disease in pediatric patients. *Journal of Pediatric Gastroenterology and Nutrition* 52(2), 170–174. DOI: 10.1097/MPG.0b013e3181ef37ba.

Lefrançois, L.H., Pujol, C., Bodier, C.C., Teixeira-Gomez, A.P., Drobecq, H. *et al.* (2011) Characterization of the *Mycobacterium avium* subsp. *paratuberculosis* laminin-binding/histone-like protein (Lbp/Hlp) which reacts with sera from patients with Crohn's disease. *Microbes and Infection* 13(6), 585–594. DOI: 10.1016/j.micinf.2011.02.002.

Li, L., Wagner, B., Freer, H., Schilling, M., Bannantine, J.P. *et al.* (2017) Early detection of *Mycobacterium avium* subsp. *paratuberculosis* infection in cattle with multiplex-bead based immunoassays. *PLoS ONE* 12(12), e0189783. DOI: 10.1371/journal.pone.0189783.

Liverani, E., Scaioli, E., Cardamone, C., Dal Monte, P. and Belluzzi, A. (2014) *Mycobacterium avium* subspecies *paratuberculosis* in the etiology of Crohn's disease. *cause or epiphenomenon? World Journal of Gastroenterology* 20(36), 13060–13070.

Lorencova, A., Babak, V., Kralova, A. and Borilova, G. (2019) Survival of *Mycobacterium avium* subsp. *paratuberculosis* in raw fermented sausages during production and storage. *Meat Science* 155, 20–26. DOI: 10.1016/j.meatsci.2019.04.023.

Lu, Y., Li, X., Liu, S., Zhang, Y. and Zhang, D. (2018) Toll-like receptors and inflammatory bowel disease. *Frontiers in Immunology* 9, 72. DOI: 10.3389/fimmu.2018.00072.

MacMicking, J.D., Taylor, G.A. and McKinney, J.D. (2003) Immune control of tuberculosis by IFN-gamma-inducible LRG-47. *Science* 302(5645), 654–659. DOI: 10.1126/science.1088063.

Magin, W.S., Van Kruiningen, H.J. and Colombel, J.-F. (2013) Immunohistochemical search for viral and bacterial antigens in Crohn's disease. *Journal of Crohn's and Colitis* 7(2), 161–166. DOI: 10.1016/j.crohns.2012.03.021.

Mendoza, J.L., San-Pedro, A., Culebras, E., Cies, R., Taxonera, C. (2010) High prevalence of viable *Mycobacterium avium* subspecies *paratuberculosis* in Crohn's disease. *World Journal of Gastroenterology* 16(36), 4558–4563. DOI: 10.3748/wjg.v16.i36.4558.

Molicotti, P., Scanu, A.M., Lumbau, A., Cannas, S., Bua, A. *et al.* (2013) Molecular identification of *Mycobacterium avium* subspecies *paratuberculosis* in oral biopsies of Crohn's disease patients. *Gut Pathogens* 5(1), 18. DOI: 10.1186/1757-4749-5-18.

Momotani, E., Romona, N.M., Yoshihara, K., Momotani, Y., Hori, M. *et al.* (2012) Molecular pathogenesis of bovine paratuberculosis and human inflammatory bowel diseases. *Veterinary Immunology and Immunopathology* 148(1–2), 55–68. DOI: 10.1016/j.vetimm.2012.03.005.

Motiwala, A.S., Strother, M., Amonsin, A., Byrum, B., Naser, S.A. *et al.* (2003) Molecular epidemiology of *Mycobacterium avium* subsp. *paratuberculosis*: evidence for limited strain diversity, strain sharing, and identification of unique targets for diagnosis. *Journal of Clinical Microbiology* 41(5), 2015–2026. DOI: 10.1128/JCM.41.5.2015-2026.2003.

Nakase, H., Tamaki, H., Matsuura, M., Chiba, T. and Okazaki, K. (2011) Involvement of *Mycobacterium avium* subspecies *paratuberculosis* in TNF-α production from macrophage: possible link between MAP and immune response in Crohn's disease. *Inflammatory Bowel Diseases* 17(11), E140–E142. DOI: 10.1002/ibd.21750.

Nazareth, N., Magro, F., Appelberg, R., Silva, J., Gracio, D. *et al.* (2015a) Increased viability but decreased culturability of *Mycobacterium avium* subsp. *paratuberculosis* in macrophages from inflammatory bowel disease patients under infliximab treatment. *Medical Microbiology and Immunology* 204(6), 647–656. DOI: 10.1007/s00430-015-0393-2.

Nazareth, N., Magro, F., Machado, E., Ribeiro, T.G., Martinho, A. *et al.* (2015b) Prevalence of *Mycobacterium avium* subsp. *paratuberculosis* and *Escherichia coli* in blood samples from patients with inflammatory bowel disease. *Medical Microbiology and Immunology* 204(6), 681–692. DOI: 10.1007/s00430-015-0420-3.

Nunberg, M.Y., Werner, L., Kopylov, U., Haberman, Y., Lahad, A. *et al.* (2018) Impaired IL-10 receptor-mediated suppression in monocyte from patients with Crohn disease. *Journal of Pediatric Gastroenterology and Nutrition* 66(5), 779–784. DOI: 10.1097/MPG.0000000000001795.

Ogura, Y., Bonen, D.K., Inohara, N., Nicolae, D.L., Chen, F.F. *et al.* (2001) A frameshift mutation in NOD2 associated with susceptibility to Crohn's disease. *Nature* 411(6837), 603–606. DOI: 10.1038/35079114.

Okura, H., Toft, N. and Nielsen, S.S. (2012) Occurrence of *Mycobacterium avium* subsp. *paratuberculosis* in milk at dairy cattle farms: a systematic review and meta-analysis. *Veterinary Microbiology* 157(3-4), 253–263. DOI: 10.1016/j.vetmic.2011.12.019.

Olsen, I., Tollefsen, S., Aagaard, C., Reitan, L.J., Bannantine, J.P. *et al.* (2009) Isolation of *Mycobacterium avium* subspecies *paratuberculosis* reactive CD4 T cells from intestinal biopsies of Crohn's disease patients. *PLoS ONE* 4(5), e5641. DOI: 10.1371/journal.pone.0005641.

Olsen, I., Lundin, K.E. and Sollid, L.M. (2013) Increased frequency of intestinal CD4 + T cells reactive with mycobacteria in patients with Crohn's disease. *Scandinavian Journal of Gastroenterology* 48(11), 1278–1285. DOI: 10.3109/00365521.2013.837952.

Opstelten, J.L., Leenders, M., Dik, V.K., Chan, S.S., van Schaik, F.D. *et al.* (2016) Dairy products, dietary calcium, and risk of inflammatory bowel disease: results from a European prospective cohort investigation. *Inflammatory Bowel Diseases* 22(6), 1403–1411.

Parkes, M., Barrett, J.C., Prescott, N.J., Tremelling, M., Anderson, C.A. *et al.* (2007) Sequence variants in the autophagy gene IRGM and multiple other replicating loci contribute to Crohn's disease susceptibility. *Nature Genetics* 39(7), 830–832. DOI: 10.1038/ng2061.

Parrish, N.M., Radcliff, R.P., Brey, B.J., Anderson, J.L., Clark, D.L. *et al.* (2009) Absence of *Mycobacterium avium* subsp. *paratuberculosis* in Crohn's patients. *Inflammatory Bowel Diseases* 15(4), 558–565. DOI: 10.1002/ibd.20799.

Paustian, M.L., Zhu, X., Sreevatsan, S., Robbe-Austerman, S., Kapur, V. *et al.* (2008) Comparative genomic analysis of *Mycobacterium avium* subspecies obtained from multiple host species. *BMC Genomics* 9(1), 135. DOI: 10.1186/1471-2164-9-135.

Pierce, E.S. (2009) Possible transmission of *Mycobacterium avium* subspecies *paratuberculosis* through potable water: lessons from an urban cluster of Crohn's disease. *Gut Pathogens* 1(1), 17. DOI: 10.1186/1757-4749-1-17.

Pierce, E.S., Borowitz, S.M. and Naser, S.A. (2011) The Broad Street pump revisited: dairy farms and an ongoing outbreak of inflammatory bowel disease in Forest, Virginia. *Gut Pathogens* 3(1), 20. DOI: 10.1186/1757-4749-3-20.

Pistone, D., Marone, P., Pajoro, M., Fabbi, M., Vicari, N. *et al.* (2012) *Mycobacterium avium paratuberculosis* in Italy: commensal or emerging human pathogen? *Digestive and Liver Disease* 44(6), 461–465. DOI: 10.1016/j.dld.2011.12.022.

Qasem, A., Safavikhasraghi, M. and Naser, S.A. (2016) A single capsule formulation of RHB-104 demonstrates higher anti-microbial growth potency for effective treatment of Crohn's disease associated with *Mycobacterium avium* subspecies *paratuberculosis*. *Gut Pathogens* 8(1), 45. DOI: 10.1186/s13099-016-0127-z.

Qasem, A., Ramesh, S. and Naser, S.A. (2019) Genetic polymorphisms in tumour necrosis factor receptors (*TNFRSF1A/1B*) illustrate differential treatment response to TNFα inhibitors in patients with Crohn's disease. *BMJ Open Gastroenterology* 6(1), e000246. DOI: 10.1136/bmjgast-2018-000246.

Qual, D.A., Kaneene, J.B., Varty, T.J., Miller, R. and Thoen, C.O. (2010) Lack of association between the occurrence of Crohn's disease and occupational exposure to dairy and beef cattle herds infected with *Mycobacterium avium* subspecies *paratuberculosis*. *Journal of Dairy Science* 93(6), 2371–2376. DOI: 10.3168/jds.2009-2344.

Rangel, J.M., Sparling, P.H., Griffin, P.M., Griffin, P.M. and Swerdlow, D.L. (2005) Epidemiology of *Escherichia coli* O157:H7 outbreaks, United States, 1982–2002. *Emerging Infectious Diseases* 11(4), 603–609. DOI: 10.3201/eid1104.040739.

RedHill Biopharma (2018a) RedHill Biopharma announces positive top-line results from phase III study of RHB-104 in Crohn's disease. Available at: https://ir.redhillbio.com/news-releases/news-release-details/redhill-biopharma-announces-positive-top-line-results-phase-iii (accessed 30 May 2019).

RedHill Biopharma (2018b) RedHill Biopharma elaborates on Its announced positive top-line results from phase III study of RHB-104 in Crohn's disease. Available at: https://ir.redhillbio.com/news-releases/news-release-details/redhill-biopharma-elaborates-its-announced-positive-top-line?fbclid=IwAR2k397OF4VusXkvYS5553GPfvURrmZzpqmAKoFp5IlRZbJtpkgBe9li1xM (accessed 30 May 2019).

Rhodes, G., Henrys, P., Thomson, B.C. and Pickup, R.W. (2013) *Mycobacterium avium* subspecies *paratuberculosis* is widely distributed in British soils and waters: implications for animal and human health. *Environmental Microbiology* 15(10), 2761–2774.

Rhodes, G., Richardson, H., Hermon-Taylor, J., Weightman, A., Higham, A. *et al.* (2014) *Mycobacterium avium* subspecies *paratuberculosis*: human exposure through environmental and domestic aerosols. *Pathogens* 3(3), 577–595. DOI: 10.3390/pathogens3030577.

Ricanek, P., Lothe, S.M., Szpinda, I., Jorde, A.T., Brackmann, S. *et al.* (2010) Paucity of mycobacteria in mucosal bowel biopsies from adults and children with early inflammatory bowel disease. *Journal of Crohn's and Colitis* 4(5), 561–566. DOI: 10.1016/j.crohns.2010.05.003.

Sartor, R.B. (2005) Does *Mycobacterium avium* subspecies *paratuberculosis* cause Crohn's disease? *Gut* 54(7), 896–898. DOI: 10.1136/gut.2004.055889.

Sasikala, M., Reddy, D.N., Pratap, N., Sharma, S.K., Balkumar, R. *et al.* (2009) Absence of MAP specific IS900 sequence in intestinal biopsy tissues of India patients with Crohn's disease. *Indian Journal of Gastroenterology* 28(5), 169–174.

Savarino, E., Bertani, L., Ceccarelli, L., Bodini, G., Zingone, F. *et al.* (2019) Antimicrobial treatment with the fixed-dose antibiotic combination RHB-104 for *Mycobacterium avium* subspecies *paratuberculosis* in Crohn's disease: pharmacological and clinical implications. *Expert Opinion on Biological Therapy* 19(2), 79–88. DOI: 10.1080/14712598.2019.1561852.

Selby, W., Pavli, P., Crotty, B., Florin, T., Radford-Smith, G. *et al.* (2007) Two-year combination antibiotic therapy with clarithromycin, rifabutin, and clofazimine for Crohn's disease. *Gastroenterology* 132(7), 2313–2319. DOI: 10.1053/j.gastro.2007.03.031.

Sharp, R.C., Beg, S.A. and Naser, S.A. (2018) Role of *PTPN2/22* polymorphisms in pathophysiology of Crohn's disease. *World Journal of Gastroenterology* 24(6), 657–670. DOI: 10.3748/wjg.v24.i6.657.

Sibartie, S., Scully, P., Keohane, J., O'Neill, S., O'Mahony, J. *et al.* (2010) *Mycobacterium avium* subsp. *paratuberculosis* (MAP) as a modifying factor in Crohn's disease. *Inflammatory Bowel Diseases* 16(2), 296–304. DOI: 10.1002/ibd.21052.

Singh, S.B., Davis, A.S., Taylor, G.A. and Deretic, V. (2006) Human IRGM induces autophagy to eliminate intracellular mycobacteria. *Science* 313(5792), 1438–1441. DOI: 10.1126/science.1129577.

Singh, S.V., Sohal, J.S., Singh, P.K. and Singh, A.V. (2009) Genotype profiles of *Mycobacterium avium* subspecies *paratuberculosis* isolates recovered from animals, commercial milk, and human beings in North India. *International Journal of Infectious Diseases* 13(5), e221–e227. DOI: 10.1016/j.ijid.2008.11.022.

Singh, A.V., Singh, S.V., Singh, P.K., Sohal, J.S. and Singh, M.K. (2011a) High prevalence of *Mycobacterium avium* subspecies *paratuberculosis* ('Indian bison type') in animal attendants suffering from gastrointestinal complaints who work with goat herds endemic for Johne's disease in India. *International Journal of Infectious Diseases* 15(10), e677–e683. DOI: 10.1016/j.ijid.2011.04.013.

Singh, A.V., Singh, S.V., Verma, D.K., Yadav, R., Singh, P.K. *et al.* (2011b) Evaluation of 'Indigenous Absorbed ELISA Kit' for the estimation of seroprevalence of *Mycobacterium avium* subspecies *paratuberculosis* antibodies in human beings in North India. *ISRN Veterinary Science* 2011(2), 636038–5. DOI: 10.5402/2011/636038.

Stappenbeck, T.S., Rioux, J.D., Mizoguchi, A., Saitoh, T., Huett, A. *et al.* (2011) Crohn disease: a current perspective on genetics, autophagy and immunity. *Autophagy* 7(4), 355–374. DOI: 10.4161/auto.7.4.13074.

Timms, V.J., Hassan, K.A., Mitchell, H.M. and Neilan, B.A. (2015) Comparative genomics between human and animal associated subspecies of the *Mycobacterium avium* complex: a basis for pathogenicity. *BMC Genomics* 16(1), 695. DOI: 10.1186/s12864-015-1889-2.

Timms, V.J., Daskalopoulos, G., Mitchell, H.M. and Neilan, B.A. (2016) The association of *Mycobacterium avium* subsp. *paratuberculosis* with inflammatory bowel disease. *PLoS ONE* 11(2), e0148731. DOI: 10.1371/journal.pone.0148731.

Travassos, L.H., Carneiro, L.A.M., Ramjeet, M., Hussey, S., Kim, Y.-G. *et al.* (2010) Nod1 and Nod2 direct autophagy by recruiting ATG16L1 to the plasma membrane at the site of bacterial entry. *Nature Immunology* 11(1), 55–62. DOI: 10.1038/ni.1823.

United States Department of Agriculture (2008) Johne's disease on U.S. dairies 1991–2007. Available at: https://www.aphis.usda.gov/animal_health/nahms/dairy/downloads/dairy07/Dairy07_is_Johnes.pdf (accessed 2 June 2019).

Verdier, J., Deroche, L., Allez, M., Loy, C., Biet, F. *et al.* (2013) Specific IgG response against *Mycobacterium avium paratuberculosis* in children and adults with Crohn's disease. *PLoS ONE* 8(5), e62780. DOI: 10.1371/journal.pone.0062780.

Vinh, D.C. and Behr, M.A. (2013) Crohn's as an immune deficiency: from apparent paradox to evolving paradigm. *Expert Review of Clinical Immunology* 9(1), 17–30. DOI: 10.1586/eci.12.87.

Waddell, L.A., Rajić, A., Sargeant, J., Harris, J., Amezcua, R. *et al.* (2008) The zoonotic potential of *Mycobacterium avium* spp. *paratuberculosis*. *Canadian Journal of Public Health* 99(2), 145–155. DOI: 10.1007/BF03405464.

Waddell, L.A., Rajic, A., Stark, K.D.C. and McEwen, S.A. (2015) The zoonotic potential of *Mycobacterium avium* ssp. *paratuberculosis* : a systematic review and meta-analyses of the evidence. *Epidemiology and Infection* 143(15), 3135–3157. DOI: 10.1017/S095026881500076X.

Wagner, J., Sim, W., Bishop, R.F., Catto-Smith, A.G., Cameron, D.J.S. *et al.* (2011) *Mycobacterium avium* subspecies *paratuberculosis* in children with early-onset Crohn's disease: A longitudinal follow-up study. *Inflammatory Bowel Disesaes* 17(8), 1825–1826.

Wagner, J., Skinner, N.A., Catto-Smith, A.G., Cameron, D.J.S., Michalski, W.P. *et al.* (2013) TLR4, IL10RA, and NOD2 mutation in paediatric Crohn's disease patients: an association with *Mycobacterium avium* subspecies *paratuberculosis* and TLR4 and IL10RA expression. *Medical Microbiology and Immunology* 202(4), 267–276. DOI: 10.1007/s00430-013-0290-5.

Weiser, T.G., Forrester, J.D. and Forrester, J.A. (2019) Tactics to prevent intra-abdominal infections in general surgery. *Surgical Infections* 20(2), 139–145. DOI: 10.1089/sur.2018.282.

Wynne, J.W., Bull, T.J., Seemann, T., Bulach, D.M., Wagner, J. *et al.* (2011) Exploring the zoonotic potential of *Mycobacterium avium* subspecies *paratuberculosis* through comparative genomics. *PLoS ONE* 6(7), e22171. DOI: 10.1371/journal.pone.0022171.

Wynne, J.W., Beller, C., Boyd, V., Francis, B., Gwoźdź, J. *et al.* (2014) SNP genotyping of animal and human derived isolates of *Mycobacterium avium* subsp. *paratuberculosis*. *Veterinary Microbiology* 172(3–4), 479–485. DOI: 10.1016/j.vetmic.2014.05.024.

Xia, A., Stempak, J.M., Grist, J., Bressler, B., Silverberg, M.S. *et al.* (2014) Effect of inflammatory bowel disease therapies on immunogenicity of *Mycobacterium paratuberculosis* proteins. *Scandinavian Journal of Gastroenterology* 49(2), 157–163. DOI: 10.3109/00365521.2013.857713.

Zamani, S., Zali, M.R., Aghdaei, H.A., Sechi, L.A., Niegowska, M. *et al.* (2017) *Mycobacterium avium* subsp. *paratuberculosis* and associated risk factors for inflammatory bowel disease in Iranian patients. *Gut Pathogens* 9, 1. DOI: 10.1186/s13099-016-0151-z.

Zarei-Kordshouli, F., Geramizadeh, B. and Khodakaram-Tafti, A. (2019) Prevalence of *Mycobacterium avium* subspecies *paratuberculosis* IS900 DNA in biopsy tissues from patients with Crohn's disease: histopathological and molecular comparison with Johne's disease in Fars Province of Iran. *BMC Infectious Diseases* 19(1), 23. DOI: 10.1186/s12879-018-3619-2.

4 Genetics of Host Susceptibility to Paratuberculosis

Holly L. Neibergs* and Jennifer N. Kiser
Department of Animal Sciences, Washington State University, Pullman, Washington, USA

4.1 Introduction

In order to reduce the prevalence of *Mycobacterium avium* subspecies *paratuberculosis* (*MAP*) infection and the associated costs of the disease, a better understanding of the mechanisms behind the disease and the underlying genetic variation contributing to host susceptibility is needed. Given that the disease is currently contagious and difficult to identify, selection for animals that are less susceptible to *MAP* infection would be a valuable tool for producers to reduce disease and economic losses and enhance biosecurity. Ideally, genomic selection would provide tools that would be useful within and across breeds. The identification of causal mutations in one species provides the opportunity to determine if the same mutations are responsible for *MAP* susceptibility in other species. Most genetic studies have focused on *MAP* susceptibility in cattle, sheep, goats and deer. This chapter focuses on reviewing new research published on host susceptibility of domestic ruminants to *MAP* infection.

4.2 Evidence for Disease Susceptibility Differences in Cattle

It is well established that prevalence of *MAP* infection in cattle varies by breed and geographical location. This range of herd-level infection of *MAP* could be influenced by a variety of factors including differences between diagnostic tests, herd management and subspecies (*Bos taurus, Bos indicus*) and breeds of cattle being tested. Predominantly Zebu (*Bos indicus*) cattle have a higher number of herds positive for *MAP* (59.3%) compared with herds composed of crossbred *Bos taurus* × *Bos indicus* (31.6%) or European (*Bos taurus*) breeds (28.9%) in Brazil (Vilar *et al.*, 2015). A large Canadian dairy study tested breed susceptibility to *MAP* infection in Jersey and Guernsey cattle compared with Holstein and Brown Swiss cattle. In this study, the odds of Jersey and Guernsey cattle testing positive for *MAP* using a milk enzyme-linked immunosorbent assay (ELISA) test were 1.4 to 8.3 times higher than in Holsteins or Brown Swiss cattle (Sorge *et al.*, 2011). In addition to studies investigating differences in breed susceptibility, heritability studies also provide estimates of non-environmental factors associated with disease susceptibility. For an in-depth review of estimated herd-level and animal-level prevalence of *MAP*, please see Chapter 1 of this volume.

4.2.1 Heritability estimates

In cattle, heritability estimates for susceptibility to *MAP* infection range between 0.03 and

*Corresponding author: neibergs@wsu.edu

© CAB International 2020. *Paratuberculosis: Organism, Disease, Control, 2nd Edition* (eds M.A. Behr *et al.*)

0.23 in Holstein cattle (van Hulzen *et al.*, 2011; Küpper *et al.*, 2012; Shook *et al.*, 2012; Kiser *et al.*, 2017; Gao *et al.*, 2018a; Kirkpatrick and Lett, 2018) and between 0.08 and 0.48 in Jersey cattle (Zare *et al.*, 2014; Kiser *et al.*, 2017). These heritability estimates are in the range of those reported for susceptibility to other mycobacterial diseases such as leprosy (0.2; Wang *et al.*, 2016) as well as other bovine diseases such as bovine respiratory disease (0.13–0.21; Neibergs *et al.*, 2014). Although *MAP* infection is present within beef herds around the globe, heritability estimates for susceptibility to *MAP* in beef cattle have not been reported, as cattle studies tend to focus on *MAP* infection in dairy cattle where it is more prevalent.

4.2.2 Genome-wide association studies

Genome-wide association studies (GWAS) have been conducted to identify loci associated with *MAP* infection. The identification of loci associated with *MAP* infection provides the foundation for genomic selection of cattle with enhanced resistance, identifies host disease susceptibility mechanisms through the regulation of genes involved in the immune response to *MAP* and could identify if the mechanisms and genomic regions associated with infection differ in cattle and other species. A key limitation for all genetic studies associated with *MAP* infection is the sensitivity of detecting the infection prior to clinical manifestation. Due to poor diagnostic sensitivity, the identification of animals that are truly negative for *MAP* can be difficult and represents a phenocopy of cattle that are truly negative for disease. This limitation reduces the power to detect associations with *MAP* infection and is likely to reduce the repeatability or validation of loci across studies with different diagnostic tests for *MAP* infection. For example, even within a single study Settles *et al.* (2009) demonstrated that faecal culture was only able to detect *MAP* in 40% of *MAP* tissue culture-positive cattle, indicating that misclassification of *MAP* infection status is dependent on the diagnostic tests used.

Closely related to the sensitivity of *MAP* diagnostics is the differences in phenotypes used to test for associations in cattle. Some of the most common diagnostic methods used to detect *MAP* infection and determine phenotypes are ELISA for serum, milk or faecal samples, culturing of tissue or faeces, quantitative polymerase chain reaction (PCR) for tissue or faeces, or a combination of these methods. The loci identified as associated with susceptibility to *MAP* infection will vary by the phenotype or method of diagnosis of the disease (Settles *et al.*, 2009). For example, when the same Holstein cattle were studied, four loci were associated ($P < 5 \times 10^{-5}$) with *MAP* susceptibility when disease was determined by faecal culture, but different loci were associated ($P < 5 \times 10^{-5}$) with *MAP* infection based on culturing of tissue (Settles *et al.*, 2009). These differences underscore the importance of using a standardized phenotype when comparing results across studies.

In the past decade, GWAS have been performed in cattle to identify loci associated with *MAP* infection (Settles *et al.*, 2009; Minozzi *et al.*, 2010; Pant *et al.*, 2010; Kirkpatrick *et al.*, 2011; Zanella *et al.*, 2011; Minozzi *et al.*, 2012; van Hulzen *et al.*, 2012a; Alpay *et al.*, 2014; Kiser *et al.*, 2017; Brito *et al.*, 2018; Gao *et al.*, 2018b; Mallikarjunappa *et al.*, 2018; Table 4.1). The phenotypes for these studies varied and most of these studies were conducted in Holsteins or Jersey cattle. Few loci associated with *MAP* susceptibility have been validated or present in more than one study (Table 4.1). Three loci, one on BTA5 and two on BTA10, were associated in studies by Pant *et al.* (2010) and Mallikarjunappa *et al.* (2018), while a fourth locus on BTA16 was associated with *MAP* infection by both Settles *et al.* (2009) and Kiser *et al.* (2017) (Table 4.1). However, Pant *et al.* (2010) and Mallikarjunappa *et al.* (2018), and Settles *et al.* (2009) and Kiser *et al.* (2017) shared study populations, so these loci were not validated in completely independent cattle populations. The failure of loci to validate could potentially be due to the differences in diagnostic methods or the marker density used for genotyping. Most studies used lower-density single nucleotide polymorphism (SNP) genotyping assays panels that contain 50,000 or fewer SNPs, although a few studies have used higher-density (777,986 SNPs) SNP assays or have imputed genotypes to a whole genome sequence level. Differences in SNP density could result in associated loci being undetected if the SNPs nearest the locus are in weak linkage disequilibrium (LD) or aren't in LD with the associated SNP(s).

Table 4.1. Summary of genome-wide association studies performed in domestic ruminants to identify loci associated with *Mycobacterium avium* subsp. *paratuberculosis* (*MAP*) infection.

Species	Chromosome(s)	Studies	Breed(s)[a]	Diagnostic used[b]
Bovine	1	Settles *et al.* (2009); Pant *et al.* (2010); Zanella *et al.* (2011); Minozzi *et al.* (2012); Gao *et al.* (2018b)	HO	ELISA F, S Culture – F, T
	2	Kirkpatrick *et al.* (2011); Zanella *et al.* (2011); Alpay *et al.* (2014); Gao *et al.* (2018b)	HO	ELISA – S Culture – F, T
	3	Settles *et al.* (2009); Kirkpatrick *et al.* (2011); Kiser *et al.* (2017); Gao *et al.* (2018b)	HO	ELISA – S Culture – F, T
	4	Kirkpatrick *et al.* (2011); van Hulzen *et al.* (2012a)	HO, HOX	ELISA – M, S Culture – F
	5	Settles *et al.* (2009); Pant *et al.* (2010); Kirkpatrick *et al.* (2011); Mallikarjunappa *et al.* (2018)	HO	ELISA – M, S Culture – F, T
	6	Pant *et al.* (2010); Kirkpatrick *et al.* (2011); Zanella *et al.* (2011); Minozzi *et al.* (2012); Alpay *et al.* (2014); Gao *et al.* (2018b); Mallikarjunappa *et al.* (2018)	HO	ELISA – F, M, S Culture – F, T
	7	Settles *et al.* (2009); Pant *et al.* (2010); Kirkpatrick *et al.* (2011); Minozzi *et al.* (2012); Alpay *et al.* (2014); Gao *et al.* (2018b); Mallikarjunappa *et al.* (2018)	HO	ELISA – F, M, S Culture – F, T
	8	Minozzi *et al.* (2010); Kiser *et al.* (2017); Gao *et al.* (2018b)	HO	ELISA – S Culture – T
	9	Settles *et al.* (2009); Minozzi *et al.* (2010); Kirkpatrick *et al.* (2011)	HO	ELISA – S Culture – F, T
	10	Pant *et al.* (2010); Kirkpatrick *et al.* (2011); Kiser *et al.* (2017); Mallikarjunappa *et al.* (2018)	HO	ELISA – M, S Culture – F, T
	11	Minozzi *et al.* (2010); Pant *et al.* (2010); Gao *et al.* (2018b)	HO	ELISA – S
	12	Minozzi *et al.* (2010), 2012; Kiser *et al.* (2017)	HO	ELISA F, S Culture – F, T
	13	Kirkpatrick *et al.* (2011); Minozzi *et al.* (2012); Gao *et al.* (2018b)	HO	ELISA F, S Culture – F, T
	14	Pant *et al.* (2010); Kirkpatrick *et al.* (2011); van Hulzen *et al.* (2012a); Kiser *et al.* (2017); Mallikarjunappa *et al.* (2018)	HO, HOX	ELISA – M, S Culture – F, T
	15	Kirkpatrick *et al.* (2011); Zanella *et al.* (2011); Minozzi *et al.* (2012); Alpay *et al.* (2014); Mallikarjunappa *et al.* (2018)	HO	ELISA – M, S Culture – F, T

Continued

Table 4.1. Continued

Species	Chromosome(s)	Studies	Breed(s)[a]	Diagnostic used[b]
	16	Settles *et al.* (2009); Kirkpatrick *et al.* (2011); Minozzi *et al.* (2012); van Hulzen *et al.* (2012a); Kiser *et al.* (2017); Gao *et al.* (2018b); Mallikarjunappa *et al.* (2018)	HO, HOX	ELISA – F, M, S Culture – F, T
	17	Kirkpatrick *et al.* (2011); Alpay *et al.* (2014)	HO	ELISA – S Culture – F
	18	Kirkpatrick *et al.* (2011); van Hulzen *et al.* (2012a); Gao *et al.* (2018b)	HO, HOX	ELISA – M, S Culture – F
	19	van Hulzen *et al.* (2012a)	HOX	ELISA – M
	20	Kirkpatrick *et al.* (2011); van Hulzen *et al.* (2012a); Gao *et al.* (2018b); Mallikarjunappa *et al.* (2018)	HO, HOX	ELISA – M, S Culture – F
	21	Settles *et al.* (2009); Kirkpatrick *et al.* (2011); Minozzi *et al.* (2012); van Hulzen *et al.* (2012a); Kiser *et al.* (2017); Mallikarjunappa *et al.* (2018)	HO, HOX	ELISA – F, M, S Culture – F, T
	22	Kirkpatrick *et al.* (2011); Minozzi *et al.* (2012); Kiser *et al.* (2017); Gao *et al.* (2018b)	HO, JE	ELISA F, S Culture – F, T
	23	Settles *et al.* (2009); Kirkpatrick *et al.* (2011); Minozzi *et al.* (2012); Gao *et al.* (2018b)	HO	ELISA F, S Culture – F, T
	24	Gao *et al.* (2018b)	HO	ELISA – S
	25	Kirkpatrick *et al.* (2011); Minozzi *et al.* (2012)	HO	ELISA – F, S Culture – F, T
	26	Kirkpatrick *et al.* (2011); Minozzi *et al.* (2012); van Hulzen *et al.* (2012a)	HO, HOX	ELISA – F, M, S Culture – F, T
	27	Minozzi *et al.* (2010); van Hulzen *et al.* (2012a); Gao *et al.* (2018b)	HO, HOX	ELISA – M, S
	29	Kirkpatrick *et al.* (2011); van Hulzen *et al.* (2012a); Alpay *et al.* (2014)	HO, HOX	ELISA – M, S Culture – F
Ovine	1, 3–6, 8–9, 12–14, 17–18, 20, 22–24, 26–27	Moioli *et al.* (2016a)	SA	ELISA – S
Caprine	4–6, 8–9, 13, 17, 19, 27	Cecchi *et al.* (2017)	GA	ELISA – S

[a]Cattle breeds are abbreviated as follows: Holstein – HO, Jersey – JE. Crossbreds are denoted with an X after the main breed code. Sheep breeds are abbreviated: Sarda – SA. Goat breeds are abbreviated: Garfagnina – GA.
[b]Sample types used for testing are abbreviated as follows: faecal – F; milk – M; serum – S; Tissue – T. Diagnostic testing method abbreviations are: enzyme-linked immunosorbent assay – ELISA.

Studying multiple breeds or crossbred cattle is needed to determine if the loci associated with *MAP* infection differ by breed. A large GWAS in Canada identified loci associated with *MAP* susceptibility in Jersey and Holstein cattle (Sallam *et al.*, 2017). In this study, single-marker comparisons of Jersey and Holstein cattle populations were followed by combining *P*-values of SNP associations to account for the possibility that LD was different between breeds. Two loci were associated ($P < 5 \times 10^{-5}$) with disease susceptibility on BTA19 and BTA23 (Sallam *et al.*, 2017). These loci differed from the locus identified on BTA27 when the GWAS was conducted using a single-marker approach on the combined population (Sallam *et al.*, 2017). The same study also analysed the GWAS using Bayes C in the combined breed population with 1 megabase (Mb) windows. Nine windows explained >1% of the genetic variance, finding associations on BTA2, BTA3 (three regions), BTA6, BTA8, BTA25, BTA27 and BTA29 (Sallam *et al.*, 2017). The identification of loci associated with *MAP* infection in multiple breeds provides the opportunity to use the same markers for genomic selection for either breed or for crossbreds.

4.2.3 Candidate gene studies

Susceptibility to *MAP* infection is likely polygenic, due to the multiple loci identified in GWAS and the complex nature of the disease. Candidate gene studies are association studies limited to genes with functions that make them good candidates for harbouring a variant that predisposes cattle to *MAP* infection. Some of the most commonly studied candidate genes whose proteins are involved in *MAP* infection in cattle are solute carrier family 11 member 1 (*SLC11A1*, formerly known as *NRAMP1*), nucleotide-binding oligomerization domain containing 2 (*NOD2*, formerly known as *CARD15*) and the toll like receptor 2 (*TLR2*) and 4 (*TLR4*). The SLC11A1 protein is involved in limiting intracellular growth of bacteria, likely through its role in transporting metal ions like iron (Forbes and Gros, 2003). The gene products of *NOD2* and the *TLRs* are pathogen recognition receptors that sense the presence of bacteria, triggering an immune response (Ferwerda *et al.*, 2007). Further, NOD2, TLR2 and TLR4 receptors are involved in the cellular recognition of *MAP*. The receptors differ on where they act on the cell, as NOD2 senses bacterial proteins in the host cytosol (Caruso *et al.*, 2014) while the TLRs recognize bacterial products on the cell surface (Medzhitov, 2001). For a detailed review of candidate genes, *SLC11A1*, *NOD2*, *TLR2* and *TLR4* in ruminants see Kirkpatrick (2010), Purdie *et al.* (2011) and Singh *et al.* (2013). Multiple studies in the past decade have further elucidated the roles of these and other candidate genes in *MAP* infection in cattle (Table 4.2). Positional candidate genes more recently investigated include CD209 molecule (*CD209*, formerly *CLEC4L*) and C-type lectin domain containing 7 (*CLEC7A*).

Both *CD209* and *CLEC7A* encode C-type lectin receptors, which are involved in pathogen recognition. The *CD209* gene encodes a protein called DC-SIGN, which can bind mannosylated lipoarabomannan that is present on many virulent species of mycobacteria (Geijtenbeek *et al.*, 2003). Mutations within *CD209* (*rs4804803*) have been associated ($P = 0.006$) with enhanced resistance to *Mycobacterium tuberculosis* infection in humans (Vannberg *et al.*, 2008), cattle tuberculosis (*Mycobacterium bovis*; Yamakawa *et al.*, 2008) and subclinical or latent *MAP* infections ($P < 0.003$) in Holstein cattle diagnosed using serum ELISA and histopathology (Vázquez *et al.*, 2014; Table 4.2). Juste *et al.* (2018) supported these findings reporting that cattle could be placed into one of four groups in relation to susceptibility to *MAP* infection based on genotype combinations from a set of five SNPs, including one associated with *CD209* (*rs210748127*).

Like the CD209 protein, the CLEC7A protein has been associated ($P < 0.01$) with *M. tuberculosis* infection and interacts with TLR2 to induce pro-inflammatory responses within macrophages in response to *M. tuberculosis* infection (Yadav and Schorey, 2006). Two studies in cattle have associated mutations within *CLEC7A* with *MAP* infection (Pant *et al.*, 2014; Kumar *et al.*, 2019; Table 4.2). In a Holstein study, Pant *et al.* (2014) identified a missense mutation within *CLEC7A* (*rs41654445*) associated ($P = 0.008$) with resistance to *MAP* infection. Kumar *et al.* (2019) validated the association of *rs41654445* ($P < 0.0001$) and identified a nearby mutation within the 5′ untranslated region of *CLEC7A* (*rs110353594*) associated ($P = 0.0007$) with resistance to *MAP* infection in Holstein, Sahiwal, Tharparkar and Vrindavani crosses. The Kumar *et al.* (2019) study confirmed the importance of *CLEC7A* in host susceptibility to *MAP* infection as well as the usefulness of this marker for genomic selection.

Table 4.2. List of candidate genes investigated for association with *Mycobacterium avium* subsp. *paratuberculosis* (*MAP*) infection in domestic ruminants.

Candidate gene	Species (reeds) [a,b,c]	Diagnostic used[d]	Microsatellite/SNP[e]	Significance[f]	Citation
CLEC7A	Bovine (HO)	ELISA – S	*rs41654445*	$P = 0.008$	Pant *et al.* (2014)
	Bovine (HO, SA, TH, VRX)	ELISA – S Culture – F IFN-γ	*rs41654445* *rs110353594*	$P < 0.0001$ $P = 0.0007$	Kumar *et al.* (2019)
CD109	Ovine (SAR)	ELISA – S	XM_004011463.1/OAR8_270360.1	OAR8_270360.1 FDR = 0.08	Moioli *et al.* (2016b)
CD209	Bovine (HO)	ELISA – S	*rs208222804*	$P < 0.003$	Vázquez *et al.* (2014)
DRB3	Caprine (JAM)	ELISA – S Culture – F	*P* *T*	$P < 0.001$	Singh *et al.* (2012)
	Bovine (HO-SSX)	ELISA – S PCR – S	Arg84Gly Asp57His Phe60Tyr Val53Glu Val53Leu	1.45^ 1.94^ 0.45^ 453.7^ 453.7^	Rastislav and Mangesh (2012)
ETH10	Caprine (GA)	ELISA – S	205*	$P < 0.05$	Cecchi *et al.* (2018)
IL10RA	Bovine (HO)	ELISA – S	*ss104807643*	$P = 0.002$	Verschoor *et al.* (2010)
IFNG	Bovine (HO, JE, BR-ANX)	ELISA – S PCR – S, M, F Culture – F	SNP1-2781	$P = 0.019$	Pinedo *et al.* (2009a)
MHC	Ovine (ME)	AGID – S Culture – F, T	163/*	$P < 0.05$	Reddacliff *et al.* (2005)
NOD2	Bovine (HO, JE, BR-ANX)	ELISA – S PCR – S, M, F Culture – F	SNP2197	$P < 0.0001$	Pinedo *et al.* (2009b)
	Bovine (HO)	ELISA – S Culture – F	*rs43710290*	$P = 0.04$	Ruiz-Larrañaga *et al.* (2010)
PGLYRP1	Bovine (HO)	ELISA – S, M	c.480G>A	$P = 0.05$	Pant *et al.* (2011)

Continued

Table 4.2. Continued

Candidate gene	Species (reeds) [a,b,c]	Diagnostic used[d]	Microsatellite/SNP[e]	Significance[f]	Citation
SLC11A1	Bovine (HO, JE, BR-ANX)	ELISA – S PCR – S, M, F Culture – F	275* 279*	$P = 0.027$ $P = 0.007$	Pinedo *et al.* (2009a)
	Bovine (HO)	Culture – F	ss119336735 ss119336734	$P = 0.014$ $P = 0.032$	Ruiz-Larrañaga *et al.* (2010)
	Ovine (ME)	AGID – S Culture – F, T	162/*	$P < 0.05$	Reddacliff *et al.* (2005)
	Caprine (UNK)	ELISA – S Culture – F	B_7	$P < 0.02$	Korou *et al.* (2010)
	Caprine (AB, MA, MAX)	ELISA – S PCR – F	B_7	$P < 0.01$	Abraham *et al.* (2017)
SRCRSP05	Caprine (GA)	ELISA – S	163*	$P < 0.05$	Cecchi *et al.* (2018)
TLR1	Bovine (BC, HO, PI, PR, SI, SS, SSX)	PCR – S	Val220Met	3.549^	Mucha *et al.* (2009)
	Bovine (EARX, EABX, HO)	ELISA – S	1380AG	$P < 0.05$	Cinar *et al.* (2018)
TLR2	Bovine (BC, HO, PI, PR, SI, SS, SSX)	PCR – S	Ile680Val	Not given	Mucha *et al.* (2009)
	Bovine (HO)	ELISA Culture – F	SNP-1903	$P = 0.0096$	Koets *et al.* (2010)
	Bovine (HO, KO, SA, NAX)	ELISA	*rs55617172*	$P = 0.04$	Sadana *et al.* (2015)
TLR4	Bovine (BC, HO, PI, PR, SI, SS, SSX)	PCR – S	Asp299Asn Gly298[Arg,Trp] Gly389Ser	Not given*	Mucha *et al.* (2009)
	Bovine (HO)	ELISA – S, M	c.-226	$P = 0.095$	Sharma *et al.* (2015)

Continued

Table 4.2. Continued

Candidate gene	Species (reeds) [a,b,c]	Diagnostic used[d]	Microsatellite/SNP[e]	Significance[f]	Citation
WNT2	Bovine (HO)	ELISA – S Culture – F	*rs43390642*	$P = 0.013$	Pauciullo *et al.* (2015)

[a]Cattle breeds are abbreviated as follows: Angus – AN, Brahman – BR, Brown Carpathians – BC, East Anatolian Red – EAR, East Anatolian Black – EAB, Holstein – HO, Jersey – JE, Kosi – KO, native Indian – NI, Pinzhauer – PI, Polish Red – PR, Sahiwal – SA, Simmental – SI, Slovak Spotted – SS, Tharparkar – TH, Vrindavani – VR. Crossbreds are denoted with an X after the main breed code.
[b]Sheep breeds are abbreviated as follows: Merino – ME, Sarda – SAR.
[c]Goat breeds are abbreviated as follows: Attappady Black – AB, Garfagnina – GA, Jamunapari – JAM, Malabari – MA, unknown – UNK. Crossbreds are denoted with an X after the main breed code.
[d]Sample types used for testing are abbreviated as follows: faecal – F, milk – M, serum – S, tissue – T. Diagnostic testing method abbreviations are: agar gel immunodiffusion test – AGID, enzyme-linked immunosorbent assay – ELISA, interferon-gamma assay – IFN-γ, polymerase chain reaction – PCR.
[e]Microsatellites are denoted with an *.
[f]Odds ratios, denoted with ^, are listed instead of *P*-values when *P*-values were not listed in the referenced study.

4.2.4 Pathway and gene set analyses

Pathway analysis seeks to identify genes or molecules that interact with one another to complete a physiological function. This analysis can combine genomic association data, gene expression data and proteomic data to generate a list of genes or proteins of interest and understand how they interact in relation to chemical or biological systems. A commonly used pathway analysis software is Ingenuity Pathway Analysis (IPA; Krämer *et al.*, 2014). In IPA, pathways or networks consist of a group of molecules, genes, proteins, chemicals, etc., which have been grouped together through a highly curated database known as the IPA Knowledge Base, which examines literature to elucidate the functions and relationships of the molecules, genes, proteins and chemicals in the pathways. The goal of performing an IPA is to identify how genes or proteins of interest fit into the larger biological process through identifying pathways and regulators related to the input data.

In cattle, multiple studies have used IPA to investigate pathways associated with *MAP* infection (Motiwala *et al.*, 2006; MacHugh *et al.*, 2012; Määttänen *et al.*, 2013; David *et al.*, 2014; Shin *et al.*, 2015; Malvisi *et al.*, 2016). Most of these studies experimentally infected cattle (Määttänen *et al.*, 2013; David *et al.*, 2014) or cattle macrophages (Motiwala *et al.*, 2006; MacHugh *et al.*, 2012) with *MAP* and then performed gene expression analyses between cattle or cells that were infected and healthy. However, two studies (Shin *et al.*, 2015; Malvisi *et al.*, 2016) investigated gene expression differences between cattle naturally infected with *MAP* rather than cattle that were experimentally challenged. All of the pathway analyses studying *MAP* infection in cattle identified pathways with functions relating to the immune system. Most commonly, pathways were associated with cell death (MacHugh *et al.*, 2012; David *et al.*, 2014), lipid metabolism (Shin *et al.*, 2015; Malvisi *et al.*, 2016) or lymphocytes (MacHugh *et al.*, 2012; David *et al.*, 2014; Malvisi *et al.*, 2016).

An alternative to pathway analysis that uses gene expression data is gene set enrichment analysis (GSEA) or gene set enrichment analysis–SNP (GSEA-SNP) when SNP data from an association analysis are used in lieu of gene expression data. In GSEA-SNP, SNPs are used as proxies for genes to identify gene sets that are enriched for susceptibility to *MAP* infection and the leading edge genes that are responsible for that enrichment. Gene sets are similar to pathways that share a common function. Leading edge genes are those genes that are associated with *MAP* susceptibility within the gene set. The GSEA-SNP method aims to identify leading edge genes with modest individual effects that collectively have a large effect. A GWAS may miss modest effect genes that are identified in a GSEA-SNP. Currently, the gene sets are based on data collected on humans, mice and rats, which may have different gene functions and gene pathways than in ruminant species, although many genes have conserved functions across species.

The first GSEA-SNP investigated in cattle for paratuberculosis tested for gene sets enriched for *MAP* susceptibility using the presence of tissue infection and faecal shedding in Holstein cattle (Neibergs *et al.*, 2010). While no gene sets were associated with faecal shedding, a single gene set from the Gene Ontology (GO) database, positive regulation of cell motion (GO:0051272), was associated with *MAP* tissue infection (normalized enrichment score (NES) = 1.77) (Neibergs *et al.*, 2010). This enriched gene set contained five leading edge genes (*EDN2*, *ACTN4*, *TDGF1*, *PIK3R1* and *TGFB2*) with functions relating to the immune response that were associated with *MAP* tissue infection.

A second GSEA-SNP investigated susceptibility to *MAP* tissue infection in two Holstein populations, including a population that had been evaluated in the Neibergs *et al.* (2010) study (Kiser *et al.*, 2018). Between the two populations, Kiser *et al.* (2018) identified 13 gene sets (NES >2.5) and 117 leading edge genes that were associated with *MAP* tissue infection. The 13 leading edge genes had functions that related to nuclear factor κβ (*NFKB*). The *NFKB* protein is an essential regulator of intestinal inflammation (Wullaert *et al.*, 2011; Liu *et al.*, 2017), and has been associated with other mycobacterial diseases such as tuberculosis (Bai *et al.*, 2013) and leprosy (Wambier *et al.*, 2014).

A third GSEA-SNP by Del Corvo *et al.* (2017) investigated gene sets associated with *MAP* infection in Holsteins using serum ELISA diagnostics to determine disease presence. This study identified a single enriched gene set, embryogenesis and morphogenesis (NES = 1.58; GO:0009790), that contained eight leading edge genes (Del Corvo *et al.*, 2017). Four of these genes had functions relating to the general immune response to infection: *APLP2*, *PGM3*, *ATP5A* and *GTF2I* (Del Corvo *et al.*, 2017). There were no leading edge genes common between the three GSEA-SNP, and only two (*TGFB2* and *PIK3R1*) were common to Neibergs *et al.* (2010) and Kiser *et al.* (2017). Another potential reason for the lack of further validation between the three studies could be the differences in diagnostic tests, as Del Corvo *et al.* (2017) used a serum ELISA and both Neibergs *et al.* (2010) and Kiser *et al.* (2017) used culturing of tissue to determine the presence of *MAP*.

The identification of pathways, gene sets and leading edge genes associated with the immune system and different mycobacterial species could provide additional insight into the pathogenesis of mycobacterial diseases, new targets for treatments and the potential to better identify at-risk populations. While no GSEA-SNP has been conducted in other ruminant species, those performed in cattle have illustrated the potential of utilizing this method to reanalyse previous data to uncover more information about the disease process and discover additional candidate genes for further research.

4.3 Evidence for Disease Susceptibility Differences in Sheep

Like cattle, reported flock-level prevalence of *MAP infection varies by breed and location.* For an in-depth review of estimated flock-level prevalence of *MAP* infection in sheep from countries across continental Europe see Nielsen and Toft (2009) and Chapter 1 of this volume. A retrospective study in India compared the imported Dorset to the indigenous Nilagiri and Sandyno breeds, and reported that the native breeds were more susceptible to infection and had higher mortality rates than Dorset or Dorset crosses (Hemalatha *et al.*, 2013). A more recent study in Australia also reported variability in disease susceptibility between Merino, Poll Dorset, Border Leicester and Suffolk breeds, where the Dorset was also more resistant to *MAP* than the other breeds (Begg *et al.*, 2017).

4.3.1 Heritability estimates

Few studies have reported heritability for *MAP* infection in sheep. A single study in 2003 reported heritability estimates of 0.07 for Romney and 0.18 for Merino (Hickey *et al.*, 2003). These estimates are similar to those reported for Holstein and Jersey cattle.

4.3.2 Genome-wide association studies

In contrast to cattle, few GWAS have been published investigating susceptibility to *MAP* infection in sheep. A recent study in Sarda sheep using the Ovine SNP50K BeadChip, which used serum ELISA to detect *MAP* infection, identified 32 loci ($P < 6 \times 10^{-4}$) associated with resistance to *MAP* infection (Moioli *et al.*, 2016a; Table 4.1). These loci were located across 18 ovine chromosomes and harboured 30 positional candidate genes (Moioli *et al.*, 2016a). The positional candidate genes had common functions pertaining to transcription, metabolism and cell growth. Five of the 30 candidate genes (*CD109*, *ITFG2*, *PCP4*, *PRDM2* and *SEMA3D*) have a direct involvement in the immune system (Moioli *et al.*, 2016a).

4.3.3 Candidate gene studies

While few GWAS have been conducted in sheep, there have been multiple studies investigating candidate genes. As in cattle, the association of *SLC11A1*, *TRL2* and *TLR4* with *MAP* infection have been explored in sheep. Candidate genes investigated in sheep that haven't been investigated in cattle include *CD109* and the major histocompatibility complex (MHC) (Table 4.2).

The *CD109* gene encodes a protein that binds to and negatively regulates transforming growth factor β (*TGFB*) (Bizet *et al.*, 2012). During *MAP* infection, upregulation of TGFB interferes with macrophage activation resulting in less effective macrophage-killing mechanisms (Khalifeh and Stabel, 2004). The coding regions of *CD109* were further investigated by Moioli *et al.* (2016b) in Sarda sheep to identify missense mutations as potential causal variants in LD with SNPs previously associated with *MAP* infection. Using this method, a single missense mutation within *CD109* (XM_004011463.1) was found to be in LD ($P < 0.0001$) with the OvineSNP50 BeadChip SNP OAR8_270360.1 (LD = 0.74) previously associated with *MAP* infection. The authors then determined that LD was maintained between these two mutations across 33 other breeds ($P < 0.001$), and suggested that OAR8_270360.1 could be used to determine sheep susceptibility/resistance to *MAP* infection through genomic selection (Moioli *et al.*, 2016b).

The MHC class II molecules function to present antigens to the host immune cells. The presentation of the antigens stimulates the immune response and activates various signalling pathways (Holling *et al.*, 2004). Mutations within these molecules have been previously associated with multiple mycobacterial infections in humans including *M. tuberculosis* (Holling *et al.*, 2004), *Mycobacterum leprae* (Hashimoto *et al.*, 2002) and *Mycobacterium avium* (LeBlanc *et al.*, 2000). In sheep, a 2005 study investigated mutations within MHC molecules for association with *MAP* infection (Reddacliff *et al.*, 2005). This study identified a microsatellite within the MHC locus in a population of Merino sheep that was associated with *MAP* infection ($P < 0.05$) when infection status was determined by culturing tissues and faeces (Table 4.2). The MHC genes have also been associated with *MAP* infection in cattle, as Weiss *et al.* (2001) found that *MAP* infection resulted in downregulation of MHC molecule on the surface of infected macrophages.

4.3.4 Pathway analysis

While GSEA-SNP has not been used to identify gene sets or leading edge genes associated with susceptibility to *MAP* infection in sheep, several pathway analyses have been conducted using IPA. Gossner *et al.* (2017) performed a pathway analysis to investigate genes and pathways associated with naturally *MAP* infected Scottish Blackface/Blackface cross ewes. In this study, the transcriptome of ileocecal lymph nodes from healthy sheep were compared with those from sheep classified as having paucibacillary or multibacillary forms of *MAP* infection. Paucibacillary infection is when few mycobacteria are present within the gut and only T cells are present within the lamina propria, whereas sheep with multibacillary infections have a high level of bacteria present in the gut and macrophage and B cells are present within the lamina propria (Smeed *et al.*, 2007). The Gossner *et al.* (2017) study reported that more than 20 pathways enriched for both the paucibacillary and multibacillary forms had functions related to inflammation and tissue repair. As would be expected, sheep with multibacillary infections had pathways enriched for T cells while sheep with paucibacillary infections did not.

A second IPA also compared healthy sheep with paucibacillary or multibacillary *MAP* infection. Purdie *et al.* (2019) used a population of Merino sheep that were experimentally exposed to *MAP* and used the transcriptome from white blood cells isolated from peripheral blood. This analysis reported that the most common pathways associated for both forms of *MAP* infection were related to cell growth and proliferation, and lipid metabolism (Purdie *et al.*, 2019). Lipids are an integral component of plasma membranes and they play an important role in phagocytosis of mycobacteria by host macrophages (Gatfield and Pieters, 2000). Supportive of the finding that lipid metabolism is important to *MAP* infection was a study of Holstein and Red cattle from Australia that also identified a lipid metabolism pathway as important to *MAP* pathogenesis (Thirunavukkarasu *et al.*, 2014).

4.4 Evidence for Disease Susceptibility Differences in Goats

Herd-level prevalence in goats varies by geographical region and breed. Few studies have investigated breed differences in susceptibility to *MAP* infection among goats although a study in India identified that breeds native to semi-arid climates, such as Barbari and Jamunapari, were more likely to suffer from *MAP* infection compared with crossbred Rajasthani goats (Singh *et al.*, 2009).

4.4.1 Heritability estimates

In goats, there have been two studies that reported an estimated heritability for susceptibility to *MAP* infection. One study estimated heritability to be between 0.01 and 0.15 for susceptibility to *MAP* infection in India for the Barbari and Jamnapari breeds (Singh *et al.*, 1990). A more recent study on Dutch White dairy goats reported heritability estimates for susceptibility to *MAP* infection that ranged between 0.07 and 0.12 (van Hulzen *et al.*, 2012b).

4.4.2 Genome-wide association studies

Few GWAS have been conducted in goats to identify loci associated with susceptibility to *MAP* infection. However, a recent study identified seven loci associated ($P < 5 \times 10^{-4}$) with *MAP* susceptibility in Italian Garfagnina sheep identified when *MAP* infection status was determined using a serum ELISA (Cecchi *et al.*, 2017; Table 4.1). These seven loci were located across seven chromosomes and harboured nine positional candidate genes, three of which (*LOC102187381*, *MANEA* and *PCSK5*) had functions related to the Golgi apparatus (Cecchi *et al.*, 2017). The Golgi apparatus has an important role in *MAP* infection as it is where TLR4 is located (Hornef *et al.*, 2002). As previously mentioned, TLR4 is involved in *MAP* recognition by the host immune cells (Ferwerda *et al.*, 2007).

4.4.3 Candidate gene studies

The number of candidate gene studies performed in goats has increased over the last several years. Now, like cattle and sheep, the association of *SLC11A1* in *MAP* susceptibility to infection in goats has been investigated (Table 4.2). In addition to *SLC11A1*, researchers have also investigated the HLA class II histocompatibility antigen, DRB1-4 β chain (*DRB3*) gene. The HLA system is a gene complex that plays a key role in immunity. The *DRB3* gene is associated (odds ratio >0) with susceptibility to *MAP* infection in Holstein–Slovak crosses of cattle (Rastislav and Mangesh, 2012; Table 4.2). This gene is part of a heterodimer that has an established role in presenting mycobacterial antigens to host immune cells (Mustafa, 2000). A study in Jamunapari goats used PCR restriction fragment-length polymorphisms with the restriction endonucleases *PstI* and *TaqI* to determine variants within *DRB* (Singh *et al.*, 2012). This study used multiple diagnostic methods to determine infection status (serum ELISA, faecal culture and microscopic faecal exams) and reported that two variants (P and T) differed ($P < 0.001$) between healthy and *MAP* infected goats (Singh *et al.*, 2012). The authors reported that the recessive genotypes 'pp' and 'tt' and the recessive haplotype 'pptt' occurred at higher frequencies within *MAP*-resistant goats compared with susceptible goats ($P_{corr} < 0.001$).

Another recent study investigated associations with *MAP* infection with 12 microsatellites in a population of 48 Garfagnina goats using a serum ELISA test to detect *MAP* infection (Cecchi *et al.*, 2018). Two of the 12 microsatellites were weakly associated with *MAP*: *ETH10* on chromosome 5 ($P < 0.1$) and *SRCRSP05* ($P < 0.05$) on chromosome 21 (Cinar *et al.*, 2018).

4.4.4 Pathway analysis

Currently, no pathway analyses in goats have been published.

4.5 Evidence for Disease Susceptibility Differences in Deer

While few studies have investigated differences in *MAP* susceptibility across farmed deer species, it has been established that susceptibility to *MAP* infection differs across lines of red deer (*Cervus elaphus*) (Mackintosh *et al.*, 2012; Dobson *et al.*, 2013). Several studies have also investigated differences in susceptibility to *MAP* infection between farmed animal species that graze common pastures, including red deer, Wapiti (elk), cattle and sheep (Verdugo *et al.*, 2014; Forde *et al.*, 2015). Species comparisons have shown that red deer tend to be more susceptible to infection by *MAP*-C than *MAP*-S strains (see Chapter 6 for a full description of *MAP* strain types) (O'Brien *et al.*, 2006; Verdugo *et al.*, 2014). It has been suggested that differences in susceptibility between deer and other species may be due to intra-uterine transmission of *MAP*, which has been reported to occur in approximately 2% of ewes (Lambeth *et al.*, 2004), in 9–39% of cows (Whittington and Windsor, 2009), and in about 78% of hinds (female red deer; Thompson *et al.*, 2007).

4.5.1 Heritability estimates

Heritability estimates in domesticated red deer have been moderate, ranging between 0.16 and 0.3, which is similar to the aforementioned heritability estimates for cattle, sheep and goats (Griffin *et al.*, 2012; Mackintosh *et al.*, 2012; Johne's Disease Research Consortium, 2016). These heritability estimates support that selection of deer and other ruminant species for enhanced resistance to *MAP* infection could be useful in reducing the disease in domestic animals.

4.5.2 Genome-wide association studies and candidate gene studies

Currently, no GWAS or candidate gene studies in farmed deer have been published.

4.5.3 Pathway analysis

While no GSEA-SNP has been performed in deer, there have been studies using IPA to investigate how the genes that are differently expressed in *MAP*-infected and healthy deer interact with one another. A *MAP* challenge study by Mackintosh *et al.* (2016) compared differentially expressed genes between red deer from resistant and susceptible sire lines. After the *MAP* challenge, deer were tested throughout the 50-week study period for *MAP* antibodies using the serum Paralisa™ ELISA test. Over the duration of the study and after its conclusion, lymph node biopsies were taken for gene expression analysis. Genes that were differentially expressed in resistant deer after the *MAP* challenge were associated with pathways related to adaptive immunity. In contrast, genes that were differentially expressed in susceptible deer after a *MAP* challenge were associated with pro-inflammatory responses pathways (Mackintosh *et al.*, 2016). Of these differentially expressed genes, two (*CD209* and *SLC11A1*) have been previously investigated as candidate genes in other species (Table 4.2).

4.6 Conclusion

The studies reported in this chapter provide evidence for differences in disease prevalence between breeds and family lines in cattle, sheep, goats and deer, supporting the hypothesis that there is a genetic influence on host susceptibility to *MAP* infection. While there has been limited validation of loci across studies, the identification of candidate genes across species suggests that there are common mechanisms influencing host susceptibility. In order to reduce disease prevalence through genomic selection a standardization of phenotypes through a standardized diagnosis of paratuberculosis is needed. The continued characterization of host susceptibility to *MAP* infection across species will allow for better understanding of the biological mechanisms behind the disease and allow for development of selection tools to help decrease its prevalence in domesticated livestock.

References

Abraham, A., Naicy, T., Raghavan, K.C., Siju, J. and Aravindakshan, T. (2017) Evaluation of the association of *SLC11A1* gene polymorphism with incidence of paratuberculosis in goats. *Journal of Genetics* 96(4), 641–646. DOI: 10.1007/s12041-017-0820-9.

Alpay, F., Zare, Y., Kamalludin, M.H., Huang, X., Shi, X. *et al.* (2014) Genome-wide association study of susceptibility to infection by *Mycobacterium avium* subspecies *paratuberculosis* in Holstein cattle. *PLoS ONE* 9(12), e111704. DOI: 10.1371/journal.pone.0111704.

Bai, X., Feldman, N.E., Chmura, K., Ovrutsky, A.R., Su, W.-L. *et al.* (2013) Inhibition of nuclear factor-kappa B activation decreases survival of *Mycobacterium tuberculosis* in human macrophages. *PLoS ONE* 8(4), e61925. DOI: 10.1371/journal.pone.0061925.

Begg, D.J., Purdie, A.C., de Silva, K., Dhand, N.K., Plain, K.M. *et al.* (2017) Variation in susceptibility of different breeds of sheep to *Mycobacterium avium* subspecies *paratuberculosis* following experimental inoculation. *Veterinary Research* 48(1), 36. DOI: 10.1186/s13567-017-0440-7.

Bizet, A.A., Tran-Khanh, N., Saksena, A., Liu, K., Buschmann, M.D. *et al.* (2012) CD109-mediated degradation of TGF-β receptors and inhibition of TGF-β responses involve regulation of SMAD7 and Smurf2 localization and function. *Journal of Cellular Biochemistry* 113(1), 238–246. DOI: 10.1002/jcb.23349.

Brito, L.F., Mallikarjunappa, S., Sargolzaei, M., Koeck, A., Chesnais, J. *et al.* (2018) The genetic architecture of milk ELISA scores as an indicator of Johne's disease (paratuberculosis) in dairy cattle. *Journal of Dairy Science* 101(11), 10062–10075. DOI: 10.3168/jds.2017-14250.

Caruso, R., Warner, N., Inohara, N. and Núñez, G. (2014) NOD1 and NOD2: signaling, host defense, and inflammatory disease. *Immunity* 41(6), 898–908. DOI: 10.1016/j.immuni.2014.12.010.

Cecchi, F., Russo, C., Iamartino, D., Galiero, A., Turchi, B. *et al.* (2017) Identification of candidate genes for paratuberculosis resistance in the native Italian Garfagnina goat breed. *Tropical Animal Health and Production* 49(6), 1135–1142. DOI: 10.1007/s11250-017-1306-8.

Cecchi, F., Russo, C., Fratini, F., Turchi, B., Preziuso, G. *et al.* (2018) Preliminary association analysis of microsatellites and *Mycobacterium avium* subspecies paratuberculosis infection in the native Garfagnina goats. *Journal of Applied Animal Research* 46(1), 879–882. DOI: 10.1080/09712119.2017.1417129.

Cinar, M.U., Hizlisoy, H., Akyüz, B., Arslan, K., Aksel, E.G. *et al.* (2018) Polymorphisms in toll-like receptor (*TLR*) 1, 4, 9 and *SLC11A1* genes and their association with paratuberculosis susceptibility in Holstein and indigenous crossbred cattle in Turkey. *Journal of Genetics* 97(5), 1147–1154. DOI: 10.1007/s12041-018-1008-7.

David, J., Barkema, H.W., Mortier, R., Ghosh, S., Guan, L.L. *et al.* (2014) Gene expression profiling and putative biomarkers of calves 3 months after infection with *Mycobacterium avium* subspecies *paratuberculosis*. *Veterinary Immunology and Immunopathology* 160(1–2), 107–117. DOI: 10.1016/j.vetimm.2014.04.006.

Del Corvo, M., Luini, M., Stella, A., Pagnacco, G., Ajmone-Marsan, P. *et al.* (2017) Identification of additional loci associated with antibody response to *Mycobacterium avium* ssp. *paratuberculosis* in cattle by GSEA–SNP analysis. *Mammalian Genome* 28(11–12), 520–527. DOI: 10.1007/s00335-017-9714-7.

Dobson, B., Liggett, S., O'Brien, R. and Griffin, J.F.T. (2013) Innate immune markers that distinguish red deer (*Cervus elaphus*) selected for resistant or susceptible genotypes for Johne's disease. *Veterinary Research* 44(1), 5. DOI: 10.1186/1297-9716-44-5.

Ferwerda, G., Kullberg, B.J., de Jong, D.J., Girardin, S.E., Langenberg, D.M.L. *et al.* (2007) *Mycobacterium paratuberculosis* is recognized by Toll-like receptors and NOD2. *Journal of Leukocyte Biology* 82(4), 1011–1018. DOI: 10.1189/jlb.0307147.

Forbes, J.R. and Gros, P. (2003) Iron, manganese, and cobalt transport by Nramp1 (Slc11a1) and Nramp2 (Slc11a2) expressed at the plasma membrane. *Blood* 102(5), 1884–1892. DOI: 10.1182/blood-2003-02-0425.

Forde, T., Pruvot, M., De Buck, J. and Orsel, K. (2015) A high-morbidity outbreak of Johne's disease in game-ranched elk. *Canadian Veterinary Journal* 56, 479–483.

Gao, Y., Cao, J., Zhang, S., Zhang, Q. and Sun, D. (2018a) *Short communication*: Heritability estimates for susceptibility to *Mycobacterium avium* ssp. *paratuberculosis* infection in Chinese Holstein cattle. *Journal of Dairy Science* 101(8), 7274–7279. DOI: 10.3168/jds.2017-13264.

Gao, Y., Jiang, J., Yang, S., Cao, J., Han, B. *et al.* (2018b) Genome-wide association study of *Mycobacterium avium* subspecies *paratuberculosis* infection in Chinese Holstein. *BMC Genomics* 19(1), 972. DOI: 10.1186/s12864-018-5385-3.

Gatfield, J. and Pieters, J. (2000) Essential role for cholesterol in entry of mycobacteria into macrophages. *Science* 288(5471), 1647–1651. DOI: 10.1126/science.288.5471.1647.

Geijtenbeek, T.B.H., van Vliet, S.J., Koppel, E.A., Sanchez-Hernandez, M., Vandenbroucke-Grauls, C.M.J.E. *et al.* (2003) Mycobacteria target DC-SIGN to suppress dendritic cell function. *The Journal of Experimental Medicine* 197(1), 7–17. DOI: 10.1084/jem.20021229.

Gossner, A., Watkins, C., Chianini, F. and Hopkins, J. (2017) Pathways and genes associated with immune dysfunction in sheep paratuberculosis. *Scientific Reports* 7(1), 46695. DOI: 10.1038/srep46695.

Griffin, J.F., O'Brien, R., Liggett, S., Dobson, B. and Mackintosh, C.G. (2012) Genetic markers for resistance and susceptibility to Johne's disease: a deer model. *Proceedings of the 11th International Colloquium on Paratuberculosis. International Association for Paratuberculosis*, Sydney, Australia, p. 99.

Hashimoto, K., Maeda, Y., Kimura, H., Suzuki, K., Masuda, A. *et al.* (2002) *Mycobacterium leprae* infection in monocyte-derived dendritic cells and its influence on antigen-presenting function. *Infection and Immunity* 70(9), 5167–5176. DOI: 10.1128/IAI.70.9.5167-5176.2002.

Hemalatha, S., Roy, P., Purushothaman, V. and Iyue, M. (2013) Paratuberculosis in different breeds of sheep: a retrospective study of cases. *International Journal of Mycobacteriology* 2(3), 166–170. DOI: 10.1016/j.ijmyco.2013.07.002.

Hickey, S.M., Morris, C.A., Dobbie, J.L. and Lake, D.E. (2003) Heritability of Johne's disease and survival data from Romney and Merino sheep. *Proceedings of the New Zealand Society of Animal Production* 63, 179–182.

Holling, T.M., Schooten, E. and van Den Elsen, P.J. (2004) Function and regulation of MHC class II molecules in T-lymphocytes: of mice and men. *Human Immunology* 65(4), 282–290. DOI: 10.1016/j.humimm.2004.01.005.

Hornef, M.W., Frisan, T., Vandewalle, A., Normark, S. and Richter-Dahlfors, A. (2002) Toll-like receptor 4 resides in the Golgi apparatus and colocalizes with internalized lipopolysaccharide in intestinal epithelial cells. *The Journal of Experimental Medicine* 195(5), 559–570. DOI: 10.1084/jem.20011788.

Johne's Disease Research Consortium (2016) Milestone 4.4 DEERSelect module. Available at: https://static1.squarespace.com/static/57857740cd0f68284f643870/t/57d77d8cf5e231039ee03be0/1473740200964/Report+1-1+JDRC+research+and+development+record+2016_web.pdf

Juste, R.A., Vazquez, P., Ruiz-Larrañaga, O., Iriondo, M., Manzano, C. *et al.* (2018) Association between combinations of genetic polymorphisms and epidemiopathogenic forms of bovine paratuberculosis. *Heliyon* 4(2), e00535. DOI: 10.1016/j.heliyon.2018.e00535.

Khalifeh, M.S. and Stabel, J.R. (2004) Effects of gamma interferon, interleukin-10, and transforming growth factor β on the survival of *Mycobacterium avium* subsp. *paratuberculosis* in monocyte-derived macrophages from naturally infected cattle. *Infection and Immunity* 72(4), 1974–1982. DOI: 10.1128/IAI.72.4.1974-1982.2004.

Kirkpatrick, B.W. (2010) Genetics of host susceptibility to paratuberculosis. In: Behr, M.A. and Collins, D.M. (eds) *Paratuberculosis: Organism, Disease, Control*. CAB International, Wallingford, UK, pp. 50–59.

Kirkpatrick, B.W. and Lett, B.M. (2018) *Short communication*: Heritability of susceptibility to infection by *Mycobacterium avium* ssp. *paratuberculosis* in Holstein cattle. *Journal of Dairy Science* 101(12), 11165–11169. DOI: 10.3168/jds.2018-15021.

Kirkpatrick, B.W., Shi, X., Shook, G.E. and Collins, M.T. (2011) Whole-genome association analysis of susceptibility to paratuberculosis in Holstein cattle. *Animal Genetics* 42(2), 149–160. DOI: 10.1111/j.1365-2052.2010.02097.x.

Kiser, J.N., White, S.N., Johnson, K.A., Hoff, J.L., Taylor, J.F. *et al.* (2017) Identification of loci associated with susceptibility to *Mycobacterium avium* subspecies *paratuberculosis* (*Map*) tissue infection in cattle. *Journal of Animal Science* 95(3), 1080–1091. DOI: 10.2527/jas.2016.1152.

Kiser, J.N., Neupane, M., White, S.N. and Neibergs, H.L. (2018) Identification of genes associated with susceptibility to *Mycobacterium avium* ssp. *paratuberculosis* (MAP) tissue infection in Holstein cattle using gene set enrichment analysis–SNP. *Mammalian Genome* 29(7–8), 539–549. DOI: 10.1007/s00335-017-9725-4.

Koets, A., Santema, W., Mertens, H., Oostenrijk, D., Keestra, M. *et al.* (2010) Susceptibility to *paratuberculosis* infection in cattle is associated with single nucleotide polymorphisms in Toll-like receptor 2 which modulate immune responses against *Mycobacterium avium* subspecies *paratuberculosis*. *Preventive Veterinary Medicine* 93(4), 305–315. DOI: 10.1016/j.prevetmed.2009.11.008.

Korou, L.M., Liandris, E., Gazouli, M. and Ikonomopoulos, J. (2010) Investigation of the association of the SLC11A1 gene with resistance/sensitivity of goats (*Capra hircus*) to paratuberculosis. *Veterinary Microbiology* 144(3–4), 353–358. DOI: 10.1016/j.vetmic.2010.01.009.

Krämer, A., Green, J., Pollard, J. and Tugendreich, S. (2014) Causal analysis approaches in ingenuity pathway analysis. *Bioinformatics* 30(4), 523–530. DOI: 10.1093/bioinformatics/btt703.

Kumar, S., Kumar, S., Singh, R.V., Chauhan, A., Kumar, A. *et al.* (2019) Association of Bovine *CLEC7A* gene polymorphism with host susceptibility to paratuberculosis disease in Indian cattle. *Research in Veterinary Science* 123, 216–222. DOI: 10.1016/j.rvsc.2019.01.016.

Küpper, J., Brandt, H., Donat, K. and Erhardt, G. (2012) Heritability estimates for *Mycobacterium avium* subspecies *paratuberculosis* status of German Holstein cows tested by fecal culture. *Journal of Dairy Science* 95(5), 2734–2739. DOI: 10.3168/jds.2011-4994.

Lambeth, C., Reddacliff, L.A., Windsor, P., Abbott, K.A., McGregor, H. *et al.* (2004) Intrauterine and transmammary transmission of *Mycobacterium avium* subsp *paratuberculosis* in sheep. *Australian Veterinary Journal* 82(8), 504–508. DOI: 10.1111/j.1751-0813.2004.tb11171.x.

LeBlanc, S.B., Naik, E.G., Jacobson, L. and Kaslow, R.A. (2000) Association of DRB1*1501 with disseminated *Mycobacterium avium* complex infection in North American AIDS patients. *Tissue Antigens* 55(1), 17–23. DOI: 10.1034/j.1399-0039.2000.550103.x.

Liu, T., Zhang, L., Joo, D. and Sun, S.-C. (2017) NF-κB signaling in inflammation. *Signal Transduction and Targeted Therapy* 2(1), 170023. DOI: 10.1038/sigtrans.2017.23.

Määttänen, P., Trost, B., Scruten, E., Potter, A., Kusalik, A. *et al.* (2013) Divergent immune responses to *Mycobacterium avium* subsp. *paratuberculosis* infection correlate with kinome responses at the site of intestinal infection. *Infection and Immunity* 81(8), 2861–2872. DOI: 10.1128/IAI.00339-13.

MacHugh, D.E., Taraktsoglou, M., Killick, K.E., Nalpas, N.C., Browne, J.A. *et al.* (2012) Pan-genomic analysis of bovine monocyte-derived macrophage gene expression in response to in vitro infection with *Mycobacterium avium* subspecies *paratuberculosis*. *Veterinary Research* 43(1), 215. DOI: 10.1186/1297-9716-43-25.

Mackintosh, C.D., de Lisle, G.W., Clark, R.G. and Griffin, J.F. (2012) Heritability/susceptibility of red deer to MAP challenge. *Proceedings of the 11th International Colloquium on Paratuberculosis. International Association for Paratuberculosis*, Sydney, Australia, pp. 117–118.

Mackintosh, C.G., Griffin, J.F.T., Scott, I.C., O'Brien, R., Stanton, J.L. *et al.* (2016) SOLiD SAGE sequencing shows differential gene expression in jejunal lymph node samples of resistant and susceptible red deer (*Cervus elaphus*) challenged with *Mycobacterium avium* subsp. *paratuberculosis*. *Veterinary Immunology and Immunopathology* 169, 102–110. DOI: 10.1016/j.vetimm.2015.10.009.

Mallikarjunappa, S., Sargolzaei, M., Brito, L.F., Meade, K.G., Karrow, N.A. *et al.* (2018) *Short communication:* Uncovering quantitative trait loci associated with resistance to *Mycobacterium avium* ssp. *paratuberculosis* infection in Holstein cattle using a high-density single nucleotide polymorphism panel. *Journal of Dairy Science* 101(8), 7280–7286. DOI: 10.3168/jds.2018-14388.

Malvisi, M., Palazzo, F., Morandi, N., Lazzari, B., Williams, J.L. *et al.* (2016) Responses of bovine innate immunity to *Mycobacterium avium* subsp. *paratuberculosis* infection revealed by changes in gene expression and levels of microRNA. *PLoS ONE* 11(10), e0164461. DOI: 10.1371/journal.pone.0164461.

Medzhitov, R. (2001) Toll-like receptors and innate immunity. *Nature Reviews Immunology* 1(2), 135–145. DOI: 10.1038/35100529.

Minozzi, G., Buggiotti, L., Stella, A., Strozzi, F., Luini, M. *et al.* (2010) Genetic loci involved in antibody response to *Mycobacterium avium* ssp. *paratuberculosis* in cattle. *PLoS ONE* 5(6), e11117. DOI: 10.1371/journal.pone.0011117.

Minozzi, G., Williams, J.L., Stella, A., Strozzi, F., Luini, M. *et al.* (2012) Meta-analysis of two genome-wide association studies of bovine paratuberculosis. *PLoS ONE* 7(3), e32578. DOI: 10.1371/journal.pone.0032578.

Moioli, B., D'Andrea, S., De Grossi, L., Sezzi, E., De Sanctis, B. *et al.* (2016a) Genomic scan for identifying candidate genes for paratuberculosis resistance in sheep. *Animal Production Science* 56(7), 1046–1055. DOI: 10.1071/AN14826.

Moioli, B., De Grossi, L., Steri, R. and Pilla, F. (2016b) Identification of missense mutations in the PCP4 and CD109 genes to validate the effect of neutral genetic markers. *Czech Journal of Animal Science* 61(07), 317–325. DOI: 10.17221/30/2015-CJAS.

Motiwala, A.S., Janagama, H.K., Paustian, M.L., Zhu, X., Bannantine, J.P. *et al.* (2006) Comparative transcriptional analysis of human macrophages exposed to animal and human isolates of *Mycobacterium avium* subspecies *paratuberculosis* with diverse genotypes. *Infection and Immunity* 74(11), 6046–6056. DOI: 10.1128/IAI.00326-06.

Mucha, R., Bhide, M.R., Chakurkar, E.B., Novak, M. and Mikula, I. (2009) Toll-like receptors TLR1, TLR2 and TLR4 gene mutations and natural resistance to *Mycobacterium avium* subsp. *paratuberculosis* infection in cattle. *Veterinary Immunology and Immunopathology* 128(4), 381–388. DOI: 10.1016/j.vetimm.2008.12.007.

Mustafa, A.S. (2000) HLA-restricted immune response to mycobacterial antigens: relevance to vaccine design. *Human Immunology* 61(2), 166–171. DOI: 10.1016/S0198-8859(99)00137-8.

Neibergs, H.L., Settles, M.L., Whitlock, R.H. and Taylor, J.F. (2010) GSEA-SNP identifies genes associated with Johne's disease in cattle. *Mammalian Genome* 21(7–8), 419–425. DOI: 10.1007/s00335-010-9278-2.

Neibergs, H.L., Seabury, C.M., Wojtowicz, A.J., Wang, Z., Scraggs, E. *et al.* (2014) Susceptibility loci revealed for bovine respiratory disease complex in pre-weaned Holstein calves. *BMC Genomics* 15(1), 1164. DOI: 10.1186/1471-2164-15-1164.

Nielsen, S.S. and Toft, N. (2009) A review of prevalences of paratuberculosis in farmed animals in Europe. *Preventive Veterinary Medicine* 88(1), 1–14. DOI: 10.1016/j.prevetmed.2008.07.003.

O'Brien, R., Mackintosh, C.G., Bakker, D., Kopecna, M., Pavlik, I. *et al.* (2006) Immunological and molecular characterization of susceptibility in relationship to bacterial strain differences in *Mycobacterium avium* subsp. *paratuberculosis* infection in the red deer (*Cervus elaphus*). *Infection and Immunity* 74(6), 3530–3537. DOI: 10.1128/IAI.01688-05.

Pant, S.D., Schenkel, F.S., Verschoor, C.P., You, Q., Kelton, D.F. *et al.* (2010) A principal component regression based genome wide analysis approach reveals the presence of a novel QTL on BTA7 for MAP resistance in holstein cattle. *Genomics* 95(3), 176–182. DOI: 10.1016/j.ygeno.2010.01.001.

Pant, S.D., Verschoor, C.P., Schenkel, F.S., You, Q., Kelton, D.F. *et al.* (2011) Bovine *PGLYRP1* polymorphisms and their association with resistance to *Mycobacterium avium* ssp. *paratuberculosis*. *Animal Genetics* 42(4), 354–360. DOI: 10.1111/j.1365-2052.2010.02153.x.

Pant, S.D., Verschoor, C.P., Schenkel, F.S., You, Q., Kelton, D.F. *et al.* (2014) Bovine CLEC7A genetic variants and their association with seropositivity in Johne's disease ELISA. *Gene* 537(2), 302–307. DOI: 10.1016/j.gene.2013.12.020.

Pauciullo, A., Küpper, J., Brandt, H., Donat, K., Iannuzzi, L. *et al.* (2015) *Wingless-type MMTV integration site family member 2* (*WNT2*) gene is associated with resistance to MAP in faecal culture and antibody response in Holstein cattle. *Animal Genetics* 46(2), 122–132. DOI: 10.1111/age.12261.

Pinedo, P.J., Buergelt, C.D., Donovan, G.A., Melendez, P., Morel, L. *et al.* (2009a) Candidate gene polymorphisms (*BoIFNG*, *TLR4*, *SLC11A1*) as risk factors for paratuberculosis infection in cattle. *Preventive Veterinary Medicine* 91(2–4), 189–196. DOI: 10.1016/j.prevetmed.2009.05.020.

Pinedo, P.J., Buergelt, C.D., Donovan, G.A., Melendez, P., Morel, L. *et al.* (2009b) Association between *CARD15/NOD2* gene polymorphisms and paratuberculosis infection in cattle. *Veterinary Microbiology* 134(3–4), 346–352. DOI: 10.1016/j.vetmic.2008.09.052.

Purdie, A.C., Plain, K.M., Begg, D.J., de Silva, K. and Whittington, R.J. (2011) Candidate gene and genome-wide association studies of *Mycobacterium avium* subsp. *paratuberculosis* infection in cattle and sheep: a review. *Comparative Immunology, Microbiology and Infectious Diseases* 34(3), 197–208. DOI: 10.1016/j.cimid.2010.12.003.

Purdie, A.C., Plain, K.M., Begg, D.J., de Silva, K. and Whittington, R.J. (2019) Gene expression profiles during subclinical *Mycobacterium avium* subspecies *paratuberculosis* infection in sheep can predict disease outcome. *Scientific Reports* 9(1), 8245. DOI: 10.1038/s41598-019-44670-w.

Rastislav, M. and Mangesh, B. (2012) BoLA-DRB3 exon 2 mutations associated with paratuberculosis in cattle. *The Veterinary Journal* 192(3), 517–519. DOI: 10.1016/j.tvjl.2011.07.005.

Reddacliff, L.A., Beh, K., McGregor, H. and Whittington, R.J. (2005) A preliminary study of possible genetic influences on the susceptibility of sheep to Johne's disease. *Australian Veterinary Journal* 83(7), 435–441. DOI: 10.1111/j.1751-0813.2005.tb13087.x.

Ruiz-Larrañaga, O., Garrido, J.M., Manzano, C., Iriondo, M., Molina, E. *et al.* (2010) Identification of single nucleotide polymorphisms in the bovine solute carrier family 11 member 1 (*SLC11A1*) gene and their association with infection by *Mycobacterium avium* subspecies *paratuberculosis*. *Journal of Dairy Science* 93(4), 1713–1721. DOI: 10.3168/jds.2009-2438.

Sadana, T., Singh, R.V., Singh, S.V., Saxena, V.K., Sharma, D. *et al.* (2015) Single nucleotide polymorphism of *SLC11A1*, *IFNG* and *TLR2* genes and their association with *Mycobacterium avium* subspecies *paratuberculosis* infection in native Indian cattle population. *Indian Journal of Biotechnology* 14, 469–475.

Sallam, A.M., Zare, Y., Alpay, F., Shook, G.E., Collins, M.T. *et al.* (2017) An across-breed genome wide association analysis of susceptibility to paratuberculosis in dairy cattle. *Journal of Dairy Science* 84, 61–67.

Settles, M., Zanella, R., McKay, S.D., Schnabel, R.D., Taylor, J.F. *et al.* (2009) A whole genome association analysis identifies loci associated with *Mycobacterium avium* subsp. *paratuberculosis* infection status in US Holstein cattle. *Animal Genetics* 40(5), 655–662. DOI: 10.1111/j.1365-2052.2009.01896.x.

Sharma, B.S., Abo-Ismail, M.K., Schenkel, F.S., You, Q., Verschoor, C.P. *et al.* (2015) Association of *TLR4* polymorphisms with *Mycobacterium avium* subspecies *paratuberculosis* infection status in Canadian Holsteins. *Animal Genetics* 46(5), 560–565. DOI: 10.1111/age.12333.

Shin, M.-K., Park, H.-T., Shin, S.W., Jung, M., Im, Y.B. *et al.* (2015) Whole-blood gene-expression profiles of cows infected with *Mycobacterium avium* subsp. *paratuberculosis* reveal changes in immune response and lipid metabolism. *Journal of Microbiology and Biotechnology* 25(2), 255–267. DOI: 10.4014/jmb.1408.08059.

Shook, G.E., Chaffer, M., Wu, X.-L. and Ezra, E. (2012) Genetic parameters for paratuberculosis infection and effect of infection on production traits in Israeli Holsteins. *Animal Genetics* 43, 56–64. DOI: 10.1111/j.1365-2052.2012.02349.x.

Singh, N., Kala, S.N., Vihan, V.S. and Singh, S.V. (1990) Genetic study on the susceptibility to Johne's disease in goats. *Indian Journal of Animal Sciences* 60, 1163–1165.

Singh, P.K., Singh, S.V., Singh, A.V. and Sohal, J.S. (2009) Variability in susceptibility of different Indian goat breeds with respect to natural and experimental infection of *Mycobacterium avium* subspecies *paratuberculosis*. *Indian Journal of Small Ruminants* 15, 35–43.

Singh, P.K., Singh, S.V., Singh, M.K., Saxena, V.K., Horin, P. *et al.* (2012) Effect of genetic variation in the *MHC* Class II DRB region on resistance and susceptibility to Johne's disease in endangered Indian Jamunapari goats. *International Journal of Immunogenetics* 39(4), 314–320. DOI: 10.1111/j.1744-313X.2012.01092.x.

Singh, S.V., Dhama, K., Chaubey, K.K., Kumar, N., Singh, P.K. *et al.* (2013) Impact of host genetics on susceptibility and resistance to *Mycobacterium avium* subspecies *paratuberculosis* infection in domestic ruminants. *Pakistan Journal of Biological Sciences* 16(6), 251–266. DOI: 10.3923/pjbs.2013.251.266.

Smeed, J.A., Watkins, C.A., Rhind, S.M. and Hopkins, J. (2007) Differential cytokine gene expression profiles in the three pathological forms of sheep paratuberculosis. *BMC Veterinary Research* 3(1), 18. DOI: 10.1186/1746-6148-3-18.

Sorge, U.S., Lissemore, K., Godkin, A., Hendrick, S., Wells, S. *et al.* (2011) Associations between paratuberculosis milk ELISA result, milk production, and breed in Canadian dairy cows. *Journal of Dairy Science* 94(2), 754–761. DOI: 10.3168/jds.2010-3404.

Thirunavukkarasu, S., Plain, K.M., de Silva, K., Begg, D., Whittington, R.J. *et al.* (2014) Expression of genes associated with cholesterol and lipid metabolism identified as a novel pathway in the early pathogenesis of *Mycobacterium avium* subspecies *paratuberculosis*-infection in cattle. *Veterinary Immunology and Immunopathology* 160(3–4), 147–157. DOI: 10.1016/j.vetimm.2014.04.002.

Thompson, B.R., Clark, R.G. and Mackintosh, C.G. (2007) Intra-uterine transmission of *Mycobacterium avium* subsp *paratuberculosis* in subclinically affected red deer (*Cervus elaphus*). *New Zealand Veterinary Journal* 55(6), 308–313. DOI: 10.1080/00480169.2007.36786.

van Hulzen, K.J.E., Nielen, M., Koets, A.P., de Jong, G., van Arendonk, J.A.M. *et al.* (2011) Effect of herd prevalence on heritability estimates of antibody response to *Mycobacterium avium* subspecies *paratuberculosis*. *Journal of Dairy Science* 94(2), 992–997. DOI: 10.3168/jds.2010-3472.

van Hulzen, K.J.E., Schopen, G.C.B., van Arendonk, J.A.M., Nielen, M., Koets, A.P. *et al.* (2012a) Genome-wide association study to identify chromosomal regions associated with antibody response to *Mycobacterium avium* subspecies *paratuberculosis* in milk of Dutch Holstein-Friesians. *Journal of Dairy Science* 95(5), 2740–2748. DOI: 10.3168/jds.2011-5005.

van Hulzen, K.J.E., Koets, A.P., Nielen, M., Hoeboer, J., van Arendonk, J.A.M. *et al.* (2012b) Genetic variation for infection status as determined by a specific antibody response against *Mycobacterium avium* subspecies *paratuberculosis* in milk of Dutch dairy goats. *Journal of Dairy Science* 95(10), 6145–6151. DOI: 10.3168/jds.2012-5616.

Vannberg, F.O., Chapman, S.J., Khor, C.C., Tosh, K., Floyd, S. *et al.* (2008) *CD209* genetic polymorphism and tuberculosis disease. *PLoS ONE* 3(1), e1388. DOI: 10.1371/journal.pone.0001388.

Vázquez, P., Ruiz-Larrañaga, O., Garrido, J.M., Iriondo, M., Manzano, C. *et al.* (2014) Genetic association analysis of paratuberculosis forms in holstein-friesian cattle. *Veterinary Medicine International* 2014(3), 1–8. DOI: 10.1155/2014/321327.

Verdugo, C., Pleydell, E., Price-Carter, M., Prattley, D., Collins, D. *et al.* (2014) Molecular epidemiology of *Mycobacterium avium* subsp. *paratuberculosis* isolated from sheep, cattle and deer on New Zealand pastoral farms. *Preventive Veterinary Medicine* 117(3–4), 436–446. DOI: 10.1016/j.prevetmed.2014.09.009.

Verschoor, C.P., Pant, S.D., You, Q., Schenkel, F.S., Kelton, D.F. *et al.* (2010) Polymorphisms in the gene encoding bovine interleukin-10 receptor alpha are associated with *Mycobacterium avium* ssp. *paratuberculosis* infection status. *BMC Genetics* 11(1), 23. DOI: 10.1186/1471-2156-11-23.

Vilar, A.L.T., Santos, C.S.A.B., Pimenta, C.L.R.M., Freitas, T.D., Brasil, A.W.L. *et al.* (2015) Herd-level prevalence and associated risk factors for *Mycobacterium avium* subsp. *paratuberculosis* in cattle in the State of Paraíba, Northeastern Brazil. *Preventive Veterinary Medicine* 121(1–2), 49–55. DOI: 10.1016/j.prevetmed.2015.06.003.

Wambier, C.G., Ramalho, L.N., Frade, M.A. and Foss, N.T. (2014) NFκB activation in cutaneous lesions of leprosy is associated with development of multibacillary infection. *Journal of Inflammation Research* 7, 133–138.

Wang, Z., Sun, Y., Fu, Xi'an., Yu, G., Wang, C. *et al.* (2016) A large-scale genome-wide association and meta-analysis identified four novel susceptibility loci for leprosy. *Nature Communications* 7(1), 13760. DOI: 10.1038/ncomms13760.

Weiss, D.J., Evanson, O.A., McClenahan, D.J., Abrahamsen, M.S. and Walcheck, B.K. (2001) Regulation of expression of major histocompatibility antigens by bovine macrophages infected with *Mycobacterium avium* subsp. *paratuberculosis* or *Mycobacterium avium* subsp. *avium*. *Infection and Immunity* 69(2), 1002–1008. DOI: 10.1128/IAI.69.2.1002-1008.2001.

Whittington, R.J. and Windsor, P.A. (2009) In utero infection of cattle with *Mycobacterium avium* subsp. *paratuberculosis*: a critical review and meta-analysis. *The Veterinary Journal* 179(1), 60–69. DOI: 10.1016/j.tvjl.2007.08.023.

Wullaert, A., Bonnet, M.C. and Pasparakis, M. (2011) NF-κB in the regulation of epithelial homeostasis and inflammation. *Cell Research* 21(1), 146–158. DOI: 10.1038/cr.2010.175.

Yadav, M. and Schorey, J.S. (2006) The β-glucan receptor dectin-1 functions together with TLR2 to mediate macrophage activation by mycobacteria. *Blood* 108(9), 3168–3175. DOI: 10.1182/blood-2006-05-024406.

Yamakawa, Y., Pennelegion, C., Willcocks, S., Stalker, A., MacHugh, N. *et al.* (2008) Identification and functional characterization of a bovine orthologue to DC-SIGN. *Journal of Leukocyte Biology* 83(6), 1396–1403. DOI: 10.1189/jlb.0807523.

Zanella, R., Settles, M.L., McKay, S.D., Schnabel, R., Taylor, J. *et al.* (2011) Identification of loci associated with tolerance to Johne's disease in Holstein cattle. *Animal Genetics* 42(1), 28–38. DOI: 10.1111/j.1365-2052.2010.02076.x.

Zare, Y., Shook, G.E., Collins, M.T. and Kirkpatrick, B.W. (2014) *Short communication*: Heritability estimates for susceptibility to *Mycobacterium avium* subspecies *paratuberculosis* infection defined by ELISA and fecal culture test results in Jersey cattle. *Journal of Dairy Science* 97(7), 4562–4567. DOI: 10.3168/jds.2013-7426.

5 *Mycobacterium avium* Complex

Christine Y. Turenne* and David C. Alexander
Department of Medical Microbiology and Infectious Diseases, University of Manitoba, Winnipeg, Manitoba, Canada

5.1 Introduction

The *Mycobacterium avium* complex (MAC) comprises several species of slow-growing mycobacteria that are prevalent in environmental, veterinary and clinical settings. The MAC includes professional pathogens of birds and livestock, opportunistic pathogens of humans, as well as organisms commonly found in soil and water. Historically, classification of MAC organisms was based on phenotypic features, including growth characteristics, source of isolation and virulence in experimental animals. Molecular taxonomy approaches have challenged traditional designations and transformed our view of the MAC. Although *Mycobacterium paratuberculosis* was once considered a separate species, current taxonomy classifies it as *M. avium* subsp. *paratuberculosis* (*MAP*), a pathogenic clone of the MAC.

5.2 *Mycobacterium avium*

5.2.1 Discovery of *Mycobacterium avium*

Avian tuberculosis is a chronic wasting disease of wild and domesticated fowl that is difficult to detect in its early phase. Advanced disease is characterized by weight loss, fatigue, ataxia, reduced egg production and ultimately death (Feldman, 1938; Thorel *et al.*, 1997). When Koch first defined the aetiology of tuberculosis, it was assumed that a single type of tubercle bacillus was responsible for all forms of the disease. However, by the early 1890s, there was evidence that the avian tubercle bacillus, originally called *Bacillus tuberculosis gallinarum*, but generally known as *M. avium*, was distinct from the human (*Mycobacterium tuberculosis*) and bovine (*Mycobacterium bovis*) types (Anonymous, 1891; Maffucci, 1892). Whereas the mammalian isolates produced colonies with a 'dry' morphology that would not grow above 41°C and were virulent in guinea pigs but not birds, avian isolates typically had a 'moist' morphology, grew at temperatures >42°C and were virulent in birds and rabbits but not guinea pigs.

By using these criteria to classify isolates from tuberculous animals, it was discovered that infections due to *M. avium* were common in swine. It had long been recognized that avian tuberculosis was contagious among birds and now it appeared that this disease could be spread to hogs. Investigations revealed that afflicted swine had usually been in contact with sick birds and, in several cases, had actually consumed offal from infected fowl (Feldman, 1939). *Mycobacterium avium* was also isolated from diseased cattle, sheep, deer, marsupials and non-human primates.

*Corresponding author: christine.y.turenne@gmail.com

© CAB International 2020. *Paratuberculosis: Organism, Disease, Control, 2nd Edition* (eds M.A. Behr *et al.*)

Although considered relatively resistant, mice, rats, squirrels, dogs, goats and horses could be infected experimentally (Feldman, 1938; Thorel *et al.*, 1997).

5.2.2 *Mycobacterium avium* and human disease

Despite the broad host range of *M. avium*, humans seemed immune to infection. Between 1901 and 1911, the British Royal Commission on Tuberculosis conducted an extensive study to address the possibility of disease transmission between animals and humans. They concluded that *M. bovis* could be transmitted to humans via infected beef and milk. In contrast, the risk posed by *M. avium*-infected eggs and fowl appeared negligible (Miller, 1911). Even so, there continued to be sporadic reports of human disease caused by *M. avium*. Most publications were methodologically flawed and unconvincing, but when Feldman conducted a critical review, he identified 13 probable cases, dating back to 1905 (Feldman, 1938). In another attempt to resolve this issue, Branch (1931) collected strains described in previous studies and subjected them to a uniform set of tests. After assessing both morphological and pathological characteristics, he determined that several isolates were genuine examples of *M. avium* and thus represented authentic cases of human infection. Of the remaining strains, some were typical of *M. tuberculosis*, whereas others did not conform to known types. Branch suggested that these atypical isolates represented novel acid-fast pathogens.

5.2.3 *Mycobacterium paratuberculosis* is a subspecies of *Mycobacterium avium*

DNA sequencing-based approaches are less subjective and more reproducible than traditional classification schemes, but results can seem at odds with phenotypic observations. Historically, the avian tubercle bacillus, the wood pigeon bacillus and the agent of Johne's disease were classified as separate species (McFadden *et al.*, 1987; Saxegaard and Baess, 1988; Yoshimura and Graham, 1988). However, genetic analysis reveals extensive homology and current nomenclature identifies these organisms as subspecies of *M. avium* (Thorel *et al.*, 1990).

Mycobacterium avium subsp. *avium* refers to the classic avian tubercle bacillus. *Mycobacterium avium* subsp. *silvaticum*, also known as the wood pigeon bacillus, is an uncommon cause of tuberculosis in wild birds and deer. The wood pigeon bacillus was traditionally distinguished by a requirement for mycobactin upon primary isolation, but this siderophore-dependence is lost upon subculture. *Mycobacterium avium* subsp. *avium* and *M. avium* subsp. *silvaticum* exhibit similar morphological characteristics and both are bird pathogens capable of causing disease in mammals; as such, their high genetic similarity (ANI >99.9%) is not entirely surprising.

Although not officially validated, the designation *M. avium* subsp. *hominissuis* is widely accepted and aims to distinguish human- and pig-derived strains from bird isolates (Mijs *et al.*, 2002). Phylogenomic analysis of isolates from Asia, Europe and North America have revealed extensive diversity within *M. avium hominissuis*, including evidence for at least five distinct lineages (Uchiya *et al.*, 2017; Yano *et al.*, 2017). Additional studies are required to capture the true global diversity of *M. avium hominissuis* and elucidate the relationship between strain genotype, phenotype and pathogenic potential.

MAP refers to the former *Mycobacterium paratuberculosis* or *Mycobacterium johnei*, the agent of paratuberculosis or Johne's disease. A thorough history of Johne's disease/paratuberculosis has been provided elsewhere (see Manning and Collins, Chapter 1, previous edition of this book), but it should be noted that in their initial description of 'pseudo-tuberculous enteritis', Johne and Frothingham postulated that the avian tubercle bacillus was responsible. By use of multilocus sequence analysis, the phylogenetic relationships of all recognized *M. avium* subspecies were better established (Turenne *et al.*, 2008). Out of the heterogeneous mix of environmental and opportunistic strains represented by *M. avium hominissuis* emerge two independently evolved pathogenic clones (Fig. 5.1.). One of these comprised the

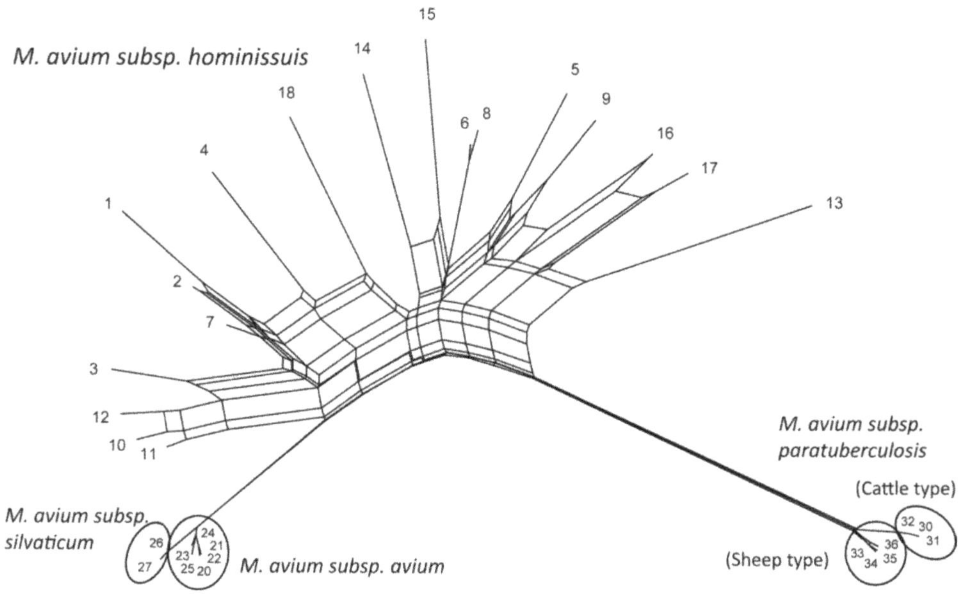

Fig. 5.1. Split network phylogeny of *Mycobacterium avium* species using the concatenated sequences of ten multilocus sequence analysis genes. (Adapted from Turenne *et al.*, 2008, with permission.)

avian subspecies (*M. avium* subsp. *avium* and *M. avium* subsp. *silvaticum*) whereas the other only included *MAP* strains, with separate phylogenetic branches observed for the sheep (S) and cattle (C) lineages. The two major sublineages of *MAP* were subsequently confirmed by whole genome sequencing (WGS) of >140 strains (Bryant *et al.*, 2016). The *MAP*-C group is traditionally associated with cattle strains, but encompasses isolates from diverse hosts, including goats, sheep, deer, bison and humans. Similarly, the *MAP*-S group, traditionally associated with sheep strains, also includes isolates from goats, deer and camels. For a thorough discussion of *MAP* genomics, see Chapter 6, this volume.

Although not formally considered a subspecies of *M. avium* at this time, genomic analysis has recently revealed that *M. lepraemurium*, the agent of murine and feline leprosy, is very closely related to *M. avium* (van Ingen *et al.*, 2018). It has been suggested that *M. lepraemurium* evolved through massive gene decay and reductive evolution from a *M. avium*-like ancestor (Benjak *et al.*, 2017).

5.3 The *Mycobacterium avium* Complex

5.3.1 *Mycobacterium intracellulare*

Through the 1940s and 1950s, numerous 'atypical' or 'anonymous' acid-fast pathogens were identified, including several responsible for serious human disease. In one lethal case of disseminated disease in a young girl, acid-fast bacteria were found in multiple organs, and a mycobacterial infection was suspected (Cuttino and McCabe, 1949). Tuberculosis, leprosy, avian tuberculosis, rat leprosy and, owing to extensive intestinal involvement, paratuberculosis were all considered, but none perfectly matched the pathological features of the case. The causative agent also seemed morphologically distinct from known mycobacteria and was ultimately named *Nocardia intracellularis*. Another atypical acid-fast pathogen, dubbed the 'Battey bacillus' because of its prevalence at the Battey State Hospital (Rome, Georgia, USA), was associated with >300 of cases of chronic pulmonary disease (Corpe, 1964). Features of Battey-type pulmonary illness (e.g. cough, weight loss and lung pathology) closely resembled those of classic tuberculosis. However, the 'Battey bacillus' was morphologically distinct from *M. tuberculosis* and harmless to guinea pigs. Epidemiological studies of the 'Battey bacillus' indicated that it was not transmissible between people. More likely it was acquired from soil (Corpe, 1964). Classification of atypical mycobacteria was a key goal of the Veterans Administration–National Tuberculosis Association Cooperative Study of Mycobacteria (Runyon, 1958). Hundreds of isolates were collected, analysed and eventually divided into four major groups (Runyon, 1958, 1965). This work revealed that the Battey bacillus and *N. intracellularis* were the same organism. They were renamed *M. intracellulare* and placed in Group III (non-photochromogens), along with *M. avium*, *M. gastri* and *M. terrae* (Runyon, 1965, 1967; Wayne, 1966).

5.3.2 The *Mycobacterium avium–intracellulare* complex

Traditionally, speciation of non-tuberculous mycobacteria was based on phenotypic characteristics such as pigmentation, growth rate, growth temperature and biochemical activities. *Mycobacterium avium* and *M. intracellulare* could be distinguished from other species of the Group III, slowly growing, non-photochromogenic mycobacteria by a positive tellurite test, and negative results for urease and Tween hydrolysis (Kent and Kubica, 1985). However, reliable differentiation of *M. avium* from *M. intracellulare* was not possible, even when using laborious procedures, such as cell wall lipid analysis, serotyping and classical animal infection experiments. In an attempt to resolve the taxonomy of Group III organisms, the International Working Group on Mycobacterial Taxonomy (IWGMT) employed a panel of 89 isolates, including 47 *M. avium* and *M. intracellulare* strains. For each culture, 292 phenotypic 'characters' were tested. Of these, 118 proved useful and improved classification of several Group III species (Meissner *et al.*, 1974), but even this numerical taxonomy approach could not reliably resolve *M. avium* and *M. intracellulare* strains. Suggestions to reclassify *M. intracellulare* as an official subspecies of *M. avium* were never adopted (Wayne, 1966; Meissner *et al.*, 1974; Runyon, 1974), but the concept of a *M. avium–intracellulare* complex (MAC) did emerge (Meissner *et al.*, 1974).

5.3.3 MAC in the molecular era

Molecular approaches to mycobacterial diagnostics and taxonomy have changed our view of the MAC. Even early nucleic acid-based typing methods, including DNA–DNA hybridization (Baess, 1983) and commercial DNA probes (Saito *et al.*, 1989, 1990), could readily differentiate *M. avium* from *M. intracellulare*. DNA sequencing-based approaches, especially 16S rRNA gene sequencing and 16S-23S internally transcribed spacer (ITS) sequencing, revealed many distinct sequevars (Frothingham and Wilson, 1993; Mijs *et al.*, 2002) and hinted at the presence of species other than *M. avium* and *M. intracellulare*. MAC now refers to a group of mycobacteria that exhibit overlapping phenotypic features and cause similar disease syndromes. Currently, MAC includes 12 validly published species: *M. avium*, *M. intracellulare*, *M. chimaera*, *M. colombiense*, *M. arosiense*, *M. vulneris*, *M. marseillense*, *M. timonense*, *M. bouchedurhonense*, *M. yongonense*, *M. paraintracellulare and M. lepraemurium* (Fig. 5.2.). Insights from WGS data are redefining the taxonomy of the MAC: using sequencing-based criteria, MAC organisms share an average nucleotide identity (ANI) >85% and exhibit high sequence identity in targets commonly used for multilocus sequence analysis, including the 16S rRNA (>99.5%), *hsp65* (>97.0%) and *rpoB* (>94%) genes (van Ingen *et al.*, 2018). Because of their extensive genomic similarity, it has been proposed that *M. yongonense* and *M. chimaera* be reduced to the rank of subspecies, i.e. *M. intracellulare* subsp. *yongonense* (Castejon *et al.*, 2018) and *M. intracellulare* subsp. *chimaera* (Nouioui *et al.*, 2018), and that *M. paraintracellulare* represents a later heterotypic synonym of *M. intracellulare* (Nouioui *et al.*, 2018). Genomic characterization of *M. bouchedurhonense* and *M. timonense* has been compromised by the absence of reference material. The strains deposited in culture collections do not match the description of the type strains of these species (Tortoli *et al.*, 2017; van Ingen *et al.*, 2018).

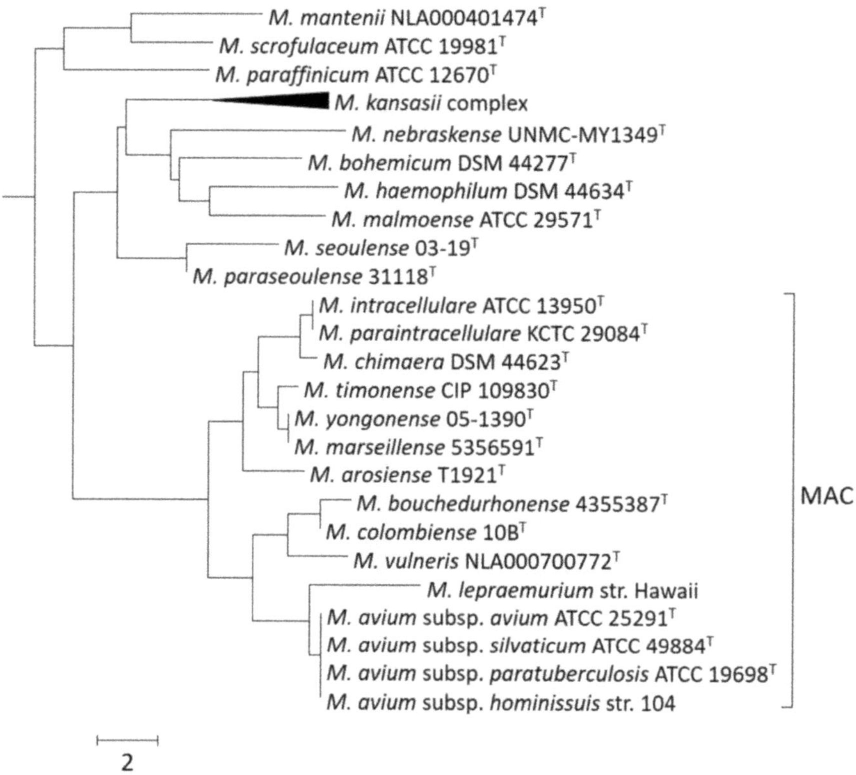

Fig. 5.2. Phylogeny reconstruction of ~1400 base pairs of the 16S rRNA gene using the neighbour-joining method and performed using the MEGA v7 software. The species shown represent the subcluster that included *Mycobacterium avium* complex (MAC) species in the context of a comprehensive 16S rRNA gene sequence alignment of all mycobacterial species. Scale represents the number of base pairs difference.

5.4 Environmental Reservoir of the *Mycobacterium avium* Complex

MAC disease, at least in humans, is not contagious. In a summary of his experiences with Battey-type pulmonary disease, Corpe (1964, p. 381) comments:

> We have never found a source case. In spite of the fact that well over 95% of the patients we see have been or are married, we have never seen either a husband or a wife also clinically ill with the disease. We have never seen two cases in the same family. This is an entirely different epidemiologic picture than is observed in *Mycobacterium tuberculosis* infections.

The epidemiology of MAC in humans differs from tuberculosis because the source of infection is different (Falkinham, 1996). MAC is typically acquired from soil or water. Despite some success in using molecular techniques to match patient specimens with isolates from environmental sources (von Reyn *et al.*, 1994; Lande *et al.*, 2019), the vast environmental reservoir of the MAC confounds such studies. These mycobacteria are abundant in diverse geographical regions, soil types, aquatic ecosystems and urban water distribution systems (Falkinham, 2002). *Mycobacterium avium* DNA was even detected in samples from the space station Mir (Kawamura *et al.*, 2001). Opportunities for exposure are so extensive that it is extraordinarily challenging to identify the time and place of infection and then, months or years later, actually isolate the causative MAC clone from that site. Such efforts are further complicated by the fact that infections can be polyclonal (Arbeit *et al.*, 1993; Wallace *et al.*, 1998). The ecological activities of the MAC are largely unknown. In aquatic systems, including swimming pools, hot tubs and municipal pipes, these mycobacteria form biofilms, which enhance resistance to disinfectants and other antimicrobial agents (Vaerewijck *et al.*, 2005). Planktonic cells sloughed from a biofilm can be aerosolized or ingested, and thus contribute to MAC infections. Aerosolized bacteria from contaminated heater-cooler units triggered an international outbreak of *M. chimaera* disease in cardiac surgery patients (van Ingen *et al.*, 2017). Not all environmental MAC are free-living. Via a process reminiscent of mammalian macrophage infection, *M. avium* can invade and replicate within protozoa (Cirillo *et al.*, 1997; Steinert *et al.*, 1998). The intracellular space is a refuge that provides the mycobacteria with nutrients and protects them from biocides (Steinert *et al.*, 1998). Experiments with tissue culture and animal models even suggest that amoeba-grown bacteria are more virulent than those propagated in standard culture media (Cirillo *et al.*, 1997). As such, MAC-infected protozoans may be an important 'environmental' reservoir for human and animal disease (Samba-Louaka *et al.*, 2018).

In birds, MAC disease (specifically *M. avium* subsp. *avium*) is considered to be contagious (Dhama *et al.*, 2011). Transmission of avium tuberculosis is similar to that of paratuberculosis. Infected animals shed large amounts of organism that contaminate the environment and can then be inhaled or ingested by healthy animals. The risk of exposure increases if hygiene practices are inadequate and/or where numerous animals are closely confined, e.g. zoo aviaries.

5.5 Diagnostics for MAC

The following section focuses on diagnostic tools and classification schemes that are typically employed for members of the MAC. How these tests do, and do not, aid in the identification of *MAP* is reviewed. More detailed information on the specific isolation and identification of *MAP* is discussed in greater detail elsewhere (see Chapter 18, this volume).

5.5.1 Culture and traditional methods for classification of MAC

Definitive diagnosis of a mycobacterial infection requires culture of the organism from a clinical specimen, followed by identification using established techniques. Both liquid and solid (agar- or egg-based) media can be used for mycobacterial culture. However, no single medium or growth condition will permit the successful isolation of all mycobacteria and therefore protocols may vary between laboratories. *Mycobacterium intracellulare* and the classical *M. avium* strains can grow on any standard mycobacterial media, with or without 10% CO_2. While the MAC grows well at 37°C incubation, *M. avium* strains may grow best at an increased temperature of 40–42°C (Kent and Kubica, 1985). MAC requires >7 days for growth and 3–4 weeks to reach maturity. Mycobacterial cultures are typically kept up to 6–8 weeks before being considered negative. However, these standard conditions are insufficient for routine isolation of *MAP*. The organism's extremely slow growth rate and requirement for mycobactin supplementation preclude detection. Even when present in immense quantities, it can take several months to detect *MAP* in the clinical setting (Richter *et al.*, 2002).

Traditionally, speciation of non-tuberculous mycobacteria was based on phenotypic characteristics such as pigmentation, growth rate, growth temperature and biochemical activities. The MAC is considered non-pigmented, although some strains may present with bright yellow pigmentation (e.g. *M. vulneris* and *M. arosiense*) and ageing cultures may adopt yellow hues. The MAC can also present with various colony morphologies (i.e. smooth or rough) and grow under wide ranges of temperature and pH. The MAC is typically differentiated from other species of the Group III, slowly growing, non-photochromogenic mycobacteria by a positive tellurite test and negative results for urease and Tween hydrolysis (Kent and Kubica, 1985). As discussed above, phenotypic distinction of MAC species is difficult and, owing to their similarity in clinical settings, identification to the complex level was generally considered sufficient. Notably, the classification algorithms used in clinical microbiology laboratories do not include *MAP*, because it is commonly not considered as a human pathogen.

For decades, serotyping was used for classification of MAC strains. Together, *M. avium* and *M. intracellulare* comprise 28 different serovars (Saito *et al.*, 1990; Wayne *et al.*, 1993). Serotyping relies on the presence of serovar-specific glycopeptidolipids (GPLs) (Brennan *et al.*, 1978; Brennan and Goren, 1979). *MAP* isolates do not produce GPLs and therefore cannot be serotyped. However, this is not a diagnostically useful characteristic since GPL mutants of *M. avium* (Belisle *et al.*, 1993) and other non-serotypeable MAC clones (De Smet *et al.*, 1996) are also encountered. Skilled technicians can also use high-performance liquid chromatography (HPLC) of mycolic acids to successfully separate some MAC species (Butler *et al.*, 1992), but differentiation of *MAP* from other *M. avium* subspecies is not possible (Dei *et al.*, 1999).

Matrix-associated laser desorption/ionization–time-of-flight mass spectrometry (MALDI-TOF MS) has become a routine method for bacterial identification. This method can distinguish *M. avium* from *M. intracellulare*, but commercial MALDI-TOF databases do not contain profiles for all validated species, nor are all species and subspecies of MAC typically evaluated in published studies. Despite promising work on the differentiation of *M. chimaera* from *M. intracellulare* (Pranada *et al.*, 2017), MALDI-TOF does not yet allow reliable identification of most MAC organisms at the species level (Brown-Elliott *et al.*, 2019). Similarly, the clinical databases available for commercial MALDI-TOF systems do not allow reliable differentiation of *MAP* from other subspecies of *M. avium*, but subspecies-level identification may be possible via creation and optimization of an in-house database (Ricchi *et al.*, 2017).

5.5.2 DNA sequencing for the identification of MAC

DNA-based analysis, including WGS, confirms that the MAC includes more than *M. avium* and *M. intracellulare*. Most validated MAC species can be identified by targeted sequencing of the full-length 16S rRNA gene. The exceptions are *M. intracellulare* and *M. paraintracellulare*, and *M. marseillense* and *M. yongonense* (Tortoli *et al.*, 2017). Additional targets, including ITSs, *hsp65* and *rpoB* allow further differentiation of MAC species and subspecies. Within *M. avium*, eight ITS sequevars have been reported. Because *MAP* strains typically present with the Mav-A sequevar, ITSs cannot be used to distinguish them from other *M. avium* subspecies (Turenne *et al.*, 2006). Similarly, the 441 base pair (BP) region at the 5′ end of the *hsp65* gene that is widely used for speciation of mycobacteria (Telenti *et al.*, 1993) does not effectively differentiate *M. avium* subspecies. In contrast, the 3′ 'tail end' of the *hsp65* gene can simultaneously identify species as well as host-associated subtypes (Turenne *et al.*, 2006). *MAP* is represented by two sequevars, one for each of the *MAP*-C and *Map*-S lineages. Another *hsp65* sequevar encompasses both of the bird-associated subspecies (i.e. subsp. *avium* and *silvaticum* together). *rpoB* sequencing is also capable of distinguishing among subspecies of *M. avium*, including *MAP* (Higgins *et al.*, 2011).

5.5.3 Other molecular assays for detection or identification of MAC

A scheme based on three insertion sequences, IS*1245*, IS*901* and IS*900*, has been used to define the subspecies of *M. avium*: IS*901* is only present in avian strains (*M. avium* subsp. *avium* and *M. avium* subsp. *silvaticum*); IS*900* is specific for *MAP*; IS*1245* is absent from *MAP* but present in all other subspecies (Ellingson *et al.*, 2000; Bartos *et al.*, 2006). Although not commercially available, hybridization-based methods (e.g. restriction fragment-length polymorphisms) that target these elements can be used for subspecies identification as well as strain typing and surveillance. However, polymerase chain reaction (PCR)-based detection of these insertion sequences should be used with caution and verified via DNA sequencing, since similar insertion elements exist throughout the MAC and in non-MAC organisms (Turenne *et al.*, 2007).

A non-sequencing-based method for reliable differentiation of *M. avium* subsp. *silvaticum* and *M. avium* subsp. *avium* has only recently been described. This high-resolution melt method can detect differences as small as a single nucleotide polymorphism (Rónai *et al.*, 2015).

5.5.4 Commercial assays for the identification of MAC

Genetic variation is the basis for a number of commercial assays currently available for the detection and/or speciation of mycobacteria. These tests offer rapid turnaround time and greater accuracy than conventional methods and thus contribute to improved patient care. Some of the first, introduced in the early 1990s and still used today, are the AccuProbe® Culture Identification Tests (Hologic) (Saito *et al.*, 1989). Currently six tests are available for mycobacteria. Each targets ribosomal RNA and permits identification from a positive culture. In addition to individual tests for the *M. tuberculosis* complex, *M. kansasii*, *M. gordonae*, *M. intracellulare* and *M. avium*, there is a MAC test that can identify any MAC organism to the complex level (Lebrun *et al.*, 1992; Viljanen *et al.*, 1993). However, on rare occasions the MAC test may cross-react with other mycobacteria (Tortoli *et al.*, 2010). Because the target of AccuProbe is ribosomal RNA, this assay cannot distinguish *MAP* from other *M. avium* subspecies. *MAP* generates a positive reaction using the MAC AccuProbe assay (Richter *et al.*, 2002).

Reverse hybridization line probe assays (LPA) are also commercially available. In addition to other non-tuberculous mycobacteria (NTM), the Inno-LiPA® MYCOBACTERIA v2 (Fujirebio) can identify *M. avium*, two subsets of *M. intracellulare* and 'MAIS complex' (Tortoli *et al.*, 2003; Lebrun *et al.*, 2005). However, the target is the ITS region, which does not permit resolution of *MAP*. Hain Lifescience markets several LPAs for mycobacteria. These detect species-specific 23S rDNA sequences. The GenoType® Mycobacteria Direct can be used for identification directly from clinical specimens, whereas the GenoType® Mycobacterium CM (Common Mycobacteria), AS (Additional Species) and NTM-DR (which can specifically identify *M. chimaera*) tests require positive cultures. The Speed-oligo Mycobacteria assay (Vircell) is the latest LPA that includes *M. avium* and *M. intracellulare*. It targets both the 16S rRNA gene and adjacent ITSs but performance evaluations are limited to date. Again, since these LPAs target the ribosomal operon, none has the capacity to differentiate *MAP* from other *M. avium*.

5.6 Concluding Remarks

The MAC includes organisms of diverse pathogenic potential. Molecular typing methods have reaffirmed the existence of the complex but also revealed differences between species, subspecies and strains. Proper classification of isolates is essential to a thorough understanding of MAC epidemiology and pathogenesis. Despite its close taxonomic proximity to other *M. avium* organisms, *MAP* stands out as an important pathogen with unique phenotypic and genetic characteristics.

References

Anonymous (1891) The tuberculosis congress. *The Lancet* 138(3547), 463–464. DOI: 10.1016/S0140-6736(02)02313-9.

Arbeit, R.D., Slutsky, A., Barber, T.W., Maslow, J.N., Niemczyk, S. *et al.* (1993) Genetic diversity among strains of *Mycobacterium avium* causing monoclonal and polyclonal bacteremia in patients with AIDS. *Journal of Infectious Diseases* 167(6), 1384–1390. DOI: 10.1093/infdis/167.6.1384.

Baess, I. (1983) Deoxyribonucleic acid relationships between different serovars of *Mycobacterium avium*, *Mycobacterium intracellulare* and *Mycobacterium scrofulaceum*. *Acta Pathologica Microbiologica Scandinavica Series B: Microbiology* 91B(1–6), 201–203. DOI: 10.1111/j.1699-0463.1983.tb00033.x.

Bartos, M., Hlozek, P., Svastova, P., Dvorska, L., Bull, T. *et al.* (2006) Identification of members of *Mycobacterium avium* species by Accu-Probes, serotyping, and single IS*900*, IS*901*, IS*1245* and

IS*901*-flanking region PCR with internal standards. *Journal of Microbiological Methods* 64(3), 333–345. DOI: 10.1016/j.mimet.2005.05.009.

Belisle, J.T., Klaczkiewicz, K., Brennan, P.J., Jacobs, W.R., Jr. and Inamine, J.M. (1993) Rough morphological variants of *Mycobacterium avium*. Characterization of genomic deletions resulting in the loss of glycopeptidolipid expression. *The Journal of Biological Chemistry* 268, 10517–10523.

Benjak, A., Honap, T.P., Avanzi, C., Becerril-Villanueva, E., Estrada-García, I. *et al.* (2017) Insights from the genome sequence of *Mycobacterium lepraemurium*: massive gene decay and reductive evolution. *MBio* 8(5), e01283–17. DOI: 10.1128/mBio.01283-17.

Branch, A. (1931) Avian tubercle Bacillus infection with special reference to mammals and to man. *Archives of Pathology* 12, 253–274.

Brennan, P.J. and Goren, M.B. (1979) Structural studies on the type-specific antigens and lipids of the *Mycobacterium avium–Mycobacterium intracellulare–Mycobacterium scrofulaceum* serocomplex: *Mycobacterium intracellulare* serotype 9. *The Journal of Biological Chemistry* 254, 4205–4211.

Brennan, P.J., Souhrada, M., Ullom, B., McClatchy, J.K. and Goren, M.B. (1978) Identification of atypical mycobacteria by thin-layer chromatography of their surface antigens. *Journal of Clinical Microbiology* 8, 374–379.

Brown-Elliott, B.A., Fritsche, T.R., Olson, B.J., Vasireddy, S., Vasireddy, R. *et al.* (2019) Comparison of two commercial matrix-assisted laser desorption/ionization-time of flight mass spectrometry (MALDI-TOF MS) systems for identification of nontuberculous mycobacteria. *American Journal of Clinical Pathology* 152(4), 527–536. DOI: 10.1093/ajcp/aqz073.

Bryant, J.M., Thibault, V.C., Smith, D.G.E., McLuckie, J., Heron, I. *et al.* (2016) Phylogenomic exploration of the relationships between strains of *Mycobacterium avium* subspecies *paratuberculosis*. *BMC Genomics* 17(1), 79. DOI: 10.1186/s12864-015-2234-5.

Butler, W.R., Thibert, L. and Kilburn, J.O. (1992) Identification of *Mycobacterium avium* complex strains and some similar species by high-performance liquid chromatography. *Journal of Clinical Microbiology* 30(10), 2698–2704. DOI: 10.1128/JCM.30.10.2698-2704.1992.

Castejon, M., Menéndez, M.C., Comas, I., Vicente, A. and Garcia, M.J. (2018) Whole-genome sequence analysis of the *Mycobacterium avium* complex and proposal of the transfer of *Mycobacterium yongonense* to *Mycobacterium intracellulare* subsp. *yongonense* subsp. nov. *International Journal of Systematic and Evolutionary Microbiology* 68(6), 1998–2005. DOI: 10.1099/ijsem.0.002767.

Cirillo, J.D., Falkow, S., Tompkins, L.S. and Bermudez, L.E. (1997) Interaction of *Mycobacterium avium* with environmental amoebae enhances virulence. *Infection and Immunity* 65(9), 3759–3767. DOI: 10.1128/IAI.65.9.3759-3767.1997.

Corpe, R.F. (1964) Clinical aspects, medical and surgical, in the management of Battey-type pulmonary disease. *Diseases of the Chest* 45(4), 380–382. DOI: 10.1378/chest.45.4.380.

Cuttino, J.T. and McCabe, A.M. (1949) Pure granulomatous nocardiosis: a new fungus disease distinguished by intracellular parasitism. *American Journal of Clinical Pathology* 25, 1–34.

De Smet, K.A.L., Hellyer, T.J., Khan, A.W., Brown, I.N. and Ivanyi, J. (1996) Genetic and serovar typing of clinical isolates of the *Mycobacterium avium–intracellulare* complex. *Tubercle and Lung Disease* 77(1), 71–76. DOI: 10.1016/S0962-8479(96)90079-9.

Dei, R., Tortoli, E., Bartoloni, A., Simonetti, M.T. and Lillini, E. (1999) HPLC does not differentiate *Mycobacterium paratuberculosis* from *Mycobacterium avium*. *Veterinary Microbiology* 65(3), 209–213. DOI: 10.1016/S0378-1135(98)00292-2.

Dhama, K., Mahendran, M., Tiwari, R., Dayal Singh, S., Kumar, D. *et al.* (2011) Tuberculosis in birds: insights into the *Mycobacterium avium* infections. *Veterinary Medicine International* 2011(1), 712369–14. DOI: 10.4061/2011/712369.

Ellingson, J.L.E., Stabel, J.R., Bishai, W.R., Frothingham, R. and Miller, J.M. (2000) Evaluation of the accuracy and reproducibility of a practical PCR panel assay for rapid detection and differentiation of *Mycobacterium avium* subspecies. *Molecular and Cellular Probes* 14(3), 153–161. DOI: 10.1006/mcpr.2000.0299.

Falkinham, J.O. (1996) Epidemiology of infection by nontuberculous mycobacteria. *Clinical Microbiology Reviews* 9(2), 177–215. DOI: 10.1128/CMR.9.2.177.

Falkinham, J.O. (2002) Nontuberculous mycobacteria in the environment. *Clinics in Chest Medicine* 23(3), 529–551. DOI: 10.1016/S0272-5231(02)00014-X.

Feldman, W.H. (1938) *Avian Tuberculosis Infections*. The William & Wilkins Co., Baltimore, Maryland.

Feldman, W.H. (1939) Types of tubercle bacilli in lesions of garbage-fed swine. *American Journal of Public Health and the Nations Health* 29(11), 1231–1238. DOI: 10.2105/AJPH.29.11.1231.

Frothingham, R. and Wilson, K.H. (1993) Sequence-based differentiation of strains in the *Mycobacterium avium* complex. *Journal of Bacteriology* 175(10), 2818–2825. DOI: 10.1128/JB.175.10.2818-2825.1993.

Higgins, J., Camp, P., Farrell, D., Bravo, D., Pate, M. *et al.* (2011) Identification of *Mycobacterium* spp. of veterinary importance using *rpoB* gene sequencing. *BMC Veterinary Research* 7(1), 77. DOI: 10.1186/1746-6148-7-77.

Kawamura, Y., Li, Y., Liu, H., Huang, X., Li, Z. *et al.* (2001) Bacterial population in Russian space station 'Mir'. *Microbiology and Immunology* 45(12), 819–828. DOI: 10.1111/j.1348-0421.2001.tb01321.x.

Kent, P.T. and Kubica, G.P. (1985) *Public Health Mycobacteriology: a Guide for the Level III Laboratory*. US Department of Health and Human Services, Centers for Disease Control, Atlanta, Georgia.

Lande, L., Alexander, D.C., Wallace, R.J., Kwait, R., Iakhiaeva, E. *et al.* (2019) *Mycobacterium avium* in community and household water, suburban Philadelphia, Pennsylvania, USA, 2010 – 2012. *Emerging Infectious Diseases* 25(3), 473–481. DOI: 10.3201/eid2503.180336.

Lebrun, L., Espinasse, F., Poveda, J.D. and Vincent-Levy-Frebault, V. (1992) Evaluation of nonradioactive DNA probes for identification of mycobacteria. *Journal of Clinical Microbiology* 30(9), 2476–2478. DOI: 10.1128/JCM.30.9.2476-2478.1992.

Lebrun, L., Weill, F.-X., Lafendi, L., Houriez, F., Casanova, F. *et al.* (2005) Use of the INNO-LiPA-MYCOBACTERIA assay (version 2) for identification of *Mycobacterium avium–Mycobacterium intracellulare–Mycobacterium scrofulaceum* complex isolates. *Journal of Clinical Microbiology* 43(6), 2567–2574. DOI: 10.1128/JCM.43.6.2567-2574.2005.

Maffucci, A. (1892) Die Hühnertuberculose. *Zeitschrift für Hygiene und Infektionskrankheiten* 11(1), 445–486. DOI: 10.1007/BF02284309.

McFadden, J.J., Butcher, P.D., Thompson, J., Chiodini, R. and Hermon-Taylor, J. (1987) The use of DNA probes identifying restriction-fragment-length polymorphisms to examine the *Mycobacterium avium* complex. *Molecular Microbiology* 1(1), 283–291. DOI: 10.1111/j.1365-2958.1987.tb01934.x.

Meissner, G., Schroder, K.H., Amadio, G.E., Anz, W., Chaparas, S. *et al.* (1974) A co-operative numerical analysis of nonscoto- and non-photochromogenic slowly growing mycobacteria. *Journal of General Microbiology* 83(2), 207–235. DOI: 10.1099/00221287-83-2-207.

Mijs, W., de Haas, P., Rossau, R., Van der Laan, T., Rigouts, L. *et al.* (2002) Molecular evidence to support a proposal to reserve the designation *Mycobacterium avium* subsp. *avium* for bird-type isolates and '*M. avium* subsp. *hominissuis*' for the human/porcine type of *M. avium*. *International Journal of Systematic and Evolutionary Microbiology* 52, 1505–1518.

Miller, J. (1911) The report of the Royal Commission on Tuberculosis: a critical survey. *British Journal of Tuberculosis* 5(4), 282–290. DOI: 10.1016/S0366-0850(11)80048-7.

Nouioui, I., Carro, L., García-López, M., Meier-Kolthoff, J.P., Woyke, T. *et al.* (2018) Genome-based taxonomic classification of the phylum *Actinobacteria*. *Frontiers in Microbiology* 9, 2007. DOI: 10.3389/fmicb.2018.02007.

Pranada, A.B., Witt, E., Bienia, M., Kostrzewa, M. and Timke, M. (2017) Accurate differentiation of *Mycobacterium chimaera* from *Mycobacterium intracellulare* by MALDI-TOF MS analysis. *Journal of Medical Microbiology* 66(5), 670–677. DOI: 10.1099/jmm.0.000469.

Ricchi, M., Mazzarelli, A., Piscini, A., Di Caro, A., Cannas, A. *et al.* (2017) Exploring MALDI-TOF MS approach for a rapid identification of *Mycobacterium avium* ssp. *paratuberculosis* field isolates. *Journal of Applied Microbiology* 122(3), 568–577. DOI: 10.1111/jam.13357.

Richter, E., Wessling, J., Lügering, N., Domschke, W. and Rüsch-Gerdes, S. (2002) *Mycobacterium avium* subsp. *paratuberculosis* infection in a patient with HIV, Germany. *Emerging Infectious Diseases* 8(7), 729–731. DOI: 10.3201/eid0807.010388.

Rónai, Z., Csivincsik, Á. and Dán, Á. (2015) Molecular identification of *Mycobacterium avium* subsp. *silvaticum* by duplex high-resolution melt analysis and subspecies-specific real-time PCR. *Journal of Clinical Microbiology* 53(5), 1582–1587. DOI: 10.1128/JCM.03556-14.

Runyon, E.H. (1958) Mycobacteria encountered in clinical laboratories. *Leprosy Briefs* 9, 21–23.

Runyon, E.H. (1965) Pathogenic mycobacteria. *Advances in Tuberculosis Research* 14, 235–287.

Runyon, E.H. (1967) Mycobacterium intracellulare. *American Review of Respiratory Disease* 95, 861–865.

Runyon, E.H. (1974) Ten mycobacterial pathogens. *Tubercle* 55(3), 235–240. DOI: 10.1016/0041-3879(74)90051-8.

Saito, H., Tomioka, H., Sato, K., Tasaka, H., Tsukamura, M. *et al.* (1989) Identification and partial characterization of *Mycobacterium avium* and *Mycobacterium intracellulare* by using DNA probes. *Journal of Clinical Microbiology* 27(5), 994–997. DOI: 10.1128/JCM.27.5.994-997.1989.

Saito, H., Tomioka, H., Sato, K., Tasaka, H. and Dawson, D.J. (1990) Identification of various serovar strains of *Mycobacterium avium* complex by using DNA probes specific for *Mycobacterium avium* and *Mycobacterium intracellulare*. *Journal of Clinical Microbiology* 28(8), 1694–1697. DOI: 10.1128/JCM.28.8.1694-1697.1990.

Samba-Louaka, A., Robino, E., Cochard, T., Branger, M., Delafont, V. *et al.* (2018) Environmental *Mycobacterium avium* subsp. *paratuberculosis* hosted by free-living amoebae. *Frontiers in Cellular and Infection Microbiology* 8, 28. DOI: 10.3389/fcimb.2018.00028.

Saxegaard, F. and Baess, I. (1988) Relationship between *Mycobacterium avium*, *Mycobacterium paratuberculosis* and 'wood pigeon mycobacteria'. Determinations by DNA–DNA hybridization. *APMIS: Acta Pathologica, Microbiologica, et Immunologica Scandinavica* 96, 37–42.

Steinert, M., Birkness, K., White, E., Fields, B. and Quinn, F. (1998) *Mycobacterium avium* bacilli grow saprozoically in coculture with *Acanthamoeba polyphaga* and survive within cyst walls. *Applied and Environmental Microbiology* 64(6), 2256–2261. DOI: 10.1128/AEM.64.6.2256-2261.1998.

Telenti, A., Marchesi, F., Balz, M., Bally, F., Böttger, E.C. *et al.* (1993) Rapid identification of mycobacteria to the species level by polymerase chain reaction and restriction enzyme analysis. *Journal of Clinical Microbiology* 31(2), 175–178. DOI: 10.1128/JCM.31.2.175-178.1993.

Thorel, M.F., Krichevsky, M. and Lévy-Frébault, V.V. (1990) Numerical taxonomy of mycobactin-dependent mycobacteria, emended description of *Mycobacterium avium*, and description of *Mycobacterium avium* subsp. *avium* subsp. nov., *Mycobacterium avium* subsp. *paratuberculosis* subsp. nov., and *Mycobacterium avium* subsp. *silvaticum* subsp. nov. *International Journal of Systematic Bacteriology* 40(3), 254–260. DOI: 10.1099/00207713-40-3-254.

Thorel, M.F., Huchzermeyer, H., Weiss, R. and Fontaine, J.J. (1997) *Mycobacterium avium* infections in animals. Literature review. *Veterinary Research* 28, 439–447.

Tortoli, E., Mariottini, A. and Mazzarelli, G. (2003) Evaluation of INNO-LiPA mycobacteria V2: improved reverse hybridization multiple DNA probe assay for mycobacterial identification. *Journal of Clinical Microbiology* 41(9), 4418–4420. DOI: 10.1128/JCM.41.9.4418-4420.2003.

Tortoli, E., Pecorari, M., Fabio, G., Messinò, M. and Fabio, A. (2010) Commercial DNA probes for mycobacteria incorrectly identify a number of less frequently encountered species. *Journal of Clinical Microbiology* 48(1), 307–310. DOI: 10.1128/JCM.01536-09.

Tortoli, E., Fedrizzi, T., Meehan, C.J., Trovato, A., Grottola, A. *et al.* (2017) The new phylogeny of the genus *Mycobacterium*: the old and the news. *Infection, Genetics and Evolution* 56, 19–25. DOI: 10.1016/j.meegid.2017.10.013.

Turenne, C.Y., Semret, M., Cousins, D.V., Collins, D.M. and Behr, M.A. (2006) Sequencing of hsp65 distinguishes among subsets of the *Mycobacterium avium* complex. *Journal of Clinical Microbiology* 44(2), 433–440. DOI: 10.1128/JCM.44.2.433-440.2006.

Turenne, C.Y., Wallace, R. and Behr, M.A. (2007) *Mycobacterium avium* in the postgenomic era. *Clinical Microbiology Reviews* 20(2), 205–229. DOI: 10.1128/CMR.00036-06.

Turenne, C.Y., Collins, D.M., Alexander, D.C. and Behr, M.A. (2008) *Mycobacterium avium* subsp. *paratuberculosis* and *M. avium* subsp. *avium* are independently evolved pathogenic clones of a much broader group of *M. avium* organisms. *Journal of Bacteriology* 190(7), 2479–2487. DOI: 10.1128/JB.01691-07.

Uchiya, K.I., Tomida, S., Nakagawa, T., Asahi, S., Nikai, T. *et al.* (2017) Comparative genome analyses of *Mycobacterium avium* reveal genomic features of its subspecies and strains that cause progression of pulmonary disease. *Scientific Reports* 7(1), 39750. DOI: 10.1038/srep39750.

Vaerewijck, M.J.M., Huys, G., Palomino, J.C., Swings, J. and Portaels, F. (2005) Mycobacteria in drinking water distribution systems: ecology and significance for human health. *FEMS Microbiology Reviews* 29(5), 911–934. DOI: 10.1016/j.femsre.2005.02.001.

van Ingen, J., Kohl, T.A., Kranzer, K., Hasse, B., Keller, P.M. *et al.* (2017) Global outbreak of severe *Mycobacterium chimaera* disease after cardiac surgery: a molecular epidemiological study. *The Lancet Infectious Diseases* 17(10), 1033–1041. DOI: 10.1016/S1473-3099(17)30324-9.

van Ingen, J., Turenne, C.Y., Tortoli, E., Wallace Jr, R.J. and Brown-Elliott, B.A. (2018) A definition of the *Mycobacterium avium* complex for taxonomical and clinical purposes, a review. *International Journal of Systematic and Evolutionary Microbiology* 68(11), 3666–3677. DOI: 10.1099/ijsem.0.003026.

Viljanen, M.K., Olkkonen, L. and Katila, M.L. (1993) Conventional identification characteristics, mycolate and fatty acid composition, and clinical significance of MAIX AccuProbe-positive isolates of *Mycobacterium avium* complex. *Journal of Clinical Microbiology* 31(5), 1376–1378. DOI: 10.1128/JCM.31.5.1376-1378.1993.

von Reyn, C.F., Marlow, J.N., Arbeit, R.D., Barber, T.W. and Falkinham, J.O. (1994) Persistent colonisation of potable water as a source of *Mycobacterium avium* infection in AIDS. *The Lancet* 343(8906), 1137–1141. DOI: 10.1016/S0140-6736(94)90239-9.

Wallace, R. J., Zhang, Y., Brown, B.A., Dawson, D., Murphy, D.T. *et al.* (1998) Polyclonal *Mycobacterium avium* complex infections in patients with nodular bronchiectasis. *American Journal of Respiratory and Critical Care Medicine* 158(4), 1235–1244. DOI: 10.1164/ajrccm.158.4.9712098.

Wayne, L.G. (1966) Classification and identification of mycobacteria. III. Species within group III. *The American Review of Respiratory Disease* 93, 919–928.

Wayne, L.G., Good, R.C., Tsang, A., Butler, R., Dawson, D. *et al.* (1993) Serovar determination and molecular taxonomic correlation in *Mycobacterium avium*, *Mycobacterium intracellulare*, and *Mycobacterium scrofulaceum*: a cooperative study of the International Working Group on Mycobacterial Taxonomy. *International Journal of Systematic Bacteriology* 43(3), 482–489. DOI: 10.1099/00207713-43-3-482.

Yano, H., Iwamoto, T., Nishiuchi, Y., Nakajima, C., Starkova, D.A. *et al.* (2017) Population structure and local adaptation of MAC lung disease agent *Mycobacterium avium* subsp. *hominissuis*. *Genome Biology and Evolution* 9(9), 2403–2417. DOI: 10.1093/gbe/evx183.

Yoshimura, H.H. and Graham, D.Y. (1988) Nucleic acid hybridization studies of mycobactin-dependent mycobacteria. *Journal of Clinical Microbiology* 26(7), 1309–1312. DOI: 10.1128/JCM.26.7.1309-1312.1988.

6 Comparative Genomics and Genomic Epidemiology of *Mycobacterium avium* subsp. *paratuberculosis* Strains

Karen Stevenson[1]* and Christina Ahlstrom[2]
[1]*Moredun Research Institute, Penicuik, UK;* [2]*US Geological Survey Alaska Science Center, Anchorage, Alaska, USA*

6.1 Introduction

Two phenotypically distinct strains of *Mycobacterium avium* subsp. *paratuberculosis* (*MAP*) were recognized in the 1930s but it was not until the introduction of restriction endonuclease analysis in the mid-1980s that these two strains, *MAP*-C and *MAP*-S, could be distinguished genetically. Since then, a plethora of molecular typing techniques has been applied to *MAP* isolates (reviewed by Li *et al.*, 2016; Fawzy *et al.*, 2018) and a complex nomenclature for *MAP* strains has evolved. Currently, the most widely used genotyping method is mycobacterial interspersed repetitive units – variable-number tandem repeats (MIRU-VNTR). However, it has limited discriminatory power within the major lineages and does not always accurately reflect genetic relatedness since the repeat sequences are subject to homoplasy (Ahlstrom *et al.*, 2015; Bryant *et al.*, 2016). Whole genome sequencing (WGS) supplies the ultimate resolution and has revolutionized *MAP* research. It has enabled determination of single nucleotide polymorphism (SNP) level diversity, clarified phylogenetic relationships between divergent lineages and closely related strains, and spawned the development of novel genotyping methods based on informative canonical SNPs (Ahlstrom *et al.*, 2016b; Leão *et al.*, 2016). This chapter presents an overview of comparative genomics and epidemiology of *MAP* strains, and also highlights the role that WGS has played in increasing our understanding of *MAP* strain diversity.

6.2 Genomic Comparison of *MAP* Strains

The advancements in DNA sequencing technologies, particularly next-generation sequencing and development of diverse new computational tools for comparative genomics, have expedited *MAP* genome research in the past decade or so. The first complete annotated genome sequence of a *MAP*-C strain (K10) was published in 2005 (Li *et al.*, 2005), subsequently revised following optical mapping (Wu *et al.*, 2009) and resequenced (Wynne *et al.*, 2010). The first complete genome sequence of a *MAP*-S strain (Telford 9.2) was recently reported (Brauning *et al.*, 2019), although draft sequences of *MAP*-S strains S397 and JIII-386 were published by Bannantine *et al.* (2012) and Möbius *et al.* (2015), respectively.

*Corresponding author: karen.stevenson@moredun.ac.uk

© CAB International 2020. *Paratuberculosis: Organism, Disease, Control, 2nd Edition* (eds M.A. Behr *et al.*)

These genome sequences provide reference genomes for mapping unassembled sequence reads of other isolates and for comparative genomics that can quickly identify large sequence polymorphisms (LSPs) including insertions, deletions, inversions, translocations and duplications, and individual SNPs that contribute to the unique genetic profiles of *MAP* strains. Other draft genome sequences are available in public databases as well as raw sequence data for a large number of *MAP* isolates, and this resource will expand as WGS becomes cheaper over time. The salient features of the *MAP* genome were described in the previous edition of this book (Paustian *et al.*, 2010), this chapter will focus on comparative genomics between *MAP* strains and genomic diversity of *MAP* isolates.

6.2.1 Phylogenetic relationships among *MAP* strains

The first comprehensive WGS study to clarify the phylogenetic relationships between previously designated strains was published by Bryant *et al.* (2016) using a global panel of 141 *MAP* isolates (modified in Fig. 6.1) that included the two major lineages corresponding to the 'Sheep-type' or 'Type S' and the 'Cattle-type' or 'Type C' designated by Collins *et al.* (1990). Type C is synonymous with the Type II strains, and Type S with the Type I/III strains, as defined by pulsed-field gel electrophoresis (Stevenson *et al.*, 2002). Type S and Type C strains differ with respect to the ease with which they can be isolated on artificial media and their respective growth rates (see

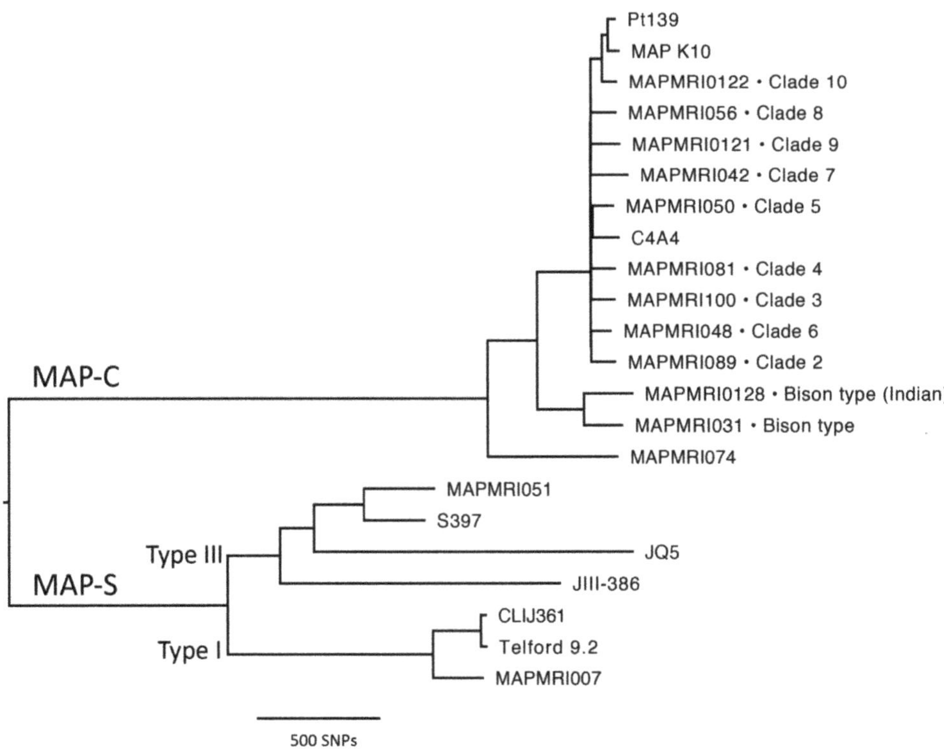

Fig. 6.1. Midpoint rooted whole genome single nucleotide polymorphism (SNP)-based phylogenetic tree of a global panel of *Mycobacterium avium* subsp. *paratuberculosis* (*MAP*) strains. Tips are annotated with strain names and, when applicable, *MAP*-C clade IDs as defined by Leão *et al.*, 2016. *MAP*-S and C lineages are labelled, as well as Type I and III *MAP*-S sub-lineages. Publicly available fastq files or assembled contigs were reference mapped to the *MAP* K10 genome and SNPs were identified using methods described by Bryant *et al.*, 2016.

Chapter 18, this volume) and also in terms of their virulence and pathogenicity (see Sections 6.2.3 and 6.4). However, since the original designation of 'Sheep-type' and 'Cattle-type' strains, it has become apparent that strain type is not always linked to host species provenance (see Section 6.3.1) and consequently this nomenclature can be confusing. Therefore, we propose that the two lineages are designated *MAP*-S and *MAP*-C and any reference to host species is avoided. These terms will be used throughout this chapter.

MAP-S is sub-divided into two sub-lineages. One sub-lineage comprises the 'Intermediate' strains (Collins *et al.*, 1990) or 'Type III' strains described by de Juan *et al.* (2005), which were hypothesized to be intermediate between Type S and Type C, a supposition subsequently not supported by phylogenomics (Fig. 6.1; Alexander *et al.*, 2009; Möbius *et al.*, 2015; Bryant *et al.*, 2016). This sub-lineage also includes the sequenced Arabian camel isolates (Ghosh *et al.*, 2012; Bryant *et al.*, 2016). The other sub-lineage corresponds to the Type I pigmented ovine isolates from the UK described by Stevenson *et al.* (2002). Pigmentation was originally thought to be a unique characteristic of Type I strains but a few pigmented isolates have since been typed as Type III (Biet *et al.*, 2012) and a non-pigmented isolate as Type I (Bryant *et al.*, 2016). Furthermore, the draft genome of a *MAP*-C pigmented isolate, *MAP* C4A4, has been published (Barbosa *et al.*, 2017). No genetic polymorphisms or the presence or absence of any particular gene that could be unique to pigmented *MAP* isolates have been identified by genomic comparisons to date.

MAP-C also has a sub-lineage that encompasses the group of isolates designated as 'Bison type' or 'Type B'. This strain type was first described by Whittington *et al.* (2001a) and was defined by the number of copies of C or T at base pair position 223 in the insertion sequence IS*1311*, which could differentiate between isolates from US bison (*Bison bison*) and other cattle isolates. More recently, a new genotype designated 'Indian Bison type' has been described, differentiated from other Type *MAP*-C strains by deletion of base pairs at positions 64 and 65 in IS*1311* (Sohal *et al.*, 2013). A divergent *MAP*-C isolate from the Netherlands (MAPMRI074) was included in the global panel of isolates sequenced by Bryant *et al.* (2016), but has not been further characterized to date. This suggests that there may be further undiscovered genetic diversity even within the *MAP*-C lineage for which a far greater number of isolates have been sequenced.

6.2.2 Genotypic differences among *MAP* strains

The two major *MAP* lineages can be distinguished by genetic polymorphisms, in particular by the presence/absence of LSPs, which can be the result of either insertions or deletions. These have been determined by microarray comparisons and *in silico* analysis of WGS data from *MAP*-C and S isolates and *Mycobacterium avium* subsp. *hominissuis* strain 104. The evolution and phylogeny of *MAP* from *M. avium* is covered elsewhere (see Chapter 5, this volume).

MAP-S strains are characterized by the presence of four clusters of open reading frames (ORFs) that are absent from *MAP* K-10 and most of which are present in the *M. avium hominissuis* genome. These clusters have been consistently detected in all *MAP*-S strains analysed to date (Semret *et al.*, 2006; Paustian *et al.*, 2008; Alexander *et al.*, 2009; Castellanos *et al.*, 2009; Bannantine *et al.*, 2012; Möbius *et al.*, 2015) and details are given in Table 6.1. Comparative genomics has revealed other polymorphisms absent from *MAP*-C strains that may reflect genetic diversity within *MAP*-S strains but alternatively could be due to sequencing errors or discrepancies between the methodologies used (Alexander *et al.*, 2009; Castellanos *et al.*, 2009; Möbius *et al.*, 2015). In the converse direction, *MAP*-C strains are differentiated from *MAP*-S strains by the presence of two clusters of ORFs that are consistently absent from *MAP*-S strains (Table 6.2).

Möbius *et al.* (2015) report fewer than 200 SNPs among three *MAP*-C strains compared with about 1000 among three *MAP*-S strains, consistent with a higher heterogeneity within the *MAP*-S lineage (in part due to the large number of SNP differences between Type I and III strains) and high similarity among *MAP*-C strains.

Table 6.1. Gene clusters in *MAP*-S but not *MAP*-C strains.

Open reading frames (ORFs)[a]	Size (kb)	Included designated regions	Encoded proteins/pathways of particular interest
MAP_{S}_15870–16180	34.37	LSP^{S}I (Möbius *et al.*, 2015) LSP^{S}1 (Bannantine *et al.*, 2012) INDEL10 (Castellanos *et al.*, 2009) LSP^{A} 4-II (Semret *et al.*, 2006; Alexander *et al.*, 2009) MAV-7 (Wu *et al.*, 2006) PIG-RDA 20 (Dohmann *et al.*, 2003)	Truncates MAP_2178 involved in mycobactin synthesis TetR transcriptional regulator, PPE proteins, HspR protein, PapA2 protein, ABC-2 type transporter, IS*1311*
MAP_{S}_46170–46350	16.39	LSP^{S}II (Möbius *et al.*, 2015) LSP^{S}2, LSP^{S}4 (Bannantine *et al.*, 2012) INDEL16 (Castellanos *et al.*, 2009) LSP^{A}18 (Semret *et al.*, 2006; Alexander *et al.*, 2009) MAV-24 (Wu *et al.*, 2006) PIG-RDA 10 (Dohmann *et al.*, 2003)	Lipid metabolism
MAP_{S}_17580–17700	16.01	LSP^{S}III (Möbius *et al.*, 2015) LSP^{S}5, LSP^{S}7 (Bannantine *et al.*, 2012) INDEL3 (Castellanos *et al.*, 2009) GPL (Paustian *et al.*, 2008; Alexander *et al.*, 2009) MAV-17 (Wu *et al.*, 2006)	Glycopeptidolipid biosynthesis
MAP_{S}_20550–20770	21.3	LSP^{S}IV (Möbius *et al.*, 2015) MAV-14 (Alexander *et al.*, 2009; Wu *et al.*, 2006) INDEL5 (Castellanos *et al.*, 2009) PIG-RDA 30 (Dohmann *et al.*, 2003)	Cytochrome P450, SecD protein, aryl-sulfatase Lipid and energy metabolism

[a]ORFs as designated by Bannantine *et al.*, 2012.

Table 6.2. Gene clusters in *MAP*-C but not in *MAP*-S strains.

Open reading frames (ORFs)	Included designated regions	Genes/pathway of particular interest
MAP_1491 to MAP_1484 c	LSP^{A}20 (Alexander *et al.*, 2009) INDEL6 (Castellanos *et al.*, 2009)	Pyruvate dehydrogenase complex
MAP_1728 c to MAP_1744	INDEL7 (Castellanos *et al.*, 2009) Del-2 (Marsh *et al.*, 2006; Alexander *et al.*, 2009) MAV-14 (Wu *et al.*, 2006) RDA3 (Marsh and Whittington, 2005)	*yfnB* (predicted hydrolase)[a] *mmpS* and *mmpL* genes[ab] MAP_1729 c & MAP-1730 (hypothetical proteins)[a] *fabG3* (lipid biosynthesis) *acg, devS*[b]

[a]Transcription upregulated in monocyte-derived macrophages (MDMs) in study by Zhu *et al.* (2008).
[b]Also identified by representational difference analysis (Marsh and Whittington, 2005).

6.2.3 Insights into the virulence and pathogenicity of *MAP* strains

Comparative genomics provides insights into the virulence and pathogenicity of *MAP* strains. Analysis of strain-specific polymorphisms reveals presence/absence of genes that may influence virulence and pathogenic traits as detailed in Table 6.3. Many of the lineage-specific polymorphisms encode hypothetical proteins of unknown function. Targeted studies will now be required to determine the functional impact of these natural polymorphisms between strains. Additionally, comparative genomics of genetically similar strains can provide insights into virulence and pathogenicity. A study of *MAP*-C vaccine strains by Bull *et al.* (2013) identified genomic variations associated with attenuation that were probably derived in a classical manner by selective subculture on different media. Twenty-five *MAP*-specific gene deletions were identified of which at least one (MAP3730; S-adenosylmethionine-dependent methyltransferase with homologues to tellurite resistance genes) could be linked to a phenotypic change that would decrease its virulence and persistence in the host. Timms *et al.* (2015) performed a comparison of *MAP*-C strains isolated from humans and animals, which suggested that although the genomes are closely related, there may be key differences in virulence factors, including the PE/PPE (proline-glutamate/proline-proline-glutamate motif) genes, mammalian cell entry (*mce*) operons and the mycobactin cluster.

6.3 Genomic Epidemiology of *MAP*

6.3.1 Host preference

There appear to be epidemiological trends associated with *MAP*-C and *MAP*-S strains with respect to transmission, host preference and susceptibility to infection. However, the results of many past epidemiological studies need to be interpreted with caution since they often employed media that would not support growth of all *MAP* strains. This could easily result in a microbiological bias in these reports, in particular an underrepresentation of *MAP*-S strains. Furthermore, many studies did not use molecular typing techniques that differentiated all *MAP* sub-lineages or strains.

MAP-S strains have been isolated predominantly, but not exclusively, from sheep and goats, suggesting a preference for these host species. Genetically distinct Type III *MAP*-S isolates have also been found in Arabian camels, though only a limited number of isolates have been genetically characterized to date (Ghosh *et al.*, 2012). Reports of the occurrence of *MAP*-S strains in wildlife are relatively uncommon. One *MAP*-S isolate was obtained from a fallow deer (*Dama dama*) (Machackova *et al.*, 2004) and another from a house mouse (*Mus musculus*) (Florou *et al.*, 2008). *MAP*-S isolates were also recovered from tissue samples of kangaroos (*Macropus fuliginosus fuliginosus*) and wallabies (*Macropus eugenii decres*) grazing with infected sheep on Kangaroo Island, though faecal shedding of *MAP* by the macropods was not observed (Cleland *et al.*, 2010).

MAP-C isolates have a broad host range and are commonly isolated from both domesticated and wildlife species, including non-ruminants (Table 15.1 Fox *et al.*, Chapter 15, this volume), although clinical disease has only been observed in ruminants, camelids, rabbits and hares. *MAP*-C is by far the most common *MAP* lineage isolated from cattle. *MAP* isolates from human Crohn's disease patients that have been typed have all been classified as *MAP*-C (Whittington *et al.*, 2000; Bull *et al.*, 2003; Ghadiali *et al.*, 2004; Wu *et al.*, 2006; Griffiths *et al.*, 2008; Paustian *et al.*, 2008; Singh *et al.*, 2009; Wynne *et al.*, 2011; Bannantine *et al.*, 2014; Timms *et al.*, 2015) and appear to be most closely related to isolates from cattle or bison in the same geographical region (Bryant *et al.*, 2016). As with the other types, Type B isolates do not show specificity for bison (*Bison bison*) and have also been isolated from cattle, sheep, goat, buffalo, bison, hog deer, rabbits, blue bull and humans in India (Sohal *et al.*, 2013), and bison and dairy cattle in the USA and Canada (Bryant *et al.*, 2016; Ahlstrom *et al.*, 2016b). This again brings into question the suitability of nomenclature based on host species provenance.

Table 6.3. *Mycobacterium avium* subsp. *paratuberculosis* (*MAP*) strain-specific gene clusters that may be associated with differences in virulence and pathogenic traits.

Gene/large sequence polymorphism (LSP)	Putative role in virulence/pathogenesis	References
MAV_1993 (LSP[A]4-II)	HspR protein. Global regulator of heat shock gene expression. Represses *acr2* involved in virulence and pathogenesis of *Mycobacterium tuberculosis*	Stewart *et al.* (2003, 2005)
MAV_1998 MAV_2006 (LSP[A]4-II)	PPE family proteins. Elicit increased humoral and cell-mediated response in infected host	Tundup *et al.* (2008)
MAV_2005 (LSP[A]4-II)	PapA2 protein. Essential for biosynthesis of *M. tuberculosis* virulence factor sulfolipid-1	Kumar *et al.* (2007)
MAV_3258–MAV_3270 (GPL) MAP_S_17680–17710	Glycopeptidolipids. Promote macrophage activation in a TLR2- and MyD88-dependent manner. Complex cluster with three configurations	Schorey and Sweet (2008); Alexander *et al.* (2009); Möbius *et al.* (2015)
MAV_2984 (MAV-14)	Cytochrome P450. Possible involvement in basic cellular processes and virulence	McLean *et al.* (2006)
MAV_2989 (MAV-14)	Aryl-sulfatase. May modulate pathogen–host interactions	Mougous *et al.* (2002)
MAP_1728c–MAP_1744	MmpL proteins: involved in fatty acid transport, associated with cell surface characteristics, biofilm formation and virulence. MmpS proteins: involved in intracellular survival and *in vivo* growth	Recht and Kolter (2001); Domenech *et al.* (2005); Lamichhane *et al.* (2005); Marsh and Whittington (2005)
MAP_1740c (Del-2)	DevS protein. Essential for *in vivo* growth of *M. tuberculosis* and induced during hypoxia	Sherman *et al.* (2001); Sassetti and Rubin (2003)
MAP_1741c (Del-2)	Upregulated during responses to heat shock and hypoxia in *M. tuberculosis*	Sherman *et al.* (2001); Stewart *et al.* (2002)
MAP_1743c (Del-2)	Acg. Associated with detoxification of nitroaromatic compounds in macrophages and granulomas in *M. tuberculosis*	Sherman *et al.* (2001); Purkayastha *et al.* (2002)
MAP_2704 (INDEL11)	Deleted in pigmented S strains. Haemolysin III homologue. Virulence factor for systemic infections of humans with isolates of *M. avium* complex	Maslow *et al.* (1999); Castellanos *et al.* (2009)
MAV_4125 MAV_4126 (INDEL12) MAP_S_39450–39500	*mce* genes involved in initiation of infection through cell entry and granuloma formation	Gioffré *et al.* (2005); Senaratne *et al.* (2008); Castellanos *et al.* (2009); Möbius *et al.* (2015)
MACPPE43	Possible human-associated PPE. Protein found only in human-associated strain of *MAP*, *M. avium* complex and *M. tuberculosis*.	Timms *et al.* (2015)

Continued

Table 6.3. Continued

Gene/large sequence polymorphism (LSP)	Putative role in virulence/pathogenesis	References
MAP_2827	*ideR*. Regulates expression of genes involved in iron acquisition and storage	Janagama *et al.* (2009), 2010)
MAP2325	Upregulated under iron-limiting conditions. Orthologue in *M. tuberculosis* (eis) important for survival in macrophages	

6.3.2 Transmission

Cross-species transmission of *MAP*-S strains has been observed in natural settings. There are some reports of cattle naturally infected with *MAP*-S strains, including a pigmented *MAP* strain (Watt, 1954), and these infections have typically been associated with bullfighting breeds (de Juan *et al.*, 2006) or cases where there has been direct or indirect contact of calves with sheep infected with *MAP*-S strains (Whittington *et al.*, 2001b). The most convincing demonstration of natural cross-species transmission of *MAP*-S strains between sheep and cattle occurred in Iceland. Paratuberculosis was introduced in 1938 via infected Karakul sheep from Germany and in 1944 spread to the local cattle population, in which it subsequently became endemic (Fridriksdottir *et al.*, 2000). The Icelandic strains were classified retrospectively as *MAP*-S (Whittington *et al.*, 2001b). *MAP*-S strains have also been isolated from farmed red deer in New Zealand (de Lisle *et al.*, 1993), but such isolations appear to be rare (Verdugo *et al.*, 2014) and experimental infection of deer calves has suggested that deer are less susceptible to *MAP*-S than to *MAP*-C strains (O'Brien *et al.*, 2006). Calves experimentally infected with either pigmented or Icelandic ovine strains developed clinical paratuberculosis (Taylor, 1953) and the pigmented strain isolated from cattle by Watt (1954) also could be transmitted experimentally to sheep. However, experimental infections typically involve high doses of *MAP* and may not accurately reflect *in vivo* transmission of the different strains.

Data are accumulating regarding the geographical distribution of *MAP* strains, which has probably been influenced by many factors, including animal movements, strain virulence and farm management systems. *MAP*-S strains are prevalent in Australian sheep and, despite the fact that *MAP*-C strains have been isolated from Australian cattle, they are rarely, if ever, isolated from Australian sheep (Whittington *et al.*, 2000; Windsor, 2015). This is in contrast to Europe and New Zealand, where both *MAP*-S and C strains are isolated from sheep (Stevenson *et al.*, 2009; Verdugo *et al.*, 2014). This difference is possibly linked to differences in management practices between some parts of Europe and Australia, the scale of the farming practices, and relative proportions of sheep and cattle co- or sequentially grazing. *MAP*-S pigmented isolates have been isolated most commonly from animals in the UK and the Faroe Islands (Taylor, 1951; Stevenson *et al.*, 2002). Pigmented strains have also been reported in Spain (Sevilla *et al.*, 2007) and Morocco (Benazzi *et al.*, 1996) but are relatively uncommon in these settings. The first confirmed *MAP*-C pigmented strain (C4A4) was recently isolated from goat faeces in Azores, Portugal (Barbosa *et al.*, 2017). In North India, Type B *MAP* isolates are most commonly isolated, though there are some regional differences, with non-Type B isolates more common in New Delhi and Agra (Singh *et al.*, 2009).

Within- and cross-species transmission of specific *MAP* strains is difficult to conclusively determine in non-experimental settings. Even with the highest resolution molecular typing tools, epidemiological data are needed to support inference regarding transmission events. This is exemplified by the low mean nucleotide substitution rate in the *MAP* genome, which was estimated to be below 0.25 substitutions per genome per year, with an upper 95% confidence limit of less than 0.5 substitutions per genome per year (Bryant *et al.*, 2016). It is important to consider this low substitution rate when inferring epidemiological connections based on genomic data alone, as just a small number of SNP differences could represent more than a decade of divergence. Nevertheless, closely related isolates are likely epidemiologically linked in the somewhat recent past and supporting epidemiological data could help unravel transmission dynamics.

WGS has been particularly useful in identifying the relative number of SNP differences between isolates to derive epidemiological inferences at different spatial scales (Ahlstrom *et al.*, 2016a; Bryant *et al.*, 2016). Sequencing of 141 diverse *MAP* isolates revealed the global diversity, though the authors found no evidence of strong geographical clustering (Bryant *et al.*, 2016). At a national and provincial scale, WGS of 182 *MAP* isolates from Canada indicated that Canadian isolates represented a subset of the known global diversity, and similarly demonstrated weak evidence of geographical clustering within Canada (Ahlstrom *et al.*, 2016a). However, a polymerase chain reaction (PCR) assay designed to target five SNPs that differentiated isolates into four strains relevant to Canadian dairy herds was used to screen 602 *MAP* isolates from Canada (Ahlstrom *et al.*, 2016b) and indicated that some strains appeared to be overrepresented in certain provinces. Comprehensive within-herd level SNP diversity has not been reported to date, though sequencing of multiple isolates per herd has confirmed the co-occurrence of genetically unrelated *MAP* isolates as well as clonal isolates within a single herd and between herds (Ahlstrom *et al.*, 2015; Davidson *et al.*, 2016).

Human-origin *MAP* isolates, isolated from patients with Crohn's disease, ulcerative colitis and non-inflammatory bowel disease controls, were found to show high genomic similarity to each other as well as to an isolate from a bovine host with clinical paratuberculosis from the same region in Australia (Wynne *et al.*, 2011). A SNP genotyping approach targeting SNPs unique to isolates derived from human, bovine and ovine hosts was used to screen 52 isolates predominantly from Australia (Wynne *et al.*, 2014). The authors found that some human-origin *MAP* isolates were highly similar to those derived from bovine hosts, whereas others were more divergent. These studies provide evidence for potential zoonotic transmission of *MAP*.

The endemic nature of *MAP* in many countries challenges efforts aimed at understanding transmission dynamics, as multiple strains co-circulate in most regions. Molecular targets previously used to estimate the diversity and genetic relationship of *MAP* isolates are often not discriminatory enough, not biologically relevant or insufficiently stable. Indeed, the reliability of traditionally used genotyping schemes to accurately reflect genetic similarity and phylogenetic relationships of *MAP* strains has been assessed using WGS (Ahlstrom *et al.*, 2015; Bryant *et al.*, 2016). For example, MIRU-VNTR typing and IS*1311* typing were each found to have limitations in their ability to differentiate strains. As such, the limitations of genotyping tools should be considered when interpreting studies based on those tools, and further comparison using WGS is warranted to determine if currently used molecular markers are appropriate to differentiate strains on an epidemiologically relevant scale.

Molecular tools have nevertheless been, and continue to be, instrumental in supplementing epidemiological data to infer inter- and intraspecies transmission of *MAP* among domestic ruminants (Verdugo *et al.*, 2014; Marquetoux *et al.*, 2016), between domestic ruminants and both ruminant and non-ruminant wildlife (Stevenson *et al.*, 2009; Fritsch *et al.*, 2012) and among wildlife (Forde *et al.*, 2012). Within a herd, short-sequence repeat (SSR) typing of *MAP* indicated that supershedders may play an important role in spreading *MAP* to herd mates (Pradhan *et al.*, 2011). Within an individual animal, SSR typing has suggested mixed infection within an individual cow, which was later confirmed by WGS (Davidson *et al.*, 2016). For more complete descriptions on the molecular epidemiology of *MAP*, see reviews (Stevenson, 2015; Li *et al.*, 2016; Fawzy *et al.*, 2018).

6.4 Comparison of the Virulence and Pathogenicity of *MAP* Strains

MAP-C, *MAP*-S and different strains within these lineages appear to differ in virulence and may be associated with different immunopathological forms of paratuberculosis in different host species. Key determinants of virulence are facilitators of adhesion, invasion, survival and colonization of host cells.

The heparin-binding haemagglutinin adhesin mediates the binding of mycobacteria to epithelial cells and fibroblasts early in infection and may play a role in dissemination. Lefrançois *et al.* (2013) determined that the C-terminal regions of the *hbhA* gene differ between *MAP*-C and *MAP*-S strains and that this correlates with their different affinities for heparin and consequently their adherence properties. There is some evidence that different *MAP* strains have different capacities for entry and survival in macrophages, although it is difficult to tease out the effects of bacterial virulence factors from host responses as the interplay of both will determine the outcome of infection. Intracellular studies undertaken by Janagama *et al.* (2006) and Gollnick *et al.* (2007) showed that *MAP*-C strains persisted in relatively higher numbers in bovine monocyte-derived macrophages (MDMs) when compared with a *MAP*-S strain. However, Abendaño *et al.* (2013) showed that *MAP*-S and C strains from sheep and goats persisted within bovine macrophages in lower numbers than those from cattle, bison, deer and wild boar regardless of *MAP* lineage. A more recent study by Alonso-Hearn *et al.* (2019) analysed infection data of both bovine and ovine MDMs with genotypically distinct *MAP* isolates, including *MAP*-C, *MAP*-S and Type B strains, from six different host species. Overall, *MAP* isolates from goat and sheep persisted or grew minimally within macrophages, whereas isolates from cattle and non-domesticated animals grew to higher numbers, suggesting that these isolates may be more virulent or better adapted to infect domestic ruminants. Furthermore, they revealed that bovine macrophages were more efficient in internalizing bovine *MAP* isolates and ovine macrophages were more efficient in internalizing ovine *MAP* isolates.

Studies have been performed to determine the genetic responses of different *MAP* strains to macrophages. Zhu *et al.* (2008) undertook transcriptional analysis of different *MAP* strains in MDMs using selective capture of transcribed sequences. Despite variations in the genes identified, the different *MAP* strains responded in a generally similar fashion to oxidative stress, to metabolic and nutritional starvation, in cell survival and in upregulating genes involved in cell wall biosynthesis. However, transcription of MAP_1728 (YfnB), MAP_1738 (MmpL5), MAP_1729c and MAP_1730 (hypothetical proteins with unknown function) was upregulated only in *MAP*-C strains, consistent with their absence from *MAP*-S strains examined to date (Tables 6.2 and 6.3). Similarly, transcriptional and proteomic profiling of *MAP*-C and *MAP*-S strains under iron-replete or -depleted conditions revealed that different strains utilize different metabolic pathways under different conditions, likely due to strain-specific polymorphisms in the promoter of the iron storage gene *bfrA* (Janagama *et al.*, 2010).

A few studies have presented evidence that different *MAP* strains may play a role in polarizing the host immune responses, which may determine the different disease pathologies observed. Janagama *et al.* (2006) investigated cytokine responses to different *MAP* strains in a bovine MDM system using real-time PCR assays, and Motiwala *et al.* (2006) performed a global-scale transcriptional analysis of human macrophages (THP-1 cells) exposed to different *MAP* strains. Both studies reported that *MAP*-C strains induced anti-inflammatory and anti-apoptotic pathways in the host cells without causing major alterations in the transcription of proinflammatory genes, which would favour bacterial survival and persistence. This pattern of gene expression was found to be the same for bovine, bison and human *MAP*-C strains, as defined by SSR analysis, although the magnitude of the responses varied. In contrast, ovine *MAP*-S strains representing distinct SSR genotypes significantly upregulated proinflammatory genes. Proinflammatory responses are generally associated with protection and elimination of mycobacteria, so this gene expression profile may help to explain why *MAP*-S strains rarely cause disease in bovine hosts. However, Borrmann *et al.* (2011) found that both *MAP*-C and *MAP*-S strains upregulated proinflammatory IL-1β genes in human macrophages, and Abendaño *et al.* (2014) demonstrated common responses of ovine MDMs to infection with *MAP*-C, *MAP*-S and Type B isolates from cattle, sheep, goat, deer and wild boar regardless of *MAP* lineage or host of origin. The results of these studies should be treated with caution since only a few *MAP* strains were investigated.

Protein expression profiling studies have been undertaken to investigate the responses of isolates representing different *MAP* lineages to conditions mimicking the host environment such as oxidative and nitrosative stress, and the stressors of temperature flux, hypoxia and nutrient starvation. These studies have only employed one *MAP*-S and one *MAP*-C strain but observed lineage-specific differential expression of a number of proteins: 10 and 9 in response to oxidative and nitrosative stress, respectively (Kawaji *et al.*, 2010); 27 to temperature flux (Gumber and Whittington, 2009); 21 and 26 to hypoxia and starvation, respectively (Gumber *et al.*, 2009). Janagama *et al.* (2010) performed protein profiling of the two lineages under iron-depleted and -replete conditions and revealed that under iron-replete conditions, ribosomal proteins, bacterioferritin, mycobacterial heme and utilization and degrader proteins were upregulated but under iron-depleted conditions aconitase, succinate dehydrogenases and superoxide dismutase were downregulated in a *MAP*-C strain only. A study by Hughes *et al.* (2012) analysed seven *MAP*-S (Type I) and 18 *MAP*-C strains and identified 17 and 16 proteins specific for *MAP*-S and *MAP*-C, respectively, which were not dependent on growth phase and were also exhibited in *MAP* isolated directly from pathogenic lesions. None of these differentially expressed proteins correlated with ORFs within the strain-specific genomic polymorphisms as detailed in Tables 6.1 and 6.2.

Verna *et al.* (2007) investigated strain-specific differences in the pathology of disease in sheep. Infection with different *MAP*-C strains resulted in a common pattern, characterized by focal lesions, mainly in the mesenteric lymph nodes, as well as the presence of fibrous tissue and occasionally necrosis and numerous Langhans giant cells in the granulomas. Infection with a *MAP*-S strain induced more severe lesions, occurring mainly in the intestinal lymphoid tissue, and there was a conspicuous absence of necrosis, fibrosis and giant cells. Lesions induced by the *MAP*-S strain were more severe than those induced by *MAP*-C strains, which suggests that *MAP*-C strains had a slow, localized development in the early stages of infection. The development of giant cells may be linked to *MAP* strain rather than host determinants, since giant cells are a feature of natural cases of bovine paratuberculosis and leporine paratuberculosis (Beard *et al.*, 2001) but not ovine paratuberculosis caused by *MAP*-S strains.

Experimental infection of lambs with *MAP*-C and *MAP*-S strains revealed specific antibody and IFN-γ production was significantly higher in animals infected with *MAP*-C strains, whereas no consistent IFN-γ responses were observed in animals infected with *MAP*-S strains (Fernández *et al.*, 2014). Furthermore, sheep infected with passaged *MAP*-C strains showed more severe and diffuse disease in the small intestine than those infected with passaged *MAP*-S strains, though lesions induced by *MAP*-C strains showed a regressive character, unlike those induced by the *MAP*-S strains. These results suggest that sheep naturally infected with *MAP*-C strains may be able to recover from the infection.

Different *MAP*-C strains have also been found to elicit different immune responses *in vitro* and *in vivo*. Differential immune responses in peripheral cells, in the proliferative responses and in cell functionality at the local level were found between two different Argentinian *MAP*-C strains in experimentally infected calves (Colavecchia *et al.*, 2016). Similarly, differences in virulence, immune response and protection of four different Argentinian *MAP*-C strains (as determined by MIRU-VNTR and SSR typing) were tested in a murine model, in which strains demonstrated different levels of virulence (Moyano *et al.*, 2018). Less virulent strains failed to induce a significant production of the proinflammatory cytokine IFN-γ, whereas a virulent strain established infection along with a proinflammatory immune response. A low virulent strain failed to provide protection from challenge, whereas the virulent strain, in its live and inactivated form, significantly reduced the bacterial count in infected mice. A study by Möbius *et al.* (2017) also provided weak evidence of differences in virulence among different *MAP*-C strains.

6.5 Concluding Thoughts

With the cost of WGS falling, more researchers are conducting in-depth genome comparison and epidemiological studies using this technology. WGS analysis of isolates with defined phenotype, virulence or pathogenicity and functional/virulence studies of genomically divergent strains should provide further insights into these distinct traits and genomic polymorphisms. More targeted studies are required to determine the functional impact of detected polymorphisms, particularly where these encode hypothetical proteins.

References

Abendaño, N., Sevilla, I.A., Prieto, J.M., Garrido, J.M., Juste, R.A. *et al.* (2013) *Mycobacterium avium* subspecies *paratuberculosis* isolates from sheep and goats show reduced persistence in bovine macrophages than cattle, bison, deer and wild boar strains regardless of genotype. *Veterinary Microbiology* 163(3–4), 325–334. DOI: 10.1016/j.vetmic.2012.12.042.

Abendaño, N., Tyukalova, L., Barandika, J.F., Balseiro, A., Sevilla, I.A. *et al.* (2014) *Mycobacterium avium* subsp. *paratuberculosis* isolates induce *in vitro* granuloma formation and show successful survival phenotype, common anti-inflammatory and antiapoptotic responses within ovine macrophages regardless of genotype or host of origin. *PLoS ONE* 9(8), e104238. DOI: 10.1371/journal.pone.0104238.

Ahlstrom, C., Barkema, H.W., Stevenson, K., Zadoks, R.N., Biek, R. *et al.* (2015) Limitations of variable number of tandem repeat typing identified through whole genome sequencing of *Mycobacterium avium* subsp. *paratuberculosis* on a national and herd level. *BMC Genomics* 16(1), 161. DOI: 10.1186/s12864-015-1387-6.

Ahlstrom, C., Barkema, H.W., Stevenson, K., Zadoks, R.N., Biek, R. *et al.* (2016a) Genome-wide diversity and phylogeography of *Mycobacterium avium* subsp. *paratuberculosis* in Canadian dairy cattle. *PLoS ONE* 11(2), e0149017. DOI: 10.1371/journal.pone.0149017.

Ahlstrom, C., Barkema, H.W. and De Buck, J. (2016b) Relative frequency of 4 major strain types of *Mycobacterium avium* ssp. *paratuberculosis* in Canadian dairy herds using a novel single nucleotide polymorphism-based polymerase chain reaction. *Journal of Dairy Science* 99(10), 8297–8303. DOI: 10.3168/jds.2016-11397.

Alexander, D.C., Turenne, C.Y. and Behr, M.A. (2009) Insertion and deletion events that define the pathogen *Mycobacterium avium* subsp. *paratuberculosis*. *Journal of Bacteriology* 191(3), 1018–1025. DOI: 10.1128/JB.01340-08.

Alonso-Hearn, M., Magombedze, G., Abendaño, N., Landin, M. and Juste, R.A. (2019) Deciphering the virulence of *Mycobacterium avium* subsp. *paratuberculosis* isolates in animal macrophages using mathematical models. *Journal of Theoretical Biology* 468, 82–91. DOI: 10.1016/j.jtbi.2019.01.040.

Bannantine, J.P., Wu, C.W., Hsu, C., Zhou, S., Schwartz, D.C. *et al.* (2012) Genome sequencing of ovine isolates of *Mycobacterium avium* subspecies *paratuberculosis* offers insights into host association. *BMC Genomics* 13(1), 89. DOI: 10.1186/1471-2164-13-89.

Bannantine, J.P., Li, L., Mwangi, M., Cote, R., Raygoza Garay, J.A. *et al.* (2014) Complete genome sequence of *Mycobacterium avium* subsp. *paratuberculosis*, isolated from human breast milk. *Genome Announcements* 2(1), e01252–13. DOI: 10.1128/genomeA.01252-13.

Barbosa, P., Leão, C., Usié, A., Amaro, A., Botelho, A. *et al.* (2017) Draft genome sequence of a rare pigmented *Mycobacterium avium* subsp. *paratuberculosis* Type C strain. *Genome Announcements* 5(41), e01066–17. DOI: 10.1128/genomeA.01066-17.

Beard, P.M., Rhind, S.M., Buxton, D., Daniels, M.J., Henderson, D. *et al.* (2001) Natural paratuberculosis infection in rabbits in Scotland. *Journal of Comparative Pathology* 124(4), 290–299. DOI: 10.1053/jcpa.2001.0466.

Benazzi, S., Hamidi, M. and Schliesser, T. (1996) Paratuberculosis in sheep flocks in Morocco: a serological, microscopical and cultural survey. *Journal of Veterinary Medicine, Series B* 43(1–10), 213–219. DOI: 10.1111/j.1439-0450.1996.tb00308.x.

Biet, F., Sevilla, I.A., Cochard, T., Lefrançois, L.H., Garrido, J.M. *et al.* (2012) Inter- and intra-subtype genotypic differences that differentiate *Mycobacterium avium* subspecies *paratuberculosis* strains. *BMC Microbiology* 12(1), 264. DOI: 10.1186/1471-2180-12-264.

Borrmann, E., Möbius, P., Diller, R. and Köhler, H. (2011) Divergent cytokine responses of macrophages to *Mycobacterium avium* subsp. *paratuberculosis* strains of Types II and III in a standardized *in vitro* model. *Veterinary Microbiology* 152(1–2), 101–111. DOI: 10.1016/j.vetmic.2011.04.002.

Brauning, R., Plain, K., Gautam, M., Russell, T., Correa, C.C. *et al.* (2019) Complete genome sequence of the Telford Type S strain of *Mycobacterium avium* subsp. *paratuberculosis*. *Microbiology Resource Announcements* 8(11), e00004–00019. DOI: 10.1128/MRA.00004-19.

Bryant, J.M., Thibault, V.C., Smith, D.G.E., McLuckie, J., Heron, I. *et al.* (2016) Phylogenomic exploration of the relationships between strains of *Mycobacterium avium* subspecies *paratuberculosis*. *BMC Genomics* 17(1), 79. DOI: 10.1186/s12864-015-2234-5.

Bull, T.J., Sidi-Boumedine, K., McMinn, E.J., Stevenson, K., Pickup, R. *et al.* (2003) Mycobacterial interspersed repetitive units (MIRU) differentiate *Mycobacterium avium* subspecies *paratuberculosis* from other species of the *Mycobacterium avium* complex. *Molecular and Cellular Probes* 17(4), 157–164. DOI: 10.1016/S0890-8508(03)00047-1.

Bull, T.J., Schock, A., Sharp, J.M., Greene, M., McKendrick, I.J. *et al.* (2013) Genomic variations associated with attenuation in *Mycobacterium avium* subsp. *paratuberculosis* vaccine strains. *BMC Microbiology* 13(1), 11. DOI: 10.1186/1471-2180-13-11.

Castellanos, E., Aranaz, A., Gould, K.A., Linedale, R., Stevenson, K. *et al.* (2009) Discovery of stable and variable differences in the *Mycobacterium avium* subsp. *paratuberculosis* Types I, II, and III genomes by pan-genome microarray analysis. *Applied and Environmental Microbiology* 75(3), 676–686. DOI: 10.1128/AEM.01683-08.

Cleland, P.C., Lehmann, D.R., Phillips, P.H., Cousins, D.V., Reddacliff, L.A. *et al.* (2010) A survey to detect the presence of *Mycobacterium avium* subspecies *paratuberculosis* in Kangaroo Island macropods. *Veterinary Microbiology* 145(3–4), 339–346. DOI: 10.1016/j.vetmic.2010.03.021.

Colavecchia, S.B., Fernández, B., Jolly, A., Minatel, L., Hajos, S.E. *et al.* (2016) Immunological findings associated with Argentinean strains of *Mycobacterium avium* subsp. *paratuberculosis* in bovine models. *Veterinary Immunology and Immunopathology* 176, 28–33. DOI: 10.1016/j.vetimm.2016.04.010.

Collins, D.M., Gabric, D.M. and de Lisle, G.W. (1990) Identification of two groups of *Mycobacterium paratuberculosis* strains by restriction endonuclease analysis and DNA hybridization. *Journal of Clinical Microbiology* 28(7), 1591–1596. DOI: 10.1128/JCM.28.7.1591-1596.1990.

Davidson, F.W., Ahlstrom, C., De Buck, J., Whitney, H.G. and Tahlan, K. (2016) Examination of *Mycobacterium avium* subspecies *paratuberculosis* mixed genotype infections in dairy animals using a whole genome sequencing approach. *PeerJ* 4(2), e2793. DOI: 10.7717/peerj.2793.

de Juan, L., Mateos, A., Domínguez, L., Sharp, J.M. and Stevenson, K. (2005) Genetic diversity of *Mycobacterium avium* subspecies *paratuberculosis* isolates from goats detected by pulsed-field gel electrophoresis. *Veterinary Microbiology* 106(3–4), 249–257. DOI: 10.1016/j.vetmic.2004.12.013.

de Juan, L., Álvarez, J., Aranaz, A., Rodríguez, A., Romero, B. *et al.* (2006) Molecular epidemiology of Types I/III strains of *Mycobacterium avium* subspecies *paratuberculosis* isolated from goats and cattle. *Veterinary Microbiology* 115(1–3), 102–110. DOI: 10.1016/j.vetmic.2006.01.008.

de Lisle, G.W., Yates, G.F. and Collins, D.M. (1993) Paratuberculosis in farmed deer: case reports and DNA characterization of isolates of *Mycobacterium paratuberculosis*. *Journal of Veterinary Diagnostic Investigation* 5(4), 567–571. DOI: 10.1177/104063879300500411.

Dohmann, K., Strommenger, B., Stevenson, K., de Juan, L., Stratmann, J. *et al.* (2003) Characterization of genetic differences between *Mycobacterium avium* subsp. *paratuberculosis* Type I and Type II isolates. *Journal of Clinical Microbiology* 41(11), 5215–5223. DOI: 10.1128/JCM.41.11.5215-5223.2003.

Domenech, P., Reed, M.B. and Barry, C.E. (2005) Contribution of the *Mycobacterium tuberculosis* MmpL protein family to virulence and drug resistance. *Infection and Immunity* 73(6), 3492–3501. DOI: 10.1128/IAI.73.6.3492-3501.2005.

Fawzy, A., Zschöck, M., Ewers, C. and Eisenberg, T. (2018) Genotyping methods and molecular epidemiology of *Mycobacterium avium* subsp. *paratuberculosis* (MAP). *International Journal of Veterinary Science and Medicine* 6(2), 258–264. DOI: 10.1016/j.ijvsm.2018.08.001.

Fernández, M., Benavides, J., Sevilla, I.A., Fuertes, M., Castaño, P. *et al.* (2014) Experimental infection of lambs with C and S-type strains of *Mycobacterium avium* subspecies *paratuberculosis*: immunological and pathological findings. *Veterinary Research* 45(1), 5. DOI: 10.1186/1297-9716-45-5.

Florou, M., Leontides, L., Kostoulas, P., Billinis, C., Sofia, M. *et al.* (2008) Isolation of *Mycobacterium avium* subspecies *paratuberculosis* from non-ruminant wildlife living in the sheds and on the pastures of Greek sheep and goats. *Epidemiology and Infection* 136(5), 644–652. DOI: 10.1017/S095026880700893X.

Forde, T., Kutz, S., Buck, J.D., Warren, A., Ruckstuhl, K. *et al.* (2012) Occurrence, diagnosis, and strain typing of *Mycobacterium avium* subspecies *paratuberculosis* infection in Rocky Mountain bighorn sheep (*Ovis canadensis canadensis*) in southwestern Alberta. *Journal of Wildlife Diseases* 48(1), 1–11. DOI: 10.7589/0090-3558-48.1.1.

Fridriksdottir, V., Gunnarsson, E., Sigurdarson, S. and Gudmundsdottir, K.B. (2000) Paratuberculosis in Iceland: epidemiology and control measures, past and present. *Veterinary Microbiology* 77(3–4), 263–267. DOI: 10.1016/S0378-1135(00)00311-4.

Fritsch, I., Luyven, G., Köhler, H., Lutz, W. and Möbius, P. (2012) Suspicion of *Mycobacterium avium* subsp. *paratuberculosis* transmission between cattle and wild-living red deer (*Cervus elaphus*) by multitarget genotyping. *Applied and Environmental Microbiology* 78(4), 1132–1139. DOI: 10.1128/AEM.06812-11.

Ghadiali, A.H., Strother, M., Naser, S.A., Manning, E.J.B. and Sreevatsan, S. (2004) *Mycobacterium avium* subsp. *paratuberculosis* strains isolated from Crohn's disease patients and animal species exhibit similar polymorphic locus patterns. *Journal of Clinical Microbiology* 42(11), 5345–5348. DOI: 10.1128/JCM.42.11.5345-5348.2004.

Ghosh, P., Hsu, C., Alyamani, E.J., Shehata, M.M., Al-Dubaib, M.A. *et al.* (2012) Genome-wide analysis of the emerging infection with *Mycobacterium avium* subspecies *paratuberculosis* in the Arabian camels (*Camelus dromedarius*). *PLoS ONE* 7(2), e31947. DOI: 10.1371/journal.pone.0031947.

Gioffré, A., Infante, E., Aguilar, D., Santangelo, M.P., Klepp, L. *et al.* (2005) Mutation in *mce* operons attenuates *Mycobacterium tuberculosis* virulence. *Microbes and Infection* 7(3), 325–334. DOI: 10.1016/j.micinf.2004.11.007.

Gollnick, N.S., Mitchell, R.M., Baumgart, M., Janagama, H.K., Sreevatsan, S. *et al.* (2007) Survival of *Mycobacterium avium* subsp. *paratuberculosis* in bovine monocyte-derived macrophages is not affected by host infection status but depends on the infecting bacterial genotype. *Veterinary Immunology and Immunopathology* 120(3–4), 93–105. DOI: 10.1016/j.vetimm.2007.07.017.

Griffiths, T.A., Rioux, K. and De Buck, J. (2008) Sequence polymorphisms in a surface PPE protein distinguish types I, II, and III of *Mycobacterium avium* subsp. *paratuberculosis*. *Journal of Clinical Microbiology* 46(4), 1207–1212. DOI: 10.1128/JCM.02031-07.

Gumber, S. and Whittington, R.J. (2009) Analysis of the growth pattern, survival and proteome of *Mycobacterium avium* subsp. *paratuberculosis* following exposure to heat. *Veterinary Microbiology* 136(1–2), 82–90. DOI: 10.1016/j.vetmic.2008.10.003.

Gumber, S., Taylor, D.L., Marsh, I.B. and Whittington, R.J. (2009) Growth pattern and partial proteome of *Mycobacterium avium* subsp. *paratuberculosis* during the stress response to hypoxia and nutrient starvation. *Veterinary Microbiology* 133(4), 344–357. DOI: 10.1016/j.vetmic.2008.07.021.

Hughes, V., Garcia-Sanchez, A., Smith, S., McLean, K., Lainson, A. *et al.* (2012) Proteome-determined type-specific proteins of *Mycobacterium avium* subspecies *paratuberculosis*. *Veterinary Microbiology* 158(1-2), 153–162. DOI: 10.1016/j.vetmic.2012.01.032.

Janagama, H., il Jeong, K., Kapur, V., Coussens, P. and Sreevatsan, S. (2006) Cytokine responses of bovine macrophages to diverse clinical *Mycobacterium avium* subsp. *paratuberculosis* strains. *BMC Microbiology* 6(1), 10. DOI: 10.1186/1471-2180-6-10.

Janagama, H.K., Senthilkumar, T.M.A., Bannantine, J.P., Rodriguez, G.M., Smith, I. *et al.* (2009) Identification and functional characterization of the iron-dependent regulator (IdeR) of *Mycobacterium avium* subsp. *paratuberculosis*. *Microbiology* 155(Pt 11), 3683–3690. DOI: 10.1099/mic.0.031948-0.

Janagama, H.K., Bannantine, J.P., Kugadas, A., Jagtap, P., Higgins, L. *et al.* (2010) Iron-sparing response of *Mycobacterium avium* subsp. *paratuberculosis* is strain dependent. *BMC Microbiology* 10(1), 268. DOI: 10.1186/1471-2180-10-268.

Kawaji, S., Zhong, L. and Whittington, R.J. (2010) Partial proteome of *Mycobacterium avium* subsp. *paratuberculosis* under oxidative and nitrosative stress. *Veterinary Microbiology* 145(3–4), 252–264. DOI: 10.1016/j.vetmic.2010.03.025.

Kumar, P., Schelle, M.W., Jain, M., Lin, F.L., Petzold, C.J. *et al.* (2007) PapA1 and PapA2 are acyltransferases essential for the biosynthesis of the *Mycobacterium tuberculosis* virulence factor sulfolipid-1. *Proceedings of the National Academy of Sciences* 104(27), 11221–11226. DOI: 10.1073/pnas.0611649104.

Lamichhane, G., Tyagi, S. and Bishai, W.R. (2005) Designer arrays for defined mutant analysis to detect genes essential for survival of *Mycobacterium tuberculosis* in mouse lungs. *Infection and Immunity* 73(4), 2533–2540. DOI: 10.1128/IAI.73.4.2533-2540.2005.

Leão, C., Goldstone, R.J., Bryant, J., McLuckie, J., Inácio, J. *et al.* (2016) Novel single nucleotide polymorphism-based assay for genotyping *Mycobacterium avium* subsp. *paratuberculosis*. *Journal of Clinical Microbiology* 54(3), 556–564. DOI: 10.1128/JCM.01958-15.

Lefrançois, L.H., Bodier, C.C., Cochard, T., Canepa, S., Raze, D. *et al.* (2013) Novel feature of *Mycobacterium avium* subsp. *paratuberculosis*, highlighted by characterization of the heparin-binding hemagglutinin adhesin. *Journal of Bacteriology* 195(21), 4844–4853. DOI: 10.1128/JB.00671-13.

Li, L., Bannantine, J.P., Zhang, Q., Amonsin, A., May, B.J. *et al.* (2005) The complete genome sequence of *Mycobacterium avium* subspecies *paratuberculosis*. *Proceedings of the National Academy of Sciences* 102(35), 12344–12349. DOI: 10.1073/pnas.0505662102.

Li, L., Katani, R., Schilling, M. and Kapur, V. (2016) Molecular epidemiology of *Mycobacterium avium* subsp. *paratuberculosis* on dairy farms. *Annual Review of Animal Biosciences* 4(1), 155–176. DOI: 10.1146/annurev-animal-021815-111304.

Machackova, M., Svastova, P., Lamka, J., Parmova, I., Liska, V. *et al.* (2004) Paratuberculosis in farmed and free-living wild ruminants in the Czech Republic (1999–2001). *Veterinary Microbiology* 101(4), 225–234. DOI: 10.1016/j.vetmic.2004.04.001.

Marquetoux, N., Heuer, C., Wilson, P., Ridler, A. and Stevenson, M. (2016) Merging DNA typing and network analysis to assess the transmission of paratuberculosis between farms. *Preventive Veterinary Medicine* 134, 113–121. DOI: 10.1016/j.prevetmed.2016.09.014.

Marsh, I.B. and Whittington, R.J. (2005) Deletion of an *mmpL* gene and multiple associated genes from the genome of the S strain of *Mycobacterium avium* subsp. *paratuberculosis* identified by representational difference analysis and *in silico* analysis. *Molecular and Cellular Probes* 19(6), 371–384. DOI: 10.1016/j.mcp.2005.06.005.

Marsh, I.B., Bannantine, J.P., Paustian, M.L., Tizard, M.L., Kapur, V. *et al.* (2006) Genomic comparison of *Mycobacterium avium* subsp. *paratuberculosis* sheep and cattle strains by microarray hybridization. *Journal of Bacteriology* 188(6), 2290–2293. DOI: 10.1128/JB.188.6.2290-2293.2006.

Maslow, J.N., Dawson, D., Carlin, E.A. and Holland, S.M. (1999) Hemolysin as a virulence factor for systemic infection with isolates of *Mycobacterium avium* complex. *Journal of Clinical Microbiology* 37(2), 445–446. DOI: 10.1128/JCM.37.2.445-446.1999.

McLean, K.J., Clift, D., Lewis, D.G., Sabri, M., Balding, P.R. *et al.* (2006) The preponderance of P450s in the *Mycobacterium tuberculosis* genome. *Trends in Microbiology* 14(5), 220–228. DOI: 10.1016/j.tim.2006.03.002.

Möbius, P., Hölzer, M., Felder, M., Nordsiek, G., Groth, M. *et al.* (2015) Comprehensive insights in the *Mycobacterium avium* subsp. *paratuberculosis* genome using new WGS data of sheep strain JIII-386 from Germany. *Genome Biology and Evolution* 7(9), 2585–2601. DOI: 10.1093/gbe/evv154.

Möbius, P., Liebler-Tenorio, E., Hölzer, M. and Köhler, H. (2017) Evaluation of associations between genotypes of *Mycobacterium avium* subsp. *paratuberculsis* and presence of intestinal lesions characteristic of paratuberculosis. *Veterinary Microbiology* 201, 188–194. DOI: 10.1016/j.vetmic.2017.01.026.

Motiwala, A.S., Janagama, H.K., Paustian, M.L., Zhu, X., Bannantine, J.P. *et al.* (2006) Comparative transcriptional analysis of human macrophages exposed to animal and human isolates of *Mycobacterium avium* subspecies *paratuberculosis* with diverse genotypes. *Infection and Immunity* 74(11), 6046–6056. DOI: 10.1128/IAI.00326-06.

Mougous, J.D., Green, R.E., Williams, S.J., Brenner, S.E. and Bertozzi, C.R. (2002) Sulfotransferases and sulfatases in mycobacteria. *Chemistry & Biology* 9(7), 767–776. DOI: 10.1016/S1074-5521(02)00175-8.

Moyano, R.D., Traveria, G.E., Alvarado Pinedo, M.F., Mon, M.L., Olivieri, M.A.C. *et al.* (2018) Protection efficacy of Argentinian isolates of *Mycobacterium avium* subsp. *paratuberculosis* with different genotypes and virulence in a murine model. *Research in Veterinary Science* 121, 4–11. DOI: 10.1016/j.rvsc.2018.09.009.

O'Brien, R., Mackintosh, C.G., Bakker, D., Kopecna, M., Pavlik, I. *et al.* (2006) Immunological and molecular characterization of susceptibility in relationship to bacterial strain differences in *Mycobacterium avium* subsp. *paratuberculosis* infection in the red deer (*Cervus elaphus*). *Infection and Immunity* 74(6), 3530–3537. DOI: 10.1128/IAI.01688-05.

Paustian, M.L., Zhu, X., Sreevatsan, S., Robbe-Austerman, S., Kapur, V. *et al.* (2008) Comparative genomic analysis of *Mycobacterium avium* subspecies obtained from multiple host species. *BMC Genomics* 9(1), 135. DOI: 10.1186/1471-2164-9-135.

Paustian, M.L., Bannantine, J.P. and Kapur, V. (2010) *Mycobacterium avium* subsp. *paratuberculosis* genome. In: Behr, M.A. and Collins, D.M. (eds) *Paratuberculosis Organism, Disease, Control*. CAB International, Wallingford, UK, pp. 73–82.

Pradhan, A.K., Mitchell, R.M., Kramer, A.J., Zurakowski, M.J., Fyock, T.L. *et al.* (2011) Molecular epidemiology of *Mycobacterium avium* subsp. *paratuberculosis* in a longitudinal study of three dairy herds. *Journal of Clinical Microbiology* 49(3), 893–901. DOI: 10.1128/JCM.01107-10.

Purkayastha, A., McCue, L.A. and McDonough, K.A. (2002) Identification of a *Mycobacterium tuberculosis* putative classical nitroreductase gene whose expression is coregulated with that of the *acr* gene within macrophages, in standing versus shaking cultures, and under low oxygen conditions. *Infection and Immunity* 70(3), 1518–1529. DOI: 10.1128/IAI.70.3.1518-1529.2002.

Recht, J. and Kolter, R. (2001) Glycopeptidolipid acetylation affects sliding motility and biofilm formation in *Mycobacterium smegmatis*. *Journal of Bacteriology* 183(19), 5718–5724. DOI: 10.1128/JB.183.19.5718-5724.2001.

Sassetti, C.M. and Rubin, E.J. (2003) Genetic requirements for mycobacterial survival during infection. *Proceedings of the National Academy of Sciences* 100(22), 12989–12994. DOI: 10.1073/pnas.2134250100.

Schorey, J.S. and Sweet, L. (2008) The mycobacterial glycopeptidolipids: structure, function, and their role in pathogenesis. *Glycobiology* 18(11), 832–841. DOI: 10.1093/glycob/cwn076.

Semret, M., Turenne, C.Y., de Haas, P., Collins, D.M. and Behr, M.A. (2006) Differentiating host-associated variants of *Mycobacterium avium* by PCR for detection of large sequence polymorphisms. *Journal of Clinical Microbiology* 44(3), 881–887. DOI: 10.1128/JCM.44.3.881-887.2006.

Senaratne, R.H., Sidders, B., Sequeira, P., Saunders, G., Dunphy, K. *et al.* (2008) *Mycobacterium tuberculosis* strains disrupted in *mce3* and *mce4* operons are attenuated in mice. *Journal of Medical Microbiology* 57(2), 164–170. DOI: 10.1099/jmm.0.47454-0.

Sevilla, I., Garrido, J.M., Geijo, M. and Juste, R.A. (2007) Pulsed-field gel electrophoresis profile homogeneity of *Mycobacterium avium* subsp. *paratuberculosis* isolates from cattle and heterogeneity of those from sheep and goats. *BMC Microbiology* 7(1), 18. DOI: 10.1186/1471-2180-7-18.

Sherman, D.R., Voskuil, M., Schnappinger, D., Liao, R., Harrell, M.I. *et al.* (2001) Regulation of the *Mycobacterium tuberculosis* hypoxic response gene encoding alpha-crystallin. *Proceedings of the National Academy of Sciences* 98(13), 7534–7539. DOI: 10.1073/pnas.121172498.

Singh, S.V., Sohal, J.S., Singh, P.K. and Singh, A.V. (2009) Genotype profiles of *Mycobacterium avium* subspecies *paratuberculosis* isolates recovered from animals, commercial milk, and human beings in North India. *International Journal of Infectious Diseases* 13(5), e221–e227. DOI: 10.1016/j.ijid.2008.11.022.

Sohal, J.S., Singh, S.V., Singh, P.K., Singh, A.V. and Kumar, N. (2013) A new marker IS*1311* L2 PCR-REA for identification of 'Indian Bison' type *Mycobacterium avium* subspecies *paratuberculosis*. *Indian Journal of Biotechnology* 12, 204–207.

Stevenson, K. (2015) Genetic diversity of *Mycobacterium avium* subspecies *paratuberculosis* and the influence of strain type on infection and pathogenesis: a review. *Veterinary Research* 46(1), 64. DOI: 10.1186/s13567-015-0203-2.

Stevenson, K., Hughes, V.M., de Juan, L., Inglis, N.F., Wright, F. *et al.* (2002) Molecular characterization of pigmented and nonpigmented isolates of *Mycobacterium avium* subsp. *paratuberculosis*. *Journal of Clinical Microbiology* 40(5), 1798–1804. DOI: 10.1128/JCM.40.5.1798-1804.2002.

Stevenson, K., Álvarez, J., Bakker, D., Biet, F., de Juan, L. *et al.* (2009) Occurrence of *Mycobacterium avium* subspecies *paratuberculosis* across host species and European countries with evidence for transmission between wildlife and domestic ruminants. *BMC Microbiology* 9(1), 212. DOI: 10.1186/1471-2180-9-212.

Stewart, G.R., Wernisch, L., Stabler, R., Mangan, J.A., Hinds, J. *et al.* (2002) Dissection of the heat-shock response in *Mycobacterium tuberculosis* using mutants and microarrays. *Microbiology* 148(Pt 10), 3129–3138. DOI: 10.1099/00221287-148-10-3129.

Stewart, G.R., Robertson, B.D. and Young, D.B. (2003) Tuberculosis: a problem with persistence. *Nature Reviews Microbiology* 1(2), 97–105. DOI: 10.1038/nrmicro749.

Stewart, G.R., Newton, S.M., Wilkinson, K.A., Humphreys, I.R., Murphy, H.N. *et al.* (2005) The stress-responsive chaperone α-crystallin 2 is required for pathogenesis of *Mycobacterium tuberculosis*. *Molecular Microbiology* 55(4), 1127–1137. DOI: 10.1111/j.1365-2958.2004.04450.x.

Taylor, A.W. (1951) Varieties *Mycobacterium johnei* isolated from sheep. *The Journal of Pathology and Bacteriology* 63(2), 333–336. DOI: 10.1002/path.1700630217.

Taylor, A.W. (1953) The experimental infection of cattle with varieties of *Mycobacterium johnei* isolated from sheep. *Journal of Comparative Pathology and Therapeutics* 63, 368–373. DOI: 10.1016/S0368-1742(53)80038-X.

Timms, V.J., Hassan, K.A., Mitchell, H.M. and Neilan, B.A. (2015) Comparative genomics between human and animal associated subspecies of the *Mycobacterium avium* complex: a basis for pathogenicity. *BMC Genomics* 16(1), 695. DOI: 10.1186/s12864-015-1889-2.

Tundup, S., Pathak, N., Ramanadham, M., Mukhopadhyay, S., Murthy, K.J.R. *et al.* (2008) The co-operonic PE25/PPE41 protein complex of *Mycobacterium tuberculosis* elicits increased humoral and cell mediated immune response. *PLoS ONE* 3(10), e3586. DOI: 10.1371/journal.pone.0003586.

Verdugo, C., Pleydell, E., Price-Carter, M., Prattley, D., Collins, D. *et al.* (2014) Molecular epidemiology of *Mycobacterium avium* subsp. *paratuberculosis* isolated from sheep, cattle and deer on New Zealand pastoral farms. *Preventive Veterinary Medicine* 117(3–4), 436–446. DOI: 10.1016/j.prevetmed.2014.09.009.

Verna, A.E., García-Pariente, C., Muñoz, M., Moreno, O., García-Marin, J.F. *et al.* (2007) Variation in the immuno-pathological responses of lambs after experimental infection with different strains of *Mycobacterium avium* subsp. *paratuberculosis*. *Zoonoses and Public Health* 54(6–7), 243–252. DOI: 10.1111/j.1863-2378.2007.01058.x.

Watt, J.A.A. (1954) Johne's disease in a bovine associated with the pigmented strain of *Mycobacterium johnei*. *The Veterinary Record* 66, 387.

Whittington, R.J., Hope, A.F., Marshall, D.J., Taragel, C.A. and Marsh, I. (2000) Molecular epidemiology of *Mycobacterium avium* subsp.*paratuberculosis*: IS900 restriction fragment length polymorphism and IS*1311* polymorphism analyses of isolates from animals and a human in Australia. *Journal of Clinical Microbiology* 38(9), 3240–3248. DOI: 10.1128/JCM.38.9.3240-3248.2000.

Whittington, R.J., Marsh, I.B. and Whitlock, R.H. (2001a) Typing of IS*1311* polymorphisms confirms that bison (*Bison bison*) with paratuberculosis in Montana are infected with a strain of *Mycobacterium avium* subsp. *paratuberculosis* distinct from that occurring in cattle and other domesticated livestock. *Molecular and Cellular Probes* 15(3), 139–145. DOI: 10.1006/mcpr.2001.0346.

Whittington, R.J., Taragel, C.A., Ottaway, S., Marsh, I., Seaman, J. *et al.* (2001b) Molecular epidemiological confirmation and circumstances of occurrence of sheep (S) strains of *Mycobacterium avium* subsp. *paratuberculosis* in cases of paratuberculosis in cattle in Australia and sheep and cattle in Iceland. *Veterinary Microbiology* 79(4), 311–322. DOI: 10.1016/S0378-1135(00)00364-3.

Windsor, P.A. (2015) Paratuberculosis in sheep and goats. *Veterinary Microbiology* 181(1–2), 161–169. DOI: 10.1016/j.vetmic.2015.07.019.

Wu, C.W., Glasner, J., Collins, M., Naser, S. and Talaat, A.M. (2006) Whole-genome plasticity among *Mycobacterium avium* subspecies: insights from comparative genomic hybridizations. *Journal of Bacteriology* 188(2), 711–723. DOI: 10.1128/JB.188.2.711-723.2006.

Wu, C.W., Schramm, T.M., Zhou, S., Schwartz, D.C. and Talaat, A.M. (2009) Optical mapping of the *Mycobacterium* avium subspecies *paratuberculosis* genome. *BMC Genomics* 10(1), 25. DOI: 10.1186/1471-2164-10-25.

Wynne, J.W., Seemann, T., Bulach, D.M., Coutts, S.A., Talaat, A.M. *et al.* (2010) Resequencing the *Mycobacterium avium* subsp. *paratuberculosis* K10 genome: improved annotation and revised genome sequence. *Journal of Bacteriology* 192(23), 6319–6320. DOI: 10.1128/JB.00972-10.

Wynne, J.W., Bull, T.J., Seemann, T., Bulach, D.M., Wagner, J. *et al.* (2011) Exploring the zoonotic potential of *Mycobacterium avium* subspecies *paratuberculosis* through comparative genomics. *PLoS ONE* 6(7), e22171. DOI: 10.1371/journal.pone.0022171.

Wynne, J.W., Beller, C., Boyd, V., Francis, B., Gwoździ, J. *et al.* (2014) SNP genotyping of animal and human derived isolates of *Mycobacterium avium* subsp. *paratuberculosis*. *Veterinary Microbiology* 172(3–4), 479–485. DOI: 10.1016/j.vetmic.2014.05.024.

Zhu, X., Tu, Z.J., Coussens, P.M., Kapur, V., Janagama, H. *et al.* (2008) Transcriptional analysis of diverse strains *Mycobacterium avium* subspecies *paratuberculosis* in primary bovine monocyte derived macrophages. *Microbes and Infection* 10(12–13), 1274–1282. DOI: 10.1016/j.micinf.2008.07.025.

7 Molecular Genetics of *Mycobacterium avium* subsp. *paratuberculosis*

Govardhan Rathnaiah[1]*, Fernanda M. Shoyama[2], Evan P. Brenner[2], Denise K. Zinniel[1], John P. Bannantine[3], Srinand Sreevatsan[2], Ofelia Chacon[4] and Raúl G. Barletta[1]

[1]University of Nebraska, Lincoln, Nebraska, USA; [2]Michigan State University, East Lansing, Michigan, USA; [3]National Animal Disease Center, USDA-Agricultural Research Service, Ames, Iowa, USA; [4]Medical Science Magnet, School District of Palm Beach, Palm Beach Gardens, Florida, USA

7.1 Introduction

The study of *Mycobacterium avium* subsp. *paratuberculosis* (*MAP*) pathogenesis and the identification and functional characterization of virulence determinants requires the use of molecular genetic approaches. This task is challenging since *MAP* is one of the slowest-growing mycobacterial species, requiring 3 weeks or more to observe colonies on Middlebrook agar supplemented with mycobactin J. In the previous edition of this book, we described the early development of genetic systems to introduce specific gene mutations to define pathogenic determinants. It is now possible to inactivate any non-essential gene by allelic exchange or to generate libraries of transposon mutants for genome-wide analysis. The identification of replicating vectors also allows the introduction of the wild-type gene into mutant strains to fulfil molecular Koch's postulates (Falkow, 2004). In this chapter, after briefly reviewing the molecular tools developed prior to 2009, we will focus on the main advances in the field to finetune the strategies for the genetic manipulation of *MAP*.

7.2 Early Developments

Progress made during the period of 1993–2008 provided key molecular tools for development of genetic systems based on transformation and transfection, and the use of reporter genes (e.g. β-galactosidase, firefly and bacterial luciferases, green fluorescent protein). One important use of these expression systems was the determination of consensus promoter sequence. It was found that this consensus for many *MAP* genes involved the hexanucleotide sequences 5′-TGMCGT-3′ and 5′-CGGCCS-3′ centred approximately 35 and 10 base pairs (bp) upstream from the transcription startpoints. These sequences do not correspond to the consensus hexanucleotides of *Escherichia coli* promoters (Bannantine *et al.*, 1997). In addition, a novel promoter upstream sequence was identified for genes up- or downregulated by the iron repressor IdeR corresponding to the consensus 5′-TWAGGTWAGSCTWACCTWA-3′, where W=A/T and S=G/C (Janagama *et al.*, 2009). It was shown that IdeR binds to the −10 box adjacent to the transcriptional start site repressing the expression of iron storage genes (e.g. *bfrA*) (Gold

*Corresponding author: gopichowvet@gmail.com

© CAB International 2020. *Paratuberculosis: Organism, Disease, Control, 2nd Edition* (eds M.A. Behr *et al.*)

et al., 2001). Upon iron starvation, IdeR is released from the promoter region allowing for the expression of iron acquisition genes (e.g. *bfd*).

The systems described above, along with the completion of the *MAP* sequencing project (Li *et al.*, 2005), led to the development of the functional genomics through generation and characterization of mutants via transposon mutagenesis, gene replacement and complementation strategies. These pioneering studies described in the previous edition of this book (Chacon and Barletta, 2010) are summarized in Table 7.1, while in the rest of the chapter we will focus on further developments in this line of research from 2009 to the present. These advances have significantly increased the understanding of *MAP* molecular pathogenesis to the level of progress made for more tractable mycobacteria.

7.3 Genome

From an epidemiological perspective, whole genome sequencing (WGS) provides the greatest resolution of strain differences in *MAP*. With the availability of benchtop next-generation sequencing platforms, it is now possible to quickly track within-herd and between-herd transmissions at higher resolutions than achieved using short sequence repeat (SSR) (Amonsin *et al.*, 2004) or variable number tandem repeats (VNTR) typing methods (Castellanos *et al.*, 2010; Douarre *et al.*, 2011). This approach has been demonstrated to be effective for defining the role of wildlife for *Mycobacterium bovis* transmission in New Zealand (Price-Carter *et al.*, 2018) and the USA (Salvador *et al.*, 2019). Although *MAP* WGS of strains has lagged well behind its *M. bovis* counterpart, there are now additional *MAP* genome sequences isolated from divergent hosts to serve as a foundation for comparisons.

The first genome sequence of *MAP* has been available for about 15 years (Li *et al.*, 2005). This bovine isolate, termed K-10 for the 10th bacterial colony cultured from a Holstein cow on a Wisconsin dairy farm, is 4.8 Mb in length and 69% GC content. More genomes of *MAP* have since followed including additional bovine isolates (Amin *et al.*, 2015; Möbius *et al.*, 2017), ovine *MAP*-S strains in the USA and Australia (Bannantine *et al.*, 2012; Brauning *et al.*, 2019), camel *MAP*-S strains (Ghosh *et al.*, 2012) and human *MAP*-C strains (Wynne *et al.*, 2011; Bannantine *et al.*, 2014). Most of the genetic diversity resides in insertions/deletions termed large sequence polymorphisms or LSPs (Semret *et al.*, 2005). These regions vary from 5 kb to >100 kb. Another source of diversity appears in single nucleotide polymorphisms (SNPs), which have been catalogued in detail using the aforementioned SSR and VNTR assays. Interestingly, *MAP* lacks a genomic region identified in *Mycobacterium avium* 109 as a pathogenicity island important for macrophage and amoeba infection (Danelishvili *et al.*, 2007).

7.4 Transposon Mutagenesis: *Himar1* vs. Tn*5367* and Tn*5370*

The thickness and impermeability of the cell wall makes it difficult to introduce plasmid vectors into *MAP*. Therefore, the shuttle plasmid vectors described in Table 7.1 yield low transformation efficiencies. To overcome this challenge recombinant shuttle phasmids were developed to introduce allelic exchange substrates and transposons into mycobacteria via the phage receptor to allow for efficient transfer of genetic material through the cell wall. These phasmids are also thermosensitive (e.g. phAE87) for the downstream recombination and selection processes that ultimately lead to gene inactivation and elimination of phage specific sequences (Chacon and Barletta, 2010).

Two types of transposons have been extensively used to generate random mutant banks: (i) IS*1096*- and (ii) *Himar1*-derived transposons (Fig. 7.1). Tn*5367* (GenBank Accession KM232614) was the first transposon widely used in mycobacteria including *MAP* (Bardarov *et al.*, 1997; Shin *et al.*, 2006; Rathnaiah *et al.*, 2014). This transposon carries a kanamycin-resistant drug marker (*aph*), and the IS*1096* transposase (*tnpA*) and resolvase (*tnpR*). This transposon has the transposase within the transposed element (Bardarov *et al.*, 1997) resulting in a relatively high transposition frequency (~1.0×10^{-5}), at least in *Mycobacterium smegmatis* (Cirillo *et al.*, 1991). Tn*5367* transposes by a cut and paste mechanism (McAdam *et al.*, 2002). This element

Table 7.1. Previously reported genetic systems and strategies used in *Mycobacterium avium* subsp. *paratuberculosis* (*MAP*) mutagenesis and gene expression studies (1993–2008).

	Component	Description	Examples
Genetic systems	Shuttle plasmids	*Escherichia coli–Mycobacterium* replicative vectors used to introduce extrachromosomal copies of genes via transformation	pMV261 and pMV262 (Stover *et al.*, 1991; Foley-Thomas *et al.*, 1995) pSMT1 (Rosseels *et al.*, 2006)
	Integrating plasmids (see Section 7.5 for use in *MAP*)	Integrase and L5 mycobacteriophage *attP* site carrying vectors used to introduce DNA into the mycobacterial chromosome via transformation at the *attB* locus	pMH94 (Lee *et al.*, 1991) pMV361 (Stover *et al.*, 1991; Hatfult and Sarkis, 1993) pMV306 (Hahn *et al.*, 2005)
	Phages and shuttle phasmids	Chimeras of a thermosensitive mycobacteriophage and an *E. coli* cosmid used to introduce DNA into the *MAP* chromosome via transduction	phAE87 vector and phAE94 carrying Tn*5367* (Bardarov *et al.*, 1997; Harris *et al.*, 1999)
	Reporter genes	Firefly and bacterial luciferase genes introduced into *MAP* via shuttle phasmids or plasmids	phAE39 (Foley-Thomas *et al.*, 1995) phAE85 (Sasahara *et al.*, 2004) pSMT1 (Rosseels *et al.*, 2006) pYUB180 (Williams *et al.*, 1999)
		Green fluorescent protein gene introduced into *MAP* via shuttle plasmids	pWES4 (Parker and Bermudez, 1997; Harris *et al.*, 2002; Park *et al.*, 2008)
		β-galactosidase promoterless truncated gene introduced via shuttle plasmid. Transcriptional/translational fusions used to identify *MAP* promoter consensus sequences and determine the strength of promoters and ribosomal binding sites	pYUB76 (Barletta *et al.*, 1992; Bannantine *et al.*, 1997)

Continued

Table 7.1. Continued

	Component	Description	Examples
Genetic manipulation strategies	Transposon mutagenesis	Transposon Tn*5367* introduced via the thermosensitive phasmid phAE87 or a shuttle plasmid used to generate and screen comprehensive mutants libraries	*MAP* strain K-10 (Harris *et al.*, 1999; Harris and Barletta, 2001) *MAP* strains 989 and TMC1613 (Cavaignac *et al.*, 2000) *MAP* strain ATCC® 19,698 (Shin *et al.*, 2006; Alonso-Hearn *et al.*, 2008)
	Gene replacement	Substrates for homologous recombination introduced via shuttle thermosensitive phasmids (phAE87) and plasmids used to inactivate specific *MAP* genes	*pknG*, *relA*, *lsr2* *MAP* K-10 mutants (Park *et al.*, 2008) *lipN* *MAP* K-10 mutant (Wu *et al.*, 2007)
	Complementation	Wild-type gene introduced into mutant strains via a multi-copy shuttle plasmid carrying apramycin-resistant marker	pMAV261-*MAP*_3464 and pMAV261-*MAP*_3464, *MAP*_3465 and *MAP*_3466 (Alonso-Hearn *et al.*, 2008)

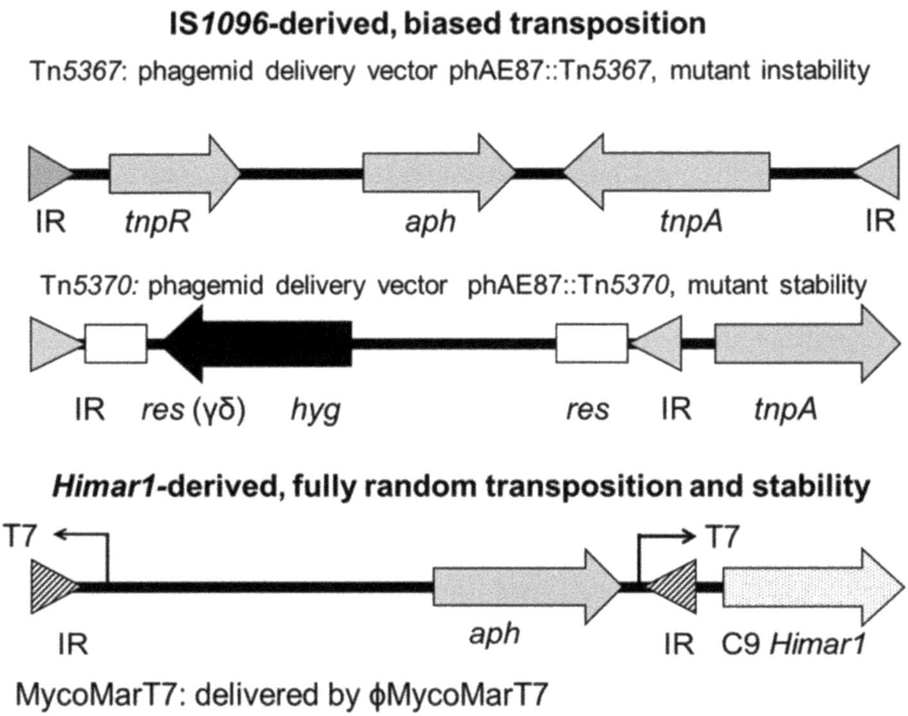

Fig. 7.1. Structure of mycobacterial transposons utilized to generate random mutant libraries. Inverted repeat (IR) for Tn*5367* or Tn*5370* (filled triangle), or *Himar1*-derived transposon (striped triangle); *tnpR*, IS*1096* resolvase; *aph*, kanamycin-resistant gene; *tnpA*, IS*1096* transposase; *res*, resolution site for transposon γ-δ; *hyg*, hygromycin resistant gene; and C9 *Himar1*, *Himar1* transposase. (Figure reproduced from Frontiers in Veterinary Science; Rathnaiah *et al.*, 2017.)

may also insert into another gene creating a new mutation and either a wild-type or mutated gene, depending on precise or imprecise excisions, in the location of the original insertion. Thus, the corresponding *MAP* mutants may not be stable enough to conduct long-term *in vivo* studies where bacilli multiply to large numbers. An improved IS*1096*-derived transposon, Tn*5370* (GenBank Accession KM232615), was constructed by removing open reading frames (ORFs) dispensable for transposition (e.g. *tnpR*) (McAdam *et al.*, 2002). Moreover, the transposase was engineered into the phasmid outside of the transposable element. Thus, Tn*5370*, as opposed to Tn*5367*, generates stable mutants upon transposition. This transposon has an additional advantage of possessing a hygromycin (*hyg*) marker outflanked by the resolution sites (*res*) of the γ-δ resolvase (Wiater and Grindley, 1988). This construction allows for the excision of the drug marker and the generation of unmarked transposon insertion mutants. However, none of these transposons inserts in a fully random manner in mycobacteria (McAdam *et al.*, 2002; Shin *et al.*, 2006).

Himar1-derived transposons have also been used to generate random mutant libraries in *Mycobacterium tuberculosis* and more recently in *MAP* (Sassetti *et al.*, 2001; Scandurra *et al.*, 2009; Rathnaiah *et al.*, 2014; Wang *et al.*, 2014). The *Himar1* transposon (MycoMarT7; GenBank Accession AF411123) was engineered into the shuttle phasmid phAE87 (Table 7.1). This vector carries the highly active C9 *Himar1* transposase outside the inverted repeats, thus yielding stable mutants. In addition, this vector carries the *aph* drug marker and a T7 promoter that reads outward to facilitate identification of insertion sites by *in vitro* transcription and complementary DNA (cDNA) synthesis (Sassetti *et al.*, 2001). The major advantage of the *Himar1* derivative is the almost fully random recognition sequence (5′-TA-3′) vs the recognition sequence of IS*1096*-derived transposons (5′-NNPy(A/T)A(A/T)NN-3′) showed experimentally to have a significant transposition bias in both *M. tuberculosis* and *MAP* (Shin *et al.*, 2006; Rathnaiah *et al.*, 2014).

Most of the *MAP* transposon mutant libraries were made with Tn*5367* (Harris *et al.*, 1999; Cavaignac *et al.*, 2000; Shin *et al.*, 2006; Rathnaiah *et al.*, 2014) and fewer libraries have been constructed using the *Himar1*-derived transposon MycoMarT7 (Scandurra *et al.*, 2009; Wang *et al.*, 2014; Rathnaiah *et al.*, 2016). A recent study performed a comparative analysis of MycoMarT7 and Tn*5367* transposon insertion sites in *MAP* (Rathnaiah *et al.*, 2016). They correlated the analysis of transposon recognition sites with actual insertions obtained in mutagenesis experiments to determine potential transposition biases. The analysis indicated that the distribution of IS*1096* insertion sites within ORFs is sparser, with 710 out of a total 4350 (16%) potential coding sequences containing no recognition sites. The corresponding mutants in those ORFs will not be observed in the mutant libraries irrespective of whether the gene is essential or non-essential in the conditions studied. In contrast, there are only 37 coding sequences not targeted by the *Himar1* transposase. Thus, Tn*5367* or Tn*5370* may not be used for a comprehensive analysis of virulence determinants and gene essentiality in *MAP*. Most importantly this study also revealed transposition biases that are independent of the transposon site recognition sequences. Both transposons have loci-dependent biases, with Tn*5367* being even more skewed. Most notably, insertions into categories of loci with fewer recognition sites seem underrepresented compared with the expected numbers of total sites. Moreover, Tn*5367* has an increased predilection for insertion within intergenic sites. These biases may explain the relatively high number of mutants with interesting phenotypes previously reported with insertions within intergenic regions (Rathnaiah *et al.*, 2014). These loci-dependent biases may lead to an underestimation of the number of independent mutants required to generate a comprehensive mutant library. This indicates that a collection of at least 400,000 mutants would be required for a representative *MAP Himar1* library. This number exceeds the 100,000 estimated for *M. tuberculosis* (Bardarov *et al.*, 1997). For Tn*5367*, considering its predilection for intergenic regions and loci biases, a comprehensive library may need a collection of ~250,000 mutants rather than 13,500 as previously estimated based on random transposition (Harris and Barletta, 2001).

Himar1 transposons have been used in *MAP* to isolate 111 mutants attenuated in bovine monocyte-derived macrophages from a random screen of 2290 (Scandurra *et al.*, 2009). Though a sensible genetic approach was used, the screening method is labour intensive to screen a representative mariner transposon library. More recently, a larger study was performed that collected 90,000 *Himar1* mutants but identified insertions in approximately 12,000 out of 52,000 (~25% saturation density) recognition sites (Wang *et al.*, 2014). The authors attributed this low number of recognition sites represented to bottle-necking effects during DNA sample preparation or the sequencing stage. Loci-dependent transposition biases may also contribute to this effect. None the less, this *Himar1* library was used to screen for mutants capable of conditional growth without mycobactin (Wang *et al.*, 2016). Based on this property, they identified a potential mycobactin-independent iron uptake system on the *MAP*-specific genomic island LSPP15. Step-by-step procedures to generate a *MAP* transposon mutant library are detailed elsewhere (Bannantine *et al.*, 2019).

7.5 Isolation of Deletion Mutants and Complementation Studies

Researchers in the field have continued using shuttle phasmids to introduce specific mutations in target genes. Scandurra *et al.* (2010) inactivated *ppiA* (MAP_0011) encoding a peptidyl-prolyl cis-trans isomerase in a clinical *MAP* isolate from New Zealand cattle strain WAg915 (Scandurra *et al.*, 2010). The role of the alternative sigma factor, SigH, was elucidated by the construction of a knockout mutant (Ghosh *et al.*, 2013). Using this methodology, deletion mutants of *leuD* (MAP_3025c) essential for leucine biosynthesis; *mpt64* (MAP_3290c) encoding an immunogenic protein; and *secA2* (MAP_1534), a preprotein translocase subunit, were generated (Chen *et al.*, 2012; Faisal *et al.*,

2013). Likewise, a functionally uncharacterized gene (*mptD*; MAP_3733c) was shown to encode a factor required for metabolic adaptation and persistence of *MAP* in the host (Meissner *et al.*, 2014). Both manuscripts reported mutant complementation difficulties due to inefficient complementation or the complex organization of the operon where the gene was located, respectively. Indeed this effect seems to be dependent upon which gene is complemented as it was successfully accomplished for the alternative global gene regulator SigL with the use of the integrating vector pMV306 (Ghosh *et al.*, 2014). Thus, our recommendation would be to attempt complementation using an integrating vector. If this is unsuccessful, a multi-copy vector may be used. In both cases, it is most important to maximize transformation efficiency for the particular mutant strain, as this may be different from the wild-type strain. For this approach, there are three markers that have been used successfully in *MAP*: kanamycin, hygromycin and apramycin (Table 7.1). A method to unmark the original mutation may ameliorate this problem, allowing the use of one selection marker for the complementation step. This procedure has been applied to *M. bovis*, *M. smegmatis* and *M. tuberculosis*, but the slow growth of *MAP* could make its application more challenging (Bardarov *et al.*, 2002).

7.6 Developing Technologies: Clustered Regularly Interspaced Short Palindromic Repeats Interference

Clustered regularly interspaced short palindromic repeats interference (CRISPRi) is a modification of the original CRISPR/Cas9 (CRISPR-associated protein 9) system, designed with a goal of gene silencing instead of knockouts. For this approach, the *Cas9* gene is modified to render its product catalytically dead (dCas9), eliminating endonuclease activity (Qi *et al.*, 2013; Adli, 2018). Lacking nuclease activity, the complex otherwise behaves identically to the wild-type. The short guiding RNA (sgRNA) directs the dCas9 to the complementary DNA and pairs with it, placing dCas9 over the target strand, where it remains unable to cleave. The target site is then silenced through simple physical hindrance, with dCas9 remaining bound so long as conditions permit sgRNA/DNA pairing. To date, there are two primary dCas9 CRISPRi systems developed for *Mycobacterium*: codon-optimized from *Streptococcus pyogenes* (Choudhary *et al.*, 2015; Singh *et al.*, 2016) and one from *Streptococcus thermophilus* (Rock *et al.*, 2017). While both systems share the advantage of being anhydrotetracycline (ATc) inducible, fully reversible at any stage and minimize toxicity from protein overexpression, the more traditional *S. pyogenes* dCas9 shows less robust and specific gene knockdowns than the *S. thermophilus* dCas9.

This section will focus on the Rock *et al.* protocol (Rock *et al.*, 2017). When using the CRISPRi system, the desired repression strength is initially selected through a list of a protospacer adjacent motifs (PAMs) with repression levels for a gene target ranging from 2.7- to 216.7-fold. The broad library of PAMs identified allows flexible target selection across mycobacterial genomes while minimizing the difficulties and risks with gene knockouts. The demonstration of efficacy in both saprophytic *M. smegmatis* and pathogenic *M. tuberculosis* establishes its function across dissimilar mycobacteria. As shown in Fig. 7.2, to disrupt transcription initiation through the promoter region, either the template (T) or non-template (NT) strand can be targeted. However, when silencing through the ORF or 5′-UTR, part of the sgRNA must pair with the NT strand. Here, we will assume a PAM is selected within the ORF. For the designed sgRNA to pair with the NT strand, a PAM is sought in the T strand. After the preferred seven nucleotide (nt) PAMs are identified, the complementary sgRNA sequence itself is pulled from the 20–25 nt directly upstream. The 5′ end of this sequence should end in a G or A, so if the 20th nucleotide is a C or T, one additional base can be moved to the 5′ direction as necessary. In this scenario, the guiding portion of the sgRNA would be identical to the template strand sequence immediately at the 5′ end of but not including the PAM sequence. This 20–25 nt sequence serves as the forward primer, and the reverse complement as its reverse. 5′-GGGA-3′ is added at the 5′ end of the forward primer, and 5′-CAAA-3′ added to the 5′ end of the reverse primer to allow cloning into the *Bsm*BI-digested vector and ligation into the dCas9 handle sequence therein. This allows the chosen pairing sequence to

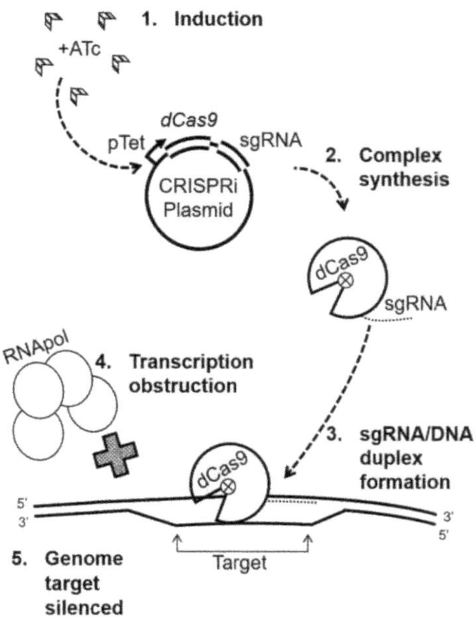

Fig. 7.2. Schematic steps in the clustered regularly interspaced short palindromic repeats interference (CRISPRi) system developed for *Mycobacterium* from *Streptococcus thermophilus* dCas9. (1) anhydrotetracycline (ATc) attaches to the tetracycline (Tet) repressor regulated optimized promoter (pTet) to induce expression of dCas9 from CRISPRi plasmid with a mycobacterial integrating vector (Table 7.1) as a backbone; (2) formation of the dCas9 complex; (3) dCas9 complex is directed to the target DNA sequence; (4) duplex formation prevents transcription initiation (RNA polymerase) or elongation (steric hindrance); (5) the gene is silenced (downregulated).

be integrated into the dCas9 protein as a full sgRNA. Transformants are selected by kanamycin resistance, and activation of the system occurs by addition of ATc. Overexpression of dCas9 is known to show toxicity to the bacterial host (Rock *et al.*, 2017). As such, Rock *et al.* designed two distinct plasmids for mycobacterial use – pLJR962 (optimized for *M. smegmatis*) and pLJR965 (optimized for *M. tuberculosis*). Each vector responds differently to ATc pools in the bacterial host, leading to differential induction and system expression. Our Michigan State Laboratory has successfully cloned pLJR965 into *MAP*. However, ATc handling in the *MAP* system and therefore the optimum vector has not yet been determined. Inducible and tunable gene silencing in *MAP* without genome modification offers promise for studying gene function and pathways in this stubborn pathogen, as its slow-growing and fastidious nature have made traditional gene knockouts comparatively difficult.

By adopting the CRISPRi system, researchers can explore the classical reductionist analysis on the effects of a single gene silenced on a host, to the broader systems approach studying the interplay of one silenced gene between pathways and the overall bacterial lifecycle. This approach still maintains the original genomic context and with minimal upstream and off-target effects. Additionally, the system's inducibility and tunable repression allows a much more nuanced view of biological effects beyond the on/off binary. For essential genes where knockouts were previously impossible, CRISPRi allows investigation by either inducing only in permissible conditions or growth phases, or by graded knockdown rather than outright depletion. The CRISPRi knockdown system offers several advantages as discussed here, but its simplicity and low cost combined with its potential for comparatively rapid implementation make this technique very attractive, especially for researchers working with *MAP*.

7.7 Concluding Remarks

Progress made in the past 10 years was important to advance the manipulation of the *MAP* genome applying the technologies developed for *M. smegmatis* and *M. tuberculosis*. Still it may be difficult to implement some of these approaches due to the extreme slow growth of *MAP* and the need for special nutritional requirements such as mycobactin J or the use of low pH media to increase iron availability. Indeed, *MAP* has a mutation in the *mbtA* gene necessary for the biosynthesis of this siderophore (Li *et al.*, 2005). Direct gene silencing and or transposon mutagenesis are critical molecular tools to begin to understand the functions of genes. This is particularly important for *MAP*, which has a higher number of genes encoding hypothetical proteins than many other bacterial genomes. More research must be conducted in this area if we are to better understand *MAP* biology. Thus,

MAP researchers must be resilient in applying novel technologies to this fastidious microorganism and we expect that significant progress will be made in the upcoming years. For example, application of the Hidden Markov Model to determine gene essentiality, single-cell genetic analysis and genome barcoding strategies would be expected in the near future (DeJesus *et al.*, 2017; Martin *et al.*, 2017; Rego *et al.*, 2017).

Acknowledgements

Support for Dr Barletta's research was by the USDA National Institute of Food and Agriculture (NIFA) Research Initiative Competitive Grant no. 2013-67015-21239, Animal Health (NEB 39-162) and Hatch/Multistate (NEB 39-168) projects. Portions of Dr Bannantine's research were also funded by Grant no. 2013-67015-21239 and the USDA-Agricultural Research Service. Research in Dr Sreevatsan's laboratory was funded by NIFA.

References

Adli, M. (2018) The CRISPR tool kit for genome editing and beyond. *Nature Communications* 9(1), 1911. DOI: 10.1038/s41467-018-04252-2.

Alonso-Hearn, M., Patel, D., Danelishvili, L., Meunier-Goddik, L. and Bermudez, L.E. (2008) The *Mycobacterium avium* subsp. *paratuberculosis* MAP3464 gene encodes an oxidoreductase involved in invasion of bovine epithelial cells through the activation of host cell Cdc42. *Infection and Immunity* 76(1), 170–178. DOI: 10.1128/IAI.01913-06.

Amin, A.S., Hsu, C.-Y., Darwish, S.F., Ghosh, P., AbdEl-Fatah, E.M. *et al.* (2015) Ecology and genomic features of infection with *Mycobacterium avium* subspecies *paratuberculosis* in Egypt. *Microbiology* 161(4), 807–818. DOI: 10.1099/mic.0.000051.

Amonsin, A., Li, L.L., Zhang, Q., Bannantine, J.P., Motiwala, A.S. *et al.* (2004) Multilocus short sequence repeat sequencing approach for differentiating among *Mycobacterium avium* subsp. *paratuberculosis* strains. *Journal of Clinical Microbiology* 42(4), 1694–1702. DOI: 10.1128/JCM.42.4.1694-1702.2004.

Bannantine, J.P., Barletta, R.G., Thoen, C.O. and Andrews, R.E. (1997) Identification of *Mycobacterium paratuberculosis* gene expression signals. *Microbiology* 143(3), 921–928. DOI: 10.1099/00221287-143-3-921.

Bannantine, J.P., Wu, C.W., Hsu, C., Zhou, S., Schwartz, D.C. *et al.* (2012) Genome sequencing of ovine isolates of *Mycobacterium avium* subspecies *paratuberculosis* offers insights into host association. *BMC Genomics* 13(1), 89. DOI: 10.1186/1471-2164-13-89.

Bannantine, J.P., Li, L., Mwangi, M., Cote, R., Raygoza Garay, J.A. *et al.* (2014) Complete genome sequence of *Mycobacterium avium* subsp. *paratuberculosis*, isolated from human breast milk. *Genome Announcements* 2(1). DOI: 10.1128/genomeA.01252-13.

Bannantine, J.P., Zinniel, D.K. and Barletta, R.G. (2019) Transposon mutagenesis in *Mycobacterium avium* subspecies *paratuberculosis*. *Methods in Molecular Biology* 2016, 117–125.

Bardarov, S., Kriakov, J., Carriere, C., Yu, S., Vaamonde, C. *et al.* (1997) Conditionally replicating mycobacteriophages: a system for transposon delivery to *Mycobacterium tuberculosis*. *Proceedings of the National Academy of Sciences* 94(20), 10961–10966. DOI: 10.1073/pnas.94.20.10961.

Bardarov, S., Bardarov, S., Pavelka, M.S., Sambandamurthy, V., Larsen, M. *et al.* (2002) Specialized transduction: an efficient method for generating marked and unmarked targeted gene disruptions in *Mycobacterium tuberculosis*, *M. bovis* BCG and *M. smegmatis*. *Microbiology* 148(10), 3007–3017. DOI: 10.1099/00221287-148-10-3007.

Barletta, R.G., Kim, D.D., Snapper, S.B., Bloom, B.R. and Jacobs, W.R. (1992) Identification of expression signals of the mycobacteriophages Bxb1, L1 and TM4 using the Escherichia-Mycobacterium shuttle plasmids pYUB75 and pYUB76 designed to create translational fusions to the lacZ gene. *Journal of General Microbiology* 138(1), 23–30. DOI: 10.1099/00221287-138-1-23.

Brauning, R., Plain, K., Gautam, M., Russell, T., Correa, C.C. *et al.* (2019) Complete genome sequence of the Telford Type S strain of *Mycobacterium avium* subsp. *paratuberculosis*. *Microbiology Resource Announcements* 8(11). DOI: 10.1128/MRA.00004-19.

Castellanos, E., Romero, B., Rodríguez, S., de Juan, L., Bezos, J. *et al.* (2010) Molecular characterization of *Mycobacterium avium* subspecies *paratuberculosis* types II and III isolates by a combination of MIRU-VNTR loci. *Veterinary Microbiology* 144(1–2), 118–126. DOI: 10.1016/j.vetmic.2009.12.028.

Cavaignac, S.M., White, S.J., de Lisle, G.W. and Collins, D.M. (2000) Construction and screening of *Mycobacterium paratuberculosis* insertional mutant libraries. *Archives of Microbiology* 173(3), 229–231. DOI: 10.1007/s002039900132.

Chacon, O. and Barletta, R.G. (2010) Molecular genetics of *Mycobacterium avium* subspecies *paratuberculosis*. In: Behr, M.A. and Collins, D.M. (eds) *Paratuberculosis: Organism, Disease, Control*. CAB International, Wallingford, UK.

Chen, J.-W., Faisal, S.M., Chandra, S., McDonough, S.P., Moreira, M.A.S. *et al.* (2012) Immunogenicity and protective efficacy of the *Mycobacterium avium* subsp. *paratuberculosis* attenuated mutants against challenge in a mouse model. *Vaccine* 30(19), 3015–3025. DOI: 10.1016/j.vaccine.2011.11.029.

Choudhary, E., Thakur, P., Pareek, M. and Agarwal, N. (2015) Gene silencing by CRISPR interference in mycobacteria. *Nature Communications* 6(1), 6267. DOI: 10.1038/ncomms7267.

Cirillo, J.D., Barletta, R.G., Bloom, B.R. and Jacobs, W.R. (1991) A novel transposon trap for mycobacteria: isolation and characterization of IS1096. *Journal of Bacteriology* 173(24), 7772–7780. DOI: 10.1128/JB.173.24.7772-7780.1991.

Danelishvili, L., Wu, M., Stang, B., Harriff, M., Cirillo, S.L.G. *et al.* (2007) Identification of *Mycobacterium avium* pathogenicity island important for macrophage and amoeba infection. *Proceedings of the National Academy of Sciences* 104(26), 11038–11043. DOI: 10.1073/pnas.0610746104.

DeJesus, M.A., Gerrick, E.R., Xu, W., Park, S.W., Long, J.E. *et al.* (2017) Comprehensive essentiality analysis of the *Mycobacterium tuberculosis* genome via saturating transposon mutagenesis. *mBio* 8(1). DOI: 10.1128/mBio.02133-16.

Douarre, P.E., Cashman, W., Buckley, J., Coffey, A. and O'Mahony, J. (2011) Molecular characterization of *Mycobacterium avium* subsp. *paratuberculosis* using multi-locus short sequence repeat (MLSSR) and mycobacterial interspersed repetitive units-variable number tandem repeat (MIRU-VNTR) typing methods. *Veterinary Microbiology* 149(3–4), 482–487. DOI: 10.1016/j.vetmic.2010.12.001.

Faisal, S.M., Chen, J.-W., Yan, F., Chen, T.-T., Useh, N.M. *et al.* (2013) Evaluation of a *Mycobacterium avium* subsp. *paratuberculosis leuD* mutant as a vaccine candidate against challenge in a caprine model. *Clinical and Vaccine Immunology* 20(4), 572–581. DOI: 10.1128/CVI.00653-12.

Falkow, S. (2004) Molecular Koch's postulates applied to bacterial pathogenicity--a personal recollection 15 years later. *Nature Reviews Microbiology* 2(1), 67–72. DOI: 10.1038/nrmicro799.

Foley-Thomas, E.M., Whipple, D.L., Bermudez, L.E. and Barletta, R.G. (1995) Phage infection, transfection and transformation of *Mycobacterium avium* complex and *Mycobacterium paratuberculosis*. *Microbiology* 141(5), 1173–1181. DOI: 10.1099/13500872-141-5-1173.

Ghosh, P., Hsu, C., Alyamani, E.J., Shehata, M.M., Al-Dubaib, M.A. *et al.* (2012) Genome-wide analysis of the emerging infection with *Mycobacterium avium* subspecies *paratuberculosis* in the Arabian camels (Camelus dromedarius). *PLoS ONE* 7(2), e31947. DOI: 10.1371/journal.pone.0031947.

Ghosh, P., Wu, C.W. and Talaat, A.M. (2013) Key role for the alternative sigma factor, SigH, in the intracellular life of *Mycobacterium avium* subsp. *paratuberculosis* during macrophage stress. *Infection and Immunity* 81(6), 2242–2257. DOI: 10.1128/IAI.01273-12.

Ghosh, P., Steinberg, H. and Talaat, A.M. (2014) Virulence and immunity orchestrated by the global gene regulator *sigL* in *Mycobacterium avium* subsp. *paratuberculosis*. *Infection and Immunity* 82(7), 3066–3075. DOI: 10.1128/IAI.00001-14.

Gold, B., Rodriguez, G.M., Marras, S.A.E., Pentecost, M. and Smith, I. (2001) The *Mycobacterium tuberculosis* IdeR is a dual functional regulator that controls transcription of genes involved in iron acquisition, iron storage and survival in macrophages. *Molecular Microbiology* 42(3), 851–865. DOI: 10.1046/j.1365-2958.2001.02684.x.

Hahn, M.-Y., Raman, S., Anaya, M. and Husson, R.N. (2005) The *Mycobacterium tuberculosis* extracytoplasmic-function sigma factor SigL regulates polyketide synthases and secreted or membrane proteins and is required for virulence. *Journal of Bacteriology* 187(20), 7062–7071. DOI: 10.1128/JB.187.20.7062-7071.2005.

Harris, N.B. and Barletta, R.G. (2001) *Mycobacterium avium* subsp. *paratuberculosis* in veterinary medicine. *Clinical Microbiology Reviews* 14(3), 489–512. DOI: 10.1128/CMR.14.3.489-512.2001.

Harris, N.B., Feng, Z., Liu, X., Cirillo, S.L., Cirillo, J.D. *et al.* (1999) Development of a transposon mutagenesis system for *Mycobacterium avium* subsp. *paratuberculosis*. *FEMS Microbiology Letters* 175(1), 21–26. DOI: 10.1111/j.1574-6968.1999.tb13597.x.

Harris, N.B., Zinniel, D.K., Hsieh, M.K., Cirillo, J.D. and Barletta, R.G. (2002) Cell sorting of formalin-treated pathogenic *Mycobacterium paratuberculosis* expressing GFP. *BioTechniques* 32(3), 522–527. DOI: 10.2144/02323st03.

Hatfult, G.F. and Sarkis, G.J. (1993) DNA sequence, structure and gene expression of mycobacteriophage L5: a phage system for mycobacterial genetics. *Molecular Microbiology* 7(3), 395–405. DOI: 10.1111/j.1365-2958.1993.tb01131.x.

Janagama, H.K., Senthilkumar, T.M.A., Bannantine, J.P., Rodriguez, G.M., Smith, I. *et al.* (2009) Identification and functional characterization of the iron-dependent regulator (IdeR) of *Mycobacterium avium* subsp. *paratuberculosis*. *Microbiology* 155(Pt 11), 3683–3690. DOI: 10.1099/mic.0.031948-0.

Lee, M.H., Pascopella, L., Jacobs, W.R. and Hatfull, G.F. (1991) Site-specific integration of mycobacteriophage L5: integration-proficient vectors for *Mycobacterium smegmatis*, *Mycobacterium tuberculosis*, and Bacille Calmette-Guerin. *Proceedings of the National Academy of Sciences* 88(8), 3111–3115. DOI: 10.1073/pnas.88.8.3111.

Li, L., Bannantine, J.P., Zhang, Q., Amonsin, A., May, B.J. *et al.* (2005) The complete genome sequence of *Mycobacterium avium* subspecies *paratuberculosis*. *Proceedings of the National Academy of Sciences* 102(35), 12344–12349. DOI: 10.1073/pnas.0505662102.

Martin, C.J., Cadena, A.M., Leung, V.W., Lin, P.L., Maiello, P. *et al.* (2017) Digitally barcoding *Mycobacterium tuberculosis* reveals *in vivo* infection dynamics in the Macaque model of tuberculosis. *mBio* 8(3). DOI: 10.1128/mBio.00312-17.

McAdam, R.A., Quan, S., Smith, D.A., Bardarov, S., Betts, J.C. *et al.* (2002) Characterization of a *Mycobacterium tuberculosis* H37Rv transposon library reveals insertions in 351 ORFs and mutants with altered virulence. *Microbiology* 148, 2975–2986.

Meissner, T., Eckelt, E., Basler, T., Meens, J., Heinzmann, J. *et al.* (2014) The *Mycobacterium avium* ssp. *paratuberculosis* specific mptD gene is required for maintenance of the metabolic homeostasis necessary for full virulence in mouse infections. *Frontiers in Cellular and Infection Microbiology* 4, 110. DOI: 10.3389/fcimb.2014.00110.

Möbius, P., Nordsiek, G., Hölzer, M., Jarek, M., Marz, M. *et al.* (2017) Complete genome sequence of JII-1961, a bovine *Mycobacterium avium* subsp. *paratuberculosis* field isolate from Germany. *Genome Announcements* 5(34). DOI: 10.1128/genomeA.00870-17.

Park, K.T., Dahl, J.L., Bannantine, J.P., Barletta, R.G., Ahn, J. *et al.* (2008) Demonstration of allelic exchange in the slow-growing bacterium *Mycobacterium avium* subsp. *paratuberculosis*, and generation of mutants with deletions at the pknG, relA, and LSR2 loci. *Applied and Environmental Microbiology* 74(6), 1687–1695. DOI: 10.1128/AEM.01208-07.

Parker, A.E. and Bermudez, L.E. (1997) Expression of the green fluorescent protein (GFP) in mycobacterium avium as a tool to study the interaction between Mycobacteria and host cells. *Microbial Pathogenesis* 22(4), 193–198. DOI: 10.1006/mpat.1996.0106.

Price-Carter, M., Brauning, R., de Lisle, G.W., Livingstone, P., Neill, M. *et al.* (2018) Whole genome sequencing for determining the source of *Mycobacterium bovis* infections in livestock herds and wildlife in New Zealand. *Frontiers in Veterinary Science* 5, 272. DOI: 10.3389/fvets.2018.00272.

Qi, L. S., Larson, M. H., Gilbert, L. A., Doudna, J. A., Weissman, J. S. *et al.* (2013) Repurposing CRISPR as an RNA-guided platform for sequence-specific control of gene expression. *Cell* 152(5), 1173–1183. DOI: 10.1016/j.cell.2013.02.022.

Rathnaiah, G., Lamont, E.A., Harris, N.B., Fenton, R.J., Zinniel, D.K. *et al.* (2014) Generation and screening of a comprehensive *Mycobacterium avium* subsp. *paratuberculosis* transposon mutant bank. *Frontiers in Cellular and Infection Microbiology* 4(88), 144. DOI: 10.3389/fcimb.2014.00144.

Rathnaiah, G., Bannantine, J.P., Bayles, D.O., Zinniel, D.K., Stabel, J.R. *et al.* (2016) Analysis of *Mycobacterium avium* subsp. *paratuberculosis* mutant libraries reveals loci-dependent transposition biases and strategies for novel mutant discovery. *Microbiology* 162(4), 633–641. DOI: 10.1099/mic.0.000258.

Rathnaiah, G., Zinniel, D.K., Bannantine, J.P., Stabel, J.R., Gröhn, Y.T. *et al.* (2017) Pathogenesis, molecular genetics, and genomics of *Mycobacterium avium* subsp. *paratuberculosis*, the etiologic agent of Johne's disease. *Frontiers in Veterinary Science* 4, 187. DOI: 10.3389/fvets.2017.00187.

Rego, E.H., Audette, R.E. and Rubin, E.J. (2017) Deletion of a mycobacterial divisome factor collapses single-cell phenotypic heterogeneity. *Nature* 546(7656), 153–157. DOI: 10.1038/nature22361.

Rock, J.M., Hopkins, F.F., Chavez, A., Diallo, M., Chase, M.R. *et al.* (2017) Programmable transcriptional repression in mycobacteria using an orthogonal CRISPR interference platform. *Nature Microbiology* 2(4), 16274. DOI: 10.1038/nmicrobiol.2016.274.

Rosseels, V., Roupie, V., Zinniel, D., Barletta, R.G. and Huygen, K. (2006) Development of luminescent *Mycobacterium avium* subsp. *paratuberculosis* for rapid screening of vaccine candidates in mice. *Infection and Immunity* 74(6), 3684–3686. DOI: 10.1128/IAI.01521-05.

Salvador, L.C.M., O'Brien, D.J., Cosgrove, M.K., Stuber, T.P., Schooley, A.M. *et al.* (2019) Disease management at the wildlife-livestock interface: using whole-genome sequencing to study the role of elk in *Mycobacterium bovis* transmission in Michigan, USA. *Molecular Ecology* 28(9), 2192–2205. DOI: 10.1111/mec.15061.

Sasahara, K.C., Gray, M.J., Shin, S.J. and Boor, K.J. (2004) Detection of viable *Mycobacterium avium* subsp. *paratuberculosis* using luciferase reporter systems. *Foodborne Pathogens and Disease* 1(4), 258–266. DOI: 10.1089/fpd.2004.1.258.

Sassetti, C.M., Boyd, D.H. and Rubin, E.J. (2001) Comprehensive identification of conditionally essential genes in mycobacteria. *Proceedings of the National Academy of Sciences* 98(22), 12712–12717. DOI: 10.1073/pnas.231275498.

Scandurra, G.M., Young, M., de Lisle, G.W. and Collins, D.M. (2009) A bovine macrophage screening system for identifying attenuated transposon mutants of *Mycobacterium avium* subsp. *paratuberculosis* with vaccine potential. *Journal of Microbiological Methods* 77(1), 58–62. DOI: 10.1016/j.mimet.2009.01.005.

Scandurra, G.M., de Lisle, G.W., Cavaignac, S.M., Young, M., Kawakami, R.P. *et al.* (2010) Assessment of live candidate vaccines for paratuberculosis in animal models and macrophages. *Infection and Immunity* 78(3), 1383–1389. DOI: 10.1128/IAI.01020-09.

Semret, M., Alexander, D.C., Turenne, C.Y., de Haas, P., Overduin, P. *et al.* (2005) Genomic polymorphisms for *Mycobacterium avium* subsp. *paratuberculosis* diagnostics. *Journal of Clinical Microbiology* 43(8), 3704–3712. DOI: 10.1128/JCM.43.8.3704-3712.2005.

Shin, S.J., Wu, C.W., Steinberg, H. and Talaat, A.M. (2006) Identification of novel virulence determinants in *Mycobacterium paratuberculosis* by screening a library of insertional mutants. *Infection and Immunity* 74(7), 3825–3833. DOI: 10.1128/IAI.01742-05.

Singh, A.K., Carette, X., Potluri, L.-P., Sharp, J.D., Xu, R. *et al.* (2016) Investigating essential gene function in *Mycobacterium tuberculosis* using an efficient CRISPR interference system. *Nucleic Acids Research* 44(18), e143. DOI: 10.1093/nar/gkw625.

Stover, C.K., de la Cruz, V.F., Fuerst, T.R., Burlein, J.E., Benson, L.A. *et al.* (1991) New use of BCG for recombinant vaccines. *Nature* 351(6326), 456–460. DOI: 10.1038/351456a0.

Wang, J., Pritchard, J.R., Kreitmann, L., Montpetit, A. and Behr, M.A. (2014) Disruption of *Mycobacterium avium* subsp. *paratuberculosis*-specific genes impairs in vivo fitness. *BMC Genomics* 15(1), 415. DOI: 10.1186/1471-2164-15-415.

Wang, J., Moolji, J., Dufort, A., Staffa, A., Domenech, P. *et al.* (2016) Iron acquisition in *Mycobacterium avium* subsp. *paratuberculosis*. *Journal of Bacteriology* 198(5), 857–866. DOI: 10.1128/JB.00922-15.

Wiater, L.A. and Grindley, N.D. (1988) Gamma delta transposase and integration host factor bind cooperatively at both ends of gamma delta. *The EMBO Journal* 7(6), 1907–1911. DOI: 10.1002/j.1460-2075.1988.tb03024.x.

Williams, S.L., Harris, N.B. and Barletta, R.G. (1999) Development of a firefly luciferase-based assay for determining antimicrobial susceptibility of *Mycobacterium avium* subsp. *paratuberculosis*. *Journal of Clinical Microbiology* 37(2), 304–309. DOI: 10.1128/JCM.37.2.304-309.1999.

Wu, C.W., Schmoller, S.K., Shin, S.J. and Talaat, A.M. (2007) Defining the stressome of *Mycobacterium avium* subsp. *paratuberculosis* in vitro and in naturally infected cows. *Journal of Bacteriology* 189(21), 7877–7886. DOI: 10.1128/JB.00780-07.

Wynne, J.W., Bull, T.J., Seemann, T., Bulach, D.M., Wagner, J. *et al.* (2011) Exploring the zoonotic potential of *Mycobacterium avium* subspecies *paratuberculosis* through comparative genomics. *PLoS ONE* 6(7), e22171. DOI: 10.1371/journal.pone.0022171.

8 Proteins and Antigens of *Mycobacterium avium* subsp. *paratuberculosis*

John P. Bannantine[1]* and Vivek Kapur[2]
[1]USDA-ARS, National Animal Disease Center, Ames, Iowa, USA; [2]Penn State University, University Park, Pennsylvania, USA

8.1 Introduction

Currently, no single diagnostic test can detect *Mycobacterium avium* subspecies *paratuberculosis* (*MAP*) infection at every stage of paratuberculosis. Furthermore, severe obstacles to disease control include the lack of early diagnosis and effective vaccines. *MAP* proteomic preparations are critical for diagnostic and vaccine applications. Killed whole-cell extracts or modified live versions of *MAP* are used as vaccine formulations (Bastida and Juste, 2011) and sonicated fractionated whole-cell extracts or cell-free supernatants, comprising secreted proteins, are used in diagnostic tests such as the interferon gamma (IFN-γ) assay or enzyme-linked immunosorbent assay (ELISA). Failure of antigen-based diagnostic accuracy is due, in part, to the inconsistency of the cell-free preps, termed purified protein derivative (PPD), and presence of many cross-reactive proteins from other closely related mycobacteria. While the failure of protective antigens is more difficult to pinpoint, it can be partially attributed to non-specific targets as well as the difficulty in measuring correlates of protection. However, a few *in vitro* screening methods have recently been introduced to triage vaccine candidates prior to expensive animal trials (Abdellrazeq *et al.*, 2018; Pooley *et al.*, 2018). In the USA, vaccination is not administered to dairy herds and thus paratuberculosis control efforts have been limited to herd management and serological testing. This situation is different for sheep and goats in Australia and other countries that do have paratuberculosis vaccine programmes (Windsor, 2015).

With the goal of aiding vaccine and diagnostic test advances, this chapter will focus primarily on *MAP* proteomic discoveries made since 2008 unless a historical perspective is needed. A few new antigens have come to light as a result of these recent studies and they will be discussed further. Another line of proteomic research that is emerging involves host biomarker discovery. A good example of this is the transthyretin and alpha haemoglobin proteins identified and purified from the serum of infected sheep (Zhong *et al.*, 2011). However, these topics are beyond the scope of this pathogen-focused chapter. Finally, many studies use the whole cell as an antigen for specific assays and these types of studies are not discussed herein. Only when the *MAP* proteome is fractionated or if a specific, defined antigen produced by the bacterium is studied, then it is mentioned herein.

*Corresponding author: john.bannantine@ars.usda.gov

© CAB International 2020. *Paratuberculosis: Organism, Disease, Control, 2nd Edition* (eds M.A. Behr *et al.*)

8.2 Study of *MAP* Proteins

The genome of *MAP* has been sequenced and available for 15 years (Li *et al.*, 2005). A dozen more genomes of *MAP* have since followed including bovine isolates (Amin *et al.*, 2015; Möbius *et al.*, 2017), ovine isolates in the USA and Australia (Bannantine *et al.*, 2012; Brauning *et al.*, 2019), camel isolates (Ghosh *et al.*, 2012) and human isolates (Wynne *et al.*, 2011; Bannantine *et al.*, 2014b). These genome sequences are discussed in depth in Chapter 6 of this book. With these genome sequences in hand, the initial strategy to identify diagnostic antigens was to perform a comparative genomics analysis and select those genes that are present only in *MAP*. Then those *MAP*-specific genes were expressed and analysed for antigenicity. However, this strategy yielded a limited number of genes due to the very high sequence identity between *MAP* and other members of the *M. avium* subspecies. Most of the unique genes resided in large-sequence polymorphisms (Semret *et al.*, 2005). Unfortunately, while some of these *MAP*-specific gene products did show immunoreactivity (Paustian *et al.*, 2004; Dernivoix *et al.*, 2017), none of these ultimately were demonstrated as a discriminating antigen that could consistently distinguish infected from non-infected animals. This finding meant that the specificity part of the equation could not be satisfied simply because the sensitivity part of the equation was low for that group of proteins. Researchers have acknowledged this fact and either looked for strong antigens that are not necessarily unique to *MAP*, or looked deeper for *MAP*-specific epitopes within otherwise conserved mycobacterial proteins. An example of this is found in the gene designated MAP_1025 (encoding an RDD family protein), which does contain an epitope that only exists in *MAP* strains (Bannantine *et al.*, 2011). The specificity of this seven-amino acid epitope is due to a single non-synonymous nucleotide polymorphism that occurs only in *MAP*. Other such examples likely exist, but their diagnostic utilities have yet to be discovered.

Current strategies for identifying diagnostic antigens are more global on the whole proteome scale as opposed to focusing on just a few *MAP*-specific proteins. Hughes *et al.* examined proteome expression differences between *MAP* and *M. avium* subspecies *avium* to find 32 *MAP* proteins that did not appear to be expressed by *M. avium avium* in the same *in vitro* culture conditions (Hughes *et al.*, 2008). However, when two of these proteins were incorporated into an IFN-γ release assay, they could not distinguish between *MAP* and *M. avium avium* infected cows (Hughes *et al.*, 2017). Still another approach is to examine lipid antigens of *MAP*. There are known differences in the lipid profiles of *MAP* and other *M. avium* subspecies, and the characterized *MAP*-specific lipopeptides are antigenic (Biet *et al.*, 2008; Mitachi *et al.*, 2016).

The study of *MAP* proteins should not be solely diagnostic focused. There is a lot of interesting biology associated with this pathogen that could yield clues to its pathogenicity. However, even with additional genome sequences and their more recent annotations, *MAP* still has an abundance of proteins with unknown function, which are termed hypothetical proteins. Approximately 70.4% of the *MAP* genome is annotated as hypothetical (unknown function, no similarity) or conserved hypothetical proteins (unknown function, but similar to other proteins in a database). This is considerably more than the 25.9% hypothetical proteins assigned for the first bacterial genome sequenced, *Haemophilus influenzae*, which sets the standard among the best-characterized bacteria (Shahbaaz *et al.*, 2013). More biochemical and proteomic studies are needed to functionally characterize the *MAP* proteins categorized as hypothetical.

8.2.1 Functional characterization of *MAP* proteins

Other than the obvious BLAST (Basic Local Alignment Search Tool) search approach (Altschul *et al.*, 1997) which is already used to annotate genomes, one way to begin functionally characterizing proteins is to identify how they interact with their environment. Proteins do not act in isolation, but interact with each other to perform specific functions. Cell development and division, transcription and translation, transport and metabolism can be defined by the activity of protein complexes. Knowing which proteins are assembled in these complexes provides critical clues to the function of these proteins. Non-denaturing (native) gel electrophoresis enables the separation and analysis of protein complexes on a proteome-wide scale. Seven *MAP* membrane protein complexes have been isolated using this method and their subunits defined (Leite *et al.*, 2015). Complexes II and VII each had nine subunit proteins identified by mass spectrometry. In complex I, the major membrane protein (MAP_2121c) was present along with a cysteine desulfurase (MAP_2120c), an enzyme that removes sulfur from cysteine for the biosynthesis of co-factors. Although this major membrane protein has long been known as a virulence protein (Bannantine *et al.*, 2003) and antigen (Triccas *et al.*, 1996), its association with cysteine desulfurase may suggest a biosynthetic function as well. Other proteins present within complex II (linocin and a peroxidase) were found to have strong interactions predicted by String analysis (Szklarczyk *et al.*, 2015). These interactions should be confirmed by Far western blot analysis (Wu *et al.*, 2007) or the older two hybrid strategies (Fields and Sternglanz, 1994). One *MAP* protein, encoded by MAP_1203, has been shown to interact with bovine epithelial cell proteins using the Far western blot method (Everman *et al.*, 2018). Initially annotated as a hypothetical protein, we now know it is surface located, promotes invasion and interacts in a specific way with the host.

Protein location on the bacterial cell is another potential clue to understanding function. For example, proteins in the cell wall are unlikely to affect DNA repair or regulate transcription, but rather might be involved in export or transport functions. The predicted locations of *MAP* proteins are shown in Table 8.1 and it is noteworthy that 1562 proteins could not be located based on this bioinformatic topology analysis (Yu *et al.*, 2010). Extracellular or secreted proteins have been considered antigenic (Lanigan *et al.*, 2007; Shin *et al.*, 2010) and *MAP* has 82 such proteins. The majority of proteins (1857) were located in the cytoplasm (Table 8.1). However, a surface exposed protein that is a strong antigen is likely the ideal diagnostic antigen. An understanding of protein, glycoprotein or lipoprotein topology within the cell wall and whether it protrudes from the outer cell wall layer to become surface exposed is generally lacking. A major effort directed at this goal of defining surface exposure of *MAP* proteins was by He and De Buck (2010). They performed a trypsin shaving method on the cell wall of intact, viable bacteria and identified 38 proteins that are surface exposed. These proteins are at the front line for adhesion, invasion, cell division or perhaps other interactions with the host cells during infection.

Table 8.1. Predicted location of *Mycobacterium avium* subsp. *paratuberculosis (MAP)* proteins based on PSORTb 3.0 analysis.

Location	No. of proteins[a]	% of proteome	Ave score[b]
Cytoplasm	1857	42.2	9.28
Extracellular	82	1.9	9.62
Membrane	833	18.9	9.66
Periplasm	60	1.4	9.69
Outer membrane	7	0.1	9.67
Unknown	1562	35.5	2.29

[a]Total number of proteins in the *MAP* proteome is 4401.
[b]PSORTb final prediction score ranges between 1 (low probability) to 10 (high probability). Score must be above 7.5 to make a location prediction (Yu *et al.*, 2010).

8.2.2 *MAP* lipoproteins

Of the *MAP* proteins that are known, 63 are annotated as lipoproteins, which are considered fertile ground for strong antigen discovery based on studies with *Mycobacterium tuberculosis* (Becker and Sander, 2016). The outer layers of the mycobacterial cell envelope are composed of at least three types of non-covalently associated glycolipids interlaced in a carbohydrate matrix. The glycolipid types include phenolic glycolipids, glycopeptidolipids (GPLs) and lipooligosaccharides. Interestingly, *MAP* does not produce GPLs like the other *M. avium* subspecies. Instead, it produces unique lipopeptides, which include Para-LP-01, a lipopentapeptide in the bovine strains (Eckstein *et al.*, 2006) and a lipotripeptide in the ovine strains (Bannantine *et al.*, 2017b). The Para-LP-01, also known as L5P, elicits an antibody response in cattle and sheep (Biet *et al.*, 2008; Thirunavukkarasu *et al.*, 2013) as well as humans (Verdier *et al.*, 2013), but not a cell-mediated immune response (Souriau *et al.*, 2017). In addition, the abundance of this lipid molecule decreases during infection (Everman *et al.*, 2015) and this, along with other documented lipids (Alonso-Hearn *et al.*, 2010) and fatty acid changes (Alonso-Hearn *et al.*, 2017), may contribute to the severe inflammation of the bovine intestine.

GPLs and lipopeptides are synthesized by non-ribosomal peptide synthetases that are encoded by large genes with a modular organization. Transposon mutagenesis of one non-ribosomal peptide synthetase, *pstA* (MAP_1242), changed the 2-D lipid profile of *MAP* as analysed by thin-layer chromatography (Wu *et al.*, 2009). Characterization of the missing lipid in the *pstA* mutant suggested it was a lipotripeptide since this gene encodes three amino acid modules, which differentiates it from *M. avium* (encoding two modules) and *Mycobacterium smegmatis* (encoding four modules). Phenotypically, the *pstA* knockout mutant could not produce an extracellular matrix important for biofilm formation and was not able to transverse the intestine as efficiently as the wild type. As mentioned above, another lipotripeptide, termed L3P, was discovered uniquely in ovine strains of *MAP*. The non-ribosomal peptide portion of this molecule is encoded by MAP_1420 and its structure was characterized in detail to reveal the tripeptide Phe-*N*-Methyl-Val-Ala attached to a lipid moiety (Bannantine *et al.*, 2017b). Still another *MAP* lipopeptide, isolated from an ethanol extract of *MAP* (Eda *et al.*, 2006), has been biochemically characterized, shown to be immunogenic and defined to five amino acids (Mitachi *et al.*, 2016). Given that the genes required for the biosynthesis of these unique peptide–lipid cell wall structures are 20 kb in length, and thus take enormous amounts of energy to replicate, it is hypothesized that they must provide critical roles. The *MAP* lipoarabinomannan molecule has largely been ignored this past decade except for a vaccination study (Jolly *et al.*, 2016) and an examination of its immunostimulatory properties on bovine macrophages (Souza *et al.*, 2013).

8.2.3 *MAP* virulence proteins

A number of virulence proteins have been proposed despite the lack of proteome-wide screens for this category of proteins. Transcriptional regulators, such as the sigma factors SigH (MAP3324c) and SigL (MAP4201) result in attenuation when they are deleted, leading investigators to infer that they may control the expression of virulence genes (Ghosh *et al.*, 2013, Ghosh *et al.*, 2014). SigH in particular is important for *MAP* survival in IFN-γ activated bovine macrophages (Ghosh *et al.*, 2013). Likewise, the LuxR transcriptional regulator (MAP1875c) is expressed at higher levels when *MAP* is exposed to milk (Alonso-Hearn *et al.*, 2010). This protein has an ATPase domain and it regulates four lipid biosynthesis genes and two invasion genes along with five other upregulated genes. Despite these notable studies, very few *MAP* virulence genes have been examined in depth and even less have been confirmed by knockout mutation experiments, which are not trivial in *MAP* (Park *et al.*, 2008; Scandurra *et al.*, 2010). None the less, a handful of knockout mutations have shown attenuation in either macrophages or animal models (Scandurra *et al.*, 2009; Bannantine *et al.*, 2014a; Ghosh *et al.*, 2014; Meissner *et al.*, 2014).

Several *MAP* virulence proteins are involved in binding and/or invasion of bovine epithelial cells, which represent the initial infection event in paratuberculosis. In addition to the 10 *MAP* proteins discussed previously (Bannantine and Bermudez, 2013), MAP_1203 also plays a role in epithelial cell invasion (Everman *et al.*, 2018). Transcription of this gene is increased 28-fold when *MAP* is exposed to milk. The gene product contains a signal sequence that guides the protein into the *MAP* cell wall where it interacts specifically with two host proteins, dihydropyrimidinase-related protein 2 and glyceraldehyde 3-phosphate dehydrogenase, at the epithelial cell membrane. Other virulence proteins include a membrane serine protease (MAP_0403), which is upregulated in cultured macrophages and epithelial cells as well as under *in vitro* acid conditions (Kugadas *et al.*, 2016). Thus MAP_0403 imparts acid stress tolerance to *MAP*, but was not detected in oxidative or nitrogen stress conditions (Kawaji *et al.*, 2010). And finally, the *mptD* gene (MAP_3733 c) present on the *MAP*-specific large-sequence polymorphism 14 is important for survival in the early macrophage environment and is necessary for virulence in the mouse model (Meissner *et al.*, 2014).

8.3 *MAP* Antigens

The vast majority of *MAP* protein studies in the literature are diagnostic-focused. This focus is well placed given the tremendous challenge in diagnosing paratuberculosis and the need for improved tests that may encourage national control efforts and inform the animal producer on management strategies. Two of the most prominent challenges involve false positive results and low sensitivity. Sera from animals exposed to environmental or ubiquitous mycobacteria, such as *M. avium*, will cross-react with *MAP* antigens, giving a false positive reaction. ELISA test sensitivities have lagged behind specificity making detection of early infection nearly impossible. This may be due to the disease characteristics, which are marked by a low antibody response early after infection such that diagnostic sensitivities in pre-clinical stages will likely be low. Herd-level testing with the goal of national control of paratuberculosis recently estimated the serum ELISA as at least 60% sensitive and 100% specific among Irish dairy herds (Sergeant *et al.*, 2019). Although this sensitivity value has improved over the years, the search for an antigen that can be used to detect infection before disease signs appear would be of considerable value. Two commercial ELISAs used for another chronic disease of small ruminants caused by *Mycoplasma agalactiae* show test sensitivities of 84%, while specificity was 95.7% for a whole-cell antigen with slightly lower sensitivity for a recombinant antigen ELISA (Poumarat *et al.*, 2012). However, these test values were obtained using well-defined serum samples, rather than field sampling dairy herds. Furthermore, 9 of 10 commercially available ELISA tests used to detect the bovine parasitic disease, neosporosis, are all above 95% sensitivity and specificity (Alvarez-García *et al.*, 2013). Examples from these animal diseases suggests that *MAP* ELISA tests could be improved dramatically. Is *MAP* really that immunologically stealthy? Especially as it relates to an antibody response? Perhaps not as sets of three or four recombinant *MAP* proteins can yield increased ELISA sensitivities (>85%) and specificities (97%) (Leroy *et al.*, 2007; Li *et al.*, 2017a), but these are not commercially available yet. Thus, *MAP* serological detection via current paratuberculosis ELISA tests has set a low bar for researchers to jump over. There are some promising new *MAP* antigens discussed below that may be developed into a commercial test.

8.3.1 Multi-antigen studies

Historically, researchers have investigated single *MAP* antigens using unclear rationale for why that particular antigen was selected for analysis. Furthermore, these antigens were not benchmarked against other *MAP* proteins or a commercial ELISA test. However, more benchmarked multi-antigen studies are now being reported. This more comprehensive method was needed and hence systematic approaches to screen all the antigens produced by *MAP* were conducted. Prior to genome sequencing, this was done by screening an expression library constructed in *Escherichia coli* (Bannantine and Stabel, 2001; Willemsen *et al.*, 2006). Here the *E. coli* host is forced to express randomly sheared DNA fragments from *MAP*. Provided the expression library contains all of the DNA fragments that constitute the entire genome, this expression library should theoretically represent the entire proteome. However, many of the cloned fragments are not present in the correct orientation or are not in frame with the promoter designed to express the cloned fragment. Moreover, *E. coli* expresses *MAP* proteins at widely varying levels, many of which are not expressed at all due to toxicity (Bannantine *et al.*, 2010). None the less, this system yielded some solid antigen candidates. Most notable was the study by Willemsen *et al.* (2006), which identified MAP_2609 and MAP_2942c by phage library screening with serum from a clinical cow. These same antigens were later identified via a protein array/ELISA combination screen with serum from cows at various stages of paratuberculosis (Bannantine *et al.*, 2017a; Li *et al.*, 2017a).

More recently, the *M. tuberculosis* protein array has been used as a screening tool to identify antigens diagnostic for paratuberculosis (Bannantine *et al.*, 2017a). Although not as closely related to *MAP* as members of the MAC complex, *M. tuberculosis* still has over half of its proteins with greater than 75% identity to *MAP*. Candidate antigens were further tested by ELISA using well-characterized serum samples that encompass all paratuberculosis stages. Rv2878c (MAP_2942c), a disulfide oxidoreductase, emerged as the strongest antigen during clinical disease. However, 10 *M. tuberculosis* antigens were identified that appear to be detected at early stages of paratuberculosis. This pilot study was then expanded to test 180 well-characterized serum samples from cows in all disease stages (Li *et al.*, 2017a), adding more significantly reactive antigen candidates while confirming those discovered in the pilot study (Bannantine *et al.*, 2017a). Interestingly, the Li *et al.* (2017a) study also revealed 27 antigens detected in early disease cows that were faecal culture-negative and ELISA negative but had high exposure to *MAP*, suggestive of the animal being infected, but not yet shedding. This result increases the possibility of developing an ELISA test with the needed sensitivity to detect early *MAP* infection in a dairy herd.

Another 2-D sodium dodecyl sulfate–polyacrylamide gel electrophoresis (SDS-PAGE) separation and immunoblot study with sera from control and infected cattle was conducted to identify proteins reactive only with the infected group. Six proteins were excised from the 2-D gels and identified by mass spectrometry including Hsp65 and MMP (MAP_2121c) (Piras *et al.*, 2015).

Other formats, such as a multiplex bead-based immunoassay, have been used on serum and milk samples. Beads were coated with six antigens identified from the Li *et al.* (2017a) study and used to test 180 serum and 90 milk samples in this bead assay format (Li *et al.*, 2017b). The results showed that the multiplex assay has higher sensitivity and specificity than the same proteins used in an ELISA format. Also, when using milk samples, areas under the curve (AUCs) of the recombinant protein multiplex assay were higher compared with the commercial IDEXX ELISA (Li *et al.*, 2017b). Combining these antigens helped increase the sensitivity further, but then specificity was reduced. These results are promising and suggest improvements to antigen-based tests for paratuberculosis are on the horizon.

8.3.2 Johnin purified protein derivative

The *MAP* purified protein derivative (PPD), also known as Johnin, is a complex secreted protein preparation that appears to vary greatly between preparations. Therefore, this antigen is very difficult to standardize for diagnostic use. It is used in two diagnostic tests, the single intradermal skin test to assay for delay-type hypersensitivity and the IFN-γ test. Both tests measure a component of the cell-mediated immune response. Although this has been considered a great antigen for these tests, its lack of standardization resulted in widely variable test results. Each PPD batch needs be checked for potency, which is performed using a 72-h guinea pig assay using a method comparable with that for tuberculin (Frankena *et al.*, 2018). In an effort to develop a reproducible reagent for these tests, proteomic analyses by 2-D SDS-PAGE were conducted on *MAP* culture filtrates and 125 proteins were initially identified (Leroy *et al.*, 2007). More recently, Wynne *et al.* (2012) characterized five PPD preps and showed that while the proteins present in each prep were remarkably consistent, they varied in terms of abundance within each prep. Time of culture harvest, ranging from 20–34 weeks, also had a significant influence on which proteins were present in the final PPD. Finally, a few proteins were identified that were consistently more abundant in the Johnin PPD preps when compared with the avium PPD preps. In separate studies, 194 total PPD proteins were identified with only 10 shared among all three preps analysed (Capsel *et al.*, 2016) and 156 PPD proteins were mentioned by Santema *et al.* (2009), but only 25 were listed in that publication. In comparing the proteins identified among these three proteomic PPD studies (Santema *et al.*, 2009; Wynne *et al.*, 2012; Capsel *et al.*, 2016), it was discovered that a core of 15 *MAP* proteins were identified from all three studies (Fig. 8.1, details in Table 8.2). These 15 proteins likely represent those in highest abundance in PPD extracts. Although these proteins are theoretically secreted based on the way the preps are made, these 15 proteins include membrane, cytoplasmic and periplasmic proteins, as well as those that could not be predicted by the PSORTb software (Yu *et al.*, 2010). It is likely that PPD comprises at least 300 proteins, thus we still have a long way to go in characterizing this useful, but hard to prepare antigen.

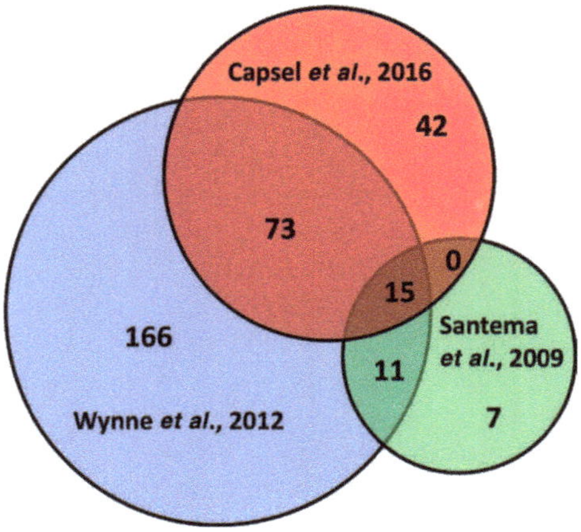

Fig. 8.1. Venn diagram of *Mycobacteria avium* subsp. *paratuberculosis* (*MAP*) proteins in three published studies of *MAP* purified protein derivative. (From Santema *et al.*, 2009; Wynne *et al.*, 2012 and Capsel *et al.*, 2016.)

8.3.3 Antigens tested in the IFN-γ assay

Cattle develop a proinflammatory T helper 1 (Th1) response to *MAP* early in the course of paratuberculosis. IFN-γ release assays measure T-cell production of IFN-γ in response to antigen exposure thus providing an assessment of cell-mediated immunity. Measurement of IFN-γ production has shown promise for paratuberculosis diagnosis at early stages of infection and is an indicator of the Th1 immune response, but the specificity is low due to extensive cross-reactions with closely related mycobacteria. Whole blood from naturally infected cows with subclinical disease shows significantly higher IFN-γ secretion than blood from healthy control cows after stimulation with the T-cell mitogen, concanavalin A, or a whole-cell extract of *MAP* (Stabel, 1996). Historically, *MAP* PPDs have been used as the stimulating antigen for the paratuberculosis IFN-γ assay. A *MAP* whole-cell extract has been compared with PPDs, but PPDs appeared to be a stronger IFN-γ stimulator (Robbe-Austerman *et al.*, 2006). Five mycobacterial lipids have also been tested this way and none can detect *MAP*-infected goats (Souriau *et al.*, 2017) or sheep (Thirunavukkarasu *et al.*, 2013). Conversely, young calves and uninfected cows often respond to PPDs used in the IFN-γ assay without having any evidence of infection (Mcdonald *et al.*, 1999) and there exist wildly fluctuating results when the same animals are serially tested (Huda *et al.*, 2003; Dernivoix *et al.*, 2017; Souriau *et al.*, 2017). Another limitation

Table 8.2. *Mycobacteria avium* subsp. *paratuberculosis* (MAP) proteins common to three proteomic purified protein derivative (PPD) studies.

Locus tag	Gene	aa length	PSORTb location	Description
MAP_0151 c	-	151	Unknown	Transcriptional regulator TetR
MAP_0494	-	318	Unknown	Putative oxidoreductase
MAP_1138 c	lprG	239	Membrane	Lipoprotein
MAP_1339	-	148	Unknown	Hypothetical protein
MAP_1595	bfrA	160	Cytoplasmic	Bacterioferritin
MAP_1609 c	fbpB	331	Unknown	Antigen 85B, mycolyltransferase
MAP_1889 c	wag31	261	Cytoplasmic	Antigen 84
MAP_1997	acpM	116	Cytoplasmic	Acyl carrier protein
MAP_2451 c	atpD	403	Membrane	ATP synthase [beta] chain
MAP_2677 c	-	133	Unknown	Putative oxidoreductase
MAP_3362 c	sahH	497	Cytoplasmic	Adenosylhomocysteinase
MAP_3531 c	fbpC2	353	Periplasmic	Antigen 85C, mycolytransferase
MAP_3840	dnaK	624	Cytoplasmic	70 kD heat shock protein
MAP_3936	groEL2	542	Cytoplasmic	60 kD chaperonin 2
MAP_4143	tuf	397	Cytoplasmic	Elongation factor EF-Tu

of this assay is the need for fresh whole blood or peripheral blood mononuclear cells (PBMCs) to conduct the antigen stimulation. Samples that have been kept for longer than 12 h show a pronounced IFN-γ decrease in response to antigens (Robbe-Austerman *et al.*, 2006). Some have tried to lengthen the usefulness of the samples out to 24 h before starting the antigen stimulation (Buza *et al.*, 2004; Jungersen *et al.*, 2012; Mikkelsen *et al.*, 2012). Not all of these difficulties can be removed simply by changing the antigen, but the goal of improved early detection and increased specificity has motivated researchers to test purified recombinant antigens in an effort to circumvent the problems associated with PPDs and whole-cell extracts.

Recently, six recombinant antigens have been examined in the IFN-γ assay using a 20-h PBMC stimulation from animals thoroughly characterized with a PPD-IFN-γ assay at the study's outset (Dernivoix *et al.*, 2017). None of these recombinant proteins showed higher sensitivity than the PPD, although three proteins were able to distinguish *MAP*-infected from *M. bovis*-infected animals. In a separate study, the stress-induced proteins, MAP_2698 c and MAP_3567, were examined by this format for sheep vaccinated with the same antigens. A 48-h stimulation resulted in IFN-γ responses, but varied depending on the adjuvant used (Gurung *et al.*, 2014b). Two other recombinant proteins, MAP_0268 c and MAP_3651 c, have been used to stimulate IFN-γ responses in sheep (Hughes *et al.*, 2013) and cattle (Hughes *et al.*, 2017) and have shown promise in both. Another study examined 16 recombinant proteins in goats (Souriau *et al.*, 2017). The proteins encoded by MAP_3651 c and MAP_1050 c emerged as the most promising candidates for detecting infected shedding goats but could not detect infected non-shedding goats. Also, use of these recombinant antigens was discouraged in adult goats (>2 years old) as infection predictions dropped below acceptable levels. A final study examined 15 recombinant proteins in a variety of time and blood preservation conditions (Mikkelsen *et al.*, 2012). Thus, only 41 antigens have been evaluated in this way these past 10 years and none appears promising enough for commercial development. Perhaps a high-throughput method should be engineered to examine more recombinant antigens for this assay.

In general, it appears that good recombinant antigen candidates for serological detection of *MAP* infection are not good candidates for IFN-γ stimulation assays. Several recombinant proteins have been tested in both assays (MAP_1589 c AhpC, for example), and thus far none works well in both. Also, those that appear to be good candidates in the IFN-γ assay are not listed among the good serological antigens and vice versa. Furthermore, it appears that recombinant proteins still cannot stimulate IFN-γ production in PBMCs from infected cows better than PPDs and this has been the main problem with using recombinant proteins in this assay. They aren't as strong an antigen as the PPDs even though they would be easier and more consistent to prepare. As mentioned in Section 8.3.2 PPDs contain hundreds of proteins that stimulate a wide variety of T cells, which a single, or even a dozen, recombinant proteins cannot. Furthermore, many studies have reported high variability in recombinant protein IFN-γ assays for paratuberculosis. This is partly due to the age of the animal, and partly the need for fresh PBMCs to conduct the assay and the differences in time used to stimulate the PBMCs, which can vary from 16–72 h. These obstacles do not mean researchers should stop looking for a strong antigen to incorporate in a marketable paratuberculosis IFN-γ assay, but the approach must be clever and antigen solid before investing too much in this line of research.

8.3.4 Cell-mediated immune (CMI) stimulating antigens of *MAP*

An early diagnostic antigen will more likely generate a measurable cell-mediated immune response than a humoral response based on what we know about progression of immune parameters to this disease. Several studies have looked beyond the proinflammatory cytokine, IFN-γ, as just one measure of a CMI response. IFN-γ is easy to measure, hence the focus on this assay for early *MAP* detection. However, measuring any type of cellular-based immunity accurately is difficult due to biological variation, standardization and technical complexity, which is why the *MAP* IFN-γ assay itself has shown high variability. To characterize a CMI response, the circulating immune cells and their subsets should be quantified by flow cytometry. This approach can identify dendritic cell maturation and activation, which drives the Th1 response. The skin test measures delayed type hypersensitivity through antigen-specific memory T cells, which creates a local inflammation at the injection site from recruitment of other immune cells. T-cell activation can be measured either by changes in surface markers (CD25, CD45RO, CD69) or by measuring intracellular cytokine levels. *MAP* proteins have played a role in all of these CMI responses.

While *MAP* itself has been shown to activate resident dendritic cells (DCs) many times, *MAP* effector proteins that cause DC activation have also been recently discovered. Antigens with this capability are well positioned as good vaccine candidates since DCs provide an important link between innate and adaptive immunity. MAP_1948 (Byun *et al.*, 2012), MAP_1981c (Kim *et al.*, 2018b), MAP_2541c (Kim *et al.*, 2018a), MAP_1305 (Lee *et al.*, 2014) and MAP_1569 (Lee *et al.*, 2009; Noh *et al.*, 2012) have all been shown to promote DC activation. Most of these same proteins have been shown to selectively activate DC through TLR-4. Dendritic maturation can go on to promote cytotoxic lymphocyte killing. Both MAP_2121c and MAP_1569 promote cytotoxic T cell activity (Noh *et al.*, 2012; Abdellrazeq *et al.*, 2018). Many other studies have examined how the intact *Mycobacterium* promotes CMI responses, but those are beyond the scope of this antigen-proteomic-focused chapter.

8.3.5 Antigens tested by ELISA

Milk has become a popular sample in recent years with the development of milk ELISA testing, but defined antigens have yet to be tested on milk samples except for one study (Li *et al.*, 2017b). Recently, a sandwich ELISA was developed with a monoclonal antibody to the Apa protein (MAP_1569) and used on faecal eluates to detect that antigen in paratuberculosis-positive cows (de Souza *et al.*, 2018), and this type of approach has been used to measure *MAP*-specific IgA in the faeces before (Begg *et al.*, 2015). But the most common sample for ELISA historically is serum.

The search for defined antigens to incorporate into an ELISA test began in 1991 with the 85C antigen and a bacterioferritin biochemically purified from a *MAP* culture filtrate (Sugden *et al.*, 1991). These antigens each showed 86 and 100% sensitivity, with the 400-kDa bacterioferritin antigen detecting all 22 infected animals in that study. Lipid antigens were readily purified and the ELISA format tested with some degree of success as well. Then molecular biology was applied to *MAP* and recombinant proteins were expressed in *E. coli* and tested.

Among *MAP* recombinant antigens incorporated into ELISA within the past 10 years, PtpA (encoded by MAP_1985), has been studied extensively. This protein is a tyrosine phosphatase that is secreted early during infection of the host. It is reported to be antigenic in humans (Slavin *et al.*, 2018), sheep (Gurung *et al.*, 2014a), and displayed a 90% ELISA sensitivity and 91% specificity in cattle (Bach *et al.*, 2018). The Bach *et al.* (2018) study also showed that the PtpA ELISA outperformed a commercial ELISA for early detection of *MAP* antibodies in a 55-animal study. Once validated, this could make PtpA the best recombinant antigen for use in a paratuberculosis ELISA test for cattle. PtpA was not among the 729 *M. tuberculosis* proteins that reacted to sera from paratuberculosis-infected cows, but it should be noted that only 81.6% identity exists between *MAP* PtpA and *M. tuberculosis* PtpA (Bannantine *et al.*, 2017a).

Fourteen culture filtrate proteins were shown to be immunogenic for paratuberculosis cows. Five of these were tested in an ELISA format based on strong reactivity with sera from *MAP*-immunized rabbits (Chaubey *et al.*, 2018). These five include MAP_1693c, MAP2168c, MAP_1569, MAP_3527 (PepA) and the aforementioned antigen 85C (MAP_3531c). They were mixed and used as a cocktail for coating ELISA plates and directly compared with a *MAP* whole-cell extract ELISA.

8.4 Conclusions

Researchers have now looked for vaccine and diagnostic antigens using proteomic approaches for at least 35 years. Although there are promising antigens that have been discovered and newer technologies to increase sensitivity or amplify signal, there is a real barrier getting these incorporated into marketable tests and vaccine formulations to control paratuberculosis or at least help the animal producer make informed decisions about herd management. There may never be an antigen that provides sterile immunity or predicts infection with 99% accuracy, but that shouldn't dissuade researchers from examining ways to improve upon what is currently available. Trends from the past 15 years show that paratuberculosis incidence is on the rise, with 68% of US dairy herds having at least one infected cow reported in the 2007 National Animal Health Monitoring System (NAHMS) dairy survey to over 90% in 2014 from NAHMS dairy 2014. Thus, it is critical to maintain the search for the best tools to fight this disease.

In the course of assembling this chapter a few proteins appeared noticeably in many studies. These include the AhpC, FAP and DnaK proteins, among others. Their repeated appearance may perhaps suggest that they are abundant in the bacilli more than being immunogenic or useful for vaccination. This might be the case for DnaK in particular, which is present in nearly all proteomic preps and is abundantly expressed in *MAP*. FAP (a.k.a. ModD) is a rare *MAP* protein that acts as a strong B-cell antigen (Li *et al.*, 2017a; Li *et al.*, 2017b), T-cell antigen (Lee *et al.*, 2009) and an activator of dendritic cells that interacts with TLR-4 (Noh *et al.*, 2012). Therefore, perhaps these three and a handful of other antigens, such as PtpA, appear to be the best targets to focus on in the future.

And finally, despite heroic efforts by *MAP* researchers around the world, the diagnostics for paratuberculosis is still at a state where (i) we cannot confidently certify a healthy appearing animal as not infected with *MAP*, and (ii) we cannot confirm *MAP* infection in an animal that does not show overt signs of disease. While this does not seem encouraging, testing has improved to a point where herd-level infections, or lack thereof, are confidently captured. In general, paratuberculosis ELISA test sensitivity and specificity has increased over time.

8.5 Future Perspectives

What is next for *MAP* proteomic research? The genome is sequenced and proteome has been defined. We now know all the 'players' and in multiple studies, we have tested them at least from a diagnostic standpoint. In many cases, we now know which antigens not to focus on! And that, at least, is some progress. However, very few proteome fractions or subunit proteins have been tested as vaccines and thus there is excellent opportunity for future successes. Animal models have been established, but the cost of using them is a likely reason so few studies have been conducted in this area. When one looks across the chronic bacterial disease landscape, it doesn't appear as though other diseases have had much success with vaccination or diagnostics, although many excellent studies have been conducted with these specific goals. Many of these diseases boast considerably more funding than that dedicated to paratuberculosis. Therefore, researchers in this field must remain creative and visionary to solve the problem of this slow, intractable disease.

References

Abdellrazeq, G.S., Elnaggar, M.M., Bannantine, J.P., Park, K.T., Souza, C.D. *et al.* (2018) A *Mycobacterium avium* subsp. *paratuberculosis relA* deletion mutant and a 35 kDa major membrane protein elicit development of cytotoxic T lymphocytes with ability to kill intracellular bacteria. *Veterinary Research* 49(1), 53. DOI: 10.1186/s13567-018-0549-3.

Alonso-Hearn, M., Eckstein, T.M., Sommer, S. and Bermudez, L.E. (2010) A *Mycobacterium avium* subsp. *paratuberculosis LuxR* regulates cell envelope and virulence. *Innate Immunity* 16(4), 235–247. DOI: 10.1177/1753425909339811.

Alonso-Hearn, M., Abendaño, N., Ruvira, M.A., Aznar, R., Landin, M. *et al.* (2017) *Mycobacterium avium* subsp. *paratuberculosis* (Map) fatty acids profile is strain-dependent and changes upon host macrophages infection. *Frontiers in Cellular and Infection Microbiology* 7, 89. DOI: 10.3389/fcimb.2017.00089.

Altschul, S.F., Madden, T.L., Schäffer, A.A., Zhang, J., Zhang, Z. *et al.* (1997) Gapped blast and PSI-BLAST: a new generation of protein database search programs. *Nucleic Acids Research* 25(17), 3389–3402. DOI: 10.1093/nar/25.17.3389.

Alvarez-García, G., García-Culebras, A., Gutiérrez-Expósito, D., Navarro-Lozano, V., Pastor-Fernández, I. *et al.* (2013) Serological diagnosis of bovine neosporosis: a comparative study of commercially available ELISA tests. *Veterinary Parasitology* 198(1–2), 85–95. DOI: 10.1016/j.vetpar.2013.07.033.

Amin, A.S., Hsu, C.-Y., Darwish, S.F., Ghosh, P., AbdEl-Fatah, E.M. *et al.* (2015) Ecology and genomic features of infection with *Mycobacterium avium* subspecies *paratuberculosis* in Egypt. *Microbiology* 161(4), 807–818. DOI: 10.1099/mic.0.000051.

Bach, E., Raizman, E.A., Vanderwal, R., Soto, P., Chaffer, M. *et al.* (2018) Immunogenicity of PTPA secreted during *Mycobacterium avium* ssp. *paratuberculosis* infection in cattle. *Veterinary Immunology and Immunopathology* 198, 1–5. DOI: 10.1016/j.vetimm.2018.02.006.

Bannantine, J.P. and Bermudez, L.E. (2013) No holes barred: invasion of the intestinal mucosa by *Mycobacterium avium* subsp. *paratuberculosis*. *Infection and Immunity* 81(11), 3960–3965. DOI: 10.1128/IAI.00575-13.

Bannantine, J.P. and Stabel, J.R. (2001) Identification of two *Mycobacterium avium* subspecies *paratuberculosis* gene products differentially recognised by sera from rabbits immunised with live mycobacteria but not heat-killed mycobacteria. *Journal of Medical Microbiology* 50(9), 795–804. DOI: 10.1099/0022-1317-50-9-795.

Bannantine, J.P., Huntley, J.F.J., Miltner, E., Stabel, J.R. and Bermudez, L.E. (2003) The *Mycobacterium avium* subsp. *paratuberculosis* 35 kDa protein plays a role in invasion of bovine epithelial cells. *Microbiology* 149(8), 2061–2069. DOI: 10.1099/mic.0.26323-0.

Bannantine, J.P., Stabel, J.R., Bayles, D.O. and Geisbrecht, B.V. (2010) Characteristics of an extensive *Mycobacterium avium* subspecies *paratuberculosis* recombinant protein set. *Protein Expression and Purification* 72(2), 223–233. DOI: 10.1016/j.pep.2010.03.019.

Bannantine, J.P., Stabel, J.R., Lamont, E.A., Briggs, R.E. and Sreevatsan, S. (2011) Monoclonal antibodies bind a SNP-sensitive epitope that is present uniquely in *Mycobacterium avium* subspecies *paratuberculosis*. *Frontiers in Microbiology* 2, 163. DOI: 10.3389/fmicb.2011.00163.

Bannantine, J.P., Wu, C.W., Hsu, C., Zhou, S., Schwartz, D.C. *et al.* (2012) Genome sequencing of ovine isolates of *Mycobacterium avium* subspecies *paratuberculosis* offers insights into host association. *BMC Genomics* 13(1), 89. DOI: 10.1186/1471-2164-13-89.

Bannantine, J.P., Everman, J.L., Rose, S.J., Babrak, L., Katani, R. *et al.* (2014a) Evaluation of eight live attenuated vaccine candidates for protection against challenge with virulent *Mycobacterium avium* subspecies *paratuberculosis* in mice. *Frontiers in Cellular and Infection Microbiology* 4(11), 88. DOI: 10.3389/fcimb.2014.00088.

Bannantine, J.P., Li, L., Mwangi, M., Cote, R., Raygoza Garay, J.A. *et al.* (2014b) Complete genome sequence of *Mycobacterium avium* subsp. *paratuberculosis*, isolated from human breast milk. *Genome Announcements* 2(1). DOI: 10.1128/genomeA.01252-13.

Bannantine, J.P., Campo, J.J., Li, L., Randall, A., Pablo, J. *et al.* (2017a) Identification of novel seroreactive antigens in Johne's disease cattle by using the *Mycobacterium tuberculosis* protein array. *Clinical and Vaccine Immunology* 24(7), e00081-17 DOI: 10.1128/CVI.00081-17.

Bannantine, J.P., Etienne, G., Laval, F., Stabel, J.R., Lemassu, A. *et al.* (2017b) Cell wall peptidolipids of *Mycobacterium avium* : from genetic prediction to exact structure of a nonribosomal peptide. *Molecular Microbiology* 105(4), 525–539. DOI: 10.1111/mmi.13717.

Bastida, F. and Juste, R.A. (2011) Paratuberculosis control: a review with a focus on vaccination. *Journal of Immune Based Therapies and Vaccines* 9(1), 8. DOI: 10.1186/1476-8518-9-8.

Becker, K. and Sander, P. (2016) *Mycobacterium tuberculosis* lipoproteins in virulence and immunity – fighting with a double-edged sword. *FEBS Letters* 590(21), 3800–3819. DOI: 10.1002/1873-3468.12273.

Begg, D.J., de Silva, K., Plain, K.M., Purdie, A.C., Dhand, N. *et al.* (2015) Specific faecal antibody responses in sheep infected with *Mycobacterium avium* subspecies *paratuberculosis*. *Veterinary Immunology and Immunopathology* 166(3–4), 125–131. DOI: 10.1016/j.vetimm.2015.06.011.

Biet, F., Bay, S., Thibault, V.C., Euphrasie, D., Grayon, M. *et al.* (2008) Lipopentapeptide induces a strong host humoral response and distinguishes *Mycobacterium avium* subsp. *paratuberculosis* from *M. avium* subsp. *avium*. *Vaccine* 26(2), 257–268. DOI: 10.1016/j.vaccine.2007.10.059.

Brauning, R., Plain, K., Gautam, M., Russell, T., Correa, C.C. *et al.* (2019) Complete genome sequence of the Telford type S strain of *Mycobacterium avium* subsp. *paratuberculosis*. *Microbiology Resource Announcements* 8(11). DOI: 10.1128/MRA.00004-19.

Buza, J.J., Hikono, H., Mori, Y., Nagata, R., Hirayama, S. *et al.* (2004) Neutralization of interleukin-10 significantly enhances gamma interferon expression in peripheral blood by stimulation with Johnin purified protein derivative and by infection with *Mycobacterium avium* subsp. *paratuberculosis* in experimentally infected cattle with *paratuberculosis*. *Infection and Immunity* 72(4), 2425–2428. DOI: 10.1128/IAI.72.4.2425-2428.2004.

Byun, E.-H., Kim, W.S., Kim, J.-S., Won, C.-J., Choi, H.-G. *et al.* (2012) *Mycobacterium paratuberculosis* CobT activates dendritic cells via engagement of Toll-like receptor 4 resulting in Th1 cell expansion. *Journal of Biological Chemistry* 287(46), 38609–38624. DOI: 10.1074/jbc.M112.391060.

Capsel, R.T., Thoen, C.O., Reinhardt, T.A., Lippolis, J.D., Olsen, R. *et al.* (2016) Composition and potency characterization of *Mycobacterium avium* subsp. *paratuberculosis* purified protein derivatives. *PLoS ONE* 11(5), e0154685. DOI: 10.1371/journal.pone.0154685.

Chaubey, K.K., Singh, S.V. and Bhatia, A.K. (2018) Evaluation of ‘recombinant secretary antigens’ based ‘cocktail ELISA’ for the diagnosis of Johne's disease and to differentiate non-infected, infected and vaccinated goats in combination with indigenous ELISA test. *Small Ruminant Research* 165, 24–29. DOI: 10.1016/j.smallrumres.2018.06.005.

de Souza, G.D.S., Rodriguez, A.B.F., Romano, M.I., Ribeiro, E.S., Oelemann, W.M.R. *et al.* (2018) Identification of the Apa protein secreted by *Mycobacterium avium* subsp. *paratuberculosis* as a novel fecal biomarker for Johne’s disease in cattle. *Pathogens and Disease* 76(6). DOI: 10.1093/femspd/fty063.

Dernivoix, K., Roupie, V., Welby, S., Roelandt, S., Viart, S. *et al.* (2017) Field performance of six *Mycobacterium avium* subsp. *paratuberculosis* antigens in a 20 H interferon gamma release assay in Belgium. *Veterinary Immunology and Immunopathology* 189, 17–27. DOI: 10.1016/j.vetimm.2017.05.008.

Eckstein, T.M., Chandrasekaran, S., Mahapatra, S., McNeil, M.R., Chatterjee, D. *et al.* (2006) A major cell wall lipopeptide of *Mycobacterium avium* subspecies *paratuberculosis*. *Journal of Biological Chemistry* 281(8), 5209–5215. DOI: 10.1074/jbc.M512465200.

Eda, S., Bannantine, J.P., Waters, W.R., Mori, Y., Whitlock, R.H. *et al.* (2006) A highly sensitive and subspecies-specific surface antigen enzyme- linked immunosorbent assay for diagnosis of Johne’s disease. *Clinical and Vaccine Immunology* 13(8), 837–844. DOI: 10.1128/CVI.00148-06.

Everman, J.L., Eckstein, T.M., Roussey, J., Coussens, P., Bannantine, J.P. *et al.* (2015) Characterization of the inflammatory phenotype of *Mycobacterium avium* subspecies *paratuberculosis* using a novel cell culture passage model. *Microbiology* 161(7), 1420–1434. DOI: 10.1099/mic.0.000106.

Everman, J.L., Danelishvili, L., Flores, L.G. and Bermudez, L.E. (2018) MAP1203 promotes *Mycobacterium avium* subspecies *paratuberculosis* binding and invasion to bovine epithelial cells. *Frontiers in Cellular and Infection Microbiology* 8, 217. DOI: 10.3389/fcimb.2018.00217.

Fields, S. and Sternglanz, R. (1994) The two-hybrid system: an assay for protein-protein interactions. *Trends in Genetics* 10(8), 286–292. DOI: 10.1016/0168-9525(90)90012-U.

Frankena, K., Jacobs, L., van Dijk, T., Good, M., Duignan, A. *et al.* (2018) A new model to calibrate a reference standard for bovine tuberculin purified protein derivative in the target species. *Frontiers in Veterinary Science* 5, 232. DOI: 10.3389/fvets.2018.00232.

Ghosh, P., Hsu, C., Alyamani, E.J., Shehata, M.M., Al-Dubaib, M.A. *et al.* (2012) Genome-wide analysis of the emerging infection with *Mycobacterium avium* subspecies *paratuberculosis* in the Arabian camels (*Camelus dromedarius*). *PLoS ONE* 7(2), e31947. DOI: 10.1371/journal.pone.0031947.

Ghosh, P., Wu, C.W. and Talaat, A.M. (2013) Key role for the alternative sigma factor, SigH, in the intracellular life of *Mycobacterium avium* subsp. *paratuberculosis* during macrophage stress. *Infection and Immunity* 81(6), 2242–2257. DOI: 10.1128/IAI.01273-12.

Ghosh, P., Steinberg, H. and Talaat, A.M. (2014) Virulence and immunity orchestrated by the global gene regulator *sigL* in *Mycobacterium avium* subsp. *paratuberculosis*. *Infection and Immunity* 82(7), 3066–3075. DOI: 10.1128/IAI.00001-14.

Gurung, R.B., Begg, D.J., Purdie, A.C., Bach, H. and Whittington, R.J. (2014a) Immunoreactivity of protein tyrosine phosphatase A (PtpA) in sera from sheep infected with *Mycobacterium avium* subspecies *paratuberculosis*. *Veterinary Immunology and Immunopathology* 160(1–2), 129–132. DOI: 10.1016/j.vetimm.2014.03.013.

Gurung, R.B., Purdie, A.C., Whittington, R.J. and Begg, D.J. (2014b) Cellular and humoral immune responses in sheep vaccinated with candidate antigens MAP2698c and MAP3567 from *Mycobacterium avium* subspecies *paratuberculosis*. *Frontiers in Cellular and Infection Microbiology* 4, 93. DOI: 10.3389/fcimb.2014.00093.

He, Z. and De Buck, J. (2010) Localization of proteins in the cell wall of *Mycobacterium avium* subsp. *paratuberculosis* K10 by proteomic analysis. *Proteome Science* 8(1), 21. DOI: 10.1186/1477-5956-8-21.

Huda, A., Lind, P., Christoffersen, A.B. and Jungersen, G. (2003) Analysis of repeated tests for interferon-gamma (IFN-γ) response and faecal excretion for diagnosis of subclinical paratuberculosis in Danish cattle. *Veterinary Immunology and Immunopathology* 94(3–4), 95–103. DOI: 10.1016/S0165-2427(03)00063-1.

Hughes, V., Bannantine, J.P., Denham, S., Smith, S., Garcia-Sanchez, A. *et al.* (2008) Immunogenicity of proteome-determined *Mycobacterium avium* subsp. *paratuberculosis*-specific proteins in sheep with paratuberculosis. *Clinical and Vaccine Immunology* 15(12), 1824–1833. DOI: 10.1128/CVI.00099-08.

Hughes, V., Denham, S., Bannantine, J.P., Chianini, F., Kerr, K. *et al.* (2013) Interferon gamma responses to proteome-determined specific recombinant proteins: potential as diagnostic markers for ovine Johne's disease. *Veterinary Immunology and Immunopathology* 155(3), 197–204. DOI: 10.1016/j.vetimm.2013.06.015.

Hughes, V., McNair, J., Strain, S., Barry, C., McLuckie, J. *et al.* (2017) Gamma interferon responses to proteome-determined specific recombinant proteins in cattle experimentally- and naturally-infected with paratuberculosis. *Research in Veterinary Science* 114, 244–253. DOI: 10.1016/j.rvsc.2017.04.018.

Jolly, A., Lompardía, S., Hajos, S.E. and Mundo, S.L. (2016) Evidence of a pro-apoptotic effect of specific antibodies in a bovine macrophage model of infection with *Mycobacterium avium* subsp. *paratuberculosis*. *Veterinary Immunology and Immunopathology* 169, 47–53. DOI: 10.1016/j.vetimm.2015.12.001.

Jungersen, G., Mikkelsen, H. and Grell, S.N. (2012) Use of the johnin PPD interferon-gamma assay in control of bovine paratuberculosis. *Veterinary Immunology and Immunopathology* 148(1–2), 48–54. DOI: 10.1016/j.vetimm.2011.05.010.

Kawaji, S., Zhong, L. and Whittington, R.J. (2010) Partial proteome of *Mycobacterium avium* subsp. *paratuberculosis* under oxidative and nitrosative stress. *Veterinary Microbiology* 145(3–4), 252–264. DOI: 10.1016/j.vetmic.2010.03.025.

Kim, W.S., Kim, J.-S., Shin, M.-K. and Shin, S.J. (2018a) A novel Th1-type T-cell immunity-biasing effect of malate dehydrogenase derived from *Mycobacterium avium* subspecies *paratuberculosis* via the activation of dendritic cells. *Cytokine* 104, 14–22. DOI: 10.1016/j.cyto.2018.01.022.

Kim, W.S., Shin, M.-K. and Shin, S.J. (2018b) MAP1981c, a putative nucleic acid-binding protein, produced by *Mycobacterium avium* subsp. *paratuberculosis*, induces maturation of dendritic cells and Th1-polarization. *Frontiers in Cellular and Infection Microbiology* 8, 206. DOI: 10.3389/fcimb.2018.00206.

Kugadas, A., Lamont, E.A., Bannantine, J.P., Shoyama, F.M., Brenner, E. *et al.* (2016) A *Mycobacterium avium* subsp. *paratuberculosis* predicted serine protease is associated with acid stress and intraphagosomal survival. *Frontiers in Cellular and Infection Microbiology* 6(599), 85. DOI: 10.3389/fcimb.2016.00085.

Lanigan, M., Shiell, B., Beddome, G., Bruce, K., Vaughan, J. *et al.* (2007) Secreted antigens of *Mycobacterium avium* subsp. *paratuberculosis*: identity, characterisation, expression profiles and immunoreactivity. *Current Topics in Peptide & Protein Research* 8, 67–79.

Lee, J.S., Shin, S.J., Collins, M.T., Jung, I.D., Jeong, Y.-I. *et al.* (2009) *Mycobacterium avium* subsp. *paratuberculosis* fibronectin attachment protein activates dendritic cells and induces a Th1 polarization. *Infection and Immunity* 77(7), 2979–2988. DOI: 10.1128/IAI.01411-08.

Lee, S.J., Noh, K.T., Kang, T.H., Han, H.D., Shin, S.J. *et al.* (2014) The *Mycobacterium avium* subsp. *paratuberculosis* protein MAP1305 modulates dendritic cell-mediated T cell proliferation through Toll-like receptor-4. *BMB Reports* 47(2), 115–120. DOI: 10.5483/BMBRep.2014.47.2.277.

Leite, F.L., Reinhardt, T.A., Bannantine, J.P. and Stabel, J.R. (2015) Envelope protein complexes of *Mycobacterium avium* subsp. *paratuberculosis* and their antigenicity. *Veterinary Microbiology* 175(2–4), 275–285. DOI: 10.1016/j.vetmic.2014.11.009.

Leroy, B., Roupie, V., Noël-Georis, I., Rosseels, V., Walravens, K. *et al.* (2007) Antigen discovery: a post-genomic approach to paratuberculosis diagnosis. *PROTEOMICS* 7(7), 1164–1176. DOI: 10.1002/pmic.200600988.

Li, L., Bannantine, J.P., Zhang, Q., Amonsin, A., May, B.J. *et al.* (2005) The complete genome sequence of *Mycobacterium avium* subspecies *paratuberculosis*. *Proceedings of the National Academy of Sciences* 102(35), 12344–12349. DOI: 10.1073/pnas.0505662102.

Li, L., Bannantine, J.P., Campo, J.J., Randall, A., Grohn, Y.T. *et al.* (2017a) Identification of sero-reactive antigens for the early diagnosis of Johne's disease in cattle. *PLoS ONE* 12(9), e0184373. DOI: 10.1371/journal.pone.0184373.

Li, L., Wagner, B., Freer, H., Schilling, M., Bannantine, J.P. *et al.* (2017b) Early detection of *Mycobacterium avium* subsp. *paratuberculosis* infection in cattle with multiplex-bead based immunoassays. *PLoS ONE* 12(12), e0189783. DOI: 10.1371/journal.pone.0189783.

Mcdonald, W.L., Ridge, S.E., Hope, A.F. and Condron, R.J. (1999) Evaluation of diagnostic tests for Johne's disease in young cattle. *Australian Veterinary Journal* 77(2), 113–119. DOI: 10.1111/j.1751-0813.1999.tb11679.x.

Meissner, T., Eckelt, E., Basler, T., Meens, J., Heinzmann, J. *et al.* (2014) The *Mycobacterium avium* ssp. *paratuberculosis* specific *mptD* gene is required for maintenance of the metabolic homeostasis necessary for full virulence in mouse infections. *Frontiers in Cellular and Infection Microbiology* 4, 110.

Mikkelsen, H., Aagaard, C., Nielsen, S.S. and Jungersen, G. (2012) Correlation of antigen-specific IFN-γ responses of fresh blood samples from *Mycobacterium avium* subsp. *paratuberculosis* infected heifers with responses of day-old samples co-cultured with IL-12 or anti-IL-10 antibodies. *Veterinary Immunology and Immunopathology* 147(1–2), 69–76. DOI: 10.1016/j.vetimm.2012.04.002.

Mitachi, K., Sharma Gautam, L.N., Rice, J.H., Eda, K., Wadhwa, A. *et al.* (2016) Structure determination of lipopeptides from *Mycobacterium avium* subspecies *paratuberculosis* and identification of antigenic lipopeptide probes. *Analytical Biochemistry* 505, 29–35. DOI: 10.1016/j.ab.2016.04.001.

Möbius, P., Nordsiek, G., Hölzer, M., Jarek, M., Marz, M. *et al.* (2017) Complete genome sequence of JII-1961, a bovine *Mycobacterium avium* subsp. *paratuberculosis* field isolate from Germany. *Genome Announcements* 5(34), e00870–17. DOI: 10.1128/genomeA.00870-17.

Noh, K.T., Shin, S.J., Son, K.H., Jung, I.D., Kang, H.K. *et al.* (2012) The *Mycobacterium avium* subsp. *paratuberculosis* fibronectin attachment protein, a toll-like receptor 4 agonist, enhances dendritic cell-based cancer vaccine potency. *Experimental & Molecular Medicine* 44(5), 340–349. DOI: 10.3858/emm.2012.44.5.038.

Park, K.T., Dahl, J.L., Bannantine, J.P., Barletta, R.G., Ahn, J. *et al.* (2008) Demonstration of allelic exchange in the slow-growing bacterium *Mycobacterium avium* subsp. *paratuberculosis*, and generation of mutants with deletions at the *pknG*, *relA*, and *LSR2* loci. *Applied and Environmental Microbiology* 74(6), 1687–1695. DOI: 10.1128/AEM.01208-07.

Paustian, M.L., Amonsin, A., Kapur, V. and Bannantine, J.P. (2004) Characterization of novel coding sequences specific to *Mycobacterium avium* subsp. *paratuberculosis*: implications for diagnosis of Johne's disease. *Journal of Clinical Microbiology* 42(6), 2675–2681. DOI: 10.1128/JCM.42.6.2675-2681.2004.

Piras, C., Soggiu, A., Bonizzi, L., Greco, V., Ricchi, M. *et al.* (2015) Identification of immunoreactive proteins of *Mycobacterium avium* subsp. *paratuberculosis*. *PROTEOMICS* 15(4), 813–823. DOI: 10.1002/pmic.201400276.

Pooley, H.B., Plain, K.M., Purdie, A.C., Begg, D.J., Whittington, R.J. *et al.* (2018) Integrated vaccine screening system: using cellular functional capacity in vitro to assess genuine vaccine protectiveness in ruminants. *Pathogens and Disease* 76(3), 3fty029. DOI: 10.1093/femspd/fty029.

Poumarat, F., Le Grand, D., Gaurivaud, P., Gay, E., Chazel, M. *et al.* (2012) Comparative assessment of two commonly used commercial ELISA tests for the serological diagnosis of contagious agalactia of small ruminants caused by *Mycoplasma agalactiae*. *BMC Veterinary Research* 8(1), 109. DOI: 10.1186/1746-6148-8-109.

Robbe-Austerman, S., Krull, A.C. and Stabel, J.R. (2006) Time delay, temperature effects and assessment of positive controls on whole blood for the gamma interferon ELISA to detect paratuberculosis. *Journal of Veterinary Medicine Series B* 53(5), 213–217. DOI: 10.1111/j.1439-0450.2006.00944.x.

Santema, W., Overdijk, M., Barends, J., Krijgsveld, J., Rutten, V. *et al.* (2009) Searching for proteins of *Mycobacterium avium* subspecies *paratuberculosis* with diagnostic potential by comparative qualitative proteomic analysis of mycobacterial tuberculins. *Veterinary Microbiology* 138(1–2), 191–196. DOI: 10.1016/j.vetmic.2009.03.021.

Scandurra, G.M., Young, M., de Lisle, G.W. and Collins, D.M. (2009) A bovine macrophage screening system for identifying attenuated transposon mutants of *Mycobacterium avium* subsp. *paratuberculosis* with vaccine potential. *Journal of Microbiological Methods* 77(1), 58–62. DOI: 10.1016/j.mimet.2009.01.005.

Scandurra, G.M., de Lisle, G.W., Cavaignac, S.M., Young, M., Kawakami, R.P. *et al.* (2010) Assessment of live candidate vaccines for paratuberculosis in animal models and macrophages. *Infection and Immunity* 78(3), 1383–1389. DOI: 10.1128/IAI.01020-09.

Semret, M., Alexander, D.C., Turenne, C.Y., de Haas, P., Overduin, P. *et al.* (2005) Genomic polymorphisms for *Mycobacterium avium* subsp. *paratuberculosis* diagnostics. *Journal of Clinical Microbiology* 43(8), 3704–3712. DOI: 10.1128/JCM.43.8.3704-3712.2005.

Sergeant, E.S.G., McAloon, C.G., Tratalos, J.A., Citer, L.R., Graham, D.A. *et al.* (2019) Evaluation of national surveillance methods for detection of Irish dairy herds infected with *Mycobacterium avium* ssp. *paratuberculosis*. *Journal of Dairy Science* 102(3), 2525–2538. DOI: 10.3168/jds.2018-15696.

Shahbaaz, M., Hassan, M. I. and Ahmad, F. (2013) Functional annotation of conserved hypothetical proteins from *Haemophilus influenzae* Rd KW20. *PLoS ONE* 8(12), e84263. DOI: 10.1371/journal.pone.0084263.

Shin, A.-R., Kim, H.-J., Cho, S.N., Collins, M.T., Manning, E.J.B. *et al.* (2010) Identification of seroreactive proteins in the culture filtrate antigen of *Mycobacterium avium* ssp. *paratuberculosis* human isolates to sera from Crohn's disease patients. *FEMS Immunology & Medical Microbiology* 58(1), 128–137. DOI: 10.1111/j.1574-695X.2009.00617.x.

Slavin, Y.N., Bo, M., Caggiu, E., Sechi, G., Arru, G. *et al.* (2018) High levels of antibodies against PtpA and PknG secreted by *Mycobacterium avium* ssp. *paratuberculosis* are present in neuromyelitis optica spectrum disorder and multiple sclerosis patients. *Journal of Neuroimmunology* 323, 49–52. DOI: 10.1016/j.jneuroim.2018.07.007.

Souriau, A., Freret, S., Foret, B., Willemsen, P.T.J., Bakker, D. *et al.* (2017) Identification of new antigen candidates for the early diagnosis of *Mycobacterium avium* subsp. *paratuberculosis* infection in goats. *Research in Veterinary Science* 115, 278–287. DOI: 10.1016/j.rvsc.2017.05.025.

Souza, C., Davis, W.C., Eckstein, T.M., Sreevatsan, S. and Weiss, D.J. (2013) Mannosylated lipoarabinomannans from *Mycobacterium avium* subsp. *paratuberculosis* alters the inflammatory response by bovine macrophages and suppresses killing of *Mycobacterium avium* subsp. avium organisms. *PLoS ONE* 8(9), e75924. DOI: 10.1371/journal.pone.0075924.

Stabel, J.R. (1996) Production of γ-interferon by peripheral blood mononuclear cells: an important diagnostic tool for detection of subclinical paratuberculosis. *Journal of Veterinary Diagnostic Investigation* 8(3), 345–350. DOI: 10.1177/104063879600800311.

Sugden, E.A., Brooks, B.W., Young, N.M., Watson, D.C., Nielsen, K.H. *et al.* (1991) Chromatographic purification and characterization of antigens A and D from *Mycobacterium paratuberculosis* and their use in enzyme-linked immunosorbent assays for diagnosis of paratuberculosis in sheep. *Journal of Clinical Microbiology* 29(8), 1659–1664. DOI: 10.1128/JCM.29.8.1659-1664.1991.

Szklarczyk, D., Franceschini, A., Wyder, S., Forslund, K., Heller, D. *et al.* (2015) String v10: protein–protein interaction networks, integrated over the tree of life. *Nucleic Acids Research* 43(D1), D447–D452. DOI: 10.1093/nar/gku1003.

Thirunavukkarasu, S., Plain, K.M., Eckstein, T.M., de Silva, K. and Whittington, R.J. (2013) Cellular and humoral immunogenicity of *Mycobacterium avium* subsp. *paratuberculosis* specific lipopentapeptide antigens. *Research in Veterinary Science* 95(1), 123–129. DOI: 10.1016/j.rvsc.2013.03.002.

Triccas, J.A., Roche, P.W., Winter, N., Feng, C.G., Butlin, C.R. *et al.* (1996) A 35-kilodalton protein is a major target of the human immune response to *Mycobacterium leprae*. *Infection and Immunity* 64(12), 5171–5177. DOI: 10.1128/IAI.64.12.5171-5177.1996.

Verdier, J., Deroche, L., Allez, M., Loy, C., Biet, F. *et al.* (2013) Specific IgG response against *Mycobacterium avium paratuberculosis* in children and adults with Crohn's disease. *PLoS ONE* 8(5), e62780. DOI: 10.1371/journal.pone.0062780.

Willemsen, P., Westerveen, J., Dinkla, A., Bakker, D., Van Zijderveld, F.G. *et al.* (2006) Secreted antigens of *Mycobacterium avium* subspecies *paratuberculosis* as prominent immune targets. *Veterinary Microbiology* 114(3–4), 337–344. DOI: 10.1016/j.vetmic.2005.12.005.

Windsor, P.A. (2015) Paratuberculosis in sheep and goats. *Veterinary Microbiology* 181(1-2), 161–169. DOI: 10.1016/j.vetmic.2015.07.019.

Wu, C.W., Schmoller, S.K., Bannantine, J.P., Eckstein, T.M., Inamine, J.M. *et al.* (2009) A novel cell wall lipopeptide is important for biofilm formation and pathogenicity of *Mycobacterium avium* subspecies *paratuberculosis*. *Microbial Pathogenesis* 46(4), 222–230. DOI: 10.1016/j.micpath.2009.01.010.

Wu, Y., Li, Q. and Chen, X.-Z. (2007) Detecting protein–protein interactions by Far western blotting. *Nature Protocols* 2(12), 3278–3284. DOI: 10.1038/nprot.2007.459.

Wynne, J.W., Bull, T.J., Seemann, T., Bulach, D.M., Wagner, J. *et al.* (2011) Exploring the zoonotic potential of *Mycobacterium avium* subspecies *paratuberculosis* through comparative genomics. *PLoS ONE* 6(7), e22171. DOI: 10.1371/journal.pone.0022171.

Wynne, J.W., Shiell, B.J., Colgrave, M.L., Vaughan, J.A., Beddome, G. *et al.* (2012) Production and proteomic characterisation of purified protein derivative from *Mycobacterium avium* subsp. *paratuberculosis*. *Proteome Science* 10(1), 22. DOI: 10.1186/1477-5956-10-22.

Yu, N.Y., Wagner, J.R., Laird, M.R., Melli, G., Rey, S. *et al.* (2010) PSORTb 3.0: improved protein subcellular localization prediction with refined localization subcategories and predictive capabilities for all prokaryotes. *Bioinformatics* 26(13), 1608–1615. DOI: 10.1093/bioinformatics/btq249.

Zhong, L., Taylor, D., Begg, D.J. and Whittington, R.J. (2011) Biomarker discovery for ovine paratuberculosis (Johne's disease) by proteomic serum profiling. *Comparative Immunology, Microbiology and Infectious Diseases* 34(4), 315–326. DOI: 10.1016/j.cimid.2011.03.001.

9 Host–Pathogen Interactions and Intracellular Survival of *Mycobacterium avium* subsp. *paratuberculosis*

Paul Coussens[1]*, Justin L. DeKuiper[1], Fernanda M. Shoyama[1], Evan Brenner[1], Elise A. Lamont[2], Edward Kabara[1] and Srinand Sreevatsan[1]
[1]Michigan State University, East Lansing, Michigan, USA; [2]University of Minnesota, St Paul, Minnesota, USA

9.1 Introduction

In *The Art of War*, Sun Tzu states that all warfare is based on deception (Tzu, 1971, p. 66). Mycobacteria, specifically *Mycobacterium avium* subsp. *paratuberculosis* (*MAP*), are no exception to this rule. Mycobacteria represent a group of closely related acid-fast bacilli that encompass a wide range of host tropisms and diseases (Harris and Barletta, 2001; Corn *et al.*, 2005; Motiwala *et al.*, 2006; Behr and Kapur, 2008). In all cases, pathogenic mycobacteria deceive the host immune system by residing within host cells. Among mycobacteria, there are two important pathogenic complexes: the *Mycobacterium tuberculosis* complex and the *M. avium* complex (MAC) (detailed in Chapter 5, this volume). The *M. tuberculosis* complex is more readily recognized due to its implications in human health and includes the major pathogens *M. tuberculosis* and *Mycobacterium bovis*. MAC organisms, despite their genetic similarity, elicit different diseases in both animals and humans, including infections of the lung, lymph nodes, bones, skin and gastrointestinal tract (Harris and Barletta, 2001; Behr and Kapur, 2008). Historically, research on MAC organisms has been limited; however, interest in this group is rapidly increasing due to associations with opportunistic infections (*M. avium* subsp. *avium* and *M. avium* subsp. *hominissuis*) in HIV/AIDS patients and a potential aetiological agent in Crohn's disease (*MAP*) (Prohászka *et al.*, 1999; Richter *et al.*, 2002; Ghadiali *et al.*, 2004; Behr and Kapur, 2008; Bentley *et al.*, 2008; Waddell *et al.*, 2008) (see also Chapter 3, this volume). Tuberculous mycobacteria (*M. tuberculosis*/*M. bovis*) and non-tuberculous mycobacteria (*M. avium*/*MAP*) have their differences, mostly in route of infection and lesion locations. Molecularly, there is much that can be learned by comparing one species with the next. Although the various mycobacterial organisms may differ in how they scavenge iron and in growth parameters, the nature and source of infection associated inflammation, cytokines triggered in response to infection and the cells involved in early responses may all be quite similar.

9.2 Persistence: The Protracted War

Sun Tzu also writes 'There has never been a protracted war from which a country has benefited' (Tzu, 1971, p. 73). In the case of pathogens and

*Corresponding author: coussens@msu.edu

© CAB International 2020. *Paratuberculosis: Organism, Disease, Control, 2nd Edition* (eds M.A. Behr *et al.*)

their hosts, one could substitute 'host' for country and 'infection' for war. A hallmark of successful pathogens is the ability to persist within a host for an indefinite period of time. Among the mycobacteria, *MAP* is a leader in this paradigm, often persisting in the ruminant host for 2–5 years before onset of clinical disease. *MAP* succeeds by efficiently invading, replicating and laying siege to the host all while potentially surviving an innate response by the immune system (Harris and Barletta, 2001; Frie *et al.*, 2017). *MAP* employs several strategies to enter host cells and causes paratuberculosis, a chronic inflammatory disorder of the gastrointestinal tract in ruminants (Clarke, 1997; Harris and Barletta, 2001; Whittington and Sergeant, 2001; Tiwari *et al.*, 2006; Zhu *et al.*, 2008).

Paratuberculosis may be categorized into three separate states: subclinical infection, subclinical disease and clinical disease (Harris and Lammerding, 2001; Chacon *et al.*, 2004). Since *MAP* is primarily transmitted by the faecal–oral route, subclinically infected cattle are of particular concern, as shedding and spread may continue unabated until clinical signs surface (Harris and Barletta, 2001; Harris and Lammerding, 2001; Whittington *et al.*, 2001, 2004; Chacon *et al.*, 2004; Crossley *et al.*, 2005; Grewal *et al.*, 2006; Tiwari *et al.*, 2006). Once ingested, *MAP* gains entrance to subepithelial macrophages by invasion into the lamina propria via microfold cells (M cells) (Momotani *et al.*, 1988; Sigurðardóttir *et al.*, 1999, 2001, 2004; Tiwari *et al.*, 2006; Wu *et al.*, 2007a; Ponnusamy *et al.*, 2013). Recent studies using a bovine ileal-loop model suggested there is a clear early response to *MAP* but it is still unclear if early responses result from direct T-cell receptor (TCR)/Toll-like receptor (TLR) signalling, other pattern recognition receptors or traditional antigen presentation (Philip Griebel, personal communication).

Interplay between *MAP* and the macrophage may dictate disease progression and outcome; therefore, it is of extreme importance to understand early macrophage responses in order to elucidate pathogenesis. Lastly, the different effects that *MAP* has on epithelial cells during different stages of their pathogenesis and their contribution to the overall immune response is important to consider. Preliminary studies using Madin-Darby bovine kidney cells (MBDKs) suggest an additional role for epithelial cells in directing a proinflammatory T-cell response (Th17) by secreting IL-23 in the presence of *MAP*. These topics will be considered later in this chapter.

The most recognizable feature of paratuberculosis is the atypical diffuse granuloma formation, which is found in the mid- and distal segments of the small intestine (Harris and Barletta, 2001; Tiwari *et al.*, 2006). These atypical granulomatous lesions are thought to represent a late response by the host to control and limit *MAP* spread to the rest of the intestine and draining lymph nodes. However, it has been suggested that mycobacteria may take advantage of granulomas to recruit new macrophages to the site of infection and allow for mycobacterial travel through the granuloma (Davis and Ramakrishnan, 2009). Thus *MAP* may use granulomas as bridges to infect new portions of the intestine as well as other organs, including the mammary gland and mesenteric lymph node (Sweeney *et al.*, 1992a, b, Sweeney *et al.*, 2006; Patel *et al.*, 2006). Clinical signs, including malabsorption, malnutrition and decreased milk yield, result within 2–5 years of infection, which may lead to death through either a direct cause or culling (Harris and Barletta, 2001; Tiwari *et al.*, 2006).

9.3 Secret Operations: Intestinal Epithelial Cells, Macrophages and *MAP*

Sun Tzu writes that 'Secret operations are essential in war; upon them the army relies to make its every move' (Tzu, 1971, p. 149). The ability of *MAP* to infiltrate and enter host cells without overtly alarming the immune system may explain why it is so well adapted to its ruminant host (Fig. 9.1). As previously mentioned, *MAP* is spread by the faecal–oral route and gains entry to intestinal walls through the small intestinal mucosa via M cells or villous epithelial cells overlying Peyer's patches in gut-associated lymphoid tissue (Sigurðardóttir *et al.*, 1999, 2004; Whittington and Sergeant, 2001; Whittington *et al.*, 2004; Crossley *et al.*, 2005; Grewal *et al.*, 2006; Tiwari *et al.*, 2006). M cells represent a primary target for *MAP* infection, which may be

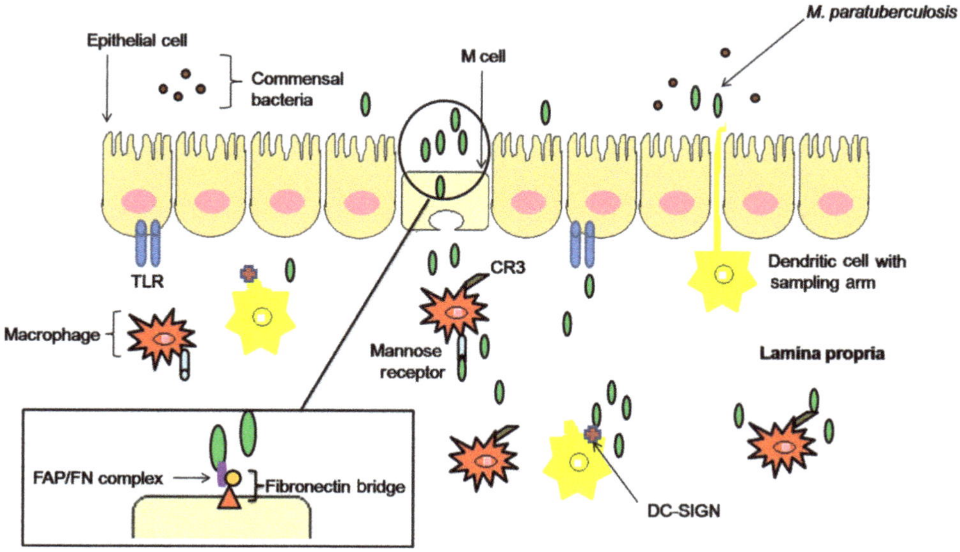

Fig. 9.1. Intestinal cell wall and macrophage invasion tactics used by *Mycobacterium avium* subsp. *paratuberculosis (MAP)*. *MAP* preferentially invades M cells by creating a fibronectin bridge and causes subsequent invasion of subepithelial macrophages. Entry into the macrophage is accomplished by ManLAM binding to CR3 and mannose receptors. *MAP* invasion into the lamina propria may also be gained through intestinal epithelial cells by an unknown mechanism. Dendritic cells may also transport *MAP* inside the lamina propria during sampling through tight junctions. *MAP* interaction with the dendritic cell receptor, DC-SIGN, may prime and promote a Th2 response. TLR, Toll-like receptor.

due to the lack of lysosomes and hydrolytic enzymes present in these cells (Miller *et al.*, 2007). Therefore, many antigenic properties of *MAP* would remain unaltered after passing through M cells. It is well established that fibronectin (FN) attachment proteins present on *MAP* facilitate FN binding of the bacterium, which in turn forms a FN bridge with β1 integrins located on intestinal epithelial cells (Sigurðardóttir *et al.*, 1999; Pieters, 2001; Secott *et al.*, 2001, 2004). More recent studies have shown that passing through epithelial cells alters *MAP*'s characteristics and changes the response of macrophages (Everman *et al.*, 2015). Preferential binding of *MAP* to M cells may be explained by the high density of β1 integrin present on the luminal surface of these cells in comparison with other cell types, such as enterocytes. Villous epithelial cell invasion is due to an unknown FN-independent mechanism (Secott *et al.*, 2001, 2004). However, *MAP* preference for M cells appears to require more than just FN–integrin interactions, since the closely related *M. avium* subsp. *avium* enters the intestinal wall via absorptive epithelial cells, despite presence of FN attachment protein genes (Secott *et al.*, 2002).

MAP primarily resides within macrophage cells; however, some studies suggest that epithelial cell processing prior to macrophage exposure may enhance efficiency of invasion into macrophages. *MAP* exposed to Mac-T cells, a mammary epithelial cell line, displayed increased invasion efficiency in subsequent infections of MDBK cells. Increased invasion efficiency by prior exposure may be due to upregulation of *MAP*3464, encoding an oxidoreductase that activates host Cdc42 and Rho internalization pathways (Patel *et al.*, 2006; Alonso-Hearn *et al.*, 2008). DNA microarray analysis of *MAP* 24h post-infection of Mac-T cells revealed upregulation of 20 *MAP* genes related to regulatory, metabolic and virulence-associated functions compared with *MAP* grown in Middlebrook 7H9 broth cultures. From this gene set, Patel *et al.* (2006) hypothesized that a 35-kDa *MAP* protein (major membrane protein, or MMP), which previously was shown to enhance invasion in epithelial cells, may be upregulated in response to prior exposure to Mac-T cells (Bannantine *et al.*, 2003). Following up, Abdellrazeq *et al.* (2018) observed cytotoxic T lymphocyte (CTL) responses to a *MAP* Δ*relA* strain and found this response targeted MMP. This CTL-mediated response was recapitulated by stimulating bovine dendritic cells *ex vivo* with MMP alone, showing killing of intracellular *MAP* and signifying the importance of MMP in infection. This group has more recently published a nanoparticle-based peptide vaccine incorporating MMP to provoke CD8 T-cell-mediated *MAP* killing (Abdellrazeq *et al.*, 2019).

Following M-cell invasion into the subepithelial dome, *MAP* may encounter dendritic cells and/or macrophages. It is well established that *MAP* interacts with intestinal dendritic cells through its cell wall glycolipid mannosylated lipoarabinomannan (ManLAM) and the dendritic cell receptor DC-SIGN (Józefowski *et al.*, 2008). *MAP* may use intestinal dendritic cell invasion as a strategic manoeuvre, since the primary function of intestinal dendritic cells is to sample and present commensal bacteria through tight junctions to the gut-associated lymphoid tissue. Thus, *MAP* would be able to overcome tight junction barriers and be directly transported to the lamina propria to interact with subepithelial macrophages. Furthermore, ManLAM–DC-SIGN interaction may prime *MAP* to direct a Th2 response, which would lead to immune subversion, as suggested for *M. tuberculosis* (Józefowski *et al.*, 2008).

Although murine monocyte-derived dendritic cells (MoDCs) are capable of phagocytizing *MAP* and secrete IL-10 after doing so (Basler *et al.*, 2013), not much is known about their interaction. Inside dendritic cells (DCs), *M. tuberculosis* vacuoles cannot access exogenous material or biosynthetic pathways and fail to replicate (Tailleux *et al.*, 2003), which is also seen with *Mycobacterium avium* (Salte *et al.*, 2011). Like macrophages however, the vacuole is arrested and does not acidify (Tailleux *et al.*, 2003). *MAP*-infected DCs also fail to mature properly when cultured in media previously used for *MAP*/MDM culture (Basler *et al.*, 2013). DC's inability to mature in the presence of *MAP* was also observed in tissues (Lei *et al.*, 2008).

As with dendritic cells, ManLAM from *MAP* is capable of interacting with macrophage cell surface receptors. The best-documented interaction is between *MAP* and the mannose receptor, which enhances macrophage phagocytosis of *MAP* (Pieters, 2001; Gatfield and Pieters, 2003; Rowe and Grant, 2006; Souza *et al.*, 2007a). Upon entry into macrophage cells, there is simultaneous replication of *MAP* and bacterial killing by the host, spurring on an initial Th1-like response (Rowe and Grant, 2006; Alonso *et al.*, 2007; Woo *et al.*, 2007). Initial killing of *MAP* may be due to a rapid phagosome acidification response from the host, allowing phagosome–lysosome fusion to occur in some cells. The end result of phagosome–lysosome fusion would presumably be destruction of *MAP* and presentation of antigens to T cells via major histocompatibility complex (MHC). However, most phagosomes containing *MAP* and other pathogenic mycobacteria fail to mature. Studies of interactions between *M. tuberculosis* ManLAM and macrophages indicate that ManLAM is indispensable for blockage of phagosome maturation (Russell *et al.*, 2002; Yates and Russell, 2005). The ability of *MAP* and other mycobacteria to inhibit phagosome–lysosome fusion is essential to prevent pathogen awareness by the host immune system, thus allowing *MAP* to hijack macrophage resources and persist unabated. An active role for *MAP* in preventing phagosome–lysosome fusion is supported by the observation that live *MAP* is able to persist within phagosomes for 15 days, while phagosome function is not interrupted following uptake of killed *MAP* (Kuehnel *et al.*, 2001). Other macrophage receptors that are important for *MAP* binding include those for complement, immunoglobulin, transferrin, scavengers and surfactant protein A (Pieters, 2001; Souza *et al.*, 2007a). *MAP* binding to one complement receptor prevents activation of an oxygen burst and is abrogated with addition of monoclonal antibodies (Sigurðardóttir *et al.*, 2004).

The efficiency with which *MAP* enters macrophage cells appears to differ with respect to *MAP* genotype, such that species-specific variation is observed (Gollnick *et al.*, 2007). Despite variation in invasion efficiency, *MAP* strains all seem to employ a similar infection protocol or modus operandi. Zhu *et al.* (2008) determined that expression patterns from three *MAP* strains of different types based on short sequence repeats showed upregulation of 27, 22 and 35 genes, respectively, when isolated from infected bovine monocyte-derived macrophages (MDMs) at 48 and 120h post-infection. Many of the genes on these lists were similar or had overlapping cellular functions. Pathway analysis categorized gene functions related to small-molecule degradation, energy metabolism, amino acid biosynthesis, lipid biosynthesis, broad regulatory functions, synthesis and modification of macromolecules, cell envelope, transport/binding proteins, virulence, antibiotic production and resistance, and conserved hypothetical proteins. All three *MAP* strains upregulated *MAP4041c* and *MAP4281*, which are suggested to play a role in protein transport and act as insertion elements, as well as genes related to lipid degradation, membrane transportation and DNA repair at 48h. Taken together, this comparative transcriptional analysis suggests that diverse *MAP* genotypes use a similar modus operandi for survival in the host.

Previous studies have investigated the ability of *MAP* strains to regulate expression of MHC molecules during macrophage infection. Although *MAP* infection of J774 murine macrophages did not affect expression of MHC class II molecules, antigen presentation decreased, which may be due to *MAP* limitation of antigen processing (Kuehnel *et al.*, 2001). However, these results conflict with those demonstrating downregulation of MHC class I and class II molecules in bovine macrophages infected with live and killed *MAP* (Weiss *et al.*, 2001). It is exciting to speculate on the potential that one or a number of hypothetical genes identified by Zhu *et al.* (2008) may be responsible for *MAP* regulation of MHC class I and class II molecules and control of antigen processing.

While immediate activation of MAPK signalling by *MAP* is of obvious importance in the response of macrophages to infection, studies focused on this cannot tell us what effect *MAP* might be having on the ability of macrophages to respond to T cells and to activate these cells to respond to infection. One of the most critical components of macrophage–T-cell interactions is engagement of CD40 on macrophages by CD154 (CD40 ligand) on activated T cells. CD40–CD154 binding is one of the major mechanisms leading to macrophage activation via T-cell interactions, and soluble CD154 can mimic many of the processes observed when T cells activate macrophages (Grewal and Flavell, 1996, Grewal and Flavell, 1998; Grewal *et al.*, 1997). CD40 is a member of the TNF receptor superfamily and is expressed on numerous cell types, including B cells, monocytes/macrophages, endothelial cells, dendritic cells, fibroblasts and vascular smooth muscle cells (Clark *et al.*, 1996; van Kooten and Banchereau, 1996). Accordingly, typically over 70% of bovine MDM stain positive for cell surface CD40 after 7 days of maturation in culture (Chiang *et al.*, 2007). In monocytes and macrophages, CD40 signalling leads to secretion of inflammatory cytokines including IL-12, chemokines including β-chemokines (Stout and Suttles, 1996; di Marzio *et al.*, 2000) and matrix metalloproteinases (Malik *et al.*, 1996); induction of inducible nitric oxide synthase (iNOS); production of nitric oxide (Tian *et al.*, 1995; van Kooten and Banchereau, 1996); enhanced cell survival; and induction of co-stimulatory molecules (Kiener *et al.*, 1995). T cells derived from CD154-deficient mice are impaired in their ability to induce macrophage effector functions (Stout and Suttles, 1996), and consequently these mice are highly susceptible to intracellular pathogens that would otherwise have been cleared by an appropriate T-cell–macrophage interaction (Soong *et al.*, 1996).

Studies on CD40 signal transduction have resulted in a complex picture of different mediators and pathways involved. Two major signalling pathways are activated downstream of CD40, which both involve activation of latent transcription factors. One pathway involves activation of the inhibitor of nuclear factor kappa B kinase complex, leading to nuclear translocation of active nuclear factor kappa B. The second mechanism is activation of the MAPK pathway, a cascade of phosphorylation events that primarily results in post-transcriptional activation of transcription factors like cyclic adenosine monophosphate (cAMP)-response element binding protein, activating transcription factor, Ets, and AP-1 (van Kooten and Banchereau, 1996, 2000). Both pathways synergize in inflammatory gene expression, including expression of IL-12p40, iNOS, IL-6, IL-8 and TNFα. Not surprisingly, CD40–CD154 signalling is a target for many intracellular pathogens. For example, Mathur *et al.* (2004)

demonstrated that *Leishmania major*, an intracellular parasite causing leishmaniasis in humans, is able to inhibit CD154–CD40-mediated IL-12p40 and iNOS gene expression in murine peritoneal macrophages. This blockade appears to involve interference with activation of two main members of the MAPK pathway, p38 and ERK1/2 (Awasthi *et al.*, 2003; Mathur *et al.*, 2004).

MAP-infected macrophages are defective in some aspects of CD40 signalling (Sommer *et al.*, 2009). In uninfected macrophages, CD40–CD154 binding results in large increases in TNFα, IL-6, IL-10, IL-8, IL-12p40 and iNOS gene expression within 6 h. In *MAP*-infected macrophages, TNFα and IL6 gene expression following CD40–CD154 binding is relatively unaffected. In contrast, *MAP*-infected macrophages fail to activate expression of IL-12p40 and iNOS gene expression. This is a critical difference, since IL-12 is a major driving force for development of an appropriate Th1-like response and production of IFNγ by T cells. In macrophages, iNOS activity and production of reactive nitrogen species is a major mechanism used to kill phagocytosed bacteria. For an intracellular bacterium such as *MAP*, limiting production of IL-12 and iNOS would ensure survival and development of an inappropriate immune response, particularly in the face of enhanced IL-10 production. It has also been suggested that failure to properly activate and/or engage T cells could lead to development of regulatory T cells, which would further reduce Th1-like immune activity against *MAP* (de Almeida *et al.*, 2008). Results of Sommer *et al.* (2009) are also consistent with observations *in vivo* (Coussens *et al.*, 2004), where *MAP*-infected intestinal tissues contained elevated levels of IL-10 but not IL-12p40 or IL-12p35. These tissues also contain elevated levels of TGFβ (Coussens *et al.*, 2004; Khalifeh and Stabel, 2004). TGFβ can be produced by regulatory Th3 cells and by regulatory γδ T cells, among others. These defects may be more important in early responses to *MAP* infection however, since despite the presence of IL-10, T regulatory cells seem to be largely absent in later stage *MAP*-infected tissues (Roussey *et al.*, 2016).

9.4 Strategic Excellence: *MAP* Gene Expression Programmes Following Host Cell Entry

Sun Tzu states 'What is of supreme importance in war is to attack the enemy's strategy' (Tzu, 1971, p. 77). The host relies on phagocytosis and phagosome maturation for infection clearance, but as noted above, *MAP* undermines these fundamental mechanisms through stress resistance and interference with host signalling.

Several sigma factors (e.g. *sigH* and *sigE*) in *MAP* were differentially co-regulated with a large number of genes, depending on the type of stressor applied (Wu *et al.*, 2007b). This was further explored by Ghosh *et al.* (2013), clarifying SigH as a stress-response sigma factor for early (<24 h) macrophage infections, and SigE induction only after 24 h. The genetic network affected was also confirmed by RNA-seq and quantitative reverse transcription polymerase chain reaction (qRT-PCR). The *MAP ΔsigH* mutant showed defects in survival in heat and diamide stress, as well as in a mouse model of infection. More revealingly, a *sigH* knockout in *M. tuberculosis* showed complete attenuation of infection in non-human primates, again supporting a conserved mycobacterial strategy for infection and the essential nature of the stress-response sigma factor SigH (Mehra *et al.*, 2012).

Wu *et al.* (2007b) showed acidic pH induced differential regulation of a large number of genes (n = 597), suggesting high sensitivity of *MAP* to acidic environments. Generally, responses to heat shock, acidity and oxidative stress were similar in *MAP* and *M. tuberculosis*, suggesting common pathways for mycobacterial defence against stressors. Subsequently, Wu *et al.* (2007b) conducted a functional analysis using a mouse model and several *MAP* mutants where differentially regulated genes were ablated. Bacterial and histopathological examinations indicated the attenuation of all gene mutants. Similarly, Vandal *et al.* (2008) utilized an existing transposon mutant library in *M. tuberculosis*, studied acid-susceptible disruptions and showed that nearly all disruptions showing acid hypersensitivity were directly linked to mycobacterial cell wall integrity. This group further explored an exception to this pattern in *rv3671c*, a then-putative gene predicted to encode a serine protease of unknown function. Disruption of this gene led to uncharacteristic death of *M. tuberculosis* in media at pH 4.5, as well as inside IFN-γ-activated macrophages. Plasmid-based complementation of *rv3671c* showed restoration of the wild-type phenotype, supporting a direct role by gene loss in acid susceptibility. Vandal *et al.* (2008) also infected mice by aerosol and observed dramatic attenuation in the *rv3671c*-disrupted strain that was absent when complemented. *Mycobacterium tuberculosis* without intact *rv3671c* yielded bacterial clearance of up to a 3.6-log reduction by 98 days, in contrast with no reduction at 1 year post-infection for wild-type strains. Gross pathology showed marked reduction in pulmonary lesions by the disrupted strain consistent with attenuation. This serine protease was labelled **myco**bacterial **a**cid **r**esistance **p**rotease (MarP), and homologues are present across *Mycobacterium*. Further research showed *ΔmarP* attenuation in zebrafish larvae for *M. marinum* (Levitte *et al.*, 2016), as well as further exploring its potential targets (Small *et al.*, 2013; Botella *et al.*, 2017). *MAP* possesses a homologue of this target (*MAP0403c* or *MAP_RS02055*), and investigation uncovered rapid *MAP marP* gene expression in bovine MDMs as well as bovine mammary epithelial (MAC-T) cells. This expression was closely tied to phagosome acidification, and if host cells were pre-treated with Bafilomycin A1, an inhibitor of phagosomal acidification, *marP* gene expression was not observed (Kugadas *et al.*, 2016). This pattern was not observed for non-pathogenic *M. smegmatis*' more distant putative *marP* homologue, supporting a conserved mechanism responsible for intracellular survival and persistence of pathogenic mycobacteria. Botella *et al.* (2017) provided evidence for peptidoglycan hydrolase RipA being a target of MarP-digestion, which ties back again to the importance of cell wall maintenance for acid stress-survival. The loss of either *marP* or *ripA* in *M. tuberculosis* led to a cell elongation phenotype where dividing cells failed to separate, but *M. smegmatis marP* was able to rescue the *M. tuberculosis ΔmarP* strain when expressed *in trans*. This supports a conserved function for MarP in proper cell division, but it is not known what other targets may exist in the *Mycobacterium* proteome.

Unlike other mycobacteria, *MAP* has special iron requirements that enhance its survival while inside the host. For optimal growth *in vitro*, *MAP* requires supplementation of the siderophore mycobactin J. Whole genome sequencing of *MAP* K-10 provided one explanation for this dependency, revealing a truncation of the *mbtA* gene. It has been suggested that this truncation impairs the production of mycobactin from the *mbtA-J* operon (Li *et al.*, 2005; Wang *et al.*, 2014b). Despite this truncation, Zhu *et al.* (2008) showed that *MAP* is still able to transcribe mycobactin synthesis genes inside macrophages. To corroborate these findings, Janagama *et al.* (2010) described the upregulation of several genes responsible for iron acquisition in infected tissues, including genes responsible for mycobactin biosynthesis.

Iron is vital to fundamental biological processes; however, high intracellular concentrations of free iron are toxic to bacteria. As such, cells have developed tightly regulated processes for intracellular metal homeostasis (Eckelt *et al.*, 2014). Bacteria control metal homeostasis by activating a set of genes regulated by metal-sensing transcription factors known as metalloregulatory proteins (Chandrangsu *et al.*, 2017). In prokaryotes, there are two major families of metalloregulators: diphtheria toxin (DtxR) and ferric uptake regulator (FUR) (Hantke, 1981). In 2009, Janagama and others identified and characterized MAP2827, an iron-dependent regulator (IdeR) in *MAP*. A member of the DtxR protein family, IdeR is involved in regulatory mechanisms to acquire, store or prevent excess accumulation of iron. The authors were able to confirm that MAP2827 was in fact IdeR and regulates genes involved in iron acquisition (*mbtB*) and iron storage (*bfrA*) (Janagama *et al.*, 2009). However, *in vitro* iron stress showed that IdeR regulation is strain dependent – while IdeR from *MAP* cattle strain K-10 regulates mycobactin synthesis and storage genes similar to IdeR from *M. tuberculosis*, IdeR from *MAP*-S strain S397 shows deficiency in iron storage function, resulting in a strain more sensitive to iron fluctuations (Janagama *et al.*, 2010).

In addition to IdeR, the *MAP* genome also encodes three putative FUR boxes, which are iron-regulated transcriptional control motifs present in a *MAP*-specific genomic island (Stratmann *et al.*, 2004). To date, there is no information about the potential roles of FUR in *MAP*. In 2009, Alexander and others confirmed that the *MAP* genome also contains another putative metal transport operon, LSP[P]15, which includes a FUR-like transcriptional regulator (Alexander *et al.*, 2009). Additionally, using transposon mutagenesis, Wang *et al.* (2016) demonstrated that this specific operon provides an alternative iron uptake system. However, a deeper investigation is necessary into how FUR in *MAP* behaves in an organism where IdeR is the protagonist. As a key virulence determinant, iron regulation in *MAP* and its role in pathogen survival and infection are important areas of research that may lead to advances in the ability to culture *MAP* and improve diagnostic techniques.

During *MAP* replication within the macrophage phagosome, host cell resources and space become limited. Therefore, *MAP* egresses from the cell by inducing apoptosis and is able to infect neighbouring macrophages within the subepithelial dome. Infected macrophages migrate into local lymphatics concurrent with disease progression, resulting in bacterial spread to regional lymph nodes, including mesenteric lymph nodes (Ayele *et al.*, 2004; Sivakumar *et al.*, 2005; Tiwari *et al.*, 2006; Zhu *et al.*, 2008). *MAP* is able to replicate within regional lymph nodes. Macroscopic lesions develop in the intestine and mesenteric lymph nodes, causing the intestinal wall to thicken and become corrugated (Sivakumar *et al.*, 2005). Infection may spread to the supramammary lymph node and mammary gland and therefore cause *MAP* contamination in colostrum and milk (Sweeney *et al.*, 1992a, b). Several studies have shown *MAP* can survive the commercial pasteurization and processing and therefore poses challenges to containment by presence in commercial calf milk replacer (Grant *et al.*, 2017), and with respect to the potential zoonotic nature of *MAP*, may pose a significant public health concern for humans susceptible to Crohn's disease (Sweeney *et al.*, 1992b; Grant, 1998).

9.5 Avoid What is Strong, Strike What is Weak: *MAP* Targets Early Host Responses to Prevent Immune Counterattack

Sun Tzu writes 'Those skilled in war subdue the enemy's army without battle ' (Tzu, 1971, p. 79). After infection, *MAP* does not attempt to fight the host immune response head-on, but instead undermines its foundation before it can begin. As with *MAP* gene and protein expression studies, technological advances have allowed us to begin unravelling the diverse effects *MAP* has on infected host cells.

Rapidly after phagocytosis, pathogenic mycobacteria induce critical signal transduction failures, impairing normal phagosomal maturation. Normally, Rab5 and phosphatidylinositol 3-phosphate (PI3P) coordinate through the Rab5 effector early endosomal antigen 1 (EEA1) for endosomal sorting and proper phagosomal maturation. Kelley and Schorey (2004) suggested the mycobacterial manipulation of phagocytosis is multifaceted and active – after artificially increasing PI3P levels in macrophages, live *M. avium hominissuis* 104 continued to impair normal phagosomal trafficking, while killed *M. avium hominissuis* was capable of being processed normally. *Mycobacterium tuberculosis* has been shown to secrete lipid phosphatase SapM and degrade PI3P, sabotaging the host process (Vergne *et al.*, 2005). *MAP* has a putative homologue of this gene (K-10: MAP_RS17645), but it has not been investigated to date. Additionally, Rab7 – required for fusion and development of the phagolysosome – is not effectively recruited despite the presence of Rab5 in infected phagosomes. In *M. bovis* BCG, this recruitment failure was proposed as a key mechanism by which mycobacterial pathogens freeze the phagosomal maturation process in the earliest stages (Via *et al.*, 1997). Later research with *M. bovis* BCG reported a *Mycobacterium*-secreted factor that inhibited Rab7-interacting lysosomal protein (RILP), interfering with Rab7 function across the entire cell and not merely in the *Mycobacterium*-containing compartment (Sun *et al.*, 2007). A similar interference was observed with live but not killed *MAP* in human monocytes (THP-1) (Keown *et al.*, 2012).

Many studies have examined gene expression patterns and intracellular signalling inside *MAP*-infected macrophages relative to uninfected macrophages. In other cases, gene expression, particularly for cytokine genes, has been examined in *MAP*-infected intestinal tissues and draining lymph nodes. Specific studies targeting intracellular signalling inside infected macrophages have been conducted. These investigations have greatly enhanced our understanding of *MAP* pathobiology, particularly when combined with *MAP* genome-sequencing efforts and the ability to genetically modify *MAP*. The effect of *MAP* on intracellular signalling in infected macrophages was highlighted in a study by Murphy *et al.* (2006), who used a bovine immune microarray. They examined differential gene expression in resting macrophages and in macrophages infected with either *MAP* or the closely related *M. avium hominissuis* 104. Although *M. avium hominissuis* can produce serious infections in immune-compromised humans, it is non-pathogenic in ruminants and is readily cleared by an efficient immune response. In general, both mycobacteria activated gene expression in macrophages 24 h post-infection (Murphy *et al.*, 2006). However, macrophage responses to *M. avium hominissuis* were consistently more robust than to *MAP*. Of particular interest, over 41% of the differentially expressed genes in *MAP*-infected cells were members of, regulators of or regulated by the mitogen-activated protein kinase (MAPK) pathway. In keeping with gene expression patterns, *M. avium hominissuis* caused a more robust activation of p38 MAPK and extracellular regulated kinase (ERK1/2), two of the major MAPK family members. This response was also longer lived in *M. avium hominissuis*-infected macrophages relative to *MAP*-infected macrophages (Murphy *et al.*, 2006). Given that activation of p38 MAPK and ERK1/2 occurred within 15 min of *MAP* or *M. avium hominissuis* infection, this response was probably mediated via Toll-like receptor signalling, supported by Weiss *et al.* (2008) (see also Weiss and Souza, 2008). The robust and sustained activation of macrophages by *M. avium hominissuis* has led to speculation that this may be one reason why infections with this organism are cleared while those with *MAP* are not. Another hypothesis is that *MAP* actively limits signalling via the MAPK pathway in infected macrophages. Similarly, another study involving systems analysis for host gene expression triggered by *MAP* and *M. avium avium* infection in a bovine jejuno-ileal loop model showed dramatic differences between the two pathogens, with *MAP* impairing the host in phagosome processing, calcium signalling, weakening the intestinal mucosal barrier and biasing the host towards an ineffectual Th2-type immune response, while the host profile for *M. avium avium* supports minimal disturbance and a transient infection (Khare *et al.*, 2016).

Several studies suggest that *MAP* subverts the ability of infected macrophages to react to normal T-cell signalling, resulting in failure of macrophages to activate and destroy *MAP*. It has been proposed that this disruption may stem from alteration of the host's ability to receive signals through the CD154–CD40 system (Sommer *et al.*, 2009). The resulting cytokine response favours an inappropriate Th2-like activity, including expression of interleukin (IL)-10, and is linked with a shift towards clinical disease (Hussain *et al.*, 2016). The failure of normal signalling between *MAP*-infected macrophages and T cells, and the absence of proper macrophage activation, are central to disease progression (Voo *et al.*, 2009). These ideas have been developed into models for immune response to *MAP* (Coussens, 2001, 2004; de Almeida *et al.*, 2008; Khare *et al.*, 2009, 2012; Arsenault *et al.*, 2014; Khare *et al.*, 2016).

In one study, it was shown that *MAP*-infected tissues contained high levels of IL-1α and TRAF1 (tumour necrosis receptor-associated factor-1) mRNA and protein (Aho *et al.*, 2003). High-level expression of TRAF1 and IL-1α was traced to *MAP*-infected macrophages infiltrating these tissues. Subsequently, it was demonstrated that *MAP* infection of MDMs *in vitro* also enhanced TRAF1 and IL-1α mRNA and protein (Chiang *et al.*, 2007). These results were significant because TRAF proteins are integral intermediates in tumour necrosis factor (TNF) receptor superfamily signalling systems, including CD40, Fas (CD95) and TNF receptors 1 and 2 (Bradley and Pober, 2001). These are all systems critical to macrophage function and activation, cytokine expression, T-cell signalling and induction of apoptosis.

This hypothesis is in line with studies demonstrating that pathogenic mycobacteria phagocytosed by macrophages pre-activated with interferon-γ (IFN-γ) or tumour necrosis factor-α (TNFα), are destroyed through normal phagosome maturation (Flynn *et al.*, 1993; Bonecini-Almeida *et al.*, 1998; Flórido *et al.*, 1999). However, treatment of macrophages with these same agents after infection fails to result in efficient destruction of mycobacteria (Denis *et al.*, 1990; Robertson and Andrew, 1991). Therefore, interfering with normal macrophage activation pathways may be a key element in persistent infection and survival of mycobacteria, including *MAP*.

Infection of macrophages with *MAP* leads to dramatic upregulation of IL-10 (Weiss *et al.*, 2005; de Almeida *et al.*, 2008). This is significant since IL-10 can severely dampen proinflammatory immune responses, which are critical to clearance of *MAP* and other intracellular infections. Activation of IL-10 gene expression in *MAP*-infected macrophages is critically dependent upon rapid signalling through p38 MAPK (Hussain *et al.*, 2016). This signal is reported to be mediated via interactions between *MAP* and Toll-like receptor 2 (Souza *et al.*, 2008; Weiss *et al.*, 2008; Weiss and Souza, 2008), and the same interactions have been noted with *M. tuberculosis* as well (Benson and Ernst, 2009). Blocking p38 MAPK signalling with specific inhibitors severely limits *MAP*-mediated increases in IL-10 mRNA (Sommer *et al.*, 2009). Inhibition of p38 MAPK also enhances the ability of macrophages to activate, acidify phagosomes and destroy *MAP* (Souza *et al.*, 2007c). Of interest, inhibition of SAPK/JNK also enhanced the ability of macrophages to destroy *MAP* (Souza *et al.*, 2006), while inhibition of ERK1/2 did not (Souza *et al.*, 2007b).

From work *in vivo*, it has been suggested that infected cattle initially develop an early and effective proinflammatory immune response to *MAP*. However, this response typically declines in cattle that progress to clinical disease, favouring a Th2-like peripheral response and antibody production without control of infection (Stabel, 2000; Coussens, 2001, 2004; Tessema *et al.*, 2001). Despite lack of a peripheral proinflammatory response, *MAP*-associated lesions continue to present a highly inflamed state. While this appears inconsistent with the loss of IFNγ and TNFα, there may be other factors at play, including perhaps a robust Th17 response and IL-23-induced inflammation. Other diverse discoveries on host–pathogen immune dynamics have been made, ranging from *MAP*-stimulated host prostaglandin E_2 production suppressing Th1 responses (Sajiki *et al.*, 2018) to *MAP*-secreted extracellular vesicles inducing cytokine releases by macrophages (Wang *et al.*, 2014a). Further complicating the issue, some question whether the Th1 to Th2 paradigm even exists or whether it is simply a product of immune system exhaustion as *MAP* confounds and defeats the host immune response (Begg *et al.*, 2011).

9.6 Alternative Host Response to *MAP* Infection

Although later stages of paratuberculosis appear to be associated with a classical Th2-like humoral immune response (Ganusov *et al.*, 2015; Koets *et al.*, 2015), the innate or early response is not as well understood. Traditionally, classical Th1 responses are thought to be a major early reaction to *MAP* infection, via the elevated expression of inflammatory cytokines IFNγ and TNFα (Begg *et al.*, 2011; Dudemaine *et al.*, 2014; Roussey *et al.*, 2014). In later stage infections, a simultaneous decrease of Th1 cytokines appears to be coupled to an increase in Th2 cytokines (Roussey *et al.*, 2014; Ganusov *et al.*, 2015; Roussey *et al.*, 2016). Recent work has also highlighted the potential importance of alternative Th17-like responses in early *M. tuberculosis* and *M. bovis* infections (Khader *et al.*, 2011; Steinbach *et al.*, 2016). The notion that *MAP* may also be activating this pattern of response is supported by macrophage and epithelial cell upregulation of mRNA encoding the Th17-driving cytokines IL-23, IL-6 and TGFβ (Dudemaine *et al.*, 2014; Everman *et al.*, 2015). Our own research has shown a mean relative percentage increase of CD4, CD8 and γδ T cells expressing the receptor for IL-23 (IL-23R) in peripheral blood mononuclear cells of *MAP* test-positive cows relative to cells from *MAP* test-negative cows. This response was inducible by *MAP in vitro* with cells from both *MAP* test-positive and test-negative cows, suggesting that IL-23R upregulation is a general feature of *MAP*-immune cell interactions. With *M. bovis* and *M. tuberculosis*, CD4+ and γδ T cells are the primary producers of the Th17 specific cytokines IL-17a and IL-22 (Steinbach *et al.*, 2016; Basile *et al.*, 2017). Th17 cell activation is a specific and early immune response to *M. bovis* and *M. tuberculosis* (Dhiman *et al.*, 2009, 2012; Tzu, 1971; Waters *et al.*, 2016). Our work has shown that while early lesions from *MAP* infected tissues express elevated mRNA encoding IL-17a, tissues exhibiting more advanced infection stages do not (Roussey *et al.*, 2016). Preliminary experimentation with isolated CD3+ T cells from cows also demonstrated a robust innate Th17 response, with production of enhanced levels of IL-17a following exposure to two strains of *MAP* or to purified protein derivative of *MAP* (PPDj). Th17 cells can exhibit a certain plasticity and may also produce IL-10 and IFNγ and be a source of these cytokines during some phases of *MAP* infection (Zielinski *et al.*, 2012). Finally, the Th17 driving cytokine IL-23 can be highly proinflammatory and prolonged expression of IL-23 can limit the effect of regulatory T cells (Ahern *et al.*, 2010; Petermann *et al.*, 2010). To this end, we are exploring the possibility that IL-23 may be one source of the continued inflammation observed in later stages of paratuberculosis.

9.7 Concluding Remarks and Future Directions

MAP is a highly successful pathogen with diverse survival mechanisms. The secrets to its unique abilities are undoubtedly encoded in its 4.8-megabase genome. Since the *MAP* genome sequence was published in 2005, little research has focused on functional analysis of this fastidious organism. Understanding the functional genome of *MAP* is essential to establish the repertoire of virulence pathways used by this organism *in vivo* or *in vitro*. Thus, future studies using systems analysis of its transcriptome and proteomes will significantly aid in improving our understanding of host–pathogen and pathogen–environment interactions of *MAP*. Also encoded in that complex genome are pathways used by *MAP* that will greatly aid in improving *in vitro* culture systems for rapid diagnosis. An additional area of emphasis for future studies in *MAP*–host interactions would be in elucidating the host innate responses in very early infection, the role vaccines might play in guiding immune responses to *MAP* and the sources of tissue inflammation characteristic of paratuberculosis. A clear understanding of *MAP*'s functional genome and its interactions with the host will also aid in establishing good animal models to study, and ultimately control this complex, but highly successful microbe.

References

Abdellrazeq, G.S., Elnaggar, M.M., Bannantine, J.P., Park, K.T., Souza, C.D. *et al.* (2018) A *Mycobacterium avium* subsp. *paratuberculosis relA* deletion mutant and a 35 kDa major membrane protein elicit development of cytotoxic T lymphocytes with ability to kill intracellular bacteria. *Veterinary Research* 49(1), 1–16. DOI: 10.1186/s13567-018-0549-3.

Abdellrazeq, G.S., Elnaggar, M.M., Bannantine, J.P., Schneider, D.A., Souza, C.D. *et al.* (2019) A peptide-based vaccine for *Mycobacterium avium* subspecies *paratuberculosis*. *Vaccine* 37(21), 2783–2790. DOI: 10.1016/j.vaccine.2019.04.040.

Ahern, P.P., Schiering, C., Buonocore, S., McGeachy, M.J., Cua, D.J. *et al.* (2010) Interleukin-23 drives intestinal inflammation through direct activity on T cells. *Immunity* 33(2), 279–288. DOI: 10.1016/j.immuni.2010.08.010.

Aho, A.D., McNulty, A.M. and Coussens, P.M. (2003) Enhanced expression of interleukin-1alpha and tumor necrosis factor receptor-associated protein 1 in ileal tissues of cattle infected with *Mycobacterium avium* subsp. *paratuberculosis*. *Infection and Immunity* 71(11), 6479–6486. DOI: 10.1128/IAI.71.11.6479-6486.2003.

Alexander, D.C., Turenne, C.Y. and Behr, M.A. (2009) Insertion and deletion events that define the pathogen *Mycobacterium avium* subsp. *paratuberculosis*. *Journal of Bacteriology* 191(3), 1018–1025. DOI: 10.1128/JB.01340-08.

Alonso, S., Pethe, K., Russell, D.G. and Purdy, G.E. (2007) Lysosomal killing of *Mycobacterium* mediated by ubiquitin-derived peptides is enhanced by autophagy. *Proceedings of the National Academy of Sciences* 104(14), 6031–6036. DOI: 10.1073/pnas.0700036104.

Alonso-Hearn, M., Patel, D., Danelishvili, L., Meunier-Goddik, L. and Bermudez, L.E. (2008) The *Mycobacterium avium* subsp. *paratuberculosis* MAP3464 gene encodes an oxidoreductase involved in invasion of bovine epithelial cells through the activation of host cell Cdc42. *Infection and Immunity* 76(1), 170–178. DOI: 10.1128/IAI.01913-06.

Arsenault, R.J., Maattanen, P., Daigle, J., Potter, A., Griebel, P. *et al.* (2014) From mouth to macrophage: mechanisms of innate immune subversion by *Mycobacterium avium* subsp. *paratuberculosis*. *Veterinary Research* 45(1), 1–15. DOI: 10.1186/1297-9716-45-54.

Awasthi, A., Mathur, R., Khan, A., Joshi, B.N., Jain, N. *et al.* (2003) CD40 signaling is impaired in *L. major*–infected macrophages and is rescued by a p38MAPK activator establishing a host-protective memory T cell response. *The Journal of Experimental Medicine* 197(8), 1037–1043. DOI: 10.1084/jem.20022033.

Ayele, W.Y., Bartos, M., Svastova, P. and Pavlik, I. (2004) Distribution of *Mycobacterium avium* subsp. *paratuberculosis* in organs of naturally infected bull-calves and breeding bulls. *Veterinary Microbiology* 103(3–4), 209–217. DOI: 10.1016/j.vetmic.2004.07.011.

Bannantine, J.P., Huntley, J.F.J., Miltner, E., Stabel, J.R. and Bermudez, L.E. (2003) The *Mycobacterium avium* subsp. *paratuberculosis* 35 kDa protein plays a role in invasion of bovine epithelial cells. *Microbiology* 149(8), 2061–2069. DOI: 10.1099/mic.0.26323-0.

Basile, J. I., Kviatcovsky, D., Romero, M.M., Balboa, L., Monteserin, J. *et al.* (2017) *Mycobacterium tuberculosis* multi-drug-resistant strain M induces IL-17+IFNγ– CD4+ T cell expansion through an IL-23 and TGF-β-dependent mechanism in patients with MDR-TB tuberculosis. *Clinical & Experimental Immunology* 187(1), 160–173.

Basler, T., Brumshagen, C., Beineke, A., Goethe, R. and Bäumer, W. (2013) *Mycobacterium avium* subspecies impair dendritic cell maturation. *Innate Immunity* 19(5), 451–461. DOI: 10.1177/1753425912470291.

Begg, D.J., de Silva, K., Carter, N., Plain, K.M., Purdie, A. *et al.* (2011) Does a Th1 over Th2 dominancy really exist in the early stages of *Mycobacterium avium* subspecies *paratuberculosis* infections? *Immunobiology* 216(7), 840–846. DOI: 10.1016/j.imbio.2010.12.004.

Behr, M.A. and Kapur, V. (2008) The evidence for *Mycobacterium paratuberculosis* in Crohn's disease. *Current Opinion in Gastroenterology* 24(1), 17–21. DOI: 10.1097/MOG.0b013e3282f1dcc4.

Benson, S.A. and Ernst, J.D. (2009) TLR2-dependent inhibition of macrophage responses to IFN-γ is mediated by distinct, gene-specific mechanisms. *PLoS ONE* 4(7), 1–9. DOI: 10.1371/journal.pone.0006329.

Bentley, R.W., Keenan, J.I., Gearry, R.B., Kennedy, M.A., Barclay, M.L. *et al.* (2008) Incidence of *Mycobacterium avium* subspecies *paratuberculosis* in a population-based cohort of patients with Crohn's disease and control subjects. *The American Journal of Gastroenterology* 103(5), 1168–1172. DOI: 10.1111/j.1572-0241.2007.01742.x.

Bonecini-Almeida, M.G., Chitale, S., Boutsikakis, I., Geng, J., Doo, H. *et al.* (1998) Induction of *in vitro* human macrophage anti-*Mycobacterium tuberculosis* activity: requirement for IFN-gamma and primed lymphocytes. *Journal of Immunology* 160, 4490–4499.

Botella, H., Vaubourgeix, J., Lee, M.H., Song, N., Xu, W. *et al.* (2017) *Mycobacterium tuberculosis* protease MarP activates a peptidoglycan hydrolase during acid stress. *The EMBO Journal* 36(4), 536–548. DOI: 10.15252/embj.201695028.

Bradley, J.R. and Pober, J.S. (2001) Tumor necrosis factor receptor-associated factors (TRAFs). *Oncogene* 20(44), 6482–6491. DOI: 10.1038/sj.onc.1204788.

Chacon, O., Bermudez, L.E. and Barletta, R.G. (2004) Johne's disease, inflammatory bowel disease, and *Mycobacterium paratuberculosis*. *Annual Review of Microbiology* 58(1), 329–363. DOI: 10.1146/annurev.micro.58.030603.123726.

Chandrangsu, P., Rensing, C. and Helmann, J.D. (2017) Metal homeostasis and resistance in bacteria. *Nature Reviews Microbiology* 15(6), 338–350. DOI: 10.1038/nrmicro.2017.15.

Chiang, S.-K., Sommer, S., Aho, A.D., Kiupel, M., Colvin, C. *et al.* (2007) Relationship between *Mycobacterium avium* subspecies *paratuberculosis*, IL-1α, and TRAF1 in primary bovine monocyte-derived macrophages. *Veterinary Immunology and Immunopathology* 116(3–4), 131–144. DOI: 10.1016/j.vetimm.2007.01.005.

Clark, L.B., Foy, T.M. and Noelle, R.J. (1996) CD40 and its ligand. *Advances in Immunology* 63, 43–78.

Clarke, C.J. (1997) The pathology and pathogenesis of paratuberculosis in ruminants and other species. *Journal of Comparative Pathology* 116(3), 217–261. DOI: 10.1016/S0021-9975(97)80001-1.

Corn, J.L., Manning, E.J.B., Sreevatsan, S. and Fischer, J.R. (2005) Isolation of *Mycobacterium avium* subsp. *paratuberculosis* from free-ranging birds and mammals on livestock premises. *Applied and Environmental Microbiology* 71(11), 6963–6967. DOI: 10.1128/AEM.71.11.6963-6967.2005.

Coussens, P. (2001) *M. avium* subspecies *paratuberculosis* and the bovine immune system. *Animal Health Research Reviews* 2, 141–161.

Coussens, P.M. (2004) Model for immune responses to *Mycobacterium avium* subspecies *paratuberculosis* in cattle. *Infection and Immunity* 72(6), 3089–3096. DOI: 10.1128/IAI.72.6.3089-3096.2004.

Coussens, P.M., Verman, N., Coussens, M.A., Elftman, M.D. and McNulty, A.M. (2004) Cytokine gene expression in peripheral blood mononuclear cells and tissues of cattle infected with *Mycobacterium avium* subsp. *paratuberculosis*: evidence for an inherent proinflammatory gene expression pattern. *Infection and Immunity* 72(3), 1409–1422. DOI: 10.1128/IAI.72.3.1409-1422.2004.

Crossley, B.M., Zagmutt-Vergara, F.J., Fyock, T.L., Whitlock, R.H. and Gardner, I.A. (2005) Fecal shedding of *Mycobacterium avium* subsp. *paratuberculosis* by dairy cows. *Veterinary Microbiology* 107(3–4), 257–263. DOI: 10.1016/j.vetmic.2005.01.017.

Davis, J.M. and Ramakrishnan, L. (2009) The role of the granuloma in expansion and dissemination of early tuberculous infection. *Cell* 136(1), 37–49. DOI: 10.1016/j.cell.2008.11.014.

de Almeida, D.E., Colvin, C.J. and Coussens, P.M. (2008) Antigen-specific regulatory T cells in bovine paratuberculosis. *Veterinary Immunology and Immunopathology* 125(3–4), 234–245. DOI: 10.1016/j.vetimm.2008.05.019.

Denis, M., Gregg, E.O. and Ghandirian, E. (1990) Cytokine modulation of *Mycobacterium tuberculosis* growth in human macrophages. *International Journal of Immunopharmacology* 12(7), 721–727. DOI: 10.1016/0192-0561(90)90034-K.

Dhiman, R., Indramohan, M., Barnes, P.F., Nayak, R.C., Paidipally, P. *et al.* (2009) IL-22 produced by human NK cells inhibits growth of *Mycobacterium tuberculosis* by enhancing phagolysosomal fusion. *The Journal of Immunology* 183(10), 6639–6645. DOI: 10.4049/jimmunol.0902587.

Dhiman, R., Periasamy, S., Barnes, P.F., Jaiswal, A.G., Paidipally, P. *et al.* (2012) NK1.1^{+}cells and IL-22 regulate vaccine-induced protective immunity against challenge with *Mycobacterium tuberculosis*. *The Journal of Immunology* 189(2), 897–905. DOI: 10.4049/jimmunol.1102833.

di Marzio, P., Mariani, R., Lui, R., Thomas, E.K. and Landau, N.R. (2000) Soluble CD40 ligand induces β-chemokine production by macrophages and resistance to HIV-1 entry. *Cytokine* 12(10), 1489–1495. DOI: 10.1006/cyto.1999.0594.

Dudemaine, P.L., Fecteau, G., Lessard, M., Labrecque, O., Roy, J.P. *et al.* (2014) Increased blood-circulating interferon-γ, interleukin-17, and osteopontin levels in bovine paratuberculosis. *Journal of Dairy Science* 97(6), 3382–3393. DOI: 10.3168/jds.2013-7059.

Eckelt, E., Jarek, M., Frömke, C., Meens, J. and Goethe, R. (2014) Identification of a lineage specific zinc responsive genomic island in *Mycobacterium avium* ssp. *paratuberculosis*. *BMC Genomics* 15(1), 1076. DOI: 10.1186/1471-2164-15-1076.

Everman, J.L., Eckstein, T.M., Roussey, J., Coussens, P., Bannantine, J.P. *et al.* (2015) Characterization of the inflammatory phenotype of *Mycobacterium avium* subspecies *paratuberculosis* using a novel cell culture passage model. *Microbiology* 161(7), 1420–1434. DOI: 10.1099/mic.0.000106.

Flórido, M., Gonçalves, A.S., Silva, R.A., Ehlers, S., Cooper, A.M. *et al.* (1999) Resistance of virulent *Mycobacterium avium* to gamma interferon-mediated antimicrobial activity suggests additional signals for induction of mycobacteriostasis. *Infection and Immunity* 67(7), 3610–3618. DOI: 10.1128/IAI.67.7.3610-3618.1999.

Flynn, J.L., Chan, J., Triebold, K.J., Dalton, D.K., Stewart, T.A. *et al.* (1993) An essential role for interferon gamma in resistance to *Mycobacterium tuberculosis* infection. *The Journal of Experimental Medicine* 178(6), 2249–2254. DOI: 10.1084/jem.178.6.2249.

Frie, M.C., Sporer, K.R.B., Kirkpatrick, B.W. and Coussens, P.M. (2017) T and B cell activation profiles from cows with and without Johne's disease in response to in vitro stimulation with *Mycobacterium avium* subspecies *paratuberculosis*. *Veterinary Immunology and Immunopathology* 193–194, 50–56. DOI: 10.1016/j.vetimm.2017.10.005.

Ganusov, V.V., Klinkenberg, D., Bakker, D. and Koets, A.P. (2015) Evaluating contribution of the cellular and humoral immune responses to the control of shedding of *Mycobacterium avium* spp. *paratuberculosis* in cattle. *Veterinary Research* 46(1), 62. DOI: 10.1186/s13567-015-0204-1.

Gatfield, J. and Pieters, J. (2003) Molecular mechanisms of host–pathogen interaction: entry and survival of mycobacteria in macrophages. *Advances in Immunology* 81, 45–96.

Ghadiali, A.H., Strother, M., Naser, S.A., Manning, E.J.B. and Sreevatsan, S. (2004) *Mycobacterium avium* subsp. *paratuberculosis* strains isolated from Crohn's disease patients and animal species exhibit similar polymorphic locus patterns. *Journal of Clinical Microbiology* 42(11), 5345–5348. DOI: 10.1128/JCM.42.11.5345-5348.2004.

Ghosh, P., Wu, C.W. and Talaat, A.M. (2013) Key role for the alternative sigma factor, SigH, in the intracellular life of *Mycobacterium avium* subsp. *paratuberculosis* during macrophage stress. *Infection and Immunity* 81(6), 2242–2257. DOI: 10.1128/IAI.01273-12.

Gollnick, N.S., Mitchell, R.M., Baumgart, M., Janagama, H.K., Sreevatsan, S. *et al.* (2007) Survival of *Mycobacterium avium* subsp. *paratuberculosis* in bovine monocyte-derived macrophages is not affected by host infection status but depends on the infecting bacterial genotype. *Veterinary Immunology and Immunopathology* 120(3–4), 93–105. DOI: 10.1016/j.vetimm.2007.07.017.

Grant, I.R. (1998) Does *Mycobacterium paratuberculosis* survive current pasteurization conditions? *Applied and Environmental Microbiology* 64, 2760–2761.

Grant, I.R., Foddai, A.C.G., Tarrant, J.C., Kunkel, B., Hartmann, F.A. *et al.* (2017) Viable *Mycobacterium avium* ssp. *paratuberculosis* isolated from calf milk replacer. *Journal of Dairy Science* 100(12), 9723–9735. DOI: 10.3168/jds.2017-13154.

Grewal, I.S. and Flavell, R.A. (1996) The role of CD40 ligand in costimulation and T-cell activation. *Immunological Reviews* 153(1), 85–106. DOI: 10.1111/j.1600-065X.1996.tb00921.x.

Grewal, I.S. and Flavell, R.A. (1998) CD40 and CD154 in cell-mediated immunity. *Annual Review of Immunology* 16(1), 111–135. DOI: 10.1146/annurev.immunol.16.1.111.

Grewal, I.S., Borrow, P., Pamer, E.G., Oldstone, M.B.A. and Flavell, R.A. (1997) The CD40–CD154 system in anti-infective host defense. *Current Opinion in Immunology* 9(4), 491–497. DOI: 10.1016/S0952-7915(97)80100-8.

Grewal, S.K., Rajeev, S., Sreevatsan, S. and Michel, F.C. (2006) Persistence of *Mycobacterium avium* subsp. *paratuberculosis* and other zoonotic pathogens during simulated composting, manure packing, and liquid storage of dairy manure. *Applied and Environmental Microbiology* 72(1), 565–574. DOI: 10.1128/AEM.72.1.565-574.2006.

Hantke, K. (1981) Regulation of ferric iron transport in Escherichia coli K12: isolation of a constitutive mutant. *Molecular and General Genetics MGG* 182(2), 288–292. DOI: 10.1007/BF00269672.

Harris, N.B. and Barletta, R.G. (2001) *Mycobacterium avium* subsp. *paratuberculosis* in veterinary medicine. *Clinical Microbiology Reviews* 14(3), 489–512. DOI: 10.1128/CMR.14.3.489-512.2001.

Harris, J.E. and Lammerding, A.M. (2001) Crohn's disease and *Mycobacterium avium* subsp. *paratuberculosis*: current issues. *Journal of Food Protection* 64(12), 2103–2110. DOI: 10.4315/0362-028X-64.12.2103.

Hussain, T., Shah, S.Z.A., Zhao, D., Sreevatsan, S. and Zhou, X. (2016) The role of IL-10 in *Mycobacterium avium* subsp. *paratuberculosis* infection. *Cell Communication and Signaling* 14(1), 1–14. DOI: 10.1186/s12964-016-0152-z.

Janagama, H.K., Senthilkumar, T.M.A., Bannantine, J.P., Rodriguez, G.M., Smith, I. *et al.* (2009) Identification and functional characterization of the iron-dependent regulator (IdeR) of *Mycobacterium avium* subsp. *paratuberculosis*. *Microbiology* 155(11), 3683–3690. DOI: 10.1099/mic.0.031948-0.

Janagama, H.K., Lamont, E.A., George, S., Bannantine, J.P., Xu, W.W. *et al.* (2010) Primary transcriptomes of *Mycobacterium avium* subsp. *paratuberculosis* reveal proprietary pathways in tissue and macrophages. *BMC Genomics* 11(1), 561. DOI: 10.1186/1471-2164-11-561.

Józefowski, S., Sobota, A. and Kwiatkowska, K. (2008) How *Mycobacterium tuberculosis* subverts host immune responses. *BioEssays* 30(10), 943–954. DOI: 10.1002/bies.20815.

Kelley, V.A. and Schorey, J.S. (2004) Modulation of cellular phosphatidylinositol 3-phosphate levels in primary macrophages affects heat-killed but not viable *Mycobacterium avium*'s transport through the phagosome maturation process. *Cellular Microbiology* 6(10), 973–985. DOI: 10.1111/j.1462-5822.2004.00415.x.

Keown, D.A., Collings, D.A. and Keenan, J.I. (2012) Uptake and persistence of *Mycobacterium avium* subsp. *paratuberculosis* in human monocytes. *Infection and Immunity* 80(11), 3768–3775. DOI: 10.1128/IAI.00534-12.

Khader, S.A., Guglani, L., Rangel-Moreno, J., Gopal, R., Junecko, B.A.F. *et al.* (2011) IL-23 is required for long-term control of *Mycobacterium tuberculosis* and B cell follicle formation in the infected lung. *The Journal of Immunology* 187(10), 5402–5407. DOI: 10.4049/jimmunol.1101377.

Khalifeh, M.S. and Stabel, J.R. (2004) Upregulation of transforming growth factor-beta and interleukin-10 in cows with clinical Johne's disease. *Veterinary Immunology and Immunopathology* 99(1–2), 39–46. DOI: 10.1016/j.vetimm.2004.01.009.

Khare, S., Nunes, J.S., Figueiredo, J.F., Lawhon, S.D., Rossetti, C.A. *et al.* (2009) Early phase morphological lesions and transcriptional responses of bovine ileum infected with *Mycobacterium avium* subsp. *paratuberculosis*. *Veterinary Pathology* 46(4), 717–728. DOI: 10.1354/vp.08-VP-0187-G-FL.

Khare, S., Lawhon, S.D., Drake, K.L., Nunes, J.E.S., Figueiredo, J.F. *et al.* (2012) Systems biology analysis of gene expression during in vivo *Mycobacterium avium paratuberculosis* enteric colonization reveals role for immune tolerance. *PLoS ONE* 7(8), e42127. DOI: 10.1371/journal.pone.0042127.

Khare, S., Drake, K.L., Lawhon, S.D., Nunes, J.E.S., Figueiredo, J.F. *et al.* (2016) Systems analysis of early host gene expression provides clues for transient *Mycobacterium avium* ssp *avium* vs. persistent *Mycobacterium avium* ssp *paratuberculosis* intestinal infections. *PLoS ONE* 11(9), e0161946–33. DOI: 10.1371/journal.pone.0161946.

Kiener, P.A., Moran-Davis, P., Rankin, B.M., Wahl, A.F., Aruffo, A. *et al.* (1995) Stimulation of CD40 with purified soluble gp39 induces proinflammatory responses in human monocytes. *Journal of Immunology* 155, 4917–4925.

Koets, A.P., Eda, S. and Sreevatsan, S. (2015) The within host dynamics of *Mycobacterium avium* ssp. *paratuberculosis* infection in cattle: where time and place matter. *Veterinary Research* 46(1), 61. DOI: 10.1186/s13567-015-0185-0.

Kuehnel, M.P., Goethe, R., Habermann, A., Mueller, E., Rohde, M. *et al.* (2001) Characterization of the intracellular survival of *Mycobacterium avium* ssp. *paratuberculosis*: phagosomal pH and fusogenicity in J774 macrophages compared with other mycobacteria. *Cellular Microbiology* 3(8), 551–566. DOI: 10.1046/j.1462-5822.2001.00139.x.

Kugadas, A., Lamont, E.A., Bannantine, J.P., Shoyama, F.M., Brenner, E. *et al.* (2016) A *Mycobacterium avium* subsp. *paratuberculosis* predicted serine protease is associated with acid stress and intraphagosomal survival. *Frontiers in Cellular and Infection Microbiology* 6(599), 1–11. DOI: 10.3389/fcimb.2016.00085.

Lei, L., Plattner, B.L. and Hostetter, J.M. (2008) Live *Mycobacterium avium* subsp. *paratuberculosis* and a killed-bacterium vaccine induce distinct subcutaneous granulomas, with unique cellular and cytokine profiles. *Clinical and Vaccine Immunology* 15(5), 783–793. DOI: 10.1128/CVI.00480-07.

Levitte, S., Adams, K. N., Berg, R. D., Cosma, C. L., Urdahl, K. B. *et al.* (2016) Mycobacterial acid tolerance enables phagolysosomal survival and establishment of tuberculous infection *in vivo*. *Cell Host & Microbe* 20(2), 250–258. DOI: 10.1016/j.chom.2016.07.007.

Li, L., Bannantine, J.P., Zhang, Q., Amonsin, A., May, B.J. *et al.* (2005) The complete genome sequence of *Mycobacterium avium* subspecies *paratuberculosis*. *Proceedings of the National Academy of Sciences* 102(35), 12344–12349. DOI: 10.1073/pnas.0505662102.

Malik, N., Greenfield, B.W., Wahl, A.F. and Kiener, P.A. (1996) Activation of human monocytes through CD40 induces matrix metalloproteinases. *Journal of Immunology* 156, 3952–3960.

Mathur, R.K., Awasthi, A., Wadhone, P., Ramanamurthy, B. and Saha, B. (2004) Reciprocal CD40 signals through p38MAPK and ERK-1/2 induce counteracting immune responses. *Nature Medicine* 10(5), 540–544. DOI: 10.1038/nm1045.

Mehra, S., Golden, N.A., Stuckey, K., Didier, P.J., Doyle, L.A. *et al.* (2012) The *Mycobacterium tuberculosis* stress response factor SigH is required for bacterial burden as well as immunopathology in primate lungs. *The Journal of Infectious Diseases* 205(8), 1203–1213. DOI: 10.1093/infdis/jis102.

Miller, H., Zhang, J., Kuolee, R., Patel, G.B. and Chen, W. (2007) Intestinal M cells: the fallible sentinels? *World Journal of Gastroenterology* 13(10), 1477–1486. DOI: 10.3748/wjg.v13.i10.1477.

Momotani, E., Whipple, D.L., Thiermann, A.B. and Cheville, N.F. (1988) Role of M cells and macrophages in the entrance of *Mycobacterium paratuberculosis* into domes of ileal Peyer's patches in calves. *Veterinary Pathology* 25(2), 131–137. DOI: 10.1177/030098588802500205.

Motiwala, A.S., Li, L., Kapur, V. and Sreevatsan, S. (2006) Current understanding of the genetic diversity of *Mycobacterium avium* subsp. *paratuberculosis*. *Microbes and Infection* 8(5), 1406–1418. DOI: 10.1016/j.micinf.2005.12.003.

Murphy, J.T., Sommer, S., Kabara, E.A., Verman, N., Kuelbs, M.A. *et al.* (2006) Gene expression profiling of monocyte-derived macrophages following infection with *Mycobacterium avium* subspecies *avium* and *Mycobacterium avium* subspecies *paratuberculosis*. *Physiological Genomics* 28(1), 67–75. DOI: 10.1152/physiolgenomics.00098.2006.

Patel, D., Danelishvili, L., Yamazaki, Y., Alonso, M., Paustian, M.L. *et al.* (2006) The ability of *Mycobacterium avium* subsp. *paratuberculosis* to enter bovine epithelial cells is influenced by preexposure to a hyperosmolar environment and intracellular passage in bovine mammary epithelial cells. *Infection and Immunity* 74(5), 2849–2855. DOI: 10.1128/IAI.74.5.2849-2855.2006.

Petermann, F., Rothhammer, V., Claussen, M.C., Haas, J.D., Blanco, L.R. *et al.* (2010) γδ T cells enhance autoimmunity by restraining regulatory T cell responses via an interleukin-23-dependent mechanism. *Immunity* 33(3), 351–363. DOI: 10.1016/j.immuni.2010.08.013.

Pieters, J. (2001) Entry and survival of pathogenic mycobacteria in macrophages. *Microbes and Infection* 3(3), 249–255. DOI: 10.1016/S1286-4579(01)01376-4.

Ponnusamy, D., Periasamy, S., Tripathi, B.N. and Pal, A. (2013) *Mycobacterium avium* subsp. *paratuberculosis* invades through M cells and enterocytes across ileal and jejunal mucosa of lambs. *Research in Veterinary Science* 94(2), 306–312. DOI: 10.1016/j.rvsc.2012.09.023.

Prohászka, Z., Daha, M.R., Süsal, C., Daniel, V., Szlávik, J. *et al.* (1999) C1Q autoantibodies in HIV infection: correlation to elevated levels of autoantibodies against 60-kDa heat-shock proteins. *Clinical Immunology* 90(2), 247–255. DOI: 10.1006/clim.1998.4620.

Richter, E., Wessling, J., Lügering, N., Domschke, W. and Rüsch-Gerdes, S. (2002) *Mycobacterium avium* subsp. *paratuberculosis* infection in a patient with HIV, Germany. *Emerging Infectious Diseases* 8(7), 729–731. DOI: 10.3201/eid0807.010388.

Robertson, A.K. and Andrew, P.W. (1991) Interferon gamma fails to activate human monocyte-derived macrophages to kill or inhibit the replication of a non-pathogenic mycobacterial species. *Microbial Pathogenesis* 11(4), 283–288. DOI: 10.1016/0882-4010(91)90032-6.

Roussey, J.A., Steibel, J.P. and Coussens, P.M. (2014) Regulatory T cell activity and signs of T cell unresponsiveness in bovine paratuberculosis. *Frontiers in Veterinary Science* 1, 20. DOI: 10.3389/fvets.2014.00020.

Roussey, J.A., Oliveira, L.J., Langohr, I.M., Sledge, D.G. and Coussens, P.M. (2016) Regulatory T cells and immune profiling in Johne's disease lesions. *Veterinary Immunology and Immunopathology* 181, 39–50. DOI: 10.1016/j.vetimm.2016.03.008.

Rowe, M.T. and Grant, I.R. (2006) *Mycobacterium avium* ssp. *paratuberculosis* and its potential survival tactics. *Letters in Applied Microbiology* 42(4), 305–311. DOI: 10.1111/j.1472-765X.2006.01873.x.

Russell, D.G., Mwandumba, H.C. and Rhoades, E.E. (2002) *Mycobacterium* and the coat of many lipids. *The Journal of Cell Biology* 158(3), 421–426. DOI: 10.1083/jcb.200205034.

Sajiki, Y., Konnai, S., Okagawa, T., Nishimori, A., Maekawa, N. *et al.* (2018) Prostaglandin E_2 induction suppresses the Th1 immune responses in cattle with Johne's disease. *Infection and Immunity* 86(5), 5. DOI: 10.1128/IAI.00910-17.

Salte, T., Pathak, S., Wentzel-Larsen, T. and Asjö, B. (2011) Increased intracellular growth of *Mycobacterium avium* in HIV-1 exposed monocyte-derived dendritic cells. *Microbes and Infection* 13(3), 276–283. DOI: 10.1016/j.micinf.2010.11.001.

Secott, T.E., Lin, T.L. and Wu, C.C. (2001) Fibronectin attachment protein homologue mediates fibronectin binding by *Mycobacterium avium* subsp. *paratuberculosis*. *Infection and Immunity* 69(4), 2075–2082. DOI: 10.1128/IAI.69.4.2075-2082.2001.

Secott, T.E., Lin, T.L. and Wu, C.C. (2002) Fibronectin attachment protein is necessary for efficient attachment and invasion of epithelial cells by *Mycobacterium avium* subsp. *paratuberculosis*. *Infection and Immunity* 70(5), 2670–2675. DOI: 10.1128/IAI.70.5.2670-2675.2002.

Secott, T.E., Lin, T.L. and Wu, C.C. (2004) *Mycobacterium avium* subsp. *paratuberculosis* fibronectin attachment protein facilitates M-cell targeting and invasion through a fibronectin bridge with host integrins. *Infection and Immunity* 72(7), 3724–3732. DOI: 10.1128/IAI.72.7.3724-3732.2004.

Sigurðardóttir, O.G., Press, C.M., Saxegaard, F. and Evensen, O. (1999) Bacterial isolation, immunological response, and histopathological lesions during the early subclinical phase of experimental infection of goat kids with *Mycobacterium avium* subsp. *paratuberculosis*. *Veterinary Pathology* 36(6), 542–550. DOI: 10.1354/vp.36-6-542.

Sigurðardóttir, O.G., Press, C.M. and Evensen, O. (2001) Uptake of *Mycobacterium avium* subsp. *paratuberculosis* through the distal small intestinal mucosa in goats: an ultrastructural study. *Veterinary Pathology* 38(2), 184–189. DOI: 10.1354/vp.38-2-184.

Sigurðardóttir, O.G., Valheim, M. and Press, C.M. (2004) Establishment of *Mycobacterium avium* subsp. *paratuberculosis* infection in the intestine of ruminants. *Advanced Drug Delivery Reviews* 56(6), 819–834. DOI: 10.1016/j.addr.2003.10.032.

Sivakumar, P., Tripathi, B.N. and Singh, N. (2005) Detection of *Mycobacterium avium* subsp. *paratuberculosis* in intestinal and lymph node tissues of water buffaloes (*Bubalus bubalis*) by PCR and bacterial culture. *Veterinary Microbiology* 108(3–4), 263–270. DOI: 10.1016/j.vetmic.2005.04.002.

Small, J.L., O'Donoghue, A.J., Boritsch, E.C., Tsodikov, O.V., Knudsen, G.M. *et al.* (2013) Substrate specificity of MarP, a periplasmic protease required for resistance to acid and oxidative stress in *Mycobacterium tuberculosis*. *Journal of Biological Chemistry* 288(18), 12489–12499. DOI: 10.1074/jbc.M113.456541.

Sommer, S., Pudrith, C.B., Colvin, C.J. and Coussens, P.M. (2009) *Mycobacterium avium* subspecies *paratuberculosis* suppresses expression of IL-12p40 and iNOS genes induced by signalling through CD40 in bovine monocyte-derived macrophages. *Veterinary Immunology and Immunopathology* 128(1–3), 44–52. DOI: 10.1016/j.vetimm.2008.10.294.

Soong, L., Xu, J.C., Grewal, I.S., Kima, P., Sun, J. *et al.* (1996) Disruption of CD40–CD40 ligand interactions results in an enhanced susceptibility to *Leishmania amazonensis* infection. *Immunity* 4(3), 263–273. DOI: 10.1016/S1074-7613(00)80434-3.

Souza, C.D., Evanson, O.A. and Weiss, D.J. (2006) Regulation by Jun N-terminal kinase/stress activated protein kinase of cytokine expression in *Mycobacterium avium* subsp *paratuberculosis*-infected bovine monocytes. *American Journal of Veterinary Research* 67(10), 1760–1765. DOI: 10.2460/ajvr.67.10.1760.

Souza, C.D., Evanson, O.A., Sreevatsan, S. and Weiss, D.J. (2007a) Cell membrane receptors on bovine mononuclear phagocytes involved in phagocytosis of *Mycobacterium avium* subsp. *paratuberculosis*. *American Journal of Veterinary Research* 68(9), 975–980. DOI: 10.2460/ajvr.68.9.975.

Souza, C.D., Evanson, O.A. and Weiss, D.J. (2007b) Role of the MAPK(ERK) pathway in regulation of cytokine expression by *Mycobacterium avium* subsp. *paratuberculosis*-exposed bovine monocytes. *American Journal of Veterinary Research* 68(6), 625–630. DOI: 10.2460/ajvr.68.6.625.

Souza, C.D., Evanson, O.A. and Weiss, D.J. (2007c) Role of the mitogen-activated protein kinase pathway in the differential response of bovine monocytes to *Mycobacterium avium* subsp. *paratuberculosis* and *Mycobacterium avium* subsp. *avium*. *Microbes and Infection* 9(14–15), 1545–1552. DOI: 10.1016/j.micinf.2007.08.008.

Souza, C.D., Evanson, O.A. and Weiss, D.J. (2008) Role of cell membrane receptors in the suppression of monocyte anti-microbial activity against *Mycobacterium avium* subsp. *paratuberculosis*. *Microbial Pathogenesis* 44(3), 215–223. DOI: 10.1016/j.micpath.2007.09.006.

Stabel, J.R. (2000) Transitions in immune responses to *Mycobacterium paratuberculosis*. *Veterinary Microbiology* 77(3–4), 465–473. DOI: 10.1016/S0378-1135(00)00331-X.

Steinbach, S., Vordermeier, H.M. and Jones, G.J. (2016) CD4+ and γδ T cells are the main producers of IL-22 and IL-17A in lymphocytes from mycobacterium bovis-infected cattle. *Scientific Reports* 6(1), 29990. DOI: 10.1038/srep29990.

Stout, R.D. and Suttles, J. (1996) The many roles of CD40 in cell-mediated inflammatory responses. *Immunology Today* 17(10), 487–492. DOI: 10.1016/0167-5699(96)10060-I.

Stratmann, J., Strommenger, B., Goethe, R., Dohmann, K., Gerlach, G.-F. *et al.* (2004) A 38-kilobase pathogenicity island specific for *Mycobacterium avium* subsp. *paratuberculosis* encodes cell surface proteins expressed in the host. *Infection and Immunity* 72(3), 1265–1274. DOI: 10.1128/IAI.72.3.1265-1274.2004.

Sun, J., Deghmane, A.-E., Soualhine, H., Hong, T., Bucci, C. *et al.* (2007) *Mycobacterium bovis* BCG disrupts the interaction of Rab7 with RILP contributing to inhibition of phagosome maturation. *Journal of Leukocyte Biology* 82(6), 1437–1445. DOI: 10.1189/jlb.10.1189.

Sweeney, R.W., Whitlock, R.H., Hamir, A.N., Rosenberger, A.E. and Herr, S.A. (1992a) Isolation of *Mycobacterium paratuberculosis* after oral inoculation in uninfected cattle. *American Journal of Veterinary Research* 53(8), 1312–1314.

Sweeney, R.W., Whitlock, R.H. and Rosenberger, A.E. (1992b) *Mycobacterium paratuberculosis* cultured from milk and supramammary lymph nodes of infected asymptomatic cows. *Journal of Clinical Microbiology* 30(1), 166–171. DOI: 10.1128/JCM.30.1.166-171.1992.

Sweeney, R.W., Uzonna, J., Whitlock, R.H., Habecker, P.L., Chilton, P. *et al.* (2006) Tissue predilection sites and effect of dose on *Mycobacterium avium* subsp. *paratuberculosis* organism recovery in a short-term bovine experimental oral infection model. *Research in Veterinary Science* 80(3), 253–259. DOI: 10.1016/j.rvsc.2005.07.007.

Tailleux, L., Neyrolles, O., Honoré-Bouakline, S., Perret, E., Sanchez, F. *et al.* (2003) Constrained intracellular survival of *Mycobacterium tuberculosis* in human dendritic cells. *The Journal of Immunology* 170(4), 1939–1948. DOI: 10.4049/jimmunol.170.4.1939.

Tessema, M.Z., Koets, A.P., Rutten, V.P. and Gruys, E. (2001) How does *Mycobacterium avium* subsp. *paratuberculosis* resist intracellular degradation? *The Veterinary Quarterly* 23(4), 153–162. DOI: 10.1080/01652176.2001.9695105.

Tian, L., Noelle, R.J. and Lawrence, D.A. (1995) Activated T cells enhance nitric oxide production by murine splenic macrophages through gp39 and LFA-1. *European Journal of Immunology* 25(1), 306–309. DOI: 10.1002/eji.1830250152.

Tiwari, A., VanLeeuwen, J.A., McKenna, S.L.B., Keefe, G.P. and Barkema, H.W. (2006) Johne's disease in Canada. Part I: clinical symptoms, pathophysiology, diagnosis, and prevalence in dairy herds. *The Canadian Veterinary Journal = La Revue Veterinaire Canadienne* 47(9), 874–882.

Tzu, S. (1971) *The Art of War, translated by Griffith, S.* Oxford University Press, Oxford, UK. DOI: 10.3389/fmicb.2016.00898.

van Kooten, C. and Banchereau, J. (1996) CD40–CD40 ligand: a multifunctional receptor–ligand pair. *Advances in Immunology* 61, 1–77. DOI: 10.1016/s0065-2776(08)60865-2.

van Kooten, C. and Banchereau, J. (2000) CD40–CD40 ligand. *Journal of Leukocyte Biology* 67(1), 2–17. DOI: 10.1002/jlb.67.1.2.

Vandal, O.H., Pierini, L.M., Schnappinger, D., Nathan, C.F. and Ehrt, S. (2008) A membrane protein preserves intrabacterial pH in intraphagosomal *Mycobacterium tuberculosis*. *Nature Medicine* 14(8), 849–854. DOI: 10.1038/nm.1795.

Vergne, I., Chua, J., Lee, H.-H., Lucas, M., Belisle, J. *et al.* (2005) Mechanism of phagolysosome biogenesis block by viable *Mycobacterium tuberculosis*. *Proceedings of the National Academy of Sciences* 102(11), 4033–4038. DOI: 10.1073/pnas.0409716102.

Via, L.E., Deretic, D., Ulmer, R.J., Hibler, N.S., Huber, L.A. *et al.* (1997) Arrest of mycobacterial phagosome maturation is caused by a block in vesicle fusion between stages controlled by Rab5 and Rab7. *Journal of Biological Chemistry* 272(20), 13326–13331. DOI: 10.1074/jbc.272.20.13326.

Voo, K.S., Wang, Y.-H., Santori, F.R., Boggiano, C., Wang, Y.-H. *et al.* (2009) Identification of IL-17-producing FOXP3$^+$ regulatory T cells in humans. *Proceedings of the National Academy of Sciences* 106(12), 4793–4798. DOI: 10.1073/pnas.0900408106.

Waddell, L.A., Rajić, A., Sargeant, J., Harris, J., Amezcua, R. *et al.* (2008) The zoonotic potential of *Mycobacterium avium* spp. *paratuberculosis*: a systematic review. *Canadian Journal of Public Health* 99(2), 145–155. DOI: 10.1007/BF03405464.

Wang, J., Chen, C., Xie, P., Pan, Y., Tan, Y. *et al.* (2014a) Proteomic analysis and immune properties of exosomes released by macrophages infected with *Mycobacterium avium*. *Microbes and Infection* 16(4), 283–291. DOI: 10.1016/j.micinf.2013.12.001.

Wang, J., Pritchard, J.R., Kreitmann, L., Montpetit, A. and Behr, M.A. (2014b) Disruption of *Mycobacterium avium* subsp. *paratuberculosis*-specific genes impairs in vivo fitness. *BMC Genomics* 15(1), 415. DOI: 10.1186/1471-2164-15-415.

Wang, J., Moolji, J., Dufort, A., Staffa, A., Domenech, P. *et al.* (2016) Iron acquisition in *Mycobacterium avium* subsp. *paratuberculosis*. *Journal of Bacteriology* 198(5), 857–866. DOI: 10.1128/JB.00922-15.

Waters, W.R., Maggioli, M.F., Palmer, M.V., Thacker, T.C., McGill, J.L. *et al.* (2016) Interleukin-17A as a biomarker for bovine tuberculosis. *Clinical and Vaccine Immunology* 23(2), 168–180. DOI: 10.1128/CVI.00637-15.

Weiss, D.J. and Souza, C.D. (2008) Review paper: modulation of mononuclear phagocyte function by *Mycobacterium avium* subsp. *paratuberculosis*. *Veterinary Pathology* 45(6), 829–841. DOI: 10.1354/vp.45-6-829.

Weiss, D.J., Evanson, O.A., McClenahan, D.J., Abrahamsen, M.S. and Walcheck, B.K. (2001) Regulation of expression of major histocompatibility antigens by bovine macrophages infected with *Mycobacterium avium* subsp. *paratuberculosis* or *Mycobacterium avium* subsp. *avium*. *Infection and Immunity* 69(2), 1002–1008. DOI: 10.1128/IAI.69.2.1002-1008.2001.

Weiss, D.J., Evanson, O.A. and Souza, C.D. (2005) Expression of interleukin-10 and suppressor of cytokine signaling-3 associated with susceptibility of cattle to infection with *Mycobacterium avium* subsp *paratuberculosis*. *American Journal of Veterinary Research* 66(7), 1114–1120. DOI: 10.2460/ajvr.2005.66.1114.

Weiss, D.J., Souza, C.D., Evanson, O.A., Sanders, M. and Rutherford, M. (2008) Bovine monocyte TLR2 receptors differentially regulate the intracellular fate of *Mycobacterium avium* subsp. *paratuberculosis* and *Mycobacterium avium* subsp. *avium*. *Journal of Leukocyte Biology* 83(1), 48–55. DOI: 10.1189/jlb.0707490.

Whittington, R.J. and Sergeant, E.S. (2001) Progress towards understanding the spread, detection and control of *Mycobacterium avium* subsp *paratuberculosis* in animal populations. *Australian Veterinary Journal* 79(4), 267–278. DOI: 10.1111/j.1751-0813.2001.tb11980.x.

Whittington, R.J., Lloyd, J.B. and Reddacliff, L.A. (2001) Recovery of *Mycobacterium avium* subsp. *paratuberculosis* from nematode larvae cultured from the faeces of sheep with Johne's disease. *Veterinary Microbiology* 81(3), 273–279. DOI: 10.1016/S0378-1135(01)00345-5.

Whittington, R.J., Marshall, D.J., Nicholls, P.J., Marsh, I.B. and Reddacliff, L.A. (2004) Survival and dormancy of *Mycobacterium avium* subsp. *paratuberculosis* in the environment. *Applied and Environmental Microbiology* 70(5), 2989–3004. DOI: 10.1128/AEM.70.5.2989-3004.2004.

Woo, S.-R., Heintz, J.A., Albrecht, R., Barletta, R.G. and Czuprynski, C.J. (2007) Life and death in bovine monocytes: the fate of *Mycobacterium avium* subsp. *paratuberculosis*. *Microbial Pathogenesis* 43(2–3), 106–113. DOI: 10.1016/j.micpath.2007.04.004.

Wu, C.W., Livesey, M., Schmoller, S.K., Manning, E.J.B., Steinberg, H. *et al.* (2007a) Invasion and persistence of *Mycobacterium avium* subsp. *paratuberculosis* during early stages of Johne's disease in calves. *Infection and Immunity* 75(5), 2110–2119. DOI: 10.1128/IAI.01739-06.

Wu, C.W., Schmoller, S.K., Shin, S.J. and Talaat, A.M. (2007b) Defining the stressome of *Mycobacterium avium* subsp. *paratuberculosis in vitro* and in naturally infected cows. *Journal of Bacteriology* 189(21), 7877–7886. DOI: 10.1128/JB.00780-07.

Yates, R.M. and Russell, D.G. (2005) Phagosome maturation proceeds independently of stimulation of Toll-like receptors 2 and 4. *Immunity* 23(4), 409–417. DOI: 10.1016/j.immuni.2005.09.007.

Zhu, X., Tu, Z.J., Coussens, P.M., Kapur, V., Janagama, H. *et al.* (2008) Transcriptional analysis of diverse strains *Mycobacterium avium* subspecies *paratuberculosis* in primary bovine monocyte derived macrophages. *Microbes and Infection* 10(12–13), 1274–1282. DOI: 10.1016/j.micinf.2008.07.025.

Zielinski, C.E., Mele, F., Aschenbrenner, D., Jarrossay, D., Ronchi, F. *et al.* (2012) Pathogen-induced human Th17 cells produce IFN-γ or IL-10 and are regulated by IL-1β. *Nature* 484(7395), 514–518. DOI: 10.1038/nature10957.

10 Drug Susceptibility Testing and Antimicrobial Resistance in *Mycobacterium avium* subsp. *paratuberculosis*

Jaryd R. Sullivan* and Marcel A. Behr
McGill University, Montreal, Quebec, Canada

10.1 Introduction

There is currently a limited role for antimicrobials in the management of infection and disease caused by *Mycobacterium avium* subsp. *paratuberculosis* (*MAP*). Yet, one can envision several potential roles, spanning from treatment of infected animals to prevention of onward spread to management of infection in spillover hosts, including humans. In order to select antimicrobial agents, whether existent or novel, a first step is determining which agents can inhibit the growth of *MAP in vitro*. This is based on the premise that, while an agent that is active *in vitro* may or may not be active *in vivo*, an agent that does not manifest activity *in vitro* is unlikely to achieve activity *in vivo*. As a foundation for such studies, it is critical to have a robust laboratory approach to assess activity. This test is often called an antimicrobial susceptibility test (AST), also known in the mycobacteria field as drug susceptibility testing (DST) (see Box 10.1).

This chapter will emphasize DST in the context of mycobacteria while focusing on the slow-growing organism *MAP*. First, a general overview of DST on mycobacteria and how DST is carried out in clinical microbiology labs will be provided. Second, reasons for establishing standardized DST in *MAP* will be discussed. Finally, new methods and current progress on antimicrobial treatment of *MAP* will be highlighted.

10.2 Drug Susceptibility Testing Definition and Description

The field of DST is rife with technical jargon that merits a brief overview. Box 10.1 defines common DST vocabulary found throughout the chapter.

AST is the cornerstone of the practice of infectious diseases and the foundation of many multidisciplinary laboratories, namely clinical laboratories, national reference centres, hospital laboratories and basic research groups. The outcome of AST on a microorganism is a determination whether the microbe is susceptible or resistant to an antimicrobial agent. A physician may be interested in susceptibility results to assist in correct antimicrobial administration, while infection control programmes may emphasize antimicrobial resistance to predict the spread of resistance. The Clinical Laboratory Standards Institute (CLSI) defines resistance as 'a decrease in susceptibility compared to a sample of drug-naïve wild strains' (Desmond *et al.*, 2011). In technical

*Corresponding author: jaryd.sullivan@mail.mcgill.ca

© CAB International 2020. *Paratuberculosis: Organism, Disease, Control, 2nd Edition* (eds M.A. Behr *et al.*)

Box 10.1. Vocabulary.

Antibiotic: A chemical compound within the subset of antimicrobials that inhibits the growth of bacteria.

Antibiotic resistance: The innate or acquired ability of bacteria to resist the effects of an antibiotic. Resistance is typically concerning at clinically relevant antibiotic concentrations. This concept is largely determined *in vitro* and is distinct from therapeutic failure, wherein an antibiotic fails to eliminate the infection.

Antimicrobial: A synthetic or naturally derived compound that inhibits the growth of bacteria, viruses or fungi.

Antimicrobial susceptibility testing (AST): Laboratory testing on microorganisms to determine resistance and susceptibility levels to antimicrobials.

CLSI: The Clinical and Laboratory Standards Institute is a US-based organization that develops clinical and laboratory standards, practices and guidelines.

Critical concentration: The lowest concentration of drug that inhibits 95% of wild strains that have never been exposed to drugs.

Drug susceptibility testing (DST): Subset of AST used in the field of mycobacteriology.

EUCAST: The European Committee on Antimicrobial Testing is a European-based society that develops laboratory standards for clinical microbiology and infectious diseases.

Inoculum: Volume/concentration of microorganism used in an AST assay.

Indeterminate (I): A microorganism that is deemed neither susceptible nor resistant according to established minimum inhibitory concentration (MIC) boundaries. These MICs do not suggest withholding the drug but emphasize appropriate adjustment of the dose. NB Some use 'I' to indicate Intermediate, but this use of I is discouraged by the CLSI because it implies a strength phenotype only, and excludes technical errors.

Isolate: A microbe isolated from a patient sample.

Microorganism: Categorical term encompassing bacteria, fungi and viruses. Commonly shortened to microbe.

Minimum inhibitory concentration (MIC): The lowest concentration of an antimicrobial that inhibits microbial growth. Typically denoted as MIC_{90} for inhibition of 90% of growth.

Resistance pattern: Summary of antimicrobial testing results for an isolate.

Resistant (R): A patient isolate *should not* be susceptible to dosage normally recommended to treat that type of infection and species.

Susceptible (S): A microorganism that is not resistant to antimicrobials at clinically relevant concentrations. A patient isolate *should* be susceptible to dosage normally recommended to treat that type of infection and species.

terms, a microorganism is considered to be resistant if greater than 0.1% of the inoculum grows through the critical concentration of the drug. In accordance, EUCAST defines resistance with a similar definition but with greater emphasis on standard dosing regimen. For example, EUCAST categorizes a microorganism as resistant when 'there is high likelihood of therapeutic failure even when there is increased exposure' (Kahlmeter, 2019, p. 17). Therefore, strains are drug resistant even with a modified dose and dosing interval or parental administration.

Mycobacteria DST is a variant of AST where resistance is established if the antimicrobial critical concentration permits more than 1.0% of inoculum growth compared with drug-naïve wild strains (Desmond *et al.*, 2011). *Mycobacterium tuberculosis* DST has been shown to predict clinical outcomes in trials and meta-analyses, so it would appear that the 99% threshold is useful (Kim, 2005; Lew *et al.*, 2008). Whether these modified standards apply to *MAP* and other non-tuberculous mycobacteria (NTM) – is unknown. Determining the resistance pattern could help guide appropriate treatment options, identify a possible mechanism for failed antimicrobial therapy and maintain surveillance for emerging drug resistance in populations (see Box 10.2, for example).

Traditionally, DST examines whether a particular drug can inhibit the growth or metabolism of a bacterium *in vitro*. Classical DST relies on macroscopic observation of growth in drug and drug-free media through colony-forming unit (CFU) counts on agar or optical density

Box 10.2. S vs R vs I.

If organisms are highly resistant to an antibiotic, it is unlikely that the antibiotic can provide clinical benefit, as seen in Panel A (Fig. 10.1). However, the MIC distribution may follow a distribution closer to that seen in Panel B, where organisms that are not S have an MIC that is increased, but to a lesser extent. In the example of *M. tuberculosis*, resistance to isoniazid is a heterogeneous phenotype. Resistance due to *katG* deletions cannot be overcome, as isoniazid is a pro-drug, for which bacterial KatG is required for activation. Resistance attributed to promoter mutations in *inhA* leads to so-called 'low-level resistance'. Here, the provision of an MIC result that is one to two dilutions higher than the standard threshold, coupled with a genetic test detecting the mutation, can inform the clinician that the resistance can be overcome with a higher dose of the antibiotic. Hence, results from DST should be interpreted appropriately with biological context and possible mechanism of action of the drug – DST means more than just reading S vs R in a microbiology report.

(OD_{600nm}) readout in liquid culture. Indeed, these methods are slow and dated for modern medicine, however, they remain the gold standard according to the CLSI guidelines.

Researchers and clinical labs have developed methodologies for DST that measure the metabolism of bacteria rather than visible growth as a means to circumvent the issue of slow-growing mycobacteria. Resazurin Microtiter Assay (REMA) and Bactiter Glo™ are two assays that monitor metabolism through intracellular redox potential and ATP synthesis, respectively (Niles *et al.*, 2008; Necchi *et al.*, 2018). Additionally, the Mycobacteria Growth Indicator Tube (MGIT) design was modified from the original BACTEC™ blood culture system from BD Diagnostic Systems to culture bacteria from patient samples (Automated Blood Culture System, 2014). Each MGIT™ tube contains an inorganic dye, which fluoresces upon exposure to CO_2 produced from metabolically active mycobacteria. MGIT™ culture systems could be described as the backbone of clinical mycobacteriology labs in many countries due to the ease of use and automated system.

The gold standard for *M. tuberculosis* frontline drug DST remains the agar proportion method for isoniazid (INH), rifampicin (RIF) and

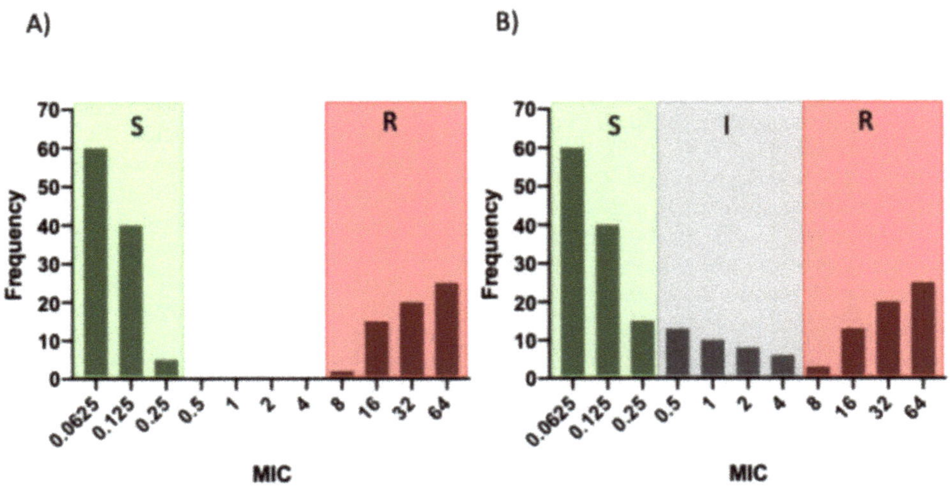

Fig. 10.1. Histogram of hypothetical minimum inhibitory concentration (MIC) distribution for a microorganism. (A) Categorical distribution. (B) Graded distribution.

ethambutol (EMB), and the radiometric proportion method for pyrazinamide (PZA) – despite the arrival of new methods with shorter incubation times. In light of these standards, the CLSI recognizes growth on solid media is time consuming and recommends broth culture with a shorter incubation time for industrialized countries. The Centres for Disease Control and Prevention also recommends broth culture for DST. The goal is to report results of primary DST within 15–30 days of receiving the patient sample. Unfortunately, countries lacking the facilities for automated culture systems must rely on standard agar growth.

10.3 Drug Susceptibility Testing for *MAP*, Why Bother?

Traditionally, *MAP* is a recognized pathogen in ruminants, and is well established as the causative agent of paratuberculosis. Unfortunately, there is no industry standard treatment for paratuberculosis. Veterinarians and cattle farmers rely on infection control programmes detailed elsewhere (Chapter 21, this volume). However, establishing DST in the context of *MAP* infections could have purpose in infectious disease medicine.

In human medicine, the hypothesized role of *MAP* as the aetiological agent of Crohn's disease (see Chapter 3, this volume) suggests the potential value of DST in managing human infection and disease (Fredricks and Relman, 1996; Chamberlin *et al.*, 2007a; Kuenstner *et al.*, 2017; Oken *et al.*, 2017; Zarei-Kordshouli *et al.*, 2019). Should new evidence emerge to support a mycobacterial link to Crohn's disease, there will be a need to develop standardized DST methods with the goal of guiding personalized therapies. These DST methods would also help promote a drug discovery pipeline for novel anti-*MAP* therapies.

10.4 *MAP* Drug Susceptibility Testing: Where Are We in 2019?

One of the key advances related to DST was the development of a reliable method to quantify *MAP* burden using a modified MGIT™ system. This nonradiometric BACTEC™ MGIT™ 960 system, which detects mycobacterial species commonly found in patient samples, was adapted for use with a culture medium specific for *MAP* (MGIT™ Para-TB medium). When comparing the new BACTEC™ MGIT™ 960 against the original BACTEC™ 460 system to enumerate *MAP*, Sung and Collins reported that the new system reduced the time to detection (TTD) from samples with various starting inocula (Sung *et al.*, 2007). The opportunity to culture *MAP* in less time may lend itself useful for *MAP* DST because it partially addresses the issue of antibiotic stability during the long period of a DST. One could incubate *MAP* and an antimycobacterial agent in the MGIT™ Para-TB medium to determine efficacy and MIC (Sung *et al.*, 2007).

In 2008, Beckler and colleagues undertook the challenge of establishing a correlation between gene mutations and phenotypic antibiotic resistance (Beckler *et al.*, 2008). They proposed that the most direct route to *MAP* DST is a molecular typing method rather than relying on the slow-growing nature of *MAP* (doubling time >24 h). Frontline anti-TB drugs RIF and rifabutin (RFB), with the well-established corresponding target RpoB, were used to illustrate correlation between single nucleotide polymorphisms (SNPs) and clinical significance. A molecular-based polymerase chain reaction (PCR) method was designed to probe the 81-bp rifampicin resistance-determining region (RRDR) in *rpoB* of the ATCC *MAP* strain, numerous human and cattle isolates, and three external species. *MAP* strains that harboured mutations in the RRDR displayed a marked increase in MIC to RIF (Table 10.1). Additionally, differing effects on MIC to RIF were observed depending on the mutations detected. The positive control was a strain of *MAP* cultured in increasing concentrations of RIF to artificially induce RIF resistance. The MIC to RIF increased 30× when leucine^{481} was substituted for proline. The test isolate UCF4 had a modest increase in the MIC to RIF when asparagine^{762} was substituted for histidine (Lin *et al.*, 2017). The study only identified one mutation from artificial culturing conditions; however, if a panel of common mutations were identified with measured MIC then mutations in *rpoB* may serve as a predictor for treatment success with RIF. Indeed, this methodology serves as the underlying principle of the commercial GeneXpert MTB/RIF assay sold by Cepheid. Molecular DST may be an appealing

Table 10.1. *rpoB* mutations and clinical significance. (Adapted from Beckler *et al.*, 2008.)

Microorganism	Strain	Rifampicin minimum inhibitory concentration (mg/l)	% Inhibited at 1 mg/l of RIF	*rpoB* gene	ΔAmino acid
MAP	ATCC 43544	≤1.0	91 ± 0.71	WT	None
MAP	UCF3	≤1.0	94 ± 2.83	WT	None
MAP	UCF4	≤2.5	79 ± 7.07	A2284C	N762H
MAP	UCF5	≤1.0	92.5 ± 0.71	WT	None
MAP	UCF7	≤1.0	88 ± 5.66	WT	None
MAP	UCF8	≤1.0	90 ± 1.41	WT	None
MAP	185	≤4.0	61 ± 7.07		None
MAP	Cow2013	≤1.0	95 ± 1.41	Silent	None
MAP	Cow5	≤1.0	96.5 ± 0.71	WT	None
MAP	61a	≤1.0	90 ± 3.54	WT	None
				WT	
UCF-RIF16r	Modified UCF5	≥30	32 ± 1.41	T1442C	L481P

MAP, *Mycobacterium avium* subsp. *paratuberculosis*; WT, Wild type.

alternative to classical DST, but further studies would be required to evaluate the DST prediction of clinical outcomes for *MAP*. Molecular typing results are categorical (wild type vs mutant) yet the clinical outcome is determined by the phenotype of the bacteria, which is a quantitative evaluation (see Box 10.2). Moreover, molecular DST loses sensitivity when mutations fall outside of the target region, illustrated by the clinical isolate MAP185. This isolate contained a silent mutation in the RRDR (i.e. no amino acid substitution in RpoB) yet still manifested an increased MIC to RIF, suggesting that RIF resistance in *MAP*, as in *M. tuberculosis* and many other bacterial species, may result from mutations outside of the RRDR.

According to a review by Gisbert and Panés in 2009, up to 40% of naïve Crohn's disease patients fail both traditional and biological therapies, specifically tumour necrosis factor (TNF) monoclonal antibody inhibitor infliximab (Gisbert and Panés, 2009). Authors of an *Expert Opinion on Biological Therapy* article propose that the 40% of patients who display primary refractoriness should be investigated for *MAP* infection and might benefit from treatment with antimicrobials for mycobacteria upon confirmation (Savarino *et al.*, 2019). Several randomized controlled trials (Borody *et al.*, 2002; Selby *et al.*, 2007; Alcedo *et al.*, 2016) and case series (Chamberlin *et al.*, 2007b; Alcedo *et al.*, 2016) reported promising results from Crohn's disease patients on anti-*MAP* regimens and complete healing of Crohn's disease ulcers over a 6-month period when RIF and clarithromycin (CLA) were used as a dual therapy, respectively. Building upon these findings, RedHill Biopharma created a triple antibiotic combination therapy in one pill that they call RHB-104: 95 mg CLA (0.63%), 10 mg clofazimine (CLO) (0.067%) and 45 mg RIF (0.30%), with an intended adult dose of five pills twice daily (Alcedo *et al.*, 2016). RHB-104 was tested on 16 clinical *MAP* isolates from Crohn's disease patients, 10 *M. avium* strains and nine other mycobacteria species. Bacteria were cultured in BD BACTEC™ MGIT™ Para-TB medium with corresponding drug dosages and combinations. The data suggest RHB-104 is a potent *MAP* antibiotic combination therapy with MIC of 0.25 μg/ml for susceptible isolates and MIC of 4–10 μg/ml for resistant isolates. Each antibiotic in the regimen was effective against *MAP;* however, when dual combinations were tested it appeared that CLA was the primary contributor to the overall anti-*MAP* activity. It has been argued that successful *MAP* therapies require ribosomal inhibitors, like CLA, to eliminate spheroplast forms of *MAP* that might potentiate the pro-inflammatory immune response, although the biology of spheroplasts in *MAP* infection is unresolved (see Chapter 3, this volume; Monif, 2018). The study reported synergy between the CLA-CLO-RIF antibiotic regimen; however, synergy tests were performed without appropriate dose–response and isobole regression analysis. Interested readers are encouraged to explore a theoretical approach to drug synergy (Tallarida, 2011). Recently, RHB-104 has been tested in a phase III, multicentre, randomized controlled trial in moderately to severely active Crohn's disease patients – published results are awaited later in 2019 or 2020.[1] Patients were stratified for concomitant use of anti-TNF agents and treated with either RHB-104 or placebo as an add-on to standard of care for 52 weeks in an open-label regimen. Primary outcome was remission after 16 weeks with RHB-104. Secondary outcomes include time of response, response duration, time to remission and remission duration. Early data suggest RHB-104 is a potentially new treatment for Crohn's disease.[2] Remission rates and safety analysis in patients with anti-TNF use indicate that RHB-104 may be effective and safe as an adjunct treatment to standard of care therapy for Crohn's disease. Whether the data provide support for the idea that *MAP* (or an alternative microbial target that responds to the same antibiotics) acts as an aetiological agent of Crohn's disease in some of these patients will be determined by the microbiological investigations embedded in the trial.

In 2011, Fecteau *et al.* postulated that similar trivalent metal ions to iron, like gallium, could interfere with iron homeostasis and cause bacterial cell death (Fecteau *et al.*, 2011a). They evaluated the efficacy of gallium nitrate as a growth inhibitor against *MAP* isolates from cattle, bison, alpaca and humans. They reported MIC_{90} values ranging from 200 to 700 μM. The authors abandoned gallium nitrate studies when it failed to provide sufficient protection to calves from *MAP* infection (Fecteau *et al.*, 2011b). The same authors turned to gallium maltolate, a metal-organic coordination complex with improved lipid solubility, as a potential *MAP* growth inhibitor and compared antimicrobial activities with the original metal salt, gallium nitrate. MIC_{90} values for two bovine isolates were reported as the following: 190 μM (164–220 95% CI) and 140 μM (102–194 95% CI) for gallium nitrate, and 94 μM (74–122) (95% CI) and 132 μM (99–179) (95% CI) for gallium maltolate (Fecteau *et al.*, 2014). Gallium maltolate displayed a slight increase in antimicrobial activity with greater lipophilicity. Despite similarities to iron, gallium appears to be a poor antimicrobial agent whether as a salt (nitrate) or coordination complex (maltolate).

Most antimicrobial agents tested against *MAP* have been small inorganic molecules (see gallium study) or repurposed TB drugs. In 2017, one of the first studies by Goněc *et al.* (2017) examined the antimicrobial activity of novel, fully synthetic n-arylpiperazines containing an ethane-1,2-diyl linker. Compounds 8a-c and 8e-h exhibited MIC_{90} >500 μM. Notwithstanding adequate lipophilic properties, the n-arylpiperazine series with ethane-1,2-diyl linker appeared to be a poor chemical scaffold against *MAP*. Although compound 8d exhibited the lowest MIC_{90} value (53 μM) reported by Goněc *et al.* (2017), compound 8d may have highlighted a unique starting scaffold for anti-*MAP* activity. It may be fruitful to probe chemical space around compound 8d with structure–activity-relationship experiments/analysis/studies.

Perhaps the most promising study of *MAP* drug discovery was from 2018. Quinolones have been recognized as a prominent class of fully synthetic antimicrobials since 1962 when Lesher *et al.* made their serendipitous discovery of nalidixic acid (Andriole, 2005). Then in 1980, the class of quinolones was revolutionized by the addition of fluorine, and the first generation of fluoroquinolones was born. The effective use of fluoroquinolones prompted medicinal chemists to probe derivatives like quinolines. Indeed, the most impactful study on drug discovery in the field of mycobacteria in recent years was the revealing of bedaquiline, a diarylquinoline antibiotic approved for the treatment of multidrug-resistant TB in 2012. This novel quinoline backbone created sparks in the medicinal chemistry community, and Lougheed and collaborators discovered anti-TB activity from primaquine (PQ), an 8-aminoquinoline anti-malarial drug (Lougheed *et al.*, 2009). Pavic *et al* synthesized four novel series consisting of amides, ureas, semicarbazides or bis-ureas as the linkers with the main primaquine pharmacophore and examined their antimycobacterial activity on *MAP* (Pavić *et al.*, 2018). They observed that hydroxyl, halogen and trifluoromethyl substituted benzene rings belonging to the urea series were universally active against all mycobacteria tested, while amides exhibited weak activity. The semicarbazide series showed equipotent activity to the parent compound PQ. Four promising candidates were discovered in the bis-urea series. In the end, compounds 2 m, 4 p and 4 u (*MAP* MIC: 2 μg/ml, 2 μg/ml, 4 μg/ml, respectively) were selected as the most potent antimycobacterial agents and well tolerated in an L6 cytotoxicity cell model (cytotoxicity: 150 μg/ml, 30 μg/ml, 81.6 μg/ml). The MIC against *MAP* and L6 cytotoxicity of PQ is 32 μg/ml and 27.1 ug/ml, respectively. Thus, Pavić *et al.* (2018) were able to build upon the primaquine pharmacophore and improve the activity of the parent compound while minimizing cytotoxic effects.

10.5 Concluding Remarks

Since the 2010 iteration of this book, only 250 studies have been published with '*Mycobacterium paratuberculosis* drug susceptibility testing' included in the PubMed Central search terms; in contrast, '*Mycobacterium tuberculosis* drug susceptibility testing' returned 8296 hits at the time of writing. It is therefore not surprising that we have yet to establish anti-*MAP* regimens either from pre-existing TB drugs or novel drug discovery efforts. Fortunately, progress has been made since the publication of the landmark article by Crohn, Ginzburg and Oppenheimer in 1932 (Crohn *et al.*, 1932). Modified media for culturing *MAP* has been validated by Sung and Collins in 2007, improving TTD. Consequently, RedHill Biopharma formulated a triple antibiotic regimen with CLA, CLO and RIF, and evaluated its potency *in vitro* against *MAP* while taking advantage of the BD BACTEC™ MGIT™ Para-TB system previously validated. Preliminary results that have been released by RedHill Biopharma suggest that RHB-104 is the first therapy to provide long-term benefit to patients with Crohn's disease. Previous clinical trial data only supported short-term benefit from antimycobacterial therapy and lacked microbial specificity (Behr and Kapur, 2008). It is known from meta-analyses that Crohn's disease patients given antibiotics improved their condition better than placebo control groups (Rahimi *et al.*, 2006; Feller *et al.*, 2007; Khan *et al.*, 2011; Davis *et al.*, 2017). Regrettably, most clinical trials examined Crohn's disease therapy outcomes on patients unconfirmed for *MAP* and, therefore, efficacy of the therapy towards *MAP* infection could not be concluded. In an *Expert Review in Gastroenterology and Hepatology* article by Davis *et al.* in 2017, the importance of evaluating the efficacy of all current and experimental antimycobacterial agents against *MAP in vitro* before designing future trials to supplement previous clinical trial findings was emphasized.

Other groups have taken an interest in developing anti-*MAP* therapies by expanding the search through non-specific mycobacterial chemical space. In other words, targeting *MAP* with general antimicrobial compounds. Some approaches have yet to produce promising results, while others are gaining traction. For example, disrupting iron homeostasis with metal ion mimic gallium proved to be unsuccessful according to Fecteau and colleagues. In addition, the piperazine backbone, recognized in antihelminth medication, failed to elicit strong anti-*MAP* activity as demonstrated by Goněc *et al.* (2017). The most recent drug discovery effort towards *MAP* came from Pavić *et al.* (2018) when they repurposed the antimalarial drug primaquine for *MAP*. Not only did the primaquine series display activity against *MAP* in their study, but select compounds in the urea and bis-urea primaquine series were also active against *M. tuberculosis* and clinical isolates of *M. avium* complex. Nevertheless, novel antimicrobials must pass the rigorous trials of therapy development. First, development must begin with *in vitro* DST and progress to *ex vivo* and/or animal model studies. Lastly, candidate therapies must demonstrate benefit in human clinical trials.

Perhaps the most direct route to establishing DST and discovering antimicrobials for *MAP* is through the approaches used for the study of *M. avium*. While working with *MAP* may be more challenging than *M. avium*, due to its slow and fastidious growth, the CLSI has provided recommendations for DST methods of other fastidious NTM, such as *M. haemophilum* (CLSI, 2011). An agar disc elution protocol is available through the CLSI manual.[3] One could envision repurposing antibiotics commonly used for *M. avium* infections for *MAP*. Moreover, success with repurposed antibiotics against *MAP* may indicate the necessity for multi-drug regimens in the future.

The fields of veterinary and human medicine have made substantial contributions towards *in vitro* DST and antimicrobial development for *MAP*. One of the critical factors in establishing DST is method validation and laboratory accreditation. A key contribution to *MAP* DST could be defined *MAP* mutants with known resistance phenotypes for distribution to clinical microbiology labs and reference centres. Such strains could provide a reference point for labs and assays to determine susceptibility vs resistance. Without appropriate controls, DST results are of uncertain significance and, by extension, have unknown capacity to predict clinical outcomes in veterinary and human medicine. Absence of reliable DST methods is expected to hinder efforts to identify new compounds with anti-*MAP* activity, further reinforcing the notion that there is a limited role for antimicrobials in the management of *MAP* infections.

Notes

[1] N.B. There are no macrophage or pre-clinical model data to substantiate the *in vitro* efficacy as RHB-104 is three pre-existing tuberculosis drugs repurposed for *MAP* infection.

[2] Trial is stated to finish October 2019.

[3] N.B. This is a non-standardized method recommended by the CLSI.

References

Alcedo, K.P., Thanigachalam, S. and Naser, S.A. (2016) RHB-104 triple antibiotics combination in culture is bactericidal and should be effective for treatment of Crohn's disease associated with *Mycobacterium paratuberculosis*. *Gut Pathogens* 8(1). DOI: 10.1186/s13099-016-0115-3.

Andriole, V.T. (2005) The quinolones: past, present, and future. *Clinical Infectious Diseases* 41(Suppl. 2), S113–S119. DOI: 10.1086/428051.

Automated Blood Culture System (2014) *NEWSLETTER Best Practices in Blood Culture Collection Fluorescent Technology in the BD BACTEC™*. Available at: www.bmsd.com.my (accessed 9 July 2019).

Beckler, D.R., Elwasila, S., Ghobrial, G., Valentine, J.F. and Naser, S.A. (2008) Correlation between rpoB gene mutation in *Mycobacterium avium* subspecies *paratuberculosis* and clinical rifabutin and rifampicin resistance for treatment of Crohn's disease. *World Journal of Gastroenterology* 14(17), 2723–2730. DOI: 10.3748/wjg.14.2723.

Behr, M.A. and Kapur, V. (2008) The evidence for *Mycobacterium paratuberculosis* in Crohn's disease. *Current Opinion in Gastroenterology* 24(1), 17–21. DOI: 10.1097/MOG.0b013e3282f1dcc4.

Borody, T.J., Leis, S., Warren, E.F. and Surace, R. (2002) Treatment of severe Crohn's disease using antimycobacterial triple therapy – approaching a cure? *Digestive and Liver Disease* 34(1), 29–38. DOI: 10.1016/S1590-8658(02)80056-1.

Chamberlin, W., Borody, T. and Naser, S. (2007a) MAP-associated Crohn's disease map, Koch's postulates, causality and Crohn's disease. *Digestive and Liver Disease* 39(8), 792–794. DOI: 10.1016/j.dld.2007.05.012.

Chamberlin, W., Ghobrial, G., Chehtane, M. and Naser, S.A. (2007b) Successful treatment of a Crohn's disease patient infected with bacteremic *Mycobacterium paratuberculosis*. *The American Journal of Gastroenterology* 102(3), 689–691. DOI: 10.1111/j.1572-0241.2007.01040_7.x.

CLSI (2011) Appendix J. agar disk elution method for *Mycobacterium haemophilum*: susceptibility testing of mycobacteria, *Nocardia* spp., and other aerobic actinomycetes. Available at: www.clsi.org (accessed 27 November 2019).

Crohn, B.B., Ginzburg, L., Oppenheimer, G.D. and Ilietis, R. (1932) Regional ileitis. *Journal of the American Medical Association* 99(16), 1323. DOI: 10.1001/jama.1932.02740680019005.

Davis, W.C., Kuenstner, J.T. and Singh, S.V. (2017) Resolution of Crohn's (Johne's) disease with antibiotics: what are the next steps? *Expert Review of Gastroenterology & Hepatology* 11(5), 393–396. DOI: 10.1080/17474124.2017.1300529.

Desmond, E., Woods, G.L., Brown-Elliott, B.A., Hall, G.S. and Conville, P.S. (2011) Susceptibility testing of mycobacteria, nocardiae and other aerobic actinomycetes. In: *Susceptibility Testing of Mycobacteria, Nocardiae, and Other Aerobic Actinomycetes, Nocardiae and other aerobic actinomycetes*, 2. Clinical and Laboratory Standards Institute, Wayne, Pennsylvania.

Fecteau, M.-E., Fyock, T.L., McAdams, S.C., Boston, R.C., Whitlock, R.H. *et al.* (2011a) Evaluation of the in vitro activity of gallium nitrate against *Mycobacterium avium* subsp *paratuberculosis*. *American Journal of Veterinary Research* 72(9), 1243–1246. DOI: 10.2460/ajvr.72.9.1243.

Fecteau, M.-E., Whitlock, R.H., Fyock, T.L., McAdams, S.C., Boston, R.C. *et al.* (2011b) Antimicrobial activity of gallium nitrate against *Mycobacterium avium* subsp. *paratuberculosis* in neonatal calves. *Journal of Veterinary Internal Medicine* 25(5), 1152–1155. DOI: 10.1111/j.1939-1676.2011.0768.x.

Fecteau, M.-E., Aceto, H.W., Bernstein, L.R. and Sweeney, R.W. (2014) Comparison of the antimicrobial activities of gallium nitrate and gallium maltolate against *Mycobacterium avium* subsp. *paratuberculosis in vitro*. *The Veterinary Journal* 202(1), 195–197. DOI: 10.1016/j.tvjl.2014.06.023.

Feller, M., Huwiler, K., Stephan, R., Altpeter, E., Shang, A. *et al.* (2007) *Mycobacterium avium* subspecies *paratuberculosis* and Crohn's disease: a systematic review and meta-analysis. *The Lancet Infectious Diseases* 7(9), 607–613. DOI: 10.1016/S1473-3099(07)70211-6.

Fredricks, D.N. and Relman, D.A. (1996) Sequence-based identification of microbial pathogens: a reconsideration of Koch's postulates. *Clinical Microbiology Reviews* 9(1), 18–33. DOI: 10.1128/CMR.9.1.18.

Gisbert, J.P. and Panés, J. (2009) Loss of response and requirement of infliximab dose intensification in Crohn's disease: a review. *The American Journal of Gastroenterology* 104(3), 760–767. DOI: 10.1038/ajg.2008.88.

Goněc, T., Malík, I., Csöllei, J., Jampílek, J., Stolaříková, J. *et al.* (2017) Synthesis and in vitro antimycobacterial activity of novel N-Arylpiperazines containing an Ethane-1,2-diyl connecting chain. *Molecules* 22(12), 2100. DOI: 10.3390/molecules22122100.

Kahlmeter, G. (2019) Gunnar Kahlmeter and the EUCAST Steering Committee. Redefining susceptibility testing categories S, I and R. Available at: www.eucast.org (accessed 9 August 2019).

Khan, K.J., Ullman, T.A., Ford, A.C., Abreu, M.T., Abadir, A. *et al.* (2011) Antibiotic therapy in inflammatory bowel disease: a systematic review and meta-analysis. *American Journal of Gastroenterology* 106(4), 661–673. DOI: 10.1038/ajg.2011.72.

Kim, S.J. (2005) Drug-susceptibility testing in tuberculosis: methods and reliability of results. *European Respiratory Journal* 25(3), 564–569. DOI: 10.1183/09031936.05.00111304.

Kuenstner, J.T., Naser, S., Chamberlin, W., Borody, T., Graham, D.Y. *et al.* (2017) The consensus from the *Mycobacterium avium* ssp. *paratuberculosis* (MAP) conference 2017. *Frontiers in Public Health* 5(208). DOI: 10.3389/fpubh.2017.00208.

Lew, W., Pai, M., Oxlade, O., Martin, D. and Menzies, D. (2008) Initial drug resistance and tuberculosis treatment outcomes: systematic review and meta-analysis. *Annals of Internal Medicine* 149(2), 123–134. Available at: http://www.embase.com/search/results?subaction=viewrecord&from=export&id=L352008599 DOI: 10.7326/0003-4819-149-2-200807150-00008.

Lin, W., Mandal, S., Degen, D., Liu, Y., Ebright, Y.W. *et al.* (2017) Structural basis of *Mycobacterium tuberculosis* transcription and transcription inhibition. *Molecular Cell* 66(2), 169–179. DOI: 10.1016/j.molcel.2017.03.001.

Lougheed, K.E.A., Taylor, D.L., Osborne, S.A., Bryans, J.S. and Buxton, R.S. (2009) New anti-tuberculosis agents amongst known drugs. *Tuberculosis* 89(5), 364–370. DOI: 10.1016/j.tube.2009.07.002.

Monif, G.R.G. (2018) Understanding therapeutic concepts in Crohn's disease. *Clinical Medicine Insights: Gastroenterology* 11, 117955221881516. DOI: 10.1177/1179552218815169.

Necchi, F., Saul, A. and Rondini, S. (2018) Setup of luminescence-based serum bactericidal assay against *Salmonella paratyphi* a. *Journal of Immunological Methods* 461, 117–121. DOI: 10.1016/j.jim.2018.06.025.

Niles, A.L., Moravec, R.A. and Riss, T.L. (2008) Update on *in vitro* cytotoxicity assays for drug development. *Expert Opinion on Drug Discovery* 3(6), 655–669. DOI: 10.1517/17460441.3.6.655.

Oken, H.A., Saleeb, P.G., Redfield, R.R. and Schimpff, S.C. (2017) Is *Mycobacterium avium paratuberculosis* the trigger in the Crohn's disease spectrum? *Open Forum Infectious Diseases* 4(3), 1–3. DOI: 10.1093/ofid/ofx104.

Pavić, K., Perković, I., Pospíšilová, Šárka., Machado, M., Fontinha, D. *et al.* (2018) Primaquine hybrids as promising antimycobacterial and antimalarial agents. *European Journal of Medicinal Chemistry* 143, 769–779. DOI: 10.1016/j.ejmech.2017.11.083.

Rahimi, R., Nikfar, S., Rezaie, A. and Abdollahi, M. (2006) A meta-analysis of broad-spectrum antibiotic therapy in patients with active Crohn's disease. *Clinical Therapeutics* 28(12), 1983–1988. DOI: 10.1016/j.clinthera.2006.12.012.

Savarino, E., Bertani, L., Ceccarelli, L., Bodini, G., Zingone, F. *et al.* (2019) Antimicrobial treatment with the fixed-dose antibiotic combination RHB-104 for *Mycobacterium avium* subspecies *paratuberculosis* in Crohn's disease: pharmacological and clinical implications. *Expert Opinion on Biological Therapy* 19(2), 79–88. DOI: 10.1080/14712598.2019.1561852.

Selby, W., Pavli, P., Crotty, B., Florin, T., Radford-Smith, G. *et al.* (2007) Two-year combination antibiotic therapy with clarithromycin, rifabutin, and clofazimine for Crohn's disease. *Gastroenterology* 132(7), 2313–2319. DOI: 10.1053/j.gastro.2007.03.031.

Sung, J.S., Jun, H.H., Manning, E.J.B. and Collins, M.T. (2007) Rapid and reliable method for quantification of *Mycobacterium paratuberculosis* by use of the BACTEC MGIT 960 system. *Journal of Clinical Microbiology* 45(6), 1941–1948.

Tallarida, R.J. (2011) Quantitative methods for assessing drug synergism. *Genes & Cancer* 2(11), 1003–1008. DOI: 10.1177/1947601912440575.

Zarei-Kordshouli, F., Geramizadeh, B. and Khodakaram-Tafti, A. (2019) Prevalence of *Mycobacterium avium* subspecies *paratuberculosis* IS 900 DNA in biopsy tissues from patients with Crohn's disease: histopathological and molecular comparison with Johne's disease in Fars province of Iran. *BMC Infectious Diseases* 19(1). DOI: 10.1186/s12879-018-3619-2.

11 Paratuberculosis in Cattle

Marie-Eve Fecteau*
University of Pennsylvania, Kennett Square, Pennsylvania, USA

11.1 Introduction: Prevalence of Paratuberculosis in Cattle

Paratuberculosis is widely distributed throughout the world in many domesticated and wild ruminant species. Because of the long incubation period of paratuberculosis in cattle (between 2 and 7 years), the majority of infected cattle (around 95%) are considered to be in the subclinical stages of the disease, with less than 5% of infected cattle displaying clinical signs of illness. The term Johne's disease (JD) is often used interchangeably with paratuberculosis but should only be used to refer to the clinical syndrome of diarrhoea and weight loss that results from *Mycobacterium avium* subspecies *paratuberculosis* (*MAP*) infection in cattle.

Studies have been conducted in the United States as well as in other countries to estimate *MAP* prevalence in cattle. Faecal cultures, environmental cultures and antibody-detection methods have been the most common diagnostic methods used to determine the prevalence of infection. Estimates of *MAP* prevalence are complicated by several factors including the low sensitivity of available diagnostic tests for subclinical animals, infected cattle in the subclinical phase may be culled for reduced fertility or production and that cattle in the clinical phase of the disease are usually culled before a final diagnosis of *MAP* infection is made (Fecteau, 2018). In the US National Animal Health Monitoring System's (NAHMS) Dairy 2007 study, the apparent herd-level prevalence of *MAP* was 70.4% (National Animal Health Monitoring System Dairy, 2007). A follow-up analysis of the 2007 NAHMS data, using a Bayesian method, which adjusts for sensitivity and specificity, determined that the true US dairy herd-level prevalence of *MAP* was actually 91.1% (Lombard *et al.*, 2013). Within herds, the most recent estimate was 027.3% of cattle, with a mean of 5.5% (National Animal Health Monitoring System, 2002). High within-herd prevalence is usually associated with higher *MAP* faecal shedding amongst positive animals, *MAP* faecal shedding at a younger age, shorter incubation period and higher numbers of animals in the clinical phase of the disease (Lombard, 2011). Compared with dairy cattle, the herd prevalence of *MAP* in beef cattle is lower, with one study estimating *MAP* prevalence at 7.9% of US herds (Lombard, 2011). This difference in prevalence between beef and dairy is worldwide, and could be explained by differences in breed, environmental conditions, feeding practices and calf-rearing practices (Roussel, 2011).

11.2 Transmission of Paratuberculosis in Cattle

11.2.1 Routes of transmission

The vast majority of new infections occur via the faecal–oral route, although other routes,

*mfecteau@vet.upenn.edu

© CAB International 2020. *Paratuberculosis: Organism, Disease, Control, 2nd Edition* (eds M.A. Behr *et al.*)

including the vertical route, are possible (Seitz *et al.*, 1989). Calves born to seropositive dams were shown to be 6.6 times more likely to be seropositive compared with calves born to seronegative dams (Aly and Thurmond, 2005). A meta-analysis estimated that 9% of fetuses born from subclinically infected dams and 39% born from clinically affected dams were infected *in utero* (Whittington and Windsor, 2009). These findings indicate that the likelihood of fetal infection is dependent on the severity of the dam's infection. As the infection status of the dam represents a major risk factor for the newborn calf, it is not advised to keep a clinical cow within the herd in the hopes of obtaining the calf prior to culling, as the cow will continue to shed *MAP* organisms into the environment and the calf has a good chance of being infected.

MAP has been isolated from uterine flush fluids of infected cattle, and *MAP* organisms have been shown to adhere to embryos *in vitro* (Rohde and Shulaw, 1990). Therefore, an embryo obtained from an infected cow could, in theory, result in an infected fetus; however, oocytes and embryos processed according to current embryo transfer recommendations are unlikely to result in infected calves (Kruip *et al.*, 2003). Semen from clinically or subclinically infected bulls can be infected with *MAP* (Ayele *et al.*, 2004; Buergelt *et al.*, 2004). However, keeping in mind some of the key features of the pathogenesis of *MAP*, such as preferred route of infection (with oral being the main route of transmission) and age susceptibility (breeding-age cattle are less likely to become infected), the probability of transmission of *MAP* via infected semen should be considered low. Additionally, bulls kept in commercial bull studs are typically rigorously tested for *MAP*. Herd sires (dairy and beef) should be tested annually to ensure their negative status. In 2004, the European Food and Safety Authority Scientific Panel on Animal Health and Welfare (AHAW) publication addressed the risk of transmission of *MAP* via bovine semen (European Food and Safety Authority Scientific Panel on Animal Health and Welfare, 2004). More specifically, the working group asked about (i) the probability that a bull selected and used at semen collection centres in the European Union (EU) is infected with *MAP*, and (ii) the probability that the use of semen from that bull results in at least one infected animal. The AHAW panel concluded that given the current lack of sufficiently sensitive diagnostic test methods for subclinically infected animals, the likelihood of detecting *MAP* infection in a bull at 12–18 months of age (the typical age at which a bull would enter a semen collection centre) is very low, and that the selection of bulls should be based on considering only bull calves from *MAP*-negative cows within herds with low test-prevalence based on a repeat testing regimen. The panel also proposed that because of the lack of data on what represents an infective dose of *MAP*-infected semen, whether used for artificial insemination or *in vitro* fertilization, a probability of infection can't be established. They however concluded that infection of reproductive age cattle with *MAP*-infected semen is an event of low probability.

Most infections with *MAP* occur in the early neonatal period and are often associated with the calf coming into contact with manure when ingesting colostrum from the dam. Multiple-use maternity pens can serve as focal points to spread the infection to the neonates. *MAP* may also be passed directly through the colostrum and milk of cattle in the later stages of infection (Sweeney *et al.*, 1992a; Streeter *et al.*, 1995). It has been demonstrated that colostrum from known infected cows is a much greater risk to spread *MAP* to calves compared with colostrum from known negative dams (Nielsen *et al.*, 2008). The practice of feeding pooled colostrum or waste milk may help spread the infection from infected adults to many calves in the herd and should be discouraged. It is also important to note that *MAP* can survive pasteurization temperatures (Chiodini and Hermon-Taylor, 1993); therefore, colostrum and milk from known infected cows, even pasteurized, should not be used. Following birth, physical separation of the dairy calves to calf hutches, or better yet, to another property such as a heifer-raising facility, reduces the risk of *MAP* transmission to the replacement stock.

11.2.2 Susceptibility to Infection

Resistance to *MAP* infection increases with age (Windsor and Whittington, 2010). Possible explanations for the increased susceptibility of young calves may include: (i) the 'openness' of the newborn gut, which allows the calf to absorb immunoglobulins in colostrum could represent a permissive barrier for *MAP* organisms; (ii) the immaturity of the calf's immune system when compared with that of an adult cow (Sweeney, 2011); and (iii) the presence of the oesophageal groove, which allows the ingested *MAP* (typically via milk feeding) to bypass the rumen and be directed to the susceptible small intestine. Age-related resistance to infection can however be overcome by pressure of infection. Replacement stock and adult cattle that are exposed to *MAP*-contaminated forages, water and extremely contaminated environments can get infected with *MAP*. For example, the practice of offering feed refusals from adult cattle to younger heifers has been shown to be a significant risk factor to spread *MAP* (Rossiter and Burhans, 1996). Additionally, dairy herds using an intensive grazing system (also known as a leader–follower system) may have an increased risk of transmission to the replacement stock, due to the continuous low-level exposure to *MAP* organisms on pasture (Fecteau *et al.*, 2010). As noted above, unless massive and repeated doses of *MAP* are consumed, this is likely to be a relatively low risk for adult cows but could represent an important risk to younger replacement stock. One could also argue that cattle that become infected later in age may not start shedding *MAP* and develop clinical signs of JD before they leave the herd for other unrelated reasons.

Susceptibility to infection with *MAP* can also be genetically influenced. It was suggested that certain breeds of cattle (Guernsey, Jersey and *Bos indicus* breeds) and certain family lines may have a genetic predisposition to becoming infected with *MAP* (Chiodini *et al.*, 1984; Roussel *et al.*, 2005; Elzo *et al.*, 2006; Osterstock *et al.*, 2008; Kirkpatrick and Lett, 2018; Kiser *et al.*, 2018). Genetic and host susceptibility to *MAP* infection in cattle is discussed in more detail in Chapter 4 in this volume.

11.3 Pathogenesis and Stages of *MAP* Infection in Cattle

MAP infection has been divided into four stages, depending on the severity of clinical signs, the potential for shedding organisms into the environment and the ease with which the disease may be detected using current laboratory methods. As noted in Chapter 1 and the preface, an alternative, more condensed classification has been proposed in 2017, but the classical staging is presented here. It has been proposed that, for every cow with advanced JD that was born in the herd (on that farm), it is likely that 15–25 others are infected (Whitlock, 1992). Only 25–30% of these infected animals will be detectable by even the most sensitive molecular testing techniques (Whitlock, 2009). The clinical animal is the 'tip of the iceberg'. As an example, consider a herd with 100 adult cattle and 100 replacement stock. Two cows born on the farm several years earlier develop clinical signs, with weight loss and diarrhoea. It is likely that 30–50 other cattle are infected but less than 30% of these will be detectable by faecal culture and/or polymerase chain reaction (PCR) or antibody detection methodology. An outline of these stages, and their relative importance within an infected herd, is presented in Table 11.1.

Table 11.1. The 'iceberg effect' with different stages of *Mycobacterium avium* subsp. *paratuberculosis* (*MAP*) infection.

Stage	Type of infection	No. of animals
Stage IV	Advanced clinical disease	1 animal
Stage III	Clinical disease	1–2 cattle
Stage II	Inapparent carrier	6–8 cattle
Stage I	Silent infection	15–25 cattle

11.3.1 Entry of the organism

Early studies suggested that 10^3 bacilli were infectious (Gilmour *et al.*, 1965) and that, with a concentration of 10^6–10^8 *MAP* colony-forming units (CFU) per gram of faeces, only a few milligrams of manure ingested by a young calf would be infectious (Jørgensen, 1982; Whittington *et al.*, 2000). Another review suggested 50–1000 CFU as infective for young calves (Chiodini, 1996). Experimental studies have shown that a 1.5×10^6 CFU/dose given orally at 21 and 22 days of age reliably induced infection in multiple tissues, yet at a low level (Sweeney *et al.*,

2006). Higher doses at younger ages resulted in greater tissue infection. Following oral ingestion, most evidence points to the ileum, more specifically, the M cells, as the main portal of entry (Momotani *et al.*, 1988). Following entry through the M cells, *MAP* organisms are released on the submucosal side of the intestinal epithelium where they are captured by macrophages. The time required for intestinal translocation from the mucosa to adjacent lymph nodes may be as short as 1 h (Wu *et al.*, 2007). Moreover, studies in tissue cultures have demonstrated that *MAP* may affect the formation of tight junctions in the intestinal mucosa providing a mechanism for increased permeability (Bannantine and Bermudez, 2013).

From this point on, whether an animal remains infected with *MAP* depends on its immune response. If the macrophages are successful at killing the phagocytosed *MAP*, the host has a chance of fighting and clearing *MAP*. But because *MAP* organisms have the ability to survive within macrophages, most likely by preventing maturation and acidification of the phagocytic vacuole (Hostetter *et al.*, 2003), the animal may remain infected and start the long incubation period that is typical of *MAP* infections.

11.3.2 Stage I: 'Silent' infection

Once infection occurs, the organism proliferates slowly in the jejunal and ileal mucosa, and spreads to the regional lymph nodes (Clarke, 1997). This silent infection phase (also known as the 'eclipse phase') usually lasts for a minimum of 2 years and up to 10 years.

Stage I-infected cattle (typically replacement stock) show no outward clinical signs of infection, and there is no appreciable effect on growth or production. Cattle in this phase of infection have no detectable serum antibodies to *MAP* and may shed *MAP* in their faeces but at a level below levels of detection using current methods, including culture and PCR, hence the name 'eclipse phase'. At post-mortem examination, the organisms in the tissues may not be visible on microscopic examination but may be detectable by culture of multiple intestinal tissues, suggesting that widespread dissemination occurs early in the disease development (Sweeney *et al.*, 2006; Stabel *et al.*, 2009).

11.3.3 Stage II: The infection progresses

Cattle enter stage II disease with higher concentrations of *MAP* in their intestinal tissues. Although these animals still do not manifest weight loss or diarrhoea, they have histological changes consistent with the intestinal granulomatous inflammatory response characteristic of paratuberculosis. Cattle in stage II may have an altered immune response with increased interferon gamma response by sensitized T cells to specific mitogens and/or increased antibody response to *MAP* (Bassey and Collins, 1997). Animals in stage II may shed *MAP* in their manure, contaminating the environment and serving as sources of infection to other animals on the farm.

The rate of disease progression through stage II is highly variable and is most likely influenced by a wide range of factors that may include: age at initial exposure to *MAP*, the dose of *MAP* at initial exposure, the frequency of re-exposure over time, genetic factors of both the host and the organism, environmental factors, nutritional factors, production effects and a variety of stressors. In addition to the progression of the intestinal infection, dissemination of *MAP* organisms to other organs such as the uterus and mammary glands may now occur (Sweeney *et al.*, 1992a, b; Streeter *et al.*, 1995). Although these animals still do not show signs of JD, studies have shown that subclinically infected cattle have a reduced milk production and reduced reproductive efficiency when compared with uninfected animals (Benedictus *et al.*, 1987; Lombard *et al.*, 2005; Gonda *et al.*, 2007). Those differences, however, may be mild enough that they may not be detectable to the producer or veterinarian.

11.3.4 Stage III: Clinical disease begins

Animals at this stage have gradual weight loss and loose manure despite a normal appetite and vital signs. Milk production and reproductive efficiency are affected. Nearly all animals at stage III are positive for *MAP* organisms on faecal culture or faecal PCR and have increased antibodies

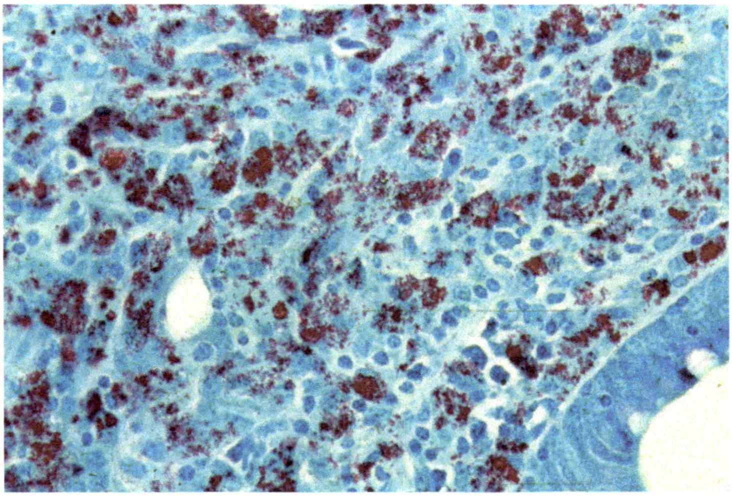

Fig. 11.1. Ziehl–Neelsen stain of ileum from a cow with advanced paratuberculosis. Copious fuchsin-staining organisms throughout the tissue indicate a multibacillary *MAP* infection. (Reproduced with the kind permission of Michael Collins.)

detectable by enzyme-linked immunosorbent assay (ELISA) test or agar gel immunodiffusion (AGID) test (Sweeney, 2011).

At this stage, the *MAP* population within the intestinal mucosal cells is very high (Fig. 11.1). The normal absorptive capacity of the bowel is abrogated, resulting in weight loss associated with a protein-losing enteropathy (Sweeney, 2011). The proliferation of reactive lymphocytes, epithelioid macrophages and giant cells results in infused blunted villi with decreased absorptive capacity. A thickened intestinal mucosa and associated lymphadenopathy along a substantial portion of the small intestinal tract is characteristic of the disease progression (Fig. 11.2). The infection becomes disseminated with *MAP* detectable in several extra-intestinal sites. These cows are at higher risk of transmitting the organism *in utero* and have a higher frequency of *MAP* isolated from the milk. Cattle in stage III shed high concentrations of *MAP* in their faeces, contaminating the environment.

11.3.5 Stage IV: Advanced clinical disease

Animals in stage IV of the disease are weak, emaciated and usually have chronic, profuse diarrhoea (Fig. 11.3). Intermandibular oedema or bottle jaw is characteristic of this phase of the disease. Animals can progress quickly from stage II to stage IV, sometimes within a few weeks, but a more gradual progression is more typical. Once the diarrhoea is profuse and hypoproteinemia occurs, the animal's condition deteriorates rapidly, often in a matter of days. Most animals are sent to slaughter for salvage at this point. Otherwise, death occurs as a result of dehydration and cachexia.

11.4 Diagnosis

Several diagnostic methods are available for the detection of *MAP* infection in cattle, and each have their relative merits and applications. They are classically divided into two categories: (i) immune-based tests that detect the animal's immune response to *MAP*, and (ii) organism detection tests that detect *MAP* organisms or *MAP* bacterial DNA. Specific diagnostic assays, including culture (see Chapter 18, this volume), PCR (see PChapter 19, this volume) and immunological assays (see Chapter 20, this volume), are considered elsewhere.

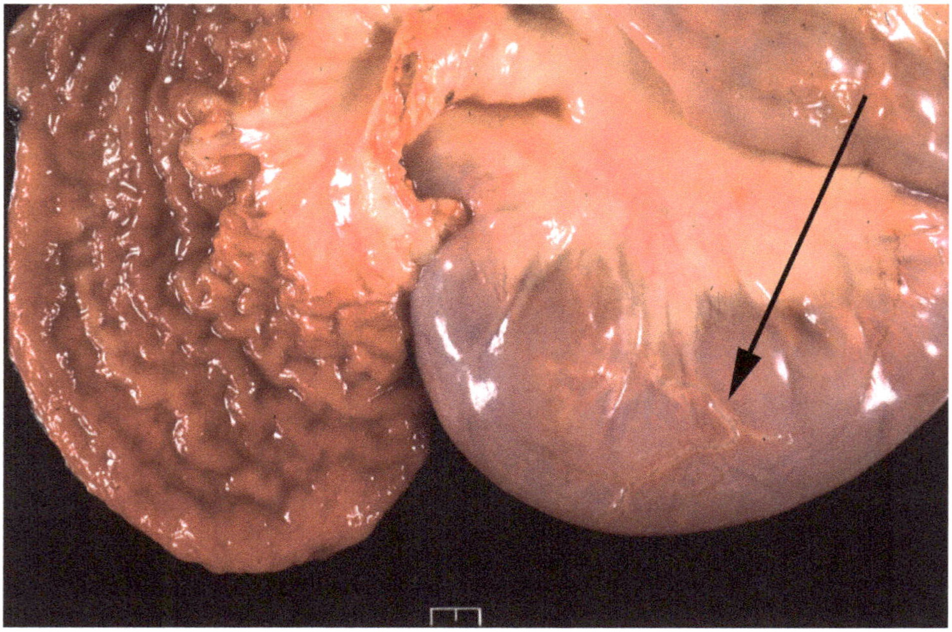

Fig. 11.2. Photograph of ileum manifesting lesions typical of paratuberculosis. To the left, the intestinal wall is thickened due to infiltration by leucocytes and epithelioid cells. To the right, there is mesenteric oedema, with a dilated mesenteric lymphatic channel, shown by the arrow. (Reproduced with the kind permission of Michael Collins.)

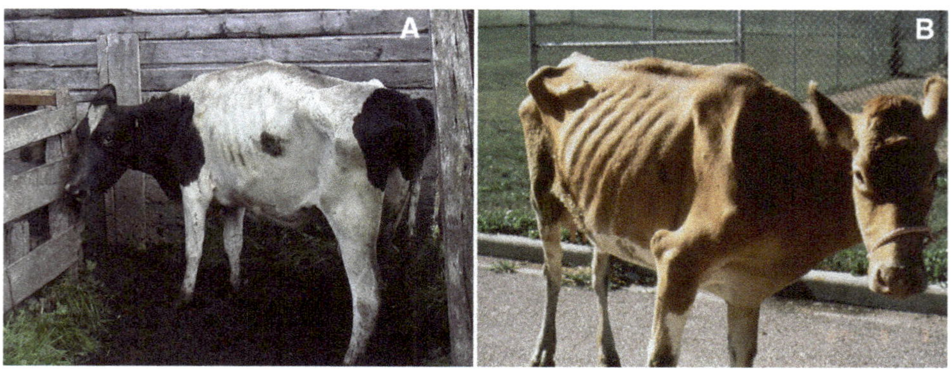

Fig. 11.3. Advanced paratuberculosis in Holstein (a) and Guernsey (b) cows. Signs of disease include weight loss, seen as marked reduction of muscle mass and visible skeletal bones such as ribs, pelvis and shoulder girdle. (Reproduced with the kind permission of Michael Collins.)

11.4.1 Testing strategies

Diagnosis of infection in cattle depends in part on whether diagnostic efforts are aimed at detecting infection in an individual animal or a herd. A consensus statement regarding testing recommendations for US cattle was published in 2006 (Collins *et al.*, 2006). The main purpose of the report was to provide simple, practical and cost-effective recommendations for cattle herds. Those recommendations were summarized and adapted by Collins in a review chapter on the diagnosis of paratuberculosis (Collins, 2011). Another consensus statement on paratuberculosis issued by the American College of Veterinary Internal Medicine and published in 2012 covered diagnostic test recommendations for various testing purposes (Sweeney *et al.*, 2012).

11.4.2 Diagnostic testing in a known-infected herd

In a known-positive herd, the main goal is to decrease environmental *MAP* load and therefore decrease the risk of exposure to susceptible animals. Determination of individual cow's *MAP* infection status will facilitate cow and calf management around calving (i.e. segregation of calving areas, colostrum management), and culling of high-positive cows. In commercial herds with high prevalence (>5%), individual ELISA on milk or serum is recommended, keeping in mind to interpret results quantitatively, based on S/P ratios (Wells *et al.*, 2002a). In low-prevalence herds (<5%), organism detection tests (culture or PCR) on pooled or individual faecal samples is recommended based on the low predictive value of ELISA-positive in those herds (Collins, 2011).

11.4.3 Diagnostic testing to determine herd *MAP* status (surveillance)

A producer or practitioner may elect to determine the *MAP* status of a given herd as part of, or outside the boundaries, of a voluntary JD control programme. Specific guidelines have been put in place for participation in control programmes at several national governing bodies. The reader should defer to those guidelines for more specifics on what the programmes entail in the respective countries. Commonly employed strategies aiming at determining the *MAP* status of a herd often include environmental testing, testing of pooled faecal samples via culture or PCR, testing of individual animals via faecal culture/PCR or ELISA and bulk tank milk ELISA testing. Individual animal testing can be done on each animal in the herd (very expensive), a random statistical subset of animals, or from specifically targeted animals based on age or suspicion of disease.

Environmental testing

Composite environmental faecal samples can serve as a good proxy to detect herd infection. Multiple studies have evaluated the use of faecal samples from high-traffic areas to estimate *MAP* prevalence in dairy herds (Raizman *et al.*, 2004; Berghaus *et al.*, 2006; Lombard *et al.*, 2006; Pillars *et al.*, 2009; Aly *et al.*, 2010; Lavers *et al.*, 2013). In those studies, sensitivity for the detection of positive herds was high (70% or greater). Expectedly, low-prevalence herds will be more susceptible to misclassification of *MAP* herd status when relying solely on composite environmental faecal sample testing. Refinement of the environmental manure sampling to include composite or pooled manure samples from high cow-traffic areas, manure storage areas and manure from pens/lots representing all cow groups has further increased diagnostic sensitivity to detect herd infection (Kalis *et al.*, 2000; Lombard *et al.*, 2006).

Faecal pooling

The main goal of faecal pooling is to maximize the economic benefit of testing without sacrificing diagnostic sensitivity (Wells *et al.*, 2002b). As expected, the sensitivity of pooled faecal cultures strongly depends on pool size and the shedding level of the cows included in the pool. Evaluation of pooled faecal samples (1:5) in a range of infected dairy herds detected at least 88% of samples that contained at least one animal with moderate to high shedding (Wells *et al.*, 2003). Optimal pool size depends on both prevalence and herd size. Samples size can vary from three samples per pool for a 500-cow herd with low prevalence to five samples per pool for a 1000-cow herd with high prevalence (van Schaik *et al.*, 2003). In another study that assessed faecal pooling strategies in low-prevalence herds using a simulation analysis, the weighted mean herd probabilities of detecting a positive herd ranged between 52 and 99.3%, with the lowest probability for pools of 10 with one positive cow in the herd and the highest probability for pools of 5 with five positive cows in the herd (McKenna *et al.*, 2018).

11.4.4 Diagnostic testing for eradication

The choice of test is dictated by herd prevalence and by how quickly and aggressively a producer wants to eradicate *MAP* from his herd. In general, the fastest approaches are also the most expensive to the producer. For example, whole-herd individual faecal culture or PCR followed by culling of positive animals has a better chance of quickly eradicating *MAP* from a seedstock herd but may be costly, while a less expensive (and less aggressive) approach might involve pooled faecal culture or serum/milk ELISA.

11.4.5 Diagnostic testing for individual animals

Diagnostic testing of an individual or group of animals that are not showing signs of paratuberculosis (pre-purchase screening) is difficult, especially in animals that are <36 months of age, as most infected animals are still in the incubation phase and therefore neither shedding *MAP* at detectable levels nor producing antibodies to *MAP*. A producer would be better advised to acquire information on the *MAP* status of the herd of origin, and buy cattle only from certified-negative herds, or better yet, never purchase cattle from outside sources.

11.5 Concluding Thoughts

Paratuberculosis remains one of the most important diseases of cattle worldwide. In cattle, the disease is debilitating and is characterized by weight loss and chronic diarrhoea in the later stages of infection. Cattle in the subclinical stages of the disease often show decreased milk production and are at higher risk for development of other common production diseases such as mastitis. Paratuberculosis is difficult to control because of the long incubation period (2–7 years), the absence of clinical signs until advanced stages of infection and the lack of reliable diagnostic methods in the preclinical stages of infection. It is generally accepted that calves become infected very early in life. Therefore, control programmes are based on preventing transmission from adult cattle shedding *MAP* organisms in faeces to young replacement stock on the farm. Control programmes specifically focus on calving-pen management, milk- and colostrum-feeding practices and rearing of young stock.

The lack of reliable diagnostic tests for the preclinical stages of the disease remains an important problem and is an important area of research. There are no drugs currently approved for the prevention or treatment of paratuberculosis in cattle. Vaccination, available on a limited basis in the USA, may reduce the incidence of clinical disease, and to a lesser extent the prevalence of infection, but vaccinates are not fully protected from infection and may still shed *MAP*. Research is needed to identify better vaccines, to prevent infection during the neonatal period.

References

Aly, S.S. and Thurmond, M.C. (2005) Evaluation of *Mycobacterium avium* subsp. *paratuberculosis* infection of dairy cows attributable to infection status of the dam. *Journal of the American Veterinary Medical Association* 227(3), 450–454. DOI: 10.2460/javma.2005.227.450.

Aly, S.S., Mangold, B.L., Whitlock, R.H., Sweeney, R.W., Anderson, R.J. *et al.* (2010) Correlation between Herrold egg yolk medium culture and real-time quantitative polymerase chain reaction results for *Mycobacterium avium* subspecies *paratuberculosis* in pooled fecal and environmental samples. *Journal of Veterinary Diagnostic Investigation* 22(5), 677–683. DOI: 10.1177/104063871002200501.

Ayele, W.Y., Bartos, M., Svastova, P. and Pavlik, I. (2004) Distribution of *Mycobacterium avium* subsp. *paratuberculosis* in organs of naturally infected bull-calves and breeding bulls. *Veterinary Microbiology* 103(3–4), 209–217. DOI: 10.1016/j.vetmic.2004.07.011.

Bannantine, J.P. and Bermudez, L.E. (2013) No holes barred: invasion of the intestinal mucosa by *Mycobacterium avium* subsp. *paratuberculosis*. *Infection and Immunity* 81(11), 3960–3965. DOI: 10.1128/IAI.00575-13.

Bassey, E.O. and Collins, M.T. (1997) Study of T-lymphocyte subsets of healthy and *Mycobacterium avium* subsp. *paratuberculosis*-infected cattle. *Infection and Immunity* 65(11), 4869–4872. DOI: 10.1128/IAI.65.11.4869-4872.1997.

Benedictus, G., Dijkhuizen, A.A. and Stelwagen, J. (1987) Economic losses due to paratuberculosis in dairy cattle. *Veterinary Record* 121(7), 142–146. DOI: 10.1136/vr.121.7.142.

Berghaus, R.D., Farver, T.B., Anderson, R.J., Jaravata, C.C. and Gardner, I.A. (2006) Environmental sampling for detection of *Mycobacterium avium* ssp. *paratuberculosis* on large California dairies. *Journal of Dairy Science* 89(3), 963–970. DOI: 10.3168/jds.S0022-0302(06)72161-0.

Buergelt, D.C., Donovan, G.A. and Williams, J.E. (2004) Identification of *Mycobacterium avium* subspecies *paratuberculosis* by polymerase chain reaction in blood and semen of a bull with clinical paratuberculosis. *International Journal of Applied Research in Veterinary Medicine* 2, 130–134.

Chiodini, R.J. (1996) Immunology: resistance to paratuberculosis. *Veterinary Clinics of North America: Food Animal Practice* 12(2), 313–343. DOI: 10.1016/S0749-0720(15)30409-6.

Chiodini, R.J. and Hermon-Taylor, J. (1993) The thermal resistance of *Mycobacterium paratuberculosis* in raw milk under conditions simulating pasteurization. *Journal of Veterinary Diagnostic Investigation* 5(4), 629–631. DOI: 10.1177/104063879300500424.

Chiodini, R.J., Van Kruiningen, H.J. and Merkal, R.S. (1984) Ruminant paratuberculosis (Johne's disease): the current status and future prospects. *The Cornell Veterinarian* 74(3), 218–262.

Clarke, C.J. (1997) The pathology and pathogenesis of paratuberculosis in ruminants and other species. *Journal of Comparative Pathology* 116(3), 217–261. DOI: 10.1016/S0021-9975(97)80001-1.

Collins, M.T. (2011) Diagnosis of paratuberculosis. *The Veterinary Clinics of North America. Food Animal Practice* 27(3), 581–591. DOI: 10.1016/j.cvfa.2011.07.013.

Collins, M.T., Gardner, I.A., Garry, F.B., Roussel, A.J. and Wells, S.J. (2006) Consensus recommendations on diagnostic testing for the detection of paratuberculosis in cattle in the United States. *Journal of the American Veterinary Medical Association* 229(12), 1912–1919. DOI: 10.2460/javma.229.12.1912.

Elzo, M.A., Rae, D.O., Lanhart, S.E., Wasdin, J.G., Dixon, W.P. *et al.* (2006) Factors associated with ELISA scores for paratuberculosis in an Angus-Brahman multibreed herd of beef cattle. *Journal of Animal Science* 84(1), 41–48. DOI: 10.2527/2006.84141x.

European Food and Safety Authority Scientific Panel on Animal Health and Welfare (2004) The risk of transmission of *Mycobacterium avium* subsp. *paratuberculosis* via bovine semen. *EFSA Journal* 110, 1–59.

Fecteau, M.E. (2018) Paratuberculosis in cattle. *Veterinary Clinics of North America: Food Animal Practice* 34, 209–222.

Fecteau, M.-E., Whitlock, R.H., Buergelt, C.D. and Sweeney, R.W. (2010) Exposure of young dairy cattle to *Mycobacterium avium* subsp. *paratuberculosis* (MAP) through intensive grazing of contaminated pastures in a herd positive for Johne's disease. *The Canadian Veterinary Journal* 51(2), 198–200.

Gilmour, N.J., Nisbet, D.I. and Brotherston, J.G. (1965) Experimental oral infection of calves with *Mycobacterium johnei*. *Journal of Comparative Pathology* 75(3), 281–286. DOI: 10.1016/0021-9975(65)90033-2.

Gonda, M.G., Chang, Y.M., Shook, G.E., Collins, M.T. and Kirkpatrick, B.W. (2007) Effect of *Mycobacterium paratuberculosis* infection on production, reproduction, and health traits in US Holsteins. *Preventive Veterinary Medicine* 80(2–3), 103–119. DOI: 10.1016/j.prevetmed.2007.01.011.

Hostetter, J., Steadham, E., Haynes, J., Bailey, T. and Cheville, N. (2003) Phagosomal maturation and intracellular survival of *Mycobacterium avium* subspecies *paratuberculosis* in J774 cells. *Comparative Immunology, Microbiology and Infectious Diseases* 26(4), 269–283. DOI: 10.1016/S0147-9571(02)00070-X.

Jørgensen, J.B. (1982) An improved medium for culture of *Mycobacterium paratuberculosis* from bovine faeces. *Acta Veterinaria Scandinavica* 23(3), 325–335.

Kalis, C.H., Hesselink, J.W., Barkema, H.W. and Collins, M.T. (2000) Culture of strategically pooled bovine fecal samples as a method to screen herds for paratuberculosis. *Journal of Veterinary Diagnostic Investigation* 12(6), 547–551. DOI: 10.1177/104063870001200609.

Kirkpatrick, B.W. and Lett, B.M. (2018) Short communication: heritability of susceptibility to infection by *Mycobacterium avium* ssp. *paratuberculosis* in Holstein cattle. *Journal of Dairy Science* 101(12), 11165–11169. DOI: 10.3168/jds.2018-15021.

Kiser, J.N., Neupane, M., White, S.N. and Neibergs, H.L. (2018) Identification of genes associated with susceptibility to *Mycobacterium avium* ssp. *paratuberculosis* (MAP) tissue infection in Holstein cattle using gene set enrichment analysis-SNP. *Mammalian Genome* 29(7–8), 539–549. DOI: 10.1007/s00335-017-9725-4.

Kruip, T.A.M., Muskens, J., van Roermund, H.J.W., Bakker, D. and Stockhofe-Zurwieden, N. (2003) Lack of association of *Mycobacterium avium* subsp. *paratuberculosis* with oocytes and embryos from moderate shedders of the pathogen. *Theriogenology* 59(7), 1651–1660. DOI: 10.1016/S0093-691X(02)01208-6.

Lavers, C.J., McKenna, S.L.B., Dohoo, I.R., Barkema, H.W. and Keefe, G.P. (2013) Evaluation of environmental fecal culture for *Mycobacterium avium* subspecies *paratuberculosis* detection in dairy herds and association with apparent within-herd prevalence. *The Canadian Veterinary Journal* 54(11), 1053–1060.

Lombard, J.E. (2011) Epidemiology and economics of paratuberculosis. *The Veterinary Clinics of North America. Food Animal Practice* 27(3), 525–535. DOI: 10.1016/j.cvfa.2011.07.012.

Lombard, J.E., Garry, F.B., McCluskey, B.J. and Wagner, B.A. (2005) Risk of removal and effects on milk production associated with paratuberculosis status in dairy cows. *Journal of the American Veterinary Medical Association* 227(12), 1975–1981. DOI: 10.2460/javma.2005.227.1975.

Lombard, J.E., Wagner, B.A., Smith, R.L., McCluskey, B.J., Harris, B.N. *et al.* (2006) Evaluation of environmental sampling and culture to determine *Mycobacterium avium* subspecies *paratuberculosis* distribution and herd infection status on US dairy operations. *Journal of Dairy Science* 89(11), 4163–4171. DOI: 10.3168/jds.S0022-0302(06)72461-4.

Lombard, J.E., Gardner, I.A., Jafarzadeh, S.R., Fossler, C.P., Harris, B. *et al.* (2013) Herd-level prevalence of *Mycobacterium avium* subsp. *paratuberculosis* infection in United States dairy herds in 2007. *Preventive Veterinary Medicine* 108(2–3), 234–238. DOI: 10.1016/j.prevetmed.2012.08.006.

McKenna, S.L.B., Ritter, C., Dohoo, I., Keefe, G.P. and Barkema, H.W. (2018) Comparison of fecal pooling strategies for detection of *Mycobacterium avium* ssp. *paratuberculosis* in cattle. *Journal of Dairy Science* 101(8), 7463–7470. DOI: 10.3168/jds.2018-14458.

Momotani, E., Whipple, D.L., Thiermann, A.B. and Cheville, N.F. (1988) Role of M cells and macrophages in the entrance of *Mycobacterium paratuberculosis* into domes of ileal Peyer's patches in calves. *Veterinary Pathology* 25(2), 131–137. DOI: 10.1177/030098588802500205.

National Animal Health Monitoring System (2002) Part I: Reference of Dairy Health and Management in the United States. USDA-APHIS-VS, CEAH, Fort Collins, Colorado, #N400.1203. Available at: https://www.aphis.usda.gov/animal_health/nahms/dairy/downloads/dairy07/Dairy07_dr_PartI.pdf (accessed 6 June 2019).

National Animal Health Monitoring System Dairy (2007) Johne's disease on US dairies, 1991–2007. USDA-APHIS-VS, CEAH, Fort Collins, Colorado, #N521.0408. Available at: https://www.aphis.usda.gov/animal_health/nahms/dairy/downloads/dairy07/Dairy07_is_Johnes.pdf (accessed 6 June 2019).

Nielsen, S.S., Bjerre, H. and Toft, N. (2008) Colostrum and milk as risk factors for infection with *Mycobacterium avium* subspecies *paratuberculosis* in dairy cattle. *Journal of Dairy Science* 91(12), 4610–4615. DOI: 10.3168/jds.2008-1272.

Osterstock, J.B., Fosgate, G.T., Cohen, N.D., Derr, J.N. and Roussel, A.J. (2008) Familial and herd-level associations with paratuberculosis enzyme-linked immunosorbent assay status in beef cattle. *Journal of Animal Science* 86(8), 1977–1983. DOI: 10.2527/jas.2008-0910.

Pillars, R.B., Grooms, D.L., Woltanski, J.A. and Blair, E. (2009) Prevalence of Michigan dairy herds infected with *Mycobacterium avium* subspecies *paratuberculosis* as determined by environmental sampling. *Preventive Veterinary Medicine* 89(3–4), 191–196. DOI: 10.1016/j.prevetmed.2009.02.022.

Raizman, E.A., Wells, S.J., Godden, S.M., Bey, R.F., Oakes, M.J. *et al.* (2004) The distribution of *Mycobacterium avium* ssp. *paratuberculosis* in the environment surrounding Minnesota dairy farms. *Journal of Dairy Science* 87(9), 2959–2966. DOI: 10.3168/jds.S0022-0302(04)73427-X.

Rohde, R.F. and Shulaw, W.P. (1990) Isolation of *Mycobacterium paratuberculosis* from the uterine flush fluids of cows with clinical paratuberculosis. *Journal of the American Veterinary Medical Association* 197(11), 1482–1483.

Rossiter, C.A. and Burhans, W. (1996) Farm specific approach to paratuberculosis (Johne's disease) control. *Veterinary Clinics of North America: Food Animal Practice* 12, 383–415.

Roussel, A.J. (2011) Control of paratuberculosis in beef cattle. *The Veterinary Clinics of North America. Food Animal Practice* 27(3), 593–598. DOI: 10.1016/j.cvfa.2011.07.005.

Roussel, A.J., Libal, M.C., Whitlock, R.L., Hairgrove, T.B., Barling, K.S. *et al.* (2005) Prevalence of and risk factors for paratuberculosis in purebred beef cattle. *Journal of the American Veterinary Medical Association* 226(5), 773–778. DOI: 10.2460/javma.2005.226.773.

Seitz, S.E., Heider, L.E., Heuston, W.D., Bech-Nielsen, S., Rings, D.M. *et al.* (1989) Bovine fetal infection with *Mycobacterium paratuberculosis*. *Journal of the American Veterinary Medical Association* 194(10), 1423–1426.

Stabel, J.R., Palmer, M.V., Harris, B., Plattner, B., Hostetter, J. *et al.* (2009) Pathogenesis of *Mycobacterium avium* subsp. *paratuberculosis* in neonatal calves after oral or intraperitoneal experimental infection. *Veterinary Microbiology* 136(3–4), 306–313. DOI: 10.1016/j.vetmic.2008.11.025.

Streeter, R.N., Hoffsis, G.F., Bech-Nielsen, S., Shulaw, W.P. and Rings, D.M. (1995) Isolation of *Mycobacterium paratuberculosis* from colostrum and milk of subclinically infected cows. *American Journal of Veterinary Research* 56(10), 1322–1324.

Sweeney, R.W. (2011) Pathogenesis of paratuberculosis. *Veterinary Clinics of North America: Food Animal Practice* 12, 537–546.

Sweeney, R.W., Whitlock, R.H. and Rosenberger, A.E. (1992a) *Mycobacterium paratuberculosis* isolated from fetuses of infected cows not manifesting signs of the disease. *American Journal of Veterinary Research* 53(4), 477–480.

Sweeney, R.W., Whitlock, R.H. and Rosenberger, A.E. (1992b) *Mycobacterium paratuberculosis* cultured from milk and supramammary lymph nodes of infected asymptomatic cows. *Journal of Clinical Microbiology* 30(1), 166–171. DOI: 10.1128/JCM.30.1.166-171.1992.

Sweeney, R.W., Uzonna, J., Whitlock, R.H., Habecker, P.L., Chilton, P. *et al.* (2006) Tissue predilection sites and effect of dose on *Mycobacterium avium* subs. *paratuberculosis* organism recovery in a short-term bovine experimental oral infection model. *Research in Veterinary Science* 80(3), 253–259. DOI: 10.1016/j.rvsc.2005.07.007.

Sweeney, R.W., Collins, M.T., Koets, A.P., McGuirk, S.M. and Roussel, A.J. (2012) Paratuberculosis (Johne's disease) in cattle and other susceptible species. *Journal of Veterinary Internal Medicine* 26(6), 1239–1250. DOI: 10.1111/j.1939-1676.2012.01019.x.

van Schaik, G., Stehman, S.M., Schukken, Y.H., Rossiter, C.R. and Shin, S.J. (2003) Pooled fecal culture sampling for *Mycobacterium avium* subsp. *paratuberculosis* at different herd sizes and prevalence. *Journal of Veterinary Diagnostic Investigation* 15(3), 233–241. DOI: 10.1177/104063870301500304.

Wells, S.J., Whitlock, R.H., Wagner, B.A., Collins, J., Garry, F. *et al.* (2002a) Sensitivity of ELISA test strategies for herd-level detection of *Mycobacterium avium* subsp. *paratuberculosis* in dairy cattle herds. *Journal of the American Veterinary Medical Association* 220, 1053–1057.

Wells, S.J., Whitlock, R.H., Lindeman, C.J. and Fyock, T. (2002b) Evaluation of bacteriologic culture of pooled fecal samples for detection of *Mycobacterium paratuberculosis*. *American Journal of Veterinary Research* 63(8), 1207–1211. DOI: 10.2460/ajvr.2002.63.1207.

Wells, S.J., Godden, S.M., Lindeman, C.J. and Collins, J.E. (2003) Evaluation of bacteriologic culture of individual and pooled fecal samples for detection of *Mycobacterium paratuberculosis* in dairy cattle herds. *Journal of the American Veterinary Medical Association* 223(7), 1022–1025. DOI: 10.2460/javma.2003.223.1022.

Whitlock, R.H. (1992) Diarrhea in cattle. In: Anderson, N.V. (ed.) *Veterinary Gastroenterology*, 2nd edn. Lea & Febiger, Philadelphia, Pennsylvania, p. 783.

Whitlock, R.H. (2009) Johne's disease. In: Smith, B.P. (ed.) *Large Animal Internal Medicine*, 4th edn. Mosby Elsevier, St Louis, Missouri, pp. 881–887.

Whittington, R.J. and Windsor, P.A. (2009) *In utero* infection of cattle with *Mycobacterium avium* subsp. *paratuberculosis*: a critical review and meta-analysis. *The Veterinary Journal* 179(1), 60–69. DOI: 10.1016/j.tvjl.2007.08.023.

Whittington, R.J., Fell, S., Walker, D., McAllister, S., Marsh, I. *et al.* (2000) Use of pooled fecal culture for sensitive and economic detection of *Mycobacterium avium* subsp. *paratuberculosis* infection in flocks of sheep. *Journal of Clinical Microbiology* 38(7), 2550–2556. DOI: 10.1128/JCM.38.7.2550-2556.2000.

Windsor, P.A. and Whittington, R.J. (2010) Evidence for age susceptibility of cattle to Johne's disease. *The Veterinary Journal* 184(1), 37–44. DOI: 10.1016/j.tvjl.2009.01.007.

Wu, C.W., Livesey, M., Schmoller, S.K., Manning, E.J.B., Steinberg, H. *et al.* (2007) Invasion and persistence of *Mycobacterium avium* subsp. *paratuberculosis* during early stages of Johne's disease in calves. *Infection and Immunity* 75(5), 2110–2119. DOI: 10.1128/IAI.01739-06.

12 Paratuberculosis in Sheep

Douglas Begg* and Richard Whittington
University of Sydney, Sydney, Australia

12.1 Introduction

The prevalence of *Mycobacterium avium* subsp. *paratuberculosis* (*MAP*) in domestic sheep worldwide is often unknown as the disease is 'non-notifiable' in many countries, even when it may be notifiable in cattle (Whittington *et al.*, 2019). Studying *MAP* in sheep has had a relatively low priority compared with *MAP* in cattle or the more politically and therefore strategically important mycobacterial diseases such as tuberculosis. There are huge populations of sheep at risk globally; in a recent survey, 50% of countries had more than 1 million sheep, half of these more than 10 million sheep and there were reports of paratuberculosis in North and South America, Australasia, the Middle East, Asia, Africa and Europe (Whittington *et al.*, 2019).

The difficulty of determining the extent of disease has been compounded by the lack of diagnostic tests with high levels of sensitivity and specificity, and the high costs involved with identifying the disease compared with the low value of a sheep. However, the prevalence of *MAP* appears to differ markedly between countries and geographical regions within countries (Sergeant, 2001; Sevilla *et al.*, 2005). Published data on the prevalence of affected sheep flocks range from 2–32% between different countries (Bakker *et al.*, 2000; Sergeant, 2001; Nielsen and Toft, 2009; Kumthekar *et al.*, 2013; Gurung *et al.*, 2018; Ma *et al.*, 2019). In Spain, the estimated prevalence within sheep flocks is 8–10% (Sevilla *et al.*, 2002). In Australia, *MAP* is thought to infect 2.4–4.4% of the flocks, based on data from abattoir surveillance. Dairy sheep flocks in Canada and Cyprus have had estimates of 66.8 and 60.8% flocks infected, although these findings may be limited to the farming enterprise examined (Liapi *et al.*, 2011; Bauman *et al.*, 2016b). In New South Wales, Australia, 6–10% of the flocks had ovine paratuberculosis, while in Western Australia less than 1% of the flocks were thought to be infected (Sergeant, 2001; Sergeant and Baldock, 2002). Within-flock prevalence has not been examined in detail but mortalities of 5–15% per year have been seen in high-prevalence flocks (Reddacliff *et al.*, 2006). In 2019, almost half of 11 countries with estimates of the prevalence of *MAP* in sheep based on objective criteria had greater than 10% of flocks affected, while almost one in five countries had greater than 40% flock-level prevalence (Whittington *et al.*, 2019).

12.2 Host and Pathogen Characteristics That Define Ovine Paratuberculosis

12.2.1 Breed, age and sex of sheep

An experimental infection study using Merino, Suffolk–Merino cross, Border Leicester and Poll

*Corresponding author: d.begg@usyd.edu.au

© CAB International 2020. *Paratuberculosis: Organism, Disease, Control, 2nd Edition* (eds M.A. Behr *et al.*)

Dorset breeds has shown genetic differences in susceptibility to *MAP* when controlling for the *MAP* strain, dose of infection and age at inoculation (Begg *et al.*, 2017). Merino and Suffolk–Merino cross animals were more susceptible to disease. An on-farm study of various breeds of sheep with natural *MAP* infection had similar findings: Dorset animals were less susceptible than the other breeds present (Hemalatha *et al.*, 2013). Both on-farm and experimental studies have shown that all ages of sheep can become infected with *MAP*, although animals infected at a younger age are more likely to have more faecal shedding and more severe lesions (McGregor *et al.*, 2012; Delgado *et al.*, 2013). Hemalatha *et al.* (2013) indicated that females were more susceptible to *MAP*, but as the numbers of animals for each sex was not provided and the two groups were housed separately, further research is required to confirm this finding.

12.2.2 The strain of *MAP*

The strain of *MAP* that infects sheep will vary with the predominant strain type in the geographical region and whether or not sheep cohabit with other species (Begg and Whittington, 2008; Sohal *et al.*, 2019). It is becoming clear from experimental infections that the *MAP* strain can have a significant impact on infection rates, the severity of pathology and the number of clinically affected animals (Verna *et al.*, 2007). Sheep develop more severe lesions when experimentally infected with *MAP*-S strains rather than *MAP*-C strains (Fernández *et al.*, 2014). Different isolates of *MAP*-C inoculated into sheep can also result in different pathological and immunological outcomes (Verna *et al.*, 2007; Fernández *et al.*, 2014). These factors may be responsible for the differences in disease outcome and prevalence between flocks in different geographical regions or countries. Quarantine authorities in endemic areas should consider excluding animals that harbour particular strains of *MAP* that are not already found in that area. *MAP* strains are described in more detail elsewhere (see Chapter 6, this volume).

12.3 Characteristics of Infection and Disease

12.3.1 Clinical signs and clinical pathology

The clinical signs of paratuberculosis in sheep are limited to chronic weight loss, which may occur from 2 years of age, with most animals succumbing to disease at 3–5 years of age. Oedema may occur occasionally. In advanced cases there may be hypoalbuminaemia and hypocalcaemia (Jones and Kay, 1996).

Most sheep that die of paratuberculosis have normal faecal pellets. Diarrhoea is not considered to be a feature of paratuberculosis in small ruminants, except in the terminal stages of disease. In a study of 50 sheep with clinical paratuberculosis, most were emaciated, half had normal faecal pellets, 30% had soft-formed faeces and 20% had severe diarrhoea (Carrigan and Seaman, 1990).

12.3.2 Pathology – gross and microscopic lesions

Advanced cases of paratuberculosis in sheep typically have thickening of the mucosa of the terminal ileum; the wall may be oedematous and the mucosa thrown up into transverse ridges; there may be cording of the subserosal lymphatics, and these are clearly visible and palpable. The caecum and colon may also be involved. In the cases described by Carrigan and Seaman (1990), lymphadenomegaly was present in 38% of the sheep. Histologically, there was moderate to severe granulomatous enteritis (94% of sheep), typhlitis (74%) and colitis (14%); the rectum was involved in 2%; lymphocytic infiltrates were also present in most sheep.

In tissue sections from the same animals, 88% of sheep had abundant acid-fast bacilli (AFB) and 12% had few AFB (Carrigan and Seaman, 1990). This dichotomy led others to describe two distinct forms of disease in sheep: paucibacillary and multibacillary (Clarke and Little, 1996). Clarke *et al.* (1996) used a lesion grading system: sheep with a mean of 0–10 AFB per macrophage were called paucibacillary, and those with >10 were multibacillary.

Animals in both groups were emaciated and had carcass oedema; the multibacillary animals were more likely to have detectable gross lesions in the intestine and associated lymph nodes. Histologically, the paucibacillary group tended to have a lymphocytic infiltrate with fewer macrophages compared with the multibacillary group, in which macrophages dominated the infiltrate (Clarke and Little, 1996). A more detailed lesion-grading system with five categories (1, 2, 3a, 3b, 3c) was proposed by Pérez *et al.* (1996). Mycobacteria were not visible in most sections from sheep with Type 1 (mild) and Type 3c (severe) lesions but were abundant in Type 3b. Although the authors were far from certain, an ordinal progression of lesions from Type 1 to Type 2 and beyond was implied. However, the time of infection and the influence of a successful immune response in hindering progression and AFB numbers were factors that could not be accounted for (Pérez *et al.*, 1996). Regression of lesions and recovery from infection has been reported, but this has only been observed in an animal with mild paucibacillary lesions (Dennis *et al.*, 2011). Lesions in some sheep can be arrested at a mild stage, but in others the progression from mild to severe lesions can be quite rapid (Dennis *et al.*, 2011).

The prevalence of the various types of pathology within an infected flock is unclear due to biases in sampling. The progression of lesions is also unknown. It is clear, however, that the first lesions develop in the ileal Peyer's patches and then spread to the surrounding mucosa (Pérez *et al.*, 1996).

12.3.3 Route of infection and transmission between individuals

The most common route of infection is faecal–oral. Infection can also be spread by intrauterine and transmammary transmission (Lambeth *et al.*, 2004; Verin *et al.*, 2016). The intratonsillar route of infection may also play a role (Begg *et al.*, 2005), although this is unlikely to be of significance in the natural infection process, mainly because the large number of bacteria that are ingested and swallowed will heavily outweigh those that may become lodged in the tonsil. Experimental infection data have shown that sheep infected by alternative routes develop altered immunological response profiles compared with those infected orally (Begg *et al.*, 2005). Lambs infected via the intratonsillar route compared with oral-challenged animals had increased interferon gamma (IFN-γ), lymphocyte proliferation and antibody levels. This may partly explain the large variation in immune responses recorded in groups of naturally infected sheep. This has implications for the design of experimental infection models and the outcomes observed from this type of research.

Ileal Peyer's patches in sheep, unlike those in humans, show prenatal maturation, with antigen-independent lymphopoiesis and a rate of lymphocyte production greater than in the thymus. The ileal Peyer's patches in sheep are unique in that they are required for the development of B cells and like the thymus they involute with age (Landsverk *et al.*, 1991). At birth, the Peyer's patches are the single biggest lymphoid tissue, accounting for 1.2% of the lamb's body weight. By 6 weeks of age, the ileal Peyer's patches of a lamb will extend 2.5 m along the terminal ileum. From 12 weeks of age, the ileal Peyer's patches start to involute, with only a few remaining by 18 months of age (Reynolds and Morris, 1983). The ileal Peyer's patch of the neonatal ruminant is considered to be a primary target for *MAP* infection (Sigurðardóttir *et al.*, 2001). The lymphoid follicles of ileal Peyer's patches contain large numbers of B cells but few T cells (Landsverk *et al.*, 1991). While jejunal Peyer's patches persist in adult animals, they contain clusters of B cells and have numerous CD4+ T cells in the associated lymphoid follicle (Landsverk *et al.*, 1991). In sheep, the ileal and jejunal Peyer's patches appear to have different functions. The jejunal Peyer's patches are essential for producing mucosal responses, while the ileal Peyer's patches seed the systemic immune organs with B cells (Mutwiri *et al.*, 1999). An intestinal loop model in lambs has shown that higher numbers of *MAP* translocate across the intestinal wall from sections of ileum than jejunum (Ponnusamy *et al.*, 2013).

It is generally accepted that the entrance of *MAP* through the intestinal wall is via M cells overlying the Peyer's patches from where the bacteria are engulfed by macrophages (Momotani *et al.*, 1988). Enterocytes may also allow *MAP* translocation, but less so than M cells (Ponnusamy *et al.*, 2013). After this, very little is known about the host–pathogen interaction. There is a period of clinical latency between when the animal is infected, typically as a lamb, and the clinical signs of disease. During the time immediately after oral challenge, most of the bacteria are passively shed into the environment (Reddacliff and Whittington, 2003). Presumably, organisms taken into the intestines are in such low numbers that animals may remain unaffected for months or years. The number of bacteria lodged in the intestines and associated lymph nodes can be below the level of detection of tissue culture for months after challenge (Reddacliff and Whittington, 2003; Begg *et al.*, 2005). It is this latency period that creates such a problem for the early diagnosis of ovine paratuberculosis.

12.3.4 Immunopathobiology

MAP dissemination and propagation are possibly due to decreased cellular immunity allowing infection to develop into clinical disease (Stabel, 2000, 2006). Containment and restriction of replication of the bacteria within the macrophage are critical, as these cells can tolerate low numbers of bacteria, whereas higher numbers of *MAP* are cytotoxic and result in apoptosis (Merkal *et al.*, 1968; Bannantine and Stabel, 2002). As bacterial numbers increase within the gut wall, the number of intracellular bacteria sloughed off into the lumen will increase. This results in larger numbers of bacteria within faecal material and an increased chance of detection by faecal culture. It is therefore unsurprising that sheep with multibacillary lesions are more likely to shed *MAP* in faeces than animals with paucibacillary lesions (Whittington *et al.*, 2000a). A sheep with multibacillary disease will shed on average 10^8 bacteria per gram of faeces or up to 10^{10} bacteria per day (Whittington *et al.*, 2000b).

Major histocompatibilty complex (MHC) processing and antigen presentation may be downregulated in *MAP* infections (Berger and Griffin, 2006). Production and expression of MHC class I and II molecules may be reduced in *MAP*-infected macrophages (Alzuherri *et al.*, 1997; Berger and Griffin, 2006). Expression of lymphocyte function-associated antigen, a molecule involved with cell-to-cell interactions, is also reduced in *MAP*-infected macrophages (Alzuherri *et al.*, 1997).

Other innate pathways that are activated during a *MAP* infection include the pattern recognition receptors (PRR). These include the Toll-like receptor family (TLR), involved in recognition of binding of pathogens by macrophages and other cells of the innate immune system. Once these receptors are engaged by microbial ligands, they initiate the innate and adaptive immune response mechanisms. Recent papers have indicated that several PRR could be upregulated during *MAP* infection, including TLR1, TLR5, TLR6, TLR8, NOD2 (alias CARD15), dectin-1 and dectin-2 (Nalubamba *et al.*, 2008; Taylor *et al.*, 2008; Plain *et al.*, 2010).

Paucibacillary and multibacillary disease states in sheep correlate broadly with the predominant pathway of immunity (Clarke, 1997). Sheep with paucibacillary lesions are likely to have an associated cell-mediated immune (CMI) response, with large numbers of lymphocytes at the site of disease (Clarke *et al.*, 1996). Lesions found in the small intestine show an increase in the numbers of CD4+ T cells and gamma delta ($\gamma\delta$) T cells, and the ratio of CD4+ to CD8+ is >1 (Little *et al.*, 1996; Reddacliff *et al.*, 2004). Increased antigen-specific lymphoproliferative responses can be seen from blood and gut cells isolated from sheep with paucibacillary disease (Kurade *et al.*, 2004; Kurade and Tripathi, 2008). Increased levels of T helper-1 (Th1) cytokines are produced within the intestinal tissues by leucocytes in animals with paucibacillary lesions. The predominant cytokines are IFN-γ, interleukin 2 (IL-2) and IL-12 (Clarke *et al.*, 1996; Burrells *et al.*, 1999; Smeed *et al.*, 2007; Rossi *et al.*, 2009). Th17 cells may also play a role in the development of paucibacillary lesions, with upregulation of the master regulator T-cell transcription factors TBX21, RORC2 and RORA being observed (Nicol *et al.*, 2016). CMI responses, as measured by lymphocyte transformation and IFN-γ assays, can be variable along the chain of mesenteric lymph nodes. The lowest level of reactivity is found in leucocytes in the intestinal lamina propria (Burrells *et al.*, 1998; Begg and Griffin, 2005; Begg *et al.*, 2005).

In contrast to paucibacillary lesions, multibacillary lesions typically contain larger numbers of macrophages, and sheep with these lesions typically have a strong Th2 (humoral antibody) response. Animals with multibacillary lesions have a significantly reduced lymphocyte proliferation response (Burrells *et al.*, 1998; Kurade and Tripathi, 2008). The number of CD4+ T cells in the lesion decreases and the ratio of CD4+ to CD8+ T cells changes from >1 to <1 (Clarke *et al.*, 1996; Clarke and Little, 1996). The reduction in CD4+ T cells may result in a reduction of the Th1 CMI response and a lack of macrophage activation and killing (Navarro *et al.*, 1998; Koets *et al.*, 2002). An increase of Type 2 cytokine production is seen in multibacillary lesions; typically the cytokines that increase are the interleukins IL-10, IL-6, IL-8 and IL-18, transforming growth factor-beta 1 (TGF-β1) and tumour necrosis factor alpha (TNF-α) (Alzuherri *et al.*, 1996; Smeed *et al.*, 2007; Muñoz *et al.*, 2009; Rossi *et al.*, 2009). There is a corresponding decrease in the production of IL-2 (Clarke *et al.*, 1996; Burrells *et al.*, 1999). Most studies report a decrease of IFN-γ production in multibacillary or clinically affected animals (Clarke *et al.*, 1996; Burrells *et al.*, 1999; Smeed *et al.*, 2007). Interestingly, increased IL-1α expression in ileal tissues, which has been observed in bovine paratuberculosis, has not been observed in ovine paratuberculosis (Smeed *et al.*, 2007). The development of multibacillary disease in sheep does not appear to require a shift from Th1 to Th2 type immunity but rather a decline in immune function; for example, the functional ability to produce IFN-γ (Begg *et al.*, 2011; Gossner *et al.*, 2017).

Researchers are now examining an important and often overlooked third group of sheep: animals that appear to contain or overcome *MAP* infection (Dennis *et al.*, 2011). These sheep may be defined as not having a significant disease (no histological lesions) or as animals identified as tissue-culture negative after experimental infection (Whittington *et al.*, 2017). These animals have increased expression of the TLR9 and TNF-α genes, and a downregulation of IL-18 (Smeed *et al.*, 2007; Nalubamba *et al.*, 2008). On a cellular level, CD4+, CD8+, CD25+ T cells and B-cell populations are increased in the blood, and CD4+, CD25+ T cells and B cells increased in the gut-associated lymph nodes (Begg and Griffin, 2005). Interestingly, the number of B cells in the blood and tissues of the resistant animals was increased relative to diseased and control animals, indicating that the Th2 humoral response may play a role in immunity to *MAP* infection. Production of antibodies in the gut may modulate the initial infection in resilient animals, with IgA and IgG responses being detected in this cohort of animals (Begg *et al.*, 2015). Early IL-10 and IFN-γ production post-infection is also associated with resilience to disease (de Silva *et al.*, 2018).

Gamma delta (γδ) T cells may also play a role in *MAP* infections. In young ruminants the γδ T cells make up a large proportion of the T cells, unlike in humans and mice, where alpha beta T cells (CD4+ and CD8+ T cells) predominate (Mackay *et al.*, 1989; Hein and Griebel, 2003). The high numbers of γδ T cells found in the ruminant gastrointestinal tract (Chiodini, 1996) may have a role in the regulation of mucosal immunity to invasion by enteric pathogens (Ferguson, 1990). There is an increase in γδ T cells in the gut-associated lymphoid tissues of lambs exposed to *MAP* (Beard *et al.*, 2000). In multibacillary lesions in *MAP*-infected sheep, the numbers of CD4+ T cells are decreased while the number of γδ T cells stays the same (Little *et al.*, 1996).

12.4 Microbiological Diagnosis

An ideal diagnostic assay would be one that detects responses early after infection and is 100% sensitive and specific. All available diagnostic methods for paratuberculosis in domestic livestock tend to underestimate true disease prevalence owing to difficulties in identifying subclinically infected animals. Ante-mortem microbial tests for *MAP* infection presently include faecal culture and polymerase chain reaction (PCR). Immunodiagnostic tests include the antibody enzyme-linked immunosorbant assay (ELISA), the agar gel immunodiffusion (AGID) assay, the lymphocyte proliferation or transformation assay, intradermal skin testing and the IFN-γ assay. All of the above assays have associated drawbacks in terms of specificity and/or sensitivity. Post-mortem tests include histopathology, and culture or PCR for *MAP* from tissues.

12.4.1 Culture

Cultivation of *MAP* from intestinal tissues is thought to be the gold standard in the diagnosis of ovine paratuberculosis. Prior to 1998, the detection of the sheep strain of *MAP* was unreliable, as available culture media were not appropriate to support its growth. The addition of egg yolk and Mycobactin J to Middlebrook 7H9 broth or 7H10 or 7H11 agar solved this problem (Whittington *et al.*, 1998, Whittington *et al.*, 1999, 2013).

Where there is subclinical infection with *MAP*, faecal shedding can be intermittent, which lowers the sensitivity of culture. However, sheep with multibacillary lesions shed large numbers of *MAP* that are detectable in culture, even when diluted 100-fold with faeces from uninfected sheep (Whittington *et al.*, 2000a). The sensitivity of pooled faecal culture for flock-level detection is 92%, assuming a sample size of 350 sheep pooled and cultured in lots of 50. Pooled faecal culture has been adopted as the laboratory method of choice for detecting *MAP*-infected sheep flocks in Australia (Sergeant, 2001). *MAP* culture is described in more detail elsewhere (see Chapter 18, this volume).

12.4.2 PCR

In recent years, the sensitivities of PCR-based assays for the detection of *MAP* from both faecal and tissue samples have greatly improved. This is primarily due to improved DNA extraction techniques and the use of quantitative or real-time PCR assays. The sensitivity of PCR for faecal samples was reported to be equivalent to that of culture with a high specificity due to primers designed to avoid detection of environmental bacteria (Kawaji *et al.*, 2007). The sensitivity of the PCR assays appears to be as good as if not better than faecal culture (Plain *et al.*, 2014; Bauman *et al.*, 2016a; Sonawane and Tripathi, 2016; Arsenault *et al.*, 2019), although validation studies for the assays following the World Organisation for Animal Health (OIE) and other guidelines (Gardner *et al.*, 2011) are few and far between for sheep, for example Bauman *et al.* (2016a) and Plain *et al.* (2014) An extensive review of the benefits and disadvantages of PCR is given elsewhere (see Chapter 19, this volume).

Diagnosis by PCR and culture for faecal samples are dependent on the stage of infection: late-stage disease is typically characterized by persistent heavy shedding in most affected animals. In the first months after exposure, intermittent or transient shedding can occur. Detection of this early shedding does not indicate the final outcome of the exposure (disease or resilience) (Begg *et al.*, 2015), although animals that shed higher numbers of *MAP* are more likely to develop disease (de Silva *et al.*, 2013).

12.5 Immunodiagnosis

Immunodiagnostic tests such as ELISA, AGID, intradermal skin testing, lymphocyte transformation and IFN-γ assays generally have poor diagnostic sensitivity and can be significantly affected by the stage of infection (Gumber *et al.*, 2006). The sensitivity of the antibody ELISA to detect *MAP* infection appears to depend on the type of kit used. One study of the Parachek™ ELISA reported a sensitivity of 16% at a specificity of 97.5% (Robbe-Austerman *et al.*, 2006a). Another antibody ELISA (Institut Pourquier, now Idexx) had a sensitivity of 34.9% and a specificity of 98.8% (Gumber *et al.*, 2006). The AGID has been found to have an even lower sensitivity than the ELISA (Gumber *et al.*, 2006; Robbe-Austerman *et al.*, 2006a). Again, the sensitivity of the IFN-γ assay is low, ranging from 40–75% with specificities ranging from 88–98% (Stewart *et al.*, 2002; Robbe-Austerman *et al.*, 2006b). The sensitivities of the intradermal skin test and the lymphocyte proliferation (or transformation) assays have not been examined in detail, with different studies using various cut-off points for diagnosis of infection (Williams *et al.*, 1985; Reddacliff and Whittington, 2003; Robbe-Austerman *et al.*, 2006a). Validation of the diagnostic assays by defining specificity and sensitivity against the gold standard tests of tissue or faecal culture/PCR is becoming more common. If validation of the assays is not done, then researchers and clinicians are left with a poor understanding of the assay's operating parameters and how they can use them under different circumstances to diagnose disease.

12.6 Spread of Infection (Epidemiology)

The behaviour of *MAP* in populations of sheep is perhaps best understood through two case studies where the organism was introduced into regions where it was previously unknown: Iceland and Australia. Common features include: (i) the cryptic nature of infection, enabling spread and evasion of normal quarantine practices; (ii) the slow progression of infection in the population; and (iii) the eventual emergence of paratuberculosis as a very serious disease.

Sheep were introduced to Iceland in the 9th century and remained free of *MAP* until 20 Karakul sheep were imported from Germany in 1933. In fact, three slowly developing infectious diseases evaded quarantine controls due to their long incubation periods: paratuberculosis, maedi/visna and jaagsiekte. Paratuberculosis was diagnosed 5 years later, and 440 farms were infected in the subsequent 20 years. In 1940, the disease appeared in cattle that were grazed with sheep, but fewer cows than sheep were affected. It later occurred in goats and a reindeer (Fridriksdottir *et al.*, 2000). Mortality rates per annum in sheep averaged 8–9% but were as high as 40% on some farms. Sheep and dairy cattle grazed together on pastures during the summer and were housed together during the colder months. Thus, there was probably a very high level of cross-species *MAP* contamination. Control measures based on fencing and zoning based on prevalence levels, along with test and cull, failed to impede spread. Destocking of all sheep in some zones was then tried, but the infection was probably maintained in cattle, as healthy sheep introduced 1 year after the destocking programme succumbed to paratuberculosis. Finally, vaccination of lambs was introduced, and this is still practised to prevent clinical disease. Paratuberculosis in Iceland is caused by *MAP*-S, which could not be routinely cultured; it was identified using molecular techniques from archives of histological paraffin blocks (Whittington *et al.*, 2001). This strain of *MAP* appears to be highly virulent for sheep but of low virulence for cattle, as, despite close contact with sheep, the clinical disease is rarely seen in cattle in Iceland; cattle have never been vaccinated in Iceland (Fridriksdottir *et al.*, 2000).

Australia has experienced a similar ‘slow epizootic’ of paratuberculosis in sheep. The most likely scenario for Australia involved importation of infected sheep from New Zealand prior to 1958 or in the 1970s, as live sheep were not imported in the intervening period (Sergeant, 2001). The first case of ovine paratuberculosis was diagnosed in New South Wales in 1980 (Seaman *et al.*, 1981), years after its likely introduction. The disease was most unlikely to have been widespread in sheep flocks at that time as there was a strong system of passive surveillance in Australia (provided by regional veterinary laboratories) at no cost to farmers. For this reason, it is likely that the rate of spread of paratuberculosis increased exponentially after about 1980, associated with the pattern of trade of live sheep in Australia. By the mid-1990s, the disease was of significant economic impact on farms where it had been present for some years. Economic losses of 6.4–8.5% in the gross income margin have been recorded on farms where ovine paratuberculosis mortality ranged from 6.2–7.8% (Bush *et al.*, 2006). By 2000, there were 823 known infected flocks in New South Wales, Victoria, Tasmania and South Australia (Sergeant, 2001). Thousands of other flocks were suspected to be infected, based on purchase of sheep from infected farms and shared farm boundaries. Western Australia, which was geographically isolated, was the last state to notify infection; at the time of detection in 2004, infection was deemed likely to have been present for 7 years or longer; seven flocks with over 82,000 sheep were infected and 144 more were under suspicion (Sunderman, 2004).

A national disease control programme commenced in Australia in 1999; it employed pooled faecal culture and abattoir surveillance to determine infection and set up an assurance programme to identify and protect flocks that did not have the disease (Sergeant, 2001). A killed vaccine was shown to be effective to prevent clinical signs of paratuberculosis in sheep (Reddacliff *et al.*, 2006), and this is now used widely. Vaccination, in combination with biosecurity and risk management, is now relied on to reduce the rate of spread of the disease.

Strains of *MAP* from sheep, cattle and many other animals can be clearly divided on genetic grounds into two lineages called *MAP*-C and *MAP*-S. A more complete description of these groups, their alternative designations and strain characterization is given elsewhere (see Chapter 6, this volume). The *MAP*-S, which was responsible for paratuberculosis in sheep in Iceland and Australia, is also prevalent in New Zealand. In other countries it may be displaced by *MAP*-C, where either sheep are uncommon or cattle are prevalent. Sheep are also susceptible to the *MAP*-C strain and it has been a common finding in sheep in Europe (de Juan *et al.*, 2005; Sevilla *et al.*, 2008; Florou *et al.*, 2009). The difficulty of cultivation of *MAP*-S strains probably leads to underestimation of its distribution and abundance. Cattle, goats and deer can become infected with *MAP*-S strains (O'Brien *et al.*, 2006; Mackintosh *et al.*, 2007; Moloney and Whittington, 2008; Sevilla *et al.*, 2008), but cattle and deer appear to be more resistant to infection and clinical disease associated with *MAP*-S. In Australia, cattle have become infected as calves if exposed to heavily infected sheep, but it is an uncommon infection (Whittington *et al.*, 2001; Moloney and Whittington, 2008).

A detailed investigation on one infected farm in Australia revealed a slow rate of transmission and clustering of infection within age classes of sheep for 7 years. Susceptibility of young sheep to lower levels of *MAP* and the long incubation period before faecal shedding may explain this (Rast and Whittington, 2005). Once contamination rates build up, this pattern may be lost as sheep of any age become infected. This results in increased prevalence rates, with infected sheep possibly shedding *MAP* sooner after infection, but this is yet to be proven. One reason for different levels of prevalence between affected flocks is variation in stocking rate during lambing (Dhand *et al.*, 2007). A higher stocking rate leads to higher levels of contamination, causing a greater risk of exposure of lambs, and this was associated with higher levels of infection (Dhand *et al.*, 2007).

12.7 Concluding Remarks

Ovine paratuberculosis is a worldwide issue that is overlooked in many countries. There are several reasons for this. One is the lower value of individual sheep compared with cattle, and the second is a greater focus on other mycobacterial infections (particularly *Mycobacterium bovis*) in other species. Arguably this follows on from the confirmed zoonotic nature of *M. bovis* and the ambiguous classification of *MAP* as a zoonotic agent, and the lack of international leadership by OIE to develop guidelines for control of paratuberculosis (Whittington *et al.*, 2019). Sadly it is the developing world that has the greatest potential loss ahead of it if control is not taken more seriously by the developed world.

The spread of the paratuberculosis in sheep has been helped by the slow development of clinical signs, the slow rate of transmission between animals and the low sensitivity of diagnostic tests, especially in the earlier stages of infection and disease. Overall there has been an increase in understanding of the issues surrounding *MAP* infection in sheep. For instance, while previous research predominately focused on diseased animals in comparison with uninfected sheep, the examination of sheep that have overcome a challenge with *MAP* may provide a better understanding of the immune processes implicated in resistance to the disease. Improvements in the focus of current and future research may lead to increased understanding of the disease process and over time may result in improved diagnosis and better control measures.

References

Alzuherri, H.M., Woodall, C.J. and Clarke, C.J. (1996) Increased intestinal TNF-α, IL-1β and IL-6 expression in ovine paratuberculosis. *Veterinary Immunology and Immunopathology* 49(4), 331–345. DOI: 10.1016/0165-2427(95)05477-4.

Alzuherri, H.M., Little, D. and Clarke, C.J. (1997) Altered intestinal macrophage phenotype in ovine paratuberculosis. *Research in Veterinary Science* 63(2), 139–143. DOI: 10.1016/S0034-5288(97)90007-2.

Arsenault, J., Singh Sohal, J., Leboeuf, A., Hélie, P., Fecteau, G. *et al.* (2019) Validation of an in-house real-time PCR fecal assay and comparison with two commercial assays for the antemortem detection of *Mycobacterium avium* subsp. *paratuberculosis* infection in culled sheep. *Journal of Veterinary Diagnostic Investigation* 31(1), 58–68. DOI: 10.1177/1040638718810744.

Bakker, D., Willemsen, P.T.J. and van Zijderveld, F.G. (2000) Paratuberculosis recognized as a problem at last: a review. *Veterinary Quarterly* 22(4), 200–204. DOI: 10.1080/01652176.2000.9695058.

Bannantine, J.P. and Stabel, J.R. (2002) Killing of *Mycobacterium avium* subspecies *paratuberculosis* within macrophages. *BMC Microbiology* 2(1), 2. DOI: 10.1186/1471-2180-2-2.

Bauman, C.A., Jones-Bitton, A., Jansen, J., Kelton, D. and Menzies, P. (2016a) Evaluation of fecal culture and fecal RT-PCR to detect *Mycobacterium avium* ssp. *paratuberculosis* fecal shedding in dairy goats and dairy sheep using latent class Bayesian modeling. *BMC Veterinary Research* 12(1), 212. DOI: 10.1186/s12917-016-0814-5.

Bauman, C.A., Jones-Bitton, A., Menzies, P., Toft, N., Jansen, J. *et al.* (2016b) Prevalence of paratuberculosis in the dairy goat and dairy sheep industries in Ontario, Canada. *The Canadian Veterinary Journal* 57, 169–175.

Beard, P.M., Rhind, S.M., Sinclair, M.C., Wildblood, L.A., Stevenson, K. *et al.* (2000) Modulation of γδ T cells and CD1 in *Mycobacterium avium* subsp. *paratuberculosis* infection. *Veterinary Immunology and Immunopathology* 77(3–4), 311–319. DOI: 10.1016/S0165-2427(00)00238-5.

Begg, D.J. and Whittington, R.J. (2008) Experimental animal infection models for Johne's disease, an infectious enteropathy caused by *Mycobacterium avium* subsp. *paratuberculosis*. *The Veterinary Journal* 176(2), 129–145. DOI: 10.1016/j.tvjl.2007.02.022.

Begg, D.J., O'Brien, R., Mackintosh, C.G. and Griffin, J.F.T. (2005) Experimental infection model for Johne's disease in sheep. *Infection and Immunity* 73(9), 5603–5611. DOI: 10.1128/IAI.73.9.5603-5611.2005.

Begg, D.J., de Silva, K., Carter, N., Plain, K.M., Purdie, A. *et al.* (2011) Does a Th1 over Th2 dominancy really exist in the early stages of *Mycobacterium avium* subspecies *paratuberculosis* infections? *Immunobiology* 216(7), 840–846. DOI: 10.1016/j.imbio.2010.12.004.

Begg, D.J., de Silva, K., Plain, K.M., Purdie, A.C., Dhand, N. *et al.* (2015) Specific faecal antibody responses in sheep infected with *Mycobacterium avium* subspecies *paratuberculosis*. *Veterinary Immunology and Immunopathology* 166(3–4), 125–131. DOI: 10.1016/j.vetimm.2015.06.011.

Begg, D.J., Purdie, A.C., de Silva, K., Dhand, N.K., Plain, K.M. *et al.* (2017) Variation in susceptibility of different breeds of sheep to *Mycobacterium avium* subspecies *paratuberculosis* following experimental inoculation. *Veterinary Research* 48(1), 36. DOI: 10.1186/s13567-017-0440-7.

Begg, D.J. and Griffin, J.F.T. (2005) Vaccination of sheep against *M. paratuberculosis*: immune parameters and protective efficacy. *Vaccine* 23(42), 4999–5008. DOI: 10.1016/j.vaccine.2005.05.031.

Berger, S.T. and Griffin, F.T. (2006) A comparison of ovine monocyte-derived macrophage function following infection with *Mycobacterium avium* ssp. *avium* and *Mycobacterium avium* ssp. *paratuberculosis*. *Immunology and Cell Biology* 84(4), 349–356. DOI: 10.1111/j.1440-1711.2006.01431.x.

Burrells, C., Clarke, C.J., Colston, A., Kay, J.M., Porter, J. *et al.* (1998) A study of immunological responses of sheep clinically-affected with paratuberculosis (Johne's disease): the relationship of blood, mesenteric lymph node and intestinal lymphocyte responses to gross and microscopic pathology. *Veterinary Immunology and Immunopathology* 66, 3–4.

Burrells, C., Clarke, C.J., Colston, A., Kay, J.M., Porter, J. *et al.* (1999) Interferon-gamma and interleukin-2 release by lymphocytes derived from the blood, mesenteric lymph nodes and intestines of normal sheep and those affected with paratuberculosis (Johne's disease). *Veterinary Immunology and Immunopathology* 68(2–4), 139–148. DOI: 10.1016/S0165-2427(99)00022-7.

Bush, R.D., Windsor, P.A. and Toribio, J.A. (2006) Losses of adult sheep due to ovine Johne's disease in 12 infected flocks over a 3-year period. *Australian Veterinary Journal* 84(7), 246–253. DOI: 10.1111/j.1751-0813.2006.00001.x.

Carrigan, M.J. and Seaman, J.T. (1990) The pathology of Johne's disease in sheep. *Australian Veterinary Journal* 67(2), 47–50. DOI: 10.1111/j.1751-0813.1990.tb07693.x.

Chiodini, R.J. (1996) Immunology: resistance to paratuberculosis. *Veterinary Clinics of North America: Food Animal Practice* 12, 313–343.

Clarke, C.J. (1997) The pathology and pathogenesis of paratuberculosis in ruminants and other species. *Journal of Comparative Pathology* 116(3), 217–261. DOI: 10.1016/S0021-9975(97)80001-1.

Clarke, C.J. and Little, D. (1996) The pathology of ovine paratuberculosis: gross and histological changes in the intestine and other tissues. *Journal of Comparative Pathology* 114(4), 419–437. DOI: 10.1016/S0021-9975(96)80017-X.

Clarke, C.J., Colston, A., Little, D., Kay, J., Alzuherri, H.M. *et al.* (1996) The immune response in paratuberculosis infection of small ruminants. *Veterinary Immunology and Immunopathology* 54(1–4), 321–324. DOI: 10.1016/S0165-2427(96)05666-8.

de Juan, L., Mateos, A., Domínguez, L., Sharp, J.M. and Stevenson, K. (2005) Genetic diversity of *Mycobacterium avium* subspecies *paratuberculosis* isolates from goats detected by pulsed-field gel electrophoresis. *Veterinary Microbiology* 106(3–4), 249–257. DOI: 10.1016/j.vetmic.2004.12.013.

de Silva, K., Begg, D.J., Plain, K.M., Purdie, A.C., Kawaji, S. *et al.* (2013) Can early host responses to mycobacterial infection predict eventual disease outcomes? *Preventive Veterinary Medicine* 112(3–4), 203–212. DOI: 10.1016/j.prevetmed.2013.08.006.

de Silva, K., Plain, K., Purdie, A., Begg, D. and Whittington, R. (2018) Defining resilience to mycobacterial disease: characteristics of survivors of ovine paratuberculosis. *Veterinary Immunology and Immunopathology* 195, 56–64. DOI: 10.1016/j.vetimm.2017.11.008.

Delgado, L., Marín, J.F.G., Muñoz, M., Benavides, J., Juste, R.A. *et al.* (2013) Pathological findings in young and adult sheep following experimental infection with 2 different doses of *Mycobacterium avium* subspecies *paratuberculosis*. *Veterinary Pathology* 50(5), 857–866. DOI: 10.1177/0300985813476066.

Dennis, M.M., Reddacliff, L.A. and Whittington, R.J. (2011) Longitudinal study of clinicopathological features of Johne's disease in sheep naturally exposed to *Mycobacterium avium* subspecies *paratuberculosis*. *Veterinary Pathology* 48(3), 565–575. DOI: 10.1177/0300985810375049.

Dhand, N.K., Eppleston, J., Whittington, R.J. and Toribio, J.-A.L.M.L. (2007) Risk factors for ovine Johne's disease in infected sheep flocks in Australia. *Preventive Veterinary Medicine* 82(1–2), 51–71. DOI: 10.1016/j.prevetmed.2007.05.007.

Ferguson, A. (1990) Mucosal immunology. *Immunology Today* 11, 1–3. DOI: 10.1016/0167-5699(90)90002-Q.

Fernández, M., Benavides, J., Sevilla, I.A., Fuertes, M., Castaño, P. *et al.* (2014) Experimental infection of lambs with C and S-type strains of *Mycobacterium avium* subspecies *paratuberculosis*: immunological and pathological findings. *Veterinary Research* 45(1), 5. DOI: 10.1186/1297-9716-45-5.

Florou, M., Leontides, L., Kostoulas, P., Billinis, C. and Sofia, M. (2009) Strain-specific sensitivity estimates of *Mycobacterium avium* subsp. *paratuberculosis* culture in Greek sheep and goats. *Zoonoses and Public Health* 56(1), 49–52. DOI: 10.1111/j.1863-2378.2008.01179.x.

Fridriksdottir, V., Gunnarsson, E., Sigurdarson, S. and Gudmundsdottir, K.B. (2000) Paratuberculosis in Iceland: epidemiology and control measures, past and present. *Veterinary Microbiology* 77(3–4), 263–267. DOI: 10.1016/S0378-1135(00)00311-4.

Gardner, I.A., Nielsen, S.S., Whittington, R.J., Collins, M.T., Bakker, D. *et al.* (2011) Consensus-based reporting standards for diagnostic test accuracy studies for paratuberculosis in ruminants. *Preventive Veterinary Medicine* 101(1–2), 18–34. DOI: 10.1016/j.prevetmed.2011.04.002.

Gossner, A., Watkins, C., Chianini, F. and Hopkins, J. (2017) Pathways and genes associated with immune dysfunction in sheep paratuberculosis. *Scientific Reports* 7(1), 46695. DOI: 10.1038/srep46695.

Gumber, S., Eamens, G. and Whittington, R. (2006) Evaluation of a Pourquier ELISA kit in relation to agar gel immunodiffusion (AGID) test for assessment of the humoral immune response in sheep and goats with and without *Mycobacterium paratuberculosis* infection. *Veterinary Microbiology* 115(1–3), 91–101. DOI: 10.1016/j.vetmic.2006.01.003.

Gurung, R.B., Begg, D.J. and Whittington, R.J. (2018) A national serosurvey to determine the prevalence of paratuberculosis in cattle in Bhutan following detection of clinical cases. *Veterinary Medicine and Science* 4(4), 288–295. DOI: 10.1002/vms3.114.

Hein, W.R. and Griebel, P.J. (2003) A road less travelled: large animal models in immunological research. *Nature Reviews Immunology* 3(1), 79–84. DOI: 10.1038/nri977.

Hemalatha, S., Roy, P., Purushothaman, V. and Iyue, M. (2013) Paratuberculosis in different breeds of sheep: a retrospective study of cases. *International Journal of Mycobacteriology* 2(3), 166–170. DOI: 10.1016/j.ijmyco.2013.07.002.

Jones, D.G. and Kay, J.M. (1996) Serum biochemistry and the diagnosis of Johne's disease (paratuberculosis) in sheep. *Veterinary Record* 139(20), 498–499. DOI: 10.1136/vr.139.20.498.

Kawaji, S., Taylor, D.L., Mori, Y. and Whittington, R.J. (2007) Detection of *Mycobacterium avium* subsp. *paratuberculosis* in ovine faeces by direct quantitative PCR has similar or greater sensitivity compared to radiometric culture. *Veterinary Microbiology* 125(1–2), 36–48. DOI: 10.1016/j.vetmic.2007.05.002.

Koets, A., Rutten, V., Hoek, A., van Mil, F., Muller, K. *et al.* (2002) Progressive bovine paratuberculosis is associated with local loss of CD4+ T cells, increased frequency of gamma delta T cells, and related changes in T-cell function. *Infection and Immunity* 70(7), 3856–3864. DOI: 10.1128/IAI.70.7.3856-3864.2002.

Kumthekar, S., Manning, E.J.B., Ghosh, P., Tiwari, K., Sharma, R.N. *et al.* (2013) *Mycobacterium avium* subspecies *paratuberculosis* confirmed following serological surveillance of small ruminants in Grenada, West Indies. *Journal of Veterinary Diagnostic Investigation* 25(4), 527–530. DOI: 10.1177/1040638713490688.

Kurade, N.P. and Tripathi, B.N. (2008) Lymphoproliferative response and its relationship with histological lesions in experimental ovine paratuberculosis and its diagnostic implications. *Veterinary Research Communications* 32(1), 107–119. DOI: 10.1007/s11259-007-9008-8.

Kurade, N.P., Tripathi, B.N., Rajukumar, K. and Parihar, N.S. (2004) Sequential development of histologic lesions and their relationship with bacterial isolation, fecal shedding, and immune responses during progressive stages of experimental infection of lambs with *Mycobacterium avium* subsp. *paratuberculosis*. *Veterinary Pathology* 41(4), 378–387. DOI: 10.1354/vp.41-4-378.

Lambeth, C., Reddacliff, L.A., Windsor, P., Abbott, K.A., Mcgregor, H. *et al.* (2004) Intrauterine and transmammary transmission of *Mycobacterium avium* subsp *paratuberculosis* in sheep. *Australian Veterinary Journal* 82(8), 504–508. DOI: 10.1111/j.1751-0813.2004.tb11171.x.

Landsverk, T., Halleraker, M., Aleksandersen, M., McClure, S., Hein, W. *et al.* (1991) The intestinal habitat for organized lymphoid tissues in ruminants; comparative aspects of structure, function and development. *Veterinary Immunology and Immunopathology* 28(1), 1–16. DOI: 10.1016/0165-2427(91)90038-E.

Liapi, M., Leontides, L., Kostoulas, P., Botsaris, G., Iacovou, Y. *et al.* (2011) Bayesian estimation of the true prevalence of *Mycobacterium avium* subsp. *paratuberculosis* infection in Cypriot dairy sheep and goat flocks. *Small Ruminant Research* 95(2–3), 174–178. DOI: 10.1016/j.smallrumres.2010.09.010.

Little, D., Alzuherri, H.M. and Clarke, C.J. (1996) Phenotypic characterisation of intestinal lymphocytes in ovine paratuberculosis by immunohistochemistry. *Veterinary Immunology and Immunopathology* 55(1–3), 175–187. DOI: 10.1016/S0165-2427(96)05716-9.

Ma, J.-G., Tian, A.-L., Zheng, W.-B., Zou, Y., Zhang, Y.-T. *et al.* (2019) First report of bovine viral diarrhea virus and *Mycobacterium avium* subspecies *paratuberculosis* infection in Tibetan sheep (*Ovis aries*) in Tibetan Plateau, China. *Tropical Animal Health and Production* 51(3), 719–722. DOI: 10.1007/s11250-018-1718-0.

Mackay, C.R., Beya, M.-F. and Matzinger, P. (1989) γ/δ T cells express a unique surface molecule appearing late during thymic development. *European Journal of Immunology* 19(8), 1477–1483. DOI: 10.1002/eji.1830190820.

Mackintosh, C.G., Labes, R.E., Clark, R.G., de Lisle, G.W. and Griffin, J.F.T. (2007) Experimental infections in young red deer *(Cervus elaphus)* with a bovine and an ovine strain of *Mycobacterium avium* subsp *paratuberculosis*. *New Zealand Veterinary Journal* 55(1), 23–29. DOI: 10.1080/00480169.2007.36731.

McGregor, H., Dhand, N.K., Dhungyel, O.P. and Whittington, R.J. (2012) Transmission of *Mycobacterium avium* subsp. *paratuberculosis*: dose-response and age-based susceptibility in a sheep model. *Preventive Veterinary Medicine* 107(1–2), 76–84. DOI: 10.1016/j.prevetmed.2012.05.014.

Merkal, R.S., Kluge, J.P., Monlux, W.S., Larsen, A.B., Kopecky, K.E. *et al.* (1968) Experimental paratuberculosis in sheep after oral, intratracheal, or intravenous inoculation: histochemical localization of hydrolase activities. *American Journal of Veterinary Research* 29, 985–994.

Moloney, B.J. and Whittington, R.J. (2008) Cross species transmission of ovine Johne's disease from sheep to cattle: an estimate of prevalence in exposed susceptible cattle. *Australian Veterinary Journal* 86(4), 117–123. DOI: 10.1111/j.1751-0813.2008.00272.x.

Momotani, E., Whipple, D.L., Thiermann, A.B. and Cheville, N.F. (1988) Role of M cells and macrophages in the entrance of *Mycobacterium paratuberculosis* into domes of ileal Peyer's patches in calves. *Veterinary Pathology* 25(2), 131–137. DOI: 10.1177/030098588802500205.

Muñoz, M., Delgado, L., Verna, A., Benavides, J., García-Pariente, C. *et al.* (2009) Expression of transforming growth factor-beta 1 (TGF-β1) in different types of granulomatous lesions in bovine and ovine paratuberculosis. *Comparative Immunology, Microbiology and Infectious Diseases* 32(3), 239–252. DOI: 10.1016/j.cimid.2007.11.009.

Mutwiri, G., Watts, T., Lew, L., Beskorwayne, T., Papp, Z. *et al.* (1999) Ileal and jejunal Peyer's patches play distinct roles in mucosal immunity of sheep. *Immunology* 97(3), 455–461. DOI: 10.1046/j.1365-2567.1999.00791.x.

Nalubamba, K., Smeed, J., Gossner, A., Watkins, C., Dalziel, R. *et al.* (2008) Differential expression of pattern recognition receptors in the three pathological forms of sheep paratuberculosis. *Microbes and Infection* 10(6), 598–604. DOI: 10.1016/j.micinf.2008.02.005.

Navarro, J.A., Ramis, G., Seva, J., Pallarés, F.J. and Sánchez, J. (1998) Changes in lymphocyte subsets in the intestine and mesenteric lymph nodes in caprine paratuberculosis. *Journal of Comparative Pathology* 118(2), 109–121. DOI: 10.1016/S0021-9975(98)80003-0.

Nicol, L., Wilkie, H., Gossner, A., Watkins, C., Dalziel, R. *et al.* (2016) Variations in T cell transcription factor gene structure and expression associated with the two disease forms of sheep paratuberculosis. *Veterinary Research* 47(1), 83. DOI: 10.1186/s13567-016-0368-3.

Nielsen, S.S. and Toft, N. (2009) A review of prevalences of paratuberculosis in farmed animals in Europe. *Preventive Veterinary Medicine* 88(1), 1–14. DOI: 10.1016/j.prevetmed.2008.07.003.

O'Brien, R., Mackintosh, C.G., Bakker, D., Kopecna, M., Pavlik, I. *et al.* (2006) Immunological and molecular characterization of susceptibility in relationship to bacterial strain differences in *Mycobacterium avium* subsp. *paratuberculosis* infection in the red deer (*Cervus elaphus*). *Infection and Immunity* 74(6), 3530–3537. DOI: 10.1128/IAI.01688-05.

Pérez, V., Marín, J.F.G. and Badiola, J.J. (1996) Description and classification of different types of lesion associated with natural paratuberculosis infection in sheep. *Journal of Comparative Pathology* 114(2), 107–122. DOI: 10.1016/S0021-9975(96)80001-6.

Plain, K.M., Purdie, A.C., Begg, D.J., de Silva, K. and Whittington, R.J. (2010) Toll-like receptor (TLR)6 and TLR1 differentiation in gene expression studies of Johne's disease. *Veterinary Immunology and Immunopathology* 137(1–2), 142–148. DOI: 10.1016/j.vetimm.2010.04.002.

Plain, K.M., Marsh, I.B., Waldron, A.M., Galea, F., Whittington, A.-M. *et al.* (2014) High-throughput direct fecal PCR assay for detection of *Mycobacterium avium* subsp. *paratuberculosis* in sheep and cattle. *Journal of Clinical Microbiology* 52(3), 745–757. DOI: 10.1128/JCM.03233-13.

Ponnusamy, D., Periasamy, S., Tripathi, B.N. and Pal, A. (2013) *Mycobacterium avium* subsp. *paratuberculosis* invades through M cells and enterocytes across ileal and jejunal mucosa of lambs. *Research in Veterinary Science* 94(2), 306–312. DOI: 10.1016/j.rvsc.2012.09.023.

Rast, L. and Whittington, R.J. (2005) Longitudinal study of the spread of ovine Johne's disease in a sheep flock in southeastern New South Wales. *Australian Veterinary Journal* 83(4), 227–232. DOI: 10.1111/j.1751-0813.2005.tb11658.x.

Reddacliff, L.A. and Whittington, R.J. (2003) Experimental infection of weaner sheep with S strain *Mycobacterium avium* subsp. *paratuberculosis*. *Veterinary Microbiology* 96(3), 247–258. DOI: 10.1016/j.vetmic.2003.07.004.

Reddacliff, L.A., McClure, S.J. and Whittington, R.J. (2004) Immunoperoxidase studies of cell mediated immune effector cell populations in early *Mycobacterium avium* subsp. *paratuberculosis* infection in sheep. *Veterinary Immunology and Immunopathology* 97(3–4), 149–162. DOI: 10.1016/j.vetimm.2003.09.001.

Reddacliff, L., Eppleston, J., Windsor, P., Whittington, R. and Jones, S. (2006) Efficacy of a killed vaccine for the control of paratuberculosis in Australian sheep flocks. *Veterinary Microbiology* 115(1–3), 77–90. DOI: 10.1016/j.vetmic.2005.12.021.

Reynolds, J.D. and Morris, B. (1983) The evolution and involution of Peyer's patches in fetal and postnatal sheep. *European Journal of Immunology* 13(8), 627–635. DOI: 10.1002/eji.1830130805.

Robbe-Austerman, S., Gardner, I.A., Thomsen, B.V., Morrical, D.G., Martin, B.M. *et al.* (2006a) Sensitivity and specificity of the agar-gel-immunodiffusion test, ELISA and the skin test for detection of paratuberculosis in United States Midwest sheep populations. *Veterinary Research* 37(4), 553–564. DOI: 10.1051/vetres:2006018.

Robbe-Austerman, S., Stabel, J.R. and Palmer, M.V. (2006b) Evaluation of the gamma interferon ELISA in sheep subclinically infected with *Mycobacterium avium* subspecies *Paratuberculosis* using a whole-cell sonicate or a Johnin purified-protein derivative. *Journal of Veterinary Diagnostic Investigation* 18(2), 189–194. DOI: 10.1177/104063870601800208.

Rossi, G., Nigro, G., Tattoli, I., Vincenzetti, S., Mariani, P. *et al.* (2009) Adhesion molecules and cytokine profile in ileal tissue of sheep infected with *Mycobacterium avium* subsp. *paratuberculosis*. *Microbes and Infection* 11(6–7), 698–706. DOI: 10.1016/j.micinf.2009.04.006.

Seaman, J.T., Gardner, I.A. and Dent, C.H.R. (1981) Johne's disease in sheep. *Australian Veterinary Journal* 57(2), 102–103. DOI: 10.1111/j.1751-0813.1981.tb00465.x.

Sergeant, E.S.G. (2001) Ovine Johne's disease in Australia - the first 20 years. *Australian Veterinary Journal* 79(7), 484–491. DOI: 10.1111/j.1751-0813.2001.tb13021.x.

Sergeant, E.S.G. and Baldock, F.C. (2002) The estimated prevalence of Johne's disease infected sheep flocks in Australia. *Australian Veterinary Journal* 80(12), 762–768. DOI: 10.1111/j.1751-0813.2002.tb11348.x.

Sevilla, I., Aduriz, G., Garrido, J., Geijo, M. V. and Juste, R.A. (2002) A preliminary survey on the prevalance of paratuberculosis in dairy cattle in Spain by bulk milk PCR. In: Juste, R. A., Geijo, M. V. and Garrido, J. (eds) *Seventh International Colloquium on Paratuberculosis*. International Association for Paratuberculosis, Bilbao, Spain, pp. 332–336.

Sevilla, I., Singh, S. V., Garrido, J.M., Aduriz, G., Rodriguez, S. *et al.* (2005) Molecular typing of *Mycobacterium avium* subspecies *paratuberculosis* strains from different hosts and regions. *Revue Scientifique et Technique* 24, 1061–1066.

Sevilla, I., Li, L., Amonsin, A., Garrido, J.M., Geijo, M.V. *et al.* (2008) Comparative analysis of *Mycobacterium avium* subsp. *paratuberculosis* isolates from cattle, sheep and goats by short sequence repeat and pulsed-field gel electrophoresis typing. *BMC Microbiology* 8(1), 204. DOI: 10.1186/1471-2180-8-204.

Sigurðardóttir, Ó. G.., Press, C.M. and Evensen, O. (2001) Uptake of *Mycobacterium avium* subsp. *paratuberculosis* through the distal small intestinal mucosa in goats: an ultrastructural study. *Veterinary Pathology* 38(2), 184–189. DOI: 10.1354/vp.38-2-184.

Smeed, J.A., Watkins, C.A., Rhind, S.M. and Hopkins, J. (2007) Differential cytokine gene expression profiles in the three pathological forms of sheep paratuberculosis. *BMC Veterinary Research* 3(1), 18. DOI: 10.1186/1746-6148-3-18.

Sohal, J.S., Arsenault, J., Leboeuf, A., Helie, P., Buczinski, S. *et al.* (2019) Molecular characterization of *Mycobacterium avium* subspecies *paratuberculosis* C-type and S-type isolated from sheep and goats by using a combination of MIRU-VNTR loci. *Canadian Journal of Veterinary Research* 83, 160–167.

Sonawane, G.G. and Tripathi, B.N. (2016) Comparative evaluation of diagnostic tests for the detection of *Mycobacterium avium* subsp. *paratuberculosis* in the tissues of sheep affected with distinct pathology of paratuberculosis. *International Journal of Mycobacteriology* 5(Suppl 1), S88–S89. DOI: 10.1016/j.ijmyco.2016.09.005.

Stabel, J. (2000) Transitions in immune responses to Mycobacterium paratuberculosis. *Veterinary Microbiology* 77(3–4), 465–473. DOI: 10.1016/S0378-1135(00)00331-X.

Stabel, J.R. (2006) Host responses to *Mycobacterium avium* subsp. *paratuberculosis* : a complex arsenal. *Animal Health Research Reviews* 7(1–2), 61–70. DOI: 10.1017/S1466252307001168.

Stewart, D.J., Stiles, P.L., Whittington, R.J., Lambeth, C., Windsor, P.A. *et al.* (2002) Validation of the interferon-g for diagnosis of ovine Johne's disease: sensitivity and specificity field trials. *Seventh International Colloquium on Paratuberculosis* 148–153.

Sunderman, F. (2004) *Update on the OJD program in Western Australia. Ovine Johne's Disease – an Update of Australian Research.* Meat & Livestock Australia, Sydney.

Taylor, D.L., Zhong, L., Begg, D.J., de Silva, K. and Whittington, R.J. (2008) Toll-like receptor genes are differentially expressed at the sites of infection during the progression of Johne's disease in outbred sheep. *Veterinary Immunology and Immunopathology* 124(1–2), 132–151. DOI: 10.1016/j.vetimm.2008.02.021.

Verin, R., Perroni, M., Rossi, G., De Grossi, L., Botta, R. *et al.* (2016) Paratuberculosis in sheep: histochemical, immunohistochemical and in situ hybridization evidence of in utero and milk transmission. *Research in Veterinary Science* 106, 173–179. DOI: 10.1016/j.rvsc.2016.04.006.

Verna, A.E., Garcia-Pariente, C., Muñoz, M., Moreno, O., García-Marin, J.F. *et al.* (2007) Variation in the immuno-pathological responses of lambs after experimental infection with different strains of *Mycobacterium avium* subsp. *paratuberculosis*. *Zoonoses and Public Health* 54(6–7), 243–252. DOI: 10.1111/j.1863-2378.2007.01058.x.

Whittington, R.J., Marsh, I., Turner, M.J., McAllister, S., Choy, E. *et al.* (1998) Rapid detection of Mycobacterium paratuberculosis in clinical samples from ruminants and in spiked environmental samples by modified BACTEC 12B radiometric culture and direct confirmation by IS900 PCR. *Journal of Clinical Microbiology* 36(3), 701–707. DOI: 10.1128/JCM.36.3.701-707.1998.

Whittington, R.J., Marsh, I., McAllister, S., Turner, M.J., Marshall, D.J. *et al.* (1999) Evaluation of modified BACTEC 12B radiometric medium and solid media for culture of *Mycobacterium avium* subsp. *paratuberculosis* from sheep. *Journal of Clinical Microbiology* 37(4), 1077–1083. DOI: 10.1128/JCM.37.4.1077-1083.1999.

Whittington, R.J., Fell, S., Walker, D., McAllister, S., Marsh, I. *et al.* (2000a) Use of pooled fecal culture for sensitive and economic detection of *Mycobacterium avium* subsp. *paratuberculosis* infection in flocks of sheep. *Journal of Clinical Microbiology* 38(7), 2550–2556. DOI: 10.1128/JCM.38.7.2550-2556.2000.

Whittington, R.J., Reddacliff, L.A., Marsh, I., Mcallister, S. and Saunders, V. (2000b) Temporal patterns and quantification of excretion of *Mycobacterium avium* subsp *paratuberculosis* in sheep with Johne's disease. *Australian Veterinary Journal* 78(1), 34–37. DOI: 10.1111/j.1751-0813.2000.tb10355.x.

Whittington, R.J., Taragel, C.A., Ottaway, S., Marsh, I., Seaman, J. *et al.* (2001) Molecular epidemiological confirmation and circumstances of occurrence of sheep (S) strains of *Mycobacterium avium* subsp. *paratuberculosis* in cases of paratuberculosis in cattle in Australia and sheep and cattle in Iceland. *Veterinary Microbiology* 79(4), 311–322. DOI: 10.1016/S0378-1135(00)00364-3.

Whittington, R.J., Whittington, A.-M., Waldron, A., Begg, D.J., de Silva, K. *et al.* (2013) Development and validation of a liquid medium (M7H9C) for routine culture of *Mycobacterium avium* subsp. *paratuberculosis* to replace modified BACTEC 12B medium. *Journal of Clinical Microbiology* 51(12), 3993–4000. DOI: 10.1128/JCM.01373-13.

Whittington, R.J., Begg, D.J., de Silva, K., Purdie, A.C., Dhand, N.K. *et al.* (2017) Case definition terminology for paratuberculosis (Johne's disease). *BMC Veterinary Research* 13(1), 328. DOI: 10.1186/s12917-017-1254-6.

Whittington, R., Donat, K., Weber, M.F., Kelton, D., Nielsen, S.S. *et al.* (2019) Control of paratuberculosis: who, why and how. A review of 48 countries. *BMC Veterinary Research* 15(1), 198. DOI: 10.1186/s12917-019-1943-4.

Williams, E.S., Demartini, J.C. and Snyder, S.P. (1985) Lymphocyte blastogenesis, complement fixation, and fecal culture as diagnostic tests for paratuberculosis in North American wild ruminants and domestic sheep. *American Journal of Veterinary Research* 46, 2317–2321.

13 Paratuberculosis in Goats

Kari R. Lybeck*, Girum T. Tessema, Annette H. Kampen, Berit Djønne and Angelika Agdestein
Norwegian Veterinary Institute, Oslo, Norway

13.1 Introduction

Mycobacterium avium subsp. *paratuberculosis* (*MAP*) has been detected in goats in most goat-rearing countries, including Austria, Norway, Portugal, Spain, Greece, the UK, South Africa, New Zealand and Australia (Stehman, 1996; Nielsen and Toft, 2009). Prevalence and control programmes for paratuberculosis in ruminants, including goats, in different countries have recently been reviewed (Whittington *et al.*, 2019). The disease in goats resembles that in sheep in many respects (see Chapter 12, this volume). The prevalence of infection in goats is difficult to estimate in any region or country because of the uncertainty of the diagnosis and the fact that reporting of cases often depends on the conduction of specific surveys or eradication programmes. There are also large variations in the estimates from studies that have used different study designs and sampling frames (Nielsen and Toft, 2009).

Paratuberculosis in goats is known to cause considerable economic losses (Sardaro *et al.*, 2017) due to reduced feed efficiency, reduced milk production and increased culling of infected animals. Moreover, subclinical *MAP* infection has been shown to result in reduced fertility of goats (Kostoulas *et al.*, 2006). Together, these considerations indicate that paratuberculosis in goats leads to decreased revenues and increased costs, including the costs of diagnostic testing and implementing control measures (de Juan *et al.*, 2005).

13.2 Host and Pathogen Factors in Paratuberculosis

13.2.1 Animal risk factors

In cattle, age of the animal at exposure, breed resistance, genetic susceptibility and other diseases have been mentioned as risk factors for development of paratuberculosis (Radostits *et al.*, 2007). Younger animals require a lower infective dose than do older animals, and adult animals are unlikely to get infected unless there is extreme environmental contamination (McKenna *et al.*, 2006). This is generally accepted for other ruminants as well, but few experimental infections have been conducted in goats (Hines *et al.*, 2007a; Begg and Whittington, 2008). However, based on experimental animal infections, goats have been found to be more susceptible to paratuberculosis than cattle and sheep (Stewart *et al.*, 2007). Resistance or sensitivity to *MAP* infection can have a genetic basis, and recent research suggests caprine candidate genes for paratuberculosis resistance (Korou *et al.*, 2010; Taka *et al.*,

*Corresponding author: kari.lybeck@vetinst.no

© CAB International 2020. *Paratuberculosis: Organism, Disease, Control, 2nd Edition* (eds M.A. Behr *et al.*)

2013, 2015; Abraham *et al.*, 2017; Cecchi *et al.*, 2017).

13.2.2 Environmental and management risk factors

Conditions predisposing to *MAP* infection include poor hygiene that exposes young kids to manure from older animals and contaminated milk. Inadequate feeding and low body condition increase the susceptibility to infection (Radostits *et al.*, 2007). Experiments have shown that infected goat kids can excrete the bacteria in their faeces and thereby infect other kids (Stewart *et al.*, 2006). In goats, *MAP* seroprevalence has been associated with a herd size of over 200 animals, the presence of foreign breeds and a high replacement rate (Mainar-Jaime and Vázquez-Boland, 1998). Recently, Freitas *et al.* (2015) and Angelidou *et al.* (2014a) have described further risk factors, e.g. water sources and the time spent by kids with their mother.

Cross-species infections are suspected to occur between cattle and goats (Holstad *et al.*, 2005). Contact with wildlife might pose a risk of *MAP* transmission to domestic animals (Daniels *et al.*, 2003). Surveys indicate that large number of rabbits and access of wildlife to feed supplies are factors that increase the likelihood of a farm having paratuberculosis (Daniels *et al.*, 2002).

13.2.3 Pathogen risk factors

MAP isolates have been classified into two main lineages based on culture characteristics and molecular characterization: *MAP*-S (also called Type S or Type I/Type III) and *MAP*-C (also called Type C or Type II) (see Chapter 6, this volume). Goats can be infected by both lineages (de Juan *et al.*, 2005), and the isolates of *MAP* obtained from goats manifest genetic variability by a number of different molecular methods (Pillai *et al.*, 2001; Djønne *et al.*, 2005; Sevilla *et al.*, 2007, 2008). Phenotypic differences between these strains of *MAP* have been described, with one group suggesting that isolates from goats have lower pathogenicity for cattle (Saxegaard, 1990). This is supported by findings in macrophages in a study by Abendaño *et al.* (2013). The genetic basis for this observation is, however, not known. In contrast, a study by Gollnick *et al.* (2007) did not find any differences in the ability of different *MAP* strains to survive in bovine monocyte-derived macrophages. At present, there does not appear to be a strain of *MAP* specific to goats, and the importance of strain variability for infection and disease in goats requires further study.

13.3 Characteristics of Infection and Disease

13.3.1 Clinical signs and clinical pathology

Infected animals can be categorized into four groups according to clinical symptoms, faecal shedding of bacteria and immunological response: (i) silent infection; (ii) subclinical disease; (iii) clinical disease; and (iv) advanced clinical disease (Whitlock and Buergelt, 1996).

The stage of silent infection has been characterized by lack of clinical signs, including no effect on body weight gain or body condition. Experimentally infected goat kids have, however, shown reduced weight gain compared with uninfected kids from 1 week after infection and throughout a 6-month study period (Malone *et al.*, 2013). There is usually neither bacterial shedding nor detectable circulating antibodies, but cellular immune responses may be detectable by the interferon gamma (IFN-γ) response test.

During the stage of subclinical disease, there are still no clinical signs of paratuberculosis. However, the animals may shed low numbers of bacteria in faeces and there may be evidence of both cellular and humoral immune responses. During the third stage of infection, referred to as clinical disease, the only consistent finding is weight loss despite apparently normal food intake (Stehman, 1996). Unlike what is observed in cattle, but in common with sheep, diarrhoea is rarely seen in goats (Manning and Collins, 2001). In this state, bacteria are typically found in faeces and animals usually have antibodies against *MAP*. Most animals, if not culled, go on to stage four.

In advanced clinical disease, animals develop a flaky skin and a poor coat, and eventually progressive emaciation, dehydration, anaemia with submandibular oedema and depression

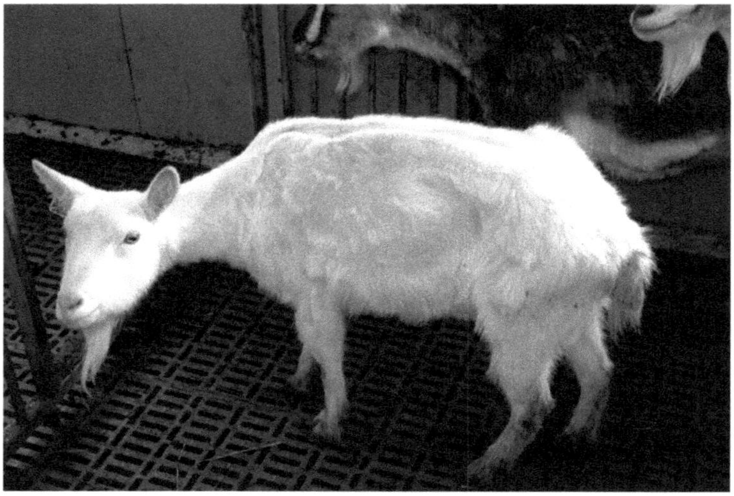

Fig. 13.1. Clinical paratuberculosis in a goat. (Reproduced with the kind permission of N. Leine.)

(Fig. 13.1). At this stage of the infection, diarrhoea, or more usually a clumping of faeces, can be seen (Stehman, 1996).

13.3.2 Pathology – gross and microscopic lesions

Macroscopic lesions are primarily seen in the intestine and in the draining lymph nodes. Intestinal lesions can be segmental or diffuse and are most commonly found in the jejunum, ileum and ileocaecal valve, but can occur throughout the whole length of the intestinal tract (Corpa *et al.*, 2000a; Storset *et al.*, 2001; Olsen *et al.*, 2002; Valheim *et al.*, 2002; Lybeck *et al.*, 2013; Krüger *et al.*, 2015b). Thickening and folding of the mucosa with transverse folds (Fig. 13.2), ulcerations, and dilated and thickened serosal and mesenteric lymphatic vessels are seen. The mesenteric lymph nodes are pale, swollen and oedematous. In goats, nodular foci of caseation and mineralization may be present both in the mucosa and in the lymph nodes

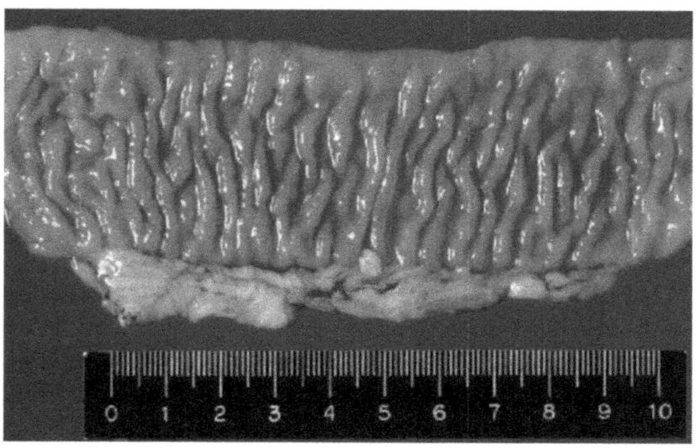

Fig. 13.2. Affected jejunum in a *Mycobacterium avium* subsp. *paratuberculosis*-infected goat. Pathological changes observed in this specimen are thickening of the mucosa with transverse folds. (Reproduced with the kind permission of Ó.G. Sigurðardóttir.)

(Corpa *et al.*, 2000a; Lybeck *et al.*, 2013; Krüger *et al.*, 2015b).

Histopathological lesions have been detected from 2–4 months after experimental infection. Infection by *MAP* results in granulomatous lesions with infiltration of lymphocytes, multinucleated giant cells and macrophages. In the earlier stages, lesions are frequently associated with gut-associated lymphoid tissue (GALT) of the jejunum, ileum and ileocaecal valve, while in more advanced stages a transmural enteritis may be seen (Sigurðardóttir *et al.*, 1999; Corpa *et al.*, 2000a; Valheim *et al.*, 2002; Munjal *et al.*, 2005; Lybeck *et al.*, 2013; Krüger *et al.*, 2015b).

The spectrum of lesions can resemble that seen in *Mycobacterium leprae* infections. In the tuberculoid form, there are focal aggregates of macrophages surrounded by large numbers of lymphocytes, with few or no acid-fast bacilli. This form is associated with strong cell-mediated immune responses. The lepromatous form is associated with strong humoral immune responses, and the lesions consist of diffuse infiltration of macrophages containing large numbers of acid-fast bacilli. Between these extremes of the spectrum are the so-called intermediary forms and 'borderline forms' (Corpa *et al.*, 2000a; Lybeck *et al.*, 2013; Krüger *et al.*, 2015a). Additionally, goats with severe diffuse lesions may still have strong cellular immune responses (Lybeck *et al.*, 2013). Subclinically infected animals will usually have focal tuberculoid lesions, while individuals with clinical symptoms may present with either the borderline tuberculoid form with only a few bacilli or the borderline lepromatous form with multiple bacilli (Sigurðardóttir *et al.*, 1999).

13.4 Route of Infection and Transmission Between Individuals

Animals are most commonly infected through the faecal–oral route: by ingestion of contaminated milk or food products or by accidental ingestion of the organism from contaminated surfaces (Harris and Barletta, 2001). *MAP* can be excreted in colostrum and milk (Streeter *et al.*, 1995), and *MAP* has been detected by polymerase chain reaction (PCR) in raw goat milk (Djønne *et al.*, 2003). *In utero* infection has been shown to occur in goats (Alinovi *et al.*, 2009). Spreading of *MAP* between flocks is usually due to trading of animals with unknown infection, but spread due to contaminated faeces on pastures also occurs (Radostits *et al.*, 2007).

13.5 Pathogenesis

The ingested bacteria enter the intestinal wall through the small intestinal mucosa. The point of entry in experimentally infected goats has been shown to be through M cells found in the follicle-associated epithelium lining the domes of the Peyer's patches and also through enterocytes in areas without Peyer's patches (Sigurðardóttir *et al.*, 2001, 2005). The bacteria are subsequently phagocytosed by subepithelial macrophages, and *MAP* is initially found in the intestinal and intestinal lymph node macrophages. Within macrophages, *MAP* resist intracellular degradation and slowly replicate, stimulating inflammatory and immunological responses (Olsen *et al.*, 2002).

Infection with *MAP* is usually characterized by an initial clinically silent period. Eventually, macrophages may lyse and facilitate spread of *MAP* (Koets *et al.*, 2015). Granulomatous arteritis has been detected in the intestine of goats and might contribute to the spread of *MAP* (Krüger *et al.*, 2015b). New infectious foci could also develop following dissemination from, for example, areas of caseous necrosis (Krüger *et al.*, 2015a, b; Lybeck *et al.*, 2013).

Most animals infected with *MAP* are able to control the infection. However, subclinically infected animals will probably remain infected for the rest of their life, and some may intermittently shed the bacteria in faeces, contributing to spread of the infection. Whether infected animals can clear the infection or not remains unclear. The bacteria could not be cultured from tissue lesions of goats that previously had been faecal shedders and IFN-γ positive, and a reduction in lesions and bacterial load was seen at 6–12 months compared with 3 months of age in some animals (Köhler *et al.*, 2015). In other studies, some individuals of experimentally infected goats where a cell-mediated immune response had been detected were found not to display lesions or harbour *MAP* bacteria (Sigurðardóttir *et al.*, 1999; Storset *et al.*, 2001). Still, it cannot be excluded that these animals were infected, but that the numbers of bacteria were too low to be detected.

13.6 Immune Response

Invasion of macrophages activates the host immune system to begin a series of attacks against *MAP*-infected macrophages. This further leads to activation of T cells, including CD4+ T helper cells and cytolytic CD8+ cells (Radostits *et al.*, 2007).

Numerous γδ T cells have been detected in early lesions (Krüger *et al.*, 2015a), but these cells were low in numbers at later stages (Valheim *et al.*, 2004; Lybeck *et al.*, 2013). This could imply an important role for γδ T cells in the early immune response.

A CD4+ T-helper type-1 response with IFN-γ production is regarded as essential since it further activates macrophages (Stabel, 2006). This cellular immune response has generally been associated with controlled infection in most animals. In line with this, granulomatous lesions with few bacteria showed large numbers of CD4+ T cells (Krüger *et al.*, 2015a). In goats exposed to *MAP* under natural conditions, IFN-γ responses were detected earlier in non-shedders than in shedders, but the strongest responses were later found in goats shedding *MAP* (Mercier *et al.*, 2016; Souriau *et al.*, 2017). This suggests that an early IFN-γ response is important for controlling the infection. The IFN-γ response may wane with progressive infection (Singh *et al.*, 2013), but this cellular response may also persist during later stages (Lybeck *et al.*, 2011; Köhler *et al.*, 2015). Animals that are unable to control the disease will usually develop a humoral immune response along with increased shedding of bacteria in faeces, eventually followed by the onset of clinical symptoms (Olsen *et al.*, 2002).

Several infectious foci in the intestine will develop, and there will be an ongoing battle between the host and the bacteria in these various foci. The local immune responses in the foci may differ, and both humoral and cellular immune responses can sometimes be present at the same time. It has previously been suggested that infected animals are able to completely recover from paratuberculosis infection. However, a rise in the IFN-γ response of experimentally infected goats after an initial drop, could indicate that reactivation of bacteria in small foci is possible (Storset *et al.*, 2001).

The timing of both pro- and anti-inflammatory responses are likely to be critical for the outcome of infection. Studies in calves and sheep have suggested that the anti-inflammatory cytokine IL-10 might reduce severity of disease and increase protective immunity (Subharat *et al.*, 2012; de Silva *et al.*, 2013), but high IL-10 has also been associated with progressive or severe *MAP* infection in goats (Lybeck *et al.*, 2013; Singh *et al.*, 2013). Recently, RNA sequencing of *MAP* antigen stimulated peripheral blood mononuclear cells (PBMC) has been performed. Among the findings were repression of proinflammatory responses by downregulation of IFN-γ and IL-17 related responses, in addition to upregulation of IL-18 binding protein in infected compared with non-infected animals 1 month post-infection (Berry *et al.*, 2018). Such new tools could help increase our knowledge on protective immune responses to *MAP*.

13.7 Diagnosis

The clinical diagnosis of paratuberculosis is challenging in goats; symptoms are vague and non-specific, as numerous other diseases present with weight loss. Goats with paratuberculosis showed complete blood cell count and protein electrophoresis values within reference intervals, though some differences in protein electrophoresis were observed between healthy goats and goats with paratuberculosis (Bonelli *et al.*, 2017). The disease can be diagnosed by pathology, microbiology (culture and PCR) and immunological methods. None of these methods is very sensitive, especially during the early stages of infection. The sensitivity and specificity of diagnostic tests for *MAP* infections vary significantly, but formal comparison of the different tests that have been reported is difficult. The main reasons are variations in study design, test components and target conditions (Nielsen and Toft, 2008). The timing and intensity of faecal shedding and immune response also vary substantially between individuals (Köhler *et al.*, 2015). Generally, the choice of laboratory diagnostic methods can vary according to the purpose of testing and the prevalence of *MAP* (Collins, 2011; Bauman *et al.*, 2019).

Volatile organic compounds have been investigated as biomarkers of *MAP* infection, and could potentially be developed into new non-invasive diagnostic tests. Volatile organic compounds have been analysed in both exhaled breath and headspace of faeces, and significant differences have been detected between infected and non-infected goats (Purkhart *et al.*, 2011; Bergmann *et al.*, 2015).

13.7.1 Culture-based diagnosis

Cultivation of *MAP* is a highly specific method, but it is costly and generally requires from 8–16 weeks of incubation to produce visible colonies on solid media (de Juan *et al.*, 2006). Using liquid culture media can significantly shorten the incubation period (Bauman *et al.*, 2016). The sensitivity of culture is difficult to determine, as this method itself is considered to be the 'gold standard' method for detection of *MAP*.

The first detection of faecal shedding has varied considerably from 1.5 months and up to 1–2 years after infection (Storset *et al.*, 2001; Stewart *et al.*, 2007; Köhler *et al.*, 2015). Intermittent faecal shedding is common, and the number of bacteria in faeces may be below detection level. Vaccination also reduces faecal shedding of *MAP*, interfering with diagnosis (Rosseels and Huygen, 2008; Hines *et al.*, 2014; Mercier *et al.*, 2014). Bacteriological culture is effectively 100% specific, as slow-growing, mycobactin-dependent, acid-fast bacteria that harbour the IS*900* element are identified as *MAP* (Olsen *et al.*, 2002).

As goats can be infected by different types of *MAP* strains, different media and an extended incubation period up to 8 months is suggested for solid media to detect *MAP* in new areas or flocks (de Juan *et al.*, 2006). Extending of the incubation period in liquid media up to 240 days showed increase in the sensitivity of *MAP* detection in goats (Bauman *et al.*, 2016). Different selective and non-selective media containing mycobactin can be used to culture *MAP* from goats, including Löwenstein–Jensen, Herrold's egg yolk medium, Middlebrook 7H11 (M7H11) or Dubos medium (Saxegaard, 1985; de Juan *et al.*, 2006). Middlebrook 7H11 supplemented with mycobactin j, OADC (Oleic Albumin Dextrose Catalase) and newborn calf serum is particularly suitable for the primary isolation of both *MAP*-C and *MAP*-S from small ruminants (Dimareli-Malli *et al.*, 2013). A study showed the combination of M7H11 and Herrold's egg yolk medium supplemented with mycobactin J and sodium pyruvate allowed the detection of all *MAP* isolates (Dimareli-Malli *et al.*, 2013).

Pooled faecal culture can be used for herd diagnosis if the animals are moderate or high shedders of *MAP*, but this would be unsuitable in herds with only a few low shedders (Eamens *et al.*, 2007). Tissue samples may also be collected at necropsy for culture. In goats, it is essential to collect mesenteric lymph nodes, and different sections of both ileum and jejunum to achieve the best possible sensitivity (Lybeck *et al.*, 2013; Krüger *et al.*, 2015b; Chaturvedi *et al.*, 2017).

13.7.2 PCR-based methods

Different methods based on IS*900* PCR have been used to detect *MAP* from different samples of naturally and experimentally infected goats, including faeces (Ikonomopoulos *et al.*, 2007), milk (Djønne *et al.*, 2003) and intestinal tissue (Whittington *et al.*, 1999). The sensitivity and specificity of PCR vary between different methods (see Chapter 19, this volume). *MAP*-S strains are very fastidious to culture and goats with a history of contact with *MAP*-positive sheep should be analysed using PCR (Collins, 2011).

13.7.3 Immunological methods

MAP antigens can be detected in paraffin-embedded tissue sections from goats by immunohistochemistry (IHC), and this method seems to be more sensitive than staining by the Ziehl–Neelsen technique (Thoresen *et al.*, 1994). IHC however, seems to be less sensitive than bacterial culture in detecting *MAP* in tissues (Krüger *et al.*, 2015b).

Cell-mediated immune response detected by the *MAP*-specific IFN-γ test can be useful to monitor the paratuberculosis status of non-vaccinated goat herds. The method seems to give fewer false-positive reactions in young goats compared with young cattle (Storset *et al.*, 2005; Lybeck *et al.*, 2011). IFN-γ responses can be found from about 2 months of age or 2 months after experimental infection (Storset *et al.*, 2001; Stewart *et al.*, 2007; Mercier *et al.*, 2016), but is often delayed for another 6–15 months or longer (Lybeck *et al.*, 2011; Köhler *et al.*, 2015; Mercier *et al.*, 2016). Cellular immune responses are generally believed to occur prior to faecal shedding and antibody responses, but several exceptions have been seen (Storset *et al.*, 2001; Stewart *et al.*, 2007; Lybeck *et al.*, 2011; Köhler *et al.*, 2015; Mercier *et al.*, 2016). Consequently, this test might not detect infected animals before they start shedding *MAP*. In addition, cattle studies have shown that the IFN-γ test cannot determine if an animal is infected, only that it has been exposed to *MAP* (Jungersen *et al.*, 2012).

The detection of a delayed-type hypersensitivity reaction to johnin has been successfully applied in goats that are naturally infected by *MAP* (Tripathi *et al.*, 2006), although the johnin skin test appears far less sensitive than faecal PCR (Rawther *et al.*, 2012). Based on studies in cattle, the skin test results are considered to be unreliable (Kalis *et al.*, 2003).

In contrast, antibody-detection-based tests for *MAP* infection are inexpensive and fast. Enzyme-linked immunosorbent assay (ELISA), agar gel immunodiffusion and complement fixation can be used in paratuberculosis control programmes for the goat industries without a major risk of generating large proportions of false-positive test results. Of these methods, the ELISA is most sensitive for detection of infected goats (Whittington *et al.*, 2003; Gumber *et al.*, 2006). All three methods show good specificity, although serological cross-reactions between *MAP* infections and other goat pathogens have been reported. False-positive reactions can occur due to cross-reacting antibodies produced against *Corynebacterium pseudotuberculosis* (Manning *et al.*, 2007) and *Mycobacterium bovis* (Álvarez *et al.*, 2008).

Commercial bovine paratuberculosis ELISAs have also been used to test goat milk. ELISA testing of goat milk samples appears to offer a useful, low-cost alternative for detection of goats with paratuberculosis that have progressed to the stage of shedding *MAP* in their faeces (Salgado *et al.*, 2007). Conflicting results exist regarding whether testing of milk is less sensitive than testing of serum for presence of *MAP* antibodies (Salgado *et al.*, 2007; Angelidou *et al.*, 2014b). In one study, testing of bulk tank milk with ELISA demonstrated comparable sensitivity with detection of *MAP* antibodies in individual milk or serum, but was less sensitive than faecal culture and PCR (Bauman *et al.*, 2019).

There is a continuous search for more specific and sensitive antigens for use in both cell-mediated and antibody tests, such as peptides or recombinant proteins from *MAP*. The combined use of several such antigens could increase test sensitivity (Casey *et al.*, 2011; Souriau *et al.*, 2017).

13.8 Treatment

Paratuberculosis is rarely treated in any species. In one study, goats treated with a combination of streptomycin, isoniazid and rifampin had a favourable clinical outcome, but extensive granulomatous inflammatory foci were detected in various organs at necropsy and *MAP* was isolated from lymph nodes (Slocombe, 1982). Consequently, antibiotics are not used to treat paratuberculosis in goats.

13.9 Control

Control of paratuberculosis is challenging and should not be underestimated. Different control measures can be applied, varying according to type of herd, animal health regulations and economical conditions (Juste and Perez, 2011; Whittington *et al.*, 2019). Control programmes are rarely successful in eliminating the infection from a herd or flock in the short term without a comprehensive and costly programme. Successful prevention and control depend on animal health authorities and livestock industries acquiring a good understanding of the nature and epidemiology of infection, and of the application of tools for diagnosis and control (Kennedy and Benedictus, 2001).

13.9.1 Environment and management

Management is considered to be the most useful tool for controlling paratuberculosis within domestic livestock herds (Harris and Barletta, 2001). In particular populations in endemic areas, paratuberculosis has been eradicated by different sanitizing approaches in goat herds (Lindheim and Sølverød, 2014; TINE Rådgiving, Helsetjenesten for geit, 2016; Gavin *et al.*, 2018). Faecal shedders should be removed, and flocks should not allow introduction of goats with a higher probability of paratuberculosis infection than their own. Care of newborn kids, separation from older animals, feeding of paratuberculosis-free colostrum and milk, a clean environment and adequate feeding of growing kids are among the most important management tools to control paratuberculosis within a flock. This was recently shown in a successful Norwegian sanitation programme (Lindheim and Sølverød, 2014, 2016; TINE Rådgiving, Helsetjenesten for geit, 2016), in which goat kids were removed from their dams at birth, fed paratuberculosis free cow's milk and reared in a presumed *MAP*-free environment.

13.9.2 Vaccination

Currently available paratuberculosis vaccines for goats are inactivated whole-cell vaccines (Corpa *et al.*, 2000b). Live vaccines have previously been used (Saxegaard and Fodstad, 1985).

Existing vaccines can delay the onset of clinical symptoms, improve body weight, and reduce mortality and faecal shedding of bacteria, but they do not protect against infection. Currently used vaccines can also result in severe injection site reactions (Rosseels and Huygen, 2008; Bastida and Juste, 2011). A number of different vaccines have been tested, with variable protection, in order to achieve improved protection. These include vaccines based on cell wall-deficient *MAP* bacteria (Hines *et al.*, 2007b), recombinant antigen vaccines (Kathaperumal *et al.*, 2009), live attenuated mutant vaccines (Park *et al.*, 2011; Faisal *et al.*, 2013a; Hines *et al.*, 2014; Shippy *et al.*, 2017) and attenuated *Salmonella* strains expressing a fusion product of *MAP* antigens (Faisal *et al.*, 2013b). So far, none of these has reached commercialization.

Vaccination is generally recommended in young animals, but vaccination of adult goats in infected flocks can reduce the number of animals developing clinical disease (Corpa *et al.*, 2000c). Vaccinated animals develop antibodies that can interfere with existing serodiagnostic tests for paratuberculosis and they can become reactive in the tuberculin skin test, used for the control of bovine tuberculosis (Hines *et al.*, 2007b; Rosseels and Huygen, 2008).

13.10 Concluding Thoughts

To further improve control of caprine paratuberculosis, continuous focus on management and prevention of transmission are important factors. Development of sensitive and validated diagnostic assays, especially in the early subclinical stage of infection, is also needed for efficient control. Thus, future research priorities should be diagnostic tests, including more knowledge on volatile organic compounds and other *MAP* biomarkers. In addition, more knowledge on protective immunity and pathogenesis in goats is needed to be able to develop improved vaccines.

References

Abendaño, N., Sevilla, I.A., Prieto, J.M., Garrido, J.M., Juste, R.A. *et al.* (2013) *Mycobacterium avium* subspecies *paratuberculosis* isolates from sheep and goats show reduced persistence in bovine macrophages than cattle, bison, deer and wild boar strains regardless of genotype. *Veterinary Microbiology* 163(3–4), 325–334. DOI: 10.1016/j.vetmic.2012.12.042.

Abraham, A., Naicy, T., Raghavan, K.C., Siju, J. and Aravindakshan, T. (2017) Evaluation of the association of SLC11A1 gene polymorphism with incidence of paratuberculosis in goats. *Journal of Genetics* 96(4), 641–646. DOI: 10.1007/s12041-017-0820-9.

Alinovi, C.A., Wu, C.C. and Lin, T.L. (2009) In utero *Mycobacterium avium* subspecies *paratuberculosis* infection of a pygmy goat. *Veterinary Record* 164(9), 276–277. DOI: 10.1136/vr.164.9.276.

Álvarez, J., de Juan, L., Bezos, J., Romero, B., Sáez, J.L. *et al.* (2008) Interference of paratuberculosis with the diagnosis of tuberculosis in a goat flock with a natural mixed infection. *Veterinary Microbiology* 128(1–2), 72–80. DOI: 10.1016/j.vetmic.2007.08.034.

Angelidou, E., Kostoulas, P. and Leontides, L. (2014a) Flock-level factors associated with the risk of *Mycobacterium avium* subsp. *paratuberculosis* (*MAP*) infection in Greek dairy goat flocks. *Preventive Veterinary Medicine* 117(1), 233–241. DOI: 10.1016/j.prevetmed.2014.09.002.

Angelidou, E., Kostoulas, P. and Leontides, L. (2014b) Bayesian validation of a serum and milk ELISA for antibodies against *Mycobacterium avium* subspecies *paratuberculosis* in Greek dairy goats across lactation. *Journal of Dairy Science* 97(2), 819–828. DOI: 10.3168/jds.2013-7218.

Bastida, F. and Juste, R.A. (2011) Paratuberculosis control: a review with a focus on vaccination. *Journal of Immune Based Therapies and Vaccines* 9(1), 8. DOI: 10.1186/1476-8518-9-8.

Bauman, C.A., Jones-Bitton, A., Jansen, J., Kelton, D. and Menzies, P. (2016) Evaluation of fecal culture and fecal RT-PCR to detect *Mycobacterium avium* ssp. *paratuberculosis* fecal shedding in dairy goats and dairy sheep using latent class Bayesian modeling. *BMC Veterinary Research* 12(1), 212. DOI: 10.1186/s12917-016-0814-5.

Bauman, C.A., Jones-Bitton, A., Jansen, J., Kelton, D. and Menzies, P. (2019) Evaluation of bulk tank milk PCR and bulk tank milk modified ELISA tests for the detection of paratuberculosis at the herd level in goat and sheep dairies in Ontario, Canada. *Journal of Dairy Science* 102(1), 511–520. DOI: 10.3168/jds.2018-15020.

Begg, D.J. and Whittington, R.J. (2008) Experimental animal infection models for Johne's disease, an infectious enteropathy caused by *Mycobacterium avium* subsp. *paratuberculosis*. *The Veterinary Journal* 176(2), 129–145. DOI: 10.1016/j.tvjl.2007.02.022.

Bergmann, A., Trefz, P., Fischer, S., Klepik, K., Walter, G. *et al.* (2015) In vivo volatile organic compound signatures of *Mycobacterium avium* subsp. *paratuberculosis*. *PLoS ONE* 10(4), e0123980. DOI: 10.1371/journal.pone.0123980.

Berry, A., Wu, C.W., Venturino, A.J. and Talaat, A.M. (2018) Biomarkers for early stages of Johne's disease infection and immunization in goats. *Frontiers in Microbiology* 9, 2284. DOI: 10.3389/fmicb.2018.02284.

Bonelli, F., Fratini, F., Turchi, B., Cantile, C., Ebani, V.V. *et al.* (2017) Evaluation of clinical pathology parameters in fecal PCR-positive or PCR-negative goats for Johne's disease. *Tropical Animal Health and Production* 49(7), 1489–1493. DOI: 10.1007/s11250-017-1351-3.

Casey, J.L., Sanalla, A.M., Tamvakis, D., Thalmann, C., Carroll, E.L. *et al.* (2011) Peptides specific for *Mycobacterium avium* subspecies *paratuberculosis* infection: diagnostic potential. *Protein Engineering Design and Selection* 24(8), 589–596. DOI: 10.1093/protein/gzr026.

Cecchi, F., Russo, C., Iamartino, D., Galiero, A., Turchi, B. *et al.* (2017) Identification of candidate genes for paratuberculosis resistance in the native Italian Garfagnina goat breed. *Tropical Animal Health and Production* 49(6), 1135–1142. DOI: 10.1007/s11250-017-1306-8.

Chaturvedi, S., Singh, S. V., Srivastava, A.K., Gangwar, N.K., Kumar, N. *et al.* (2017) Comparative evaluation of FAT, IS*900* PCR and microscopy vis-a-vis histopathology for the detection of *Mycobacterium avium* subsp. *paratuberculosis* infection in tissues of goats naturally died in herds endemic for Johne's disease. *The Indian Journal of Animal Sciences* 87, 21–29.

Collins, M.T. (2011) Diagnosis of paratuberculosis. *Veterinary Clinics of North America: Food Animal Practice* 27(vi), 581–591.

Corpa, J.M., Garrido, J., García Marín, J.F. and Pérez, V. (2000a) Classification of lesions observed in natural cases of paratuberculosis in goats. *Journal of Comparative Pathology* 122(4), 255–265. DOI: 10.1053/jcpa.1999.0368.

Corpa, J.M., Pérez, V. and García Marín, J.F. (2000b) Differences in the immune responses in lambs and kids vaccinated against paratuberculosis, according to the age of vaccination. *Veterinary Microbiology* 77(3–4), 475–485. DOI: 10.1016/S0378-1135(00)00332-1.

Corpa, J.M., Pérez, V., Garcia Marin, J.F. Sánchez, M.A. and Marín, J.F. (2000c) Control of paratuberculosis (Johne's disease) in goats by vaccination of adult animals. *Veterinary Record* 146(7), 195–196. DOI: 10.1136/vr.146.7.195.

Daniels, M.J., Hutchings, M.R., Allcroft, D.J., McKendrick, I.J. and Greig, A. (2002) Risk factors for Johne's disease in Scotland – the results of a survey of farmers. *Veterinary Record* 150(5), 135–139. DOI: 10.1136/vr.150.5.135.

Daniels, M.J., Hutchings, M.R., Beard, P.M., Henderson, D., Greig, A. *et al.* (2003) Do non-ruminant wildlife pose a risk of paratuberculosis to domestic livestock and vice versa in Scotland? *Journal of Wildlife Diseases* 39(1), 10–15. DOI: 10.7589/0090-3558-39.1.10.

de Juan, L., Mateos, A., Domínguez, L., Sharp, J.M. and Stevenson, K. (2005) Genetic diversity of *Mycobacterium avium* subspecies *paratuberculosis* isolates from goats detected by pulsed-field gel electrophoresis. *Veterinary Microbiology* 106(3–4), 249–257. DOI: 10.1016/j.vetmic.2004.12.013.

de Juan, L., Alvarez, J., Romero, B., Bezos, J., Castellanos, E. *et al.* (2006) Comparison of four different culture media for isolation and growth of type II and type I/III *Mycobacterium avium* subsp. *paratuberculosis* strains isolated from cattle and goats. *Applied and Environmental Microbiology* 72(9), 5927–5932. DOI: 10.1128/AEM.00451-06.

de Silva, K., Begg, D.J., Plain, K.M., Purdie, A.C., Kawaji, S. *et al.* (2013) Can early host responses to mycobacterial infection predict eventual disease outcomes? *Preventive Veterinary Medicine* 112(3–4), 203–212. DOI: 10.1016/j.prevetmed.2013.08.006.

Dimareli-Malli, Z., Mazaraki, K., Stevenson, K., Tsakos, P., Zdragas, A. *et al.* (2013) Culture phenotypes and molecular characterization of *Mycobacterium avium* subsp. *paratuberculosis* isolates from small ruminants. *Research in Veterinary Science* 95(1), 49–53. DOI: 10.1016/j.rvsc.2013.03.010.

Djønne, B., Jensen, M.R., Grant, I.R. and Holstad, G. (2003) Detection by immunomagnetic PCR of *Mycobacterium avium* subsp. *paratuberculosis* in milk from dairy goats in Norway. *Veterinary Microbiology* 92(1–2), 135–143. DOI: 10.1016/S0378-1135(02)00355-3.

Djønne, B., Pavlik, I., Svastova, P., Bartos, M. and Holstad, G. (2005) IS*900* restriction fragment length polymorphism (RFLP) analysis of *Mycobacterium avium* subsp. *paratuberculosis* isolates from goats and cattle in Norway. *Acta Veterinaria Scandinavica* 46(1), 13–18. DOI: 10.1186/1751-0147-46-13.

Eamens, G.J., Walker, D.M., Porter, N.S. and Fell, S.A. (2007) Pooled faecal culture for the detection of *Mycobacterium avium* subsp. *paratuberculosis* in goats. *Australian Veterinary Journal* 85(6), 243–251. DOI: 10.1111/j.1751-0813.2007.00160.x.

Faisal, S.M., Chen, J.-W., Yan, F., Chen, T.-T., Useh, N.M. *et al.* (2013a) Evaluation of a *Mycobacterium avium* subsp. *paratuberculosis* leuD mutant as a vaccine candidate against challenge in a caprine model. *Clinical and Vaccine Immunology* 20(4), 572–581. DOI: 10.1128/CVI.00653-12.

Faisal, S.M., Yan, F., Chen, T.-T., Useh, N.M., Guo, S. *et al.* (2013b) Evaluation of a *Salmonella* vectored vaccine expressing *Mycobacterium avium* subsp. *paratuberculosis* antigens against challenge in a goat model. *PLoS ONE* 8(8), e70171. DOI: 10.1371/journal.pone.0070171.

Freitas, T.D., Azevedo, S.S., Silva, M.L.C.R., Garino Júnior, F., Santos, C.D.S.A.B. *et al.* (2015) Epidemiological characterization and risk factors associated with *Mycobacterium avium* subsp. *paratuberculosis* infection in dairy goats in the Brazilian semiarid region. *Semina: Ciências Agrárias* 36(1), 267–275. DOI: 10.5433/1679-0359.2015v36n1p267.

Gavin, W.G., Porter, C.A., Hawkins, N., Schofield, M.J. and Pollock, J.M. (2018) Johne's disease: a successful eradication programme in a dairy goat herd. *Veterinary Record* 182(17), 483–483. DOI: 10.1136/vr.104507.

Gollnick, N.S., Mitchell, R.M., Baumgart, M., Janagama, H.K., Sreevatsan, S. *et al.* (2007) Survival of *Mycobacterium avium* subsp. *paratuberculosis* in bovine monocyte-derived macrophages is not affected by host infection status but depends on the infecting bacterial genotype. *Veterinary Immunology and Immunopathology* 120(3–4), 93–105. DOI: 10.1016/j.vetimm.2007.07.017.

Gumber, S., Eamens, G. and Whittington, R. (2006) Evaluation of a Pourquier ELISA kit in relation to agar gel immunodiffusion (AGID) test for assessment of the humoral immune response in sheep and goats with and without *Mycobacterium paratuberculosis* infection. *Veterinary Microbiology* 115(1–3), 91–101. DOI: 10.1016/j.vetmic.2006.01.003.

Harris, N.B. and Barletta, R.G. (2001) *Mycobacterium avium* subsp. *paratuberculosis* in veterinary medicine. *Clinical Microbiology Reviews* 14(3), 489–512. DOI: 10.1128/CMR.14.3.489-512.2001.

Hines, M.E., Stabel, J.R., Sweeney, R.W., Griffin, F., Talaat, A.M. *et al.* (2007a) Experimental challenge models for Johne's disease: a review and proposed international guidelines. *Veterinary Microbiology* 122(3–4), 197–222. DOI: 10.1016/j.vetmic.2007.03.009.

Hines, M.E., Stiver, S., Giri, D., Whittington, L., Watson, C. *et al.* (2007b) Efficacy of spheroplastic and cell-wall competent vaccines for *Mycobacterium avium* subsp. *paratuberculosis* in experimentally-challenged baby goats. *Veterinary Microbiology* 120(3–4), 261–283. DOI: 10.1016/j.vetmic.2006.10.030.

Hines, M.E., Turnquist, S.E., Ilha, M.R.S., Rajeev, S., Jones, A.L. *et al.* (2014) Evaluation of novel oral vaccine candidates and validation of a caprine model of Johne's disease. *Frontiers in Cellular and Infection Microbiology* 4, 26. DOI: 10.3389/fcimb.2014.00026.

Holstad, G., Sigurðardóttir, Ó.G., Storset, A.K., Tharaldsen, J., Nyberg, O. *et al.* (2005) Description of the infection status in a Norwegian cattle herd naturally infected by *Mycobacterium avium* subsp. *paratuberculosis*. *Acta Veterinaria Scandinavica* 46(1), 45–56. DOI: 10.1186/1751-0147-46-45.

Ikonomopoulos, J., Balaskas, C., Kantzoura, B., Fragiadaki, E., Pavlik, I. *et al.* (2007) Comparative evaluation of positive tests to *Mycobacterium avium* subsp. *paratuberculosis* in clinically healthy sheep and goats in south-west Greece using molecular techniques, serology, and culture. *The Veterinary Journal* 174(2), 337–343. DOI: 10.1016/j.tvjl.2006.09.004.

Jungersen, G., Mikkelsen, H. and Grell, S.N. (2012) Use of the johnin PPD interferon-gamma assay in control of bovine paratuberculosis. *Veterinary Immunology and Immunopathology* 148(1–2), 48–54. DOI: 10.1016/j.vetimm.2011.05.010.

Juste, R.A. and Perez, V. (2011) Control of paratuberculosis in sheep and goats. *Veterinary Clinics of North America: Food Animal Practice* 27(1), 127–138. DOI: 10.1016/j.cvfa.2010.10.020.

Kalis, C.H.J., Collins, M.T., Hesselink, J.W. and Barkema, H.W. (2003) Specificity of two tests for the early diagnosis of bovine paratuberculosis based on cell-mediated immunity: the Johnin skin test and the gamma interferon assay. *Veterinary Microbiology* 97(1–2), 73–86. DOI: 10.1016/j.vetmic.2003.07.003.

Kathaperumal, K., Kumanan, V., McDonough, S., Chen, L.H., Park, S.-U. *et al.* (2009) Evaluation of immune responses and protective efficacy in a goat model following immunization with a cocktail of recombinant antigens and a polyprotein of *Mycobacterium avium* subsp. *paratuberculosis*. *Vaccine* 27(1), 123–135. DOI: 10.1016/j.vaccine.2008.10.019.

Kennedy, D.J. and Benedictus, G. (2001) Control of *Mycobacterium avium* subsp. *paratuberculosis* infection in agricultural species. *Revue Scientifique et Technique de l'OIE* 20(1), 151–179. DOI: 10.20506/rst.20.1.1274.

Koets, A.P., Eda, S. and Sreevatsan, S. (2015) The within host dynamics of *Mycobacterium avium* ssp. *paratuberculosis* infection in cattle: where time and place matter. *Veterinary Research* 46(1), 61. DOI: 10.1186/s13567-015-0185-0.

Köhler, H., Soschinka, A., Meyer, M., Kather, A., Reinhold, P. *et al.* (2015) Characterization of a caprine model for the subclinical initial phase of *Mycobacterium avium* subsp. *paratuberculosis* infection. *BMC Veterinary Research* 11(1), 74. DOI: 10.1186/s12917-015-0381-1.

Korou, L.M., Liandris, E., Gazouli, M. and Ikonomopoulos, J. (2010) Investigation of the association of the SLC11A1 gene with resistance/sensitivity of goats (*Capra hircus*) to paratuberculosis. *Veterinary Microbiology* 144(3-4), 353–358. DOI: 10.1016/j.vetmic.2010.01.009.

Kostoulas, P., Leontides, L., Billinis, C., Amiridis, G.S. and Florou, M. (2006) The association of subclinical paratuberculosis with the fertility of Greek dairy ewes and goats varies with parity. *Preventive Veterinary Medicine* 74(2-3), 226–238. DOI: 10.1016/j.prevetmed.2005.12.001.

Krüger, C., Köhler, H. and Liebler-Tenorio, E.M. (2015a) Cellular composition of granulomatous lesions in gut-associated lymphoid tissues of goats during the first year after experimental infection with *Mycobacterium avium* subsp. *paratuberculosis*. *Veterinary Immunology and Immunopathology* 163(1-2), 33–45. DOI: 10.1016/j.vetimm.2014.11.002.

Krüger, C., Köhler, H. and Liebler-Tenorio, E.M. (2015b) Sequential development of lesions 3, 6, 9, and 12 months after experimental infection of goat kids with *Mycobacterium avium* subsp *paratuberculosis*. *Veterinary Pathology* 52(2), 276–290. DOI: 10.1177/0300985814533804.

Lindheim, D. and Sølverød, L. (2014) *The Norwegian Healthier Goat Project. Proceedings from the 4th ParaTB Forum*. Parma, Italy, pp. 65–66.

Lindheim, D. and Sølverød, L. (2016) The Norwegian Healthier Goats Project. In: Kukovics, S. (ed.) *Sustainable Goat Breeding and Goat Farming in Central and Eastern European Countries, European Regional Conference on Goats 7–13 April 2014*. Food and Agriculture Organization of the United Nations (FAO), Rome, Italy, pp. 209–213.

Lybeck, K.R., Storset, A.K., Djønne, B., Valheim, M. and Olsen, I. (2011) Faecal shedding detected earlier than immune responses in goats naturally infected with *Mycobacterium avium* subsp. *paratuberculosis*. *Research in Veterinary Science* 91(1), 32–39. DOI: 10.1016/j.rvsc.2010.08.012.

Lybeck, K.R., Løvoll, M., Johansen, T.B., Olsen, I., Storset, A.K. *et al.* (2013) Intestinal strictures, fibrous adhesions and high local interleukin-10 levels in goats infected naturally with *Mycobacterium avium* subsp. *paratuberculosis*. *Journal of Comparative Pathology* 148(2–3), 157–172. DOI: 10.1016/j.jcpa.2012.05.007.

Mainar-Jaime, R.C. and Vázquez-Boland, J.A. (1998) Factors associated with seroprevalence to *Mycobacterium paratuberculosis* in small-ruminant farms in the Madrid region (Spain). *Preventive Veterinary Medicine* 34(4), 317–327. DOI: 10.1016/S0167-5877(97)00091-3.

Malone, A.N., Fletcher, D.M., Vogt, M.B., Meyer, S.K., Hess, A.M. *et al.* (2013) Early weight development of goats experimentally infected with *Mycobacterium avium* subsp. *paratuberculosis*. *PLoS ONE* 8(12), e84049. DOI: 10.1371/journal.pone.0084049.

Manning, E.J.B. and Collins, M.T. (2001) *Mycobacterium avium* subsp. *paratuberculosis*: pathogen, pathogenesis and diagnosis. *Revue Scientifique et Technique de l'OIE* 20(1), 133–150. DOI: 10.20506/rst.20.1.1275.

Manning, E.J.B., Cushing, H.F., Hietala, S. and Wolf, C.B. (2007) Impact of *Corynebacterium pseudotuberculosis* infection on serologic surveillance for Johne's disease in goats. *Journal of Veterinary Diagnostic Investigation* 19(2), 187–190. DOI: 10.1177/104063870701900209.

McKenna, S.L., Keefe, G.P., Tiwari, A., VanLeeuwen, J. and Barkema, H.W. (2006) Johne's disease in Canada part II: disease impacts, risk factors, and control programs for dairy producers. *Canadian Veterinary Journal* 47, 1089–1099.

Mercier, P., Brémaud, I. and Gautier, M.-P. (2014) Vaccination of kids under one month of age with a killed vaccine and reduction in the frequency of faecal shedding of *Mycobacterium* avium *subspecies paratuberculosis*. *Small Ruminant Research* 121(2–3), 425–433. DOI: 10.1016/j.smallrumres.2014.09.007.

Mercier, P., Freret, S., Laroucau, K., Gautier, M.-P., Brémaud, I. *et al.* (2016) A longitudinal study of the *Mycobacterium avium* subspecies paratuberculosis infection status in young goats and their mothers. *Veterinary Microbiology* 195, 9–16. DOI: 10.1016/j.vetmic.2016.08.021.

Munjal, S.K., Tripathi, B.N. and Paliwal, O.P. (2005) Progressive immunopathological changes during early stages of experimental infection of goats with *Mycobacterium avium* subspecies *paratuberculosis*. *Veterinary Pathology* 42(4), 427–436. DOI: 10.1354/vp.42-4-427.

Nielsen, S.S. and Toft, N. (2008) Ante mortem diagnosis of paratuberculosis: a review of accuracies of ELISA, interferon-γ assay and faecal culture techniques. *Veterinary Microbiology* 129(3–4), 217–235. DOI: 10.1016/j.vetmic.2007.12.011.

Nielsen, S.S. and Toft, N. (2009) A review of prevalences of paratuberculosis in farmed animals in Europe. *Preventive Veterinary Medicine* 88(1), 1–14. DOI: 10.1016/j.prevetmed.2008.07.003.

Olsen, I., Sigurðardóttir, Ó.G.. and Djønne, B. (2002) Paratuberculosis with special reference to cattle a review. *Veterinary Quarterly* 24(1), 12–28. DOI: 10.1080/01652176.2002.9695120.

Park, K.T., Allen, A.J., Bannantine, J.P., Seo, K.S., Hamilton, M.J. *et al.* (2011) Evaluation of two mutants of *Mycobacterium avium* subsp. *paratuberculosis* as candidates for a live attenuated vaccine for Johne's disease. *Vaccine* 29(29–30), 4709–4719. DOI: 10.1016/j.vaccine.2011.04.090.

Pillai, S.R., Jayarao, B.M., Gummo, J.D., Hue, E.C., Tiwari, D. *et al.* (2001) Identification and subtyping of *Mycobacterium avium* subsp. *paratuberculosis* and *Mycobacterium avium* subsp. *avium* by randomly amplified polymorphic DNA. *Veterinary Microbiology* 79(3), 275–284. DOI: 10.1016/S0378-1135(00)00358-8.

Purkhart, R., Köhler, H., Liebler-Tenorio, E., Meyer, M., Becher, G. *et al.* (2011) Chronic intestinal *Mycobacteria* infection: discrimination via VOC analysis in exhaled breath and headspace of feces using differential ion mobility spectrometry. *Journal of Breath Research* 5(2), 027103. DOI: 10.1088/1752-7155/5/2/027103.

Radostits, O.M., Gay, C.C., Hinchcliff, K.W. and Constable, P.D. (2007) Diseases associated with bacteria. In: Rodenhuis, J. and Ball, E. (eds) *Veterinary Medicine: a Textbook of the Diseases of Cattle, Horses, Sheep, Pigs and Goats*, 10th edn. Saunders Elsevier, Philadelphia, Pennsylvania, pp. 1017–1044.

Rawther, S.S., Saseendranath, M.R., Nair, G.P.K., Tresamol, P.V., Pillai, U.N. *et al.* (2012) Diagnosis of paratuberculosis in goats by cell mediated immune response, conventional and molecular diagnostic techniques. *Tropical Animal Health and Production* 44(4), 911–914. DOI: 10.1007/s11250-011-9986-y.

Rosseels, V. and Huygen, K. (2008) Vaccination against paratuberculosis. *Expert Review of Vaccines* 7(6), 817–832. DOI: 10.1586/14760584.7.6.817.

Salgado, M., Kruze, J. and Collins, M.T. (2007) Diagnosis of paratuberculosis by fecal culture and ELISA on milk and serum samples in two types of Chilean dairy goat herds. *Journal of Veterinary Diagnostic Investigation* 19(1), 99–102. DOI: 10.1177/104063870701900117.

Sardaro, R., Pieragostini, E., Rubino, G. and Petazzi, F. (2017) Impact of *Mycobacterium avium* subspecies *paratuberculosis* on profit efficiency in semi-extensive dairy sheep and goat farms of Apulia, southern Italy. *Preventive Veterinary Medicine* 136, 56–64. DOI: 10.1016/j.prevetmed.2016.11.013.

Saxegaard, F. (1985) Isolation of *Mycobacterium paratuberculosis* from intestinal mucosa and mesenteric lymph nodes of goats by use of selective Dubos medium. *Journal of Clinical Microbiology* 22(2), 312–313. DOI: 10.1128/JCM.22.2.312-313.1985.

Saxegaard, F. (1990) Experimental infection of calves with an apparently specific goat-pathogenic strain of *Mycobacterium paratuberculosis*. *Journal of Comparative Pathology* 102(2), 149–156. DOI: 10.1016/S0021-9975(08)80120-X.

Saxegaard, F. and Fodstad, F. (1985) Control of paratuberculosis (Johne's disease) in goats by vaccination. *Veterinary Record* 116(16), 439–441. DOI: 10.1136/vr.116.16.439.

Sevilla, I., Garrido, J.M., Geijo, M. and Juste, R.A. (2007) Pulsed-field gel electrophoresis profile homogeneity of *Mycobacterium avium* subsp. *paratuberculosis* isolates from cattle and heterogeneity of those from sheep and goats. *BMC Microbiology* 7(1), 18. DOI: 10.1186/1471-2180-7-18.

Sevilla, I., Li, L., Amonsin, A., Garrido, J.M., Geijo, M.V. *et al.* (2008) Comparative analysis of *Mycobacterium avium* subsp. *paratuberculosis* isolates from cattle, sheep and goats by short sequence repeat and pulsed-field gel electrophoresis typing. *BMC Microbiology* 8(1), 204. DOI: 10.1186/1471-2180-8-204.

Shippy, D.C., Lemke, J.J., Berry, A., Nelson, K., Hines, M.E. *et al.* (2017) Superior protection from live-attenuated vaccines directed against Johne's disease. *Clinical and Vaccine Immunology* 24(1), e00478-16, pii. DOI: 10.1128/CVI.00478-16.

Sigurðardóttir, Ó.G., Press, C.M., Saxegaard, F. and Evensen, Ø. (1999) Bacterial isolation, immunological response, and histopathological lesions during the early subclinical phase of experimental infection of goat kids with *Mycobacterium avium* subsp. *paratuberculosis*. *Veterinary Pathology* 36(6), 542–550. DOI: 10.1354/vp.36-6-542.

Sigurðardóttir, Ó.G., Press, C.M. and Evensen, Ø. (2001) Uptake of *Mycobacterium avium* subsp. *paratuberculosis* through the distal small intestinal mucosa in goats: an ultrastructural study. *Veterinary Pathology* 38(2), 184–189. DOI: 10.1354/vp.38-2-184.

Sigurðardóttir, O.G., Bakke-McKellep, A.M., Djønne, B. and Evensen, Ø. (2005) *Mycobacterium avium* subsp. *paratuberculosis* enters the small intestinal mucosa of goat kids in areas with and without Peyer's patches as demonstrated with the everted sleeve method. *Comparative Immunology, Microbiology and Infectious Diseases* 28(3), 223–230. DOI: 10.1016/j.cimid.2005.01.004.

Singh, P.K., Singh, S.V., Saxena, V.K., Singh, M.K., Singh, A.V. *et al.* (2013) Expression profiles of different cytokine genes in peripheral blood mononuclear cells of goats infected experimentally with native strain of *Mycobacterium avium* subsp. *paratuberculosis*. *Animal Biotechnology* 24(3), 187–197. DOI: 10.1080/10495398.2012.762008.

Slocombe, R.F. (1982) Combined streptomycin–isoniazid–rifampin therapy in the treatment of Johne's disease in a goat. *Canadian Veterinary Journal* 23, 160–163.

Souriau, A., Freret, S., Foret, B., Willemsen, P.T.J., Bakker, D. *et al.* (2017) Identification of new antigen candidates for the early diagnosis of *Mycobacterium avium* subsp. *paratuberculosis* infection in goats. *Research in Veterinary Science* 115, 278–287. DOI: 10.1016/j.rvsc.2017.05.025.

Stabel, J.R. (2006) Host responses to *Mycobacterium avium* subsp. *paratuberculosis*: a complex arsenal. *Animal Health Research Reviews* 7(1–2), 61–70. DOI: 10.1017/S1466252307001168.

Stehman, S.M. (1996) Paratuberculosis in small ruminants, deer, and South American camelids. *Veterinary Clinics of North America: Food Animal Practice* 12(2), 441–455. DOI: 10.1016/S0749-0720(15)30416-3.

Stewart, D.J., Vaughan, J.A., Stiles, P.L., Noske, P.J., Tizard, M.L.V. *et al.* (2006) A long-term study in Angora goats experimentally infected with *Mycobacterium avium* subsp. *paratuberculosis*: clinical disease, faecal culture and immunological studies. *Veterinary Microbiology* 113(1–2), 13–24. DOI: 10.1016/j.vetmic.2005.09.015.

Stewart, D., Vaughan, J., Stiles, P., Noske, P., Tizard, M. *et al.* (2007) A long-term bacteriological and immunological study in Holstein–Friesian cattle experimentally infected with *Mycobacterium avium* subsp. *paratuberculosis* and necropsy culture results for Holstein–Friesian cattle, merino sheep and Angora goats. *Veterinary Microbiology* 122(1–2), 83–96. DOI: 10.1016/j.vetmic.2006.12.030.

Storset, A.K., Hasvold, H.J., Valheim, M., Brun-Hansen, H., Berntsen, G. *et al.* (2001) Subclinical paratuberculosis in goats following experimental infection. An immunological and microbiological study. *Veterinary Immunology and Immunopathology* 80, 271–287.

Storset, A.K., Berg, I. and Djønne, B. (2005) Evaluation of the gamma interferon test for diagnosis of paratuberculosis in goats. *Veterinary Immunology and Immunopathology* 107(1–2), 87–94. DOI: 10.1016/j.vetimm.2005.03.015.

Streeter, R.N., Hoffsis, G.F., Bech-Nielsen, S., Shulaw, W.P. and Rings, M. (1995) Isolation of *Mycobacterium paratuberculosis* from colostrum and milk of subclinically infected cows. *American Journal of Veterinary Research* 56, 1322–1324.

Subharat, S., Shu, D., Neil Wedlock, D., Price-Carter, M., de Lisle, G.W. *et al.* (2012) Immune responses associated with progression and control of infection in calves experimentally challenged with *Mycobacterium avium* subsp. *paratuberculosis*. *Veterinary Immunology and Immunopathology* 149(3–4), 225–236. DOI: 10.1016/j.vetimm.2012.07.005.

Taka, S., Liandris, E., Gazouli, M., Sotirakoglou, K., Theodoropoulos, G. *et al.* (2013) In vitro expression of the SLC11A1 gene in goat monocyte-derived macrophages challenged with *Mycobacterium avium* subsp *paratuberculosis*. *Infection, Genetics and Evolution* 17, 8–15. DOI: 10.1016/j.meegid.2013.03.033.

Taka, S., Gazouli, M., Sotirakoglou, K., Liandris, E., Andreadou, M. *et al.* (2015) Functional analysis of 3'UTR polymorphisms in the caprine SLC11A1 gene and its association with the *Mycobacterium avium* subsp. *paratuberculosis* infection. *Veterinary Immunology and Immunopathology* 167(1–2), 75–79. DOI: 10.1016/j.vetimm.2015.06.004.

Thoresen, O.F., Falk, K. and Evensen, Ø. (1994) Comparison of immunohistochemistry, acid-fast staining, and cultivation for detection of *Mycobacterium paratuberculosis* in goats. *Journal of Veterinary Diagnostic Investigation* 6(2), 195–199. DOI: 10.1177/104063879400600210.

TINE Rådgiving, Helsetjenesten for geit (2016) Syk – friskere – friskest. Sluttrapport prosjekt Friskere geiter 2001-2015. [Healthier Goats Project, final report 2001-2015, No]. Ås: Helsetjenesten for Geit; 2016. Available at: http://geithelse.tine.no/friskere-geiter/prosjektrapport-friskere-geiter

Tripathi, B.N., Periasamy, S., Paliwal, O.P. and Singh, N. (2006) Comparison of IS*900* tissue PCR, bacterial culture, johnin and serological tests for diagnosis of naturally occurring paratuberculosis in goats. *Veterinary Microbiology* 116(1-3), 129–137. DOI: 10.1016/j.vetmic.2006.03.017.

Valheim, M., Storset, A.K., Aleksersen, M., Brun-Hansen, H. and Press, C.M. (2002) Lesions in subclinical paratuberculosis of goats are associated with persistent gut-associated lymphoid tissue. *Journal of Comparative Pathology* 127(2–3), 194–202. DOI: 10.1053/jcpa.2002.0581.

Valheim, M., Sigurðardóttir, Ó.G.., Storset, A.K., Aune, L.G. and Press, C.M. (2004) Characterization of macrophages and occurrence of T cells in intestinal lesions of subclinical paratuberculosis in goats. *Journal of Comparative Pathology* 131(2–3), 221–232. DOI: 10.1016/j.jcpa.2004.04.004.

Whitlock, R.H. and Buergelt, C. (1996) Preclinical and clinical manifestations of paratuberculosis (including pathology). *Veterinary Clinics of North America: Food Animal Practice* 12(2), 345–356. DOI: 10.1016/S0749-0720(15)30410-2.

Whittington, R.J., Reddacliff, L., Marsh, I. and Saunders, V. (1999) Detection of *Mycobacterium avium* subsp. *paratuberculosis* in formalin-fixed paraffin-embedded intestinal tissue by IS900 polymerase chain reaction. *Australian Veterinary Journal* 77(6), 392–397. DOI: 10.1111/j.1751-0813.1999.tb10315.x.

Whittington, R.J., Eamens, G.J. and Cousins, D.V. (2003) Specificity of absorbed ELISA and agar gel immuno-diffusion tests for paratuberculosis in goats with observations about use of these tests in infected goats. *Australian Veterinary Journal* 81(1–2), 71–75. DOI: 10.1111/j.1751-0813.2003.tb11437.x.

Whittington, R., Donat, K., Weber, M.F., Kelton, D., Nielsen, S.S. *et al.* (2019) Control of paratuberculosis: who, why and how. A review of 48 countries. *BMC Veterinary Research* 15(1), 198. DOI: 10.1186/s12917-019-1943-4.

14 Paratuberculosis in Deer, Camelids and Other Ruminants

Rory O'Brien[1]*, Colin Mackintosh[2] and Frank Griffin[3]
[1]DRL, Invermay Agricultural Centre, Mosgiel, New Zealand; [2]AgResearch, Invermay Agricultural Centre, Mosgiel, New Zealand; [3]University of Otago, Dunedin, New Zealand

14.1 Paratuberculosis in Wild and Farmed Deer

14.1.1 Wild and captive deer

There have been a few isolated reports of paratuberculosis in deer prior to 1970, but since then it has been reported frequently in a range of deer species in the wild and in parks, zoos and enclosures in a number of countries (Temple *et al.*, 1979; Pacetti *et al.*, 1994). This includes reindeer/caribou (*Rangifer tarandus*) in Russia, northern Canada and Greenland (Strogov, 1973; Forde *et al.*, 2012), axis deer (*Axis axis*) in California (Riemann *et al.*, 1979), Tule elk (*Cervus canadensis nannodes*) and white-tailed deer (*Odocoileus virginianus*) in the USA (Jessup *et al.*, 1981; Chiodini and Van Kruiningen, 1983), fallow (*Dama dama*) and red deer (*Cervus elaphus*) in Spain (Marco *et al.*, 2002; Reyes-García *et al.*, 2008), red, roe (*Capreolus capreolus*) and fallow deer in the Czech Republic (Pavlik *et al.*, 2000; Machackova *et al.*, 2004; Kopecna *et al.*, 2008), red deer in the western Alps of Italy (Nebbia *et al.*, 2000) and Austria (Glawischnig *et al.*, 2006), red and roe deer in Italy (Robino *et al.*, 2008; Galiero *et al.*, 2018), and red, roe, fallow and sika deer (*Cervus nippon*) in Scotland (Sharp *et al.*, 1997; Simpson *et al.*, 2012). A serological survey of free-ranging cervids in Norway (Tryland *et al.*, 2004) showed evidence of paratuberculosis in moose (*Alces alces*), red deer, roe deer and semi-domesticated reindeer, but no evidence was seen in wild reindeer. Paratuberculosis has been diagnosed in wild red deer (*Cervus elaphus*) in Germany (Fritsch *et al.*, 2012), where it is suspected that disease may have been transmitted from domestic cattle to wild deer.

14.1.2 Farmed deer

Since the early 1980s, paratuberculosis has been diagnosed on deer farms in Europe, Asia, North and South America and Oceania. Reported cases include farmed deer in the UK (Gilmour, 1988; Fawcett *et al.*, 1995), Ireland (Power *et al.*, 1993), Denmark (Jorgensen and Jorgensen, 1987), Germany (Commichau, 1982), Belgium (Godfroid *et al.*, 2000), New Zealand (Gumbrell, 1986; de Lisle *et al.*, 1993), Canada (Starke, 1991), Hungary (Pavlik *et al.*, 1994; Machackova *et al.*, 2004), the USA (Manning *et al.*, 1998), Argentina (Mereb *et al.*,

*Corresponding author: rory@drl.net.nz

© CAB International 2020. *Paratuberculosis: Organism, Disease, Control, 2nd Edition* (eds M.A. Behr *et al.*)

1994) and Australia (Kennedy and Allworth, 2000).

New Zealand is unique in that it has a population of approximately 1 million farmed European red deer (*Cervus elaphus*) and a smaller herd of North American elk (*Cervus elaphus nelsoni/manitobensis*), a farming practice not commonly employed internationally. The first case of paratuberculosis in farmed red deer in New Zealand was diagnosed in 1979 (Gumbrell, 1986) and paratuberculosis has been the subject of investigation and research since the mid-1990s. Passive surveillance, principally by the examination of suspect tuberculous lesions identified in deer slaughter plants, subsequently resulted in *Mycobacterium avium* subsp. *paratuberculosis* (*MAP*) being identified in over 600 farmed deer on 300 properties (de Lisle *et al.*, 2003). The study suggested that the herd prevalence at that time was approximately 6%. Serological surveillance for paratuberculosis on 627 of New Zealand's deer farms has been carried out since 2000, using blood samples submitted as part of the national tuberculosis eradication scheme (Griffin *et al.*, 2003, 2005). The findings from this project suggested a national prevalence of herd infection of around 63% (Griffin *et al.*, 2006). A recent non-random sample of 115 New Zealand deer herds of unknown infection status found that 43% had pooled faecal samples culture-positive for *MAP* (Glossop *et al.*, 2006). Since 2007 abattoir surveillance and recording of enlarged mesenteric lymph nodes suggestive of paratuberculosis has been in place in New Zealand involving >95% of deer farms. While low prevalence outbreaks of clinical paratuberculosis occurred commonly prior to 2010, regular herd testing with ELISA and culling of reactors since then has reduced clinical outbreaks of paratuberculosis in New Zealand deer herds to a very low level.

14.1.3 Disease

The pathogenesis of *MAP* infection in cervids is different from that observed in other ruminant species. Deer appear to be innately susceptible to *MAP* infection and show clinical evidence of disease and severe pathology at an earlier age than other ruminants. Clinical disease has been reported in fawns and yearlings of white-tailed deer, sika and fallow deer (Temple *et al.*, 1979), red deer (Gilmour and Nyange, 1989; Machackova *et al.*, 2004; Mackintosh *et al.*, 2004) and North American elk (Manning *et al.*, 1998). In spite of this, only a small proportion of deer exposed to *MAP* develop clinical disease. Two clinical syndromes have been recognized in heavily infected herds of red deer: sporadic cases in adult deer, which may affect <2% per annum, and outbreaks in 8–27-month-old deer, which generally affect <5% of a group but in severe cases may involve up to 20% (Mackintosh *et al.*, 2004). Adult deer tend to show chronic loss of condition and weight, similar to the disease in sheep and cattle. The clinical disease in young deer tends to be much more acute, with rapid loss of weight and muscle mass over a period of a few weeks rather than months. Affected deer usually have rough or 'moth-eaten' coats and have faecal staining around the tail and hocks as a result of soft or liquid faeces. Blood samples taken for clinical pathology tests typically show low total protein and low serum albumin concentrations, and there may be elevated concentrations of acute inflammatory proteins such as fibrinogen and haptoglobin. Cervids typically mount a vigorous immune response to *MAP* infection characterized by high titres of antibody (Griffin *et al.*, 2005) and are capable of shedding high titres of *MAP* organisms into the environment as the disease progresses from the paucibacillary to multibacillary state (O'Brien *et al.*, 2013). Subclinical paratuberculosis may result in poor growth rates in yearlings, reduced calving percentages in hinds, reduced velvet production in stags, interference with bovine tuberculosis (TB) testing and TB-like lesions in gut lymph nodes at slaughter (Mackintosh, 2008).

14.1.4 Pathology

Post-mortem examination of clinically affected deer typically reveals greatly enlarged jejunal and ileocaecal lymph nodes, occasionally with white or cream-coloured caseous lesions. There may be some gross thickening of the jejunum and terminal ileum, but this is usually less obvious than that seen in sheep and cattle. However, the lymphatic drainage vessels from the jejunum to adjacent lymph nodes are often notably thickened, and the omentum may be oedematous and in some cases may be adherent to the affected jejunum and lymph nodes (Gilmour and Nyange, 1989; Mackintosh *et al.*, 2004). Subclinically affected deer may also have bovine TB-like caseous lesions in the mesenteric lymph nodes and occasionally these can be found in the retropharyngeal lymph nodes. These paratuberculosis lesions can be indistinguishable from those caused by *Mycobacterium bovis*.

Histopathological examination of gut and lymph node lesions typically shows extensive areas of invasion of affected lymph nodes by macrophages laden with acid-fast organisms (AFOs) (multibacillary disease). In these cases, there are often giant cells, caseation and foci of calcification. The ileocaecal valve may show loss of villous structure and a mixed cellular infiltrate and contain numerous AFOs, and similar lesions are also commonly found proximally in the ileum and/or jejunum. Some deer with less severe disease may show milder lesions and these may have few visible AFOs (paucibacillary disease) (Gilmour and Nyange, 1989; Power *et al.*, 1993; Mackintosh *et al.*, 2004). Histopathological changes due to infection by *MAP*, *M. bovis* and *M. avium* subsp. *avium* can be indistinguishable, and a presumptive microbial diagnosis needs confirmation by culture or polymerase chain reaction (PCR). A scoring system has been developed to assist in the objective scoring of the severity of lesions due to paratuberculosis in deer (Mackintosh *et al.*, 2007) and for histopathological grading of disease severity (Clark *et al.*, 2010, 2011).

14.1.5 Epidemiology

Although *MAP* strains of both *MAP*-C (Cattle-type or Type C) and *MAP*-S (Sheep-type or Type S) have been isolated from deer, *MAP*-C appears to be both more common and more pathogenic and is responsible for the outbreaks of serious disease in both young, farmed red deer (O'Brien *et al.*, 2006; Verdugo *et al.*, 2014) and in wild red deer (Galiero *et al.*, 2018). A more detailed discussion of the different strain types of *MAP* is given elsewhere (see Chapter 6, this volume). Clinical disease developed in 5 out of 16 deer, 21–38 weeks after heavy oral experimental challenge of 4-month-old deer with a *MAP*-C strain, while the *MAP*-S strain was relatively non-pathogenic (Mackintosh *et al.*, 2007). Over 95% of isolates of *MAP* from farmed deer in New Zealand have been *MAP*-C (de Lisle *et al.*, 2006).

Young deer appear to be more susceptible to clinical disease than older animals. In an experimental challenge study, 30 3-month-old weaners, 20 yearlings and 20 adult red deer were all dosed with 4×10^9 colony-forming units of a *MAP*-C strain of *MAP*. Ten of the weaner animals developed clinical disease in the year following challenge while none of the yearlings or adults was similarly affected (Mackintosh *et al.*, 2010).

Intrauterine transmission from dam to fetus has been reported in red deer. van Kooten *et al.* (2006) reported that in a sample of 9 late-stage pregnant hinds clinically affected by paratuberculosis, 9 out of 10 (90%) of their fetuses were BACTEC culture-positive for *MAP*. In another study, *MAP* was isolated from 78% of fetuses of a group of 18 subclinically infected red deer hinds (Thompson *et al.*, 2007). In contrast, *MAP* was isolated from only 39% of fetuses from clinically affected dairy cows and 9% of fetuses from subclinically infected dairy cows (Whittington and Windsor, 2007). In sheep, intrauterine transmission is thought to occur in <10% of infected ewes. *MAP*-infected colostrum and milk may play a role in pseudo-vertical transmission.

As with sheep and cattle, young deer also become infected via the oral route from contaminated pasture and water. On farms, deer-to-deer transmission appears the most common route, although cattle-to-deer and sheep-to-deer transmission may occur. The role of wildlife is not known. Resistance to paratuberculosis has a genetic component, which may be related to resistance against bovine TB (Mackintosh *et al.*, 2000) and similarly heritable differential outcomes to experimental *MAP* challenge have also been reported (Robinson *et al.*, 2008, Robinson *et al.*, 2011; Mackintosh *et al.*, 2011; Marfell *et al.*, 2013; Dobson *et al.*, 2013).

14.1.6 Diagnosis

A presumptive diagnosis of paratuberculosis can be made in clinically affected deer based on typical clinical signs. Gross lesions at necropsy and histopathological examination of lesions and tissues can be highly suggestive but may not distinguish between lesions caused by *MAP*, *M. bovis* and *M. avium* subsp. *avium*. The isolation of *MAP* by culture from faeces in living deer or from tissues post-mortem was generally regarded as the diagnostic 'gold standard' but latterly has been supplanted by quantitative faecal PCR tests, which can specifically and quantitatively identify *MAP* in faeces and tissues. While still used in some areas culture has now been largely superseded and quantitative PCR recognized as the routine ante-mortem diagnostic test of choice for direct detection and quantitation of *MAP* bacteria in faeces from suspect animals (O'Brien *et al.*, 2011, O'Brien *et al.*, 2013, O'Brien *et al.*, 2014).

Diagnosis of subclinically infected deer is more difficult. The chronic nature of infection and the relatively low incidence of disease on many deer farms can result in unnoticed *MAP* infection and an underestimate of the effects on production. Similarly, the disease may go unnoticed in wild populations of deer. Thus, *MAP* infection represents an example of a chronic bacterial disease that behaves more like a parasitic disease than a typical acute bacterial infection.

Antibody tests and cell-mediated immunity tests may provide supporting evidence of infection. Distinct features of these tests as they pertain to paratuberculosis in deer are discussed below, while a more complete description of the tests and their uses is given elsewhere (see Chapter 20, this volume).

Commercial enzyme-linked immunosorbent antibody assays (ELISAs) (e.g. IDEXX, Portland, Maine, USA) have been used to test deer (e.g. white-tailed deer in North America) but, because they rely on cross-reactivity with cervine antibodies, may have limited or no capacity to detect antibodies from species other than for which they were developed (Pruvot *et al.*, 2013). Thus, commercial tests must be recalibrated and their performance evaluated for each deer species that is being tested. An IgG_1 antibody ELISA (Paralisa; Griffin *et al.*, 2005) was developed specifically for deer, which incorporates two antigens, Johne's protoplasmic antigen and purified protein derivative johnin, and the responses are read in parallel so that if either is positive the test is considered positive. The Paralisa test is reported to have an estimated sensitivity of >80% and an estimated specificity of >98% (Griffin *et al.*, 2005) although these parameters have been reported to be significantly lower in very young animals, as might be expected (Stringer *et al.*, 2013a). Similarly, an ethanol-treated protoplasmic preparation obtained from a fallow deer *MAP* isolate has been used in Spain to diagnose paratuberculosis in fallow deer, with a sensitivity of 88.0% in animals with overt pathology and 62.5% in animals with latent infection (Prieto *et al.*, 2014). There is no commercially available gamma interferon assay marketed for use in deer since Cervigam® was withdrawn; some laboratories use an in-house interferon gamma assay.

14.1.7 Prevention and control, including vaccination

Ideally, deer farmers should prevent the introduction of *MAP* on to their farm by acquiring only uninfected livestock originating from test-negative herds, maintaining a closed herd and avoiding cross-grazing with other livestock species. Use of artificial insemination to bring in new bloodlines and control of wildlife on the property may also be worthwhile. However, these measures are not as easy as they may appear. Establishing the infection status of a herd or source of uninfected deer is very difficult, and avoiding cross-grazing, surface water runoff from neighbouring paddocks, and controlling wildlife such as rabbits, rodents, roe deer and mustelids may be very difficult. As well as only sourcing deer from 'low-risk' farms, farmers should only purchase deer that are negative to the most sensitive blood tests available or to an individual or pooled faecal PCR test.

In the face of a recognized infection, the aim should be to reduce the level of infection in order to eliminate clinical disease and minimize subclinical effects on productivity. Because of limitations in the sensitivity of current tests, it is unrealistic to try to eradicate paratuberculosis from a deer herd. On affected farms, the real cost of paratuberculosis should be estimated by adding some or all of the following production losses: outbreaks of mortality in 8–27-month-old deer, sporadic losses of mixed-age hinds and stags, reduced growth rates in calves and yearlings, increased non-pregnancy rates in hinds, suboptimal antler growth in stags and interference with TB testing. Control measures to reduce these costs are identification and culling of clinically affected deer, as well as culling infected hinds, because their calves will almost invariably be infected either *in utero* or shortly after birth.

In addition, a number of vaccines have been used for decades, including in deer, where they provide partial protection by reducing faecal shedding (Harris and Barletta, 2001). In Scotland, vaccination was used in deer on a farm that experienced a severe outbreak of paratuberculosis; although infection still occurred, there was a marked reduction in the incidence of clinical disease (Fawcett *et al.*, 1995). There have been a small number of efficacy studies in deer of various live attenuated and killed *MAP* vaccines (Mackintosh *et al.*, 2003, 2005, 2008), which showed that oil-adjuvanted killed vaccines gave some protection against clinical disease but did not prevent infection. However, vaccination produced some interference with TB diagnosis, resulting in false-positive reactions to the bovine tuberculin skin test. Because of this interference, it is recommended that these vaccines are not used in breeding animals on deer farms in New Zealand; they may be cost-effective in young deer raised for slaughter as meat animals, as these animals would not need to be tested for TB. Vaccination with a killed *MAP* strain 316F vaccine (Silirum) in a randomized clinical trial in young deer in naturally exposed herds significantly reduced paratuberculosis incidence by 60% by approximately 12 months of age, but did not reduce the proportion shedding (Stringer *et al.*, 2013b).

14.2 Paratuberculosis in Camelids and Other Ruminants

Apart from its presence in cattle, sheep, goats and deer, paratuberculosis has also been diagnosed in a wide range of other free-ranging and domesticated ruminants. These include moose (*Alces alces*) (Soltys *et al.*, 1967), bison (*Bison bison*) (Buergelt *et al.*, 2000), riverine buffalo (*Bubalus bubalis*) (Yadav *et al.*, 2008), bighorn sheep (*Ovis canadensis*) (Williams *et al.*, 1983), mouflon (*Ovis orientalis orientalis*) (Pavlik *et al.*, 2000) and various antelope species. Paratuberculosis has also been described in free-ranging and captive camelids, including camels, llamas and alpacas.

14.2.1 Bison

Approximately 200,000 bison inhabit the USA, with most on ranches or in zoos, or in federal- and state-controlled areas. Paratuberculosis was first described in bison in a breeding herd of 2800 animals. Clinically affected bison showed severe weight loss, failure to shed winter hair and poor body condition (Buergelt *et al.*, 2000). Necropsy revealed mucosal corrugation in the distal small intestine and enlarged mesenteric lymph nodes. Particular strains of *MAP*, which express a distinct IS*1311* genotype, have been isolated from North American bison (*Bison bison*) (Whittington *et al.*, 2001) and Indian bison (Sohal *et al.*, 2013; see Chapter 6, this volume). These bison strains are a subset of *MAP*-C, are not exclusive to bison and have been reported to have the capacity to act as an important zoonotic pathogen for humans in India (Chaubey *et al.*, 2017).

14.2.2 Riverine buffalo

The world population of riverine buffalo is over 160 million and they are spread over 129 countries, with most being in India, which has nearly 100 million. Paratuberculosis has been reported in riverine buffalo in India, where the prevalence of infection was over 40% in some populations (Yadav *et al.*, 2008). Gross lesions typical of paratuberculosis occur naturally in around 5% of domesticated water buffalo in India (Sivakumar *et al.*, 2006), although it has been reported that water buffalo are comparatively resistant to paratuberculosis (Mukherjee and Lahiri, 1966). Clinical cases of paratuberculosis have been diagnosed in riverine buffalo in Brazil (Dalto *et al.*, 2012) and low prevalence

(0–3%) infection was reported in Italy (Lillini *et al.*, 1999) and Pakistan (Rehman *et al.*, 2018), where lighter, non-pregnant females had a higher prevalence of disease.

14.2.3 Camelids

Paratuberculosis has been reported widely in many populations of camels in Asia, the Middle East, Africa and the Russia (Ivanov and Skalinskii, 1957; Poddubskii *et al.*, 1962; Burgemeister *et al.*, 1975; Feldman *et al.*, 1981; Ovdienko *et al.*, 1985; Gameel *et al.*, 1994; Wernery and Kinne, 2007). There has also been one report from a zoo in North America (Amand, 1974). The disease in camels may have a more rapid course than in cattle, with death occurring after 4–6 weeks' illness (Wernery *et al.*, 2007). In Bactrian camels (*Camelus bactrianus*), the disease was most severe in 3–5-year-old animals (Ovdienko *et al.*, 1985). Paratuberculosis causes scouring and wasting disease in Arabian dromedary camels (*Camelus dromedarius*) and is found widely throughout camel herds in the Kingdom of Saudi Arabia (Salem *et al.*, 2019). Lesions have been documented in the ileum, colon, rectum, mesenteric lymph nodes, liver, mediastinal lymph nodes and spleen. The camel may be unique in its high incidence of liver involvement (Alharbi *et al.*, 2012). Camels have been found to be infected with *MAP*-S strains (Ghosh *et al.*, 2012).

Paratuberculosis was first reported in llamas (*Lama glama*) in North America (Belknap *et al.*, 1994), but a review in 2000 considered it was uncommon in llamas and alpacas (*Lama pacos*) in the USA (Miller *et al.*, 2000). Paratuberculosis has also been occasionally reported in alpacas in Australia (Ridge *et al.*, 1995; Harkin, 1998), the UK (Davis *et al.*, 1998) and New Zealand (Gibson and Varney, 2007). Clinical disease has occurred in 12–24-month-old alpacas as well as in adults, with signs of diarrhoea, weight loss and hypoproteinaemia. As with other species, necropsy showed emaciation, thickened intestines and enlarged mesenteric lymph nodes (Belknap *et al.*, 1994).

A study conducted in the USA on 84 llamas and 16 alpacas in a research herd believed to be free of paratuberculosis infection was designed to test the specificity of modified antibody-based diagnostic assays (Kramsky *et al.*, 2000; Miller *et al.*, 2000). The specificity of the ELISAs ranged from 48–98% and the agar gel immunodiffusion assay had a specificity of 100% on these samples, leading to the recommendation that seropositive results require culture confirmation. A modified commercial bovine ELISA appears to give a good balance of sensitivity (67%) and specificity (99%) (Kramsky *et al.*, 2000).

Paratuberculosis can be managed and eliminated by combining stringent management with frequent testing and culling, or by combining vaccination (where permitted) with management of faecal–oral transmission (Stehman, 1996). Management of faecal–oral transmission is particularly important to prevent exposure of young stock to the infection. In Australia, a market assurance programme, AlpacaMAP, has been developed for the alpaca industry and it is an integral part of the Australian National Johne's Disease Program. This programme provides a herd classification scheme to assure alpaca breeders and their clients that participating herds have been objectively assessed as having a low risk of *MAP* infection. Details may be obtained from the following website: https://www.animalhealthaustralia.com.au/what-we-do/endemic-disease/maps/alpacamap/.

References

Alharbi, K.B., Al-Swailem, A., Al-Dubaib, M.A., Al-Yamani, E., Al-Naeem, A. *et al.* (2012) Pathology and molecular diagnosis of camels. *Tropical Animal Health and Production* 44, 173–177.

Amand, W.B. (1974) Paratuberculosis in a dromedary camel. *American Association of Zoo Veterinarians Annual Proceedings 1974*, pp. 150–153.

Belknap, E.B., Getzy, D.M., Johnson, L.W., Ellis, R.P., Thompson, G.L. *et al.* (1994) *Mycobacterium paratuberculosis* infection in two llamas. *Journal of the American Veterinary Medical Association* 204, 1805–1808.

Buergelt, C.D., Layton, A.W., Ginn, P.E., Taylor, M., King, J.M. *et al.* (2000) The pathology of spontaneous paratuberculosis in the North American bison (*Bison bison*). *Veterinary Pathology* 37(5), 428–438. DOI: 10.1354/vp.37-5-428.

Burgemeister, R., Leyk, W. and Gossler, R. (1975) Studies on the occurrence of parasites and bacterial and viral infections in southern Tunisian dromedaries. *Deutsche Tierarztliche Wochenschrift* 82, 352–354.

Chaubey, K.K., Singh, S.V., Gupta, S., Singh, M., Sohal, J.S. *et al.* (2017) *Mycobacterium avium* subspecies *paratuberculosis* – an important food borne pathogen of high public health significance with special reference to India: an update. *Veterinary Quarterly* 37(1), 282–299. DOI: 10.1080/01652176.2017.1397301.

Chiodini, R. and Van Kruiningen, H. (1983) Eastern white-tailed deer as a reservoir of ruminant paratuberculosis. *Journal of American Veterinary Medical Association* 182, 168–169.

Clark, R.G., Griffin, J.F.T. and Mackintosh, C.G. (2010) Johne's disease caused by *Mycobacterium avium* subsp. *paratuberculosis* infection in red deer (*Cervus elaphus*): an histopathological grading system, and comparison of paucibacillary and multibacillary disease. *New Zealand Veterinary Journal* 58(2), 90–97. DOI: 10.1080/00480169.2010.65263.

Clark, R.G., Griffin, J.F.T. and Mackintosh, C.G. (2011) Modification to histopathological lesion severity score in red deer (*Cervus elaphus*) affected by Johne's disease. *New Zealand Veterinary Journal* 59(5), 261–262. DOI: 10.1080/00480169.2011.596261.

Commichau, C. (1982) Demonstration of paratuberculosis in fallow deer (*Dama dama*). *Berliner und Münchener Tierarztliche Wochenschrift* 95, 248–250.

Dalto, A.C., Bandarra, P.M., Pavarini, S.P., Boabaid, F.M., de Bitencourt, A.P.G. *et al.* (2012) Clinical and pathological insights into Johne's disease in buffaloes. *Tropical Animal Health and Production* 44(8), 1899–1904. DOI: 10.1007/s11250-012-0154-9.

Davis, R., Keeble, E., Wright, A. and Morgan, K.L. (1998) South American camelids in the United Kingdom: population statistics, mortality rates and causes of death. *Veterinary Record* 142(7), 162–166. DOI: 10.1136/vr.142.7.162.

de Lisle, G.W., Yates, G.F. and Collins, D.M. (1993) Paratuberculosis in farmed deer: case reports and DNA characterization of isolates of *Mycobacterium paratuberculosis*. *Journal of Veterinary Diagnostic Investigation* 5(4), 567–571. DOI: 10.1177/104063879300500411.

de Lisle, G.W., Yates, G.F. and Montgomery, R.H. (2003) The emergence of *Mycobacterium paratuberculosis* in farmed deer in New Zealand — a review of 619 cases. *New Zealand Veterinary Journal* 51(2), 58–62. DOI: 10.1080/00480169.2003.36341.

de Lisle, G.W., Cannon, M.C., Yates, G.F. and Collins, D.M. (2006) Use of a polymerase chain reaction to subtype *Mycobacterium avium* subspecies *paratuberculosis*, an increasingly important pathogen from farmed deer in New Zealand. *New Zealand Veterinary Journal* 54(4), 195–197. DOI: 10.1080/00480169.2006.36695.

Dobson, B., Liggett, S., O'Brien, R. and Griffin, J.F.T. (2013) Innate immune markers that distinguish red deer (*Cervus elaphus*) selected for resistant or susceptible genotypes for Johne's disease. *Veterinary Research* 44(1), 5. DOI: 10.1186/1297-9716-44-5.

Fawcett, A., Goddard, P., McKelvey, W., Buxton, D., Reid, H. *et al.* (1995) Johne's disease in a herd of farmed red deer. *Veterinary Record* 136(7), 165–169. DOI: 10.1136/vr.136.7.165.

Feldman, B.F., Keen, C.L., Kaneko, J.J. and Parver, T.B. (1981) Husbandry and diseases of camels. *Tierarztliche Praxis* 9, 389–402.

Forde, T., Orsel, K., Buck, J.D., Côté, S.D., Cuyler, C. *et al.* (2012) Detection of *Mycobacterium avium* subspecies *paratuberculosis* in several herds of Arctic Caribou (*Rangifer tarandus ssp.*). *Journal of Wildlife Diseases* 48(4), 918–924. DOI: 10.7589/2011-09-261.

Fritsch, I., Luyven, G., Köhler, H., Lutz, W. and Möbius, P. (2012) Suspicion of *Mycobacterium avium* subsp. *paratuberculosis* transmission between cattle and wild-living red deer (*Cervus elaphus*) by multitarget genotyping. *Applied and Environmental Microbiology* 78(4), 1132–1139. DOI: 10.1128/AEM.06812-11.

Galiero, A., Leo, S., Garbarino, C., Arrigoni, N., Russo, S. *et al.* (2018) *Mycobacterium avium* subsp. *paratuberculosis* isolated from wild red deer (*Cervus elaphus*) in northern Italy. *Veterinary Microbiology* 217, 167–172. DOI: 10.1016/j.vetmic.2018.03.015.

Gameel, A.A., Ali, A.S., Razig, S.A., Brown, J., Alhendi, S.A. *et al.* (1994) A clinical–pathological study on spontaneous paratuberculosis in camels (*Camelus dromedarius*) in Saudi Arabia. *Pakistan Veterinary Journal* 14, 15–19.
Ghosh, P., Hsu, C., Alyamani, E.J., Shehata, M.M., Al-Dubaib, M.A. *et al.* (2012) Genome-wide analysis of the emerging infection with *Mycobacterium avium* subspecies *paratuberculosis* in the Arabian camels (*Camelus dromedarius*). *PLoS ONE* 7(2), e31947. DOI: 10.1371/journal.pone.0031947.
Gibson, I. and Varney, K. (2007) Quarterly review of diagnostic cases – January to March 2007. *Surveillance* 34, 28–31.
Gilmour, N.J.L. (1988) Paratuberculosis. In: Reid, H.W. (ed.) *The Management and Health of Farmed Deer*. Kluwer Academic Publishers, Dordrecht/Boston/London, pp. 113–119.
Gilmour, N. and Nyange, J. (1989) Paratuberculosis (Johne's disease) in deer. *In Practice* 11(5), 193–196. DOI: 10.1136/inpract.11.5.193.
Glawischnig, W., Steineck, T. and Spergser, J. (2006) Infections caused by *Mycobacterium avium* subspecies *avium*, *hominissuis*, and *paratuberculosis* in free-ranging red deer (*Cervus elaphus hippelaphus*) in Austria, 2001–2004. *Journal of Wildlife Diseases* 42(4), 724–731. DOI: 10.7589/0090-3558-42.4.724.
Glossop, J., Wilson, P., Heuer, C., West, D. and Mackintosh, C. (2006) JD in deer: practical steps towards taking control. In: Wilson, P. (ed.) *New Zealand Veterinary Association: Deer Branch 2006*. New Zealand Veterinary Association, Wellington, New Zealand, pp. 1–5.
Godfroid, J., Boelaert, F., Heier, A., Clavareau, C., Wellemans, V. (2000) First evidence of Johne's disease in farmed red deer (*Cervus elaphus*) in Belgium. *Veterinary Microbiology* 77(3-4), 283–290. DOI: 10.1016/S0378-1135(00)00313-8.
Griffin, J.F.T., O'Brien, R., Spittle, E., Liggett, S., Cooper, M. *et al.* (2003) Immunodiagnostic tests in the management of Johne's disease in deer herds. *Proceedings of a Deer Course for Veterinarians, No. 20, Deer Branch of the New Zealand Veterinary Association*, New Zealand Veterinary Association, Wellington, New Zealand, pp. 99–102.
Griffin, J.F.T., Spittle, E., Rodgers, C.R., Liggett, S., Cooper, M. *et al.* (2005) Immunoglobulin G_1 enzyme-linked immunosorbent assay for diagnosis of Johne's disease in red deer (*Cervus elaphus*). *Clinical and Diagnostic Laboratory Immunology* 12, 1401–1409.
Griffin, J.F.T., Spittle, E., Mackintosh, C.G., Rodgers, C.R., Liggett, S. *et al.* (2006) Serodiagnosis of Johne's disease in red deer (*Cervus elaphus*). In: Manning, E.J.B. and Nielsen, S.S. (eds) *Proceedings of the 8th International Colloquium on Paratuberculosis*. International Society of Paratuberculosis, Madison, Wisconsin, USA, pp. 496–500.
Gumbrell, R. (1986) Johne's disease in deer. *Surveillance* 13, 15–16.
Harkin, J.T. (1998) Johne's disease and Australian camelids. *Australian Camel Veterinary Association News* 14, 6–11.
Harris, N.B. and Barletta, R.G. (2001) *Mycobacterium avium* subsp. *paratuberculosis* in veterinary medicine. *Clinical Microbiology Reviews* 14(3), 489–512. DOI: 10.1128/CMR.14.3.489-512.2001.
Ivanov, B.G. and Skalinskii, E.I. (1957) Pathological changes in paratuberculosis of camels. *Trudy Vsesoyuznogo Instituta Eksperimental'noi Veterinarii* 20, 186–206.
Jessup, D., Abbas, B. and Behymer, D. (1981) Paratuberculosis in Tule elk in California. *Journal American Veterinary Medicine Association* 179, 1252–1254.
Jorgensen, J.B. and Jorgensen, R.J. (1987) Paratuberculosis in two red deer. *Dansk Veterinaertidsskrift* 70, 322–324.
Kennedy, D. and Allworth, M.B. (2000) Progress in national control and assurance programs for bovine Johne's disease in Australia. *Veterinary Microbiology* 77(3–4), 443–451. DOI: 10.1016/S0378-1135(00)00329-1.
Kopecna, M., Parmova, I., Dvorska-Bartosova, L., Moravkova, M., Babak, V. *et al.* (2008) Distribution and transmission of *Mycobacterium avium* subspecies *paratuberculosis* in farmed deer (*Cervus elaphus*) studied by faecal culture, serology and IS*900* RFLP examinations. *Veterinari Medicina* 53, 510–523.
Kramsky, J., Miller, D.S., Hope, A. and Collins, M.T. (2000) Modification of a bovine ELISA to detect camelid antibodies to *Mycobacterium paratuberculosis*. *Veterinary Microbiology* 77(3-4), 333–337. DOI: 10.1016/S0378-1135(00)00318-7.
Lillini, E., Gamberale, F. and Di Guardo, G. (1999) *Mycobacterium paratuberculosis* infection in a water buffalo (*Bubalus bubalis*) from central Italy. In: Manning, E.J.B. and Collins, M.T. (eds) *Proceedings of the 6th International Colloquium on Paratuberculosis*. International Society of Paratuberculosis, Madison, Wisconsin, p. 254.

Machackova, M., Svastova, P., Lamka, J., Parmova, I., Liska, V. *et al.* (2004) Paratuberculosis in farmed and free-living wild ruminants in the Czech Republic (1999–2001). *Veterinary Microbiology* 101(4), 225–234. DOI: 10.1016/j.vetmic.2004.04.001.

Mackintosh, C. (2008) Review: epidemiology of Johne's disease in farmed red deer (*Cervus elaphus*) in New Zealand. In: Sumner, R. (ed.) *Proceedings of the New Zealand Society of Animal Production*. New Zealand Society of Animal Production, Hamilton, New Zealand, pp. 117–121.

Mackintosh, C.G., Qureshi, T., Waldrup, K., Labes, R.E., Dodds, K.G. *et al.* (2000) Genetic resistance to experimental infection with *Mycobacterium bovis* in red deer (*Cervus elaphus*). *Infection and Immunity* 68(3), 1620–1625. DOI: 10.1128/IAI.68.3.1620-1625.2000.

Mackintosh, C.G., Labes, R.E., de Lisle, G.W. and Griffin, J.F.T. (2003) Experimental challenge trial of novel Johne's disease vaccine. In: Wilson, P.R. (ed.) *Proceedings of a Deer Course for Veterinarians, Deer Branch of the New Zealand Veterinary Association*. New Zealand Veterinary Association, Wellington, New Zealand, pp. 91–98.

Mackintosh, C.G., de Lisle, G.W., Collins, D.M. and Griffin, J.F.T. (2004) Mycobacterial diseases of deer. *New Zealand Veterinary Journal* 52(4), 163–174. DOI: 10.1080/00480169.2004.36424.

Mackintosh, C.G., Labes, R.E. and Griffin, J.F.T. (2005) The effect of Johne's vaccination on tuberculin testing in farmed red deer (*Cervus elaphus*). *New Zealand Veterinary Journal* 53(4), 216–222. DOI: 10.1080/00480169.2005.36549.

Mackintosh, C.G., Labes, R.E., Clark, R.G., de Lisle, G.W. and Griffin, J.F.T. (2007) Experimental infections in young red deer (*Cervus elaphus)* with a bovine and an ovine strain of *Mycobacterium avium* subsp *paratuberculosis*. *New Zealand Veterinary Journal* 55(1), 23–29. DOI: 10.1080/00480169.2007.36731.

Mackintosh, C.G., Labes, R.E., Thompson, B.R., Clark, R.G., de Lisle, G.W. *et al.* (2008) Efficacy, immune responses and side-effects of vaccines against Johne's disease in young red deer (*Cervus elaphus*) experimentally challenged with *Mycobacterium avium* subsp. *paratuberculosis*. *New Zealand Veterinary Journal* 56(1), 1–9. DOI: 10.1080/00480169.2008.36797.

Mackintosh, C.G., Clark, R.G., Thompson, B., Tolentino, B., Griffin, J.F.T. *et al.* (2010) Age susceptibility of red deer (*Cervus elaphus*) to paratuberculosis. *Veterinary Microbiology* 143(2–4), 255–261. DOI: 10.1016/j.vetmic.2009.11.014.

Mackintosh, C.G., Clark, R.G., Tolentino, B., de Lisle, G.W., Liggett, S. *et al.* (2011) Immunological and pathological responses of red deer resistant or susceptible genotypes, to experimental challenge with *Mycobacterium avium* subsp. *paratuberculosis*. *Veterinary Immunology and Immunopathology* 143(1–2), 131–142. DOI: 10.1016/j.vetimm.2011.06.035.

Manning, E.J.B., Steinberg, H., Rossow, K., Ruth, G.R. and Collins, M.T. (1998) Epizootic of paratuberculosis in farmed elk (*Cervus elaphus*). *Journal of the American Veterinary Medical Association* 213, 1320–1322.

Marco, I., Ruiz, M., Juste, R., Garrido, J.M. and Lavin, S. (2002) Paratuberculosis in free-ranging fallow deer in Spain. *Journal of Wildlife Diseases* 38(3), 629–632. DOI: 10.7589/0090-3558-38.3.629.

Marfell, B.J., O'Brien, R. and Griffin, J.F.T. (2013) Global gene expression profiling of monocyte-derived macrophages from red deer (*Cervus elaphus*) genotypically resistant or susceptible to *Mycobacterium avium* subspecies *paratuberculosis* infection. *Developmental & Comparative Immunology* 40(2), 210–217. DOI: 10.1016/j.dci.2013.02.004.

Mereb, G.C., Bedotti, D.O., Suarez, V.H., Busetti, M.R., Moreira, A.R. *et al.* (1994) Paratuberculosis in red deer. *Veterinaria Argentina* 11, 107–112.

Miller, D.S., Collins, M.T., Smith, B.B., Anderson, P.R., Kramsky, J. *et al.* (2000) Specificity of four serologic assays for *Mycobacterium avium* ssp *paratuberculosis* in llamas and alpacas: a single herd study. *Journal of Veterinary Diagnostic Investigation* 12(4), 345–353. DOI: 10.1177/104063870001200408.

Mukherjee, A. and Lahiri, A. (1966) Investigation of Johne's disease in buffaloes. *Indian Veterinary Journal* 37, 349–353.

Nebbia, P., Robino, P., Ferroglio, E., Rossi, L., Meneguz, G. *et al.* (2000) Paratuberculosis in red deer (*Cervus elaphus hippelaphus*) in the Western Alps. *Veterinary Research Communications* 24(7), 435–443. DOI: 10.1023/A:1006482520336.

O'Brien, R., Mackintosh, C.G., Bakker, D., Kopecna, M., Pavlik, I. *et al.* (2006) Immunological and molecular characterization of susceptibility in relationship to bacterial strain differences in *Mycobacterium avium* subsp. *paratuberculosis* infection in the red deer (*Cervus elaphus*). *Infection and Immunity* 74(6), 3530–3537. DOI: 10.1128/IAI.01688-05.

O'Brien, R., Hughes, A. and Griffin, F. (2011) Quantitative PCR as an ancillary, ante-mortem diagnostic test for Johne's disease in deer: a comparison between bacteriologic culture, serological reactivity and faecal shedding. In: Wilson, P.R. (ed.) *Proceedings of the Deer Branch of the New Zealand Veterinary Association*. 28. New Zealand Veterinary Association, Wellington, New Zealand, pp. 121–127.

O'Brien, R., Hughes, A., Liggett, S. and Griffin, F. (2013) Composite testing for *ante-mortem* diagnosis of Johne's disease in farmed New Zealand deer: correlations between bacteriological culture, histopathology, serological reactivity and faecal shedding as determined by quantitative PCR. *BMC Veterinary Research* 9(1), 72. DOI: 10.1186/1746-6148-9-72.

O'Brien, R., Liggett, S. and F, G.. and (2014) Serological assay followed up by quantitative PCR; a synergistic combo for antemortem diagnosis of Johne's disease in farmed New Zealand deer. *Proceedings of the 12th International Colloquium on Paratuberculosis*, International Society of Paratuberculosis Inc., Madison, Wisconsin, USA, p. p93.

Ovdienko, N.P., Khon, F.K., Sharov, V.A. and Yakusheva, O.V. (1985) *Diagnosis of Paratuberculosis in Camels*. 4. Veterinariya, Moscow, USSR, pp. 65–68.

Pacetti, A., Belletti, G., Gabbi, M. and Mutinelli, F. (1994) Paratuberculosis in red deer. *Obiettivie-Documenti-Veterinari* 15, 67–70.

Pavlik, I., Pavlas, M. and Bejckova, L. (1994) Incidence, economic importance and diagnosis of paratuberculosis. *Veterinarni Medicina* 39, 451–496.

Pavlik, I., Bartl, J., Dvorska, L., Svastova, P., du Maine, R. (2000) Epidemiology of paratuberculosis in wild ruminants studied by restriction fragment length polymorphism in the Czech Republic during the period 1995–1998. *Veterinary Microbiology* 77(3–4), 231–251. DOI: 10.1016/S0378-1135(00)00309-6.

Poddubskii, I.V., Shchurevskii, V.E. and Alikaeva, A.P. (1962) Paratuberculosis among farm animals in the Soviet Union. *Trudy Vsesoyuznogo Instituta Eksperimental'noi Veterinarii* 26, 115–134.

Power, S., Haagsma, J. and Smyth, D. (1993) Paratuberculosis in farmed red deer (*Cervus elaphus*) in Ireland. *Veterinary Record* 132(9), 213–216. DOI: 10.1136/vr.132.9.213.

Prieto, J.M., Balseiro, A., Casais, R., Abendaño, N., Fitzgerald, L.E. *et al.* (2014) Sensitive and specific enzyme-linked immunosorbent assay for detecting serum antibodies against *Mycobacterium avium* subsp. *paratuberculosis* in fallow deer. *Clinical and Vaccine Immunology* 21(8), 1077–1085. DOI: 10.1128/CVI.00159-14.

Pruvot, M., Forde, T.L., Steele, J., Kutz, S.J., De Buck, J. *et al.* (2013) The modification and evaluation of an ELISA test for the surveillance of *Mycobacterium avium* subsp. *paratuberculosis* infection in wild ruminants. *BMC Veterinary Research* 9(1), 5. DOI: 10.1186/1746-6148-9-5.

Rehman, A.U., Javed, M.T., Aslam, M.S., Kahn, M.N., Hussain, S.M. *et al.* (2018) Prevalence of paratuberculosis in water buffaloes on public livestock farms of Punjab, Pakistan. *Veterinaria Italiana* 54, 287–292.

Reyes-García, R., Pérez-de-la-Lastra, J.M., Vicente, J., Ruiz-Fons, F., Garrido, J.M. *et al.* (2008) Large-Scale ELISA testing of Spanish red deer for paratuberculosis. *Veterinary Immunology and Immunopathology* 124(1-2), 75–81. DOI: 10.1016/j.vetimm.2008.01.032.

Ridge, S.E., Harkin, J.T., Badman, R.T., Mellor, A.M. and Larsen, J.W.A. (1995) Johne's disease in alpacas (*Lama pacos*) in Australia. *Australian Veterinary Journal* 72(4), 150–153. DOI: 10.1111/j.1751-0813.1995.tb15040.x.

Riemann, H., Zaman, M.R., Ruppanner, R., Aalund, O., Jorgensen, J.B. *et al.* (1979) Paratuberculosis in cattle and free-living exotic deer. *Journal of the American Veterinary Medical Association* 174, 841–843.

Robino, P., Nebbia, P., Tramuta, C., Martinet, M., Ferroglio, E. *et al.* (2008) Identification of *Mycobacterium avium* subsp. *paratuberculosis* in wild cervids (*Cervus elaphus hippelaphus* and *Capreolus capreolus*) from northwestern Italy. *European Journal of Wildlife Research* 54(2), 357–360. DOI: 10.1007/s10344-007-0141-z.

Robinson, M., O'Brien, R., Mackintosh, C. and Griffin, F. (2008) Differential immune responses of red deer (*Cervus elaphus*) following experimental challenge with *Mycobacterium avium* subsp. *paratuberculosis*. *Clinical and Vaccine Immunology* 15(6), 963–969. DOI: 10.1128/CVI.00031-08.

Robinson, M.W., O'Brien, R., Mackintosh, C.G., Clark, R.G. and Griffin, J.F.T. (2011) Immunoregulatory cytokines are associated with protection from immunopathology following *Mycobacterium avium* subspecies *paratuberculosis* infection in red deer. *Infection and Immunity* 79(5), 2089–2097. DOI: 10.1128/IAI.00779-10.

Salem, M.M., El-Deeb, W.M., Zaghawa, A.A., Housawi, F.M. and Alluwaimi, A.M. (2019) Investigation of *Mycobacterium paratuberculosis* in Arabian dromedary camels (*Camelus dromedarius*). *Veterinary World* 12, 218–223.

Sharp, J.M., Stevenson, K., Challans, J.A., Ramage, C., Hitchcock, D. *et al.* (1997) Mycobacterial infections of free-living deer in Scotland. In: Chiodini, R.J., Hines II, M.E. and Collins, M.T. (eds) *Proceedings of the Fifth International Colloquium on Paratuberculosis*. International Association for Paratuberculosis Inc., Madison, Wisconsin, pp. 180–182.

Simpson, J.W., Morrison, L. and Stevenson, K. (2012) Johne's disease. *Journal of the British Deer Society* 16, 42–46.

Sivakumar, P., Tripathi, B.N., Singh, N. and Sharma, N.V. (2006) Pathology of naturally occurring paratuberculosis in water buffalo. *Veterinary Pathology* 43, 455–462.

Sohal, J.S., Singh, S.V., Singh, P.K., Singh, A.V. and Kumar, N. (2013) A new marker IS*1311* L2 PCR-REA for identification of 'Indian Bison' type *Mycobacterium avium* subspecies *paratuberculosis*. *Indian Journal of Biotechnology* 12, 204–207.

Soltys, M.A., Andress, G.E. and Fletch, A.L. (1967) Johne's disease in a moose (*Alces alces*). *Bulletin of the Wildlife Disease Association* 3(4), 183–184. DOI: 10.7589/0090-3558-3.4.183.

Starke, R.K.A. (1991) Paratuberculosis (Johne's disease) in a captive wapiti. In: Renecker, L.A. and Hudson, R.J. (eds) *Wildlife Production: Conservation and Sustainable Development*. University of Alaska, Fairbanks, Alaska, pp. 435–437.

Stehman, S.M. (1996) Paratuberculosis in small ruminants, deer, and South American camelids. *Veterinary Clinics of North America: Food Animal Practice* 12(2), 441–455. DOI: 10.1016/S0749-0720(15)30416-3.

Stringer, L.A., Jones, G., Jewell, C.P., Noble, A.D., Heuer, C. *et al.* (2013a) Bayesian estimation of the sensitivity and specificity of individual fecal culture and Paralisa to detect *Mycobacterium avium* subspecies *paratuberculosis* infection in young farmed deer. *Journal of Veterinary Diagnostic Investigation* 25(6), 759–764. DOI: 10.1177/1040638713505587.

Stringer, L.A., Wilson, P.R., Heuer, C. and Mackintosh, C.G. (2013b) A randomised controlled trial of Silirum vaccine for control of paratuberculosis in farmed red deer. *Veterinary Record* 173(22), 551. DOI: 10.1136/vr.101799.

Strogov, A.K. (1973) [Paratuberculosis in reindeer]. In: *Maloizvestnye Zaraznye Bolezni Zhivotnykh [Little-known Contagious Diseases of Animals]*, 2nd edn. Izdatel'stvo Kolos, USSR, pp. 231–236.

Temple, R.M.S., Muscoplat, C.C., Thoen, C.O., Himes, E.M. and Johnson, D.W. (1979) Observations on diagnostic tests for paratuberculosis in a deer herd. *Journal of the American Veterinary Medical Association*, 914–915.

Thompson, B.R., Clark, R.G. and Mackintosh, C.G. (2007) Intra-uterine transmission of *Mycobacterium avium* subsp. *paratuberculosis* in subclinically affected red deer (*Cervus elaphus*). *New Zealand Veterinary Journal* 55(6), 308–313. DOI: 10.1080/00480169.2007.36786.

Tryland, M., Olsen, I., Vikøren, T., Handeland, K., Arnemo, J.M. *et al.* (2004) Serologic survey for antibodies against *Mycobacterium avium* subsp. *paratuberculosis* in free-ranging cervids from Norway. *Journal of Wildlife Diseases* 40(1), 32–41. DOI: 10.7589/0090-3558-40.1.32.

van Kooten, H.C.J., Mackintosh, C.G. and Koets, A.P. (2006) Intra-uterine transmission of paratuberculosis (Johne's disease) in farmed red deer. *New Zealand Veterinary Journal* 54(1), 16–20. DOI: 10.1080/00480169.2006.36598.

Verdugo, C., Pleydell, E., Price-Carter, M., Prattley, D., Collins, D. *et al.* (2014) Molecular epidemiology of *Mycobacterium avium* subsp. *paratuberculosis* isolated from sheep, cattle and deer on New Zealand pastoral farms. *Preventive Veterinary Medicine* 117(3–4), 436–446. DOI: 10.1016/j.prevetmed.2014.09.009.

Wernery, U. and Kinne, J. (2007) An approach to the diagnosis of camel diseases: clinical pathology and post-mortem criteria. In: Gahlot, T.K. (ed.) *Proceedings of the International Camel Conference 'Recent Trends in Camelid Research and Future Strategies for Saving Camels'*. Rajasthan, India, pp. 1–18.

Wernery, U., Kinne, J., Jahans, K., Vordermeier, H., Esfandiari, J. *et al.* (2007) Tuberculosis outbreak in a dromedary racing herd and rapid serological detection of infected camels. *Veterinary Microbiology* 122(1–2), 108–115. DOI: 10.1016/j.vetmic.2007.01.012.

Whittington, R.J. and Windsor, P.A. (2007) *In utero* infection of cattle with *Mycobacterium avium* subsp. *paratuberculosis*. In: Nielsen, S.S. (ed.) *Proceedings of the 9th International Colloquium on Paratuberculosis*. International Society of Paratuberculosis, Inc., Madison, Wisconsin, p. 287. Available at: http://www.paratuberculosis.net/proceedings/proc9.pdf

Whittington, R.J., Marsh, I.B. and Whitlock, R.H. (2001) Typing of IS 1311 polymorphisms confirms that bison (*Bison bison*) with paratuberculosis in Montana are infected with a strain of *Mycobacterium avium* subsp. *paratuberculosis* distinct from that occurring in cattle and other domesticated livestock. *Molecular and Cellular Probes* 15(3), 139–145. DOI: 10.1006/mcpr.2001.0346.

Williams, E.S., Snyder, S.P. and Martin, K.L. (1983) Pathology of spontaneous and experimental infection of North American wild ruminants with *Mycobacterium paratuberculosis*. *Veterinary Pathology* 20(3), 274–290. DOI: 10.1177/030098588302000304.

Yadav, D., Singh, S.V., Singh, A.V., Sevilla, I., Juste, R.A. *et al*. (2008) Pathogenic 'bison-type' *Mycobacterium avium* subspecies *paratuberculosis* genotype characterized from riverine buffalo (*Bubalus bubalis*) in North India. *Comparative Immunology, Microbiology and Infectious Diseases* 31(4), 373–387. DOI: 10.1016/j.cimid.2007.06.007.

15 Infection of Non-Ruminant Wildlife by *Mycobacterium avium* subsp. *paratuberculosis*

Naomi J. Fox[1]*, Lesley A. Smith[1], Karen Stevenson[2], Ross S. Davidson[1], Glenn Marion[3] and Michael R. Hutchings[1]

[1]SRUC, Edinburgh, UK; [2]Moredun Research Institute, Penicuik, UK; [3]Biomathematics Statistics Scotland, Edinburgh, UK

15.1 Introduction

Paratuberculosis has historically been considered an infection of ruminants. First indications of a wider host range date to the 1970s, but it was not until the late 1990s that non-ruminant wildlife hosts of paratuberculosis were confirmed and further investigated in relation to their significance in the epidemiology of livestock infection. Numerous theoretical epidemiological studies and practical disease control programmes have demonstrated the added difficulties that a wildlife host can bring to livestock disease control. Ultimately, if the infection can persist in the wildlife host population in isolation (i.e. no infection pressure from other host species) for extended periods of time and there is a viable route of transmission from the wildlife host to the livestock host, then the wildlife species must be included in any disease control strategy. In this chapter we consider the role of non-ruminant wildlife in the epidemiology of paratuberculosis.

15.2 Host Infection

15.2.1 Known host range

Mycobacterium avium subsp. *paratuberculosis* (*MAP*) has been identified in a broad range of non-ruminant wildlife species. Suspected isolation of *MAP* was first reported from a European brown hare (*Lepus europaeus*) in England in 1977 (Matthews and Sargent, 1977), although the organism responsible was not fully characterized. Lesions attributed to paratuberculosis were subsequently described in a wild rabbit (*Oryctolagus cuniculus*) from Scotland (Angus, 1990), and presence of *MAP* has since been confirmed in rabbits (Greig *et al.*, 1997, 1999; Beard *et al.*, 2001a) by a polymerase chain reaction (PCR) assay based on the species-specific IS*900* insertion sequence (Vary *et al.*, 1990). Following detection of *MAP* in rabbits, studies were extended to investigate other wildlife species associated with infected farms in the same region. *MAP* was isolated from foxes (*Vulpes vulpes*) and stoats (*Mustela erminea*) (Beard *et al.*,

*Corresponding author: Naomi.Fox@sruc.ac.uk

© CAB International 2020. *Paratuberculosis: Organism, Disease, Control, 2nd Edition* (eds M.A. Behr *et al.*)

1999), then subsequently from weasels (*Mustela nivalis*), badgers (*Meles meles*), wood mice (*Apodemus sylvaticus*), rats (*Rattus norvegicus*), brown hares (*Lepus europaeus*), jackdaws (*Corvus monedula*), rooks (*Corvus frugilegus*) and crows (*Corvus corone*) (Beard *et al.*, 2001a). Following these studies researchers have looked for *MAP* in non-ruminant species around the world, and *MAP* has been isolated from a diverse range of non-ruminant species, from marsupials in New Zealand (Nugent *et al.*, 2011) to birds in the USA (Corn *et al.*, 2005). *MAP* has also been detected in the intestinal tracts of a number of invertebrates including cockroaches (Fischer *et al.*, 2003a), earthworms (Fischer *et al.*, 2003b) and a number of diptera species (Fischer *et al.*, 2001). Table 15.1 contains a complete list of the known non-ruminant wildlife species in which *MAP* has been detected. However, a species should not be considered a competent reservoir host based solely on isolation of the bacteria.

The *MAP* strains that infect non-ruminants are principally of the 'Cattle-type' (*MAP*-C). However, 'Sheep-type' strains (*MAP*-S) are occasionally isolated from wildlife, with reports of *MAP*-S being isolated from the house mouse (*Mus musculus*) (Florou *et al.*, 2008), western grey kangaroo (*Macropus fuliginosus*) and tammar wallabies (*Macropus eugenii decres*) (Cleland *et al.*, 2010). More detailed descriptions of the different types of *MAP* are given elsewhere (see Chapter 6, this volume).

15.2.2 Pathology

In wild ruminants, gross lesions and clinical signs have been reported as similar to those in infected cattle and sheep, where the disease is ultimately fatal (Williams *et al.*, 1979; Buergelt *et al.*, 2000). In contrast, macroscopic lesions are rare in non-ruminant wildlife. To date, reports of clinical cases of paratuberculosis in non-ruminant species are limited.

The histopathology of non-ruminant *MAP* infections has been studied most in naturally infected rabbits, in which both severe and mild lesions have been observed in the intestines (Greig *et al.*, 1997; Beard *et al.*, 2001b). Severe lesions are characterized by large numbers of infiltrating epithelioid macrophages and giant cells, also a prominent feature of early bovine paratuberculosis and some forms of ovine paratuberculosis. Acid-fast staining reveals numerous intracellular organisms (acid-fast organisms, or AFOs) that can be bacillary or coccobacillary. In rabbits with severe intestinal lesions, similar histopathological changes are apparent in the mesenteric lymph nodes and gut-associated lymphoid tissue. Some rabbits have been found to exhibit multiple foci of chronic granulomatous inflammation in the liver. However, no observable clinical cases of *MAP* infection have been found in rabbits and they are considered asymptomatic.

Disease-induced mortality has been observed in primates, with clinical and pathological features similar to those in ruminant hosts (McClure *et al.*, 1987; Zwick *et al.*, 2002). In addition to high mortality rates, high prevalence was reported within an infected colony of stumptail macaques (*Macaca arctoides*) (76%, $n = 38$), with up to 10^8 colony-forming units (CFU)/g intestinal tissue in clinical hosts, and high shedding (up to 2×10^6 CFU/g faeces) in non-clinical hosts (McClure *et al.*, 1987).

The pathology of *MAP* infections in other non-ruminants is much more subtle. In foxes, weasels and stoats, small numbers of single, large macrophage-like cells or granulomata consisting of 10 or fewer cells have been observed in the mesenteric lymph nodes and mucosa-associated lymphoid tissue of the gut (Beard *et al.*, 2001a). Only small numbers of AFOs have been detected in the cytoplasm of these macrophage-like cells. No histopathological lesions have been observed in the intestines or liver of these host species. Similar mild histopathological lesions have also been observed in a crow and a wood mouse. Beard *et al.* (2001b) observed cells containing fewer than five AFOs scattered throughout the lamina propria of a crow intestine. Multiple granulomata were observed in the liver but did not contain AFOs. In the wood mouse, macrophage-like cells containing AFOs were observed as both single cells and small granulomata within the cortex of the mesenteric lymph node and small intestine.

15.2.3 Prevalence and excretion rates

The contribution made by a species to the amount of *MAP* in the environment is a function

Table 15.1. Known range of non-ruminant wildlife species in which *MAP* has been described.

Class	Order	Species common name	Species Latin name	Reference
Mammalia	Rodents	Rat	*Rattus norvegicus*	Beard *et al.*, 2001a
		Wood mouse	*Apodemus sylvaticus*	Beard *et al.*, 2001a
		House mouse	*Mus musculus*	Florou *et al.*, 2008
		Black rat	*Rattus rattus*	Florou *et al.*, 2008
		Hispid cotton rat	*Sigmodon hispidus*	Corn *et al.*, 2005
		Bank vole	*Myodes glareolus*	Beard *et al.*, 2001a
		Common vole	*Microtus arvalis*	Anderson *et al.*, 2007
	Lagomorphs	Rabbit	*Oryctolagus cuniculus*	Greig *et al.*, 1999
		Brown hare	*Lepus europaeus*	Matthews and Sargent, 1977
		Eastern cottontail	*Sylvilagus floridanus*	Corn *et al.*, 2005
	Carnivora	Stoat	*Mustela erminea*	Beard *et al.*, 2001a
		Weasel	*Mustela nivalis*	Beard *et al.*, 2001a
		Badger	*Meles meles*	Beard *et al.*, 2001a
		Eurasian otter	*Lutra lutra*	Matos *et al.*, 2013
		Feral ferrets	*Mustela putorius furo*	de Lisle *et al.*, 2003
		Fox	*Vulpes vulpes*	Beard *et al.*, 2001a
		Brown bear	*Ursus arctos*	Kopecna *et al.*, 2006
		Raccoon	*Procyon lotor*	Corn *et al.*, 2005
		Striped skunk	*Mephitis mephitis*	Corn *et al.*, 2005
		Coyote	*Canis latrans*	Anderson *et al.*, 2007
		Feral cat	*Felis catus*	Palmer *et al.*, 2005
	Eulipotyphla	South-eastern shrew	*Sorex longirostris*	Corn *et al.*, 2005
		Lesser white-toothed shrew	*Crocidura suaveolens*	Kopecna *et al.*, 2008
		European hedgehog	*Erinaceus europaeus*	Nugent *et al.*, 2011
	Primates	Mandrill	*Mandrillus sphinx*	Zwick *et al.*, 2002
		Stumptail macaque	*Macaca arctoides*	McClure *et al.*, 1987
	Artiodactyls	Wild boar	*Sus scrofa*	Alvarez *et al.*, 2005
	Cingulata	Armadillo	*Dasypus novemcinctus*	Corn *et al.*, 2005

Continued

Table 15.1. Continued

Class	Order	Species common name	Species Latin name	Reference
Mammalia (infraclass marsupiala)	Didelphimorphia	Opossum	*Didelphis virginiana*	Corn *et al*., 2005
	Diprotodontia	Western grey kangaroo	*Macropus fuliginosus fuliginosus*	Cleland *et al*., 2010
		Tammar wallaby	*Macropus eugenii decres*	Cleland *et al*., 2010
		Brushtail possum	*Trichosurus vulpecula*	Nugent *et al*., 2011
Aves	Passeriformes	Crow	*Corvus corone*	Beard *et al*., 2001a
		Rook	*Corvus frugilegus*	Beard *et al*., 2001a
		Jackdaw	*Corvus monedula*	Beard *et al*., 2001a
		House sparrow	*Passer domesticus*	Corn *et al*., 2005
		European starling	*Sturnus vulgaris*	Corn *et al*., 2005
		Savi's warbler	*Locustella luscinioides*	Gronesova *et al*., 2008
	Charadriiformes	Common snipe	*Gallinago gallinago*	Corn *et al*., 2005
		Black-headed gull	*Larus ridibundus*	Gronesova *et al*., 2008
		European curlew	*Numenius arquata*	Gronesova *et al*., 2008
		Ruff	*Philomachus pugnax*	Gronesova *et al*., 2008
	Cuculiformes	Common cuckoo	*Cuculus canorus*	Gronesova *et al*., 2008
	Anseriformes	Paradise shelduck	*Tadorna variegata*	Nugent *et al*., 2011
Insecta	Diptera	Dungfly	*Scatophaga* spp.	Fischer *et al*., 2001
		Blowfly	*Calliphora vicina*	Fischer *et al*., 2001
		Blowfly	*Lucilia caesar*	Fischer *et al*., 2001
	Blatodia	Oriental cockroach	*Blatta orientalis*	Fischer *et al*., 2003a
Clitellata	Haplotaxida	Earthworm	*Lumbricus* spp.	Fischer *et al*., 2003b

of both the numbers of infected animals (prevalence × population size) and the rates of *MAP* excretion. In the UK, prevalences of *MAP* are generally higher in carnivores such as the fox compared with prey species such as lagomorphs and rodents. With an average prevalence of 85%, the fox might be considered a useful indicator species of on-farm infection (Beard *et al.*, 2001a). Of the prey species from the Beard *et al.* (2001a) study, prevalences were highest in hare (17%) and less than 10% in rats and mice. While prevalences in carnivores tend to be higher than those in prey species, prevalences in rabbits on a single farm in the UK can be as high as 79% (Shaughnessy *et al.*, 2013).

High prevalences have been found in rabbits (26%, $n = 113$), possums (25%, $n = 73$), and hedgehogs (36% $n = 42$) in New Zealand, and it has been postulated that these species could contribute to the persistence of *MAP* at the wildlife–livestock interface (Nugent *et al.*, 2011). High prevalence rates and levels of infection in non-ruminant wildlife as found in the UK and New Zealand have not been reported in Europe or the USA. The prevalence of *MAP* in wildlife in the USA ranged from 1.7–25%, although the sample sizes for those with a prevalence of 25% were small ($n = 4$) (Corn *et al.*, 2005). Most species with a prevalence of greater than 10% were predators or scavengers (including armadillo, feral cat, opossum and raccoon) (Corn *et al.*, 2005), which follows the patterns of prevalence found in the UK. Faecal samples were culture-positive from raccoons, armadillos, an opossum and a feral cat, suggesting that these animals shed the bacteria in their faeces (Corn *et al.*, 2005) and therefore have the potential to play a part in the onward transmission of the disease.

Published data on *MAP* shedding rates in non-ruminant wildlife are limited, and primarily restricted to lagomorph hosts. The mean number of CFU from infected rabbit faeces was $7.6 \times 10^5 \pm 5.2 \times 10^5$ CFU/g (Daniels *et al.*, 2003a), which is lower than the 10^8 CFU/g reported in faeces from clinically affected cattle (Cranwell, 1997; Whittington *et al.*, 2000). Infected rabbits may also shed *MAP* in their urine ($n = 2/17$), although the levels of shedding are thought to be far lower than those in faeces (Daniels *et al.*, 2003a). However, as rabbits are asymptomatic, the excretion rates in rabbit faeces is a mean across different levels of infection and as such cannot be compared with the clinically affected cattle excretion rates that are often cited in the literature. Although not directly quantified, pathological comparisons suggest that the shedding rates of other non-ruminant wildlife would be expected to be far lower than for rabbits. A relative estimate of the input of *MAP* on to pasture suggested that sheep and cattle potentially contributed 4 and 125 times more organisms per hectare per day, respectively, than rabbits. None the less, rabbits were estimated to contribute $>10^6$ CFU of *MAP* per hectare per day (Daniels *et al.*, 2003a).

15.3 Epidemiology

15.3.1 Spatial and temporal patterns of infection in host populations and the environment

Clustering of hosts and pathogens in the environment can lead to hot spots of disease at local, regional, national and international levels. Scotland contains regional hot spots of *MAP* in rabbits and this mirrors the distribution in cattle (Greig *et al.*, 1999). *MAP*-infected rabbits, and the associated risk of interspecies transmission, are highly clustered in the environment (Greig *et al.*, 1999). However, this is mostly due to the clustered distribution of the general rabbit population in the environment. In addition to spatial clustering, there is also temporal variation in infection levels. For example, within a farm with an overall *MAP* infection prevalence of 40%, the temporal distribution of infection in rabbits followed a cyclical pattern, with a peak in spring of 55.4% and a low in summer of 19.4% (Judge *et al.*, 2005a).

15.3.2 Inter- and intraspecies routes of transmission within wildlife communities

The faecal–oral route is generally seen as the major route of horizontal transmission in ruminant species, although vertical and pseudovertical (via the ingestion of colostrum or milk contaminated with faeces) have also been shown to occur (Sweeney, 1996). Few studies have examined intra- and interspecific routes of transmission within wildlife communities. *MAP* has been isolated from the testes, uterus, placenta and fetuses of rabbits, suggesting the possibility of both sexual and vertical transmission (Judge *et al.*, 2006). *MAP* has also been isolated from the milk of lactating rabbits. However, as rabbits wean their young on their own cecotropes and faeces, the role of milk vs faecal–oral transmission may be difficult to separate. In a population of rabbits with 42% infected adult females, field studies suggest that vertical, i.e. transplacental and/or pseudovertical, transmission can occur in 14% of offspring entering the population at 1 month of age, and this equates to a probability of infection via this route of 0.326 (Judge *et al.*, 2006). In the same population, an analysis of the prevalence of infection as a function of age estimated that the probability of infection via horizontal transmission (including interspecies transmission) was as high as 3.7% per month. Given an average 18-month lifespan of a wild rabbit, this estimate suggests a high probability of acquiring *MAP* infection in an affected population (Judge *et al.*, 2006).

Carnivores have relatively little direct contact with contaminated pastures or farm buildings, and it is therefore more likely that they become infected with *MAP* through ingestion of infected prey. *MAP* has been isolated from the mesenteric lymph nodes and intestinal tissue of infected rabbits and rats (Greig *et al.*, 1997, 1999; Florou *et al.*, 2008), and carnivores often ingest entire rodents and large enough proportions of rabbit carcasses to include lymph nodes and intestinal tissue. Lagomorphs and small rodents form the majority of the diet of foxes, stoats and weasels in Scotland (Harris and Lloyd, 1991; King, 1991a, b; Leckie *et al.*, 1998), and coyotes and raccoons are also known to feed on rodents and lagomorphs. This may explain the relatively high numbers of PCR positive samples from these species (Anderson *et al.*, 2007). As detailed above, the prevalence of infection in carnivore species is generally higher than that in prey species. In one study, 62% of the predators (fox, stoat and weasel) sampled were infected compared with 10% of the prey species (rabbits, rats and mice). Transmission of pathogens through predation has been reported and is explained by carnivores' higher trophic level in the food chain (Zarnke *et al.*, 2000). At least part of the diet of the predators and opportunistic animals such as the opossum and a large part of the diet of corvids may consist of scavenging potentially infected prey carcasses (Mason and MacDonald, 1995; Anderson *et al.*, 2007). Therefore, scavenging may constitute an additional mode of paratuberculosis transmission.

15.3.3 Potential livestock-to-wildlife routes of transmission

The association between *MAP* in non-ruminant wild species and paratuberculosis infections on farms suggests that interspecies transmission may occur between livestock and non-ruminant wildlife. Evidence for interspecies transmission has been provided by molecular typing of cattle and rabbit *MAP* isolates, which has shown that the same strains infect both species (Greig *et al.*, 1999; Stevenson *et al.*, 2009). Interspecies transmission has also been demonstrated experimentally in the case of rabbits infected with a cattle isolate of *MAP* (Mokresh and Butler, 1990) and calves infected with a rabbit isolate (Beard *et al.*, 2001c). Given the high rates of shedding by clinically infected cattle and sheep (Cranwell, 1997; Whittington *et al.*, 2000), it is likely that wild herbivores such as rabbits or hares sharing pasture with infected cattle become infected through the ingestion of contaminated grass.

Based on a sample of 20 farms across Scotland, Greig *et al.* (1999) found a statistically significant relationship between farms with an ongoing or recent paratuberculosis problem in cattle and the presence of similar strains of paratuberculosis in cattle and rabbits, providing further support for transmission between the species. In the case of *MAP* infection of granivorous or omnivorous rodents, exposure is more likely to occur in contaminated buildings than pasture (Beard *et al.*, 2001a). Indeed, in one study, rodents from which *MAP* was isolated were all caught in buildings that housed cattle, whereas those captured in adjacent field margins or woodland were negative on culture (M.R. Hutchings, personal communication).

15.3.4 Potential wildlife-to-livestock routes of transmission

There are two main ways sheep and cattle can potentially ingest faecal matter from infected wildlife: grazing contaminated pasture and consuming contaminated farm-stored feed. Grazing livestock tend to avoid pastures contaminated with their own and other species' faeces through: (i) the selection of non-contaminated sward patches; (ii) reduced bite rate when forced to graze contaminated sward patches; and (iii) reduced grazing depth when grazing contaminated swards (Hutchings *et al.*, 1998; Smith *et al.*, 2009b). However, the only known species for which livestock do not use faeces-avoidance behaviour to limit exposure to parasites and pathogens is the rabbit. Sheep and cattle do not avoid rabbit faeces and graze contaminated pastures as they do non-contaminated pastures (Daniels *et al.*, 2001; Judge *et al.*, 2005b; Smith *et al.*, 2009a, b), so the number of rabbit faecal pellets ingested by calves is directly proportional to the level of contamination of the pasture. Given a density of 10–59 rabbits/ha for Scotland, the number of rabbit faecal pellets present at any one time on a grazing pasture is between 30,657 and 290,752 pellets/ha (Daniels *et al.*, 2003a). Cattle ingest 1.3% of the faecal pellets, or between 389 and 3693 pellets/ha grazed. If one-third of such pellets harbour live *MAP* organisms, one calf could ingest 140–1329 infected faecal pellets/ha grazed. Sheep ingest an average of 0.5% of the faecal pellets, giving the potential to ingest between 146 and 1384 faecal pellets/ha grazed, of which 53–498 may be infected. Consequently, of the known non-ruminant wildlife host range in the UK, rabbits represent the greatest *MAP* risk to sympatric livestock due to the combination of a relatively high prevalence of infection, high population density of rabbits with access to livestock pasture and the lack of avoidance of rabbit faeces by grazing livestock. This risk is compounded by the ability of rabbit isolates of *MAP* to infect calves, which has been demonstrated experimentally (Beard *et al.*, 2001c).

Livestock are also susceptible to infection through feed. Wildlife excreta, particularly from rodents and birds, can contaminate feed during on-farm storage, after which faeces are likely to be ingested when the feed is given to livestock. Levels of wildlife faecal contamination during on-farm winter feed storage have been quantified in feed stores on four farms in the east of Scotland by Daniels *et al.* (2003b). They reported a mean of 79.9 rodent faeces (95% confidence interval (CI): 37.5–165.9) and 24.9 bird faeces (95% CI: 14.3–41.7) deposited per m^2 of stored feed per month. It was estimated that individual cattle and sheep could encounter 1626 and 814 rabbit faeces, respectively, over the winter. Similarly, infected rodents were found to be associated with livestock premises in Wisconsin (Corn *et al.*, 2005; Anderson *et al.*, 2007), and rodents were trapped in sheds housing goats and sheep in Greece (Florou *et al.*, 2008). It is hypothesized that they became infected through scavenging livestock feed from the floor of the sheds where it has become contaminated with livestock faeces (Florou *et al.*, 2008). Once infected it is likely that the rodents would contribute to contaminating the livestock feed and potentially contribute to the cycle of infection in the host community.

Livestock that are fed supplementary feedstuffs are often unable to avoid contaminated feed. When presented with contaminated feed such as concentrates, cattle and sheep have some limited ability to actively avoid rat faecal pellets, but this is not the case for mouse faeces (Daniels and Hutchings, 2001). Moreover, the discriminatory power of livestock is limited, and group feeding that increases competition between animals further increases the rates of ingestion of rodent faeces (Daniels and Hutchings, 2001). Given such low levels of avoidance, even with extremely low probabilities of *MAP* transmission, infection may be maintained in livestock populations (Daniels *et al.*, 2003b).

The importance of carnivores as paratuberculosis hosts at the wildlife–livestock interface is more likely to be indirect, since livestock are unlikely to come into contact with large amounts of their faeces. Such species could still play a role in disease dynamics as infected carnivores and corvids could theoretically transport *MAP* over far greater distances than most prey species. For example, in the east of Scotland the home range size of rabbits is 6.3 ha (Hulbert *et al.*, 1996; Daniels *et al.*, 2003c) but is 4000 ha for foxes (Harris and Lloyd, 1991).

15.3.5 Persistence of infection in non-ruminant wildlife

For a wildlife species to be considered a reservoir host, the pathogen must be capable of being permanently maintained within that host population. A longitudinal study of an island population investigated if *MAP* infection levels would be maintained in a rabbit population following the implementation of paratuberculosis control in sympatric cattle (Fox *et al.*, 2018). It was found that applying *MAP* control measures in cattle led to a reduction in *MAP* seroprevalence in the cattle, however the same trend in prevalence was not echoed in the rabbit population (Fig. 15.1). This is the first study to show that efforts to control paratuberculosis in cattle do not reduce *MAP* prevalence in sympatric rabbits (Fox *et al.*, 2018). This finding adds to mounting evidence indicating that if *MAP* became established in rabbit populations, they would represent a persistent and widespread

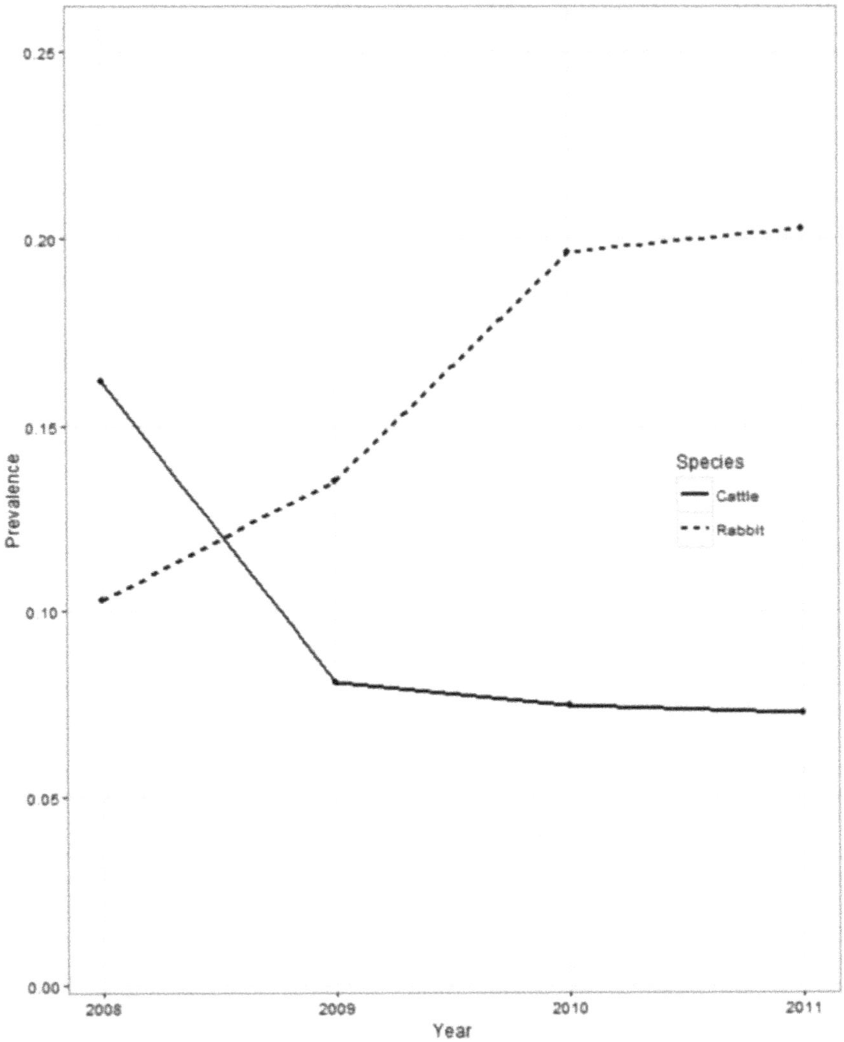

Fig. 15.1. Prevalence of paratuberculosis enzyme-linked immunosorbent assay (ELISA)-positive cattle and *Mycobacterium avium* subsp. *paratuberculosis* (*MAP*) culture positive rabbits from 2008 to 2011, following the implementation of a paratuberculosis control programme on all cattle farms from 2008. Composite means across 15 cattle study farms.

source of infection, potentially impeding control strategies in livestock. This supports the findings of Shaughnessy *et al.* (2013), who demonstrated that farms that responded well to *MAP* control and had low levels of *MAP* in sympatric rabbits could achieve effective paratuberculosis control in their herds, while farms with high *MAP* levels in rabbits were unable to control paratuberculosis in cattle despite implementing the same control measures. This suggests that the risks from rabbits translates to additional difficulties for the livestock industry to achieve control.

The findings of these field studies support the predictions of a mathematical model exploring the maintenance of *MAP* within rabbit populations. Due to the difficulties in experimentally demonstrating the long-term persistence of an infection in a host species, especially a slow-growing pathogen such as *MAP*, mathematical modelling is often used to synthesize known information on the host demography and pathogen transmission to predict if the infection is stable within a host population. Judge *et al.* (2007) developed a mathematical model and applied it to combinations of vertical (including pseudovertical) and horizontal rates of transmission, which covered a broad range of plausible values. This exploration of parameter space accounted for uncertainties in the value of R0 in wild populations. The long-run averages of the equilibrium prevalences were used to characterize disease persistence for each combination of horizontal and vertical/pseudovertical transmission rates.

This approach predicted that paratuberculosis would persist in rabbit populations at all values of the horizontal and vertical transmission parameters in the range estimated from the field data presented above, and in many cases at all values within 95% confidence intervals around this range. Indeed, the prevalence of *MAP* infection in the study site from which the transmission rates were estimated was 39.7%, falling within the prevalence ranges produced by the model. The persistence of *MAP* infection in rabbit populations suggests that they may act as a reservoir of infection (i.e. self-sustaining) for sympatric livestock. This may have significant implications for paratuberculosis control. For example, destocking is often considered as a management measure for disease control in livestock populations and was included in the National Ovine Johne's Disease Control and Evaluation Program in Australia (Sergeant, 2001; Whittington *et al.*, 2003). However, this type of strategy fails to account for the presence of a wildlife reservoir and may therefore be unsuccessful if wildlife is infected. Furthermore, the possibility of the disease spreading from infected livestock farms to previously uninfected livestock areas through the dispersal of infected wildlife needs to be considered in control strategies. For the UK and other countries with high rabbit populations, it would seem unrealistic to eradicate rabbits at the scale necessary to prevent spillover infection in current hot spots of disease (Judge *et al.*, 2005b). Consequently, management of disease should probably be restricted to rabbit population management (e.g. keeping population densities low), managing livestock–wildlife interactions (e.g. keeping animals, especially the most susceptible animals, away from areas with high rabbit densities) and good livestock management practices in terms of husbandry and movement between herds.

15.4 Control of Paratuberculosis in Wildlife

Reduction in wildlife populations is a common method for the control of livestock infections that have wildlife hosts; its success is dependent on the characteristics of the infection, as well as on the spatial and social structure of the wildlife host. Few studies have considered wildlife population reduction to control paratuberculosis. However, a modelling study suggests that unrealistically high rabbit culls (>95% population reduction) are needed if infection is to be eradicated from local rabbit populations with a single population reduction event, of either individuals or social groups (Davidson *et al.*, 2009). Repeated annual culls are more effective at reducing the prevalence of infection in rabbit populations and eradicating infection. However, annual population reductions of >40% are required over extended periods of time (many years). Thus, this model predicts that *MAP* will persist in rabbit populations without significant and prolonged effort to achieve control (Davidson *et al.*, 2009). Currently, *MAP*-infected rabbit populations may be considered relatively uncommon in the UK and around the world. However, given the ubiquitous distribution of the European rabbit in the UK and New Zealand and the likely persistence of infection in rabbit populations, the use of repeated control strategies over extended time periods may be advisable. Where possible these should be conducted across farms at the landscape scale via landowner cooperation. In settings where other non-ruminant hosts are documented to be infected with *MAP*, similar considerations may apply.

15.5 Conclusions

The high prevalence of *MAP* in some non-ruminant wildlife species and their interaction with susceptible ruminant livestock raise the possibility that they play an active role in the current epidemiology of the disease in livestock and can present a significant challenge to control efforts. The risks of transmission from wildlife to livestock have been frequently raised but are difficult to prove in the field, mainly due to the long incubation period of the disease and the difficulty in excluding other potential sources of infection (Williams *et al.*, 1979; Chiodini and Van Kruiningen, 1983; Greig *et al.*, 1997; Ferroglio *et al.*, 2000; Beard *et al.*, 2001a, b). Nevertheless, a growing body of evidence implicates the European rabbit as a persistent risk of infection to sympatric livestock. Paratuberculosis in domestic livestock populations is a notoriously difficult disease to control (Stabel, 1998). The participation of non-ruminant wildlife carriers of *MAP* may partially account for these difficulties. Further research is required to quantify the role of rabbits in the epidemiology of *MAP* infection. For example, the force of infection of *MAP* from wildlife to domestic livestock and vice versa is currently unknown. This knowledge is needed if we are to improve our approach to *MAP* control and fully appreciate the role of non-ruminant wildlife species in the epidemiology of paratuberculosis.

References

Alvarez, J., De Juan, L., Briones, V., Romero, B., Aranaz, A. *et al.* (2005) *Mycobacterium avium* subspecies *paratuberculosis* in fallow deer and wild boar in Spain. *Veterinary Record* 156(7), 212–213. DOI: 10.1136/vr.156.7.212.

Anderson, J.L., Meece, J.K., Koziczkowski, J.J., Clark, D.L., Radcliff, R.P. *et al.* (2007) *Mycobacterium avium* subsp. *paratuberculosis* in scavenging mammals in Wisconsin. *Journal of Wildlife Diseases* 43(2), 302–308. DOI: 10.7589/0090-3558-43.2.302.

Angus, K.W. (1990) Intestinal lesions resembling paratuberculosis in a wild rabbit (*Oryctolagus cuniculus*). *Journal of Comparative Pathology* 103(1), 101–105. DOI: 10.1016/S0021-9975(08)80140-5.

Beard, P.M., Henderson, D., Daniels, M.J., Pirie, A., Buxton, D. *et al.* (1999) Evidence of paratuberculosis in fox (*Vulpes vulpes*) and stoat (*Mustela erminea*). *Veterinary Record* 145(21), 612–613. DOI: 10.1136/vr.145.21.612.

Beard, P.M., Daniels, M.J., Henderson, D., Pirie, A., Rudge, K. *et al.* (2001a) Paratuberculosis infection of nonruminant wildlife in Scotland. *Journal of Clinical Microbiology* 39(4), 1517–1521. DOI: 10.1128/JCM.39.4.1517-1521.2001.

Beard, P.M., Rhind, S.M., Buxton, D., Daniels, M.J., Henderson, D. *et al.* (2001b) Natural paratuberculosis infection in rabbits in Scotland. *Journal of Comparative Pathology* 124(4), 290–299. DOI: 10.1053/jcpa.2001.0466.

Beard, P.M., Stevenson, K., Pirie, A., Rudge, K., Buxton, D. *et al.* (2001c) Experimental paratuberculosis in calves following inoculation with a rabbit isolate of *Mycobacterium avium* subsp. *paratuberculosis*. *Journal of Clinical Microbiology* 39(9), 3080–3084. DOI: 10.1128/JCM.39.9.3080-3084.2001.

Buergelt, C.D., Layton, A.W., Ginn, P.E., Taylor, M., King, J.M. *et al.* (2000) The pathology of spontaneous paratuberculosis in the North American bison (*Bison bison*). *Veterinary Pathology* 37(5), 428–438. DOI: 10.1354/vp.37-5-428.

Chiodini, R.J. and Van Kruiningen, H.J. (1983) Eastern white-tailed deer as a reservoir of ruminant paratuberculosis. *Journal of the American Veterinary Medical Association* 182(2), 168–169.

Cleland, P.C., Lehmann, D.R., Phillips, P.H., Cousins, D.V., Reddacliff, L.A. *et al.* (2010) A survey to detect the presence of *Mycobacterium avium* subspecies *paratuberculosis* in Kangaroo Island macropods. *Veterinary Microbiology* 145(3-4), 339–346. DOI: 10.1016/j.vetmic.2010.03.021.

Corn, J.L., Manning, E.J.B., Sreevatsan, S. and Fischer, J.R. (2005) Isolation of *Mycobacterium avium* subsp. *paratuberculosis* from free-ranging birds and mammals on livestock premises. *Applied and Environmental Microbiology* 71(11), 6963–6967. DOI: 10.1128/AEM.71.11.6963-6967.2005.

Cranwell, M.P. (1997) Johne's disease: past and present. *Cattle Practice* 5, 217–223.

Daniels, M.J. and Hutchings, M.R. (2001) The response of cattle and sheep to feed contaminated with rodent faeces. *The Veterinary Journal* 162(3), 211–218. DOI: 10.1053/tvjl.2000.0552.

Daniels, M.J., Ball, N., Hutchings, M.R. and Greig, A. (2001) The grazing response of cattle to pasture contaminated with rabbit faeces and the implications for the transmission of paratuberculosis. *The Veterinary Journal* 161(3), 306–313. DOI: 10.1053/tvjl.2000.0550.

Daniels, M.J., Henderson, D., Greig, A., Stevenson, K., Sharp, J.M. *et al.* (2003a) The potential role of wild rabbits *Oryctolagus cuniculus* in the epidemiology of paratuberculosis in domestic ruminants. *Epidemiology and Infection* 130(3), 553–559. DOI: 10.1017/S0950268803008471.

Daniels, M.J., Hutchings, M.R. and Greig, A. (2003b) The risk of disease transmission to livestock posed by contamination of farm stored feed by wildlife excreta. *Epidemiology and Infection* 130(3), 561–568. DOI: 10.1017/S0950268803008483.

Daniels, M.J., Lees, J.D., Hutchings, M.R. and Greig, A. (2003c) The ranging behaviour and habitat use of rabbits on farmland and their potential role in the epidemiology of paratuberculosis. *The Veterinary Journal* 165(3), 248–257. DOI: 10.1016/S1090-0233(02)00238-1.

Davidson, R.S., Marion, G., White, P.C.L. and Hutchings, M.R. (2009) Use of host population reduction to control wildlife infection: rabbits and paratuberculosis. *Epidemiology and Infection* 137(1), 131–138. DOI: 10.1017/S0950268808000642.

de Lisle, G., Yates, G.F., Cavaignac, S.M., Collins, D.M., Paterson, B.M. *et al.* (2003) *Mycobacterium avium* subsp. *paratuberculosis* in feral ferrets – a potential reservoir of Johne's disease. In: Juste, A.R., Geijo, M.V. and Garrido, J.M. (eds) *Proceedings of the Seventh International Colloquium on Paratuberculosis*. International Association for Paratuberculosis, Madison, Wisconsin, pp. 361–362.

Ferroglio, E., Nebbia, P., Robino, P., Rossi, L. and Rosati, S. (2000) *Mycobacterium paratuberculosis* infection in two free-ranging Alpine ibex. *Revue Scientifique et Technique de l'OIE* 19(3), 859–862. DOI: 10.20506/rst.19.3.1260.

Fischer, O., Mátlová, L., Dvorská, L., Svástová, P., Bartl, J. *et al.* (2001) Diptera as vectors of mycobacterial infections in cattle and pigs. *Medical and Veterinary Entomology* 15(2), 208–211. DOI: 10.1046/j.1365-2915.2001.00292.x.

Fischer, O.A., Matlova, L., Dvorska, L., Svastova, P. and Pavlik, I. (2003a) Nymphs of the Oriental cockroach (*Blatta orientalis*) as passive vectors of causal agents of avian tuberculosis and paratuberculosis. *Medical and Veterinary Entomology* 17(2), 145–150. DOI: 10.1046/j.1365-2915.2003.00417.x.

Fischer, O.A., Matlova, L., Bartl, J., Dvorska, L., Svastova, P. *et al.* (2003b) Earthworms (*Oligochaeta, Lumbricidae*) and mycobacteria. *Veterinary Microbiology* 91(4), 325–338. DOI: 10.1016/S0378-1135(02)00302-4.

Florou, M., Leontides, L., Kostoulas, P., Billinis, C., Sofia, M. *et al.* (2008) Isolation of *Mycobacterium avium* subspecies *paratuberculosis* from non-ruminant wildlife living in the sheds and on the pastures of Greek sheep and goats. *Epidemiology and Infection* 136(5), 644–652. DOI: 10.1017/S095026880700893X.

Fox, N.J., Caldow, G.L., Liebeschuetz, H., Stevenson, K. and Hutchings, M.R. (2018) Counterintuitive increase in observed *Mycobacterium avium* subspecies *paratuberculosis* prevalence in sympatric rabbits following the introduction of paratuberculosis control measures in cattle. *Veterinary Record* 182(22), 634–634. DOI: 10.1136/vr.104638.

Greig, A., Stevenson, K., Perez, V., Pirie, A.A., Grant, J.M. *et al.* (1997) Paratuberculosis in wild rabbits (*Oryctolagus cuniculus*). *Veterinary Record* 140(6), 141–143. DOI: 10.1136/vr.140.6.141.

Greig, A., Stevenson, K., Henderson, D., Perez, V., Hughes, V. *et al.* (1999) Epidemiological study of paratuberculosis in wild rabbits in Scotland. *Journal of Clinical Microbiology* 37(6), 1746–1751. DOI: 10.1128/JCM.37.6.1746-1751.1999.

Gronesova, P., Ficova, M., Mizakova, A., Kabat, P., Trnka, A. *et al.* (2008) Prevalence of avian influenza viruses, *Borrelia garinii, Mycobacterium avium*, and *Mycobacterium avium* subsp. *paratuberculosis* in waterfowl and terrestrial birds in Slovakia, 2006. *Avian Pathology* 37(5), 537–543. DOI: 10.1080/03079450802356953.

Harris, S. and Lloyd, H.G. (1991) Fox (*Vulpes vulpes*). In: Corbet, G.B. and Harris, S. (eds) *A Handbook of British Mammals*. Blackwell Scientific Press, Oxford, UK, pp. 351–367.

Hulbert, I.A.R., Iason, G.R., Elston, D.A. and Racey, P.A. (1996) Home-range sizes in a stratified upland landscape of two lagomorphs with different feeding strategies. *The Journal of Applied Ecology* 33(6), 1479–1488. DOI: 10.2307/2404786.

Hutchings, M.R., Kyriazakis, I., Anderson, D.H., Gordon, I.J. and Coop, R.L. (1998) Behavioural strategies used by parasitized and non-parasitized sheep to avoid ingestion of gastro-intestinal nematodes associated with faeces. *Animal Science* 67(1), 97–106. DOI: 10.1017/S1357729800009838.

Judge, J., Kyriazakis, I., Greig, A., Allcroft, D.J. and Hutchings, M.R. (2005a) Clustering of *Mycobacterium avium* subsp. *paratuberculosis* in rabbits and the environment: how hot is a hot spot? *Applied and Environmental Microbiology* 71(10), 6033–6038. DOI: 10.1128/AEM.71.10.6033-6038.2005.

Judge, J., Greig, A., Kyriazakis, I. and Hutchings, M.R. (2005b) Ingestion of faeces by grazing herbivores – risk of inter-species disease transmission. *Agriculture, Ecosystems & Environment* 107(2–3), 267–274. DOI: 10.1016/j.agee.2004.10.028.

Judge, J., Kyriazakis, I., Greig, A., Davidson, R.S. and Hutchings, M.R. (2006) Routes of intraspecies transmission of *Mycobacterium avium* subsp. *paratuberculosis* in rabbits (*Oryctolagus cuniculus*): a field study. *Applied and Environmental Microbiology* 72(1), 398–403. DOI: 10.1128/AEM.72.1.398-403.2006.

Judge, J., Davidson, R.S., Marion, G., White, P.C.L. and Hutchings, M.R. (2007) Persistence of *Mycobacterium avium* subspecies *paratuberculosis* in rabbits: the interplay between horizontal and vertical transmission. *Journal of Applied Ecology* 44(2), 302–311. DOI: 10.1111/j.1365-2664.2007.01282.x.

King, C.M. (1991a) Stoat (*Mustela erminea*). In: Corbet, G.B. and Harris, S. (eds) *A Handbook of British Mammals*. Blackwell Scientific Press, Oxford, UK, pp. 377–387.

King, C.M. (1991b) Weasel (*Mustela nivalis*). In: Corbet, G.B. and Harris, S. (eds) *A Handbook of British Mammals*. Blackwell Scientific Press, Oxford, UK, pp. 387–396.

Kopecna, M., Ondrus, S., Literak, I., Klimes, J., Horvathova, A. *et al.* (2006) Detection of *Mycobacterium avium* subsp. *paratuberculosis* in two brown bears in the central European Carpathians. *Journal of Wildlife Diseases* 42(3), 691–695. DOI: 10.7589/0090-3558-42.3.691.

Kopecna, M., Trcka, I., Lamka, J., Moravkova, M., Koubek, P. *et al.* (2008) The wildlife hosts of *Mycobacterium avium* subsp. *paratuberculosis* in the Czech Republic during the years 2002–2007. *Veterinární Medicína* 53(8), 420–426. DOI: 10.17221/1931-VETMED.

Leckie, F.M., Thirgood, S.J., May, R. and Redpath, S.M. (1998) Variation in the diet of red foxes on Scottish moorland in relation to prey abundance. *Ecography* 21(6), 599–604. DOI: 10.1111/j.1600-0587.1998.tb00552.x.

Mason, C.F. and MacDonald, S.M. (1995) Corvids feeding on carrion. *Bird Study* 42(3), 255–256. DOI: 10.1080/00063659509477176.

Matos, A.C., Figueira, L., Martins, M.H., Matos, M., Alvares, S, Álvares, Pinto, M.L. *et al.* (2013) Disseminated *Mycobacterium avium* subsp. *paratuberculosis* infection in two wild Eurasian otters (*Lutra lutra* L.) from Portugal. *Journal of Zoo and Wildlife Medicine: Official Publication of the American Association of Zoo Veterinarians* 44(1), 193–195. DOI: 10.1638/1042-7260-44.1.193.

Matthews, P.R. and Sargent, A. (1977) The isolation of mycobacteria from the brown hare (*Lepus europaeus*). *British Veterinary Journal* 133(4), 399–404. DOI: 10.1016/S0007-1935(17)34041-1.

McClure, H.M., Chiodini, R.J., Anderson, D.C., Swenson, R.B., Thayer, W.R. *et al.* (1987) *Mycobacterium paratuberculosis* infection in a colony of stumptail macaques (*Macaca arctoides*). *Journal of Infectious Diseases* 155(5), 1011–1019. DOI: 10.1093/infdis/155.5.1011.

Mokresh, A.H. and Butler, D.G. (1990) Granulomatous enteritis following oral inoculation of newborn rabbits with *Mycobacterium paratuberculosis* of bovine origin. *Canadian Journal of Veterinary Research* 54(3), 313–319.

Nugent, G., Whitford, E.J., Hunnam, J.C., Wilson, P.R., Cross, M.L. *et al.* (2011) *Mycobacterium avium* subsp. *paratuberculosis* infection in wildlife on three deer farms with a history of Johne's disease. *New Zealand Veterinary Journal* 59(6), 293–298. DOI: 10.1080/00480169.2011.605747.

Palmer, M.V., Stoffregen, W.C., Carpenter, J.G. and Stabel, J.R. (2005) Isolation of *Mycobacterium avium* subsp. *paratuberculosis* (*MAP*) from feral cats on a dairy farm with *MAP*-infected cattle. *Journal of Wildlife Diseases* 41(3), 629–635. DOI: 10.7589/0090-3558-41.3.629.

Sergeant, E.S. (2001) Ovine Johne's disease in Australia – the first 20 years. *Australian Veterinary Journal* 79(7), 484–491. DOI: 10.1111/j.1751-0813.2001.tb13021.x.

Shaughnessy, L.J., Smith, L.A., Evans, J., Anderson, D., Caldow, G. *et al.* (2013) High prevalence of paratuberculosis in rabbits is associated with difficulties in controlling the disease in cattle. *The Veterinary Journal* 198(1), 267–270. DOI: 10.1016/j.tvjl.2013.08.030.

Smith, L.A., Marion, G., Swain, D.L., White, P.C.L. and Hutchings, M.R. (2009a) Inter- and intra-specific exposure to parasites and pathogens via the faecal–oral route: a consequence of behaviour in a patchy environment. *Epidemiology and Infection* 137(5), 630–643. DOI: 10.1017/S0950268808001313.

Smith, L.A., White, P.C.L., Marion, G. and Hutchings, M.R. (2009b) Livestock grazing behavior and inter-versus intraspecific disease risk via the fecal–oral route. *Behavioral Ecology* 20(2), 426–432. DOI: 10.1093/beheco/arn143.

Stabel, J.R. (1998) Biosecurity and disease – Johne's disease: a hidden threat. *Journal of Dairy Science* 81, 283–288.

Stevenson, K., Alvarez, J., Bakker, D., Biet, F., de Juan, L. *et al.* (2009) Occurrence of *Mycobacterium avium* subspecies *paratuberculosis* across host species and European countries with evidence for transmission between wildlife and domestic ruminants. *BMC Microbiology* 9, 212. DOI: 10.1186/1471-2180-9-212.

Sweeney, R.W. (1996) Transmission of paratuberculosis. In: Sweeney, R.W. (ed.) *Paratuberculosis (Johne's Disease)*. 12. W.B. Saunders Company, Philadelphia, Pennsylvania, pp. 305–312. DOI: 10.1016/S0749-0720(15)30408-4.

Vary, P.H., Andersen, P.R., Green, E., Hermon-Taylor, J. and McFadden, J.J. (1990) Use of highly specific DNA probes and the polymerase chain reaction to detect *Mycobacterium paratuberculosis* in Johne's disease. *Journal of Clinical Microbiology* 28(5), 933–937. DOI: 10.1128/JCM.28.5.933-937.1990.

Whittington, R.J., Reddacliff, L.A., Marsh, I., McAllister, S. and Saunders, V. (2000) Temporal patterns and quantification of excretion of *Mycobacterium avium* subsp. *paratuberculosis* in sheep with Johne's disease. *Australian Veterinary Journal* 78(1), 34–37. DOI: 10.1111/j.1751-0813.2000.tb10355.x.

Whittington, R.J., Marsh, I.B., Taylor, P.J., Marshall, D.J., Taragel, C. *et al.* (2003) Isolation of *Mycobacterium avium* subsp. *paratuberculosis* from environmental samples collected from farms before and after destocking sheep with paratuberculosis. *Australian Veterinary Journal* 81(9), 559–563. DOI: 10.1111/j.1751-0813.2003.tb12887.x.

Williams, E.S., Spraker, T.R. and Schoonveld, G.G. (1979) Paratuberculosis (Johne's disease) in bighorn sheep and a Rocky Mountain goat in Colorado. *Journal of Wildlife Diseases* 15(2), 221–227. DOI: 10.7589/0090-3558-15.2.221.

Zarnke, R.L., Dubey, J.P., Kwok, O.C. and Ver Hoef, J.M. (2000) Serologic survey for *Toxoplasma gondii* in selected wildlife species from Alaska. *Journal of Wildlife Diseases* 36(2), 219–224. DOI: 10.7589/0090-3558-36.2.219.

Zwick, L.S., Walsh, T.F., Barbiers, R., Collins, M.T., Kinsel, M.J. *et al.* (2002) Paratuberculosis in a mandrill (*Papio sphinx*). *Journal of Veterinary Diagnostic Investigation* 14(4), 326–328. DOI: 10.1177/104063870201400409.

16 Experimental Animal Models of Paratuberculosis

Adel M. Talaat[1]*, Chia-wei Wu[1] and Murray E. Hines II[2]
[1]University of Wisconsin-Madison, Madison, Wisconsin, USA; [2]Department of Pathology, University of Georgia, Athens, Georgia, USA

16.1 Introduction

Animal models play a crucial role in understanding the pathogenesis of infection and disease, and in testing novel vaccine candidates. Historically the use of non-standardized experimental animal models for paratuberculosis research has been a major problem that has prevented the ability to establish pathogen and host factors that predispose to disease, and to successfully compare the efficacy of vaccine candidates among research groups. In 2006, paratuberculosis researchers from the USA, the European Union (EU), Australia and New Zealand were convened under the sponsorship of the Johne's Disease Integrated Project (JDIP) as a committee to critically review the paratuberculosis animal model literature and develop international standardized challenge models for cattle, sheep, goats, cervids and mice. The JDIP Animal Model Standardization Committee (AMSC) members consisted of 16 paratuberculosis researchers, most of whom had successfully completed and published *Mycobacterium avium* subsp. *paratuberculosis* (*MAP*) challenge and/or vaccine efficacy studies in various species. The intent was to develop and propose international standard guidelines for models based on animal species that would gain acceptance worldwide. Parameters essential for the development of infection and vaccine models were outlined and harmonized to provide 'best fit' paratuberculosis challenge models. The AMSC findings were published in 2007 (Hines *et al.*, 2007a). Another extensive critical review of paratuberculosis experimental infection models was also published by Begg and Whittington (2008). The conclusions of both extensive reviews were often similar. The intent of this chapter is to provide an update of the current *MAP* experimental infection and vaccine models, across multiple species.

16.2 Rodent Models

Small rodent models of paratuberculosis are usually utilized to reduce the cost associated with larger models of infection (e.g. cattle or goat models). In addition, the use of small animal models can successfully reproduce certain signs of paratuberculosis in a shorter period compared with ruminant models. Mice represent the forefront small animal model for paratuberculosis. As for many infectious agents, mice provide a convenient model to study certain aspects related to the immunopathogenesis of infection with *MAP*. Given the impressive repertoire of immunological reagents and genetically

*Corresponding author: atalaat@wisc.edu

© CAB International 2020. *Paratuberculosis: Organism, Disease, Control, 2nd Edition* (eds M.A. Behr *et al.*)

distinct breeds (unlike ruminant models), mice were utilized efficiently to test the virulence of several *MAP* mutants, dissect the host–pathogen interactions and examine several vaccine candidates directed against paratuberculosis (Mullerad *et al.*, 2002a; Shin *et al.*, 2006; Bull *et al.*, 2007; Chen *et al.*, 2008; Settles *et al.*, 2014; Ghosh *et al.*, 2015). Another small animal, the white New Zealand rabbit (Beard *et al.*, 2001a), is infrequently used. The leporid (rabbit) model is usually utilized to produce antibodies for researchers focused on paratuberculosis, but has gained more attention owing to its value in microbiota (Arrazuria *et al.*, 2016a) and vaccination studies (Arrazuria *et al.*, 2016b).

16.2.1 Key features of the murine model of paratuberculosis

As often described among researchers in the field, mouse is the 'poor man's' model for initial testing of vaccine candidates before investing more efforts and funds in more relevant models. Depending on the specific goals of the project, different protocols can be employed to examine *MAP* immune-pathogenesis or test novel vaccine candidates. In a typical vaccine study, candidate vaccines can be inoculated by the preferred route and then protection can be measured at later times (e.g. 6, 12 or 24 weeks) following challenge with a fully virulent strain of *MAP*. These studies usually combine infection and immune-pathogenesis models based on the presence of control groups that receive only vaccines or only infection with a virulent strain of *MAP*. To evaluate the development of infection, sections from the liver, spleen and intestine are usually analysed for mycobacterial burden and histological lesions. In addition, immunological assessment can be done at the time of challenge with the virulent strain, as an attempt to uncover potential correlates of protection. Generally, a significant reduction of organ load with *MAP* (at least one log) and granuloma formation scores, compared with controls, are the major parameters for evaluation (Ghosh *et al.*, 2014; Settles *et al.*, 2014). These two outcomes depend on the genetic background of the mouse and the virulence of the mycobacterial challenge strain (Tanaka *et al.*, 1994; Mullerad *et al.*, 2002a; Shin *et al.*, 2006). Some studies have used BALB/c or C57BL/6 mice, which are immunocompetent but susceptible to mycobacterial infection (Veazey *et al.*, 1995). Others have used C3H mice, which are more resistant to *MAP* infection (Tanaka *et al.*, 1994). Despite the presence of granuloma-like lesions in both susceptible and resistant breeds of mice, the number of lesions and bacterial colonization levels declined dramatically in the resistant breed. Alternatively, athymic nude mice inoculated via oral route with *MAP* resulted in bacterial shedding in the faecal materials of mice (Hamilton *et al.*, 1991) – a rare clinical sign in mice.

Another important parameter of the murine model is the dose of infection and the route of administration. Oral inoculation of 10^{11} colony-forming units (CFU)/animal produces granulomatous lesions in 58% of the animals; however, lesions are limited to the mesenteric lymph node (Veazey *et al.*, 1995). On the other hand, intra-peritoneal (IP) injection of a low dose of *MAP* (10^6 CFU) induces a small number of epithelioid granulomas, as compared with the multifocal granulomas composed of macrophages and epithelioid cells that are observed after a higher dose of infection (10^8 CFU) (Tanaka *et al.*, 2000). In our hands, an IP dose of 10^8 CFU/animal of virulent strains of *MAP* typically produces enough granulomatous responses to score by 3 weeks post-infection (WPI) and these lesions are well formed by 6 and 12 WPI (Shin *et al.*, 2006; Wu *et al.*, 2007b). In several reports, IP injection produced the highest reproducible infection rate (100%) compared with other routes of delivery (Mutwiri *et al.*, 1992). In a high-throughput format, analysis is focused on the liver and intestine of inoculated animals. Both organs are analysed for histological lesions as well as mycobacterial burden over a period of 12 weeks (Shin *et al.*, 2006). Recently, oral inoculation of immune-competent mice resulted in establishing *MAP* infection by 12 weeks post-inoculation that required repeated inoculation of high doses of *MAP* (10^9 CFU/mouse, twice) (Cooney *et al.*, 2014). More importantly, this oral infection elicited an immune profile in mice that is similar to that observed in cattle.

Despite the several advantages of the murine model for *MAP*, several shortcomings limit the utility of this model. For example, there are known anatomical, physiological and immunological differences between cattle and mice that could impact the outcome of infection. Also, some of the typical features of paratuberculosis in cattle (e.g. diarrhoea, severe intestinal lesions) cannot be reproduced in mice. None the less, the murine model for paratuberculosis may play an important role in vaccine development against paratuberculosis. Unlike the bovine, caprine or ovine models of paratuberculosis, many immunological reagents are well developed for mice, permitting a more mechanistic assessment of the histopathologic outcomes observed. Also, mice with variable genetic background are readily available allowing the dissection of host–pathogen interactions on a molecular level.

16.2.2 *MAP* infection models

Because *MAP* infection can require prolonged times from infection to clinical signs of paratuberculosis, mice offer the advantage that *MAP*-specific lesions can be observed within 3–6 weeks of infection (Tanaka *et al.*, 1994; Shin *et al.*, 2006). Using immuno-compromised mice (SCID), intestinal lesions characteristic for paratuberculosis in cattle were obtained, including luminal infiltration with mononuclear cells containing acid-fast organisms (Mutwiri *et al.*, 2001). Also, infected intestines show increased ion secretion and damage to the epithelial cell lining of intestinal villi, both potential reasons for the chronic wasting syndrome associated with paratuberculosis in infected cows. Factors associated with *MAP* invasion and attachment to intestinal cells, particularly M cells, were initially examined using the ileal loop of mice where mycobacterial FAP (fibronectin attachment protein) was analysed (Secott *et al.*, 2002, 2004). Another set of virulence factors associated with how *MAP* regulates gene networks (sigma factors) was examined in mice (Ghosh *et al.*, 2013, 2014) and showed that both *sigL* and *sigH* genes play a key role in *MAP* survival during infection. In these studies, at least 8 and 30 genes were under regulatory control of both *sigL* and *sigH*, respectively (Table 16.1).

An attractive aspect of the murine model is the ability to utilize a large number of animals in an experiment with a relatively low budget. This economy of numbers is especially marked in comparison with ruminant models. In our experience, we have been able to assay the virulence of 11 *MAP* transposon mutants in BALB/c within a relatively short time (12 weeks of infection), enabling us to identify seven virulence genes in *MAP*, including *pstA* and *papA2* (Shin *et al.*, 2006). In a subsequent experiment, we assayed six additional mutants, again in a short time frame, permitting us to document the contribution of transcriptionally regulated genes (e.g. *lipN*, *prrA*) to *MAP* survival during infection. Further validation of the attenuation of some of these mutants was successfully performed for the Δ*gcpE* mutant (Wu *et al.*, 2007a) and the Δ*pstA* mutant (Wu *et al.*, 2009) in the cattle model, as well as the Δ*lipN* mutant in the caprine model (Shippy *et al.*, 2017). Results from the ruminant models validated the strategy for testing the virulence of *MAP* mutants first in the murine model before testing in any ruminant model of infection. It will be interesting to see whether mutants attenuated in ruminants are found for which no phenotype is detected in mice. Needless to say, mice were not able to reproduce exact organ lesions associated with paratuberculosis to those observed in ruminant models.

With the advancement in murine genetics, several investigators took advantage of the available knockout strains of mice to investigate the role and type of immune responses directed against *MAP* infection. Studies of macrophages derived from mice with specific mutations on pathogen recognition receptors (e.g. TLR-2, TLR-4, NOD2) have pointed to a role for innate immunity in the recognition and immune response directed against *MAP* (Ferwerda *et al.*, 2007). For adaptive immunity, γδT cells were studied in γδT cell-knockout BALB/c mice to determine their role in granuloma formation (Tanaka *et al.*, 2000). Another study, using C57BL/6 mice, demonstrated a major role for γ cells compared with γδ cells, by quantifying tissue lesions in mice deficient for the γ T-cell receptors (Stabel and Ackermann, 2002). Reconstitution of *MAP*-infected SCID mice with spleen cells from immunocompetent BALB/c mice significantly reduced lesions caused by *MAP* infection, indicating that protection against *MAP* is dependent on cell-mediated immunity (CMI) (Mutwiri *et al.*, 2002).

Table 16.1. Murine models for *Mycobacterium avium* subsp. *paratuberculosis* (*MAP*) infection.

	Mouse breed	Intended use of model	Age at start	*MAP* strain[a]	Method of *MAP* preparation	Infection method	Dose for testing colonization	Single or multiple doses	Vaccine administration	Challenge inoculum	End point	Time for first readout	Readouts	Samples	Culturing protocol	Shortcomings	Conclusions
Chandler, 1961	Swiss White	Pathogenesis	NR[b]	Cow isolate	Cultured in Taylor's broth	IP	2, 8 or 24 mg wet wt	1	No vaccination	No challenge	52 weeks	12 weeks	Histo, tissue smear count	Liver, spleen	Taylor's media	Use cortisone, not defined inocula, no clinical signs	Cortisone is not effective, high dose → high infection rate
Chandler, 1962	C57	Pathogenesis	NR[b]	Cow isolate		IP, IV, oral	2, 8 or 24 mg wet wt	1	No vaccination	No challenge	52 weeks	12 weeks	Histo, isolation of *MAP*	Liver, spleen	Taylor's media	Use cortisone, not defined inocula, no clinical signs	Highest infection: IV > IP > oral. C57 more sensitive than Swiss White
Mutwiri *et al.*, 1992	SCID	Pathogenesis	NR[b]	Cow isolate	Infected Ileum	IP, oral	1E + 05	1	No vaccination	No challenge	26 weeks	6 weeks	Histo, CFU, body wt	Liver, spleen	Herrold'segg yolk medium (HEYM)	No clinical signs	Quick model
Adams *et al.*, 1993	Athymic BALB/c	Immunology	NR[b]	*MAP* Linda	Culture in 7 H9	Gastric tube	1E + 10	1	No vaccination	No challenge	24 weeks	6 weeks	Histo, CFU, aecal shed	Liver, spleen	7 H10	A prolonged model	Immune cells otherthan CD4[+] and CD8[+] cells are more important for defendingMAP infection.
Tanaka *et al.*, 1994	BALB/c & C3H	Pathogenesis	NR[b]	ATCC19698	Culture in 7 H9	IP	1E + 08	1	No vaccination	No challenge	32 weeks	3 weeks	Histo, tissue smear count	Liver, spleen intestine	NR[b]	No clinical signs	BALB/c (BCGs) is more sensitive to infection than C3H (BCGr)
Veazey *et al.*, 1995	C57BL/6	Pathogenesis	NR[b]	ATCC19698	Culture in 7 H9	Oral	1E + 11	1	No vaccination	No challenge	44 weeks	28 weeks	Histo, tissue smear count	Liver, spleen, intestine, lymph node	NR[b]	Low infection rate	C57/BL6 is sensitive to *MAP* (BCGs)
Tanaka *et al.*, 2000	BALB/c	Pathogenesis	NR[b]	ATCC19698	Culture in 7 H9	IP	1E + 6 or 1E + 9	1	No vaccination	No challenge	18 weeks	6 weeks	Histo, CFU count	Liver, spleen, intestine	HEYM	No clinical signs	Gamma and delta T cells are important to form granulomas
Stabel and Ackermann, 2002	C57BL/6	Immunology	NR[b]	ATCC19698 and Ben	Culture in 7 H9	IP	1E + 08	1	No vaccination	No challenge	24 weeks	12 weeks	Histo, CFU count	Liver, spleen, intestine, lymph node	HEYM	No clinical signs	Alpha and beta T cells are more important than gamma and delta cells in forming granulomas
Mullerad *et al.*, 2002a,2003	BALB/c	Immunology	5–8 weeks	None	NR[b]	NR[b]	NR[b]	NRb	rAg85, rSOD/ SC (subcutaneous)	No challenge	6 weeks	6 weeks	DTH, cytokine, prolif. assay	Spleen, footpad	NR[b]	No challenge	Ag85B and SOD are immunogenic, subunit vaccines are possible against *MAP*

Continued

Table 16.1. Continued

	Mouse breed	Intended use of model	Age at start	*MAP* strain[a]	Method of *MAP* preparation	Infection method	Dose for testing colonization	Single or multiple doses	Vaccine administration	Challenge inoculum	End point	Time for first readout	Readouts	Samples	Culturing protocol	Shortcomings	Conclusions
Rosseels *et al.*, 2006	B6 beig/beig	Immunology	NR[b]	ATCC19698	Culture in 7 H9	IV	1E+06	1	No vaccination	No challenge	10 weeks	4–5 weeks	IFN-g, CFU count	Spleen	7 H11	No challenge	Ag85 complex is T-cell immunodominant antigens
Basagoudanavar *et al.*, 2006	Swiss Albino	Immunology	6 weeks	NR[b]	NR[b]	NR[b]	NR[b]	NR[b]	r35 kDa from MAP 316F	No challenge	5 weeks	5 weeks	IFN-γ, cytokine level	Spleen	NR[b]	No challenge	r35 kDa can elicit cell-mediated immunity
Shin *et al.*, 2006	BALB/c	Pathogenesis	NR[b]	ATCC19698 and mutants	Culture in 7 H9	IP	1E+07 or 1E+08	1	No vaccination	No challenge	12 weeks	3 weeks	Histo, CFU count	Liver, spleen, intestine	7 H10	No clinical signs	Several *MAP* mutants are attenuated for mice tissue colonization
Optimal model (Hines *et al.*, 2007b)	BALB/c or C3H	Pathogenesis or vaccine	4 weeks	ATC19698 or K-10	Culture in 7 H9	IP	1E+07 or 1E+08	1	Recombinant or attenuated	ATC19698 or K-10, 1E07 CFU	12 weeks	3 weeks	Histo, CFU count and IFN-γ production or DTH	Liver, spleen, intestine, mesenteric lymph node	7 H10	No clinical signs	Identify potential attenuated mutants or protective vaccine candidates
Scandurra *et al.*, 2010	BALB/c	Vaccine	4–7 weeks	*MAP* K-10, strain 989 and mutants	Culture in 7 H9	IP	1E+08	1	Attenuated mutants, 2E+06 CFU	*MAP* K-10, 2E+08 CFU	12 weeks	3 weeks	CFU count	Liver, spleen, intestine	7 H11	No clinical signs	Limited protection was observed from mice vaccinated with one of the mutants
Chen *et al.*, 2012	C57BL/6	Vaccine	6–8 weeks	*MAP* K-10, strain 66115–98	Culture in 7 H9	IP	5E+08	1	Prime and booster, attenuated mutants, 1E+07 CFU each	Strain 66115-98, 5E+08 CFU	21 weeks	8 weeks	Histo, CFU count, splenocyte IFN-γ production, T-cell proliferation	Liver, spleen	7 H10	No clinical signs	Several *MAP* mutants are attenuated and provide protection against challenge
Bull *et al.*, 2013	C57BL/6	Genomic and virulence variations in vaccine strains	5 weeks	JD87/107, 2eUK2001, 316FUK2001, IIUK2001	Culture in 7 H11	IP	1.1E+08–10.4E+08 by microscopic count	1	No vaccination	No challenge	12 weeks	4 weeks	Histo, CFU count	Liver, spleen	7 H11	No clinical signs	The vaccine strains showed reduced survival in organs

Continued

Table 16.1. Continued

	Mouse breed	Intended use of model	Age at start	*MAP* strain[a]	Method of *MAP* preparation	Infection method	Dose for testing colonization	Single or multiple doses	Vaccine administration	Challenge inoculum	End point	Time for first readout	Readouts	Samples	Culturing protocol	Shortcomings	Conclusions
Cooney *et al.*, 2014	BALB/c	Establishment of oral infection model	4weeks	*MAP* K-10	Culture in 7 H9	Oral gavage	1E+04 1E+05, or 1E+09 for 2 days	2	No vaccination	No challenge	24 weeks	3 weeks	Histo, CFU count, cytokine profiling	Liver, spleen, intestine, mesenteric LN	7 H10	No clinical signs	High-dose infections showed disseminated infection and elicited antigen-specific IFN-γ, IL-2 and IL-10 secretion
Koc *et al.*, 2014	C57BL/6	Immunology	8–12 weeks	DSM 44135	Watson–Reid medium	IP	1E+08	1	No vaccination	No challenge	8weeks	8 weeks	CFU count, body weight, spleen weight, colonic inflammatory response	Liver, spleen	7 H10	No clinical signs	T-cell-dependent inflammatory reactions were observed in intestines after infection
Ghosh *et al.*, 2014	BALB/c, C57BL/6	Pathogenesis, vaccine	NR[b]	*MAP* K-10 and a mutant	Culture in 7 H9	SC, IP	2E+08	1	Prime and booster, attenuated mutant, 2E+06 CFU each, SC	*MAP* K-10, 7E+08 CFU, IP	18 weeks	3 weeks	Histo, CFU count, splenocyte IFN-γ and IL-17A production	Liver, spleen, intestine	7 H11	No clinical signs	The vaccine strain showed reduced colonization, caused milder inflammation in organs and provided protection against challenge
Bannantine *et al.*, 2014	C57BL/6	Vaccine	6 weeks	*MAP* K-10 and mutants	Culture in 7 H9	IP	NDc	NDc	Attenuated mutants, 1E+05 CFU	*MAP* K-10, 6.5E+06 CFU	18 weeks	12 weeks	CFU count, spleen weight, liver weight	Liver, spleen	7 H11	No clinical signs	Mutants showed various levels of protection against challenge
Meissner *et al.*, 2014	C57BL/6	Pathogenesis	8 weeks	DSM 44135	Culture in 7 H9	IP	1E+08–2E+08	1	No vaccination	No challenge	4 weeks	4 weeks	CFU count, liver granuloma count	Liver, spleen, mesentery	7 H10	No clinical signs	The *mptD* gene is required for full virulence of *MAP* in mice
Settles *et al.*, 2014	C57BL/6	Vaccine	NR[b]	*MAP* K-10 and a mutant	Culture in 7 H9	SC, IP	NDc	NDc	Prime and booster, attenuated mutant, 1E+06 CFU each, SC	*MAP* K-10, 1E+08 CFU, IP	18 weeks	12 weeks	Histo, CFU count, splenocyte IFN-γ, IL-4 and IL-17A production, Ab titre	Liver, spleen, intestine	7 H10	No clinical signs	The vaccine strains showed reduced colonization, caused milder inflammation in organs and provided protection against challenge

Continued

Table 16.1. Continued

	Mouse breed	Intended use of model	Age at start	*MAP* strain[a]	Method of *MAP* preparation	Infection method	Dose for testing colonization	Single or multiple doses	Vaccine administration	Challenge inoculum	End point	Time for first readout	Readouts	Samples	Culturing protocol	Shortcomings	Conclusions
Suwandi *et al.*, 2014	C57BL/6J	Disease model	9 weeks	DSM 44135	Watson–Reid medium	IP	1E+08	1	No vaccination	No challenge	3 weeks	1 day	Histo, CFU count, TEM, Ab titre, T-cell counts, IFN-γ and TNF-α expression	Liver, spleen, intestine, mesentery	7H10	No clinical signs, no direct evidence to prove causal relationship	*MAP* is able to infect the inflamed colonic mucosal tissue and significantly increases both mucosal and systemic signs of inflammation
Ghosh *et al.*, 2015	C57BL/6	Vaccine	NR[b]	*MAP* K-10 and a mutant	Culture in 7H9	SC, IP	2E+06	1	Prime and booster, attenuated mutant, 2E+06 CFU each, SC	*MAP* K-10, 5E+08 CFU, IP	18 weeks	6 weeks	Histo, CFU count, splenocyte IFN-γ and IL-10 production, Ab titre	Liver, spleen	7H10	No clinical signs	ΔsigH mutant vaccination reduced organ colonization and severity of inflammation after challenge
Ghosh *et al.*, 2015	C57BL/6	Vaccine	NR[b]	*MAP* K-10, JTC1285 and a mutant	Culture in 7H9	SC, IP	1E+08	1	Attenuated mutant, 1E+08 CFU, SC	JTC1285, 1E+09 CFU, IP	15 weeks	6 weeks	Histo, CFU count, splenocyte IFN-γ and IL-10 production, Ab titre	Liver, spleen	7H10	No clinical signs	ΔsigH mutant vaccination, with or without adjuvant Quil-A, reduced organ colonization and severity of inflammation after challenge

[a]Where *MAP* strain was not provided, the provenance of the infecting strain is reported.
[b]NR = not reported.
[c]ND = not determined.
CFU, colony-forming units; SCID, Severe combined immunodeficient; BCGs, *Mycobacterium bovis* bacillus Calmette-Guérin (BCG) strains; LN, Lymph node; SOD, Superoxide dismutase; DTH, Delayed-type hypersensitivity; Ab, Antibody; IP, Intraperitoneal; IV, intravenous; SC, subcutaneous.

By studying T-cell responses specific to the 35-kDa protein of *MAP*, the involvement of CMI in responding to infection with *MAP* was further documented (Basagoudanavar *et al.*, 2006). Finally, murine models provided key information on the quality of adaptive immune responses. Both Th1 and Th2 responses were observed in mice inoculated with recombinant antigens (e.g. superoxide dismutase, SOD) (Mullerad *et al.*, 2002a), opening the door for testing these antigens as vaccine candidates against *MAP* infection. In addition to paratuberculosis, the murine model is also used to investigate the potential link of *MAP* with human Crohn's disease. Although the casual relationship between the bacterium and the human disease remains debatable, *MAP* infection in mice was shown to worsen existing colitis that resembles Crohn's disease (Suwandi *et al.*, 2014).

16.2.3 *MAP* vaccine models

Traditionally, the antigenicity of *MAP*-specific novel antigens was tested first in mice as a prelude to further analysis on naturally infected cattle (Paustian *et al.*, 2004). This step-wise approach provides a paradigm for examining the functional significance of novel antigens and may also serve in the testing of novel vaccine candidates. Early on, several vaccine candidates (live-attenuated, subunit and vectored vaccines) developed against *MAP* were tested in the murine model of paratuberculosis (Table 16.1) before further testing in ruminant models. Recombinant proteins of Ag85B, SOD and thiol peroxidase (TP) of *MAP* elicited strong immune responses in mice (Mullerad *et al.*, 2002a, b, 2003). In these models, subunit vaccine inoculation triggered strong levels of cytokines (e.g. IFN-γ, TNF-α, IL-4, IL-6) only in the presence of the Ribi adjuvant. Interestingly, a polypeptide encoding three open reading frames (ORFs) of *MAP* antigens provided protective responses against challenge with a virulent strain of *MAP* (Chen *et al.*, 2008). Because DNA immunization does not require protein preparations (Huntley *et al.*, 2005; Talaat and Stemke-Hale, 2005), a library of *MAP* encoding ORFs were analysed in mice (Huntley *et al.*, 2005). A total of 26 protective antigens have been identified in such a screen. To take advantage of safety of the traditional vaccine vectors, an adenovirus-based vaccine (Ad5) encoding four fused antigens elicited strong immune responses against *MAP* using a strategy based on vaccination with DNA constructs followed by Ad5 inoculation of mice (Bull *et al.*, 2007). This vaccine candidate was later further assessed and validated in a calf model of paratuberculosis, showing increased antigen-specific $CD4^+$, $CD8^+$ IFN-γ producing T-cell populations up to 7 months post-challenge (Bull *et al.*, 2014).

Recently, more attention has been directed towards live-attenuated vaccine (LAV) candidates because of their ability to elicit a robust immunity against mycobacterial infections (Kaufmann, 2011). Some members among the group who established the AMSC standards also examined the utility of mice to screen eight LAV candidates in mice (Bannantine *et al.*, 2014) to show that not all vaccine constructs need to be persistent to offer protection against paratuberculosis. Targeting specific gene deletions in extra-cytoplasmic sigma factors of *MAP*, *sigH* and *sigL* mutants were found to induce protective immunity in mice against challenge with virulent strain of *MAP* (Ghosh *et al.*, 2013, 2014). When tested in goats, *sigH* mutant was also protective, similar to *lipN* mutant (Shippy *et al.*, 2017). In all murine vaccine testing scenarios, the murine model was utilized as a preliminary screen of vaccine candidates to eliminate non-protective constructs. Whether the mouse model faithfully serves this purpose will require comparative experiments in different hosts.

16.2.4 Leporid (rabbit) models

Rabbits can be naturally infected with *MAP* (Judge *et al.*, 2005) and provide an attractive model to study the immunopathogenesis of *MAP*. Historically, rabbits were used to generate polyclonal antibodies against *MAP* antigens. Both live and heat-killed *MAP* isolates were able to elicit strong immune responses in rabbits (Stabel *et al.*, 1996). Several proteins expressed from clones of a genomic library of *MAP* were reactive to rabbit antibodies, including the 25-kDa protein (*a.k.a* Csp1) and Pks7

(Bannantine and Stabel, 2001). One of these antigens (Csp1) was shown to be expressed during macrophage survival. Recently, rabbits were also used to generate antibodies against a specific *MAP* antigen (34 kDa) that was shown to exist in naturally infected cows (Malamo *et al.*, 2007). All of these reports suggested that rabbits can provide a valid model for generating antibodies necessary to investigate *MAP* pathogenesis. With the success of the leporid model in immune-diagnostics of *MAP*, this model was also investigated to study the pathogenesis of *MAP*. All models of *MAP* pathogenesis reported so far included oral administration of *MAP* with variable success in inducing infection in all of the animals (Mokresh and Butler, 1990; Mokresh *et al.*, 1989; Vaughan *et al.*, 2005). In one report, almost 50% of the inoculated animals did not show any sign of infection despite the high dose of infection (10^7 CFU/animal) (Mokresh *et al.*, 1989). None the less, higher inocula (10^8 CFU) resulted in a higher infection rate (~70%) of inoculated rabbits (Mokresh and Butler, 1990). Interestingly, rabbits showing signs of the disease suffered from granulomatous enteritis and diarrhoea (Vaughan *et al.*, 2005), which are hallmarks of paratuberculosis in cattle. Additionally, signs of body weight loss and clinical emaciation were also evident in a considerable number of inoculated rabbits (Vaughan *et al.*, 2005). In all of these reports, *MAP* was isolated from variable tissues, especially the intestinal tract and mesenteric lymph nodes. However, faecal shedding of *MAP* was not always detectable in infected animals (Vaughan *et al.*, 2005). To investigate factors that affect susceptibility to oral challenges of *MAP* in rabbits, dietary variations were applied 4 days prior to oral administration of *MAP*. The group given regular rabbit diet showed more signs, including low weight gain, faecal shedding, tissue colonization and histopathology, which are consistent with paratuberculosis infection, compared with groups switched to a high protein or high fibre diet (Arrazuria *et al.*, 2015). These phenomena are thought to be partly attributed to shifted gut microbiota upon *MAP* infection, as evidenced by microbiome analysis of samples of caecum and sacculus rotundus from *MAP*-infected rabbits under different dietary regimens (Arrazuria *et al.*, 2016a).

In a novel leporid model of vaccination study, heat-killed *MAP* vaccine was administrated 1 month before or after an oral *MAP* challenge (Arrazuria *et al.*, 2016b). Both groups showed a elevated humoral response against *MAP* and reduced *MAP* colonization, but vaccination after infection provided a therapeutic effect as it showed slightly stronger protection compared with vaccination before infection. Overall, rabbits provide an important small animal model to study key aspects of paratuberculosis that cannot be investigated in mice. Aspects related to the induction of diarrhoea, weight loss or intestinal colonization could be better investigated in rabbits rather than mice. In most reports, rabbits developed signs of paratuberculosis within the first year post-infection, a time period that is much shorter than calf infections. Despite all of these advantages, the use of the leporid model in *MAP* research is limited because of the limited availability of immunological reagents and the expenses associated with housing rabbits compared with mice. See Table 16.2 for key aspects associated with both the murine and leporid models of paratuberculosis.

16.3 Bovine Models

16.3.1 Long-term bovine *MAP* challenge model

Because of the chronic nature of paratuberculosis, bovine infection models that examined inoculated animals for long periods (>6 months) are common (Table 16.3). This long-term bovine model is typically used for studies of vaccination efficacy, preventive or therapeutic interventions, long-term pathogenesis, the immune response and assessment of diagnostic assays. A wide variety of *MAP* strains has been used in experimental infections (Table 16.3). Many reports identify strains only as '*M. johnei*', 'field isolate' or 'clinical isolate', while others provide more specific strain designations. Strain K-10 was selected by the AMSC as the prototype strain for bovine infection models because it is known to be pathogenic and has a well-characterized genotype. However, the laboratory passage status of the K-10 *MAP* strain is currently unknown. Other low-passage virulent strains with a similar

Table 16.2. Main features of the murine and leporid (rabbit) models of paratuberculosis.

Model	Murine	Leporid (rabbit)
Breeds	C57/BL6, BALB/c, SCID C3H, Beige, Swiss Albino	White New Zealand
Route of administration	Oral, intraperitoneal	Oral
Infectious dose	10^7–10^8 CFU/animal	10^5–10^9 CFU/animal
Sample collection	Liver, spleen, intestine, lymph nodes, sera	Liver, spleen, intestine, lymph nodes, sera
Diarrhoea	Never	Frequent
Faecal shedding	Rarely	Occasionally
Experimental endpoints	CFU count, histopathology, immunological evaluations	CFU count, histopathology, clinical signs
Shortcomings	Rare development of clinical signs of paratuberculosis	Lack of reagents, incomplete clinical signs of paratuberculosis

CFU, colony-forming units; IT, Intratonsillar; SC, Subcutaneous; ATCC, American Type Culture Collection; IFN, Interferon; LBT, Lymphocyte blastogenesis test; IL, Interleukin.

short-segment repeat genotype (e.g. 15g–5ggt genotype; Ghadiali *et al.* (2004)) or equivalent pulsed-field gel electrophoresis or amplified fragment-length polymorphism genotype may also be used (Hines *et al.*, 2007b). This allows use of various local *MAP* strains and minimizes regulatory issues concerning the import of infectious organisms.

Types of inocula that have been administered include faeces from an infected animal, intestinal mucosal scrapings, lymph node homogenates or *in vitro*-cultured *MAP* (Table 16.3). In previous studies, *MAP* has been harvested from solid media (Herrold's egg yolk, Middlebrook 7H11, Middlebrook 7H10, Taylor's, Dubos) or from broth (Middlebrook 7H9). Inocula prepared from homogenized intestinal tissue, lymphoid tissue or faeces have been used, but these are difficult to standardize and aliquot for numerous studies. To address these issues, the AMSC has suggested that the chosen strain of *MAP* should be propagated *in vitro* from a master seed stock to mid-log-phase growth. Middlebrook 7H9 broth, supplemented with Oleic AlbuminDextrose Catalase (OADC), mycobactin J and 1% glycerol, was recommended as the culture medium (Hines *et al.*, 2007b).

Storage of the inoculum at 4°C beyond 4 weeks may result in reduced viability (R. Whitlock, Philadelphia, 2008, personal communication) but storage up to 2 weeks at 4°C is generally considered acceptable. However, it is recommended that the stored inoculum should be incubated at 37°C for 2–3 h prior to administration. All samples within a study should be treated in the same manner.

Oral, intragastric and parenteral routes of inoculation have been successfully used (Table 16.3). However, the oral route is recommended by the AMSC as this most closely mimics natural exposure and allows *MAP* uptake by tonsillar tissue as well as the intestinal tract. The calf is induced to suckle from a syringe containing *MAP* mixed with a small volume of milk replacer or pasteurized milk. The mixture can also be gently expressed over the back of the tongue to induce swallowing. Administration by gastric tube is not recommended (Hines *et al.*, 2007b).

Published doses range from 10–200 mg wet weight and 10^6 to 10^{11} CFU (Table 16.3) but were not based on standardized procedures. Doses of 5×10^8 CFU given on 2 consecutive days reliably resulted in infection in calves by 12–14 weeks post-inoculation (Sweeney *et al.*, 2006). The AMSC recommends use of a standard bovine challenge dose of approximately 10^9 CFU/dose (100 mg wet weight) given on each of 2 successive days. Quantification of CFU should be by serial dilution and plating on solid medium known to support the organism without added antibiotics. Excessively large doses that produce clinical disease in cattle less than 18 months post-inoculation should be

Table 16.3. Bovine models for *Mycobacterium avium* subsp. *paratuberculosis* (*MAP*) infection.

Reference	Age at start	*MAP* strain	Dose(s)	Infection method	Study length	Clinical disease	Faecal culture	Tissue culture	Immune testing	Histopathology
Rankin, 1958	1 month	Clinical isolate (eight subcultures)	100 mg wet wt	IV	4 years	4/6	6/6	6/6	ND[a]	4/6
Rankin, 1961a	3 years	Clinical isolate (eight subcultures)	100 mg wet wt	IV	4 years	0/5	0/5	5/5	ND[a]	1/5
Rankin, 1961b	1, 3, 6 months	Clinical isolate	Unknown	Faecal–oral exposure	5 years	4/9	8/9	9/9	ND[a]	9/9
Payne and Rankin, 1961a	3 months	Clinical isolate	200 mg wet wt	Oral (milk)	Variable up to 14 months	No	ND[a]	Variable	ND[a]	Increased with period of infection
Payne and Rankin, 1961b	3 months or 3 years	Clinical isolate	200 mg wet wt	Oral (milk or water)	Variable up to 6 months	No	ND[a]	6/8 calves 1/8 cows	ND[a]	Increased with period of infection
Rankin, 1962	Variable	Clinical isolate	Unknown	Faecal–oral exposure	4 years	5/6 calves 1/7 cows	6/6 calves 4/7 cows	4/7 cows	Skin test + CFT –	6/6 calves 1/7 cows
Gilmour *et al.*, 1965	3 weeks	Clinical isolate	1×10^8; 1×10^{10} one dose × 10 weeks	Oral (tube)	Variable up to 13 months	No	ND[a]	Increased with period of infection and dose	Skin test +	Increased with period of infection and dose
Stuart, 1965	1 week	Clinical isolate	100 mg wet wt	IV	10 months	Yes	40/40	NR[b]	CFT +	18/40
Larsen *et al.*, 1978	1 month	Naturally infected cows	Unknown	Oral (natural exposure)	Up to 6 years	Yes	20/175	31/175	Skin test +	22/175
Larsen *et al.*, 1974	16 days	Clinical isolate	180 mg wet wt	Oral (milk)	5 months	No	8/8	7/8	Skin test +	8/8
Thorel *et al.*, 1984	4 weeks	Various isolates of *MAP*	1×10^9	IV	12 months	No	ND[a]	23/23	ND[a]	8/23
Krishnappa *et al.*, 1989	NR calves	Clinical isolate from mucosal scrapings	50 g one dose × 10 weeks	Oral	30 weeks	No	1/12	ND[a]	AGID + CIE +	ND[a]

Continued

Table 16.3. Continued

Reference	Age at start	*MAP* strain	Dose(s)	Infection method	Study length	Clinical disease	Faecal culture	Tissue culture	Immune testing	Histopathology
Szilágyi *et al.*, 1989	17 days	Strain 5889	1×10^8 two × 5 days with 15 days between	Oral	400 days	No	NR[b]	NR[b]	NR[b]	ND[a]
Saxegaard, 1990	4 weeks	Clinical goat isolate from tissue	10 mg wet wt; 10 doses over 10 days	Oral (milk)	Variable up to 18 months	No	ND[a]	4/4	ELISA –	No
McDonald *et al.*, 1999	2 months	Clinical isolate	2 g wet wt three doses	Oral (gastric tube)	Up to 27 months	No	2/4	1/4	IFN-γ + ELISA –	2/4
McDonald *et al.*, 1999	2 months	Clinical isolate	20 g wet wt three doses	Oral (gastric tube)	Up to 27 months	No	4/4	4/4	IFN-γ + ELISA +	4/4
Beard *et al.*, 2001b	1 week	Rabbit and bovine isolate	1×10^9 three doses one ×/week	Oral	6 months	No	5/8 – rabbit 0/4 – bovine	7/8 – rabbit 3/4 – bovine	ND[a]	3/8 – rabbit 2/4 – bovine
Waters *et al.*, 2003	2 weeks	*MAP* K-10	1.6×10^7 one × 4 weeks	IT	320 days	No	3/3	3/3	IFN-γ + ELISA +	0/3
Uzonna *et al.*, 2003	28 days	Clinical isolate	1×10^{10} two doses	Oral (milk)	49 days	No	ND[a]	15/15	IFN-γ + ELISA –	0/15
Koo *et al.*, 2004	1–2 days	NR[b]	1×10^7 seven doses	Oral	6 months	No	ND[a]	2/3 – PCR	IFN-γ + ELISA –	ND[a]
Simutis *et al.*, 2005	4 weeks	Strain 19,698	1×10^8	SC	150 days	No	0/25	1/25	IFN-γ + Skin test +	0/25
Stabel *et al.*, 2003	4 months	Clinical isolate (bovine and bison)	1×10^9 five doses	Oral (gastric tube)	6 months	No	1/6 – cattle 2/6 – bison	5/6 – cattle 6/6 – bison	IFN-γ + ELISA –	0/6 – cattle 0/6 – bison
Sweeney *et al.*, 2006	2–3 days	Clinical isolate (ATCC 700533)	2.5×10^{10} two doses	Oral (milk)	44 days	No	0/6	60/6	ND[a]	06/6
Sweeney *et al.*, 2006	21–22 days	Clinical isolate (ATCC 700533)	H – 5×10^9 M – 5×10^8 L – 1.5×10^6	Oral (milk)	44 days	No	H – 0/8 M – 0/6 L – 0/6	H – 8/8 M – 6/6 L – 6/6	ND[a]	H – 0/8 M – 0/6 L – 0/6

Continued

Table 16.3. Continued

Reference	Age at start	*MAP* strain	Dose(s)	Infection method	Study length	Clinical disease	Faecal culture	Tissue culture	Immune testing	Histopathology
Kurade *et al.*, 2004	1 month	Clinical cow faeces (high shedder)	20 g nine doses	Oral (faeces in milk)	644 days	No	8/10	ND[a]	IFN-γ + ELISA +	ND[a]
Rosseels *et al.*, 2006	2–3 weeks	Strain 19,698	1×10^8 10 doses	Oral	875 days	No	NR[b]	ND[a]	Skin test + ELISA –	ND[a]
Wu *et al.*, 2007a	3–4 weeks	*MAP* K-10, strain 19698, Δ*gcpE* mutant	1×10^7–10^8 one dose	Ileum injection	4 days to 9 months	No	0/5	5/5	Skin test – ELISA – IFN-γ + TNFα + IL12 + IL4 –	5/5
Stewart *et al.*, 2007	6 weeks 14 weeks	Bovine isolate	1–2×10^{10} 15–20 g tissue	Oral	54 months	No No	3/5 5/5	0/3 0/5	IFN-γ + ELISA	0/8
Stewart *et al.*, 2007	6 weeks 14 weeks	Ovine isolate	Four weekly doses one or two $\times 10^{10}$ 15–20 g tissue four weekly doses	Oral	35 months	No No	0/5 3/5	0/5 0/4	IFN-γ + ELISA	0/9
Kathaperumal *et al.*, 2008	8 weeks	*MAP*66115-98	1×10^7 daily doses, two groups rechallenged 8 weeks later with 5×10^8 for 5 days	Oral	23 weeks PI	No	NR[b]	13/24	IFN-γ + LBT + TNFα + IL12 + IL2 + IL4 – IL10 – FLOW	0/24
Khare *et al.*, 2009	3–4 weeks	Strain 19,698	3×10^9	Ileal and Jejunal injection	12 h	No	No	Yes	Gene expression	4/4 SEM and TEM
Stabel *et al.*, 2009	1–2 days	*MAP* K-10 and *MAP* isolate 509	4×10^{11} to 8×10^{12} over 14 days	Oral and Intraperitoneal	12 months	No	Yes and PCR	Yes and PCR	ND[a]	19/19

Continued

Table 16.3. Continued

Reference	Age at start	*MAP* strain	Dose(s)	Infection method	Study length	Clinical disease	Faecal culture	Tissue culture	Immune testing	Histopathology
Stabel and Robbe-Austerman, 2011	1–2 days	*MAP* K-10 and *MAP* isolate 509	4×10^{11} to 8×10^{12} over 14 days	Oral and Intraperitoneal	12 months	No	NR[b]	NR[b]	IFN-γ + IL-10 LBT+ FLOW NOX	NR[b]
Begg *et al.*, 2018a, b	2–4 months	*MAP* CM00/416 K-10 like	Variable doses over three trials	Oral	Up to 4.75 years	Yes	Yes and PCR	Yes	IFN-γ + ELISA	35/50

[a]ND = not determined.
[b]NR = not reported.
AGID, agar gel immunodiffusion test; CFT, complement fixation test; CIE, crossed immunoelectrophoresis; FLOW, flow cytometry; NOX, nitric oxide; PI, post-inoculation; SEM, scanning electron microscopy; TEM, transmission electron microscopy; ELISA, enzyme-linked immunosorbent assay; PCR, polymerase chain reaction.

avoided as they are not typical in naturally occurring paratuberculosis (Hines *et al.*, 2007b).

Previous studies reported challenge of cattle from 1 day of age to adulthood (Table 16.3). While studies have suggested age-associated reduction in susceptibility (Hagan, 1938; Doyle, 1953, 1956; ; Larsen *et al.*, 1975), a definitive relationship between infectious dose and age has not been established yet. Some flexibility in age of inoculation is required due to differences in experimental objectives and timing of other interventions, such as vaccination. Although some have suggested breed differences in susceptibility to infection (e.g. increased susceptibility of Jersey cattle to natural infection; Koets *et al.*, 2000), this has not been formally evaluated in an experimental infection study. However, because of the difference in clinical presentation of paratuberculosis in *Bos indicus* breeds, the AMSC recommends that *Bos taurus* breeds be used exclusively for studies of bovine paratuberculosis (Hines *et al.*, 2007b). Current experience suggests that 100 mg pelleted wet weight on 2 successive days results in demonstrable infection in calves less than 8 weeks of age (Hines *et al.*, 2007b).

Current diagnostic methods are ineffective for screening of individual calves at or before 8 weeks of age. Calves should be purchased from paratuberculosis-free farms to ensure lack of prior exposure. In the USA, herds that have achieved the equivalent of Status Level 3 or 4 in the National Voluntary Bovine Johne's Disease Status Control Program should be chosen as source herds (Hines *et al.*, 2007b; Usda-Aphis, 2010). This equates to a closed herd with no history of paratuberculosis in the previous 5 years, tested annually with at least one whole-herd negative serological test and one whole-herd negative faecal test (second lactation and older animals).

Calves should have received adequate quantities of colostrum. Milk replacer, if used, should be high quality and of animal origin (i.e. casein not soy protein) and high-quality rations should be fed. Routine vaccinations may be administered but should not be given on the same days as paratuberculosis vaccination or *MAP* administration. Anthelmintics and parasiticides should be given to all animals at normal dosages. If individual therapy is necessary (i.e. respiratory infection), an antibiotic known to have minimal effects on *MAP* (i.e. ceftiofur) should be used.

Passive (pass-through) shedding occurs as early as 12 h after oral inoculation. Detection of passive shedding by culture provides additional confirmation of inoculum viability and the sensitivity of the faecal culture method. Positive faecal culture results more than 14 days after inoculation should be considered shedding due to infection, except in very heavily contaminated environments (R. Whitlock, Philadelphia, 2008, personal communication). The AMSC recommends that animals should have faecal cultures at least monthly during the course of the study. Gross examination and culture of tissues for *MAP* as well as histopathological examination of tissues to identify acid-fast organisms and lesions characteristic of paratuberculosis should be performed in all studies. At the inoculation dose recommended, colonization of tissues can be detected by culture in most animals by 4–12 weeks after inoculation, although culture of multiple tissues (minimum of three ileum including ileocaecal valve, three jejunum, one duodenum, one spiral colon and three mesenteric lymph nodes including ileocaecal nodes) is necessary. Although tissue samples should be culture-positive by 12 weeks, investigators should not always expect to find histological lesions at this early stage. A necropsy and histopathology scoring system should be used (González *et al.*, 2005; Hines *et al.*, 2007b). The clinical status of the study animals should, at a minimum, be assessed and recorded monthly and at necropsy (Hines *et al.*, 2007b).

The method of culturing faecal and tissue samples should permit quantification (or at least semi-quantification, e.g. by counting CFU on solid medium or time to positive detection in automated liquid culture systems). Decontamination by incubation of faecal and tissue samples in 0.6% hexadecyl pyridinium chloride (HPC) for 14–16 h and 3 h, respectively, is strongly recommended (Hines *et al.*, 2007b). There is wide variation in culture methods currently used and batch-to-batch differences occur in media. All samples should be processed on medium from the same batch. Fresh samples may be processed immediately or frozen at −70°C and thawed only once, with all samples treated in the same manner.

16.3.2 Short-term bovine *MAP* challenge models

Experimental *MAP* infection models that employ direct surgical access to the ileum for administration of *MAP* and collection of intestinal samples have been described (Allen *et al.*, 2005, 2009). In general, these models are most useful for short-term study of host–pathogen interactions. The time frame of such studies can vary from a few hours to a few days or even a few weeks (Wu *et al.*, 2007a). Many parameters should be the same as for long-term challenge, including strain, inoculum preparation and quantification, storage, animal selection criteria and quality control issues. Exceptions are age of administration, dose, experimental endpoints and sample collection.

Ileal cannulation model

In this model, calves are cannulated at 8 weeks of age under general anaesthesia, using a modification of the method of Streeter *et al.* (1991), as previously described (Hines *et al.*, 2007b; Allen *et al.*, 2009). Calves are inoculated in the ileum with 10^{10} CFU of *MAP* in 20 ml phosphate-buffered saline. *MAP* is taken up by M cells and by dendritic and epithelial cells of the ileum and jejunum within 30 min (Momotani *et al.*, 1988; Sigurðardóttir *et al.*, 1999, 2001). The inoculation procedure is repeated the next day. Ileal mucosal biopsies can be obtained at various time intervals, to test for *MAP* infection and local immune responses.

Invasion/surgical model

Surgical incision and direct deposition of *MAP* into the ileum has been recently employed to establish a model for intestinal invasion to other organs within a few hours of infection (Wu *et al.*, 2007a). Although infected calves survive the surgery and repeated biopsy sampling up to 10 months following infection, the focus of this model is to examine early *MAP* intestinal interactions. The movement of *MAP* from intestine to liver, spleen or mesenteric lymph node was shown to differentiate between *MAP* strains with different virulence phenotypes.

Intestinal loop model

An alternative surgical approach is the ligated intestinal loop model with an injected inoculum of 3×10^9 CFU (~300 mg wet weight) (Momotani *et al.*, 1988; Khare *et al.*, 2009). This is only suitable for studies <12 h in duration and includes endpoints such as *MAP* detection, transcriptional gene expression and histopathology. This model is very useful for ultrastructural studies of the early host–pathogen interaction and the early host immune response as measured by cytokine gene expression.

16.4 Caprine Models

16.4.1 Long-term caprine *MAP* challenge model

The long-term caprine *MAP* challenge model is intended primarily for pathogenesis and vaccine efficacy studies; however, it could also be useful for other studies, including evaluation of diagnostic assays. No particular goat breed has been recommended and any common local breed is generally considered acceptable since no studies have been performed to investigate breed susceptibility.

The animals should undergo a selection process that includes validation/precertification of infection status. The paratuberculosis status of the herd of origin is the best method to validate infection status of young kids. Kids should be selected only from closed herds in which all adults have been tested negative by ELISA and faecal culture/PCR for at least 1 year and paratuberculosis vaccination is not practised. All kids should receive sufficient quantities of colostrum. No specific recommendations were made for rations other than they should be of high quality (Hines *et al.*, 2007b).

Goat challenge studies have successfully used *MAP* isolates from clinically infected goats, cattle and sheep, as well as humans (Table 16.3). Twenty-six to 50% of isolates from clinical cases in goats have been genotyped as *MAP*-C (also called Cattle-type) (Motiwala *et al.*, 2004; Sevilla *et al.*, 2005, 2007). A more detailed description of the different types of *MAP* is given elsewhere (see Chapter 6, this volume). Motiwala *et al.*

(2004) detected little genetic diversity between and within bovine and caprine isolates, but Sevilla *et al.* (2007) showed that caprine isolates had a high degree of genetic heterogeneity, and Mobius *et al.* (2008) found a high degree of genetic heterogeneity in isolates from German cattle.

In some studies, the inoculum consisted of *in vitro*-cultured *MAP*, while in others it was prepared from homogenized intestinal mucosal tissue scrapings from a clinically diseased animal. The use of a virulent, low-passage caprine clinical isolate cultured *in vitro* from a master seed stock, with a genotype identical or closely related to bovine strain K-10, is recommended (Hines *et al.*, 2007b).

The recommended method of inoculum preparation, quantification and storage is the same as stated above for the bovine model. The majority of studies used Middlebrook 7H9 + OADC + mycobactin J + glycerol or Tween 80 for *in vitro* cultivation. This medium with the addition of glycerol (1%), but without Tween or antibiotics, is considered to be the best choice (Hines *et al.*, 2007b). The routes of administration have varied (Table 16.4), and essentially all routes tested have been successful in establishing infection, with even the aerosol route resulting in intestinal pathology (Harding, 1957). The oral route most closely parallels natural exposure and is generally considered the best route of administration.

The size of the challenge inoculum has also varied widely (Table 16.4), ranging from 2.37 to 200 mg and 3×10^7 to 8×10^{10} CFU, and all were generally successful in establishing infection. Two consecutive daily doses of 10^9 organisms (approximately 100 mg pelleted wet weight/dose and ~200 mg total dose) should establish infection in most kids without overwhelming experimental interventions (Hines *et al.*, 2007b).

The age at which to inoculate has varied from the day of birth to 10 months of age (Table 16.4). It is not known whether age-related resistance occurs in goats. The age of administration will depend on the experimental goals and endpoints, but generally the goats should be less than 4 months of age. Clinical disease is expected to develop in a low percentage of animals at 9–10 months post-inoculation (Hines *et al.*, 2007b).

The experimental endpoints will depend on the goals of the study. The minimal AMSC recommendations for goat experiments were to determine infection status through a combination of culture, PCR and histopathology, using quantitative or semi-quantitative methods (Hines *et al.*, 2007a, b). A lesion grading system for gross and histopathological findings (Hines *et al.*, 2007a, b) should be used, with a sufficient range in values to allow statistical analysis. Recommendations for methods of faecal and tissue *MAP* culture and for sample handling are the same as for the bovine model. Like cattle, positive faecal cultures within 2 weeks or more post-inoculation should be considered to be due to infection. Faecal cultures/PCR should be performed at least monthly from all animals.

16.4.2 Short-term caprine *MAP* challenge models

In general, these models are used for short-term study of host–pathogen interactions. Many parameters should be the same as for long-term challenge, including strain, inoculum preparation and quantification, storage, animal selection criteria and quality control issues.

Intestinal loop model

An intestinal loop model, as previously described in goats by Sigurðardóttir *et al.* (2001), can be used for studying initial paratuberculosis bacterial–host interactions including bacterial attachment and internalization, bacterial localization, cytokine regulation and early bacterial gene regulation. *MAP* strain, dose, quantification, culture medium, animal selection and sample collection should be similar to the long-term challenge model (Hines *et al.*, 2007b). Strain selection and quantification, animal selection and sample collection should also be similar to those for the long-term challenge model.

Table 16.4. Caprine models for *Mycobacterium avium* subsp. *paratuberculosis* (*MAP*) infection.

Reference	Age at start	*MAP* strain	Dose(s)	Infection method	Study length	Clinical disease	Faecal culture	Tissue culture	Immune testing	Histopathology
Harding, 1957	NR[a]	NR[a]	NR[a]	IH IV IV/oral oral	9–16 months	NR[a]	ND[b]	17/24	ND[b]	23/24
Van Kruiningen *et al.*, 1986	2–12 days	Human isolate Linda	3.2×10^7 4.0×10^8 50 mg wet wt	Oral (milk)	Up to 310 days	No	1/4	4/4	ND[b]	4/4
Sigurðardóttir *et al.*, 1999	7–26 days	Clinical goat isolate (three passages)	10 mg dry wt 1×10 days	Oral (milk)	Up to 49 weeks	No	0/8	2/8	Skin test + ELISA + CFT +	3/8
Sigurðardóttir *et al.*, 2001	18–21 days	Clinical goat isolate	2.365 mg dry wt/3 ml four loops	Distal ileal ligation	1 h	No	ND[b]	ND[b]	ND[b]	*MAP* in M cells and leucocytes
Sigurðardóttir *et al.*, 2001	23–39 days	ATCC bovine strain 19,698	Three $\times 10^7$ per sleeve	Everted intestine sleeve	1 h	No	ND[b]	ND[b]	ND[b]	*MAP* uptake by M cells and enterocytes
Storset *et al.*, 2001	5–8 weeks	Clinical goat isolate (P173) two passages	10 mg three $\times$ 10 weeks	Oral (milk)	Up to 117 weeks	Yes	4/7	5/7	IFN-γ + LBT + ELISA +	6/7
Valheim *et al.*, 2002	5–8 weeks	Clinical goat isolate (P173) two passages	10 mg three $\times$ 10 weeks	Oral (milk)	Up to 117 weeks	Yes	ND[b]	5/7	ND[b]	6/7
Munjal *et al.*, 2005	5–8 weeks	Clinical goat isolate (tissue)	1×10^{10} seven $\times$ 2 days	Oral	Up to 270 days	Yes	1/10	2/10 (PCR)	LBT + ELISA + AGID +	5/10
Stewart *et al.*, 2006	5 months	Clinical bovine isolate (tissue/culture)	1×10^{10} 20 g wet wt	Oral	54 months	Yes	10/10	8	IFN-γ + ELISA +	NR[a]
Stewart *et al.*, 2006	10 months	Clinical sheep isolate (tissue/culture)	One $\times$ 4 weeks	Oral	35 months	Yes	9/10	1	IFN-γ + ELISA +	NR[a]

Continued

Table 16.4. Continued

Reference	Age at start	*MAP* strain	Dose(s)	Infection method	Study length	Clinical disease	Faecal culture	Tissue culture	Immune testing	Histopathology
Hines *et al.*, 2007b	3–7 days	Clinical goat isolate	1.5×10^9 four × alternate days	Oral	6–9 months	Yes	Yes and PCR	Yes and PCR	IFN-γ + Skin test + ELISA + AGID +	50/50
Kathaperumal *et al.*, 2009	5–10 days	*MAP* 66115-98	5×10^8 daily doses, for 7 days	Oral	38 weeks post-vaccination	No	0/25	15/25	IFN-γ + LBT + IL 10+ FLOW	12/25
Faisal *et al.*, 2013	8–10 weeks	*MAP* 66115-98 *LeuD* mutant vaccine	5×10^8 7 consecutive days	Oral	30 weeks post-vaccinatiion	No	Yes	Yes	IFN-γ + Skin test + LBT + IL2, 4,6, 8+ + IL10, 12, 17 + IL17, 18, 23 + TNF, TGF-b+ FLOW	17/18
Hines *et al.*, 2014	8 weeks	*MAP* ATCC-700535 (K-10 like)	1.0×10^9 two × alternate days	Oral	13 months post-challenge	Yes	Yes and PCR	Yes and PCR	Skin test + ELISA + AGID +	80/80
Shippy *et al.*, 2017	4 weeks	*MAP* JTC-1285 and mutant vaccines	1.0×10^9 three × days	Oral	12 months post-challenge	Yes	Yes	Yes	Skin test + ELISA + IFN-γ +	24/24

[a]NR = not reported.
[b]ND = not determined.
ELISA, enzyme-linked immunosorbent assay; PCR, polymerase chain reaction; TNF, tumour necrosis factor; TGF, tumour growth factor; CFU, colony-forming units; AGID, agar gel immunodiffusion test; CFT, complement fixation test; LBT, lymphocyte blastogenesis test; ATCC, American Type Culture Collection; IFN, Interferon; LBT, Lymphocyte blastogenesis test; FLOW, Flow cytometry; IL, Interleukin.

Everted intestine sleeve model

An everted intestine sleeve model, as described by Sigurðardóttir *et al.* (2005), is also useful for studying initial host–bacterial interactions, such as bacterial attachment and internalization, bacterial localization and early bacterial gene regulation, but may have limitations when evaluating host gene and early cytokine regulation in response to infection. Approximately 1.0-cm segments of small intestine are excised, everted, washed, maintained in tissue culture and bathed in a suspension of *MAP* for short periods of time. Strain selection and quantification, animal selection and sample collection should be similar to those for the long-term challenge model.

16.5 Ovine Models

16.5.1 Long-term ovine *MAP* challenge model

This model is used primarily for pathogenesis and vaccine efficacy studies but could also be used for evaluating diagnostic assays. Some sheep breeds, including the Merino may be more susceptible to *MAP* than others (see Chapter 12, this volume). While the Merino breed is preferred by members of the AMSC, any breed shown to be susceptible to *MAP* is acceptable (Hines *et al.*, 2007b).

Homogenized lymphoid tissue or intestinal mucosal scrapings from a clinically diseased animal appear to be the best source for reproducing the ovine infection (Table 16.4), but this method is not recommended by the AMSC as a standard for experimental challenge studies in sheep. Successful ovine experimental studies have used *MAP* isolates from cattle, sheep, wildlife and humans (Table 16.5). Until 2000, the majority of experimental studies in sheep did not use characterized ovine strains of *MAP* (*MAP*-S, see Chapter 6, this volume). In contrast, later strains have used tissue homogenates or low-passage *MAP*-S strains grown *in vitro* (Gwozdz *et al.*, 2000; Stewart *et al.*, 2004; Begg *et al.*, 2005). Because of the marked genetic diversity detected among ovine, caprine and bovine isolates (Motiwala *et al.*, 2004; Ghosh *et al.*, 2012), we recommend using a confirmed virulent ovine isolate. Therefore, to most closely reproduce natural ovine infections, any confirmed-virulent ovine clinical isolate that can be cultivated *in vitro* is considered acceptable (Hines *et al.*, 2007b). Begg and co-workers more recently have successfully used lyophilized, low-passage, pure culture, seed stock of *MAP* for challenge studies and recommend an international bank of master seed stock to be established, containing low- passage isolates representing the major S and C lineages of *MAP* for research (Begg *et al.*, 2010, 2018a, b).

Culture media used in published ovine studies have included Middlebrook 7H11 broth, Middlebrook 7H10 agar and BACTEC™ 460 medium with increased egg concentration. Middlebrook 7H11 broth is the medium preferred by the AMSC for inoculum preparation, but any medium capable of supporting the growth of ovine strains is acceptable. The quantification of organisms and handling and storage of the inoculum should be similar to those described for cattle (Hines *et al.*, 2007b).

The usual route of administration has been oral, using either saline or milk suspensions, but other protocols have also been successfully used (Table 16.5). Since the oral route most closely parallels natural exposure, it is generally considered the best route of administration. The challenge inoculum in previous studies has ranged from 15 to 200 mg wet weight of organisms, 2.6×10^1 to 2.6×10^{11} CFU and 0.65 to 80 g of macerated infected tissue from a clinical case (Table 16.4). The lowest doses have generally not been effective in establishing infection (Reddacliff *et al.*, 2004). Route and frequency of dosing have varied widely between studies (Table 16.5). Based on recent experimental evidence, three consecutive daily oral doses consisting of 10^9 organisms per dose (~100 mg pelleted wet weight/dose; 300 mg total) should be used to establish infection (Begg and Griffin, 2005).

The age at inoculation has ranged from day of birth to 10 months (Table 16.5). Experimental challenge at any time up to 4 months of age is considered appropriate. However, age at time of challenge will be influenced by experimental objectives. The AMSC made no recommendation as to when a paratuberculosis vaccine or other intervention should be administered. As for other species' models, positive faecal cultures 2 weeks or more post-inoculation should be considered to be due to infection. Faecal culture is less reliable in animals infected with sheep *MAP* strains (as compared with cattle strains), and expense is a major

Table 16.5. Ovine models for *Mycobacterium avium* subsp. *paratuberculosis* (*MAP*) infection.

Reference	Age at start	*MAP* strain	Dose(s)	Infection method	Study length	Clinical disease	Faecal culture	Tissue culture	Immune testing	Histopathology
Brotherston *et al.*, 1961	10 weeks	Sheep isolate (var. bovine)	1×10^7	IV	Up to 22 months	No	ND[a]	13/16	ND[a]	ND[a]
Brotherston *et al.*, 1961	1 week		1×10^8	Oral	53 weeks	No	ND[a]	10/16	ND[a]	ND[a]
			One × 3 weeks			No				
						Yes				
	7–10 days		1×10^8	Oral	53 weeks		ND[a]	9/9	ND[a]	ND[a]
			One × 8 weeks							
	3 weeks		$1 \times 10^{3-9}$	Oral	Up to 9 months		ND[a]	24/51	ND[a]	ND[a]
			One × 10 weeks							
Gilmour and Brotherston, 1962	8 months	Sheep isolate (var. bovine)	1×10^9	Oral	Up to 56 days	No	ND[a]	10/12	ND[a]	ND[a]
Nisbet *et al.*, 1962	1 week	Sheep isolate (var. bovine)	1×10^8	Oral	53 weeks	NR[b]	ND[a]	ND[a]	ND[a]	6/12
	7–10 days		1×10^8		53 weeks		ND[a]	ND[a]	ND[a]	7/9
			One × 8 weeks							
	3 weeks		$1 \times 10^{3-9}$		Up to 9 months		ND[a]	23/35	ND[a]	20/35
			One × 10 weeks							
	3 months		1×10^6		Up to 9 months		ND[a]		ND[a]	5/12
			One × 10 weeks					6/12		
Gilmour and Brotherston, 1962	3 months 10 months	Sheep isolate (var. bovine)	1×10^6 one × 10 weeks	Oral	Up to 18 weeks	NR[b]	ND[a]	6/9 9/9	Skin test +	ND[a]
Kluge *et al.*, 1968	3 weeks	Clinical bovine isolate (tissue)	50 mg dry wt 50 mg dry wt 200 mg dry wt	IV IT Oral	Up to 16 months	Yes	NR[b]	NR[b]	NR[b]	Yes
Merkal *et al.*, 1968a, b	3 weeks	Clinical bovine isolate (tissue)	50 mg dry wt 50 mg dry wt 200 mg dry wt	IV IT Oral (milk)	Up to 16 months	Yes	NR[b]	NR[b]	Skin test + CFT + AGID +	Yes

Continued

Table 16.5. Continued

Reference	Age at start	*MAP* strain	Dose(s)	Infection method	Study length	Clinical disease	Faecal culture	Tissue culture	Immune testing	Histopathology
Gilmour *et al.*, 1977	5 months	Sheep isolate (three passages)	1×10^9 one × 10 weeks	Oral	Up to 27 months	Yes	ND[a]	11/22	Skin test +	15/22
Williams *et al.*, 1983a, b	4–5 months	Clinical sheep isolate	50 mg wet wt	Oral	Up to 12 months	No	0/9	5/9	ND[a]	ND[a]
Dukes *et al.*, 1992	1 day	Saiga antelope (tissue)	0.325 g mucosa two doses	Oral	18 months	Yes			ND[a]	Yes
Juste *et al.*, 1994	3 months	Bovine isolate (three passages)	150 mg wet wt two doses	Oral	220 days	No	0/5	Yes	ELISA +	Yes
Bull *et al.*, 2013	1 day	Deer isolate strain JD88/107	1×10^8 one × 9 weeks	Oral	NR[b]	No	ND[a]	ND[a]	IFN-γ + LBT +	ND[a]
Begara-McGorum *et al.*, 1998	5–9 days	Deer isolate strain JD88/107	1×10^9 three × 2 days	Oral	Up to 41 days	No	1/8	3/8	IFN-γ – ELISA –	4/8
Gwozdz and Thompson, 2002	1–4 weeks	Clinical sheep isolate (tissue)	3.4×10^9	Oral (gastric tube)	108 weeks	Yes	ND[a]	1/10 – PCR	IFN-γ + ELISA + AGID +	1/10
			4.4×10^8		53 weeks			4/9–PCR	CFT +	3/9
Gwozdz *et al.*, 2000	1–2 months	Clinical sheep isolate (tissue)	4.4×10^8	Oral (gastric tube)	53 weeks	Yes	3/14 – PCR	10/14 – PCR	IFN-γ + ELISA +	9/14
Gwozdz *et al.*, 2001	1–4 weeks	Clinical sheep isolate (tissue)	3.4×10^9	Oral (gastric tube)	108 weeks	Yes	ND[a]	ND[a]	ND[a]	18/28
Reddacliff and Whittington, 2003	12–16 weeks	Sheep isolate (faeces)	2.6×10^1 1×10^4 1×10^8 three × 1 week 10 × 1 week	Oral		No	0/30	0/12 0/12 6/6	IFN-γ + Skin test + ELISA +	0/30
Kurade *et al.*, 2004	8–12 weeks	Clinical sheep isolate (tissue)	1×10^{10} eight × 3 days	Oral	Up to 330 days	Yes	3/20	7/20	LBT + ELISA +	20/20
Stewart *et al.*, 2004	6 months	Clinical bovine isolate (tissue/culture)	1×10^{10} 20 g wet wt	Oral	54 months	Yes	7/10	1/5	IFN-γ +	NR[b]
	10 months	Clinical sheep isolate (tissue/culture)	One × 4 weeks		35 months	Yes	5/10	1/5	ELISA +	

Continued

Table 16.5. Continued

Reference	Age at start	*MAP* strain	Dose(s)	Infection method	Study length	Clinical disease	Faecal culture	Tissue culture	Immune testing	Histopathology
Begg *et al.*, 2005	12 weeks	Clinical sheep isolates (JD3 – tissue and W – high- and low-passage culture)	1×10^9 four × 3 days	Oral	10 months	No	NR[b]	21/30	IFN-γ + LBT + ELISA +	17/30
			5×10^8	Oral	13 months	Yes	NR[b]	16/30		22/30
			1×10^9							
			One × 3 weeks							
			5×10^7	IT		No	NR[b]	8/12		7/12
			5×10^7	Oral						
			1×10^9	Oral	16 months			8/12		9/12
			Two × 1 month					3/12		1/12
Begg and Griffin, 2005	2.5 months	Clinical sheep isolate – JD3	5×10^8 one × 3 weeks	Oral	Up to 22 months	Yes	NR[b]	NR[b]	IFN-γ + LBT + ELISA +	23/30
Begg *et al.*, 2010	3–7 months	*MAP* strain Telford 9.2	Variable dose over six trials 6.9×10^5 to 1.4×10^{10}	Oral	6.5–23 months over 4-year period	Yes	Yes	Yes	IFN-γ + ELISA +	128/128

[a]ND = not determined.
[b]NR = not reported.
ELISA, enzyme-linked immunosorbent assay; PCR, polymerase chain reaction; IT, intrathecal; IV, intravenous; AGID, agar gel immunodiffusion test; CFT, complement fixation test; LBT, lymphocyte blastogenesis test; IFN, Interferon.

consideration in large ovine studies, particularly in the field; none the less, all animals should have periodic faecal cultures using validated methods.

Animals to be included in challenge studies should be validated as infection-free based upon the infection-free status of the foundation flock. Lambs for experimental studies should only be selected from closed flocks in which all adults have been negative on ELISA and faecal culture for at least 1 year and paratuberculosis vaccination is not practised. All lambs should receive adequate quantities of colostrum.

A combination of necropsy with histopathology and bacterial culture to determine levels of tissue colonization is recommended by the AMSC as the minimal requirement to determine infection status of experimental animals (Hines *et al.*, 2007b). The lesion grading system used for gross and histopathology findings should have a sufficient range in values to allow for statistical analysis (Hines *et al.*, 2007a, Hines *et al.*, 2007b; Juste *et al.*, 1994). Animals should be observed until clinical signs develop in a proportion of the group. The type, quantity and processing of samples collected will vary with the purpose, number of animals, goals and cost of the study. All samples should be collected, handled and processed in the same way to ensure uniformity. Ovine strains apparently grow better in liquid medium, and BACTEC™ 460 was suggested by the AMSC to be the best system by which to recover ovine strains at that time, but no single standard culture medium or method for culture of faeces or tissues was recommended. However, in 2012 the BACTEC™ 460 system was discontinued by its manufacturer and is being replaced by other non-radiometric types of liquid culture. Decontamination and culture should be performed similarly to those previously described for cattle, providing, at a minimum, semi-quantitative results (Hines *et al.*, 2007b).

16.5.2 Short-term ovine *MAP* challenge models

In general, these models are most useful for short-term study of host–pathogen interactions. Many parameters should be the same as for long-term challenge, including strain, inoculum preparation and quantification, storage, animal selection criteria and quality control issues.

Ileal cannulation model

An ileal cannulation model similar to that previously described for cattle has not been investigated in sheep. This model is considered to have only limited practical use in sheep. However, Hein and co-workers have developed a model for investigating enteric diseases where the afferent lymphatics draining the small intestine are cannulated (Hein *et al.*, 2004). This model allows longitudinal sampling of immune cells and fluids draining the small intestine, which may be useful in evaluating the immunology associated with paratuberculosis in sheep.

Intestinal loop and everted intestine sleeve models

No study describing an intestinal loop or everted intestinal sleeve model for sheep was found in our literature review. Models similar to that described previously in goats should be equally applicable in sheep for studying the initial bacterial–host interactions *in vivo*.

16.6 Cervid Models

16.6.1 Long-term cervid *MAP* challenge model

No published studies were found in our literature review where intestinal loop, everted intestine or other short-term models have been used in deer. Models similar to that described previously for goats should be equally applicable for deer in studying the initial bacterial–host interactions *in vivo*. In contrast, there is some literature on the long-term cervid challenge model, primarily used for pathogenesis and vaccine efficacy studies. This model may also be useful for evaluation of diagnostic assays and heritable resistance.

High levels of heritable resistance and susceptibility to mycobacterial infection are documented in deer (Mackintosh *et al.*, 2000). Deer appear to be a useful host for disclosing diagnostic markers by which to monitor infection (Griffin *et al.*, 2005), protective immunity or resistance to infection. The majority of studies

Table 16.6. Cervid and/or exotic ruminant species models for *Mycobacterium avium* subsp. *paratuberculosis* (*MAP*) infection.

Reference	Species	Age at start	*MAP* strain	Dose(s)	Infection method	Study length	Clinical disease	Faecal culture	Tissue culture	Immune testing	Histopathology[c]
Williams *et al.*, 1983a	Bighorn × mouflon	4–5 months	Clinical bighorn sheep isolate	50 mg wet wt	Oral	6 or 12 months	No	NR[a]	9/9	ND[b]	8/9
	Mule deer						Yes		8/8		8/8
	White-tailed deer						Yes		2/2		2/2
	Elk						No		8/8		8/8
Williams *et al.*, 1983b	Bighorn × mouflon	4–5 months	Clinical bighorn sheep isolate	50 mg wet wt	Oral	6 or 12 months	No	9/9	9/9	ND[b]	NR[a]
	Mule deer						Yes	2/8	8/8		
	White-tailed deer						Yes	0/2	2/2		
	Elk						No	0/8	8/8		
Mackintosh *et al.*, 2003	Red deer	4 months	Clinical deer isolate (bovine var.)	NR one × 4 days	Oral	12 months	Yes	NR[a]	NR[a]	Skin test + LBT + ELISA +	39/43
Stabel *et al.*, 2003	Bison	4 months	Clinical isolate (bovine and bison)	1×10^9 five doses	Oral (gastric tube)	6 months	No	2/6	6/6	IFN-γ + ELISA –	0/6
Mackintosh *et al.*, 2005	Red deer	4 months	Clinical deer isolate (tissue; bovine var.)	1×10^9	Oral	12 months	Yes	NR[a]	NR[a]	LBT +	42/42
				One × 4 days						ELISA +	
O'Brien *et al.*, 2006	Red deer	4 months	Clinical red deer isolate (tissue; bovine var.)	1×10^9	Oral	Up to 44 weeks	NR[a]	NR[a]	40/64	IFN-γ +	Lesions apparent
			Clinical sheep isolate (JD3)	1×10^7						LBT +	
				1×10^3						ELISA +	
				One × 4 days							
				1×10^7			NR[a]		11/16	IFN-γ +/–	Lesions apparent

Continued

Table 16.6. Continued

Reference	Species	Age at start	*MAP* strain	Dose(s)	Infection method	Study length	Clinical disease	Faecal culture	Tissue culture	Immune testing	Histopathology[c]
				One × 4 days						LBT +ELISA –	
Mackintosh *et al.*, 2007	Red deer	4 months	Clinical deer isolate (tissue; bovine var.) Sheep tissue isolate Ovine strain	$10^3 \times 4$ (LB) $10^7 \times 4$ (MB) $10^9 \times 4$ (HB) $10^7 \times 4$ (MO)	Oral	12 months	NR[a]	NR[a]	8/16 LB 16/16 MB 16/16 HB 8/16 MO	LBT + IFN-γ ELISA + (IgG1)	NR[a]
Palmer *et al.*, 2007	White-tailed deer	9 days	*MAP* K-10 (NADC)	1.87×10^{10}	Oral	390–658 days	Yes	Yes and PCR	Yes and PCR	IFN-γ + ELISA +	Lesions apparent 3/4
Robinson *et al.*, 2008	Red deer	6.5 months	Deer tissue isolate (bovine var.)	1×10^9 four daily doses	Oral	56 weeks	Yes	NR	NR	IFN-γ + ELISA + FLOW	Lesions apparent
Mackintosh *et al.*, 2008	Red deer	4–4.5 months	Deer tissue isolate (bovine var.)	1×10^8 four daily doses	Oral	14 months	No	21–33%	89–97%	ELISA + Skin test +	Lesions apparent (~20%)

[a]NR = not reported.
[b]ND = not determined.
[c]in the cervid model histopathology following necropsy may be used to stratify disease severity.
ELISA, enzyme-linked immunosorbent assay; PCR, polymerase chain reaction; LBT, lymphocyte blastogenesis test; FLOW, flow cytometry; IFN, Interferon; LB, Group receiving low doses of a bovine strain of MAP; MB, Group receiving medium doses of a bovine strain of MAP; HB, Group receiving high doses of a bovine strain of MAP; MO, Group receiving medium doses of an ovine strain of MAP.

performed in cervids (Table 16.6) have involved red deer (*Cervus elaphus*), which seem to be more naturally susceptible (Mackintosh *et al*., 2003) than other cervids (Williams *et al*., 1983a, b). However, white-tailed deer have also been shown to be an effective model (Palmer *et al*., 2007).

Experimentally infected deer have a broad spectrum of responses, ranging from limited infection to extremes of pathology and clinical disease. The range of pathology allows clear stratification and analysis of the impact of interventions (Mackintosh *et al*., 2007). The predominance of *MAP*-C strains in naturally infected deer suggests that cervids may provide an alternative experimental challenge model for paratuberculosis in cattle. Red deer may be naturally infected with either *MAP*-C or *MAP*-S strains (de Lisle *et al*., 2006), but they appear more susceptible and develop more severe disease with *MAP*-C strains (O'Brien *et al*., 2006). Clinical signs develop in 25–30% of red deer between 4 and 12 months post-bovine strain challenge (Mackintosh *et al*., 2005). Red deer are recommended as the model of choice for cervid challenge studies.

Strain, dose, route of inoculation and infection or disease endpoints used in cattle and goat models appear to apply equally well to cervids (Hines *et al*., 2007b). In addition, there is evidence that most of the bovine immunological reagents are useful in studies with red deer, with the known exception of TGFβ (Frank Griffin, Dunedin, 2008, personal communication). A dose of 10^3 CFU of a *MAP*-C strain produces equivalent infection and pathology to 10^6 CFU of a *MAP*-S strain (O'Brien *et al*., 2006). Thus, *MAP*-C strains are preferred, and all challenge parameters in red deer should be the same as those for the bovine and caprine models. A strain with genotype similar to the *MAP*-C strain, *MAP* K-10, at the bovine dose (10^9 organisms, 100 mg) given on 2 consecutive days was recommended as most appropriate for deer challenge studies by the AMSC (Hines *et al*., 2007b). Age of administration will vary depending on the goals of the study. Deer should be challenged at less than 3 months of age. The relatively early onset of pathology and disease in deer suggest that this model may be cost-effective and informative for study of *MAP* infection. Access to deer bloodlines with resistant or susceptible phenotypes for *M. bovis* infection may be valuable in exploring candidate genes that contribute to heritable resistance to other mycobacterial infections.

References

Adams, J., Follett, D., Hamilton, H. and Czuprynski, C. (1993) Effects of administration of anti-CD4 and anti-CD8 monoclonal antibodies on Mycobacterium paratuberculosis infection in intragastrically challenged mice. *Immunology Letters* 35(2), 183–189. DOI: 10.1016/0165-2478(93)90089-K.

Allen, A.J., Barrington, G.M., Hamilton, M.J., Stabel, J.R. and Robbe-Austerman, S. (2005) Development of a calf ileal cannulation model to facilitate testing of existing and candidate vaccines for *M. avium* subsp. *paratuberculosis*. In: Manning, E.J.B. and Nielsen, S.S. (eds) *Proceedings of the 8th International Colloquium for Paratuberculosis*. Copenhagen, Denmark, p. 148.

Allen, A.J., Park, K.T., Barrington, G.M., Lahmers, K.K., Hamilton, M.J. *et al*. (2009) Development of a bovine ileal cannulation model to study the immune response and mechanisms of pathogenesis of paratuberculosis. *Clinical and Vaccine Immunology* 16(4), 453–463. DOI: 10.1128/CVI.00347-08.

Arrazuria, R., Molina, E., Mateo-Abad, M., Arostegui, I., Garrido, J.M. *et al*. (2015) Effect of various dietary regimens on oral challenge with *Mycobacterium avium* subsp. *paratuberculosis* in a rabbit model. *Research in Veterinary Science* 101, 80–83. DOI: 10.1016/j.rvsc.2015.06.006.

Arrazuria, R., Elguezabal, N., Juste, R.A., Derakhshani, H. and Khafipour, E. (2016a) *Mycobacterium avium* subspecies *paratuberculosis* infection modifies gut microbiota under different dietary conditions in a rabbit model. *Frontiers in Microbiology* 7(130), 446. DOI: 10.3389/fmicb.2016.00446.

Arrazuria, R., Molina, E., Garrido, J.M., Pérez, V., Juste, R.A. *et al*. (2016b) Vaccination sequence effects on immunological response and tissue bacterial burden in paratuberculosis infection in a rabbit model. *Veterinary Research* 47(1), 77. DOI: 10.1186/s13567-016-0360-y.

Bannantine, J.P. and Stabel, J.R. (2001) Identification of two *Mycobacterium avium* subspecies *paratuberculosis* gene products differentially recognised by sera from rabbits immunised with live mycobacteria but not heat-killed mycobacteria. *Journal of Medical Microbiology* 50(9), 795–804. DOI: 10.1099/0022-1317-50-9-795.

Bannantine, J.P., Everman, J.L., Rose, S.J., Babrak, L., Katani, R. *et al.* (2014) Evaluation of eight live attenuated vaccine candidates for protection against challenge with virulent *Mycobacterium avium* subspecies *paratuberculosis* in mice. *Frontiers in Cellular and Infection Microbiology* 4(11), 88. DOI: 10.3389/fcimb.2014.00088.

Basagoudanavar, S.H., Goswami, P.P. and Tiwari, V. (2006) Cellular immune responses to 35 kDa recombinant antigen of *Mycobacterium avium paratuberculosis*. *Veterinary Research Communications* 30(4), 357–367. DOI: 10.1007/s11259-006-3253-0.

Beard, P.M., Rhind, S.M., Buxton, D., Daniels, M.J., Henderson, D. *et al.* (2001a) Natural paratuberculosis infection in rabbits in Scotland. *Journal of Comparative Pathology* 124(4), 290–299. DOI: 10.1053/jcpa.2001.0466.

Beard, P.M., Stevenson, K., Pirie, A., Rudge, K., Buxton, D. *et al.* (2001b) Experimental paratuberculosis in calves following inoculation with a rabbit isolate of *Mycobacterium avium* subsp. *paratuberculosis*. *Journal of Clinical Microbiology* 39(9), 3080–3084. DOI: 10.1128/JCM.39.9.3080-3084.2001.

Begara-McGorum, I., Wildblood, L.A., Clarke, C.J., Connor, K.M., Stevenson, K. *et al.* (1998) Early immunopathological events in experimental ovine paratuberculosis. *Veterinary Immunology and Immunopathology* 63(3), 265–287. DOI: 10.1016/S0165-2427(98)00107-X.

Begg, D.J. and Griffin, J.F.T. (2005) Vaccination of sheep against *M. paratuberculosis*: immune parameters and protective efficacy. *Vaccine* 23(42), 4999–5008. DOI: 10.1016/j.vaccine.2005.05.031.

Begg, D.J. and Whittington, R.J. (2008) Experimental animal infection models for Johne's disease, an infectious enteropathy caused by *Mycobacterium avium* subsp. *paratuberculosis*. *The Veterinary Journal* 176(2), 129–145. DOI: 10.1016/j.tvjl.2007.02.022.

Begg, D.J., O'Brien, R., Mackintosh, C.G. and Griffin, J.F.T. (2005) Experimental infection model for Johne's disease in sheep. *Infection and Immunity* 73(9), 5603–5611. DOI: 10.1128/IAI.73.9.5603-5611.2005.

Begg, D.J., de Silva, K., Di Fiore, L., Taylor, D.L., Bower, K. *et al.* (2010) Experimental infection model for Johne's disease using a lyophilised, pure culture, seedstock of *Mycobacterium avium* subspecies *paratuberculosis*. *Veterinary Microbiology* 141(3–4), 301–311. DOI: 10.1016/j.vetmic.2009.09.007.

Begg, D.J., de Silva, K., Di Fiore, L., Taylor, D.L., Bower, K. *et al.* (2018a) Corrigendum to 'Experimental infection model for Johne's disease using a lyophilised, pure culture, seedstock of *Mycobacterium avium* subspecies *paratuberculosis*' [Veterinary Microbiology 141 (2010) 301-311]. *Veterinary Microbiology* 224, 119. DOI: 10.1016/j.vetmic.2018.08.023.

Begg, D.J., Plain, K.M., de Silva, K., Gurung, R., Gunn, A. *et al.* (2018b) Immunopathological changes and apparent recovery from infection revealed in cattle in an experimental model of Johne's disease using a lyophilised culture of *Mycobacterium avium* subspecies *paratuberculosis*. *Veterinary Microbiology* 219, 53–62. DOI: 10.1016/j.vetmic.2018.03.029.

Brotherston, J.G., Gilmour, N.J.L. and Samuel, J.M.C.A. (1961) Quantitative studies of *Mycobacterium johnei* in the tissues of sheep. *Journal of Comparative Pathology and Therapeutics* 71, 286–299. DOI: 10.1016/S0368-1742(61)80035-0.

Bull, T.J., Gilbert, S.C., Sridhar, S., Linedale, R., Dierkes, N. *et al.* (2007) A novel multi-antigen virally vectored vaccine against *Mycobacterium avium* subspecies *paratuberculosis*. *PLoS ONE* 2(11), e1229. DOI: 10.1371/journal.pone.0001229.

Bull, T.J., Schock, A., Sharp, J.M., Greene, M., McKendrick, I.J. *et al.* (2013) Genomic variations associated with attenuation in Mycobacterium avium subsp. paratuberculosis vaccine strains. *BMC Microbiology* 13, 11. DOI: 10.1186/1471-2180-13-11.

Bull, T.J., Vrettou, C., Linedale, R., McGuinnes, C., Strain, S. *et al.* (2014) Immunity, safety and protection of an Adenovirus 5 prime - Modified Vaccinia virus Ankara boost subunit vaccine against *Mycobacterium avium* subspecies *paratuberculosis* infection in calves. *Veterinary Research* 45(1), 112. DOI: 10.1186/s13567-014-0112-9.

Chandler, R.L. (1961) Infection of laboratory animals with Mycobacterium johnei. I. infection in Swiss white mice and its modification by suramin and cortisone. *Journal of Comparative Pathology* 71, 118–130. DOI: 10.1016/s0368-1742(61)80015-5.

Chandler, R.L. (1962) Infection of laboratory animals with Mycobacterium johnei. V. Further studies on the comparative susceptibility of C57 black mice. *Journal of Comparative Pathology* 72, 198–213. DOI: 10.1016/s0368-1742(62)80024-1.

Chen, L.-H., Kathaperumal, K., Huang, C.-J., McDonough, S.P., Stehman, S. *et al.* (2008) Immune responses in mice to *Mycobacterium avium* subsp. *paratuberculosis* following vaccination with a novel 74F recombinant polyprotein. *Vaccine* 26(9), 1253–1262. DOI: 10.1016/j.vaccine.2007.12.014.

Chen, J.-W., Faisal, S.M., Chandra, S., McDonough, S.P., Moreira, M.A.S. *et al.* (2012) Immunogenicity and protective efficacy of the Mycobacterium avium subsp. paratuberculosis attenuated mutants against challenge in a mouse model. *Vaccine* 30(19), 3015–3025. DOI: 10.1016/j.vaccine.2011.11.029.

Cooney, M.A., Steele, J.L., Steinberg, H. and Talaat, A.M. (2014) A murine oral model for *Mycobacterium avium* subsp. *paratuberculosis* infection and immunomodulation with *Lactobacillus casei* ATCC 334. *Frontiers in Cellular and Infection Microbiology* 4, 11. DOI: 10.3389/fcimb.2014.00011.

de Lisle, G.W., Cannon, M.C., Yates, G.F. and Collins, D.M. (2006) Use of a polymerase chain reaction to subtype *Mycobacterium avium* subspecies *paratuberculosis*, an increasingly important pathogen from farmed deer in New Zealand. *New Zealand Veterinary Journal* 54(4), 195–197. DOI: 10.1080/00480169.2006.36695.

Doyle, T.M. (1953) Susceptibility to Johne's disease in relation to age. *The Veterinary Record* 65, 363–365.

Doyle, T.M. (1956) Johne's disease. *The Veterinary Record* 68, 869–878.

Dukes, T.W., Glover, G.J., Brooks, B.W., Duncan, J.R. and Swendrowski, M. (1992) Paratuberculosis in saiga antelope (*Saiga tatarica*) and experimental transmission to domestic sheep. *Journal of Wildlife Diseases* 28(2), 161–170. DOI: 10.7589/0090-3558-28.2.161.

Faisal, S.M., Chen, J.-W., Yan, F., Chen, T.-T., Useh, N.M. *et al.* (2013) Evaluation of a *Mycobacterium avium* subsp. *paratuberculosis* leuD mutant as a vaccine candidate against challenge in a caprine model. *Clinical and Vaccine Immunology* 20(4), 572–581. DOI: 10.1128/CVI.00653-12.

Ferwerda, G., Kullberg, B.J., de Jong, D.J., Girardin, S.E., Langenberg, D.M.L. *et al.* (2007) *Mycobacterium paratuberculosis* is recognized by Toll-like receptors and NOD2. *Journal of Leukocyte Biology* 82(4), 1011–1018. DOI: 10.1189/jlb.0307147.

Ghadiali, A.H., Strother, M., Naser, S.A., Manning, E.J.B. and Sreevatsan, S. (2004) *Mycobacterium avium* subsp. *paratuberculosis* strains isolated from Crohn's disease patients and animal species exhibit similar polymorphic locus patterns. *Journal of Clinical Microbiology* 42(11), 5345–5348. DOI: 10.1128/JCM.42.11.5345-5348.2004.

Ghosh, P., Hsu, C., Alyamani, E.J., Shehata, M.M., Al-Dubaib, M.A. *et al.* (2012) Genome-wide analysis of the emerging infection with *Mycobacterium avium* subspecies *paratuberculosis* in the Arabian camels (*Camelus dromedarius*). *PLoS ONE* 7(2), e31947. DOI: 10.1371/journal.pone.0031947.

Ghosh, P., Wu, C.-W. and Talaat, A.M. (2013) Key role for the alternative sigma factor, SigH, in the intracellular life of *Mycobacterium avium* subsp. *paratuberculosis* during macrophage stress. *Infection and Immunity* 81(6), 2242–2257. DOI: 10.1128/IAI.01273-12.

Ghosh, P., Steinberg, H. and Talaat, A.M. (2014) Virulence and immunity orchestrated by the global gene regulator *sigL* in *Mycobacterium avium* subsp. *paratuberculosis*. *Infection and Immunity* 82(7), 3066–3075. DOI: 10.1128/IAI.00001-14.

Ghosh, P., Shippy, D.C. and Talaat, A.M. (2015) Superior protection elicited by live-attenuated vaccines in the murine model of paratuberculosis. *Vaccine* 33(51), 7262–7270. DOI: 10.1016/j.vaccine.2015.10.116.

Gilmour, N.J.L. and Brotherston, J.G. (1962) Quantitative studies of *Mycobacterium johnei* in the tissues of sheep. *Journal of Comparative Pathology and Therapeutics* 72, 165–169. DOI: 10.1016/S0368-1742(62)80020-4.

Gilmour, N.J.L., Nisbet, D.I. and Brotherston, J.G. (1965) Experimental oral infection of calves with *Mycobacterium johnei*. *Journal of Comparative Pathology* 75(3), 281–286. DOI: 10.1016/0021-9975(65)90033-2.

Gilmour, N.J.L., Angus, K.W. and Mitchell, B. (1977) Intestinal infection and host response to oral administration of *Mycobacterium johnei* in sheep. *Veterinary Microbiology* 2(3), 223–235. DOI: 10.1016/0378-1135(77)90015-3.

González, J., Geijo, M.V., García-Pariente, C., Verna, A., Corpa, J.M. *et al.* (2005) Histopathological classification of lesions associated with natural paratuberculosis infection in cattle. *Journal of Comparative Pathology* 133(2–3), 184–196. DOI: 10.1016/j.jcpa.2005.04.007.

Griffin, J.F.T., Spittle, E., Rodgers, C.R., Liggett, S., Cooper, M. *et al.* (2005) Immunoglobulin G1 enzyme-linked immunosorbent assay for diagnosis of Johne's disease in red deer (*Cervus elaphus*). *Clinical and Vaccine Immunology* 12(12), 1401–1409. DOI: 10.1128/CDLI.12.12.1401-1409.2005.

Gwozdz, J.M. and Thompson, K.G. (2002) Antigen-Induced production of interferon-gamma in samples of peripheral lymph nodes from sheep experimentally inoculated with *Mycobacterium avium* subsp. *paratuberculosis*. *Veterinary Microbiology* 84(3), 243–252. DOI: 10.1016/S0378-1135(01)00462-X.

Gwozdz, J.M., Thompson, K.G., Manktelow, B.W., Murray, A. and West, D.M. (2000) Vaccination against paratuberculosis of lambs already infected experimentally with *Mycobacterium avium* subspecies *paratuberculosis*. *Australian Veterinary Journal* 78(8), 560–566. DOI: 10.1111/j.1751-0813.2000.tb11902.x.

Gwozdz, J.M., Thompson, K.G. and Manktelow, B.W. (2001) Lymphocytic neuritis of the ileum in sheep with naturally acquired and experimental paratuberculosis. *Journal of Comparative Pathology* 124(4), 317–320. DOI: 10.1053/jcpa.2001.0468.

Hagan, W.A. (1938) Age as a factor in susceptibility to Johne's disease. *The Cornell Veterinarian* 28, 34–40.

Hamilton, H.L., Cooley, A.J., Adams, J.L. and Czuprynski, C.J. (1991) *Mycobacterium paratuberculosis* monoassociated nude mice as a paratuberculosis model. *Veterinary Pathology* 28(2), 146–155. DOI: 10.1177/030098589102800207.

Harding, H.P. (1957) Experimental infection with *Mycobacterium johnei*. *Journal of Comparative Pathology and Therapeutics* 67, 37–IN11. DOI: 10.1016/S0368-1742(57)80005-8.

Hein, W.R., Barber, T., Cole, S.-A., Morrison, L. and Pernthaner, A. (2004) Long-term collection and characterization of afferent lymph from the ovine small intestine. *Journal of Immunological Methods* 293(1–2), 153–168. DOI: 10.1016/j.jim.2004.07.008.

Hines, M.E., Stiver, S., Giri, D., Whittington, L., Watson, C. *et al.* (2007a) Efficacy of spheroplastic and cell-wall competent vaccines for *Mycobacterium avium* subsp. *paratuberculosis* in experimentally-challenged baby goats. *Veterinary Microbiology* 120(3–4), 261–283. DOI: 10.1016/j.vetmic.2006.10.030.

Hines, M.E., Stabel, J.R., Sweeney, R.W., Griffin, F., Talaat, A.M. *et al.* (2007b) Experimental challenge models for Johne's disease: a review and proposed international guidelines. *Veterinary Microbiology* 122(3-4), 197–222. DOI: 10.1016/j.vetmic.2007.03.009.

Hines, M.E., Turnquist, S.E., Ilha, M.R.S., Rajeev, S., Jones, A.L. *et al.* (2014) Evaluation of novel oral vaccine candidates and validation of a caprine model of Johne's disease. *Frontiers in Cellular and Infection Microbiology* 4, 26. DOI: 10.3389/fcimb.2014.00026.

Huntley, J.F., Stabel, J.R., Paustian, M.L., Reinhardt, T.A. and Bannantine, J.P. (2005) Expression library immunization confers protection against *Mycobacterium avium* subsp. *paratuberculosis* infection. *Infection and Immunity* 73(10), 6877–6884. DOI: 10.1128/IAI.73.10.6877-6884.2005.

Judge, J., Kyriazakis, I., Greig, A., Allcroft, D.J. and Hutchings, M.R. (2005) Clustering of *Mycobacterium avium* subsp. *paratuberculosis* in rabbits and the environment: how hot is a hot spot? *Applied and Environmental Microbiology* 71(10), 6033–6038. DOI: 10.1128/AEM.71.10.6033-6038.2005.

Juste, R.A., García Marín, J.F., Peris, B., Sáez de Ocáriz, C. and Badiola, J.J. (1994) Experimental infection of vaccinated and non-vaccinated lambs with *Mycobacterium paratuberculosis*. *Journal of Comparative Pathology* 110(2), 185–194. DOI: 10.1016/S0021-9975(08)80189-2.

Kathaperumal, K., Park, S.-U., McDonough, S., Stehman, S., Akey, B. *et al.* (2008) Vaccination with recombinant *Mycobacterium avium* subsp. *paratuberculosis* proteins induces differential immune responses and protects calves against infection by oral challenge. *Vaccine* 26(13), 1652–1663. DOI: 10.1016/j.vaccine.2008.01.015.

Kathaperumal, K., Kumanan, V., McDonough, S., Chen, L.-H., Park, S.-U. *et al.* (2009) Evaluation of immune responses and protective efficacy in a goat model following immunization with a cocktail of recombinant antigens and a polyprotein of *Mycobacterium avium* subsp. *paratuberculosis*. *Vaccine* 27(1), 123–135. DOI: 10.1016/j.vaccine.2008.10.019.

Kaufmann, S.H.E. (2011) Fact and fiction in tuberculosis vaccine research: 10 years later. *The Lancet Infectious Diseases* 11(8), 633–640. DOI: 10.1016/S1473-3099(11)70146-3.

Khare, S., Nunes, J.S., Figueiredo, J.F., Lawhon, S.D., Rossetti, C.A. *et al.* (2009) Early phase morphological lesions and transcriptional responses of bovine ileum infected with *Mycobacterium avium* subsp. *paratuberculosis*. *Veterinary Pathology* 46(4), 717–728. DOI: 10.1354/vp.08-VP-0187-G-FL.

Kluge, J.P., Merkal, R.S., Monlux, W.S., Larsen, A.B., Kopecky, K.E. *et al.* (1968) Experimental paratuberculosis in sheep after oral, intratracheal, or intravenous inoculation lesions and demonstration of etiologic agent. *American Journal of Veterinary Research* 29, 953–962.

Koc, A., Bargen, I., Suwandi, A., Roderfeld, M., Tschuschner, A. *et al.* (2014) Systemic and mucosal immune reactivity upon Mycobacterium avium ssp. paratuberculosis infection in mice. *PLoS ONE* 9(4), e94624.

Koets, A.P., Adugna, G., Janss, L.L.G., van Weering, H.J., Kalis, C.H.J. *et al.* (2000) Genetic variation of susceptibility to *Mycobacterium avium* subsp. *paratuberculosis* infection in dairy cattle. *Journal of Dairy Science* 83(11), 2702–2708. DOI: 10.3168/jds.S0022-0302(00)75164-2.

Koo, H.C., Park, Y.H., Hamilton, M.J., Barrington, G.M., Davies, C.J. *et al.* (2004) Analysis of the immune response to *Mycobacterium avium* subsp. *paratuberculosis* in experimentally infected calves. *Infection and Immunity* 72(12), 6870–6883. DOI: 10.1128/IAI.72.12.6870-6883.2004.

Krishnappa, G., Jagannath, C. and Rao, B.U. (1989) The specificity of antibody response in experimental and natural bovine paratuberculosis studied by crossed immunoelectrophoresis with intermediate gel. *Veterinary Microbiology* 21(1), 67–78. DOI: 10.1016/0378-1135(89)90019-9.

Kurade, N.P., Tripathi, B.N., Rajukumar, K. and Parihar, N.S. (2004) Sequential development of histologic lesions and their relationship with bacterial isolation, fecal shedding, and immune responses during progressive stages of experimental infection of lambs with Mycobacterium avium subsp. paratuberculosis. *Veterinary Pathology* 41(4), 378–387. DOI: 10.1354/vp.41-4-378.

Larsen, A.B., Merkal, R.S. and Moon, H.W. (1974) Evaluation of a paratuberculosis vaccine given to calves before infection. *American Journal of Veterinary Research* 35, 367–369.

Larsen, A.B., Merkal, R.S. and Cutlip, R.C. (1975) Age of cattle as related to resistance to infection with *Mycobacterium paratuberculosis*. *American Journal of Veterinary Research* 36, 255–257.

Larsen, A.B., Moyle, A. I. and Himes, E.M. (1978) Experimental vaccination of cattle against paratuberculosis (Johne's disease) with killed bacterial vaccines: a controlled field study. *American Journal of Veterinary Research* 39, 65–69.

Mackintosh, C.G., Qureshi, T., Waldrup, K., Labes, R.E., Dodds, K.G. *et al.* (2000) Genetic resistance to experimental infection with *Mycobacterium bovis* in red deer (*Cervus elaphus*). *Infection and Immunity* 68(3), 1620–1625. DOI: 10.1128/IAI.68.3.1620-1625.2000.

Mackintosh, C., Labes, R., De Lisle, G. and Griffin, J. (2003) Experimental challenge trial of novel Johne's disease vaccine. *Proceedings of the Deer Branch of the New Zealand Veterinary Association*, The Deer Branch New Zealand Veterinary Association.

Mackintosh, C.G., Labes, R.E. and Griffin, J.F.T. (2005) The effect of Johne's vaccination on tuberculin testing in farmed red deer (*Cervus elaphus*). *New Zealand Veterinary Journal* 53(4), 216–222. DOI: 10.1080/00480169.2005.36549.

Mackintosh, C.G., Labes, R.E., Clark, R.G., de Lisle, G.W. and Griffin, J.F.T. (2007) Experimental infections in young red deer (*Cervus elaphus*) with a bovine and an ovine strain of *Mycobacterium avium* subsp *paratuberculosis*. *New Zealand Veterinary Journal* 55(1), 23–29. DOI: 10.1080/00480169.2007.36731.

Mackintosh, C.G., Labes, R.E., Thompson, B.R., Clark, R.G., de Lisle, G.W. *et al.* (2008) Efficacy, immune responses and side-effects of vaccines against Johne's disease in young red deer (*Cervus elaphus*) experimentally challenged with *Mycobacterium avium* subsp *paratuberculosis*. *New Zealand Veterinary Journal* 56(1), 1–9. DOI: 10.1080/00480169.2008.36797.

Malamo, M., Okazaki, K., Sakoda, Y. and Kida, H. (2007) Carboxyl terminus of the 34 kDa protein of *Mycobacterium paratuberculosis* shares homologous B-cell epitopes with *Mycobacterium avium* and *Mycobacterium intracellulare*. *The Veterinary Record* 161, 853–857.

McDonald, W.L., Ridge, S.E., Hope, A.F. and Condron, R.J. (1999) Evaluation of diagnostic tests for Johne's disease in young cattle. *Australian Veterinary Journal* 77(2), 113–119. DOI: 10.1111/j.1751-0813.1999.tb11679.x.

Meissner, T., Eckelt, E., Basler, T., Meens, J., Heinzmann, J. *et al.* (2014) The Mycobacterium avium ssp. paratuberculosis specific mptD gene is required for maintenance of the metabolic homeostasis necessary for full virulence in mouse infections. *Frontiers in cellular and infection microbiology* 4, 110. DOI: 10.3389/fcimb.2014.00110.

Merkal, R.S., Kluge, J.P., Monlux, W.S., Larsen, A.B., Kopecky, K.E. *et al.* (1968a) Experimental paratuberculosis in sheep after oral, intratracheal, or intravenous inoculation: histochemical localization of hydrolase activities. *American Journal of Veterinary Research* 29, 985–994.

Merkal, R.S., Larsen, A.B., Kopecky, K.E., Kluge, J.P., Monlux, W.S. *et al.* (1968b) Experimental paratuberculosis in sheep after oral, intratracheal, or intravenous inoculation: serologic and intradermal tests. *American Journal of Veterinary Research* 29, 963–969.

Mokresh, A.H. and Butler, D.G. (1990) Granulomatous enteritis following oral inoculation of newborn rabbits with *Mycobacterium paratuberculosis* of bovine origin. *Canadian Journal of Veterinary Research* 54, 313–319.

Mobius, P., Luyven, G., Hotzel, H. and Kohler, H. (2008) High genetic diversity among *Mycobacterium avium* subsp. *paratuberculosis* strains from German cattle herds shown by combination of IS*900* restriction fragment length polymorphism analysis and mycobacterial interspersed repetitive unit-variable-number tandem-repeat typing. *Journal of Clinical Microbiology* 46(3), 972–981. DOI: 10.1128/JCM.01801-07.

Mokresh, A.H., Czuprynski, C.J. and Butler, D.G. (1989) A rabbit model for study of *Mycobacterium paratuberculosis* infection. *Infection and Immunity* 57(12), 3798–3807. DOI: 10.1128/IAI.57.12.3798-3807.1989.

Momotani, E., Whipple, D.L., Thiermann, A.B. and Cheville, N.F. (1988) Role of M cells and macrophages in the entrance of *Mycobacterium paratuberculosis* into domes of ileal Peyer's patches in calves. *Veterinary Pathology* 25(2), 131–137. DOI: 10.1177/030098588802500205.

Motiwala, A.S., Amonsin, A., Strother, M., Manning, E.J.B., Kapur, V. *et al.* (2004) Molecular epidemiology of *Mycobacterium avium* subsp. *paratuberculosis* isolates recovered from wild animal species. *Journal of Clinical Microbiology* 42(4), 1703–1712. DOI: 10.1128/JCM.42.4.1703-1712.2004.

Mullerad, J., Hovav, A.-H., Fishman, Y., Barletta, R.G. and Bercovier, H. (2002a) Antigenicity of *Mycobacterium paratuberculosis* superoxide dismutase in mice. *FEMS Immunology & Medical Microbiology* 34(1), 81–88. DOI: 10.1111/j.1574-695X.2002.tb00606.x.

Mullerad, J., Michal, I., Fishman, Y., Hovav, A.-H., Barletta, R.G. *et al.* (2002b) The immunogenicity of *Mycobacterium paratuberculosis* 85B antigen. *Medical Microbiology and Immunology* 190(4), 179–187. DOI: 10.1007/s00430-001-0104-z.

Mullerad, J., Hovav, A.-H., Nahary, R., Fishman, Y. and Bercovier, H. (2003) Immunogenicity of a 16.7 kDa *Mycobacterium paratuberculosis* antigen. *Microbial Pathogenesis* 34(2), 81–90. DOI: 10.1016/S0882-4010(02)00209-7.

Munjal, S.K., Tripathi, B.N. and Paliwal, O.P. (2005) Progressive immunopathological changes during early stages of experimental infection of goats with *Mycobacterium avium* subspecies *paratuberculosis*. *Veterinary Pathology* 42(4), 427–436. DOI: 10.1354/vp.42-4-427.

Mutwiri, G.K., Butler, D.G., Rosendal, S. and Yager, J. (1992) Experimental infection of severe combined immunodeficient beige mice with *Mycobacterium paratuberculosis* of bovine origin. *Infection and Immunity* 60(10), 4074–4079. DOI: 10.1128/IAI.60.10.4074-4079.1992.

Mutwiri, G.K., Kosecka, U., Benjamin, M., Rosendal, S., Perdue, M. *et al.* (2001) *Mycobacterium avium* subspecies *paratuberculosis* triggers intestinal pathophysiologic changes in beige/*scid* mice. *Comparative Medicine* 51, 538–544.

Mutwiri, G.K., Rosendal, S., Kosecka, U., Yager, J.A., Perdue, M. *et al.* (2002) Adoptive transfer of BALB/c mouse splenocytes reduces lesion severity and induces intestinal pathophysiologic changes in the *Mycobacterium avium* subspecies *paratuberculosis* beige/*scid* mouse model. *Comparative Medicine* 52, 332–341.

Nisbet, D.I., Gilmour, N.J.L. and Brotherston, J.G. (1962) Quantitative studies of *Mycobacterium johnei* in tissues of sheep. *Journal of Comparative Pathology and Therapeutics* 72, 80–IN19. DOI: 10.1016/S0368-1742(62)80009-5.

O'Brien, R., Mackintosh, C.G., Bakker, D., Kopecna, M., Pavlik, I. *et al.* (2006) Immunological and molecular characterization of susceptibility in relationship to bacterial strain differences in *Mycobacterium avium* subsp. *paratuberculosis* infection in the red deer (*Cervus elaphus*). *Infection and Immunity* 74(6), 3530–3537. DOI: 10.1128/IAI.01688-05.

Palmer, M.V., Stabel, J.R., Waters, W.R., Bannantine, J.P. and Miller, J.M. (2007) Experimental infection of white-tailed deer (*Odocoileus virginianus*) with *Mycobacterium avium* subsp. *paratuberculosis*. *Journal of Wildlife Diseases* 43(4), 597–608. DOI: 10.7589/0090-3558-43.4.597.

Paustian, M.L., Amonsin, A., Kapur, V. and Bannantine, J.P. (2004) Characterization of novel coding sequences specific to *Mycobacterium avium* subsp. *paratuberculosis*: implications for diagnosis of Johne's disease. *Journal of Clinical Microbiology* 42(6), 2675–2681. DOI: 10.1128/JCM.42.6.2675-2681.2004.

Payne, J.M. and Rankin, J.D. (1961a) A comparison of the pathogenesis of experimental Johne's disease in calves and cows. *Research in Veterinary Science* 2(2), 175–179. DOI: 10.1016/S0034-5288(18)34971-3.

Payne, J.M. and Rankin, J.D. (1961b) The pathogenesis of experimental Johne's disease in calves. *Research in Veterinary Science* 2(2), 167–176. DOI: 10.1016/S0034-5288(18)34970-1.

Rankin, J.D. (1958) The experimental infection of cattle with *Mycobacterium johnei*. I. Calves inoculated intravenously. *Journal of Comparative Pathology and Therapeutics* 68, 331–IN24.

Rankin, J.D. (1961a) The experimental infection of cattle with *Mycobacterium johnei*. II. Adult cattle inoculated intravenously. *Journal of Comparative Pathology and Therapeutics* 71, 6–9.

Rankin, J.D. (1961b) The experimental infection of cattle with *Mycobacterium johnei*. III. Calves maintained in an infectious environment. *Journal of Comparative Pathology and Therapeutics* 71, 10–15.

Rankin, J.D. (1962) The experimental infection of cattle with *Mycobacterium johnei*. IV. Adult cattle maintained in an infectious environment. *Journal of Comparative Pathology and Therapeutics* 72, 113–117.

Reddacliff, L.A. and Whittington, R.J. (2003) Experimental infection of weaner sheep with S strain *Mycobacterium avium* subsp. *paratuberculosis*. *Veterinary Microbiology* 96(3), 247–258. DOI: 10.1016/j.vetmic.2003.07.004.

Reddacliff, L.A., McClure, S.J. and Whittington, R.J. (2004) Immunoperoxidase studies of cell mediated immune effector cell populations in early *Mycobacterium avium* subsp. *paratuberculosis* infection in sheep. *Veterinary Immunology and Immunopathology* 97(3–4), 149–162. DOI: 10.1016/j.vetimm.2003.09.001.

Robinson, M., O'Brien, R., Mackintosh, C. and Griffin, F. (2008) Differential immune responses of red deer (*Cervus elaphus*) following experimental challenge with *Mycobacterium avium* subsp. *paratuberculosis*. *Clinical and Vaccine Immunology* 15(6), 963–969. DOI: 10.1128/CVI.00031-08.

Rosseels, V., Marche, S., Roupie, V., Govaerts, M., Godfroid, J. *et al.* (2006) Members of the 30- to 32-kilodalton mycolyl transferase family (Ag85) from culture filtrate of *Mycobacterium avium* subsp. *paratuberculosis* are immunodominant Th1-type antigens recognized early upon infection in mice and cattle. *Infection and Immunity* 74(1), 202–212. DOI: 10.1128/IAI.74.1.202-212.2006.

Saxegaard, F. (1990) Experimental infection of calves with an apparently specific goat-pathogenic strain of *Mycobacterium paratuberculosis*. *Journal of Comparative Pathology* 102(2), 149–156. DOI: 10.1016/S0021-9975(08)80120-X.

Scandurra, G.M., de Lisle, G.W., Cavaignac, S.M., Young, M., Kawakami, R.P. *et al.* (2010) Assessment of live candidate vaccines for paratuberculosis in animal models and macrophages. *Infection and immunity* 78(3), 1383–1389. DOI: 10.1128/IAI.01020-09.

Secott, T.E., Lin, T.L. and Wu, C.C. (2002) Fibronectin attachment protein is necessary for efficient attachment and invasion of epithelial cells by *Mycobacterium avium* subsp. *paratuberculosis*. *Infection and Immunity* 70(5), 2670–2675. DOI: 10.1128/IAI.70.5.2670-2675.2002.

Secott, T.E., Lin, T.L. and Wu, C.C. (2004) *Mycobacterium avium* subsp. *paratuberculosis* fibronectin attachment protein facilitates M-cell targeting and invasion through a fibronectin bridge with host integrins. *Infection and Immunity* 72(7), 3724–3732. DOI: 10.1128/IAI.72.7.3724-3732.2004.

Settles, E.W., Kink, J.A. and Talaat, A. (2014) Attenuated strains of *Mycobacterium avium* subspecies *paratuberculosis* as vaccine candidates against Johne's disease. *Vaccine* 32(18), 2062–2069. DOI: 10.1016/j.vaccine.2014.02.010.

Sevilla, I., Singh, S.V., Garrido, J.M., Aduriz, G., Rodriguez, S. *et al.* (2005) Tipificación molecular de cepas de *Mycobacterium avium* subespecie *paratuberculosis* de diferentes huéspedes y regiones. *Revue Scientifique et Technique de l'OIE* 24(3), 1061–1066. DOI: 10.20506/rst.24.3.1634.

Sevilla, I., Garrido, J.M., Geijo, M. and Juste, R.A. (2007) Pulsed-field gel electrophoresis profile homogeneity of *Mycobacterium avium* subsp. *paratuberculosis* isolates from cattle and heterogeneity of those from sheep and goats. *BMC Microbiology* 7, 18. DOI: 10.1186/1471-2180-7-18.

Shin, S.J., Wu, C.-W., Steinberg, H. and Talaat, A.M. (2006) Identification of novel virulence determinants in *Mycobacterium paratuberculosis* by screening a library of insertional mutants. *Infection and Immunity* 74(7), 3825–3833. DOI: 10.1128/IAI.01742-05.

Shippy, D.C., Lemke, J.J., Berry, A., Nelson, K., Hines, M.E. *et al.* (2017) Superior protection from live-attenuated vaccines directed against Johne's disease. *Clinical and Vaccine Immunology* 24(1). DOI: 10.1128/CVI.00478-16.

Sigurðardóttir, O.G., Press, C.M., Saxegaard, F. and Evensen, Ø. (1999) Bacterial isolation, immunological response, and histopathological lesions during the early subclinical phase of experimental infection of goat kids with *Mycobacterium avium* subsp. *paratuberculosis*. *Veterinary Pathology* 36(6), 542–550. DOI: 10.1354/vp.36-6-542.

Sigurðardóttir, O.G., Press, C.M. and Evensen, Ø. (2001) Uptake of *Mycobacterium avium* subsp. *paratuberculosis* through the distal small intestinal mucosa in goats: an ultrastructural study. *Veterinary Pathology* 38(2), 184–189. DOI: 10.1354/vp.38-2-184.

Sigurðardóttir, O.G., Bakke-McKellep, A.M., Djønne, B. and Evensen, Ø. (2005) *Mycobacterium avium* subsp. *paratuberculosis* enters the small intestinal mucosa of goat kids in areas with and without Peyer's patches as demonstrated with the everted sleeve method. *Comparative Immunology, Microbiology and Infectious Diseases* 28(3), 223–230. DOI: 10.1016/j.cimid.2005.01.004.

Simutis, F.J., Cheville, N.F. and Jones, D.E. (2005) Investigation of antigen-specific T-cell responses and subcutaneous granuloma development during experimental sensitization of calves with *Mycobacterium avium* subsp *paratuberculosis*. *American Journal of Veterinary Research* 66(3), 474–482. DOI: 10.2460/ajvr.2005.66.474.

Stabel, J.R. and Ackermann, M.R. (2002) Temporal *Mycobacterium paratuberculosis* infection in T-cell receptor (TCR)-α and TCR-δ-deficient mice. *Veterinary Immunology and Immunopathology* 89(3–4), 127–132. DOI: 10.1016/S0165-2427(02)00167-8.

Stabel, J.R. and Robbe-Austerman, S. (2011) Early immune markers associated with *Mycobacterium avium* subsp. *paratuberculosis* infection in a neonatal calf model. *Clinical and Vaccine Immunology* 18(3), 393–405. DOI: 10.1128/CVI.00359-10.

Stabel, J.R., Ackermann, M.R. and Goff, J.P. (1996) Comparison of polyclonal antibodies to three different preparations of *Mycobacterium paratuberculosis* in immunohistochemical diagnosis of Johne's disease in cattle. *Journal of Veterinary Diagnostic Investigation* 8(4), 469–473. DOI: 10.1177/104063879600800412.

Stabel, J.R., Palmer, M.V. and Whitlock, R.H. (2003) Immune responses after oral inoculation of weanling bison or beef calves with a bison or cattle isolate of *Mycobacterium avium* subsp. *paratuberculosis*. *Journal of Wildlife Diseases* 39(3), 545–555. DOI: 10.7589/0090-3558-39.3.545.

Stabel, J.R., Palmer, M.V., Harris, B., Plattner, B., Hostetter, J. *et al.* (2009) Pathogenesis of *Mycobacterium avium* subsp. *paratuberculosis* in neonatal calves after oral or intraperitoneal experimental infection. *Veterinary Microbiology* 136(3–4), 306–313. DOI: 10.1016/j.vetmic.2008.11.025.

Stewart, D.J., Vaughan, J.A., Stiles, P.L., Noske, P.J., Tizard, M.L.V. *et al.* (2004) A long-term study in Merino sheep experimentally infected with *Mycobacterium avium* subsp. *paratuberculosis*: clinical disease, faecal culture and immunological studies. *Veterinary Microbiology* 104(3–4), 165–178. DOI: 10.1016/j.vetmic.2004.09.007.

Stewart, D.J., Vaughan, J.A., Stiles, P.L., Noske, P.J., Tizard, M.L.V. *et al.* (2006) A long-term study in Angora goats experimentally infected with *Mycobacterium avium* subsp. *paratuberculosis*: clinical disease, faecal culture and immunological studies. *Veterinary Microbiology* 113(1–2), 13–24. DOI: 10.1016/j.vetmic.2005.09.015.

Stewart, D.J., Vaughan, J.A., Stiles, P.L., Noske, P.J., Tizard, M.L.V. *et al.* (2007) A long-term bacteriological and immunological study in Holstein-Friesian cattle experimentally infected with *Mycobacterium avium* subsp. *paratuberculosis* and necropsy culture results for Holstein-Friesian cattle, Merino sheep and Angora goats. *Veterinary Microbiology* 122(1–2), 83–96. DOI: 10.1016/j.vetmic.2006.12.030.

Storset, A.K., Hasvold, H.J., Valheim, M., Brun-Hansen, H., Berntsen, G. *et al.* (2001) Subclinical paratuberculosis in goats following experimental infection. *Veterinary Immunology and Immunopathology* 80(3–4), 271–287. DOI: 10.1016/S0165-2427(01)00294-X.

Streeter, M.N., Barron, S.J., Wagner, D.G., Hibberd, C.A., Owens, F.N. *et al.* (1991) Technical note: a double L intestinal cannula for cattle. *Journal of Animal Science* 69(6), 2601. DOI: 10.2527/1991.6962601x.

Stuart, P. (1965) Vaccination against Johne's disease in cattle exposed to experimental infection. *British Veterinary Journal* 121, 289–318. DOI: 10.1016/S0007-1935(17)41102-X.

Suwandi, A., Bargen, I., Roy, B., Pils, M.C., Krey, M. *et al.* (2014) Experimental colitis is exacerbated by concomitant infection with *Mycobacterium avium* ssp. *paratuberculosis*. *Inflammatory Bowel Diseases* 20(11), 1962–1971. DOI: 10.1097/MIB.0000000000000157.

Sweeney, R.W., Uzonna, J., Whitlock, R.H., Habecker, P.L., Chilton, P. *et al.* (2006) Tissue predilection sites and effect of dose on *Mycobacterium avium* subs. *paratuberculosis* organism recovery in a short-term bovine experimental oral infection model. *Research in Veterinary Science* 80(3), 253–259. DOI: 10.1016/j.rvsc.2005.07.007.

Szilágyi, M., Körmendy, B., Suri, A., Tuboly, S. and Nagy, G. (1989) Experimental paratuberculosis (Johne's disease) – studies on biochemical parameters in cattle. *Archiv fur experimentelle Veterinarmedizin* 43(3), 463–470.

Talaat, A.M. and Stemke-Hale, K. (2005) Expression library immunization: a road map for discovery of vaccines against infectious diseases. *Infection and Immunity* 73(11), 7089–7098. DOI: 10.1128/IAI.73.11.7089-7098.2005.

Tanaka, S., Sato, M., Taniguchi, T. and Yokomizo, Y. (1994) Histopathological and morphometrical comparison of granulomatous lesions in BALB/c and C3H/HeJ mice inoculated with *Mycobacterium paratuberculosis*. *Journal of Comparative Pathology* 110(4), 381–388. DOI: 10.1016/S0021-9975(08)80315-5.

Tanaka, S., Itohara, S., Sato, M., Taniguchi, T. and Yokomizo, Y. (2000) Reduced formation of granulomata in gamma(delta) T cell knockout BALB/c mice inoculated with *Mycobacterium avium* subsp. *paratuberculosis*. *Veterinary Pathology* 37(5), 415–421. DOI: 10.1354/vp.37-5-415.

Thorel, M.F., Pardon, P., Irgens, K., Marly, J. and Lechopier, P. (1984) Paratuberculose expérimentale: pouvoir pathogène chez le veau de souches de Mycobactéries mycobactine-dépendantes. *Annales de Recherches Veterinaires* 15, 365–374.

Usda-Aphis (2010) *Uniform Program Standards for the Voluntary Bovine Johne's Disease Control Program*. Usda-Aphis, Washington, DC.

Uzonna, J.E., Chilton, P., Whitlock, R.H., Habecker, P.L., Scott, P. *et al.* (2003) Efficacy of commercial and field-strain *Mycobacterium paratuberculosis* vaccinations with recombinant IL-12 in a bovine experimental infection model. *Vaccine* 21(23), 3101–3109. DOI: 10.1016/S0264-410X(03)00261-5.

Valheim, M., Storset, A.K., Aleksersen, M., Brun-Hansen, H. and C McL Press., (2002) Lesions in subclinical paratuberculosis of goats are associated with persistent gut-associated lymphoid tissue. *Journal of Comparative Pathology* 127(2–3), 194–202. DOI: 10.1053/jcpa.2002.0581.

Van Kruiningen, H.J., Chiodini, R.J., Thayer, W.R., Coutu, J.A., Merkal, R.S. *et al.* (1986) Experimental disease in infant goats induced by a *Mycobacterium* isolated from a patient with Crohn's disease. A preliminary report. *Digestive Diseases and Sciences* 31(12), 1351–1360. DOI: 10.1007/BF01299814.

Vaughan, J.A., Lenghaus, C., Stewart, D.J., Tizard, M.L. and Michalski, W.P. (2005) Development of a Johne's disease infection model in laboratory rabbits following oral administration of *Mycobacterium avium* subspecies *paratuberculosis*. *Veterinary Microbiology* 105(3–4), 207–213. DOI: 10.1016/j.vetmic.2004.10.019.

Veazey, R.S., Horohov, D.W., Krahenbuhl, J.L., Taylor, H.W., Oliver, J.L. *et al.* (1995) Comparison of the resistance of C57BL/6 and C3H/He mice to infection with *Mycobacterium paratuberculosis*. *Veterinary Microbiology* 47(1–2), 79–87. DOI: 10.1016/0378-1135(95)00057-H.

Waters, W.R., Miller, J.M., Palmer, M.V., Stabel, J.R., Jones, D.E. *et al.* (2003) Early induction of humoral and cellular immune responses during experimental *Mycobacterium avium* subsp. *paratuberculosis* infection of calves. *Infection and Immunity* 71(9), 5130–5138. DOI: 10.1128/IAI.71.9.5130-5138.2003.

Williams, E.S., Snyder, S.P. and Martin, K.L. (1983a) Experimental infection of some North American wild ruminants and domestic sheep with *Mycobacterium paratuberculosis*: clinical and bacteriological findings. *Journal of Wildlife Diseases* 19(3), 185–191. DOI: 10.7589/0090-3558-19.3.185.

Williams, E.S., Snyder, S.P. and Martin, K.L. (1983b) Pathology of spontaneous and experimental infection of North American wild ruminants with *Mycobacterium paratuberculosis*. *Veterinary Pathology* 20(3), 274–290. DOI: 10.1177/030098588302000304.

Wu, C.W., Livesey, M., Schmoller, S.K., Manning, E.J.B., Steinberg, H. *et al.* (2007a) Invasion and persistence of *Mycobacterium avium* subsp. *paratuberculosis* during early stages of Johne's disease in calves. *Infection and Immunity* 75(5), 2110–2119. DOI: 10.1128/IAI.01739-06.

Wu, C.W., Schmoller, S.K., Shin, S.J. and Talaat, A.M. (2007b) Defining the stressome of *Mycobacterium avium* subsp. *paratuberculosis* in vitro and in naturally infected cows. *Journal of Bacteriology* 189(21), 7877–7886. DOI: 10.1128/JB.00780-07.

Wu, C.W., Schmoller, S.K., Bannantine, J.P., Eckstein, T.M., Inamine, J.M. *et al.* (2009) A novel cell wall lipopeptide is important for biofilm formation and pathogenicity of *Mycobacterium avium* subspecies *paratuberculosis*. *Microbial Pathogenesis* 46(4), 222–230. DOI: 10.1016/j.micpath.2009.01.010.

17 Immunology of Paratuberculosis Infection and Disease

Judith R. Stabel[1]*, Ad Koets[2] and Kumudika de Silva[3]
[1]USDA-ARS, Ames, Iowa, USA; [2]Wageningen Bioveterinary Research, Lelystad, The Netherlands and Faculty of Veterinary Medicine, Utrecht University, The Netherlands; [3]The University of Sydney, Sydney, Australia

17.1 Introduction

The study of host immune responses to *Mycobacterium avium* subsp. *paratuberculosis* (*MAP*) is critical to the identification and development of tools to control and manage the disease. As one of the more fastidious mycobacteria, *MAP* infection results in long periods of asymptomatic infection. Some of these animals may progress to a clinical form of the disease, whereas others will either clear the infection or remain infected but maintain a subclinical disease state. The biology of the disease is complex, influenced not only by host and environmental factors, but by the wily nature of the pathogen. In this chapter, we have included the most current studies of host immunity after infection with *MAP*, embracing earlier studies for comprehensive knowledge. The discussion also conflates studies on host immunity to closely related mycobacterial pathogens to help fill in knowledge gaps that exist for *MAP*. The co-authors hope to provide a comprehensive treatise to improve the understanding of a very complicated disease process with the anticipation that it provides focus for development of diagnostic tools and new vaccines.

17.2 Bacterial Uptake

The primary route of transmission is through ingestion of faecal matter, or milk and colostrum containing *MAP*. It is generally thought that neonates are most susceptible to infection due to their undeveloped immune system combined with the high level of exposure to *MAP*, especially if their dams are clinically infected (Larsen *et al.*, 1975; Smith *et al.*, 2015). By nature, *MAP* is an enteric pathogen, preferentially homing to regions of the small intestine to initiate the infectious process. Lymphoid aggregates such as gut-associated lymphoid tissue (GALT) line the mucosa of the intestines and are key components of the infectious process. The most critical GALT lymphoid aggregates involved in *MAP* infection are the Peyer's patches. Peyer's patches are regulators of the intestine, acting as a gateway for pathogen uptake primarily through the M cells. M cells are specialized epithelial cells whose sole function is to transport antigens from the lumen of the intestine for presentation to the immune cells in the lamina propria (Gebert *et al.*, 1996). M cells lack the typical brush-border microvilli, digestive enzymes and surface mucus associated with enterocytes, and instead have a thin glycocalyx

*Corresponding author: judy.stabel@ars.usda.gov

© CAB International 2020. *Paratuberculosis: Organism, Disease, Control, 2nd Edition* (eds M.A. Behr *et al.*)

allowing endocytosis of antigens (Featherstone, 1997). Transcytosis of antigens through the cell then occurs, engaging with macrophages or dendritic cells (DCs) on the basolateral side. The process of *MAP* uptake in the small intestine was first defined in a pioneering study using ileal loops of neonatal calves, in which it was found that *MAP* was preferentially taken up by M cells and then by subepithelial and intraepithelial macrophages within the domes of the Peyer's patches (Momotani *et al.*, 1988). Additional studies using the everted sleeve method demonstrated that uptake of *MAP* by intestinal mucosa of goat kids was not restricted to M cells but also included regional enterocytes (Sigurðardóttir *et al.*, 2004, 2005). This was further corroborated more recently wherein the presence of M cells was evaluated in a temporal study of *MAP* uptake within loops of Peyer's patch and non-Peyer's patch ileum and jejunum from lambs (Ponnusamy *et al.*, 2013). Results demonstrated that *MAP* was able to invade both M cells and enterocytes, but translocation was greater for M cells in both ileum and jejunum. Furthermore, uptake of *MAP* was more pronounced for ileal loops compared with the jejunal regions of the small intestine. Uptake of *MAP* has experimentally been shown to occur as rapidly as 30 min post-inoculation (Sigurðardóttir *et al.*, 2001), with colonization of distant organs such as liver and lymph nodes occurring within 1 h. However, it is a time-dependent process, with higher numbers of *MAP* taken up by the intestine at the more extended time points of 12, 20 and 24 h (Momotani *et al.*, 1988; Ponnusamy *et al.*, 2013).

A high density of B1 integrin receptors is noted on the apical membrane of the M cell, whereas integrin receptors are located more exclusively on the basolateral side of enterocytes, suggesting a probable mechanism for preferential uptake of *MAP* by M cells. The role of integrin receptors in the uptake of *MAP* was demonstrated after pre-opsonization of *MAP* with fibronectin increased uptake by M cells 2.6-fold compared with non-opsonized *MAP* (Secott *et al.*, 2001). Further, uptake of *MAP* by M cells was inhibited 52–73% following treatment of fibronectin-opsonized *MAP* with monoclonal antibodies directed against integrin subunits a5, aV, b1 and B3 (Secott *et al.*, 2004). In addition to integrins, other pathogen recognition receptors (PRRs) such as Toll-like receptors (TLRs) and platelet-activating factor receptor are involved in the recognition of bacterial pathogens (Tyrer *et al.*, 2006). Cell-surface PRRs such as TLR2 and TLR4 recognize microbial components such as lipoproteins (TLR2) and lipopolysaccharide (LPS) (TLR4) upon infection and act as 'gatekeepers' to the intestinal epithelium, including M cells (Chabot *et al.*, 2006). In *MAP* infection, gene expression of TLR2 and TLR4 was upregulated in the ileum and jejunum of naturally infected sheep compared with non-infected animals, with increasing TLR expression correlating with bacterial load in the tissue (Taylor *et al.*, 2008). Interestingly, divergent expression of TLR2 and TLR4 was observed on peripheral blood mononuclear cells (PBMCs) in early and late stages of infection, respectively, in sheep that ultimately progressed to multibacillary disease (Thirunavukkarasu *et al.*, 2013). In addition, blocking of TLR2 expression in RAW 264.7 macrophages upon exposure to *MAP* precluded the expression of nitric oxide, suggesting that TLR2 may be a critical factor in controlling early infection in the host. Intracellular PRRs present within epithelial cells such as TLR9 and NOD1 have also been recognized as innate recognition receptors involved in the uptake of *MAP* (Pott *et al.*, 2009). More finite studies determined that internalization of *MAP* by intestinal epithelial cells was directed by a small GTPase Rac in combination with Cdc42 (a cell division control protein), and that the internalization resulted in an upregulation of proinflammatory cytokine MIP-2 (Miltner *et al.*, 2005; Alonso-Hearn *et al.*, 2008; Pott *et al.*, 2009). This work demonstrated that activation of epithelial cells is dependent upon ligand release from the bacterial pathogen after uptake by the cell (Pott *et al.*, 2009). Alternatively, bacterial proteins such as a 35-kDa major membrane protein and CipA, a surface protein present within a putative invasion region, have been implicated as playing key roles in the uptake of *MAP* by epithelial cells (Bannantine *et al.*, 2003; Bannantine and Bermudez, 2013; Harriff *et al.*, 2009). Using an *in vitro* model of infection, greater than 6-fold upregulation of *MAP3464* gene was observed in MDBK cells within 30 min of exposure to live *MAP* (Alonso-Hearn *et al.*, 2008). The *MAP3464* gene is purported to be part of the ABC transporter family, a series of genes acting as membrane pumps in biological membranes. The composition of the cell wall of *MAP* has been implicated in the efficiency of invasion of epithelial cells (Alonso-Hearn *et al.*, 2010). A mycobacterial LuxR transcriptional regulator controlling the expression of genes such as *FadE8*, *MAP1420*, *LprL* and *MAP1203* demonstrated that envelope genes could control the entry of *MAP* into cells. Further studies with *MAP* modified with a pJAM:*MAP1203* plasmid showed that invasion of bovine MDBK cells was enhanced compared with pJAM *MAP* controls (Everman *et al.*, 2018). MAP1203 contains an arginine–glycine–aspartic acid (RGD) peptide motif that facilitates the binding and invasion of microorganisms (Everman *et al.*, 2018). The role of bacterial factors and host receptors in the initial uptake of *MAP* by epithelial cells requires further study, as it conceptually represents a point in the infection process amenable to intervention.

17.3 Innate Response to *MAP* Infection

Upon transcytosis across the intestinal epithelium, *MAP* is taken up by macrophages and DCs in the subepithelium. In general, these cells will then traffic to the lamina propria of the small intestine, the mesenteric lymph nodes and the peripheral circulation, thereby initiating an immune response to clear or control the infection (Lugton, 1999). Survivability of *MAP* within the macrophage is dependent upon the host, as *MAP* can thwart degradation by blocking maturation and preventing acidification of phagosomes (Sturgill-Koszycki *et al.*, 1994; Hostetter *et al.*, 2003). Hostetter *et al.* (2003) found that phagosomes within macrophages infected with live *MAP* expressed higher levels of transferrin, an early phagosome marker and lower levels of lysosome-associated membrane protein one, a late maturation marker, compared with phagosomes within macrophages infected with heat-killed *MAP* or *M. avium* subsp. *avium*. The arrestment of phagosome maturation has been correlated with inhibition of phagosome acidification, a process that produces bactericidal agents such as nitric oxide, reactive oxygen species (H_2O_2 and OH^-) and lysosomal hydrolases (Akaki *et al.*, 1997). The mechanism by which *MAP* is taken up by phagocytes can affect the subsequent ability of the phagocyte to effectively destroy the pathogen. The engagement of complement receptors (CR1, CR3 and CR4) for *MAP* uptake can limit macrophage activation (Hostetter *et al.*, 2005). Blocking bovine CR3 receptor by incubating activated macrophages with anti-CD18/CD11b reduced uptake of opsonized *MAP* (Hostetter *et al.*, 2005). Opsonization of *MAP* with heat-treated serum limited uptake, further demonstrating that CR is a primary means of *MAP* entry into macrophage. Perhaps even more interesting was the observation that opsonization of *MAP* lowered levels of phagosome acidification, suggestive of decreased ability to kill the pathogen (Hostetter *et al.*, 2005). More recently, pre-incubation of bovine monocyte-derived macrophages with mannosylated-lipoarabinomannan (Man-LAM) derived from *MAP* was shown to reduce killing of live *M. avium* subsp. *avium* (Souza *et al.*, 2013). Reduced killing was directly associated with lower phagosome acidification and fusion of the phagosome-lysosome, along with increased IL-10 expression. IL-10 is induced via a MAPK-p38 pathway in macrophages, a signalling pathway that has been shown to negatively impact acidification of the phagosome and killing of *MAP* within the macrophage (Souza *et al.*, 2006). Activation of the MAPK pathway in bovine monocytes upon infection with *MAP* is mediated through TLR2 (Weiss *et al.*, 2008). Blocking TLR2 on bovine monocytes resulted in increased phagosome acidification, phagosome-lysosome fusion and *MAP* survival, suggesting that TLR2 signalling is critical to macrophage bactericidal activities (Weiss *et al.*, 2008). It is clear that IL-10 plays a substantial role in the ability of the macrophages to arrest the infectious process (Hussain *et al.*, 2016). IL-10 is a negative regulator of IL-12 production by antigen-presenting cells, thereby influencing the cross-talk between macrophages and T cells and retarding the amplification of the adaptive immune response (Rahim *et al.*, 2005). Although IL-10 plays an essential role in controlling inflammatory responses of the host during infection, one contrary aspect may be a weaponization by the intracellular pathogen to defeat the macrophage. A recent review outlines a critical role for regulation of IL-10 by mycobacteria, suggesting that it is a key effector in survival of the pathogen (Abdalla *et al.*, 2016). Another cytokine that controls macrophage responses to intracellular infection is IL-1β (Lamont *et al.*, 2012). The infection of MAC-T-bovine monocyte-derived macrophages with live *MAP* demonstrated that phagosome acidification was calcium-dependent and linked to IL-1β processing by the cells. It was suggested that IL-1β plays a critical role in the recruitment of other monocyte-derived macrophages to the site of infection and that *MAP* has the ability to exploit this by inducing calcium influx in the cell. A comprehensive RNA-seq analysis of genes expressed by monocyte-derived macrophages infected with *MAP* complemented those results as they found upregulation of IL-1 and other proinflammatory genes in early infection, concomitant with expression of IL-10 and other proinflammatory mediators (Casey *et al.*, 2015). It is clear that macrophage–mycobacteria interactions have to be balanced between protection and pathology and this is not always successful in the infected host (Divangahi *et al.*, 2018). Participation of other cell types involved in adaptive immunity may be required to skew the host response to a more protective nature.

17.4 Early Infection – Macrophage–T-Cell Interaction

Macrophages activated by mycobacteria produce a number of cytokines, such as interleukin (IL)-1, tumour necrosis factor (TNF)-α and IL-12 (Wang *et al.*, 1999; Hope *et al.*, 2004). Infection with *MAP* initiates the upregulation of the aforementioned cytokines, in addition to other proinflammatory cytokines, IL-6, IL-8 and IL-10 (Adams and Czuprynski, 1994; Coussens *et al.*, 2004; Weiss *et al.*, 2004). Presentation of major histocompatibility complex (MHC) class II antigens on the surface of the macrophage, along with IL-1 secretion, results in the activation of T cells. Activated T cells produce IL-2, which aids in the clonal expansion of specific CD4+ T helper cells and CD8+ cytolyticT cells. Upon activation, the naive CD4+ T cell can differentiate into several subtypes including Th1, Th2 or Th17 cells, based upon the nature of the antigen presented and other factors in the microenvironment. Differentiation of the CD4+ T-cell population is skewed towards a Th1 T-cell subpopulation in the early stages of *MAP* infection, characterized by the secretion of the Th1-associated cytokines, IFN-γ, IL-2 and TNF-α (Burrells *et al.*, 1998; Stabel, 2000). The strong bias towards a Th1-mediated immune response in the early stages of infection is dominated by the key effector cytokine, IFN-γ. Studies have reported higher expression of IFN-γ in PBMCs of cattle infected with *MAP* (Coussens *et al.*, 2003, 2004; Khalifeh and Stabel, 2004a), and this was correlated with higher levels of IFN-γ secreted by PBMCs isolated from animals in the subclinical or early stages of infection, whether it be natural or experimental (Stabel, 2000; Waters *et al.*, 2003; Khalifeh and Stabel, 2004a). In sheep, the early appearance of robust IFN-γ responses was associated with the control of disease progression to a more advanced stage (de Silva *et al.*, 2013, 2018). IFN-γ plays a crucial role in the activation of T cells and macrophages, DC maturation, upregulation of MHC class I and II molecules, and production of reactive oxygen and nitrogen species by macrophages (Delvig *et al.*, 2002; Shankar *et al.*, 2003). CD4+ T cells appear to be the primary source of IFN-γ in mycobacterial, including *MAP*, infections, but CD8+ and γδ T cells also produce IFN-γ. Quantities of IFN-γ secreted by γδ T cells in response to *MAP* antigens appear to be lower than amounts produced by CD4+ or CD8+ T cells, but this may be antigen-dependent (Shin *et al.*, 2005). γδ T cells accumulate at the site of infection prior to CD4+ T-cell arrival, secrete significant amounts of IFN-γ (Plattner *et al.*, 2013) and induce intracellular killing of *MAP* in infected macrophages (Baquero and Plattner, 2017); in particular, the WC1.1+ subset appears to influence disease control in the early stages of infection (Albarrak *et al.*, 2018).

IFN-γ not only induces the secretion of IL-12 by APCs, resulting in Th1 induction via a paracrine pathway, but also acts to directly augment Th1 polarization via an autocrine mechanism that does not involve IL-12 (Teixeira *et al.*, 2005). Together with a downregulation of MHC class II-mediated antigen presentation pathways that mediate CD4+ T-cell responses, there is an upregulation of MHC class I gene expression indicating a switch to CD8+ T-cell responses early in *MAP* infection of cattle (Purdie *et al.*, 2012).

Similarly in sheep, a lack of protection against *MAP* infection is associated with early dysfunction of CD4+ T-cell, B-cell and γδ T-cell responses, but not CD8+ T-cell responses (de Silva *et al.*, 2015).

IL-12 and IL-18 are both important mediators of anti-mycobacterial immunity and appear to induce IFN-γ synergistically through activation of Th1 cells (Kohno *et al.*, 1997; O'Donnell *et al.*, 1999; Sugawara *et al.*, 1999). Although they share some biological properties, these two cytokines utilize different signal transduction pathways in the induction of IFN-γ, suggesting a unique regulatory process in T-cell activation (Nembrini *et al.*, 2006). Very little information has been published on the roles of IL-12 and IL-18 in *MAP* infections to date. Gene expression studies have generated variable results (Coussens *et al.*, 2004; Tanaka *et al.*, 2005; Smeed *et al.*, 2007). Interestingly, when used as an adjuvant for paratuberculosis vaccines, IL-12 has not demonstrated increased IFN-γ secretion or reduced *MAP* colonization in tissues of vaccinated animals (Uzonna *et al.*, 2003; Kathaperumal *et al.*, 2008). This is in contrast to a study that demonstrated that co-immunization with a plasmid containing IL-12 and *M. tuberculosis* Ag85B enhanced IFN-γ secretion and protection against *M. tuberculosis* infection (Triccas et *al.*, 2002). It has been suggested that induction of a strong T-cell receptor (TCR) response is responsible for the majority of IFN-γ production and that IL-12 and IL-18 play supporting roles, perhaps becoming more critical when a TCR signal is weak or misaligned (Nembrini *et al.*, 2006). Since mycobacterial cell wall components are highly antigenic, it is likely that IL-12 and IL-18 are both used by the host as compensatory mediators of IFN-γ production.

Early immune responses are pivotal in determining eventual disease outcome in *MAP* infection and, in sheep, the magnitude of early IFN-γ responses appears to be an important factor (de Silva *et al.*, 2013, 2018).

17.5 Transition From Early to Late Infection – T-Cell Subpopulations

Activated CD4+, CD8+ and γδ T-cell subpopulations are recognized sources of IFN-γ, but CD4+ T cells appear to be most highly reactive to mycobacterial antigens in the early stage of infection (Flynn *et al.*, 1993). Naturally infected cows with paratuberculosis have an increased frequency of CD4+ and CD8+ T cells in the total PBMC population compared with healthy controls, with equivalent percentages of both within the subclinical and clinical infection groups (Stabel *et al.*, 2007). Further delineation of the CD4+ subset demonstrated increased expression of CD25 and CD45RO, suggesting that these cells were highly activated and that prior exposure to *MAP* via natural infection fostered an effector–memory cell phenotype within the CD4+ subpopulation. These findings agree with an earlier study evaluating cellular reactivity to antigens over a 320-day infection period in calves experimentally infected with *MAP* (Koo *et al.*, 2004). CD4+ T cells with a memory cell phenotype (CD4+ CD45RO+) are the predominant population of cells responding to antigens in the first year of infection, with an increase in the expression of CD25 and CD26 activation markers observed within the CD4+ subpopulation. In contrast, CD8+ CD45RO+T cells have low levels of CD25 and CD26 expression initially, but after 18 months of infection the cell phenotype begins to shift towards increased expression of these activation markers on CD8+ T cells (Koo *et al.*, 2004). In a study of naturally infected cows, greater intracellular IFN-γ was noted for CD8+ CD45RO+ cells compared with CD4+ CD45RO+ cells for clinical cows after stimulation of cells with johnin purified protein derivative (PPD) (Stabel *et al.*, 2007). Higher levels of secreted and intracellular IFN-γ were also observed for CD8+ cells after stimulation of PBMC with Bacillus Calmette–Guérin (BCG)-infected DCs in cattle vaccinated with BCG (Hope *et al.*, 2004). These results suggest that, although CD4+ T cells may be critical in early responses to infection, CD8+ T cells may play a more significant role in late infection. Interestingly, a study in naturally infected cattle found that unlike CD4+ T cells and γδ T cells, CD8+ T cells did not respond to antigen stimulation *ex vivo* (Frie et *al.*, 2017). The role that CD8+ T cells fulfil may be influenced by the maturity of the host immune system or the number of exposures to a particular pathogen.

Since γδ T cells produce IFN-γ in response to mycobacteria, it is believed that they contribute to the protection of the host early in the infection process, in a way that is distinct from that of γT cells (Kaufmann, 1996). In a *MAP* challenge study, TCR-α-deficient mice had higher levels of *MAP* colonization in their tissues compared with TCR-δ-deficient mice or C57BL/6 control mice (Stabel and Ackermann, 2002). Lesions were located predominantly in the liver or the ileum, depending upon the period of infection, and lesion scores were higher for TCR-α-deficient mice. Further suggestion of a unique role for γδ T cells stems from evidence that during experimental infection with intracellular pathogens such as *Listeria* and *M. avium* subsp. *avium*, TCR-γδ-deficient mice formed atypical lesions in their tissues instead of the granulomatous lesions seen in wild-type mice (Mombaerts *et al.*, 1993; Saunders *et al.*, 1998). Similarly, Tanaka *et al.* (2000) presented evidence of reduced granulomatous lesions in TCR-γδ-deficient mice after challenge with *MAP*. These studies suggest that γδ T cells play a role in clearance of the pathogen after infection, which either harmonizes with or compensates for γT cells.

More recently, the role that Th17 cells play in cellular immunity to mycobacterial infections has come into focus. Th17 cells are characterized by cell surface expression of CD4 and the ability to secrete IL-17 and IL-22. The proinflammatory cytokine IL-17 is involved in the pathogenesis of chronic inflammatory diseases and is also secreted by a variety of innate cells including macrophages, natural killer cells and γδ T cells (Onishi and Gaffen, 2010). In natural infection with *MAP*, plasma IL-17 was lower and Th17-derived cytokine genes were downregulated in asymptomatic cows that were shedding *MAP* and were enzyme-lined immunosorbent assay (ELISA) positive (Dudemaine *et al.*, 2014; Park *et al.*, 2018). This is in contrast to the increased IL-17 expression observed in intestinal tissues of red deer with advanced disease (Robinson *et al.*, 2011). It is plausible that Th17-type immune responses become more established as disease progresses in subclinical cattle and this likely reflects the inability to control intracellular *MAP* infection.

Less studied but none the less proving to be critical checkpoints in modulation of host immunity in early and late stages of T-cell responsiveness, respectively, are the cytotoxic T-lymphocyte antigen-4 (CTLA-4) and programmed death-1 (PD-1) pathways (Buchbinder and Desai, 2016). Increased cell surface expression of CTLA-4 in effector cells leads to inhibition of T-cell proliferation and survival, with suppressive functions of regulatory T cells (Tregs) correlated to their constitutive expression of CTLA-4. PD-1 expression is characteristic of 'exhausted' T cells and is associated with chronic infections. The presence of CTLA-4+ CD4+ T cells was significantly higher in naturally infected cows in the subclinical stage of disease compared with clinical cows, suggesting an engagement in limiting T-cell effector function as disease progresses (Leite *et al.*, 2015). In later subclinical stages of bovine infection, *MAP*-specific T-cell exhaustion has been observed to be partially mediated via upregulation of PD-1 and lymphocyte activation gene-3 (*LAG-3*) (Okagawa *et al.*, 2016).

In an effort to understand the dynamic disease progression in *MAP* infection, mathematical models were developed using datasets from well-characterized naturally infected cattle (Magombedze *et al.*, 2014). Using the assumptions that Th1 responses are driven by intracellular bacteria and that Th2 responses are driven by extracellular bacteria, and that these immune responses are not cross-suppressive, this model supports previous experimental studies (Stabel, 2000) to predict that the Th1 to Th2 switch is a result of disease progression. Although such models are provocative, they are hampered by the complexity of immune responses and changes in the microenvironment at the site of infection that are not completely understood as yet.

17.6 Late Infection – T-Cell Responses

The progression of paratuberculosis from a subclinical to clinical state is associated with a switch from Th1 to Th2 immune response. The production of Th2 regulatory cytokines, IL-4, IL-5 and IL-10, supports a humoral immune response characterized by the expansion of B lymphocytes, immunoglobulin secretion and control of Th1-mediated responses. Both IL-4 and IL-10 have been noted to play specific roles in the suppression of IFN-γ production by CD4+ Th1 cells (Peleman *et al.*, 1989; Ito *et al.*, 1999). There is little evidence to suggest a specific role for IL-4 in mediating protective immunity against *MAP*, as expression of IL-4 in tissues does not appear to be associated with the infection status of the host animal (Sweeney *et al.*, 1998; Coussens *et al.*, 2004). However, two reports have documented reduced expression of IL-4 in PBMC from subclinically and clinically infected cows as compared with control cows after stimulation with *MAP* (Coussens *et al.*, 2004; Karcher *et al.*, 2008). In contrast, IL-4 secretion by PBMCs stimulated with a whole-cell sonicate of *MAP* was greater in clinical cows compared with control and subclinical cows, fitting the paradigm shift to Th2-mediated immunity in clinical disease (Karcher *et al.*, 2008).

IL-10 seems to play a more significant role in mycobacterial infections in general, and specifically during *MAP* infection. An upregulation of IL-10 in tissues, including ileum, mesenteric lymph nodes and cultured PBMCs, from naturally infected cattle and sheep with paratuberculosis has been observed (Coussens *et al.*, 2004; Khalifeh and Stabel, 2004b; Smeed *et al.*, 2007). Increased expression of IL-10 was observed in bovine monocyte-derived macrophages infected with *MAP* as compared with *M. avium* subsp. *avium* (Weiss *et al.*, 2002). Interestingly, prostaglandin E2, a proinflammatory mediator, contributes to the progression of paratuberculosis infection by suppressing Th1 responses and activating IL-10 production (Sajiki *et al.*, 2018). In both sheep and cattle, IL-10 responses can also be elevated in early stages of disease (Nagata *et al.*, 2010; de Silva *et al.*, 2011) but does not appear to be related to disease susceptibility or resistance. The role of IL-10 varies and mathematical modelling indicates that it is related to infectious burden (Magombedze *et al.*, 2015).

Further understanding of a role for IL-10 was gained through studies demonstrating that the addition of exogenous IL-10 to bovine cell cultures prior to infection with live *MAP* reduced IFN-γ secretion (Khalifeh and Stabel, 2004a). Alternatively, neutralization of IL-10 in johnin-PPD-stimulated whole blood increased IFN-γ production 23-fold in cattle with subclinical paratuberculosis (Buza *et al.*, 2004). The effects of IL-10 on IFN-γ may be mediated through IL-12, as the addition of exogenous IL-10 to human monocyte cultures reduces secretion of IL-12, whereas neutralization of IL-10 activity results in increased IL-12 secretion (Fulton *et al.*, 1998). Similarly, with bovine T cells, sensitization of a cow to mycobacteria with PPD and IL-12 resulted in an IL-12-mediated increase in IFN-γ secretion associated with decreased IL-10 expression (Tuo *et al.*, 1999). Moreover, neutralization of IL-10 activity in bovine monocyte-derived macrophage cultures resulted in increased acidification of phagosomes, increased expression of IL-12, IL-8, TNF-α and MHC class II molecules, and increased killing of *MAP* in culture (Weiss *et al.*, 2005). These studies provide evidence of the importance of interplay between Th1 and Th2 cytokines in the regulation of host responses to mycobacterial pathogens, allowing a balance to be maintained between protective and pathogenic responses.

Detailed analysis of T-cell responses at the ileal site of predilection in *MAP* infection is challenging. The development of a surgically isolated ileal loop model in newborn calves allowed study of localized immunity in the ileum during persistent *MAP* infection (Charavaryamath *et al.*, 2013). An accumulation of CD8+ T cells and γδ T cells in the lamina propria of infected intestinal segments was observed, concomitant with an increase in *MAP*-specific IFN-γ and TNF by leucocytes in the lamina propria. This is an interesting observation as a recently discovered subset of γδ T cells that co-expresses CD8γ+ is considered to play an important role in mucosal healing in inflammatory bowel disease (Kadivar *et al.*, 2016). In addition, γδ T cells lose their responsiveness to *MAP* antigen in the clinical stages of disease (Albarrak *et al.*, 2018). In ileal tissues of cattle with clinical disease, there is a significant increase in WC1+ γδ T cells expressing IL-10 and a significant increase in TGF-β expression by non-WC1+ cells compared with subclinical and non-infected cattle. However, there is no difference in expression of IFN-γ, IL-17 and TNF-α in mucosal cells, in these later stages of disease (Albarrak *et al.*, 2018). In sheep, paucibacillary lesions are linked to a Th1/Th17 response, characterized by high levels of IL-12, IL-17A and IFN-γ, while multibacillary lesions are associated with a Th2 response and high levels of IL-5 and IL-10 (Burrells *et al.*, 1998; Whittington *et al.*, 2012). However, multibacillary lesions are unlikely to be solely driven by a Th2 response (Nicol *et al.*, 2016). As disease progresses, Treg cells likely play an important role in controlling these changes (Robinson *et al.*, 2011).

17.7 Late Infection – T Regulatory Cells

A subpopulation of T cells known as T regulatory (Treg) cells further mediates host immune responses to infectious diseases. During infectious disease including mycobacterial infections Tregs play a dual role: they benefit the host by limiting immune-mediated pathology and also facilitate chronic pathogen persistence by reducing effector immunity and clearance of infection. During acute infection, the beneficial role of Tregs seems to predominate by facilitating optimal antigen presentation and lymphocyte recirculation. However, while Treg cells successfully control excessive immune responses in the host to preserve homeostasis, as a consequence this may lead to enhanced pathogen survival and in the case of mycobacterial infection, long-term pathogen persistence (Belkaid and Tarbell, 2009). Several types of Tregs are currently recognized and can be thymus-derived, peripheral tissue-derived or from unknown origin. Another distinction is natural vs adaptive Tregs. In mice, several characteristic phenotypes have been described but unique markers are lacking in non-murine species. Tregs are preferably defined by a combination of markers such as forkhead box P3 (FoxP3) transcription factor and CD25, along with demonstration of suppressive activity. The latter is important since other non-Treg populations may also express 'Treg markers' (Boer *et al.*, 2015). It has been noted that bovine CD4+ CD25+ FoxP3+ T cells lack functional suppressive activity, while γδ T-cell subsets do show a suppressive functional phenotype, suggesting that γδ T cells may represent true Tregs in cattle (Hoek *et al.*, 2009; Guzman *et al.*, 2014). Key regulatory cytokines such as IL-10 have been associated with reduced secretion of IFN-γ by PBMCs isolated from cows with subclinical paratuberculosis (de Almeida *et al.*, 2008). Further, depletion of IL-10-producing T cells from the PBMCs of infected cattle revealed the characteristic phenotype of Tregs, CD4+ CD25+. Previously, a highly significant reduction in IFN-γ production was noted with the addition of exogenous IL-10 and TGF-β to PBMC cultures, with and without the presence of live *MAP*, an effect that was independent of the infection status of the cows (Khalifeh and Stabel, 2004a). Interestingly, an increase in IL-10 secretion was observed after the addition of TGF-β to cell cultures. Similarly, exogenous IL-10 and TGF-β suppressed the production of IFN-γ by PBMCs from TB skin-test-positive individuals, with greater suppression in the presence of both IL-10 and TGF-β (Othieno *et al.*, 1999). The least suppression was noted with the addition of IL-10 alone, indicating that TGF-β mediates the suppression of IFN-γ both directly and indirectly through IL-10. A comprehensive review of Tregs in paratuberculosis suggested that additional mechanisms for immune regulation may exist, especially in cattle (Coussens *et al.*, 2012). Recent studies with *M. tuberculosis* have indicated the presence and importance of myeloid regulatory cells (MRCs), also termed monocytic myeloid-derived suppressor cells (M-MDSCs) (Dorhoi and Du Plessis, 2018). These populations include regulatory DCs, as well as regulatory and alternatively activated macrophages (M2-like macrophages). These phagocytes are capable of pathogen internalization and persistence, while exerting localized suppressive activity of T cells and induction of Treg cells. Patients with active tuberculosis or after recent TB treatment had an increased frequency of myeloid-derived suppressor cells leading to reduced proinflammatory T-cell function (du Plessis *et al.*, 2013). Immunohistochemical staining of mid-ileal tissue from cattle has also indicated a predominance of M2 anti-inflammatory macrophages coordinating with T cells in advanced stages of bovine paratuberculosis. While this may help limit inflammatory tissue damage within the host, it may also lead to a failure to control bacterial replication leading to increased bacterial burden within the tissue (Jenvey *et al.*, 2019).

17.8 Late Infection – B-Cell Responses

The role that B cells play in mycobacterial infections is not well understood and is understudied based upon the assumption that antibodies are not useful against intracellular pathogens. Although secretion of antibody is considered a main role, B cells also act as professional antigen-presenting cells and play a crucial role in T-cell activation. It has now become clear that

B-cell diversity resembles that of the T-cell subsets. B cells can be divided into subpopulations, including naïve B cells, immature B cells, plasma cells, regulatory B cells (Bregs) and memory B cells based upon lineage and differentiation markers (Moir and Fauci, 2009). Therefore, it reasonable to assume that B cells responsible for T-cell activation may critically determine the final fate and direction of the ensuing T-cell response. Evidence has been presented indicating a regulatory role for B cells in chronic inflammatory pathogenesis that is mediated through IL-10 secretion (Lampropoulou *et al.*, 2008). Upon aerosol infection with *M. tuberculosis*, B cell−/− mice had higher IL-10 production, bacterial burden and increased immunopathology in the lungs (Maglione *et al.*, 2007). Previously, a significant increase in B-cell numbers was noted in naturally infected cows demonstrating clinical signs of paratuberculosis as compared with subclinical cows or healthy controls (Waters *et al.*, 1999). In cattle, the expansion of a $CD5^{bright}$ subpopulation of B cells, suggestive of the B1a population, in the peripheral blood of cattle subclinically infected with *MAP* has been described (Stabel and Khalifeh, 2008). This finding suggests that a shift in the B-cell subpopulations is taking place during infection, which may directly or indirectly affect pathogenesis induced by inflammatory T cells. In human TB, the role of B cells is currently being reevaluated as reviewed recently (Akhbar *et al.*, 2015). They argue that in TB, antibody could potentially affect the outcome of infection in the alveolar space through opsonization, complement activation and FcR-mediated enhanced *M. tuberculosis* phagocytosis and killing. Inflammation at the alveolar level during the initial state of infection would increase serum permeability and allow the transfer of antibodies and complement into the alveolar space. B cells located at the germinal centre of lymphoid organs could play roles in the defence against infection as professional antigen-presenting cells and T-cell activators; via induction of cytokines with further T-cell expansion; and by clearing immunomodulatory *M. tuberculosis* antigens by forming immune complexes. Antibody could also affect granuloma formation through its intrinsic pro- and anti-inflammatory effects. Formation of better organized granulomas in the presence of antibody could potentially translate into reduced dissemination and local control of infection (Akhbar *et al.*, 2015). A recent study comparing active TB with latent TB infection (LTBI) in human patients revealed reduced B-cell function in LTBI patients. Moreover, these results indicated that normally functioning B cells, present in treated TB patients, are critical for optimal activation of human *M. tuberculosis*-specific T-cell immunity (Joosten *et al.*, 2016). Based upon these observations it can be hypothesized that the late or delayed activation of B cells and production of antibodies may be an intrinsic part of the pathogenic mechanisms of mycobacteria. The fact that a number of vaccination as well as antibody transfer studies indicate protective effects of antibodies also highlights the need to further study the role of B cells and antibody during mycobacterial infection (Chan *et al.*, 2014; Koets *et al.*, 2015). Beneficial effects of therapeutic heat shock protein 70 (Hsp70) vaccination in cattle with naturally acquired chronic *MAP* infection were significantly associated with the ability of the host to mount specific antibody responses, rather than with induction of specific Th1 cells (Santema *et al.*, 2013). Together these observations suggest that the interaction of B cells and T cells is an integral component in the host response to mycobacterial infection and that B cells play a role in regulating pathogenesis of infection.

17.9 Immunopathology

Granulomatous lesions in the intestine are a hallmark characteristic of *MAP* infection, and the composition of cell types within the lesions is correlated with stage of infection. The localized immune response at the site of lesions changes in accordance with lesion severity. The progression of disease from asymptomatic to clinical has been associated with decreased CD4+ T-cell and increased γδ T-cell populations in the lamina propria of the ileum, with no differences in cell frequency noted between asymptomatic infected and non-infected cattle (Koets *et al.*, 2002;Valheim *et al.*, 2004). An increase in CD8+ T cells, concomitant with reduced numbers of CD4+ T cells, was also noted in goats with clinical paratuberculosis (Navarro *et al.*, 1998). The pattern of cytokine expression in intestinal tissue is also influenced by the state of infection (Sweeney *et al.*, 1998; Lee *et al.*, 2001; Coussens *et al.*, 2004; Khalifeh and Stabel, 2004b; Tanaka *et al.*, 2005). Major findings include an upregulation of proinflammatory cytokines such as IFN-γ, IL-1, TNF-α, TGF-β, IL-5 and IL-8 in tissues from infected animals compared with non-infected controls. Studies contrasting cytokine expression in the asymptomatic and symptomatic stages of disease have demonstrated that subclinically infected animals generally have increased levels of IFN-γ and IL-18 in tissues compared with clinically infected cows (Sweeney *et al.*, 1998; Khalifeh and Stabel, 2004b; Tanaka *et al.*, 2005). In contrast, clinically infected animals have greater expression of proinflammatory cytokines, IL-1 and TNF-α; Th2 cytokines, IL-4 and IL-10; and Th3 cytokine, TGF-β (Alzuherri *et al.*, 1996; Lee *et al.*, 2001; Khalifeh and Stabel, 2004b; Tanaka *et al.*, 2005). More recently a comprehensive evaluation of cytokine gene expression in the ileal tissues of sheep that were classified into one of three groups – paucibacillary, multibacillary or asymptomatic – was performed (Smeed *et al.*, 2007). Expression of many of the proinflammatory cytokines, such as IL-1β, IL-6, IL-8, TNF-α and IL-12, increased progressively according to stage of disease, from lowest for non-infected animals to highest for multibacillary sheep. Secretion of proinflammatory cytokines may trigger a sequence of events that contributes to the formation of lesions in the tissues. Although proinflammatory cytokines are critical for the recruitment of immune cells to fight infection, protracted activation may result in tissue damage and lesion formation (Clarke, 1997). A linear increase in TGF-β expression was correlated with increased pathology in the intestinal tissue of infected sheep (Clarke, 1997), suggesting that TGF-β was responding to the need to control tissue damage potentiated by the secretion of proinflammatory cytokines (Smeed *et al.*, 2007). A study in red deer focusing also on Treg gene expression indicated that animals with clinical signs of disease and severe multibacillary infection showed reduced transcription of IL-4, FoxP3, GATA3 and TGF-β (Robinson *et al.*, 2011). The pattern of cytokine expression in target tissues of infected animals may also reflect that increased proinflammatory gene expression driven by or combined with a loss of regulatory gene expression contributes to enhanced pathology and progressive disease in late stage disease rather than protection. The role of (local) macrophages in the outcome of mycobacterial infections has recently been studied in more detail. Immunohistochemical staining of mid-ileal tissues from cattle in various stages of disease demonstrated that subclinical cows had a nearly equal ratio of host defence (M1) and resolution and repair (M2) phenotype macrophages compared with dominant M1 and M2 macrophage phenotypes in control and clinical cows, respectively (Jenvey *et al.*, 2019). Recent insights into tuberculosis granuloma pathophysiology also indicate that a delicate balance, rather than skewing in lymphocyte function and cytokine production, controls disease outcome and macrophages appear to play a critical role in maintaining the inflammatory balance (Flynn *et al.*, 2011).

17.10 Summary

The stealthy nature of *MAP* is rewarded by a chronic infection within the host, characterized by long periods of asymptomatic disease. This ensures that the disease can spread quietly and efficiently within and between groups of animals. Even animals that are able to mount robust responses to *MAP* infection are at risk of the pathogen circumventing normal control measures and eventually overtaking the host. Negative feedback mechanisms designed to control immunopathological responses may, in fact, be contributing to a loss of protective immunity in the host. Although our understanding of host responses to *MAP* has progressed in recent years, the complicated biology associated with mycobacterial pathogens continues to thwart full comprehension.

References

Abdalla, A.E., Lambert, N., Duan, X. and Xie, J. (2016) Interleukin-10 family and tuberculosis: an old story renewed. *International Journal of Biological Sciences* 12(6), 710–717. DOI: 10.7150/ijbs.13881.

Adams, J.L. and Czuprynski, C.J. (1994) Mycobacterial cell wall components induce the production of TNF-α, IL-1, and IL-6 by bovine monocytes and the murine macrophage cell line RAW 264.7. *Microbial Pathogenesis* 16(6), 401–411. DOI: 10.1006/mpat.1994.1040.

Akaki, T., Sato, K., Shimizu, T., Sano, C., Kajitani, H. *et al.* (1997) Effector molecules in expression of the antimicrobial activity of macrophages against *Mycobacterium avium* complex: roles of reactive nitrogen intermediates, reactive oxygen intermediates, and free fatty acids. *Journal of Leukocyte Biology* 62(6), 795–804. DOI: 10.1002/jlb.62.6.795.

Akhbar, J.M., Chan, J. and Casadevall, A. (2015) B cells and antibodies in the defense against *Mycobacterium tuberculosis* infection. *Immunological Reviews* 264, 167–181.

Albarrak, S.M., Waters, W.R., Stabel, J.R. and Hostetter, J.M. (2018) Evaluating the cytokine profile of the WC1+ γδ T cell subset in the ileum of cattle with the subclinical and clinical forms of *MAP* infection. *Veterinary Immunology and Immunopathology* 201, 26–31. DOI: 10.1016/j.vetimm.2018.05.003.

Alonso-Hearn, M., Patel, D., Danelishvili, L., Meunier-Goddik, L. and Bermudez, L.E. (2008) The *Mycobacterium avium* subsp. *paratuberculosis* MAP3464 gene encodes an oxidoreductase involved in invasion of bovine epithelial cells through the activation of host cell CDc42. *Infection and Immunity* 76(1), 170–178. DOI: 10.1128/IAI.01913-06.

Alonso-Hearn, M., Eckstein, T.M., Sommer, S. and Bermudez, L.E. (2010) A *Mycobacterium avium* subsp. *paratuberculosis* LuxR regulates cell envelope and virulence. *Innate Immunity* 16(4), 235–247. DOI: 10.1177/1753425909339811.

Alzuherri, H.M., Woodall, C.J. and Clarke, C.J. (1996) Increased intestinal TNF-α, IL-1β and IL-6 expression in ovine paratuberculosis. *Veterinary Immunology and Immunopathology* 49(4), 331–345. DOI: 10.1016/0165-2427(95)05477-4.

Bannantine, J.P. and Bermudez, L.E. (2013) No holes barred: invasion of the intestinal mucosa by *Mycobacterium avium* subsp. *paratuberculosis*.

Bannantine, J.P., Huntley, J.F.J., Miltner, E., Stabel, J.R. and Bermudez, L.E. (2003) The *Mycobacterium avium* subsp. *paratuberculosis* 35 kDa protein plays a role in invasion of bovine epithelial cells. *Microbiology* 149(8), 2061–2069. DOI: 10.1099/mic.0.26323-0.

Baquero, M.M. and Plattner, B.L. (2017) Bovine peripheral blood WC1 + and WC1 neg γδ T lymphocytes modulate monocyte-derived macrophage effector functions during in vitro *Mycobacterium avium* subspecies *paratuberculosis* infection. *Cellular Immunology* 315, 34–44. DOI: 10.1016/j.cellimm.2017.01.009.

Belkaid, Y. and Tarbell, K. (2009) Regulatory T cells in the control of host-microorganism interactions. *Annual Review of Immunology* 27(1), 551–589. DOI: 10.1146/annurev.immunol.021908.132723.

Boer, M.C., Joosten, S.A. and Ottenhoff, T.H.M. (2015) Regulatory T-cells at the interface between human host and pathogens in infectious diseases and vaccination. *Frontiers in Immunology* 6, 217. DOI: 10.3389/fimmu.2015.00217.

Buchbinder, E. I. and Desai, A. (2016) CTLA-4 and PD-1 pathways: similarities, differences, and implications of their inhibition. *American Journal of Clinical Oncology* 39, 98–106.

Burrells, C., Clarke, C.J., Colston, A., Kay, J.M., Porter, J. *et al.* (1998) A study of immunological responses of sheep clinically-affected with paratuberculosis (Johne's disease). The relationship of blood, mesenteric lymph node and intestinal lymphocyte responses to gross and microscopic pathology. *Veterinary Immunology and Immunopathology* 66, 343–358.

Buza, J.J., Hikono, H., Mori, Y., Nagata, R., Hirayama, S. *et al.* (2004) Neutralization of interleukin-10 significantly enhances gamma interferon expression in peripheral blood by stimulation with johnin purified protein derivative and by infection with *Mycobacterium avium* subsp. *paratuberculosis* in experimentally infected cattle with *paratuberculosis*. *Infection and Immunity* 72(4), 2425–2428. DOI: 10.1128/IAI.72.4.2425-2428.2004.

Casey, M.E., Meade, K.G., Nalpas, N.C., Taraktsoglou, M., Browne, J.A. *et al.* (2015) Analysis of the bovine monocyte-derived macrophage response to *Mycobacterium avium* subspecies *paratuberculosis* infection using RNA-seq. *Frontiers in Immunology* 6, 1–14. DOI: 10.3389/fimmu.2015.00023.

Chabot, S., Wagner, J.S., Farrant, S. and Neutra, M.R. (2006) TLRs regulate the gatekeeping functions of the intestinal follicle-associated epithelium. *The Journal of Immunology* 176(7), 4275–4283. DOI: 10.4049/jimmunol.176.7.4275.

Chan, J., Mehta, S., Bharrhan, S., Chen, Y., Achkar, J.M. *et al.* (2014) The role of B cells and humoral immunity in *Mycobacterium tuberculosis* infection. *Seminars in Immunology* 26(6), 588–600. DOI: 10.1016/j.smim.2014.10.005.

Charavaryamath, C., Gonzalez-Cano, P., Fries, P., Gomis, S., Doig, K. *et al.* (2013) Host responses to persistent *Mycobacterium avium* subspecies *paratuberculosis* infection in surgically isolated bovine ileal segments. *Clinical and Vaccine Immunology* 20(2), 156–165. DOI: 10.1128/CVI.00496-12.

Clarke, C.J. (1997) The pathology and pathogenesis of paratuberculosis in ruminants and other species. *Journal of Comparative Pathology* 116(3), 217–261. DOI: 10.1016/S0021-9975(97)80001-1.

Coussens, P.M., Colvin, C.J., Rosa, G.J.M., Perez Laspiur, J. and Elftman, M.D. (2003) Evidence for a novel gene expression program in peripheral blood mononuclear cells from *Mycobacterium avium* subsp. *paratuberculosis*-infected cattle. *Infection and Immunity* 71(11), 6487–6498. DOI: 10.1128/IAI.71.11.6487-6498.2003.

Coussens, P.M., Verman, N., Coussens, M.A., Elftman, M.D. and McNulty, A.M. (2004) Cytokine gene expression in peripheral blood mononuclear cells and tissues of cattle infected with *Mycobacterium avium* subsp. *paratuberculosis*: evidence for an inherent proinflammatory gene expression pattern. *Infection and Immunity* 72(3), 1409–1422. DOI: 10.1128/IAI.72.3.1409-1422.2004.

Coussens, P.M., Sipkovsky, S., Murphy, B., Roussey, J. and Colvin, C.J. (2012) Regulatory T cells in cattle and their potential role in bovine paratuberculosis. *Comparative Immunology, Microbiology and Infectious Diseases* 35(3), 233–239. DOI: 10.1016/j.cimid.2012.01.004.

de Almeida, D.E., Colvin, C.J. and Coussens, P.M. (2008) Antigen-specific regulatory T cells in bovine paratuberculosis. *Veterinary Immunology and Immunopathology* 125(3-4), 234–245. DOI: 10.1016/j.vetimm.2008.05.019.

de Silva, K., Begg, D. and Whittington, R. (2011) The interleukin 10 response in ovine Johne's disease. *Veterinary Immunology and Immunopathology* 139(1), 10–16. DOI: 10.1016/j.vetimm.2010.07.022.

de Silva, K., Begg, D.J., Plain, K.M., Purdie, A.C., Kawaji, S. *et al.* (2013) Can early host responses to mycobacterial infection predict eventual disease outcomes? *Preventive Veterinary Medicine* 112(3-4), 203–212. DOI: 10.1016/j.prevetmed.2013.08.006.

de Silva, K., Plain, K., J. Begg, D., C. Purdie, A. and J. Whittington, R. (2015) CD4+ T-cells, γδ T-cells and B-cells are associated with lack of vaccine protection in *Mycobacterium avium* subspecies *paratuberculosis* infection. *Vaccine* 33(1), 149–155. DOI: 10.1016/j.vaccine.2014.10.082.

de Silva, K., Plain, K., Purdie, A., Begg, D. and Whittington, R. (2018) Defining resilience to mycobacterial disease: characteristics of survivors of ovine paratuberculosis. *Veterinary Immunology and Immunopathology* 195, 56–64. DOI: 10.1016/j.vetimm.2017.11.008.

Delvig, A.A., Lee, J.J., Chrzanowska-Lightowlers, Z.M. and Robinson, J.H. (2002) TGF-beta and IFN-gamma cross-regulate antigen presentation to CD4 T cells by macrophages. *Journal of Leukocyte Biology* 72, 163–166.

Divangahi, M., Khan, N. and Kaufmann, E. (2018) Beyond killing *Mycobacterium tuberculosis*: disease tolerance. *Frontiers in Immunology* 9, 2976. DOI: 10.3389/fimmu.2018.02976.

Dorhoi, A. and Du Plessis, N. (2018) Monocytic myeloid-derived suppressor cells in chronic infections. *Frontiers in Immunology* 8, 1895. DOI: 10.3389/fimmu.2017.01895.

du Plessis, N., Loebenberg, L., Kriel, M., von Groote-Bidlingmaier, F., Ribechini, E. *et al.* (2013) Increased frequency of myeloid-derived suppressor cells during active tuberculosis and after recent *Mycobacterium tuberculosis* infection suppresses T-cell function. *American Journal of Respiratory and Critical Care Medicine* 188(6), 724–732. DOI: 10.1164/rccm.201302-0249OC.

Dudemaine, P.L., Fecteau, G., Lessard, M., Labrecque, O., Roy, J.P. *et al.* (2014) Increased blood-circulating interferon-γ, interleukin-17, and osteopontin levels in bovine paratuberculosis. *Journal of Dairy Science* 97(6), 3382–3393. DOI: 10.3168/jds.2013-7059.

Everman, J.L., Danelishvili, L., Flores, L.G. and Bermudez, L.E. (2018) MAP1203 promotes *Mycobacterium avium* subspecies *paratuberculosis* binding and invasion to bovine epithelial cells. *Frontiers in Cellular and Infection Microbiology* 8, 217–226. DOI: 10.3389/fcimb.2018.00217.

Featherstone, C. (1997) M cells: portals to the mucosal immune system. *The Lancet* 350(9086), 1230. DOI: 10.1016/S0140-6736(05)63467-8.

Flynn, J.L., Chan, J., Triebold, K.J., Dalton, D.K., Stewart, T.A. *et al.* (1993) An essential role for interferon gamma in resistance to *Mycobacterium tuberculosis* infection. *The Journal of Experimental Medicine* 178(6), 2249–2254. DOI: 10.1084/jem.178.6.2249.

Flynn, J.L., Chan, J. and Lin, P.L. (2011) Macrophages and control of granulomatous inflammation in tuberculosis. *Mucosal Immunology* 4(3), 271–278. DOI: 10.1038/mi.2011.14.

Frie, M.C., Sporer, K.R.B., Kirkpatrick, B.W. and Coussens, P.M. (2017) T and B cell activation profiles from cows with and without Johne's disease in response to in vitro stimulation with *Mycobacterium avium* subspecies *paratuberculosis*. *Veterinary Immunology and Immunopathology* 193-194, 50–56. DOI: 10.1016/j.vetimm.2017.10.005.

Fulton, S.A., Cross, J.V., Toossi, Z.T. and Boom, W.H. (1998) Regulation of interleukin-12 by interleukin-10, transforming growth factor-beta, tumor necrosis factor-α, and interferon-γ in human monocytes infected with *Mycobacterium tuberculosis* H37Ra. *The Journal of Infectious Diseases* 178(4), 1105–1114. DOI: 10.1086/515698.

Gebert, A., Rothkotter, H.J. and Pabst, R. (1996) M cells in Peyer's patches of the intestine. *International Reviews of Cytology* 167, 91–159.

Guzman, E., Hope, J., Taylor, G., Smith, A.L., Cubillos-Zapata, C. *et al.* (2014) Bovine γδ T cells are a major regulatory T cell subset. *The Journal of Immunology* 193(1), 208–222. DOI: 10.4049/jimmunol.1303398.

Harriff, M.J., Danelishvili, L., Wu, M., Wilder, C., McNamara, M. *et al.* (2009) *Mycobacterium avium* genes MAV_5138 and MAV_3679 are transcriptional regulators that play a role in invasion of epithelial cells, in part by their regulation of CipA, a putative surface protein interacting with host cell signaling pathways. *Journal of Bacteriology* 191(4), 1132–1142. DOI: 10.1128/JB.01359-07.

Hoek, A., Rutten, V.P., Kool, J., Arkesteijn, G.J., Bouwstra, R.J. *et al.* (2009) Subpopulations of bovine WC1(+) gammadelta T cells rather than CD4(+)CD25(high) Foxp3(+) T cells act as immune regulatory cells ex vivo. *Veterinary Research* 40, 6.

Hope, J.C., Thom, M.L., McCormick, P.A. and Howard, C.J. (2004) Interaction of antigen presenting cells with mycobacteria. *Veterinary Immunology and Immunopathology* 100(3–4), 187–195. DOI: 10.1016/j.vetimm.2004.04.007.

Hostetter, J., Steadham, E., Haynes, J., Bailey, T. and Cheville, N. (2003) Phagosomal maturation and intracellular survival of *Mycobacterium avium* subspecies *paratuberculosis* in J774 cells. *Comparative Immunology, Microbiology and Infectious Diseases* 26(4), 269–283. DOI: 10.1016/S0147-9571(02)00070-X.

Hostetter, J., Kagan, R. and Steadham, E. (2005) Opsonization effects on *Mycobacterium avium* subsp. *paratuberculosis*—macrophage interactions. *Clinical Diagnostics and Laboratory Investigation* 12, 793–796.

Hussain, T., Shah, S.Z.A., Zhao, D., Sreevatsan, S. and Zhou, X. (2016) The role of IL-10 in *Mycobacterium avium* subsp. *paratuberculosis* infection. *Cell Communication and Signaling* 14(1), 29–42. DOI: 10.1186/s12964-016-0152-z.

Ito, S., Ansari, P., Sakatsume, M., Dickensheets, H., Vazquez, N. *et al.* (1999) Interleukin-10 inhibits expression of both interferon α- and interferon γ- induced genes by suppressing tyrosine phosphorylation of STAT1. *Blood* 93(5), 1456–1463. DOI: 10.1182/blood.V93.5.1456.

Jenvey, C.J., Shircliff, A.L., Bannantine, J.P. and Stabel, J.R. (2019) Phenotypes of macrophages present in the intestine are impacted by stage of disease in cattle naturally infected with *Mycobacterium avium* subsp. *paratuberculosis*. *PLoS ONE* 14(5), e0217649. DOI: 10.1371/journal.pone.0217649.

Joosten, S.A., van Meijgaarden, K.E., del Nonno, F., Baiocchini, A., Petrone, L. *et al.* (2016) Patients with tuberculosis have a dysfunctional circulating B-cell compartment, which normalizes following successful treatment. *PLoS Pathogens* 12(6), e1005687. DOI: 10.1371/journal.ppat.1005687.

Kadivar, M., Petersson, J., Svensson, L. and Marsal, J. (2016) CD8αβ^{+} γδ T cells: A novel T cell subset with a potential role in inflammatory bowel disease. *The Journal of Immunology* 197(12), 4584–4592. DOI: 10.4049/jimmunol.1601146.

Karcher, E.L., Beitz, D.C. and Stabel, J.R. (2008) Modulation of cytokine gene expression and secretion during the periparturient period in dairy cows naturally infected with *Mycobacterium avium* subsp. *paratuberculosis*. *Veterinary Immunology and Immunopathology* 123(3–4), 277–288. DOI: 10.1016/j.vetimm.2008.02.006.

Kathaperumal, K., Park, S.-U., McDonough, S., Stehman, S., Akey, B. *et al.* (2008) Vaccination with recombinant *Mycobacterium avium* subsp. *paratuberculosis* proteins induces differential immune responses and protects calves against infection by oral challenge. *Vaccine* 26(13), 1652–1663. DOI: 10.1016/j.vaccine.2008.01.015.

Kaufmann, S.H. (1996) Gamma/delta and other unconventional T lymphocytes: what do they see and what do they do? *Proceedings of the National Academy of Sciences* 93(6), 2272–2279. DOI: 10.1073/pnas.93.6.2272.

Khalifeh, M.S. and Stabel, J.R. (2004a) Effects of gamma interferon, interleukin-10, and transforming growth factor beta on the survival of *Mycobacterium avium* subsp. *paratuberculosis* in monocyte-derived macrophages from naturally infected cattle. *Infection and Immunity* 72(4), 1974–1982. DOI: 10.1128/IAI.72.4.1974-1982.2004.

Khalifeh, M.S. and Stabel, J.R. (2004b) Upregulation of transforming growth factor-beta and interleukin-10 in cows with clinical Johne's disease. *Veterinary Immunology and Immunopathology* 99(1–2), 39–46. DOI: 10.1016/j.vetimm.2004.01.009.

Koets, A., Rutten, V., Hoek, A., van Mil, F., Muller, K. *et al.* (2002) Progressive bovine paratuberculosis is associated with local loss of CD4+ T cells, increased frequency of gamma delta T cells, and related changes in T-cell function. *Infection and Immunity* 70(7), 3856–3864. DOI: 10.1128/IAI.70.7.3856-3864.2002.

Koets, A.P., Eda, S. and Sreevatsan, S. (2015) The within host dynamics of *Mycobacterium avium* ssp. *paratuberculosis* infection in cattle: where time and place matter. *Veterinary Research* 46(1), 61. DOI: 10.1186/s13567-015-0185-0.

Kohno, K., Kataoka, J., Ohtsuki, T., Suemoto, Y., Okamoto, I. *et al.* (1997) IFN-gamma-inducing factor (IGIF) is a costimulatory factor on the activation of Th1 but not Th2 cells and exerts its effect independently of IL-12. *Journal of Immunology* 158, 1541–1550.

Koo, H.C., Park, Y.H., Hamilton, M.J., Barrington, G.M., Davies, C.J. *et al.* (2004) Analysis of the immune response to *Mycobacterium avium* subsp. *paratuberculosis* in experimentally infected calves. *Infection and Immunity* 72(12), 6870–6883. DOI: 10.1128/IAI.72.12.6870-6883.2004.

Lamont, E.A., O'Grady, S.M., Davis, W.C., Eckstein, T. and Sreevatsan, S. (2012) Infection with *Mycobacterium avium* subsp. *paratuberculosis* results in rapid interleukin-1β release and macrophage transepithelial migration. *Infection and Immunity* 80(9), 3225–3235. DOI: 10.1128/IAI.06322-11.

Lampropoulou, V., Hoehlig, K., Roch, T., Neves, P., Calderón Gómez, E. *et al.* (2008) TLR-activated B cells suppress T cell-mediated autoimmunity. *The Journal of Immunology* 180(7), 4763–4773. DOI: 10.4049/jimmunol.180.7.4763.

Larsen, A.B., Merkal, R.S. and Cutlip, R.C. (1975) Age of cattle as related to resistance to infection with *Mycobacterium paratuberculosis*. *American Journal of Veterinary Research* 36, 255–257.

Lee, H., Stabel, J.R. and Kehrli, M.E. (2001) Cytokine gene expression in ileal tissues of cattle infected with *Mycobacterium paratuberculosis*. *Veterinary Immunology and Immunopathology* 82(1–2), 73–85. DOI: 10.1016/S0165-2427(01)00340-3.

Leite, F.L., Eslabão, L.B., Pesch, B., Bannantine, J.P., Reinhardt, T.A. *et al.* (2015) ZAP-70, CTLA-4 and proximal T cell receptor signaling in cows infected with *Mycobacterium avium* subsp. *paratuberculosis*. *Veterinary Immunology and Immunopathology* 167(1–2), 15–21. DOI: 10.1016/j.vetimm.2015.06.017.

Lugton, I.W. (1999) Mucosa-associated lymphoid tissues as sites for uptake, carriage and excretion of tubercle bacilli and other pathogenic mycobacteria. *Immunology and Cell Biology* 77(4), 364–372. DOI: 10.1046/j.1440-1711.1999.00836.x.

Maglione, P.J., Xu, J. and Chan, J. (2007) B cells moderate inflammatory progression and enhance bacterial containment upon pulmonary challenge with *Mycobacterium tuberculosis*. *The Journal of Immunology* 178(11), 7222–7234. DOI: 10.4049/jimmunol.178.11.7222.

Magombedze, G., Eda, S. and Ganusov, V.V. (2014) Competition for antigen between Th1 and Th2 responses determines the timing of the immune response switch during *Mycobacterium avium* subspecies *paratuberculosis* infection in ruminants. *PLoS Computational Biology* 10(1), e1003414. DOI: 10.1371/journal.pcbi.1003414.

Magombedze, G., Eda, S. and Stabel, J. (2015) Predicting the role of IL-10 in the regulation of the adaptive immune responses in *Mycobacterium avium* subsp. *paratuberculosis* infections using mathematical models. *PLoS ONE* 10(11), e0141539. DOI: 10.1371/journal.pone.0141539.

Miltner, E., Daroogheh, K., Mehta, P.K., Cirillo, S.L.G., Cirillo, J.D. *et al.* (2005) Identification of *Mycobacterium avium* genes that affect invasion of the intestinal epithelium. *Infection and Immunity* 73(7), 4214–4221. DOI: 10.1128/IAI.73.7.4214-4221.2005.

Moir, S. and Fauci, A.S. (2009) B cells in HIV infection and disease. *Nature Reviews Immunology* 9(4), 235–245. DOI: 10.1038/nri2524.

Mombaerts, P., Arnoldi, J., Russ, F., Tonegawa, S. and Kaufmann, S.H.E. (1993) Different roles of αβ and γδ T cells in immunity against an intracellular bacterial pathogen. *Nature* 365(6441), 53–56. DOI: 10.1038/365053a0.

Momotani, E., Whipple, D.L., Thiermann, A.B. and Cheville, N.F. (1988) Role of M cells and macrophages in the entrance of *Mycobacterium paratuberculosis* into domes of ileal Peyer's patches in calves. *Veterinary Pathology* 25(2), 131–137. DOI: 10.1177/030098588802500205.

Nagata, R., Kawaji, S., Minakawa, Y., Wang, X., Yanaka, T. *et al.* (2010) A specific induction of interleukin-10 by the Map41 recombinant PPE antigen of *Mycobacterium avium* subsp. *paratuberculosis*. *Veterinary Immunology and Immunopathology* 135(1–2), 71–78. DOI: 10.1016/j.vetimm.2009.11.002.

Navarro, J.A., Ramis, G., Seva, J., Pallarés, F.J. and Sánchez, J. (1998) Changes in lymphocyte subsets in the intestine and mesenteric lymph nodes in caprine paratuberculosis. *Journal of Comparative Pathology* 118(2), 109–121. DOI: 10.1016/S0021-9975(98)80003-0.

Nembrini, C., Abel, B., Kopf, M. and Marsland, B.J. (2006) Strong TCR signaling, TLR ligands, and cytokine redundancies ensure robust development of type 1 effector T cells. *The Journal of Immunology* 176(12), 7180–7188. DOI: 10.4049/jimmunol.176.12.7180.

Nicol, L., Wilkie, H., Gossner, A., Watkins, C., Dalziel, R. *et al.* (2016) Variations in T cell transcription factor gene structure and expression associated with the two disease forms of sheep paratuberculosis. *Veterinary Research* 47(1), 83. DOI: 10.1186/s13567-016-0368-3.

O'Donnell, M.A., Luo, Y., Chen, X., Szilvasi, A., Hunter, S.E. *et al.* (1999) Role of IL-12 in the induction and potentiation of IFN-gamma in response to Bacillus Calmette-Guerin. *Journal of Immunology* 163, 4246–4252.

Okagawa, T., Konnai, S., Nishimori, A., Ikebuchi, R., Mizorogi, S. *et al.* (2016) Bovine immunoinhibitory receptors contribute to suppression of *Mycobacterium avium* subsp. *paratuberculosis*-specific T-cell responses. *Infection and Immunity* 84(1), 77–89. DOI: 10.1128/IAI.01014-15.

Onishi, R.M. and Gaffen, S.L. (2010) Interleukin-17 and its target genes: mechanisms of interleukin-17 function in disease. *Immunology* 129(3), 311–321. DOI: 10.1111/j.1365-2567.2009.03240.x.

Othieno, C., Hirsch, C.S., Hamilton, B.D., Wilkinson, K., Ellner, J.J. *et al.* (1999) Interaction of *Mycobacterium tuberculosis*-induced transforming growth factor β1 and interleukin-10. *Infection and Immunity* 67(11), 5730–5735. DOI: 10.1128/IAI.67.11.5730-5735.1999.

Park, H.-E., Park, H.-T., Jung, Y.H. and Yoo, H.S. (2018) Gene expression profiles of immune-regulatory genes in whole blood of cattle with a subclinical infection of *Mycobacterium avium* subsp. *paratuberculosis*. *PLoS ONE* 13(4), e0196502. DOI: 10.1371/journal.pone.0196502.

Peleman, R., Wu, J., Fargeas, C. and Delespesse, G. (1989) Recombinant interleukin 4 suppresses the production of interferon gamma by human mononuclear cells. *The Journal of Experimental Medicine* 170(5), 1751–1756. DOI: 10.1084/jem.170.5.1751.

Plattner, B.L., Huffman, E.L. and Hostetter, J.M. (2013) Gamma-delta T-cell responses during subcutaneous *Mycobacterium avium* subspecies *paratuberculosis* challenge in sensitized or naive calves using matrix biopolymers. *Veterinary Pathology* 50(4), 630–637. DOI: 10.1177/0300985812463404.

Ponnusamy, D., Periasamy, S., Tripathi, B.N. and Pal, A. (2013) *Mycobacterium avium* subsp. *paratuberculosis* invades through M cells and enterocytes across ileal and jejunal mucosa of lambs. *Research in Veterinary Science* 94(2), 306–312. DOI: 10.1016/j.rvsc.2012.09.023.

Pott, J., Basler, T., Duerr, C.U., Rohde, M., Goethe, R. *et al.* (2009) Internalization-dependent recognition of *Mycobacterium avium* ssp. *paratuberculosis* by intestinal epithelial cells. *Cellular Microbiology* 11(12), 1802–1815. DOI: 10.1111/j.1462-5822.2009.01372.x.

Purdie, A.C., Plain, K.M., Begg, D.J., de Silva, K. and Whittington, R.J. (2012) Expression of genes associated with the antigen presentation and processing pathway are consistently regulated in early *Mycobacterium avium* subsp. *paratuberculosis* infection. *Comparative Immunology, Microbiology and Infectious Diseases* 35(2), 151–162. DOI: 10.1016/j.cimid.2011.12.007.

Rahim, S.S., Khan, N., Boddupalli, C.S., Hasnain, S.E. and Mukhopadhyay, S. (2005) Interleukin-10 (IL-10) mediated suppression of IL-12 production in RAW 264.7 cells also involves c-rel transcription factor. *Immunology* 114(3), 313–321. DOI: 10.1111/j.1365-2567.2005.02107.x.

Robinson, M.W., O'Brien, R., Mackintosh, C.G., Clark, R.G. and Griffin, J.F.T. (2011) Immunoregulatory cytokines are associated with protection from immunopathology following *Mycobacterium avium* subspecies *paratuberculosis* infection in red deer. *Infection and Immunity* 79(5), 2089–2097. DOI: 10.1128/IAI.00779-10.

Sajiki, Y., Konnai, S., Okagawa, T., Nishimori, A., Maekawa, N. *et al.* (2018) Prostaglandin E2 induction suppresses the Th1 immune responses in cattle with Johne's disease. *Infection and Immunity* 86(5), e00910-17. DOI: 10.1128/IAI.00910-17.

Santema, W., Rutten, V., Segers, R., Poot, J., Hensen, S. *et al.* (2013) Postexposure subunit vaccination against chronic enteric mycobacterial infection in a natural host. *Infection and Immunity* 81(6), 1990–1995. DOI: 10.1128/IAI.01121-12.

Saunders, B.M., Frank, A.A., Cooper, A.M. and Orme, I.M. (1998) Role of γδ T cells in immunopathology of pulmonary *Mycobacterium avium* infection in mice. *Infection and Immunity* 66(11), 5508–5514. DOI: 10.1128/IAI.66.11.5508-5514.1998.

Secott, T.E., Lin, T.L. and Wu, C.C. (2001) Fibronectin attachment protein homologue mediates fibronectin binding by *Mycobacterium avium* subsp. *paratuberculosis*. *Infection and Immunity* 69(4), 2075–2082. DOI: 10.1128/IAI.69.4.2075-2082.2001.

Secott, T.E., Lin, T.L. and Wu, C.C. (2004) *Mycobacterium avium* subsp. *paratuberculosis* fibronectin attachment protein facilitates M-cell targeting and invasion through a fibronectin bridge with host integrins. *Infection and Immunity* 72(7), 3724–3732. DOI: 10.1128/IAI.72.7.3724-3732.2004.

Shankar, G., Pestano, L.A. and Bosch, M.L. (2003) Interferon-gamma added during bacillus Calmette-Guerin induced dendritic cell maturation stimulates potent Th1 immune responses. *Journal of Translational Medicine* 1(1), 7. DOI: 10.1186/1479-5876-1-7.

Shin, S.J., Chang, C.-F., Chang, C.-D., McDonough, S.P., Thompson, B. *et al.* (2005) *In vitro* cellular immune responses to recombinant antigens of *Mycobacterium avium* subsp. *paratuberculosis*. *Infection and Immunity* 73(8), 5074–5085. DOI: 10.1128/IAI.73.8.5074-5085.2005.

Sigurðardóttir, Ó.G., Press, C.M. and Evensen, Ø. (2001) Uptake of *Mycobacterium avium* subsp. *paratuberculosis* through the distal small intestinal mucosa in goats: an ultrastructural study. *Veterinary Pathology* 38(2), 184–189. DOI: 10.1354/vp.38-2-184.

Sigurðardóttir, O., Valheim, M. and Press, C.M. (2004) Establishment of *Mycobacterium avium* subsp. *paratuberculosis* infection in the intestine of ruminants. *Advanced Drug Delivery Reviews* 56(6), 819–834. DOI: 10.1016/j.addr.2003.10.032.

Sigurðardóttir, O.G., Bakke-McKellep, A.M., Djønne, B. and Evensen, O. (2005) *Mycobacterium avium* subsp. *paratuberculosis* enters the small intestinal mucosa of goat kids in areas with and without Peyer's patches as demonstrated with the everted sleeve method. *Comparative Immunology, Microbiology and Infectious Diseases* 28(3), 223–230. DOI: 10.1016/j.cimid.2005.01.004.

Smeed, J.A., Watkins, C.A., Rhind, S.M. and Hopkins, J. (2007) Differential cytokine gene expression profiles in the three pathological forms of sheep is an important component of the anti-inflammatory response in *Mycobacterium avium* subsp. *paratuberculosis*-infected bovine monocytes. *Microbial Pathogenesis* 41, 59–66.

Smith, R.L., Schukken, Y.H. and Gröhn, Y.T. (2015) A new compartmental model of *Mycobacterium avium* subsp. *paratuberculosis* infection dynamics in cattle. *Preventive Veterinary Medicine* 122(3), 298–305. DOI: 10.1016/j.prevetmed.2015.10.008.

Souza, C.D., Evanson, O.A. and Weiss, D.J. (2006) Mitogen activated protein kinase p38 pathway is an important component of the anti-inflammatory response in *Mycobacterium avium* subsp. *paratuberculosis*-infected bovine monocytes. *Microbial Pathogenesis* 41(2–3), 59–66. DOI: 10.1016/j.micpath.2006.04.002.

Souza, C., Davis, W.C., Eckstein, T.M., Sreevatsan, S. and Weiss, D.J. (2013) Mannosylated lipoarabinomannans from *Mycobacterium avium* subsp. *paratuberculosis* alters the inflammatory response by bovine macrophages and suppresses killing of *Mycobacterium avium* subsp. *avium* organisms. *PLoS ONE* 8(9), 375924. DOI: 10.1371/journal.pone.0075924.

Stabel, J.R. (2000) Transitions in immune responses to *Mycobacterium paratuberculosis*. *Veterinary Microbiology* 77(3–4), 465–473. DOI: 10.1016/S0378-1135(00)00331-X.

Stabel, J.R. and Ackermann, M.R. (2002) Temporal *Mycobacterium paratuberculosis* infection in T-cell receptor (TCR)-alpha and TCR-delta-deficient mice. *Veterinary Immunology and Immunopathology* 89(3–4), 127–132. DOI: 10.1016/S0165-2427(02)00167-8.

Stabel, J.R. and Khalifeh, M.S. (2008) Differential expression of CD5 on B lymphocytes in cattle infected with *Mycobacterium avium* subsp. *paratuberculosis*. *Veterinary Immunology and Immunopathology* 126(3–4), 211–219. DOI: 10.1016/j.vetimm.2008.07.004.

Stabel, J.R., Kimura, K. and Robbe-Austerman, S. (2007) Augmentation of secreted and intracellular gamma interferon following johnin purified protein derivative sensitization of cows naturally infected with *Mycobacterium avium* subsp. *paratuberculosis*. *Journal of Veterinary Diagnostic Investigation* 19(1), 43–51. DOI: 10.1177/104063870701900107.

Sturgill-Koszycki, S., Schlesinger, P.H., Chakraborty, P., Haddix, P.L., Collins, H.L. *et al.* (1994) Lack of acidification in *Mycobacterium* phagosomes produced by exclusion of the vesicular proton-ATPase. *Science* 263(5147), 678–681. DOI: 10.1126/science.8303277.

Sugawara, I., Yamada, H., Kaneko, H., Mizuno, S., Takeda, K. *et al.* (1999) Role of interleukin-18 (IL-18) in mycobacterial infection in IL-18-gene-disrupted mice. *Infection and Immunity* 67(5), 2585–2589. DOI: 10.1128/IAI.67.5.2585-2589.1999.

Sweeney, R.W., Jones, D.E., Habecker, P. and Scott, P. (1998) Interferon-gamma and interleukin 4 gene expression in cows infected with *Mycobacterium paratuberculosis*. *American Journal of Veterinary Research* 59(7), 842–847.

Tanaka, S., Itohara, S., Sato, M., Taniguchi, T. and Yokomizo, Y. (2000) Reduced formation of BALB/c mice inoculated with *Mycobacterium avium* subsp. *paratuberculosis*. *Veterinary Pathology* 37, 415–421.

Tanaka, S., Sato, M., Onitsuka, T., Kamata, H. and Yokomizo, Y. (2005) Inflammatory cytokine gene expression in different types of granulomatous lesions during asymptomatic stages of bovine paratuberculosis. *Veterinary Pathology* 42(5), 579–588. DOI: 10.1354/vp.42-5-579.

Taylor, D.L., Zhong, L., Begg, D.J., de Silva, K. and Whittington, R.J. (2008) Toll-like receptor genes are differentially expressed at the sites of infection during the progression of Johne's disease in outbred sheep. *Veterinary Immunology and Immunopathology* 124(1–2), 132–151. DOI: 10.1016/j.vetimm.2008.02.021.

Teixeira, L.K., Fonseca, B.P.F., Vieira-de-Abreu, A., Barboza, B.A., Robbs, B.K. *et al.* (2005) IFN-gamma production by CD8+ T cells depends on NFAT1 transcription factor and regulates Th differentiation. *The Journal of Immunology* 175(9), 5931–5939. DOI: 10.4049/jimmunol.175.9.5931.

Thirunavukkarasu, S., de Silva, K., Whittington, R.J. and Plain, K.M. (2013) In vivo and in vitro expression pattern of Toll-like receptors in *Mycobacterium avium* subspecies *paratuberculosis* infection. *Veterinary Immunology and Immunopathology* 156(1–2), 20–31. DOI: 10.1016/j.vetimm.2013.08.008.

Triccas, J.A., Sun, L., Palendira, U. and Britton, W.J. (2002) Comparative affects of plasmid-encoded interleukin 12 and interleukin 18 on the protective efficacy of DNA vaccination against *Mycobacterium tuberculosis*. *Immunology and Cell Biology* 80(4), 346–350. DOI: 10.1046/j.1440-1711.2002.01087.x.

Tuo, W., Estes, D.M. and Brown, W.C. (1999) Comparative effects of interleukin-12 and interleukin-4 on cytokine responses by antigen-stimulated memory CD4+ T cells of cattle: IL-12 enhances IFN-gamma production, whereas IL-4 has marginal effects on cytokine expression. *Journal of Interferon & Cytokine Research* 19(7), 741–749. DOI: 10.1089/107999099313587.

Tyrer, P., Foxwell, A.R., Cripps, A.W., Apicella, M.A. and Kyd, J.M. (2006) Microbial pattern recognition receptors mediate M-cell uptake of a gram-negative bacterium. *Infection and Immunity* 74(1), 625–631. DOI: 10.1128/IAI.74.1.625-631.2006.

Uzonna, J.E., Chilton, P., Whitlock, R.H., Habecker, P.L., Scott, P. *et al.* (2003) Efficacy of commercial and field-strain *Mycobacterium paratuberculosis* vaccinations with recombinant IL-12 in a bovine experimental infection model. *Vaccine* 21(23), 3101–3109. DOI: 10.1016/S0264-410X(03)00261-5.

Valheim, M., Sigurdardóttir, O.G., Storset, A.K., Aune, L.G. and Press, C.M. (2004) Characterization of macrophages and occurrence of T cells in intestinal lesions of subclinical paratuberculosis in goats. *Journal of Comparative Pathology* 131(2–3), 221–232. DOI: 10.1016/j.jcpa.2004.04.004.

Wang, J., Wakeham, J., Harkness, R. and Xing, Z. (1999) Macrophages are a significant source of type 1 cytokines during mycobacterial infection. *Journal of Clinical Investigation* 103(7), 1023–1029. DOI: 10.1172/JCI6224.

Waters, W.R., Stabel, J.R., Sacco, R.E., Harp, J.A., Pesch, B.A. *et al.* (1999) Antigen-specific B-cell unresponsiveness induced by chronic *Mycobacterium avium* subsp. *paratuberculosis* infection of cattle. *Infection and Immunity* 67(4), 1593–1598.

Waters, W.R., Miller, J.M., Palmer, M.V., Stabel, J.R., Jones, D.E. *et al.* (2003) Early induction of humoral and cellular immune responses during experimental *Mycobacterium avium* subsp. *paratuberculosis* infection of calves. *Infection and Immunity* 71(9), 5130–5138. DOI: 10.1128/IAI.71.9.5130-5138.2003.

Weiss, D.J., Evanson, O.A., Moritz, A., Deng, M.Q. and Abrahamsen, M.S. (2002) Differential responses of bovine macrophages to *Mycobacterium avium* subsp. *paratuberculosis* and *Mycobacterium avium* subsp. *avium*. *Infection and Immunity* 70(10), 5556–5561. DOI: 10.1128/IAI.70.10.5556-5561.2002.

Weiss, D.J., Evanson, O.A., Deng, M. and Abrahamsen, M.S. (2004) Sequential patterns of gene expression by bovine monocyte-derived macrophages associated with ingestion of mycobacterial organisms. *Microbial Pathogenesis* 37(4), 215–224. DOI: 10.1016/j.micpath.2004.07.001.

Weiss, D.J., Evanson, O.A., de Souza, C. and Abrahamsen, M.S. (2005) A critical role of interleukin-10 in the response of bovine macrophages to infection by *Mycobacterium avium* subsp. *paratuberculosis*. *American Journal of Veterinary Research* 66(4), 721–726. DOI: 10.2460/ajvr.2005.66.721.

Weiss, D.J., Souza, C.D., Evanson, O.A., Sanders, M. and Rutherford, M. (2008) Bovine monocyte TLR2 receptors differentially regulate the intracellular fate of *Mycobacterium avium* subsp. *paratuberculosis* and *Mycobacterium avium* subsp. *avium*. *Journal of Leukocyte Biology* 83(1), 48–55. DOI: 10.1189/jlb.0707490.

Whittington, R.J., Begg, D.J., de Silva, K., Plain, K.M. and Purdie, A.C. (2012) Comparative immunological and microbiological aspects of paratuberculosis as a model mycobacterial infection. *Veterinary Immunology and Immunopathology* 148(1–2), 29–47. DOI: 10.1016/j.vetimm.2011.03.003.

18 Cultivation of *Mycobacterium avium* subsp. *paratuberculosis*

Richard Whittington*
University of Sydney, Sydney, Australia

18.1 Introduction

Cultivation and identification of *Mycobacterium avium* subsp. *paratuberculosis* (*MAP*) remains the definitive diagnostic test for paratuberculosis. A positive culture provides information that no other test can: viable pathogen present; potentially infectious. Despite the ubiquity of rapid molecular diagnostic tests, culture is still required in a wide range of contexts. Ironically, the simplification, marketing and accessibility of molecular testing platforms has led to a decline in expertise in culture, and traditional skills such as making culture media are disappearing from laboratories.

Performing culture on clinical samples informs the case definitions that are vital in disease control (Whittington *et al.*, 2017). Culture of *MAP* allows classification of animals using these definitions, which permeate research studies on topics ranging from genome-wide association (Purdie *et al.*, 2011) through to diagnostic test validation (Alinovi *et al.*, 2009; Gardner *et al.*, 2011; Keller *et al.*, 2014; Plain *et al.*, 2014; Butot *et al.*, 2019). Culture is required to propagate the organism for experimental infection models to study the disease and to enumerate viable cells in the inoculum in such models (Begg *et al.*, 2010; Corbett *et al.*, 2018). It is the basis for production of commercial vaccines and for studies of their efficacy (Reddacliff *et al.*, 2006b; Mercier *et al.*, 2014). In food safety and public health, culture of *MAP* is central to understanding exposure, disinfection and thermal inactivation (Whittington *et al.*, 2010; Van Brandt *et al.*, 2011b; Faria *et al.*, 2014; Botsaris *et al.*, 2016; Galiero *et al.*, 2016). In medicine, culture is used to assess the susceptibility of *MAP* to antimicrobials (Krishnan *et al.*, 2009b). In epidemiology, culture informs our understanding of transmission and spread of *MAP* through data on its prevalence in livestock (Bauman *et al.*, 2016; Noll *et al.*, 2017; Selim *et al.*, 2019), its infection of wild animals (Beard *et al.*, 2001; Kumar *et al.*, 2010; Matos *et al.*, 2014; Salgado *et al.*, 2016; Galiero *et al.*, 2018), its persistence in a viable form in the environment (Eppleston *et al.*, 2014), its fate and transport to waterways, in air and even to plants (Aboagye and Rowe, 2011; Eisenberg *et al.*, 2011; Salgado *et al.*, 2011; Kaevska *et al.*, 2014).

The pathogenesis of *MAP* infection is such that subclinical paratuberculosis is always more common than clinical disease (Whittington and Sergeant, 2001; Vazquez *et al.*, 2014).While paratuberculosis can be detected at post-mortem examination or in an abattoir during meat inspection, *MAP* is present in the intestinal mucosa or mesenteric lymph nodes long before microscopic or gross lesions develop in these tissues, rendering culture more sensitive than gross pathology (Elze *et al.*, 2013; Stringer *et al.*, 2013). For diagnosis, the microscopic lesions induced by *MAP*

*richard.whittington@sydney.edu.au

© CAB International 2020. *Paratuberculosis: Organism, Disease, Control, 2nd Edition* (eds M.A. Behr *et al.*)

in the intestinal tract of ruminants are quite characteristic, but even where acid-fast bacilli (AFB) are visualized in tissues associated with the granulomatous infiltrate, specificity is not assured because other mycobacterial species sometimes infect the gut and can cause similar changes. Furthermore, it is not possible to determine from mild lesions alone whether an animal is still infected because recovery is possible (Dennis *et al.*, 2011; Begg *et al.*, 2018). Immunohistochemistry and *in situ* hybridization reactions conducted on tissue sections containing AFB are not specific for *MAP* and neither are serological tests such as enzyme-linked immunosorbent assay (ELISA). *MAP* is defined as an obligate parasite and pathogen; for this reason, its precise identification by cultivation is the benchmark for diagnosis of paratuberculosis. At herd/flock level, if *MAP* is detected there must be – or must recently have been – an infected animal somewhere nearby. Thus, it is standard practice to use culture to confirm a presumptive diagnosis in individual animals and for index cases. Furthermore, culture of samples pooled from more than one animal, or the environment, is now accepted for herd and flock surveillance. One area where the specificity of culture is yet to be widely applied is in the investigation of the proposed link between *MAP* and Crohn's disease in humans, where there is still an overreliance on molecular tests to demonstrate organism DNA.

Important aspects in the culture of *MAP* are the analytical sensitivity of the particular method, the contamination rate, the cost of the test and the time taken to report results. Unfortunately, this information is lacking for many culture protocols. Culture of *MAP* is usually used as a gold standard to evaluate other types of diagnostic test, but unless the most sensitive culture protocols are employed, sensitivity estimates for tests such as direct polymerase chain reaction (PCR), serology and histopathology will be inflated. It is a fact that not all culture protocols are equally sensitive and therefore claims in many diagnostic test evaluations are flawed.

18.2 Historical Perspective

The first report of reliable cultivation of *MAP* was published in 1912 and was both detailed and rigorous (Twort and Ingram, 1912). The study was stimulated by the challenge to grow an organism that was obviously present in the lesions of affected livestock but clearly different from *Mycobacterium bovis*, which was prevalent at the time. Twort and Ingram deduced that as tuberculosis and paratuberculosis could occur in the same cow there must be a nutrient missing from *in vitro* culture media that precluded growth of the Johne's disease bacillus. They speculated that it had lived a pathogenic existence for long enough to lose an attribute possessed by its wild ancestor but retained by the tuberculosis organism. To compensate for this loss, heat-killed human tubercle bacilli were added to a Dorset egg medium base and *MAP* then grew from clinical samples from cows with paratuberculosis. They tested many strains of *M. tuberculosis* for their ability to support the growth of *MAP* and found this property to be quite variable so they evaluated other mycobacterial species. The timothy grass bacillus – *Mcobacterium phlei* – was superior to *M. tuberculosis* and inclusion of this organism, or mycobactin derived from it or other mycobacteria, has enabled the culture of *MAP* ever since.

Twort and Ingram studied the active ingredient from *M. phlei* and found that it could be extracted in hot ethanol. They showed that inclusion of egg in culture media was beneficial, that extracted liver broth and agar media were less suitable than a clotted egg medium, that albumin was not needed for growth and that glycerol was beneficial. *MAP* grew at 28–43°C. They described the morphology of *in vitro* grown *MAP*, showed that it was non-spore forming, non-motile, acid-fast, was a strict aerobe when growing but was not killed in the absence of oxygen for 3 months at 39°C. They reported a slow rate of growth with tiny discrete colonies forming in 3–5 weeks. Cultures were not killed by light and were resistant to disinfectants.

In summary, much of the useful knowledge about cultivation of *MAP* was provided in one paper in 1912. This breakthrough enabled additional work that has been of enormous benefit, including the evaluation of alternative culture media, assessment of decontamination solutions and improvements to sample preparation protocols (Minett, 1942; Merkal *et al.*, 1982). Although progress was made prior to the molecular era, including the realization that there were different strains of *MAP* with distinct cultural requirements, substantial advances have been made since, particularly with the introduction of PCR to identify the presence and also the growth of *MAP* in cultures.

18.3 General Principles for the Cultivation of *MAP*

The cultivation of *MAP* is based on traditional methods for the culture of slowly growing mycobacteria from clinical samples (Merkal *et al.*, 1964; Gillespie, 1999). There are four critical steps, which will be described in more detail in the sections that follow:

1. Decontamination of clinical samples to destroy or suppress irrelevant, mostly rapidly growing microorganisms, which include both bacteria and fungi. These are present in vast numbers in faeces and most types of environmental samples, but there are fewer in tissues, blood or milk.
2. Prolonged incubation in an appropriate medium containing antimicrobial agents to suppress any remaining contaminants for long enough for *MAP* to emerge.
3. Recognition of *MAP* colonies on a solid medium, or observation of a particular sign of growth or reaching a prescribed incubation period endpoint in a broth medium.
4. Identification of *MAP* by phenotypic and/or genotypic means.
5. Subculture to confirm phenotype (mycobactin dependence), or for molecular typing.

18.4 Requirement for Mycobactin

Mycobactin is an essential component of media for primary culture of *MAP* in most circumstances. It is a chelating compound required for iron acquisition and storage, and demonstration of mycobactin dependence is a taxonomic criterion for identification of *MAP* (Merkal and McCullough, 1982; Thorel *et al.*, 1990). The underlying basis of the requirement for mycobactin and gene regulation of iron metabolism in *MAP* is still being determined (Janagama *et al.*, 2009; Wang *et al.*, 2016). Exceptions to the rule of mycobactin dependence for primary isolation are discussed below.

18.5 Requirement for Egg Yolk and Pyruvate

The benefits of including egg or egg yolk in culture media for isolation of *MAP* have been noted since 1912 but the reasons it is needed are still not clear (Twort and Ingram, 1912). Certainly some strains of the organism can be grown in some media without egg, while it is vital in other circumstances (Cousins *et al.*, 1995; Whittington *et al.*, 1999; Grant *et al.*, 2003). When benzalkonium chloride (BAC) or hexadecylpyridinium chloride (HPC) is used to decontaminate faecal samples, egg yolk is required in 7H10 or 7H9 media to enable growth of *MAP*; simply washing the *MAP* pellet once in water enabled growth in these media without addition of egg yolk (Whipple and Merkal, 1983). Thus egg may play a role in neutralizing disinfectants carried over into the culture medium with the inoculum (Merkal and Curran, 1974). This may explain why egg yolk was a necessary ingredient for primary isolation but not for subculture (Cousins *et al.*, 1995). There is more information about egg yolk in section 18.6.

Sodium pyruvate was reported to stimulate the growth of *MAP* and is included in both the Lowenstein Jensen medium (LJ) and the Herrold's egg yolk medium (HEYM) used by some laboratories (Merkal and Curran, 1974; Jorgensen, 1982; Kim *et al.*, 1989; Juste *et al.*, 1991; Eamens *et al.*, 2000). Pyruvate enables *MAP* colonies to grow to a larger size on LJ medium, compensating for the inhibitory effects of the antibiotics that are included in this medium (Jorgensen, 1982).

18.6 Other Chemical and Physical Aspects of Culture

Antimicrobials such as malachite green, cycloheximide, amphotericin B, vancomycin, nalidixic acid, chloramphenicol, penicillin G, polymyxin B, trimethoprim, azlocillin and ampicillin have been included in various media but thorough evaluation of their potential to inhibit growth of *MAP* is mostly lacking (Shin, 1989; Whitlock and Rosenberger, 1990; Stabel, 1997; Kalis *et al.*, 1999; Whittington *et al.*, 1999; Gumber and Whittington, 2007).

Dyes may be added to solid media to improve visibility of colonies. Brilliant green was effective (Minett, 1942). Methylene blue becomes concentrated in the colonies as they develop (Fig. 18.1) (Whittington *et al.*, 1999). Dye uptake by colonies can vary between strains and variant colonies may be observed within a culture (Parrish *et al.*, 2004). Colonies are readily observed on media containing malachite green, particularly LJ medium, which has a smooth surface. The surface of HEYM slopes is often finely pitted, which can make recognition of small colonies quite challenging despite the inclusion of the dye. Small colonies are relatively difficult to see on modified Middlebrook 7H10 and 7H11 agars, which are pale yellow due to their egg yolk content.

Polyoxyethylene sorbates (Tweens) and polyoxyethylene stearates (POES) influence the *in vitro* growth of mycobacteria, leading to alterations in replication rate, intracellular storage vacuoles, cell surface properties and colony morphology; they can be stimulatory or inhibitory depending on concentration (Stinson and Solotorovsky, 1971; Cutler *et al.*, 1987; van Boxtel *et al.*, 1990; Realini *et al.*, 1997). Tweens are a source of carbon and energy and are wetting agents, enabling dispersal of bacterial cells (Stinson and Solotorovsky, 1971). One per cent w/v Tween 80 was associated with optimal growth of *MAP* clinical isolates in Middlebrook 7H9 medium with mycobactin J (van Boxtel *et al.*, 1990) but its inclusion in similar medium in another study had equivocal effects on growth of C and S strains of *MAP* (Whittington *et al.*, 2013). POES was present in BACTEC 12B medium through the antibiotic mixture PANTA-PLUS to which it may have been added to aid dissolution of antibiotics during media preparation, but it is not present in BACTEC MGIT Para TB medium (BD). It did not appear to be required for optimal growth of *MAP*-C and S strains in Middlebrook 7H9 medium with egg yolk (Whittington *et al.*, 2013), probably because the egg yolk provides major carbon and energy sources in addition to those already included in the complex medium, and also provides the surfactant lecithin. Oleic acid, the hydrolysis product of Tween, is an additive in Middlebrook 7H11 agar and MGIT medium, but it is not required for growth of *MAP* in the presence of egg yolk because *MAP* grows equally well on Middlebrook 7H10 agar that does not contain added oleic acid (Whittington *et al.*, 1999).

Twort and Ingram, and later Merkal, investigated many fundamental properties of media and incubation conditions (Twort and Ingram, 1912; Merkal *et al.*, 1982). Shaking a liquid culture reduced growth of *MAP*, a CO_2 atmosphere was not required and pH 6 was beneficial to growth compared with a higher pH.

18.7 Resuscitation of *MAP* and Growth Stimulation

Dormancy, a physiological state in which *MAP* is viable but non-culturable, can occur during adverse environmental conditions (Whittington *et al.*, 2004). Lamont *et al.* (2012) reported that 1-year-old broth cultures of *MAP* contained a mix of vegetative cells and a morphotype resembling a spore. Conceptually it may be possible to improve culture success rates by stimulating dormant cells to resume vegetative growth. Recently, Rock *et al.* (2016) reported significant improvement in growth and recovery rate of presumed dormant *MAP* when a commercial reducing agent Oxyrase® was incorporated in M7H9C liquid medium, but there does not appear to have been any further work published. Claims have also been made about the substantial benefits of a non-disclosed commercial growth stimulant for cultivation of *MAP* (Bull *et al.*, 2016), but without proper validation data or follow-up studies it is not possible to assess these claims.

18.8 Contemporary Culture Media

Both solid and liquid media can be used to cultivate *MAP*. Solid media are sometimes cheaper, involve less instrumentation and identification of the organism can be simpler. However, they have lower analytical sensitivity. Culture media for *MAP* must include essential nutrients, often

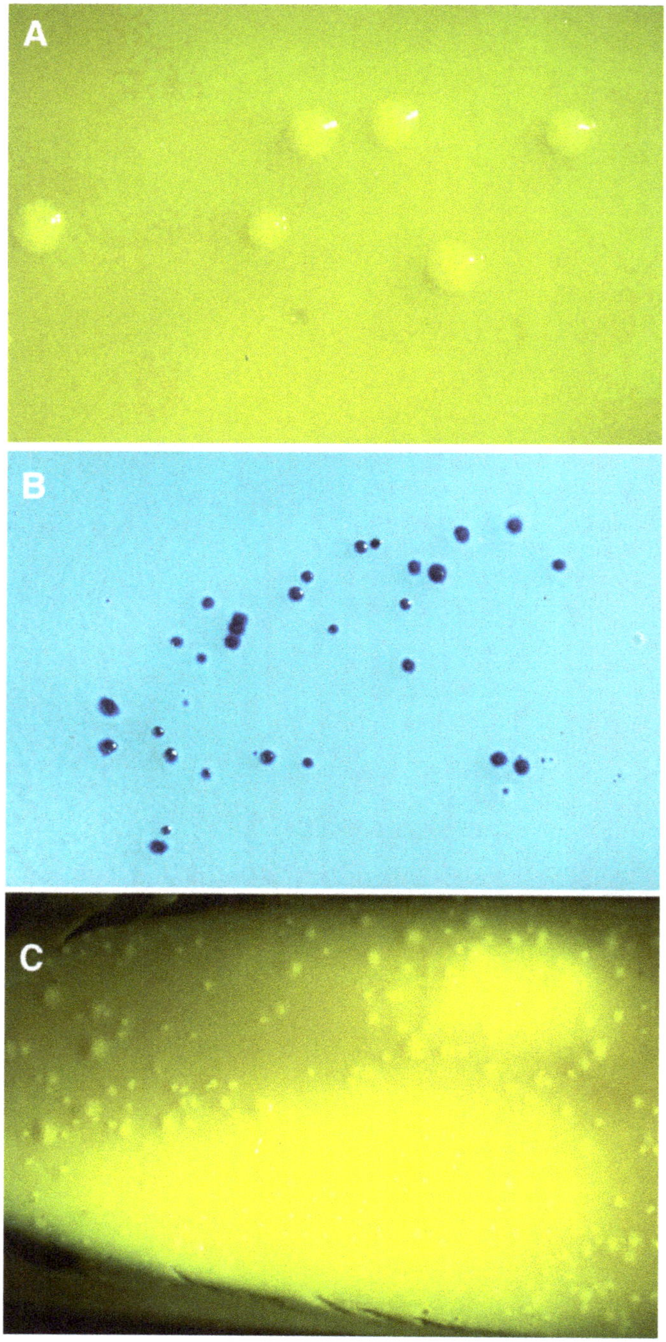

Fig. 18.1. Colonies of S strain *Mycobacterium avium* subsp. *paratuberculosis* (*MAP*) growing on modified Middlebrook 7H10 agar. Upper panel, subculture after 13 weeks' incubation; middle panel, subculture after 13 weeks' incubation with methylene blue incorporated in the agar; lower panel, primary isolate after 20 weeks' incubation.

Table 18.1. Comparison of contemporary solid media suitable for cultivation of *Mycobacterium avium* subsp. *paratuberculosis* Cattle-strains (*MAP*-C). Amounts are per litre.

Ingredient	HEYM[a]	LJ[b]
LJ base (Difco)	-	23.25 g
Sodium chloride	4.5 g	-
Beef extract	2.7 g	-
Peptone	9.0 g	-
Agar	15.3 g	-
Whole egg homogenate	-	625 ml
Egg yolk	120 ml	-
Malachite green	100 mg	250 mg
Glycerol	34 ml	7.5 ml
Sodium pyruvate	4.1 g[c]	4.0 g
Mycobactin	2.0 mg	2.0 mg
Cycloheximide	-	0.75 g
Chloramphenicol	-	0.20 g
Penicillin G	-	2 × 10^5 U

[a](Merkal and Curran, 1974; Herrold, 1931; Merkal *et al.*, 1964; Merkal, 1970). Most animal health laboratories in the USA include the antibiotics vancomycin, amphotericin B and nalidixic acid (VAN) in HEYM (Robbe-Austermann, pers comm, 2009)
[b](Kalis *et al.*, 2000).
[c]Optional.

include antimicrobials to discourage growth of contaminants and may include dyes to assist recognition of colonies.

18.8.1 Solid culture media

Many media were evaluated by early workers leading to progressive refinement of simple egg-based slants such as Dubos medium, but these have fallen out of favour and do not seem to have been used for more than 35 years (Saxegaard, 1985). Researchers eventually reached consensus that two media were most suitable for clinical microbiology: LJ in some European countries and HEYM elsewhere (Table 18.1). However, the media based on Middlebrook 7H10 or 7H11 (solid media) and 7H9 (broth) are arguably better base media for culture of all known strains of *MAP* within a reasonable incubation period (see below), but for optimal growth, egg yolk must be added to these media (Table 18.2).

18.8.2 Liquid culture media

Chemically defined liquid culture medium

Laboratory adaptation of some strains of *MAP* has enabled their propagation in the chemically defined Watson–Reid broth without mycobactin or egg (Table 18.2) (Watson, 1935; Morrison, 1965). This medium is unsuitable for cultivation of field strains of *MAP* (Merkal and Curran, 1974) but is still used for bulk culture for antigen production.

Liquid culture medium with a radiometric growth indicator

The inclusion of C^{14}-labelled palmitic acid as a carbon source in liquid medium to enable highly sensitive radiometric detection of C^{14}-labelled carbon dioxide produced through microbial respiration was a breakthrough in medical microbiology (Middlebrook *et al.*, 1977; Reggiardo and Tigertt, 1977). The growth signal is not specific, but rather triggers an examination of the broth for the pathogen of interest. In practice, the culture bottles were incubated routinely and transferred periodically to a semi-automated ion-chamber reader (BACTEC 460 machine), which sampled the gas phase after piercing a rubber seal with a sterile needle. This method was adapted for *MAP* by adding egg yolk and mycobactin to commercial BACTEC 12B medium (Becton Dickinson) (no longer available commercially), which has a Middlebrook 7H9 base (Damato *et al.*, 1987; Damato and Collins, 1990).

Liquid culture medium systems with another growth indicator

Other liquid culture systems designed for medical microbiology have been applied to cultivate *MAP* (Grant *et al.*, 2003; Ellingson *et al.*, 2004; Stich *et al.*, 2004). All utilize a proprietary medium that is inevitably based on Middlebrook 7H9 broth. The BACTEC MGIT 960 system (Becton Dickinson) relies on detection of a fluorescent

Table 18.2. Composition of the Middlebrook liquid and solid media, which are suitable for cultivation of both *Mycobacterium avium* subsp. *paratuberculosis* Cattle and Sheep strains (*MAP*-C and *MAP*-S), and the chemically defined Watson–Reid medium, which supports the growth of some laboratory adapted strains. Amounts are per litre.

Ingredient	Modified Middlebrook 7H9 broth (M7H9C/BACTEC 12B;)[a]	Modified Middlebrook 7H10 agar[b]	Modified Middlebrook 7h11 agar[b]	Watson–Reid medium[c]
BASE MEDIUM				
Casein digest	667 mg[d,e]	1.0g[e]	1.0g	-
Ammonium sulfate	333 mg	500 mg	500 mg	-
Monopotassium phosphate	667 mg	1.5g	1.5g	2g
Disodium phosphate	1.7g	1.5g	1.5g	-
Sodium citrate	67 mg	400 mg	400 mg	-
Magnesium sulfate	33 mg	25 mg	50 mg	1g
Calcium chloride	0.33 mg	0.5 mg	-	20 mg
Zinc sulfate	0.67 mg	1 mg	-	10 mg
Copper sulfate	0.67 mg	1 mg	-	-
Arginine	-	-	-	5g
L-glutamic acid	333 mg	500 mg	500 mg	-
Cobalt chloride	-	-	-	2 mg
Sodium chloride	-	-	850 mg	2g
Ferric ammonium citrate	27 mg	40 mg	40 mg	75 mg
Pyridoxine	0.67 mg	1 mg	1 mg	-
Biotin	0.33 mg	0.5 mg	0.5 mg	-
Malachite green	-	0.25 mg	1 mg	-
Bacto agar	-	15g	15g	-
ENRICHMENT	ADC/-	ADC	OADC	
Oleic acid	-	-	50 mg	-
Albumin fraction V, Bovine	1.34/3.30 g[d]	5.0g	5.0g	-
Dextrose	534 mg/-	2.0g	2.0g	10g
Catalase	0.8 mg/32,000 U[d]	3 mg	4 mg	-
ADDITIVES				
Egg yolk	167 ml[f]	250 ml	250 ml	-
Mycobactin J	0.83 mg[f]	1.25 mg	1.25 mg	-
PANTA[g]	33.3 ml[f]	50 ml	50 ml	-
Glycerol	-	5 ml	5 ml	60 ml

[a]Adapted from Whittington *et al*. (2013).
[b]Adapted from Whittington *et al*. (1999).
[c]From Merkal and Curran (1974).
[d]Albumin: 1.34g added by user to form M7H9C; 3.30g added by manufacturer to form BACTEC 12B medium. Catalase: 0.8 mg added by user to form M7H9C; 32,000 U added by manufacturer to form BACTEC 12B medium. Dextrose: added by user to form M7H9C; not added to BACTEC 12B. BACTEC 12B also contains 667 uCi C-14 palmitic acid.
[e]Added as casitone to M7H9C and 7H10 agar; already added by manufacturer to form BACTEC 12B.
[f]Added to the enriched BACTEC 12B media at a rate of 1 ml egg yolk, 5 µg (100 µl) Mycobactin J, 200 µl PANTA and 700 µl water per 4 ml vial; final volume 6 ml/vial. (Data from Difco Manual 10th Edition 1984; Becton Dickenson BACTEC 12B package insert.)
[g]Consists of polymyxin B 1000 U/ml, amphoteracin B 100 µg/ml, nalidixic acid 400 µg/ml, trimethoprim 100 µg/ml, azlocillin 100 µg/ml; PANTA PLUS includes polyoxyethylene stearate, which is not essential.

signal, which develops from an indicator at the base of the culture vial as oxygen is consumed during microbial respiration; the amount of egg yolk that can be added is limited as it interferes with measurement of the growth signal. ESP culture technology (Trek Diagnostics) relies on detection of a change in the pressure of the gas phase in a sealed culture vial. The MB/BacT system (BioMerieux) is based on a colorimetric indication of carbon dioxide production. Each of these culture formats enables early identification of microbial growth by incubating culture vials within a machine that regularly monitors the growth signal. Like the original BACTEC 460 system, these newer methods require the formal identification of *MAP* when a growth signal is detected. The MGIT system is gaining acceptance for detection of *MAP* in clinical samples (Grant *et al.*, 2003; Shin *et al.*, 2007). However, vancomycin, an ingredient included at 18 μg/ml in commercial MGIT ParaTB medium for paratuberculosis, is inhibitory to some common strains of *MAP* (Thornton *et al.*, 2002; Gumber and Whittington, 2007). Others found 100 μg/ml vancomycin to be the threshold concentration for inhibition of growth (Pribylova *et al.*, 2012).

Liquid culture media without a growth indicator

M7H9C liquid medium was developed in Australia as a direct response to the withdrawal of commercial BACTEC 12B medium from the global market. M7H9C medium was validated using *MAP*-C and *MAP*-S strains and a large number of clinical samples in Australia and subsequently in Germany (Whittington *et al.*, 2013; Schwalm *et al.*, 2018). Based on Middlebrook 7H9 medium and after supplementation, M7H9C mimics modified BACTEC 12B in composition so that prior evaluations demonstrating capacity of the latter to support the growth of diverse strains of *MAP* remain relevant (Whittington *et al.*, 2011). The shelf life of M7H9C exceeds 12 months at 7–8°C (Schwalm *et al.*, 2018). It can be made in any laboratory by appropriately experienced staff and is cost effective compared with the commercial broths with growth readouts that are designed for specific machine platforms.

A comparison of the compositions of M7H9C, BACTEC12B and MGIT ParaTB medium is provided in Table 18.3. Other liquid media based on Middlebrook 7H9 have been developed but with limited validation. An example is that of Pozzato *et al.* (2011), partially described by the authors as consisting of Middlebrook 7H9 powder (0.47%), casitone (0.1%), glycerol (0.5%), 10% OADC, egg yolk (16%), mycobactin J (1 μg/ml) and PANTA PLUS (2.5%).

18.9 Cultural Requirements of Different Strains of *MAP*

A variety of types of *MAP* occur in cattle, sheep, goats, deer and other hosts (Stevenson, 2015). Molecular genetic differences between *MAP* isolates led to the naming of so-called C (cattle) and S (sheep) strains based on the hosts of origin of the isolates (Collins *et al.*, 1990a). Other types of *MAP* have been reported: yellow-orange pigmented strains in the UK and so-called bison strains in the USA and India (see Chapter 6, this volume). Compared with *MAP*-C, the *MAP*-S, pigmented and bison strains grow less well, poorly or not at all on HEYM (Collins *et al.*, 1990a; Juste *et al.*, 1991; Whittington *et al.*, 1999; Whitlock *et al.*, 1999; Stevenson *et al.*, 2002). Furthermore, *MAP*-S strains do not grow readily in MGIT ParaTB medium due to the inclusion of vancomycin (Gumber and Whittington, 2007). Kawaji *et al.* (2014) excluded vancomycin from MGIT medium and improved the detection time for growth of a laboratory adapted *MAP*-S strain, but the *MAP*-S strain still required a prolonged incubation in this medium compared with the *MAP*-C strain.

In an attempt to understand apparent disparities in culture success between strains and countries, 35 field isolates comprising 14 different *Bst*EII IS*900*/IS*1311* genotypes and representing *MAP*-C, *MAP*-S and bison strains that originated from the USA, Spain, Northern Ireland and Australia were propagated in BACTEC 12B medium, Middlebrook 7H10 agar, LJ agar, HEYM, modified Middlebrook 7H10 agar, Middlebrook 7H11 agar and WatsonReid (WR) agars (Whittington *et al.*, 2011). Most *MAP*-S strains grew poorly on HEYM and pyruvate was inhibitory to some isolates, but

Table 18.3. Comparison of the composition of the Middlebrook-based liquid culture media M7H9C, BACTEC 12B and MGIT ParaTB.

	Amount/litre			Amount/vial		
Ingredient	M7H9C	BACTEC	MGIT	M7H9C	BACTEC	MGIT
Base-Middlebrook 7H9	3.1 g			18.6 mg	4.0 ml	7.0 ml
C-14 palmitic acid	-	667 uCi	-	-	4 uCi	-
Fluorescent indicator	-	-	13.25 ml	-	-	110 µl
Casein	667 mg[a]	667 mg[b]	48.2 mg [c]	4.0 mg	4.0 mg	0.4 mg
Bovine albumin	1340 mg[d]	3300 mg[b]	4820 mg[c]	8.0 mg	19.8 mg	40 mg
Dextrose	534 mg[d]	-	-	3.2 mg	-	-
Catalase	0.8 mg[d]	32,000 U[b]	4627 U[c]	4.8 µg	192 U	38.4 U
Oleic acid	-	-	9.64 mg[c]	-	-	0.08 mg
Egg yolk	167 ml[e]	167 ml[e]	30.1 ml[f]	1.0 ml	1.0 ml	0.25 ml
Mycobactin J	0.83 mg[e]	0.83 mg[e]	Included[g]	5 µg	5 µg	Included
Polyoxyethylene stearate	optional	133 mg[h]	-	Optional	0.80 mg [h]	-
Amphotericin B	3.33 mg[h]	3.33 mg[h]	7 mg[i]	20 µg	20 µg	60 µg
Nalidixic acid	13.3 mg[h]	13.3 mg[h]	18 mg[i]	80 µg	80 µg	150 µg
Vancomycin hydrochloride	-	-	18 mg[i]	-	-	150 µg
Polymyxin B	33,300 U[h]	33,300 U[h]	-	200 U	200 U	-
Trimethoprim	3.33 mg[h]	3.33 mg[h]	-	20 µg	20 µg	-
Azlocillin	3.33 mg[h]	3.33 mg[h]	-	20 µg	20 µg	-
Ampicillin (optional)	100 mg	100 mg	100 mg	0.6 mg	0.6 mg	0.83 mg
Final volume	1000 ml	1000 ml	1000 ml	6.0 ml	6.0 ml	8.3 ml

[a]Added by user as casitone.
[b]Already added by manufacturer to Middlebrook 7H9 broth to form BACTEC 12B medium.
[c]Added by user to each MGIT ParaTB medium vial as MGIT ParaTB supplement.
[d]Added by user as ADC, 26.7 ml per litre or 0.16 ml per vial.
[e]Added by user after dispensing medium to each M7H9C or BACTEC 12B medium vial (100 µl of 50 µg/ml mycobactin J solution, 600 µl water, 1 ml egg yolk) resulting in a final volume of 6 ml medium.
[f]Egg yolk (0.25 ml plus 0.25 ml water) added by user to each MGIT ParaTB medium vial resulting in a final volume of 8.3 ml; BD recommends egg yolk enrichment, a 50% egg yolk solution, at a rate of 0.5 ml per MGIT ParaTB medium vial.
[g]Amount not specified by the manufacturer.
[h]Added by user to each M7H9C or BACTEC 12B medium vial as 200 µl PANTA PLUS; an alternative form of antibiotics such as BD BBL MGIT PANTA, which does not contain polyoxyethylene stearate, may be used.
[i]Added by user to each MGIT ParaTB medium vial from stock antibiotic solutions.

all 35 grew on modified Middlebrook 7H10 agar. Growth was slower and less prolific on LJ agar and mycobactin J was required for growth on all media except 7H11 agar. These results were generally supported by those of Dimareli-Malli *et al.* (2013), who compared the suitability of HEYM, 7H11 and LJ agars to support growth of *MAP* from small ruminants where both *MAP*-C and *MAP*-S strains were present in the livestock population of northern Greece.

The suboptimal growth of *MAP* that has been reported historically in diverse settings, including for example reports of extremely long incubation periods and minute colony sizes, likely has been strongly influenced by the interaction between strain of *MAP* and type of culture medium. When an optimal culture medium is used (such as a suitably modified Middlebrook medium), less difference between the growth phenotypes of diverse strains of

MAP is observed. Caution should be exercised in the use of a commercial liquid culture system marketed for detection of *MAP* unless it has been proven to support the growth of all common strains of *MAP*. Otherwise it is not possible to interpret negative culture outcomes with any confidence.

18.10 Incubation Period

MAP is a very slowly growing mycobacterium; generation times in liquid media range from 1.3 to 4.4 days and are inversely proportional to the inoculum size (Lambrecht *et al.*, 1988; Kim *et al.*, 2002; Elguezabal *et al.*, 2011). The reasons for the very slow rate of growth of *MAP* compared with its close relative *M. avium* subsp. *hominissuis* are unclear (Bannantine *et al.*, 2003). When grown on the most suitable media, colonies of *MAP* typically do not appear for several weeks and sometimes not for several months. Incubation periods of 12–20 weeks are typically specified for solid media before negative culture results are reported from livestock. Occasionally, extremely long incubation periods such as 9 months on HEYM have been reported in veterinary studies (Salgado *et al.*, 2009), raising a question about the suitability of the chosen culture protocol. Growth in liquid media is more rapid than on solid media (Damato and Collins, 1990; Schwalm *et al.*, 2018), and incubation is usually terminated after 8–12 weeks. In optimal liquid media, differences in growth rate associated with the C or S genotype of *MAP* were minor, but there can be an influence of the background source of the bacteria (which can be correlated with the decontamination protocol applied to them) because *MAP*-S isolates from sheep faeces and soil grew more slowly than those from sheep tissues and subcultures (Reddacliff *et al.*, 2003a), while isolates of different genotype grew more slowly from sheep than from other ruminant hosts (Abendaño *et al.*, 2012). Furthermore, isolates from sheep with paucibacillary disease grew more slowly than those from sheep with multibacillary disease, suggesting some growth inhibition in the former (Reddacliff *et al.*, 2003a). This was analogous to the longer generation time and delayed growth of *MAP* from the faeces of cows with low- compared with high-shedding rates (Kim *et al.*, 2002). Overall, these data suggest that the environment in which *MAP* lies immediately prior to incubation in culture medium has a large impact on its subsequent *in vitro* growth.

18.11 Specific Applications of Culture

18.11.1 Faecal samples

Decontamination of samples

The first attempts to cultivate *MAP* from clinical samples utilized the methods that were available at the time for tuberculosis. These involved soaking or mixing the sample in a decontamination solution for long enough to inactivate contaminants but not so long as to completely destroy all *MAP* present. Sodium hypochlorite (previously called antiformin) was once popular (Dunkin, 1928) but trials were also undertaken with sodium hydroxide (NaOH), sulfuric acid, hydrochloric acid, phenol, benzalkonium chloride (BAC) and oxalic acid (OA) (Minett, 1942; Merkal *et al.*, 1964; Gunnarsson, 1979; Gunnarsson and Fodstad, 1979). By 1980 a combination of NaOH and OA was adopted in protocols used in northern Europe (Jorgensen, 1982). The cationic quaternary ammonium compound HPC was used first in the USA, replacing BAC (Merkal, 1984), and is now the basis for current protocols in many countries including North America, Europe and Australasia (Table 18.4). Antimicrobial dyes such as malachite green and/or one or more antibiotics – vancomycin, amphotericin B, nalidixic acid, polymyxin B, carbenicillin, trimethoprim and neomycin – have often been included in decontamination solutions (Cameron, 1956; Shin, 1989; Whitlock and Rosenberger, 1990; Kalis *et al.*, 2000; Ellingson *et al.*, 2005).

Faecal samples present a particular problem due to their high load of enteric bacteria, some of which break through decontamination routines and multiply in culture media. Common contaminants from bovine faeces in Chile

Table 18.4. Common methods for cultivation of *Mycobacterium avium* subsp. *paratuberculosis* (*MAP*) from faeces.

Method	Decontamination	Medium and inoculum	Incubation	Notes and references[a]
North American methods				
1. Sedimentation (one step-HPC)	Faeces 1 g added to 25 ml 0.75% HPC, shaken for 30 min and allowed to stand for 2 h; 4 ml of supernatant aspirated into a transfer pipette and allowed to settle overnight in the upright pipette. Drops from the pipette are used as inoculum	HEYM, sodium pyruvate 4.1 mg/ml; three slopes per sample; three drops of inoculum per tube allowed to dry on surface of loosely capped tubes for 1 week then tubes were sealed	16 weeks	Kim *et al.* (1989) Modified 7H9 liquid media may be used in place of HEYM (Collins *et al.*, 1990b); Cousins *et al.* (1995)
2. Sedimentation (two-step water-HPC)	Step 1. Faeces 1 g added to 35–40 ml water, shaken for 30 min at 22°C then allowed to settle for 30 min. Step 2. 5 ml of upper part of supernatant added to 25 ml 0.9% HPC (final conc. 0.75%), mixed and allowed to stand overnight (16–24 h) at room temperature. The sediment is used as inoculum	HEYM; four slopes per sample; one may have no MJ; 0.1 ml of inoculum per tube allowed to dry on surface of loosely capped tubes for 1 week then tubes were sealed	12 weeks; slopes were examined with a dissecting microscope	Whipple and Merkal (1983), Whipple *et al.* (1991) HPC replaced BAC in the original method of Merkal in 1982
3. Sedimentation and centrifugation (two-step double incubation centrifugation)	Step 1 as above except faeces 2–5 g, water 35 ml. Step 2 as above except 0.9% HPC made up in ½ strength BHI, incubated at 35–37°C, then centrifuged at 900 × *g* for 30 min. The pellet is resuspended in 1 ml ½ strength BHI containing VAN and incubated overnight or longer (it may be stable for 3 days at 35–37°C). This suspension is used as inoculum	HEYM; 4 slopes per sample; 0.25 ml of inoculum per tube allowed to dry on surface of loosely capped tubes for 1 week then tubes were sealed	16 weeks	Shin (1989), Whitlock and Rosenberger (1990) Inclusion of BHI and VAN was developed at Cornell University by Shin. This method is used in most countries. In Australia, 10–12 ml saline is used instead of water; incubation in VAN for 3 days; modified 7H9 liquid media or 7H10 agar may be used in place of HEYM (Whittington, 2009)

Continued

Table 18.4. Continued

Method	Decontamination	Medium and inoculum	Incubation	Notes and references[a]
4. Sedimentation and centrifugation (two-step double incubation centrifugation method, NADC modification)	Step 1 as above except faeces 1–2 g, water 35 ml. Step 2 as above except the entire 25–30 ml supernatant is removed, centrifuged 1700 × *g* 20 min, the pellet resuspended in 30 ml 0.9% HPC-BHI, mixed and incubated overnight at 37°C, centrifuged 1700 x *g* 20 min, the pellet resuspended in VAN overnight 37°C. This suspension is used as inoculum	HEYM with naladixic acid 50 µg/ml, vancomycin 50 µg/ml; four slopes per sample; 0.2 ml of inoculum per tube allowed to dry on surface of loosely capped tubes for 1 week then tubes were sealed	12 weeks; slopes were examined with a dissecting microscope	Stabel (1997)
North European culture method				
5. Three-step sedimentation centrifugation	Faeces 5 ml or approx 2 g mixed with 8 ml 4% NaOH and shaken for 15 min then centrifuged 1000 x *g* 15 min. Pellet added to 5 ml of 5 mg/ml OA and 1 mg/ml malachite green solution and mixed for 15 min then centrifuged 1000 x *g* 15 min. Pellet mixed with 4 ml of 0.5 mg/ml neomycin and 50 mg/ml amphotericin B solution and allowed to stand overnight at 20°C. The clear upper layer just above the sediment is used as inoculum	LJ Three or four drops of inoculum was allowed to dry on the surface of loosely capped tubes for 24 h then tubes were sealed	26 weeks	Jorgensen (1982); Kalis *et al*. (1999). Jorgensen's decontamination method was based on Beerworth (1967).

[a]The first reference is the earliest mention of the original method; VAN, vancomycin 100 µg/ml, amphotericin B 50 µg/ml, naladixic acid 100 µg/ml; BHI, brain heart infusion.
OA, oxalic acid; NaOH, sodium hydroxide; HPC, hexadecylpyridinium chloride; BAC, benzalkonium chloride.

were *Paenibacillus* sp., *Enterobacteriaceae* and *Pseudomonas aeruginosa* (Steuer *et al*., 2015). Most protocols require removal of the particulate matter with which many microbes associate. This is achieved by suspending faeces in water or decontamination solution followed by filtration through cheesecloth (Cameron, 1956), light centrifugation (Merkal *et al*., 1968) or more commonly by sedimentation. The filtrate or supernatant is removed, and if it is already a decontamination solution it can be inoculated directly into culture media. In contrast, a water suspension is added first to a decontamination solution and the sediment from this is inoculated (Merkal *et al*., 1968).

Concerns about the failure to inactivate contaminants that were not in a vegetative growth phase led to incubation of the faecal suspension in antimicrobial solutions made up in nutrient media, in the hope that spores would germinate and then be killed. This approach has been used for preparation of milk samples and is commonly applied to faecal samples where HPC and antibiotics are constituted in half-strength brain heart infusion (BHI) broth (Shin, 1989; Whitlock and Rosenberger, 1990; Ellingson *et al.*, 2005).

Culture protocols have evolved independently in different laboratories over the years (Table 18.4). In some countries in Europe, the NaOH-OA decontamination protocol is used in conjunction with LJ medium, while in the US and most other countries, sedimentation methods with HPC are combined with HEYM or liquid media. In a small study, 20 AFB smear-positive bovine faecal and tissue samples were processed using either NaOH-OA or HPC and inoculated on to LJ media with resulting sensitivity of 70 and 85%, respectively; all samples were positive by culture in modified BACTEC 12B medium after HPC decontamination (Collins *et al.*, 1990b). In the absence of data from a larger study it is unclear whether the decontamination protocol or medium can be interchanged without loss of sensitivity or an increase in the contamination rate. An attempt was made to compare the two most popular types of solid media – HEYM and LJ – in a large study in Denmark but using only the NaOH-OA decontamination protocol. Ironically the HEYM medium appeared to have slightly greater sensitivity (Nielsen *et al.*, 2004).

Decontamination of other types of samples is discussed below under each sample type.

Concentration of MAP from the faecal sample

To increase the analytical sensitivity of faecal culture, *MAP* may be concentrated by a variety of methods during or after decontamination of the sample. This was first tried by Merkal, who allowed coarse particles to settle then removed and centrifuged the supernatant of the faecal water suspension; the pellet was then decontaminated (Merkal *et al.*, 1964). Others modified this by centrifuging the decontamination solution (Kim *et al.*, 1989). Another modification combined both approaches by centrifuging all of the water supernatant, resuspending its pellet in decontamination solution and after incubation, centrifuging the entire solution (Stabel, 1997). In a similar approach using three decontaminants, centrifugation was undertaken after the first two steps (Kalis *et al.*, 2000). As an alternative to centrifugation, 3 μm filtration to trap *MAP* was tried (Collins *et al.*, 1990b). These approaches were all successful in concentrating *MAP* from a sample, although the penalty may be an increased contamination rate as other bacteria may also be concentrated and included in the inoculum (Eamens *et al.*, 2000). Immunomagnetic bead separation of *MAP* was ineffective for faecal samples (Mason *et al.*, 2001).

Pooled faecal culture

The pooling of faecal samples from more than one animal for culture is a logical way to reduce the cost of detection of *MAP* at herd or flock level. Initial attempts were not encouraging due to loss of sensitivity associated with the dilution effect of pooling (Vialard *et al.*, 1993). However, pooled culture was shown to be highly sensitive in sheep provided that an animal with multibacillary disease was present, as these cases shed about 10^8 *MAP*/gram of faeces (Whittington *et al.*, 2000b). Pooling rates of greater than 50 sheep per culture were clearly possible without losing sensitivity. For practical application a pooling rate of 50 was selected and was shown to provide higher flock-level sensitivity than serological examination of the same animals (Whittington *et al.*, 2000a; Sergeant *et al.*, 2002). By 2002 in Australia this test together with abattoir surveillance for gross lesions of paratuberculosis had replaced other screening tests. Research in cattle showed similar benefits, but a lower pooling rate was required, generally five (Kalis *et al.*, 2000; Wells *et al.*, 2002, 2003; Eamens *et al.*, 2007b, 2008). Both computer modelling and empiric results showed that pooled faecal culture is practical and cost effective for cattle (van Schaik *et al.*, 2003; Kalis *et al.*, 2004). The same approach is applicable in goats and probably most other species (Eamens *et al.*, 2007a). Pooled faecal culture provides aggregate results but the data can be used to estimate the prevalence of infection within a population provided that some pools yield negative results.

There are several published methods for calculating within-herd prevalence from pooled samples (Toribio and Sergeant, 2007). Apart from ensuring the thorough mixing of the pooled sample prior to selecting an aliquot to culture, there are no additional technical considerations for the laboratory. The individual samples that were used to create a pool can be stored if it is desired to determine later which individual animals contributed to the positive pooled culture result.

18.11.2 Tissue samples

MAP can be isolated readily from the intestinal tissues of infected animals and has sometimes been cultured from the tissues of humans (Chiodini, 1989) (see Chapter 3, this volume). Less effort has been devoted to the development of protocols for culture of tissues than there has been for faeces, probably because contamination rates have not caused concern (Taylor, 1950; Smith, 1953). Tissues are usually collected in bulk at post-mortem, but small surgical biopsy samples from animals and humans can also be cultured (Kirkwood *et al.*, 2009; Dennis *et al.*, 2011; Begg *et al.*, 2018). Methods described for tissues are simpler than those for faeces, usually involving only one decontamination step. Incubation of homogenized tissues overnight or for up to 3 days in HPC is sufficient to remove most contaminants from the intestinal wall or lymph nodes (Cousins *et al.*, 1995; Whittington *et al.*, 1999). Tissues may be prepared using a variety of methods. Usually fat is trimmed away, and the remaining intestine or lymph node is finely divided with scissors, ground, blended mechanically or disrupted using a stomacher machine in solutions that may include proteases to disrupt tissue structure (Merkal *et al.*, 1964; Merkal, 1973; Sweeney *et al.*, 1992; Adúriz *et al.*, 1995). The homogenate is decontaminated by suspension in a solution of antimicrobials, and the sediment is used as the inoculum. As for faeces, centrifugation of the suspension may be used to concentrate *MAP* and improve analytical sensitivity, but this tends to increase the contamination rate (Reddacliff *et al.*, 2003b). An alternative approach involved the use of the Zwitterionic detergent CB-18 to release *MAP*, combined with a cocktail of lytic enzymes to destroy contaminants; this was tried on intestinal tissues from cattle and bison but has not been compared independently with traditional methods (Thornton *et al.*, 2002).

18.11.3 Meat

MAP has been cultured from red meat in several countries using solid (HEYM, LJ) and liquid (BACTEC, MGIT and Trek) media. Alonso-Hearn *et al.* (2009) homogenized 2 g of diaphragm muscle and decontaminated it in 0.75% HPC before culture on HEYM and LJ slopes using a protocol for faecal culture. Mutharia *et al.* (2010) obtained colonies of *MAP* on HEYM after inoculation of stomacher fluid from a 5-g meat sample, without any decontamination. Others used 3-g samples of ground beef in an HPC decontamination routine and obtained isolates of *MAP* in Trek liquid media from two of 140 samples (Savi *et al.*, 2015).

To overcome potentially very low concentrations of the organism in muscle, an acid enzymatic digestion method was developed to enable processing of large samples (20 g) (Reddacliff *et al.*, 2010). This method was cumbersome, and a modification of a routine faecal culture protocol (centrifugation at 3000 × *g* for 20 min instead of 900 × *g* for 30 min, decontamination in HPC for 24 h instead of 72 h, inoculum volume 0.3 m instead of 0.1 ml) using a 2-g sample, although less sensitive, was more practical, and had a similar rate of isolation of *MAP* from field samples in BACTEC medium due to there being fewer contaminated samples (Reddacliff *et al.*, 2010).

Processed meats are another potential source of *MAP* for humans. Akineden *et al.* (2011) evaluated methods for raw sausage using spiked samples and recommended decontamination in HPC followed by culture on HEYM, but they did not evaluate liquid media.

18.11.4 Milk and processed dairy products

The shedding of *MAP* into the milk of infected cows was recognized many years ago (Alexejeff-Goloff, 1935; Taylor *et al.*, 1981), while its isolation from the milk of two Crohn's disease patients has fuelled the debate about its role in disease in humans (Naser *et al.*, 2000). There has been a great deal of research on ways to detect the organism in samples of milk from individual cows, in bulk tank milk and in pasteurized milk, including retail samples (Millar *et al.*, 1996; Grant *et al.*, 2002). Pragmatically, reliable culture from milk was vital to study the effects of pasteurization on the viability of *MAP* (Stabel *et al.*, 1997). Most culture protocols involve centrifugation of milk at about 2500 × *g* for 15 min to produce a pellet containing *MAP*; to increase the chance of detection, the cream layer – where *MAP* also partitions – may also be collected and pooled with the pellet for culture (Gao *et al.*, 2005). Sample volumes of 40–50 ml of milk are necessary as the efficiency of recovery is <20% and there may be <10 colony-forming units per 50 ml (Sweeney *et al.*, 1992; Gao *et al.*, 2005). Immunomagnetic bead separation of *MAP* from milk has also been reported but does not appear to have found wide application (Grant *et al.*, 1998).

Several groups have shown that decontamination of milk in 0.75% HPC in a one-step sedimentation protocol is appropriate to minimize contamination and is less damaging to *MAP* than double incubation methods (Sweeney *et al.*, 1992; Dundee *et al.*, 2001; Gao *et al.*, 2005). As milk ages, the population of contaminants increases so that by 8 days decontamination in HPC is ineffective (Gao *et al.*, 2005). Raw milk should be processed within 2 days. Further work using spiked samples led to recommendations that milk be decontaminated for 15 min using N-acetyl-L-cysteine (NALC) with 1.5% NaOH; recovery of viable *MAP* was greater and contamination rates were lower compared with decontamination with 0.75% HPC for 5 h (Bradner *et al.*, 2013a). These recommendations were validated by the same authors in subsequent work using natural milk samples from seven dairies (Bradner *et al.*, 2014).

Culture of *MAP* from yoghurt was achieved on 7H10 agar with antibiotics (Van Brandt *et al.*, 2011a). For hard cheese, Donaghy *et al.* (2003) found that a decontamination step inhibited all cheddar cheese starter and nonstarter microorganisms, and that Middlebrook 7H10 agar containing polymyxin, amphotericin B, nalidixic acid, trimethoprim and azlocillin (PANTA) and to a lesser degree HEYM containing vancomycin, amphotericin B and nalidixic acid (VAN) inhibited growth of these microflora when no decontamination step was employed. *MAP* was subsequently cultured from cheddar cheese on HEYM agar medium containing VAN without a decontamination step (Donaghy *et al.*, 2004). *MAP* was isolated recently from a soft goat cheese that had been decontaminated in 0.75% HPC and cultured on 7H11 agar (Galiero *et al.*, 2015).

18.11.5 Blood

Blood culture for *MAP* is not undertaken routinely and there have been no systematic studies on suitable methodologies or on the prevalence of bacteraemia. In the first study involving blood culture from livestock (Koenig *et al.*, 1993), heparinized buffy coat-plasma preparations were decontaminated in 0.75% HPC and cultured on HEYM, with positive results from one of seven cows that had advanced disease. After first optimizing the methods for liquid culture in BACTEC medium (Bower *et al.*, 2010), the occurrence of mycobacteraemia was investigated in 111 sheep. Dissemination of infection to the liver and hepatic lymph node was detected in 18 of 53 infected sheep while the bacterium was isolated from the blood of only 4 of these; prolonged incubation periods prior to growth were consistent with inhibition of growth or dormancy of *MAP* in some cultures (Bower *et al.*, 2011). The problem of contamination of blood cultures was dealt with by incubation of blood lysates in 0.75% (w/v) HPC for 72 h (Bower *et al.*, 2010, 2011). There have been at least two studies looking for *MAP* in the blood of Crohn's disease patients, with various methods, different levels of stringency in microbiological assessment and conflicting results (Naser *et al.*, 2004; Parrish *et al.*, 2008).

18.11.6 Environmental samples

MAP tends to associate with soil particles, especially organic matter and certain clays (Dhand *et al.*, 2009a, b). It tends to remain on grass and in the upper layers of soil where it is available to grazing livestock and can contaminate the runoff after heavy rain (Dhand *et al.*, 2009a; Salgado *et al.*, 2011). To determine the status of grazing livestock at herd/flock level, *MAP* can be cultured from environmental samples including soil, water, pasture, faeces collected from the environment, manure and slurry collected from drains, dust from barns, boot swabs and collection socks (Whittington *et al.*, 2003b; Khol *et al.*, 2009; Pillars *et al.*, 2009; Eisenberg *et al.*, 2011, 2013; Arango-Sabogal *et al.*, 2016a; Ritter *et al.*, 2016; Noll *et al.*, 2017). The entire approach was reviewed recently (Wolf *et al.*, 2017). Culture methods for environmental samples are usually based on those for faeces (Whittington *et al.*, 1998, 2004, 2005; Pickup *et al.*, 2005; Salgado *et al.*, 2011).

The sensitivity of culture of faecal samples collected from the environment is less than that of individual or pooled culture of faecal samples collected directly from animals; for example, see Lavers *et al.* (2013).

Environmental culture is most applicable in dairy cattle where manure collects in drains adjacent to the dairy parlour. Faeces from most cows will collect there over time and is readily available for sampling. It is the cheapest method of sampling for determination of herd infection status (Lombard *et al.*, 2006; Tavornpanich *et al.*, 2008). As the number of animals contributing to a particular sample is unknown, environmental sampling cannot be used to reliably estimate within-herd prevalence but can be useful for regional prevalence estimation (Kruze *et al.*, 2013). The method has also been applied in extensively grazed sheep but few positive cultures were obtained from pastures on affected farms (Whittington *et al.*, 2003a).

Other applications of environmental culture include the search for possible transport vectors of *MAP* such as blowflies, earthworms and parasitic nematode larvae (Whittington *et al.*, 2001; Fischer *et al.*, 2003, 2004).

Observations of *MAP* in the environment using culture revealed an important phenomenon – dormancy – whereby the organism was able to survive for an extended period in a non-cultivable state (Whittington *et al.*, 2004). Further evidence of dormancy was an extended lag phase when stressed *MAP* were inoculated into liquid media (Gumber *et al.*, 2009). *MAP* possesses genes and proteins that are known from homology searches with other mycobacteria to be dormancy-associated. It may be possible to revive dormant *MAP* cells to enhance culture success rates, but further studies are required (Rock *et al.*, 2016).

Concentration of water samples by centrifugation or filtration is required prior to culture (Aboagye and Rowe, 2011). Decontamination methods suitable for river water and municipal drinking water have received relatively little attention. Recent work to understand the interaction between decontamination protocols and culture media employed for water sediment samples has confirmed the need for further studies leading to optimized methods (Aboagye and Rowe, 2018).

18.11.7 Animal wastes, compost and manufactured pasture by-products

Stall waste consisting of bedding straw, faeces and the slurry of washings may be converted into useful products such as biogas, compost and fertilizer. A question remains about the viability of *MAP* after such industrial processes (Gobec *et al.*, 2009; Khol *et al.*, 2010; Slana *et al.*, 2011; Mazzone *et al.*, 2018; Donat *et al.*, 2019). Similar questions arise about the risk of transmission of *MAP* when straw or silage is made from pastures on which livestock have grazed or to which their manure has been applied (Katayama *et al.*, 2001; Fecteau *et al.*, 2013) and the ultimate fate of the pathogen when manure is sprayed on to grasslands is also of interest (Salgado *et al.*, 2011). Protocols based on those for faeces can be used to culture *MAP* from such samples to inform risk assessments.

18.11.8 Evaluation of antimicrobials

Treatment of *MAP* infection in animals has not been attempted very often due to the adverse economics and the practical constraints of administering drugs over a prolonged period; for a zoological collection example, see Bryant *et al.* (2012). However, there may be applications for antimicrobials in neonatal prophylaxis (Fecteau *et al.*, 2011; Ali *et al.*, 2019). Antimicrobials are also relevant in the context of Crohn's disease in humans, where individual patients may be treated and clinical trials undertaken (Selby *et al.*, 2007; Krishnan *et al.*, 2009a, b). To assess antimicrobial sensitivity of *MAP* isolates, growth inhibition assays are possible in liquid or solid media, but standards for such tests are needed (Krishnan *et al.*, 2009a, b; Fecteau *et al.*, 2014; Nowotarska *et al.*, 2017; Steuer *et al.*, 2018; Ali *et al.*, 2019).

18.11.9 *In vitro* propagation and stability of MAP

There are many applications that require bulk propagation of *MAP*. These include for production of DNA for non-PCR molecular typing methods such as restriction fragment length polymorphism (RFLP) analysis, antigen production for immunological tests such as delayed type hypersensitivity (Johnin PPD) or production of live bacterial pellets or suspensions for inclusion in vaccines and for use as inoculum in experimental animal models. There has been an implicit assumption when *MAP* is cultivated for each of these applications that the organism is genetically stable. In fact, most authors do not report the passage history (media, growth conditions, number of subcultures) of isolates used in their studies. While there have been few studies to investigate genetic changes following laboratory propagation of *MAP*, the findings have been enlightening. Kasnitz *et al.* (2013) showed that alterations can occur in IS*900* *BstE*11 RFLP profiles in some isolates of *MAP* after subculture to different media and after repeated subculture, due to factors such as activity of transposase within IS*900* and point mutations at specific recognition sequences for restriction endonucleases. Furthermore, some short sequence repeats were unstable. Bull *et al.* (2013) showed that different culture histories over decades in different laboratories was associated with substantial genome diversity in the vaccine strain *MAP* 316, including large tandem genomic duplications, deletions and transposable element mobility. This mirrored findings with BCG, which has a similar history of unregulated, multicentric, laboratory passage (Behr *et al.*, 1999). Evidence was presented recently to show that *in vitro* culture can lead to a reduction in the virulence of *MAP* (Fernández *et al.*, 2015), which is consistent with a genetic change in the organism through selection. Bryant *et al.* (2016) investigated genome stability in two *MAP* field strains that were passaged several times on a 7H11 agar medium and found mutations in genes responsible for nitrogen assimilation. Authors should record the provenance of *MAP* strains and their passage history when reporting their findings, regardless of the type of study.

18.12 Enumeration of *MAP*

Estimation of the number of *MAP* organisms that are present in a clinical sample can be used to determine the level of risk posed by livestock through contamination of the environment. In research, accurate enumeration of *MAP* is often needed, for example in experimental infection models to enable a repeatable dose of inoculum between trials, and in vaccine efficacy experiments to study faecal shedding rates between vaccinates and controls (Reddacliff *et al.*, 2006a; Begg and Whittington, 2008). This topic was recently reviewed by Marquetoux *et al.* (2019).

Direct microscopic cell counts measure the total number of cells while PCR-based methods measure total DNA levels, which can be related back to the total number of *MAP* cells. Both of these methods tend to overestimate counts by including both live and dead cells. In contrast, culture-based methods provide an estimate of the number of viable cells. Direct colony counts on HEYM slopes have often been used. Turbidimetric estimates, most probable number (MPN) estimates or time to growth are applicable to liquid cultures. As *MAP* tends to clump in suspension, counts from suspensions can be underestimates and considerable effort is required to achieve thorough dispersion of cells.

Turbidimetric estimates are only suitable for high concentrations of cells – one absorbance unit at 600 nm corresponds to a concentration of about 10^8 cells/ml (Shin *et al.*, 2007). Most probable number methods are applicable over a wide range of cell numbers but require the culture of replicates of a serial dilution through to an endpoint, which is the highest dilution before growth no longer occurs; thus they are very costly in terms of media (Whittington *et al.*, 2000b; Reddacliff *et al.*, 2003a). Estimation of count by the time to growth in radiometric BACTEC 12B medium is based on the observation that generation time is proportional to inoculum size; this is less susceptible than other methods to error through bacterial clumping, is highly sensitive, can be done with a single culture vial and is applicable to a range of sample types and strains of *MAP* (Lambrecht *et al.*, 1988; Reddacliff *et al.*, 2003a). The same approach has been used in MGIT media (Shin *et al.*, 2007; Abendaño *et al.*, 2012). Time to detection in the Trek ESP system is also related to inoculum size (Kim *et al.*, 2002). Colony counts are likely to substantially underestimate the actual count relative to MPN counts or time to growth in liquid media (Reddacliff *et al.*, 2003a), although there was close correlation between counts on HEYM, BACTEC 12B medium and MGIT medium in one study (Shin *et al.*, 2007).

To study pathogenesis and provide an indirect measure of vaccine efficacy, Pooley *et al.* (2016) developed a liquid culture-based method for enumeration of *MAP* in intracellular infection assays; it required a 2-week incubation and had a PCR readout.

18.13 Contamination Rate

An important characteristic of any *MAP* culture protocol is its capacity to prevent growth of irrelevant microbes that may be present in a clinical sample. On solid media, colonies of *MAP* may not be visible among colonies of other organisms, but sometimes *MAP* can still be detected by PCR from the surface of the slope (Secott *et al.*, 1999; Whittington, 2009). There is lack of consistency in the literature in the use of the term 'contamination', which makes it difficult to compare rates between different studies. It can mean a mixed culture of *MAP* with other organisms, a light growth of irrelevant organisms or their complete overgrowth of the medium requiring cultures to be repeated. Such cultures can simply be classified as non-interpretable (Rangel *et al.*, 2017).

The reported contamination rates in *MAP* faecal culture are extremely variable and there is still a need for a better culture protocol (Table 18.5). Contamination rates for culture on solid media in human clinical mycobacteriology are also highly variable, and in one study ranged from 0.4 to 41% (van Griethuysen *et al.*, 1996).

Contamination of faecal cultures is due to the presence of microbes that resist decontamination; not unexpectedly their prevalence can vary with the diet of livestock and their geographical location and typically is clustered within groups of samples from certain farms or localities (Whitlock *et al.*, 1989; Whittington, 2009; Rangel *et al.*, 2017). In one study of 2599 faeces from 137 farms, between 1.7 and 11% of samples among farms yielded growth of irrelevant organisms (Whittington, 2009). Culture protocols may need to be modified, for example by inclusion of additional antibiotics, in order to obtain satisfactory results for samples from some farms.

Contamination of intestinal tissue and mesenteric lymph node cultures is less problematic. The long-term rate for 2577 intestinal tissues and associated lymph nodes from sheep was <0.2% using decontamination in HPC and growth in modified BACTEC 12B medium (Whittington, 2009). The contamination rate of cultures of skeletal muscle following an acid enzymatic digestion procedure was about 8.5% (Reddacliff *et al.*, 2010).

Contamination can be a problem with milk samples too; about 40% of liquid cultures were contaminated but as for faecal samples in other studies, contaminated cultures were clustered within herds (range 4–83% of samples among 132 cows from seven herds) (Bradner *et al.*, 2014).

18.14 Loss of *MAP* During Sample Storage and Processing – Analytical Sensitivity

The care and handling of samples between the time of their collection and processing them for culture in the laboratory is important. Common

Table 18.5. Reported contamination rates of faecal cultures for *Mycobacterium avium* subsp. *paratuberculosis* (*MAP*) using contemporary culture methods.

Species	Decontamination protocol[a]	Media	No. of samples (no. of slopes per sample)	Contamination rate	Definition of contamination	Reference
Cattle	3. Sedimentation, centrifugation HPC	HEYM	4688 samples (4)	9.6% of samples 58% of slopes	Overgrowth of contaminants on ≥2 slopes	Whitlock *et al*. (1989)
Cattle	1. and 2. Sedimentation or centrifugation HPC	HEYM	131 (3)	26 and 60%	Presence of contaminants	Kim *et al*. (1989)
Cattle	1. Sedimentation, 3 μm filtered HPC	BACTEC 12B	603	3.9%	Uncertain	Collins *et al*. (1990b)
Cattle	1. Sedimentation HPC	BACTEC 12B	453	7.7%	Presence of contaminants	Cousins *et al*. (1995)
Alpaca	1. Sedimentation HPC	BACTEC 12B	137	7.3%	Presence of contaminants	Cousins *et al*. (1995)
Cattle	5. Centrifugation NaOH-OA	LJ	2989 samples (4)	0.13% of samples 7.0% of slopes	Overgrowth	Kalis *et al*. (1999)
Cattle	3. Sedimentation, centrifugation HPC	HEYM	463 samples (4)	30%	Overgrowth of contaminants on ≥1 slope	Secott *et al*. (1999)
Cattle	5. Centrifugation NaOH-OA	LJ and HEYM	2513 (4 HEYM; 2 LJ)	13 to 14% of samples	Overgrowth	Nielsen *et al*. (2004)
Sheep	3. Sedimentation, centrifugation HPC	BACTEC 12B	5066	11.7%	Presence of contaminants	Whittington (2009)

[a]The number corresponds to a method in Table 18.4.
HPC, hexadecylpyridinium chloride;

practice is to keep clinical samples out of direct sunlight and chilled during transport to the laboratory. The viability of *MAP* in samples declines during storage at −20°C (e.g. see Hahn *et al.*, 2017; Schwalm *et al.*, 2018). If there is likely to be a delay in set up of cultures, samples should be stored at −80°C where viability in faeces is many years (unpublished data). Recovery of *MAP* from faecal samples that had been thermally stressed by eight freeze thaw cycles from −20°C was greater on M7H9C liquid medium (75% of 20 samples) than on solid HEYM (25%) (Schwalm *et al.*, 2018).

There can be a dramatic loss of viable *MAP* when samples are cultured. The factors are both physical and chemical. A progressive loss of viable organisms occurs as only part of the sample material from one step is taken forward to the next step in the protocol. Decontamination solutions kill some *MAP* and some of the antimicrobials in the culture media may also have a deleterious effect on the viability and growth of *MAP*. In fact, about 99% (two logs) of the viable *MAP* organisms in a sample can be killed or lost during routine processing.

Protocols with NaOH or OA reduced the concentration of *MAP* from cattle by one to two logs in 4 h and BAC reduced it by one log; HPC at concentrations of up to 1% did not affect the viability of *MAP* over a 5-day incubation (Merkal *et al.*, 1982; Whipple and Merkal, 1983). The NaOH-OA protocol was confirmed independently to cause a two-log loss of viable *MAP* (Jorgensen, 1982). The double incubation protocol (method 3 in Table 18.4), which involves both HPC and the antibiotic mixture vancomycin, amphotericin B and nalidixic acid (VAN) caused a 2.7-log loss of the *MAP*-S strain, of which 1.3–1.7 logs loss was due to the VAN step and little due to HPC (Reddacliff *et al.*, 2003b). When these factors are taken into account several researchers have reported the analytical sensitivity of faecal culture to be no better than 10^2 viable organisms per 1–2 g of faeces (Jorgensen, 1982; Reddacliff *et al.*, 2003b). Obviously, there will be false-negative culture outcomes for samples that contain low numbers of organisms due to the loss of *MAP* during the culture process.

The situation with tissue samples is similar or worse; there was a 3.1-log loss with the HPC protocol (Reddacliff *et al.*, 2003b). There is a striking deleterious impact of HPC on *MAP* in milk: the temperature (22°C optimal) and duration of incubation (2–5 h optimal) in this decontaminant can be critical to minimize inactivation of MAP. After 5 h of incubation in HPC, losses are such that analytical sensitivity in liquid media is no better than 10^2 to 10^3 *MAP* cells per millilitre of milk (Grant *et al.*, 2003). These results have been confirmed, leading to recommendations that NALC-NaOH be used instead of HPC (Bradner *et al.*, 2013b, 2014) (see Section 18.11.4).

Conflicting data for the impact of HPC on the recovery of *MAP* from faeces, tissues and milk suggest that there might be complex interactions between this chemical, the type of substrate, the strain of *MAP* and the culture media.

18.15 Identification of *MAP* in Cultures

When colonies that consist of AFB are recognized on solid media or when growth is reported in broth culture, the next challenge is to identify *MAP*. A presumptive identification can be made based on slow growth – colonies develop after more than 2 weeks – and host tissue predilection, as cases of granulomatous enteritis associated with AFB in livestock are most likely to be paratuberculosis. Microbiologists once used guinea pig inoculation to differentiate the types of AFB isolated from cattle – if disease did not develop in the guinea pig then the organisms could not have been *M. tuberculosis* or *M. bovis* (Tamarin and Landau, 1961; Wilson and Miles, 1975). Lambs or calves then had to be inoculated with the isolate to prove its potential to cause lesions of paratuberculosis; the high cost and time required to do this meant that it was not often accomplished.

18.15.1 Colony morphology

Twort and Ingram were the first to describe *MAP* colonies. They were initially round, smooth and white then tended to heap up slightly and become dull light yellow with wrinkling of the surface; pigmentation of colonies was influenced by the colour of the *M. phlei* or egg in the medium (Twort and Ingram, 1912). Colony morphology is dependent on the medium, and addition of supplements can dramatically alter it: Tween compounds cause otherwise irregular granular colonies on 7H9 agar to appear instead as entire, smooth and domed (van Boxtel *et al.*, 1990). Colonies of a *MAP*-S strain growing on 7H10 agar reach ≤1 mm in diameter by 6 weeks and are white, circular, shiny, raised and convex (Whittington *et al.*, 1999) (Fig. 18.1).

18.15.2 Mycobactin dependency

Clinical isolates of *MAP* generally require an exogenous source of mycobactin for growth on commonly used media and this became the primary feature used to distinguish *MAP* from other mycobacteria (Morrison, 1965; Thorel *et al.*, 1990). The routine to demonstrate mycobactin dependency involves subculture of a colony from the primary slope or from broth to two media, one with and the other without mycobactin, then comparing the degree of growth on the two media (Whipple *et al.*, 1991). Demonstration of mycobactin dependency is not infallible and the result is influenced by the pH of the medium and the concentration of iron (Lambrecht and Collins, 1992). There may be sufficient carryover of mycobactin attached to the cell wall of *MAP* from the primary culture medium to new media to disguise the need for mycobactin during subculture (Lambrecht and Collins, 1992). However, this did not appear to be a practical concern in a later study on a variety of culture media (Whittington *et al.*, 2011). *MAP* cultured from the intestinal tissues of sheep in Spain were not dependent on mycobactin for primary isolation on 7H11 agar (Adúriz *et al.*, 1995) and it was not required for growth of a variety of *MAP* strains on egg-free 7H11 agar but did stimulate their growth (Whittington *et al.*, 2011). Mycobactin dependency should not be assessed on egg-free Middlebrook 7H11 agar (Whittington *et al.*, 2011) but can be undertaken on Middlebrook 7H10 agar, HEYM or LJ agar (Whittington *et al.*, 2011). To further complicate matters, some isolates of *M. avium* require mycobactin for primary culture but the dependency is lost on subculture (Matthews *et al.*, 1978; Thorel, 1984).

18.15.3 Molecular confirmation using IS*900*

The discovery of IS*900*, an insertion element thought to be specific for *MAP* (Green *et al.*, 1989) provided a molecular basis for identification, and PCR is now used alone or in combination with mycobactin dependency for the identification of *MAP* in cultures. Arguably this is sufficient for livestock on endemically infected farms but is insufficient for index cases given the potential adverse consequences of a positive diagnosis.

IS*900* is a member of a family of insertion sequences, some of which closely resemble IS*900* and so positive results can be obtained from other mycobacterial species using common probes and primers (Cousins *et al.*, 1999; Englund *et al.*, 2002; Kim *et al.*, 2002). Strategies to resolve closely related IS*900*-like insertion sequences include restriction endonuclease analysis of the PCR product, sequencing of the PCR product, use of internal probes in assays based on technology such as Taqman and high-stringency real-time PCR protocols with specific primers and the assessment of melting temperatures (Cousins *et al.*, 1999; Englund *et al.*, 2002; Kim *et al.*, 2002; Kawaji *et al.*, 2007). Because the identification of *MAP* has been shown to be less than 100% specific on several counts, and regardless of assurances from laboratories, microbiological diagnoses should be questioned if they do not make epidemiological sense. This is how the organisms that carry IS*900*-like sequences were discovered (Cousins *et al.*, 1999).

For confirmation of an index case in any species, it is critical that IS*900*-based PCR not serve as a stand-alone test for culture-confirmation. Additional microbiological data to confirm the identification of *MAP* are needed, such as sequencing of the IS*900* amplicon, identification of *MAP*-specific elements such as F57, demonstration of mycobactin dependency, colony morphology and growth rate, acid-fast staining and microscopic morphology of bacterial cells.

PCR analysis of colonies of AFB can be conducted after release of DNA from bacterial cells using simple methods such as boiling a suspension made from the colony (Whittington *et al.*, 1999). However, it can be quite difficult to obtain suitable samples for identification of *MAP* from liquid cultures such as BACTEC, because egg yolk inhibits PCR amplification of IS*900*. The first protocols for PCR confirmation required subculture to BACTEC 12B medium without egg yolk and incubation for a few weeks, which delayed the diagnosis (Cousins *et al.*, 1995). Later, a simple extraction of the BACTEC 12B broth in ethanol was developed to remove the egg yolk, followed by heating of the supernatant to lyse *MAP* cells and release DNA (Whittington *et al.*, 1998); this method was used in diagnostic laboratories in Australia. Where the liquid culture contained other types of bacteria as well as *MAP*, purification of DNA from the lysate using a silica column was shown to remove residual inhibitors of PCR (Whittington, 2009). To address the problem of identification of *MAP* from liquid cultures, Okwumabua *et al.* (2010) compared three methods for *MAP* nucleic acid extraction (magnetic bead, silica spin column and phenol-chloroform) after growth in Trek broth. The recovery of amplifiable *MAP* DNA was greatest for the magnetic bead method. After widespread adoption of M7H9C liquid medium in Australia, the manual ethanol extraction method was replaced by a semi-automated purification system based on bead beating and magnetic bead purification of nucleic acid (Plain *et al.*, 2015). Others have also tackled this problem. Simply boiling the broth to release DNA was almost as successful as a commercial column kit in one study (Sweeney *et al.*, 2006). Salgado *et al.* (2014) found that an in-house sequential sodium dodecyl sulphate/proteinase K lysis, bead beating and ethanol extraction method was superior to a commercial bead beating, proteinase K, column purification kit and a cethyltrimethylammonium bromide (CTAB) chloroform purification method, but only 12 samples were evaluated. Thirumalapura *et al.* (2018) recently evaluated three commercial products (Chelex 100 resin 6% w/v, InstaGene Matrix, Bio-Rad Laboratories; alkaline polyethylene glycol, DNAzol Direct, Molecular Research Center; DNA Extract All Reagents KitTM, Life Technologies) on *MAP* grown in Trek liquid culture medium; the latter method was superior.

All of the semi-automated methods for DNA extraction (such as magnetic bead) and those involving commercial kits are generally multistep procedures. There are interactions between machine platforms and kits. Kits from different manufacturers may not be interchangeable, their composition is often not known and components can change over time without notification. Consequently, methods need to be fully documented, properly validated and monitored closely over time using process controls.

18.15.4 Other identification approaches

Matrix-assisted laser desorption/ionization time-of-flight (MALDI-TOF) mass spectrometry (MS) is widely used in medical microbiology for rapid identification of clinical isolates, typically by analysis of bacterial colonies. Conceptually MALDI-TOF MS produces a molecular fingerprint that can be compared against a database of fingerprints from known taxa. This method has been used to differentiate *MAP* from other mycobacterial species and to differentiate strains of *MAP* (Ravva *et al.*, 2017). Similarly, volatile emissions can be collected from the headspace above a viable culture and analysed by gas chromatography mass spectrometry (GC-MS). Happily the resulting fingerprint is known as a bouquet (Nawrath *et al.*, 2012). Volatile emissions are influenced by culture media and growth conditions, but *MAP* can be differentiated from other mycobacterial species (Trefz *et al.*, 2013; Küntzel *et al.*, 2016). Küntzel *et al.* (2019) showed that there is a group of 28 volatiles common to *MAP* even when there is variation in cultivation conditions. However, specificity of these new approaches for *MAP* is unproven.

18.15.5 Effect of contaminants on the identification of MAP

Many cultures of *MAP* are mixed cultures, i.e. they also contain irrelevant microorganisms. The latter confound the identification of *MAP* in liquid culture by inhibiting IS*900* PCR, increasing the number of samples for which purification of DNA extracts is required prior to PCR (Whittington, 2009). On solid medium, contaminants can inhibit the growth of *MAP* or obscure *MAP* colonies (Secott *et al.*, 1999; Whittington, 2009). Thus contaminants complicate, delay and increase the costs of culture of *MAP*. There are several studies in which PCR has been used to reveal the presence of *MAP* in contaminated or overgrown culture media (Whittington, 2009; Arango-Sabogal *et al.*, 2016b; Rangel *et al.*, 2017). Contaminated cultures may need to be redone, either from stored sample material or following collection of replacement samples.

18.16 Comparison of Culture Methods

Routine methods should have the following characteristics: capacity to support the growth of all strains of *MAP*; high analytical sensitivity; low contamination rate; short incubation period; ease of identification of *MAP*; low overall cost; and low occupational health and environmental impact.

Not all media support the growth of all strains of *MAP*. However, suitably modified Middlebrook 7H10 agar, 7H11 agar and 7H9 broth (M7H9C and BACTEC 12B media) appear to do so within reasonable incubation periods (Whittington *et al.*, 2011). Common *MAP*-S strains grow more slowly, poorly or not at all on HEYM, LJ or in MGIT ParaTB medium (Gumber and Whittington, 2007, 2009; Abendaño *et al.*, 2012; Dimareli-Malli *et al.*, 2013; Kawaji *et al.*, 2014). HEYM selects for a narrower range of genetic types of *MAP* than does liquid culture (Cernicchiaro *et al.*, 2008; Whittington *et al.*, 2011).

The various protocols that are used for culture of *MAP* on solid media have different sensitivities (Jorgensen, 1982). Seemingly small differences between protocols can affect both sensitivity and contamination rate. For example, sensitivity for several methodological variations using HPC and HEYM ranged from 39–68% in one study (Eamens *et al.*, 2000) and varied by 10-fold in another (Stabel, 1997).

For culture of *MAP* from faeces, liquid culture methods have greater diagnostic sensitivity than protocols based on solid media, regardless of the strain of *MAP* (Table 18.6). This table shows data for paired samples cultured in both liquid and solid media. Even though the type of liquid medium that was available changed over time, the outcome was always the same. The lower sensitivity of solid media is disguised by the fact that protocols often specify inoculation of multiple (up to four) solid medium slopes. A reduction from four to three HEYM slopes reduced sensitivity by 8% (Whitlock *et al.*, 1989). The reason for greater diagnostic sensitivity of liquid medium is that it has greater analytical sensitivity; that is, it supports the growth of lower concentrations of *MAP* in an inoculum. This has been shown in limiting dilution experiments, where typically, *MAP* grows from a two- or three-log greater dilution of inoculum as a suspension in liquid media than as colonies on solid medium (Pozzato *et al.*, 2011) (and unpublished data).

The sensitivity of culture of *MAP* from milk in BACTEC 12B and MGIT media was similar (Grant *et al.*, 2003) but analytical sensitivity of HEYM (10^4 CFU/ml) was two logs less than for BACTEC 12B liquid medium (10^2 CFU/ml) (Bradner *et al.*, 2013b).

In addition to the percentage of positive cultures always being higher for liquid media, the growth rate of *MAP* is faster in liquid media than solid media (Cousins *et al.*, 1995; Pozzato *et al.*, 2011). Thus time to reporting a result is shorter with liquid media than with solid media. For example *MAP* was evident 5 weeks sooner in BACTEC 12B medium than on HEYM slopes (Damato and Collins, 1990). Comparing the MGIT ParaTB system and 7H10 agar, Kawaji *et al.* (2014) found the time to obtain a final result in faecal culture was several weeks earlier for the MGIT system than the time to confirm *MAP* colonies on the solid medium.

Growth rates on solid media depend on the type of medium. *MAP* colonies took 3 weeks longer to appear on HEYM than on 7H10 agar (Damato and Collins, 1990). Some strains grow very slowly on HEYM, a finding that has been confirmed by comparing a wide range of *MAP* isolates on a wide range of media (Whittington *et al.*, 2011). The growth of many different strains of *MAP* on LJ agar was very slow (>12 weeks to appearance of colonies) compared with 7H10 agar (3–4 weeks) (Whittington *et al.*, 2011).

Table 18.6. Table 18.6.In this Studies in which the sensitivity of culture on solid and liquid media were compared.[a] Various decontamination protocols based on hexadecylpyridinium chloride (HPC) were used. Data are the percentage of positive tests usually from paired samples. The number of slopes or broths inoculated is shown in parentheses.

			Solid media		Liquid media					
Species	No.	No. positive	HEYM	7H10	BACTEC 12B	MGIT ParaTB	TREK-ESP	MB/BacT	M7H9C	Reference
Cattle	603	75	60 (2)	-	92	-	-	-	-	Collins *et al.* (1990b)
Cattle	179	38	39–68 (3)	-	89	-	-	-	-	Eamens *et al.* (2000)
Cattle	25[b]	17	82	-	-	-	-	88	-	Stich *et al.* (2004)
Cattle	240	81	46–65 (2)	-	-	-	-	80; 94[c]	-	Motiwala *et al.* (2005)
Cattle	61	61	-	42 (2)	-	66 (2)	-	-	-	Kawaji *et al.* (2014)
Cattle	313–345	na	46	-	-	-	55%	-	-	Laurin *et al.* (2015)
Cattle	107	107	62–73 (1–3)[d]	-	-	-	-	-	84–97 (1–3)[e]	Schwalm *et al.* (2018)
Sheep	1535	202		34	98	-	-	-	-	Whittington (2009)

[a]One study where sample size was not stated was excluded from this table but showed similar trends (Pozzato *et al.*, 2011).
[b]Selected NVSL interlaboratory check test samples.
[c]Identification by: subculture; molecular detection.
[d]na, not available.
[e]Sensitivity increased with the number of slopes or broths inoculated; HEYM data after 12 weeks of incubation; M7H9C data after 6 weeks of incubation.

There are no contemporary cost–benefit analyses for any of the culture methods but incubation time and the time taken to clarify the status of contaminated cultures are costly to laboratories. In regulatory applications, costs to the users of test results include business risk associated with the delay between submission of samples and receipt of results and the costs of false-negative test results. There is demand from industry for the development of quicker culture methods.

BACTEC 12B medium is a low-level radioactive source, which creates a waste disposal problem in some jurisdictions, whereas the other liquid systems have fewer environmental and occupational health and safety constraints. All methods create substantial laboratory waste through use of disposable plastic-ware, antibiotics and other chemicals.

18.17 Quality Control

Good laboratory practice dictates that positive and negative control samples are included with each batch of samples to be processed. Although several international studies have been conducted, there are few published reports of the results of inter-laboratory quality control tests for culture of *MAP* (Sockett *et al.*, 1992; Ridge *et al.*, 2010; Heuvelink *et al.*, 2017). Programmes to evaluate the accuracy and consistency of culture test outcomes between laboratories are conducted in several countries. In the USA this is managed by the National Veterinary Services Laboratory, Ames, Iowa. In Australia, annual inter-laboratory proficiency tests for paratuberculosis are conducted under guidance from the OIE reference laboratory in Melbourne (Gwodz, 2006) and are managed by the Australian National Quality Assurance Program for veterinary laboratories; participation is required for laboratory accreditation under ISO17025 administered by the National Association of Testing Authorities. Faecal culture tests are evaluated less often than serological tests. The degree of difficulty in procuring samples to represent the full spectrum of species and levels of *MAP* burden, standardizing samples, confirming viability and shipping faecal samples in good condition to multiple laboratories is a disincentive to the regular conduct of such tests for *MAP* cultivation.

18.18 General Recommendations and Conclusions

If better control of paratuberculosis through detection of infected herds is desired, liquid culture methods, which are much more sensitive than solid medium culture, will need to be used more widely. Commercial BACTEC 12B was the only liquid medium shown to support the growth of a wide variety of *MAP* strains. Since BACTEC 12B was discontinued, a fully validated alternative has been made available to the international community. International standards for culture protocols are required, as there are many variations of protocols, which have significant effects on sensitivity and prevent the meaningful analysis of new diagnostics for which culture serves as the gold standard. Contamination of faecal cultures remains a major hurdle to timely and economical culture of *MAP* and improved decontamination protocols would be a major advance, enabling greater use of liquid culture. The ongoing widespread use of solid media for diagnostic culture masks the true prevalence of MAP as well as strain diversity, and results in exaggerated estimates of sensitivity for other diagnostic tests that are compared with culture. Failure to routinely employ liquid culture methods to more fully explore the role of *MAP* infection in Crohn's disease is inexplicable.

References

Abendaño, N., Sevilla, I., Prieto, J.M., Garrido, J.M., Juste, R.A. *et al.* (2012) Quantification of *Mycobacterium avium* subsp. *paratuberculosis* strains representing distinct genotypes and isolated from domestic and wildlife animal species by use of an automatic liquid culture system. *Journal of Clinical Microbiology* 50(8), 2609–2617. DOI: 10.1128/JCM.00441-12.

Aboagye, G. and Rowe, M.T. (2011) Occurrence of *Mycobacterium avium* subsp. *paratuberculosis* in raw water and water treatment operations for the production of potable water. *Water Research* 45(11), 3271–3278. DOI: 10.1016/j.watres.2011.03.029.

Aboagye, G. and Rowe, M.T. (2018) Optimisation of decontamination method and influence of culture media on the recovery of *Mycobacterium avium* subspecies *paratuberculosis* from spiked water sediments. *Journal of Microbiological Methods* 150, 24–28. DOI: 10.1016/j.mimet.2018.05.015.

Adúriz, J.J., Juste, R.A. and Cortabarria, N. (1995) Lack of mycobactin dependence of mycobacteria isolated on Middlebrook 7H11 from clinical cases of ovine paratuberculosis. *Veterinary Microbiology* 45(2–3), 211–217. DOI: 10.1016/0378-1135(95)00037-B.

Akineden, O., Fernandez-Silva, J.A., Weirich, S., Abdulmawjood, A. and Bulte, M. (2011) Comparison of two decontamination procedures, three culture media, and real time-PCR assay for the detection of *Mycobacterium avium* subsp *paratuberculosis* (MAP) from artificially contaminated raw sausage. *Archiv Fur Lebensmittelhygiene* 62, 150–156.

Alexejeff-Goloff, N.A. (1935) The pathogenesis and excretion of the bacilli in bovine paratuberculosis (Johne's disease). *Journal of Comparative Pathology and Therapeutics* 48, 81–82. DOI: 10.1016/S0368-1742(35)80011-8.

Ali, Z.I., Saudi, A.M., Albrecht, R. and Talaat, A.M. (2019) The inhibitory effect of nisin on *Mycobacterium avium* ssp. *paratuberculosis* and its effect on mycobacterial cell wall. *Journal of Dairy Science* 102(6), 4935–4944. DOI: 10.3168/jds.2018-16106.

Alinovi, C.A., Ward, M.P., Lin, T.L., Moore, G.E. and Wu, C.C. (2009) Real-time PCR, compared to liquid and solid culture media and ELISA, for the detection of *Mycobacterium avium* ssp. *paratuberculosis*. *Veterinary Microbiology* 136(1–2), 177–179. DOI: 10.1016/j.vetmic.2008.10.012.

Alonso-Hearn, M., Molina, E., Geijo, M., Vazquez, P., Sevilla, I. *et al.* (2009) Isolation of *Mycobacterium avium* subsp. *paratuberculosis* from muscle tissue of naturally infected cattle. *Foodborne Pathogens and Disease* 6(4), 513–518. DOI: 10.1089/fpd.2008.0226.

Arango-Sabogal, J.C., Cote, G., Pare, J., Labrecque, O., Roy, J.P. *et al.* (2016a) Detection of *Mycobacterium avium* subspecies *paratuberculosis* in tie-stall dairy herds using a standardized environmental sampling technique and targeted pooled samples. *Canadian Journal of Veterinary Research-Revue Canadienne De Recherche Veterinaire* 80, 175–182.

Arango-Sabogal, J.C., Labrecque, O., Paré, J., Fairbrother, J.-H., Roy, J.-P. *et al.* (2016b) Evaluation of a PCR assay on overgrown environmental samples cultured for *Mycobacterium avium* subsp. *paratuberculosis*. *Journal of Veterinary Diagnostic Investigation* 28(6), 638–645. DOI: 10.1177/1040638716662302.

Bannantine, J.P., Zhang, Q., Li, L.-L. and Kapur, V. (2003) Genomic homogeneity between *Mycobacterium avium* subsp. *avium* and *Mycobacterium avium* subsp. *paratuberculosis* belies their divergent growth rates. *BMC Microbiology* 3(1), 10. DOI: 10.1186/1471-2180-3-10.

Bauman, C.A., Jones-Bitton, A., Menzies, P., Toft, N., Jansen, J. *et al.* (2016) Prevalence of paratuberculosis in the dairy goat and dairy sheep industries in Ontario, Canada. *The Canadian Veterinary Journal* 57, 169–175.

Beard, P.M., Rhind, S.M., Buxton, D., Daniels, M.J., Henderson, D. *et al.* (2001) Natural paratuberculosis infection in rabbits in Scotland. *Journal of Comparative Pathology* 124(4), 290–299. DOI: 10.1053/jcpa.2001.0466.

Beerworth, W. (1967) Culture of mycobacteria from faeces of domestic animals and their significance for the epidemiology of tuberculosis (Die Zuchtung von Mycobakterien aus dem Kot der Haustiere und ihre Bedeutung fur die Epidemiologie der Tuberkulose). *Praxis der Pneumologie* 21, 189–202.

Begg, D.J. and Whittington, R.J. (2008) Experimental animal infection models for Johne's disease, an infectious enteropathy caused by *Mycobacterium avium* subsp. *paratuberculosis*. *The Veterinary Journal* 176(2), 129–145. DOI: 10.1016/j.tvjl.2007.02.022.

Begg, D.J., de Silva, K., Di Fiore, L., Taylor, D.L., Bower, K. *et al.* (2010) Experimental infection model for Johne's disease using a lyophilised, pure culture, seedstock of *Mycobacterium avium* subspecies *paratuberculosis*. *Veterinary Microbiology* 141(3–4), 301–311. DOI: 10.1016/j.vetmic.2009.09.007.

Begg, D.J., Plain, K.M., de Silva, K., Gurung, R., Gunn, A. *et al.* (2018) Immunopathological changes and apparent recovery from infection revealed in cattle in an experimental model of Johne's disease using a lyophilised culture of *Mycobacterium avium* subspecies *paratuberculosis*. *Veterinary Microbiology* 219, 53–62. DOI: 10.1016/j.vetmic.2018.03.029.

Behr, M.A., Wilson, M.A., Gill, W.P., Salmon, H., Schoolnik, G.K. (1999) Comparative genomics of BCG vaccines by whole-genome DNA microarray. *Science* 284(5419), 1520–1523. DOI: 10.1126/science.284.5419.1520.

Botsaris, G., Swift, B.M.C., Slana, I., Liapi, M., Christodoulou, M. *et al.* (2016) Detection of viable *Mycobacterium avium* subspecies *paratuberculosis* in powdered infant formula by phage-PCR and confirmed by culture. *International Journal of Food Microbiology* 216, 91–94. DOI: 10.1016/j.ijfoodmicro.2015.09.011.

Bower, K., Begg, D.J. and Whittington, R.J. (2010) Optimisation of culture of *Mycobacterium avium* subspecies *paratuberculosis* from blood samples. *Journal of Microbiological Methods* 80(1), 93–99. DOI: 10.1016/j.mimet.2009.11.005.

Bower, K.L., Begg, D.J. and Whittington, R.J. (2011) Culture of *Mycobacterium avium* subspecies *paratuberculosis* (MAP) from blood and extra-intestinal tissues in experimentally infected sheep. *Veterinary Microbiology* 147(1-2), 127–132. DOI: 10.1016/j.vetmic.2010.06.016.

Bradner, L., Robbe-Austerman, S., Beitz, D.C. and Stabel, J.R. (2013a) Chemical decontamination with N-acetyl-L-cysteine-sodium hydroxide improves recovery of viable *Mycobacterium avium* subsp. *paratuberculosis* organisms from cultured milk. *Journal of Clinical Microbiology* 51(7), 2139–2146. DOI: 10.1128/JCM.00508-13.

Bradner, L., Robbe-Austerman, S., Beitz, D.C. and Stabel, J.R. (2013b) Optimization of hexadecylpyridinium chloride decontamination for culture of *Mycobacterium avium* subsp. *paratuberculosis* from milk. *Journal of Clinical Microbiology* 51(5), 1575–1577. DOI: 10.1128/JCM.00333-13.

Bradner, L., Robbe-Austerman, S., Beitz, D.C. and Stabel, J.R. (2014) Short communication: Application of an N-acetyl-l-cysteine-NaOH decontamination method for the recovery of viable *Mycobacterium avium* subspecies *paratuberculosis* from milk of naturally infected cows. *Journal of Dairy Science* 97(6), 3694–3699. DOI: 10.3168/jds.2013-7730.

Bryant, B., Blyde, D., Eamens, G. and Whittington, R. (2012) *Mycobacterium avium* subspecies *paratuberculosis* cultured from the feces of a Southern black rhinoceros (*Diceros bicornis minor*) with diarrhea and weight loss. *Journal of Zoo and Wildlife Medicine* 43(2), 391–393. DOI: 10.1638/2010-0161.1.

Bryant, J.M., Thibault, V.C., Smith, D.G.E., McLuckie, J., Heron, I. *et al.* (2016) Phylogenomic exploration of the relationships between strains of *Mycobacterium avium* subspecies *paratuberculosis*. *BMC Genomics* 17(1). DOI: 10.1186/s12864-015-2234-5.

Bull, T.J., Schock, A., Sharp, J.M., Greene, M., McKendrick, I.J. *et al.* (2013) Genomic variations associated with attenuation in *Mycobacterium avium* subsp. *paratuberculosis* vaccine strains. *BMC Microbiology* 13(1), 11. DOI: 10.1186/1471-2180-13-11.

Bull, T.J., Munshi, T., Mikkelsen, H., Hartmann, S.B., Sorensen, M.R. *et al.* (2016) Improved culture medium (Tika) for *Mycobacterium avium* subspecies *paratuberculosis* (MAP) matches qPCR sensitivity and reveals significant proportions of non-viable MAP in lymphoid tissue of vaccinated MAP challenged animals. *Frontiers in Microbiology* 7, 2112.

Butot, S., Ricchi, M., Sevilla, I.A., Michot, L., Molina, E. *et al.* (2019) Estimation of performance characteristics of analytical methods for *Mycobacterium avium* subsp. *paratuberculosis* detection in dairy products. *Frontiers in Microbiology* 10, 11. DOI: 10.3389/fmicb.2019.00509.

Cameron, J. (1956) Isolation of *Mycobacterium johnei* from faeces. *The Journal of Pathology and Bacteriology* 71(1), 223–225. DOI: 10.1002/path.1700710128.

Cernicchiaro, N., Wells, S.J., Janagama, H. and Sreevatsan, S. (2008) Influence of type of culture medium on characterization of *Mycobacterium avium* subsp. *paratuberculosis* subtypes. *Journal of Clinical Microbiology* 46(1), 145–149. DOI: 10.1128/JCM.01769-07.

Chiodini, R.J. (1989) Crohn's disease and the mycobacterioses: a review and comparison of two disease entities. *Clinical Microbiology Reviews* 2(1), 90–117. DOI: 10.1128/CMR.2.1.90.

Collins, D.M., Gabric, D.M. and de Lisle, G.W. (1990a) Identification of two groups of *Mycobacterium paratuberculosis* strains by restriction endonuclease analysis and DNA hybridization. *Journal of Clinical Microbiology* 28(7), 1591–1596. DOI: 10.1128/JCM.28.7.1591-1596.1990.

Collins, M.T., Kenefick, K.B., Sockett, D.C., Lambrecht, R.S., McDonald, J. *et al.* (1990b) Enhanced radiometric detection of *Mycobacterium paratuberculosis* by using filter-concentrated bovine fecal specimens. *Journal of Clinical Microbiology* 28(11), 2514–2519. DOI: 10.1128/JCM.28.11.2514-2519.1990.

Corbett, C.S., Barkema, H.W. and De Buck, J. (2018) Quantifying fecal shedding of *Mycobacterium avium* ssp. *paratuberculosis* from calves after experimental infection and exposure. *Journal of Dairy Science* 101(2), 1478–1487. DOI: 10.3168/jds.2017-13544.

Cousins, D. V., Evans, R.J. and Francis, B.R. (1995) Use of BACTEC radiometric culture method and polymerase chain reaction for the rapid screening of faeces and tissues for *Mycobacterium paratuberculosis*. *Australian Veterinary Journal* 72(12), 458–462. DOI: 10.1111/j.1751-0813.1995.tb03489.x.

Cousins, D.V., Whittington, R., Marsh, I., Masters, A., Evans, R.J. *et al.* (1999) Mycobacteria distinct from *Mycobacterium avium* subsp. *paratuberculosis* isolated from the faeces of ruminants possess IS*900*-like sequences detectable by IS*900* polymerase chain reaction: implications for diagnosis. *Molecular and Cellular Probes* 13(6), 431–442. DOI: 10.1006/mcpr.1999.0275.

Cutler, R.R., Wilson, P. and Clarke, F.V. (1987) The effect of polyoxyethylene stearate (POES) on the growth of mycobacteria in radiometric 7H12 Middlebrook TB medium. *Tubercle* 68(3), 209–220. DOI: 10.1016/0041-3879(87)90057-2.

Damato, J.J. and Collins, M.T. (1990) Growth of *Mycobacterium paratuberculosis* in radiometric, Middlebrook and egg-based media. *Veterinary Microbiology* 22(1), 31–42. DOI: 10.1016/0378-1135(90)90122-C.

Damato, J.J., Knisley, C. and Collins, M.T. (1987) Characterization of *Mycobacterium paratuberculosis* by gas-liquid and thin-layer chromatography and rapid demonstration of mycobactin dependence using radiometric methods. *Journal of Clinical Microbiology* 25(12), 2380–2383. DOI: 10.1128/JCM.25.12.2380-2383.1987.

Dennis, M.M., Reddacliff, L.A. and Whittington, R.J. (2011) Longitudinal study of clinicopathological features of Johne's disease in sheep naturally exposed to *Mycobacterium avium* subspecies *paratuberculosis*. *Veterinary Pathology* 48(3), 565–575. DOI: 10.1177/0300985810375049.

Dhand, N.K., Eppleston, J., Whittington, R.J. and Toribio, J.-A.L.M.L. (2009a) Association of farm soil characteristics with ovine Johne's disease in Australia. *Preventive Veterinary Medicine* 89(1–2), 110–120. DOI: 10.1016/j.prevetmed.2009.02.017.

Dhand, N.K., Toribio, J.-A.L.M.L. and Whittington, R.J. (2009b) Adsorption of *Mycobacterium avium* subsp. *paratuberculosis* to soil particles. *Applied and Environmental Microbiology* 75(17), 5581–5585. DOI: 10.1128/AEM.00557-09.

Dimareli-Malli, Z., Mazaraki, K., Stevenson, K., Tsakos, P., Zdragas, A. *et al.* (2013) Culture phenotypes and molecular characterization of *Mycobacterium avium* subsp. *paratuberculosis* isolates from small ruminants. *Research in Veterinary Science* 95(1), 49–53. DOI: 10.1016/j.rvsc.2013.03.010.

Donaghy, J.A., Totton, N.L. and Rowe, M.T. (2003) Evaluation of culture media for the recovery of *Mycobacterium avium* subsp. *paratuberculosis* from cheddar cheese. *Letters in Applied Microbiology* 37(4), 285–291. DOI: 10.1046/j.1472-765X.2003.01392.x.

Donaghy, J.A., Totton, N.L. and Rowe, M.T. (2004) Persistence of *Mycobacterium paratuberculosis* during manufacture and ripening of cheddar cheese. *Applied and Environmental Microbiology* 70(8), 4899–4905. DOI: 10.1128/AEM.70.8.4899-4905.2004.

Donat, K., Eisenberg, S.W.F., Einax, E., Reinhold, G. and Zoche-Golob, V. (2019) Reduction of viable *Mycobacterium avium* ssp. *paratuberculosis* in slurry subjected to anaerobic digestion in biogas plants. *Journal of Dairy Science* 102(7), 6485–6494. DOI: 10.3168/jds.2018-15937.

Dundee, L., Grant, I.R., Ball, H.J. and Rowe, M.T. (2001) Comparative evaluation of four decontamination protocols for the isolation of *Mycobacterium avium* subsp. *paratuberculosis* from milk. *Letters in Applied Microbiology* 33(3), 173–177. DOI: 10.1046/j.1472-765x.2001.00979.x.

Dunkin, G.W. (1928) A diagnostic agent for the detection of Johne's disease and its method of preparation. *Journal of Comparative Pathology and Therapeutics* 41, 94–108. DOI: 10.1016/S0368-1742(28)80011-7.

Eamens, G.J., Whittington, R.J., Marsh, I.B., Turner, M.J., Saunders, V. *et al.* (2000) Comparative sensitivity of various faecal culture methods and ELISA in dairy cattle herds with endemic Johne's disease. *Veterinary Microbiology* 77(3-4), 357–367. DOI: 10.1016/S0378-1135(00)00321-7.

Eamens, G.J., Walker, D.M., Porter, N.S. and Fell, S.A. (2007a) Pooled faecal culture for the detection of *Mycobacterium avium* subsp *paratuberculosis* in goats. *Australian Veterinary Journal* 85(6), 243–251. DOI: 10.1111/j.1751-0813.2007.00160.x.

Eamens, G.J., Whittington, R.J., Turner, M.J., Austin, S.L., Fell, S.A. *et al.* (2007b) Evaluation of radiometric faecal culture and direct PCR on pooled faeces for detection of *Mycobacterium avium* subsp. *paratuberculosis* in cattle. *Veterinary Microbiology* 125(1–2), 22–35. DOI: 10.1016/j.vetmic.2007.04.043.

Eamens, G.J., Walker, D.M., Porter, N.S. and Fell, S.A. (2008) Radiometric pooled faecal culture for the detection of *Mycobacterium avium* subsp *paratuberculosis* in low-shedder cattle. *Australian Veterinary Journal* 86(7), 259–265. DOI: 10.1111/j.1751-0813.2008.00313.x.

Eisenberg, S., Nielen, M., Hoeboer, J., Bouman, M., Heederik, D. *et al.* (2011) *Mycobacterium avium* subspecies *paratuberculosis* in bioaerosols after depopulation and cleaning of two cattle barns. *Veterinary Record* 168(22), 587–587. DOI: 10.1136/vr.d1091.

Eisenberg, T., Wolter, W., Lenz, M., Schlez, K. and Zschöck, M. (2013) Boot swabs to collect environmental samples from common locations in dairy herds for *Mycobacterium avium* ssp. *paratuberculosis* (MAP) detection. *Journal of Dairy Research* 80(4), 485–489. DOI: 10.1017/S002202991300040X.

Elguezabal, N., Bastida, F., Sevilla, I.A., González, N., Molina, E. *et al.* (2011) Estimation of *Mycobacterium avium* subsp. *paratuberculosis* growth parameters: strain characterization and comparison of methods. *Applied and Environmental Microbiology* 77(24), 8615–8624. DOI: 10.1128/AEM.05818-11.

Ellingson, J.A.Y.L.E., Koziczkowski, J.J. and Anderson, J.L. (2004) Comparison of PCR prescreening to two cultivation procedures with PCR confirmation for detection of *Mycobacterium avium* subsp. *paratuberculosis* in U.S. department of agriculture fecal check test samples. *Journal of Food Protection* 67(10), 2310–2314. DOI: 10.4315/0362-028X-67.10.2310.

Ellingson, J.A.Y.L.E., Anderson, J.L., Koziczkowski, J.J., Radcliff, R.O.Y.P., Sloan, S.J. *et al.* (2005) Detection of viable *Mycobacterium avium* subsp. *paratuberculosis* in retail pasteurized whole milk by two culture methods and PCR. *Journal of Food Protection* 68(5), 966–972. DOI: 10.4315/0362-028X-68.5.966.

Elze, J., Liebler-Tenorio, E., Ziller, M. and Kohler, H. (2013) Comparison of prevalence estimation of *Mycobacterium avium* subsp. *paratuberculosis* infection by sampling slaughtered cattle with macroscopic lesions *vs.* systematic sampling. *Epidemiology and Infection* 141(7), 1536–1544. DOI: 10.1017/S0950268812002452.

Englund, S., Bölske, G. and Johansson, K.E. (2002) An IS900-like sequence found in a *Mycobacterium* sp. other than *Mycobacterium avium* subsp. *paratuberculosis*. *FEMS Microbiology Letters* 209(2), 267–271. DOI: 10.1111/j.1574-6968.2002.tb11142.x.

Eppleston, J., Begg, D.J., Dhand, N.K., Watt, B. and Whittington, R.J. (2014) Environmental survival of *Mycobacterium avium* subsp. *paratuberculosis* in different climatic zones of eastern Australia. *Applied and Environmental Microbiology* 80(8), 2337–2342. DOI: 10.1128/AEM.03630-13.

Faria, A.C.S., Schwarz, D.G.G., Carvalho, I.A., Rocha, B.B., De Carvalho Castro, K.N. *et al.* (2014) Short communication: Viable *Mycobacterium avium* subspecies *paratuberculosis* in retail artisanal Coalho cheese from Northeastern Brazil. *Journal of Dairy Science* 97(7), 4111–4114. DOI: 10.3168/jds.2013-7835.

Fecteau, M.-E., Fyock, T.L., McAdams, S.C., Boston, R.C., Whitlock, R.H. *et al.* (2011) Evaluation of the in vitro activity of gallium nitrate against *Mycobacterium avium* subsp *paratuberculosis*. *American Journal of Veterinary Research* 72(9), 1243–1246. DOI: 10.2460/ajvr.72.9.1243.

Fecteau, M.E., Hovingh, E., Whitlock, R.H. and Sweeney, R.W. (2013) Persistence of *Mycobacterium avium* subsp. *paratuberculosis* in soil, crops, and ensiled feed following manure spreading on infected dairy farms. *Canadian Veterinary Journal* 54, 1083–1085.

Fecteau, M.-E., Aceto, H.W., Bernstein, L.R. and Sweeney, R.W. (2014) Comparison of the antimicrobial activities of gallium nitrate and gallium maltolate against *Mycobacterium avium* subsp. *paratuberculosis* in vitro. *The Veterinary Journal* 202(1), 195–197. DOI: 10.1016/j.tvjl.2014.06.023.

Fernández, M., Delgado, L., Sevilla, I.A., Fuertes, M., Castaño, P. *et al.* (2015) Virulence attenuation of a *Mycobacterium avium* subspecies *paratuberculosis* S-type strain prepared from intestinal mucosa after bacterial culture. Evaluation in an experimental ovine model. *Research in Veterinary Science* 99, 180–187. DOI: 10.1016/j.rvsc.2015.02.001.

Fischer, O.A., Matlova, L., Bartl, J., Dvorska, L., Svastova, P. *et al.* (2003) Earthworms (Oligochaeta, Lumbricidae) and mycobacteria. *Veterinary Microbiology* 91(4), 325–338. DOI: 10.1016/S0378-1135(02)00302-4.

Fischer, O.A., Matlova, L., Dvorska, L., Svastova, P., Bartl, J. *et al.* (2004) Blowflies *Calliphora vicina* and *Lucilia sericata* as passive vectors of *Mycobacterium avium* subsp. *avium*, *M. a. paratuberculosis* and *M. a. hominissuis*. *Medical and Veterinary Entomology* 18(2), 116–122. DOI: 10.1111/j.0269-283X.2004.00477.x.

Galiero, A., Fratini, F., Turchi, B., Colombani, G., Nuvoloni, R. *et al.* (2015) Detection of *Mycobacterium avium* subsp. *paratuberculosis* in a sheep flock in Tuscany. *Tropical Animal Health and Production* 47(8), 1567–1571. DOI: 10.1007/s11250-015-0899-z.

Galiero, A., Fratini, F., Mataragka, A., Turchi, B., Nuvoloni, R. *et al.* (2016) Detection of *Mycobacterium avium* subsp. *paratuberculosis* in cheeses from small ruminants in Tuscany. *International Journal of Food Microbiology* 217, 195–199. DOI: 10.1016/j.ijfoodmicro.2015.10.029.

Galiero, A., Leo, S., Garbarino, C., Arrigoni, N., Russo, S. *et al.* (2018) *Mycobacterium avium* subsp. *paratuberculosis* isolated from wild red deer (*Cervus elaphus*) in Northern Italy. *Veterinary Microbiology* 217, 167–172. DOI: 10.1016/j.vetmic.2018.03.015.

Gao, A., Odumeru, J., Raymond, M. and Mutharia, L. (2005) Development of improved method for isolation of *Mycobacterium avium* subsp. *paratuberculosis* from bulk tank milk: effect of age of milk, centrifugation, and decontamination. *Canadian Journal of Veterinary Research* 69, 81–87.

Gardner, I.A., Nielsen, S.S., Whittington, R.J., Collins, M.T., Bakker, D. *et al.* (2011) Consensus-based reporting standards for diagnostic test accuracy studies for paratuberculosis in ruminants. *Preventive Veterinary Medicine* 101(1–2), 18–34. DOI: 10.1016/j.prevetmed.2011.04.002.

Gillespie, S.H. (1999) Microbiological diagnosis of granulomatous infections. In: James, D.G. and Zumla, A. (eds) *The Granulomatous Disorders*. Cambridge University Press, Cambridge, pp. 122–131.

Gobec, I., Ocepek, M., Pogacnik, M. and Dobeic, M. (2009) Inactivation of *Mycobacterium avium paratuberculolosis* in sheep manure. *Slovanian Veterinary Research* 46, 105–113.

Grant, I.R., Ball, H.J. and Rowe, M.T. (1998) Isolation of *Mycobacterium paratuberculosis* from milk by immunomagnetic separation. *Applied and Environmental Microbiology* 64(9), 3153–3158. DOI: 10.1128/AEM.64.9.3153-3158.1998.

Grant, I.R., Ball, H.J. and Rowe, M.T. (2002) Incidence of *Mycobacterium paratuberculosis* in bulk raw and commercially pasteurized cows' milk from approved dairy processing establishments in the United Kingdom. *Applied and Environmental Microbiology* 68(5), 2428–2435. DOI: 10.1128/AEM.68.5.2428-2435.2002.

Grant, I.R., Kirk, R.B., Hitchings, E. and Rowe, M.T. (2003) Comparative evaluation of the MGIT™ and BACTEC culture systems for the recovery of *Mycobacterium avium* subsp. *paratuberculosis* from milk. *Journal of Applied Microbiology* 95(1), 196–201. DOI: 10.1046/j.1365-2672.2003.02003.x.

Green, E.P., Tizard, M.L.V., Moss, M.T., Thompson, J., Winterbourne, D.J. *et al.* (1989) Sequence and characteristics or IS *900*, an insertion element identified in a human Crohn's disease isolate or *Mycobacterium paratuberculosis*. *Nucleic Acids Research* 17(22), 9063–9073. DOI: 10.1093/nar/17.22.9063.

Gumber, S. and Whittington, R.J. (2007) Comparison of BACTEC 460 and MGIT 960 systems for the culture of *Mycobacterium avium* subsp. *paratuberculosis* S strain and observations on the effect of inclusion of ampicillin in culture media to reduce contamination. *Veterinary Microbiology* 119(1), 42–52. DOI: 10.1016/j.vetmic.2006.08.009.

Gumber, S. and Whittington, R.J. (2009) Corrigendum to Comparison of BACTEC 460 and MGIT 960 systems for the culture of *Mycobacterium avium* subsp. *paratuberculosis* S strain and observations on the effect of inclusion of ampicillin in culture media to reduce contamination. *Veterinary Microbiology* 139(3–4), 423–424. DOI: 10.1016/j.vetmic.2006.08.035.

Gumber, S., Taylor, D.L., Marsh, I.B. and Whittington, R.J. (2009) Growth pattern and partial proteome of *Mycobacterium avium* subsp. *paratuberculosis* during the stress response to hypoxia and nutrient starvation. *Veterinary Microbiology* 133(4), 344–357. DOI: 10.1016/j.vetmic.2008.07.021.

Gunnarsson, E. (1979) Isolation of *Mycobacterium paratuberculosis* from sheep and cattle in Iceland. *Acta Veterinaria Scandinavica* 20, 191–199.

Gunnarsson, E. and Fodstad, F.H. (1979) Cultural and biochemical characteristics of *Mycobacterium paratuberculosis* isolated from goats in Norway. *Acta Veterinaria Scandinavica* 20, 122–134.

Gwodz, J.M. (2006) Paratuberculosis. In: Anon. (ed.) *Annual Reports of OIE Reference Laboratories and Collaborating Centres*. World Organisation for Animal Health, Paris, pp. 125–127.

Hahn, N., Failing, K., Eisenberg, T., Schlez, K., Zschöck, P.-M. *et al.* (2017) Evaluation of different diagnostic methods for the detection of *Mycobacterium avium* subsp. *paratuberculosis* in boot swabs and liquid manure samples. *BMC Veterinary Research* 13(1), 8. DOI: 10.1186/s12917-017-1173-6.

Herrold, R.D. (1931) Egg yolk agar medium for the growth of tubercle bacilli. *Journal of Infectious Diseases* 48(2), 236–241. DOI: 10.1093/infdis/48.2.236.

Heuvelink, A., Hassan, A.A., van Weering, H., van Engelen, E., Bülte, M. *et al.* (2017) An intra-laboratory cultural and real-time PCR method comparison and evaluation for the detection of subclinical paratuberculosis in dairy herds. *Folia Microbiologica* 62(3), 197–205. DOI: 10.1007/s12223-016-0488-1.

Janagama, H.K., Senthilkumar, T.M.A., Bannantine, J.P., Rodriguez, G.M., Smith, I. *et al.* (2009) Identification and functional characterization of the iron-dependent regulator (IdeR) of *Mycobacterium avium* subsp. *paratuberculosis*. *Microbiology* 155(11), 3683–3690. DOI: 10.1099/mic.0.031948-0.

Jorgensen, J.B. (1982) An improved medium for culture of *Mycobacterium paratuberculosis* from bovine faeces. *Acta Veterinaria Scandinavica* 23, 325–335.

Juste, R.A., Marco, J.C., Sáez de Ocáriz, C. and Adúriz, J.J. (1991) Comparison of different media for the isolation of small ruminant strains of *Mycobacterium paratuberculosis*. *Veterinary Microbiology* 28(4), 385–390. DOI: 10.1016/0378-1135(91)90073-O.

Kaevska, M., Lvoncik, S., Slana, I., Kulich, P. and Kralik, P. (2014) Microscopy, culture, and quantitative real-time PCR examination confirm internalization of mycobacteria in plants. *Applied and Environmental Microbiology* 80(13), 3888–3894. DOI: 10.1128/AEM.00496-14.

Kalis, C.H.J., Hesselink, J.W., Russchen, E.W., Barkema, H.W., Collins, M.T. *et al.* (1999) Factors influencing the isolation of *Mycobacterium avium* subsp. *paratuberculosis* from bovine faecal samples. *Journal of Veterinary Diagnostic Investigation* 11(4), 345–351. DOI: 10.1177/104063879901100409.

Kalis, C.H.J., Hesselink, J.W., Barkema, H.W. and Collins, M.T. (2000) Culture of strategically pooled bovine fecal samples as a method to screen herds for paratuberculosis. *Journal of Veterinary Diagnostic Investigation* 12(6), 547–551. DOI: 10.1177/104063870001200609.

Kalis, C.H.J., Collins, M.T., Barkema, H.W. and Hesselink, J.W. (2004) Certification of herds as free of *Mycobacterium paratuberculosis* infection: actual pooled faecal results versus certification model predictions. *Preventive Veterinary Medicine* 65(3–4), 189–204. DOI: 10.1016/j.prevetmed.2004.07.005.

Kasnitz, N., Köhler, H., Weigoldt, M., Gerlach, G.F. and Möbius, P. (2013) Stability of genotyping target sequences of *Mycobacterium avium* subsp. *paratuberculosis* upon cultivation on different media, in vitro- and in vivo passage, and natural infection. *Veterinary Microbiology* 167(3–4), 573–583. DOI: 10.1016/j.vetmic.2013.09.008.

Katayama, N., Tanaka, C., Fujita, T., Suzuki, T., Watanabe, S. *et al.* (2001) Effects of silage fermentation and ammonia treatment on activity of *Mycobacterium avium* subsp. *paratuberculosis*. *Grassland Science* 47, 296–299.

Kawaji, S., Taylor, D.L., Mori, Y. and Whittington, R.J. (2007) Detection of *Mycobacterium avium* subsp. *paratuberculosis* in ovine faeces by direct quantitative PCR has similar or greater sensitivity compared to radiometric culture. *Veterinary Microbiology* 125(1–2), 36–48. DOI: 10.1016/j.vetmic.2007.05.002.

Kawaji, S., Nagata, R. and Mori, Y. (2014) Detection and confirmation of *Mycobacterium avium* subsp. *paratuberculosis* in direct quantitative PCR positive fecal samples by the manual fluorescent MGIT culture system. *Journal of Veterinary Medical Science* 76(1), 65–72. DOI: 10.1292/jvms.13-0366.

Keller, S.M., Stephan, R., Kuenzler, R., Meylan, M. and Wittenbrink, M.M. (2014) Comparison of fecal culture and F57 real-time polymerase chain reaction for the detection of *Mycobacterium avium* subspecies *paratuberculosis* in Swiss cattle herds with a history of paratuberculosis. *Acta Veterinaria Scandinavica* 56(1), 68. DOI: 10.1186/s13028-014-0068-9.

Khol, J., Vill, M., Dunser, M., Geisbacher, E., Tichy, A. *et al.* (2009) Environmental faecal sampling, a new approach in diagnosis and surveillance of paratuberculosis in Austrian cattle herds. *SWeiner Tierarztilche Monatsschrift* 96, 279–285.

Khol, J.L., Beran, V., Kralik, P., Trckova, M., Pavlik, I. *et al.* (2010) Grass silage contaminated with *Mycobacterium avium* subspecies *paratuberculosis* (MAP): a possible source of paratuberculosis infection in ruminants? *Veterinární Medicína* 55(5), 225–232. DOI: 10.17221/2996-VETMED.

Kim, Y.G., Bech-Nielsen, S., Slemons, R.D. and Spangler, E. (1989) Comparison of two methods for isolation of *Mycobacterium paratuberculosis* from bovine fecal samples. *American Journal of Veterinary Research* 50, 1110–1113.

Kim, S.G., Shin, S.J., Jacobson, R.H., Miller, L.J., Harpending, P.R. *et al.* (2002) Development and application of quantitative polymerase chain reaction assay based on the ABI 7700 system (TaqMan) for detection and quantification of *Mycobacterium avium* subsp *paratuberculosis*. *Journal of Veterinary Diagnostic Investigation* 14(2), 126–131. DOI: 10.1177/104063870201400206.

Kirkwood, C.D., Wagner, J., Boniface, K., Vaughan, J., Michalski, W.P. *et al.* (2009) *Mycobacterium avium* subspecies *paratuberculosis* in children with early-onset Crohn's disease. *Inflammatory Bowel Diseases* 15(11), 1643–1655. DOI: 10.1002/ibd.20967.

Koenig, G.J., Hoffsis, G.F., Shulaw, W.P., Bech-Nielsen, S., Rings, D.M. *et al.* (1993) Isolation of *Mycobacterium paratuberculosis* from mononuclear cells in tissues, blood, and mammary glands of cows with advanced paratuberculosis. *American Journal of Veterinary Research* 54, 1441–1445.

Krishnan, M.Y., Manning, E.J.B. and Collins, M.T. (2009a) Comparison of three methods for susceptibility testing of *Mycobacterium avium* subsp. *paratuberculosis* to 11 antimicrobial drugs. *Journal of Antimicrobial Chemotherapy* 64(2), 310–316. DOI: 10.1093/jac/dkp184.

Krishnan, M.Y., Manning, E.J.B. and Collins, M.T. (2009b) Effects of interactions of antibacterial drugs with each other and with 6-mercaptopurine on in vitro growth of *Mycobacterium avium* subspecies *paratuberculosis*. *Journal of Antimicrobial Chemotherapy* 64(5), 1018–1023. DOI: 10.1093/jac/dkp339.

Kruze, J., Monti, G., Schulze, F., Mella, A. and Leiva, S. (2013) Herd-level prevalence of MAP infection in dairy herds of southern Chile determined by culture of environmental fecal samples and bulk-tank milk qPCR. *Preventive Veterinary Medicine* 111(3–4), 319–324. DOI: 10.1016/j.prevetmed.2013.05.011.

Kumar, S., Singh, S.V., Singh, A.V., Singh, P.K., Sohal, J.S. *et al.* (2010) Wildlife (*Boselaphus tragocamelus*)–small ruminant (goat and sheep) interface in the transmission of 'Bison type' genotype of *Mycobacterium avium* subspecies *paratuberculosis* in India. *Comparative Immunology, Microbiology and Infectious Diseases* 33(2), 145–159. DOI: 10.1016/j.cimid.2008.08.006.

Küntzel, A., Fischer, S., Bergmann, A., Oertel, P., Steffens, M. *et al.* (2016) Effects of biological and methodological factors on volatile organic compound patterns during cultural growth of *Mycobacterium avium* ssp . *paratuberculosis*. *Journal of Breath Research* 10(3), 14. DOI: 10.1088/1752-7155/10/3/037103.

Küntzel, A., Weber, M., Gierschner, P., Trefz, P., Miekisch, W. *et al.* (2019) Core profile of volatile organic compounds related to growth of *Mycobacterium avium* subspecies *paratuberculosis* – a comparative extract of three independent studies. *PLoS ONE* 14(8), e0221031–e0221031. DOI: 10.1371/journal.pone.0221031.

Lambrecht, R.S. and Collins, M.T. (1992) *Mycobacterium paratuberculosis* factors that influence mycobactin dependence. *Diagnostic Microbiology and Infectious Disease* 15(3), 239–246. DOI: 10.1016/0732-8893(92)90119-E.

Lambrecht, R.S., Carriere, J.F. and Collins, M.T. (1988) A model for analyzing growth kinetics of a slowly growing *Mycobacterium* sp. *Applied and Environmental Microbiology* 54(4), 910–916. DOI: 10.1128/AEM.54.4.910-916.1988.

Lamont, E.A., Bannantine, J.P., Armién, A., Ariyakumar, D.S. and Sreevatsan, S. (2012) Identification and characterization of a spore-like morphotype in chronically starved *Mycobacterium avium* subsp. *paratuberculosis* cultures. *PLoS ONE* 7(1), e30648. DOI: 10.1371/journal.pone.0030648.

Laurin, E.L., Chaffer, M., McClure, J.T., McKenna, S.L.B. and Keefe, G.P. (2015) The association of detection method, season, and lactation stage on identification of fecal shedding in *Mycobacterium avium* ssp. *paratuberculosis* infectious dairy cows. *Journal of Dairy Science* 98(1), 211–220. DOI: 10.3168/jds.2014-8406.

Lavers, C.J., Mckenna, S.L.B., Dohoo, I.R., Barkema, H.W. and Keefe, G.P. (2013) Evaluation of environmental fecal culture for *Mycobacterium avium* subspecies *paratuberculosis* detection in dairy herds and association with apparent within-herd prevalence. *Canadian Veterinary Journal* 54, 1053–1060.

Lombard, J.E., Wagner, B.A., Smith, R.L., McCluskey, B.J., Harris, B.N. *et al.* (2006) Evaluation of environmental sampling and culture to determine *Mycobacterium avium* subspecies *paratuberculosis* distribution and herd infection status on US dairy operations. *Journal of Dairy Science* 89(11), 4163–4171. DOI: 10.3168/jds.S0022-0302(06)72461-4.

Marquetoux, N., Ridler, A., Heuer, C. and Wilson, P. (2019) What counts? A review of in vitro methods for the enumeration of *Mycobacterium avium* subsp. *paratuberculosis*. *Veterinary Microbiology* 230, 265–272. DOI: 10.1016/j.vetmic.2019.02.011.

Mason, O., Marsh, I.B. and Whittington, R.J. (2001) Comparison of immunomagnetic bead separation-polymerase chain reaction and faecal culture for the detection of *Mycobacterium avium* subsp *paratuberculosis* in sheep faeces. *Australian Veterinary Journal* 79(7), 497–500. DOI: 10.1111/j.1751-0813.2001.tb13024.x.

Matos, A.C., Figueira, L., Martins, M.H., Loureiro, F., Pinto, M.L. *et al.* (2014) Survey of *Mycobacterium avium* subspecies *paratuberculosis* in road-killed wild carnivores in Portugal. *Journal of Zoo and Wildlife Medicine* 45(4), 775–781. DOI: 10.1638/2014-0010.1.

Matthews, P.R.J., Mcdiarmid, A., Collins, P. and Brown, A. (1978) The dependence of some strains of *Mycobacterium avium* on mycobactin for initial and subsequent growth. *Journal of Medical Microbiology* 11(1), 53–57. DOI: 10.1099/00222615-11-1-53.

Mazzone, P., Corneli, S., Di Paolo, A., Maresca, C., Felici, A. *et al.* (2018) Survival of *Mycobacterium avium* subsp. *paratuberculosis* in the intermediate and final digestion products of biogas plants. *Journal of Applied Microbiology* 125(1), 36–44. DOI: 10.1111/jam.13762.

Mercier, P., Brémaud, I. and Gautier, M.-P. (2014) Vaccination of kids under one month of age with a killed vaccine and reduction in the frequency of faecal shedding of *Mycobacterium avium* subspecies *paratuberculosis*. *Small Ruminant Research* 121(2–3), 425–433. DOI: 10.1016/j.smallrumres.2014.09.007.

Merkal, R.S. (1970) Diagnostic methods for detection of paratuberculosis (Johne's disease). *Proceedings of the 74th Annual Meeting of the United States Animal Health Association*, USAHA, pp. 620–623.

Merkal, R.S. (1973) Laboratory diagnosis of bovine paratuberculosis. *Journal of the American Veterinary Medical Association* 163, 1100–1102.

Merkal, R.S. (1984) Paratuberculosis: advances in cultural, serologic, and vaccination methods. *Journal of the American Veterinary Medical Association* 184, 939–943.

Merkal, R.S. and Curran, B.J. (1974) Growth and metabolic characteristics of *Mycobacterium paratuberculosis*. *Applied Microbiology* 28(2), 276–279. DOI: 10.1128/AEM.28.2.276-279.1974.

Merkal, R.S. and McCullough, W.G. (1982) A new mycobactin, mycobactin J, from *Mycobacterium paratuberculosis*. *Current Microbiology* 7(6), 333–335. DOI: 10.1007/BF01572599.

Merkal, R.S., Kopecky, K.E., Larsen, A.B. and Thurston, J.R. (1964) Improvements in the techniques for primary cultivation of *Mycobacterium paratuberculosis*. *American Journal of Veterinary Research* 25, 1290–1294.

Merkal, R.S., Larsen, A.B., Kopecky, K.E. and Ness, R.D. (1968) Comparison of examination and test methods for early detection of paratuberculous cattle. *American Journal of Veterinary Research* 29, 1533–1538.

Merkal, R.S., Lyle, P. and Whipple, D.L. (1982) Decontamination, media and culture methods for *Mycobacterium paratuberculosis*. *Proceedings of the 86th Annual Meeting of the United States Animal Health Association*, pp. 519–523.

Middlebrook, G., Reggiardo, Z. and Tigertt, W.D. (1977) Automatable radiometric detection of growth of *Mycobacterium tuberculosis* in selective media. *American Review of Respiratory Disease* 115, 1066–1069.

Millar, D., Ford, J., Sanderson, J., Withey, S., Tizard, M. *et al.* (1996) IS*900* PCR to detect *Mycobacterium paratuberculosis* in retail supplies of whole pasteurized cow's milk in England and Wales. *Applied and Environmental Microbiology* 62(9), 3446–3452. DOI: 10.1128/AEM.62.9.3446-3452.1996.

Minett, F.C. (1942) The diagnosis of Johne's disease of cattle by cultural methods. *The Journal of Pathology and Bacteriology* 54(2), 209–219. DOI: 10.1002/path.1700540208.

Morrison, N.E. (1965) Circumvention of the mycobactin requirement of *Mycobacterium paratuberculosis*. *Journal of Bacteriology* 89(3), 762–767. DOI: 10.1128/JB.89.3.762-767.1965.

Motiwala, A.S., Strother, M., Theus, N.E., Stich, R.W., Byrum, B. *et al.* (2005) Rapid detection and typing of strains of *Mycobacterium avium* subsp. *paratuberculosis* from broth cultures. *Journal of Clinical Microbiology* 43(5), 2111–2117. DOI: 10.1128/JCM.43.5.2111-2117.2005.

Mutharia, L.M., Klassen, M.D., Fairles, J., Barbut, S. and Gill, C.O. (2010) *Mycobacterium avium* subsp. *paratuberculosis* in muscle, lymphatic and organ tissues from cows with advanced Johne's disease. *International Journal of Food Microbiology* 136(3), 340–344. DOI: 10.1016/j.ijfoodmicro.2009.10.026.

Naser, S.A., Schwartz, D. and Shafran, I. (2000) Isolation of *Mycobacterium avium* subsp *paratuberculosis* from breast milk of Crohn's disease patients. *The American Journal of Gastroenterology* 95(4), 1094–1095. DOI: 10.1111/j.1572-0241.2000.01954.x.

Naser, S.A., Ghobrial, G., Romero, C. and Valentine, J.F. (2004) Culture of *Mycobacterium avium* subspecies *paratuberculosis* from the blood of patients with Crohn's disease. *The Lancet* 364(9439), 1039–1044. DOI: 10.1016/S0140-6736(04)17058-X.

Nawrath, T., Mgode, G.F., Weetjens, B., Kaufmann, S.H.E. and Schulz, S. (2012) The volatiles of pathogenic and nonpathogenic mycobacteria and related bacteria. *Beilstein Journal of Organic Chemistry* 8, 290–299. DOI: 10.3762/bjoc.8.31.

Nielsen, S.S., Kolmos, B. and Christoffersen, A.B. (2004) Comparison of contamination and growth of *Mycobacterium avium* subsp. *paratuberculosis* on two different media. *Journal of Applied Microbiology* 96(1), 149–153. DOI: 10.1046/j.1365-2672.2003.02138.x.

Noll, I., Eisenberg, T., Failing, K., Rohde, M., Schlez, K. *et al.* (2017) Herd prevalence of *Mycobacterium avium* ssp *paratuberculosis* (MAP) in Hessian dairy herds. *Berliner Und Munchener Tierarztliche Wochenschrift* 130, 13–20.

Nowotarska, S., Nowotarski, K., Grant, I., Elliott, C., Friedman, M. *et al.* (2017) Mechanisms of antimicrobial action of cinnamon and oregano oils, cinnamaldehyde, carvacrol, 2,5-dihydroxybenzaldehyde, and 2-hydroxy-5-methoxybenzaldehyde against *Mycobacterium avium* subsp. *paratuberculosis* (MAP). *Foods* 6(9), 72. DOI: 10.3390/foods6090072.

Okwumabua, O., Shull, E., O'Connor, M., Moua, T.V., Danz, T. *et al.* (2010) Comparison of three methods for extraction of *Mycobacterium avium* subspecies *paratuberculosis* DNA for polymerase chain reaction from broth-based culture systems. *Journal of Veterinary Diagnostic Investigation* 22(1), 67–69. DOI: 10.1177/104063871002200111.

Parrish, N.M., Ko, C.G., Dick, J.D., Jones, P.B. and Ellingson, J.L.E. (2004) Growth, Congo red agar colony morphotypes and antibiotic susceptibility testing of *Mycobacterium avium* subspecies *paratuberculosis*. *Clinical Medicine & Research* 2(2), 107–114. DOI: 10.3121/cmr.2.2.107.

Parrish, N.M., Radcliff, R.P., Brey, B.J., Anderson, J.L., Clark, D.L., Jr. *et al.* (2008) Absence of *Mycobacterium avium* subsp. *paratuberculosis* in Crohn's patients. *Inflammatory Bowel Disease,* 558–565.

Pickup, R.W., Rhodes, G., Arnott, S., Sidi-Boumedine, K., Bull, T.J. *et al.* (2005) *Mycobacterium avium* subsp. *paratuberculosis* in the catchment area and water of the River Taff in South Wales, United Kingdom, and its potential relationship to clustering of Crohn's disease cases in the city of Cardiff. *Applied and Environmental Microbiology* 71(4), 2130–2139. DOI: 10.1128/AEM.71.4.2130-2139.2005.

Pillars, R.B., Grooms, D.L., Woltanski, J.A. and Blair, E. (2009) Prevalence of Michigan dairy herds infected with *Mycobacterium avium* subspecies *paratuberculosis* as determined by environmental sampling. *Preventive Veterinary Medicine* 89(3–4), 191–196. DOI: 10.1016/j.prevetmed.2009.02.022.

Plain, K.M., Marsh, I.B., Waldron, A.M., Galea, F., Whittington, A.-M. *et al.* (2014) High-throughput direct fecal PCR assay for detection of *Mycobacterium avium* subsp. *paratuberculosis* in sheep and cattle. *Journal of Clinical Microbiology* 52(3), 745–757. DOI: 10.1128/JCM.03233-13.

Plain, K.M., Waldron, A.M., Begg, D.J., de Silva, K., Purdie, A.C. *et al.* (2015) Efficient, validated method for detection of mycobacterial growth in liquid culture media by use of bead beating, magnetic-particle-based nucleic acid isolation, and quantitative PCR. *Journal of Clinical Microbiology* 53(4), 1121–1128. DOI: 10.1128/JCM.03521-14.

Pooley, H.B., de Silva, K., Purdie, A.C., Begg, D.J., Whittington, R.J. *et al.* (2016) A rapid method for quantifying viable *Mycobacterium avium* subsp. *paratuberculosis* in cellular infection assays. *Applied and Environmental Microbiology* 82(18), 5553–5562. DOI: 10.1128/AEM.01668-16.

Pozzato, N., Gwozdz, J., Gastaldelli, M., Capello, K., Ben, C.D. *et al.* (2011) Evaluation of a rapid and inexpensive liquid culture system for the detection of *Mycobacterium avium* subsp. *paratuberculosis* in bovine faeces. *Journal of Microbiological Methods* 84(3), 413–417. DOI: 10.1016/j.mimet.2011.01.019.

Pribylova, R., Kubickova, L., Babak, V., Pavlik, I. and Kralik, P. (2012) Effect of short- and long-term antibiotic exposure on the viability of *Mycobacterium avium* subsp. *paratuberculosis* as measured by propidium monoazide f57 real time quantitative PCR and culture. *The Veterinary Journal* 194(3), 354–360. DOI: 10.1016/j.tvjl.2012.05.002.

Purdie, A.C., Plain, K.M., Begg, D.J., de Silva, K. and Whittington, R.J. (2011) Candidate gene and genome-wide association studies of *Mycobacterium avium* subsp. *paratuberculosis* infection in cattle and sheep: a review. *Comparative Immunology, Microbiology and Infectious Diseases* 34(3), 197–208. DOI: 10.1016/j.cimid.2010.12.003.

Rangel, S., Arango-Sabogal, J.C., Labrecque, O., Paré, J., Fairbrother, J.-H. *et al.* (2017) Evaluation of a PCR assay on overgrown individual fecal samples cultured for *Mycobacterium avium* subsp. *paratuberculosis*. *Journal of Veterinary Diagnostic Investigation* 29(6), 912–915. DOI: 10.1177/1040638717724837.

Ravva, S.V., Harden, L.A. and Sarreal, C.Z. (2017) Characterization and differentiation of *Mycobacterium avium* subsp. *paratuberculosis* from other mycobacteria using matrix assisted laser desorption/ionization time-of-flight mass spectrometry. *Frontiers in Cellular and Infection Microbiology* 7, 8. DOI: 10.3389/fcimb.2017.00297.

Realini, L., Van Der Stuyft, P., De Ridder, K., Hirschel, B. and Portaels, F. (1997) Inhibitory effects of polyoxyethylene stearate, PANTA, and neutral pH on growth of *Mycobacterium genavense* in BACTEC primary cultures. *Journal of Clinical Microbiology* 35(11), 2791–2794. DOI: 10.1128/JCM.35.11.2791-2794.1997.

Reddacliff, L.A., Nicholls, P.J., Vadali, A. and Whittington, R.J. (2003a) Use of growth indices from radiometric culture for quantification of sheep strains of *Mycobacterium avium* subsp. *paratuberculosis*. *Applied and Environmental Microbiology* 69(6), 3510–3516. DOI: 10.1128/AEM.69.6.3510-3516.2003.

Reddacliff, L.A., Vadali, A. and Whittington, R.J. (2003b) The effect of decontamination protocols on the numbers of sheep strain *Mycobacterium avium* subsp. *paratuberculosis* isolated from tissues and faeces. *Veterinary Microbiology* 95(4), 271–282. DOI: 10.1016/S0378-1135(03)00181-0.

Reddacliff, L., Eppleston, J., Windsor, P., Whittington, R. and Jones, S. (2006a) Efficacy of a killed vaccine for the control of paratuberculosis in Australian sheep flocks. *Veterinary Microbiology* 115(1–3), 77–90. DOI: 10.1016/j.vetmic.2005.12.021.

Reddacliff, L., Eppleston, J., Windsor, P., Whittington, R. and Jones, S. (2006b) Efficacy of a killed vaccine for the control of paratuberculosis in Australian sheep flocks. *Veterinary Microbiology* 115(1–3), 77–90. DOI: 10.1016/j.vetmic.2005.12.021.

Reddacliff, L.A., Marsh, I.B., Fell, S.A., Austin, S.L. and Whittington, R.J. (2010) Isolation of *Mycobacterium avium* subspecies *paratuberculosis* from muscle and peripheral lymph nodes using acid-pepsin digest prior to BACTEC culture. *Veterinary Microbiology* 145(1–2), 122–128. DOI: 10.1016/j.vetmic.2010.03.011.

Reggiardo, Z. and Tigertt, W.D. (1977) Automatable radiometric detection of growth of *Mycobacterium tuberculosis* in selective media. *American Review of Respiratory Disease* 115, 1066–1069.

Ridge, S.E., Andreata, S., Jones, K., Cantlon, K., Francis, B. *et al.* (2010) Inter-laboratory comparison of radiometric culture for *Mycobacterium avium* subsp. *paratuberculosis* using raw milk from known infected herds and individual dairy cattle in Victoria. *Australian Veterinary Journal* 88(7), 249–254. DOI: 10.1111/j.1751-0813.2010.00579.x.

Ritter, C., Wolf, R., Adams, C.L., Kelton, D.F., Pickel, C. *et al.* (2016) Short communication: Herd-level prevalence of *Mycobacterium avium* ssp. *paratuberculosis* is not associated with participation in a voluntary Alberta Johne's disease control program. *Journal of Dairy Science* 99(3), 2157–2160. DOI: 10.3168/jds.2015-10356.

Rock, K.E., Kinkead, L.C. and Secott, T.E. (2016) The biological reducing agent oxyrase improves the resuscitation of dormant *Mycobacterium smegmatis* and *Mycobacterium avium* subsp. *paratuberculosis*. *Microbiology Research* 7(1), 13–18. DOI: 10.4081/mr.2016.5695.

Salgado, M., Herthnek, D., Bölske, G., Leiva, S. and Kruze, J. (2009) First isolation of *Mycobacterium avium* subsp. *paratuberculosis* from wild guanacos (*Lama guanicoe*) on Tierra del Fuego Island. *Journal of Wildlife Diseases* 45(2), 295–301. DOI: 10.7589/0090-3558-45.2.295.

Salgado, M., Collins, M.T., Salazar, F., Kruze, J., Bölske, G. *et al.* (2011) Fate of *Mycobacterium avium* subsp. *paratuberculosis* after application of contaminated dairy cattle manure to agricultural soils. *Applied and Environmental Microbiology* 77(6), 2122–2129. DOI: 10.1128/AEM.02103-10.

Salgado, M., Verdugo, C., Heuer, C., Castillo, P. and Zamorano, P. (2014) A novel low-cost method for *Mycobacterium avium* subsp. *paratuberculosis* DNA extraction from an automated broth culture system for real-time PCR analysis. *Journal of Veterinary Science* 15(2), 233–239. DOI: 10.4142/jvs.2014.15.2.233.

Salgado, M., Sevilla, I., Rios, C., Crossley, J., Tejeda, C. *et al.* (2016) Presence of *Mycobacterium avium* subsp. *paratuberculosis* in alpacas (*Lama pacos*) inhabiting the Chilean altiplano. *Journal of Zoo and Wildlife Medicine* 47(1), 12–16. DOI: 10.1638/2012-0255.1.

Savi, R., Ricchi, M., Cammi, G., Garbarino, C., Leo, S. *et al.* (2015) Survey on the presence of *Mycobacterium avium* subsp. *paratuberculosis* in ground beef from an industrial meat plant. *Veterinary Microbiology* 177(3–4), 403–408. DOI: 10.1016/j.vetmic.2015.03.013.

Saxegaard, F. (1985) Isolation of *Mycobacterium paratuberculosis* from intestinal mucosa and mesenteric lymph nodes of goats by use of selective Dubos medium. *Journal of Clinical Microbiology* 22(2), 312–313. DOI: 10.1128/JCM.22.2.312-313.1985.

Schwalm, A.K., Obiegala, A., Pfeffer, M. and Sting, R. (2018) Enhanced sensitivity and fast turnaround time in laboratory diagnosis for bovine paratuberculosis in faecal samples. *Journal of Microbiological Methods* 152, 39–47. DOI: 10.1016/j.mimet.2018.07.010.

Secott, T.E., Ohme, A.M., Barton, K.S., Wu, C.C. and Rommel, F.A. (1999) *Mycobacterium paratuberculosis* detection in bovine feces is improved by coupling agar culture enrichment to an IS *900*-specific polymerase chain reaction assay. *Journal of Veterinary Diagnostic Investigation* 11(5), 441–447. DOI: 10.1177/104063879901100509.

Selby, W., Pavli, P., Crotty, B., Florin, T., Radford-Smith, G. *et al.* (2007) Two-year combination antibiotic therapy with clarithromycin, rifabutin, and clofazimine for Crohn's disease. *Gastroenterology* 132(7), 2313–2319. DOI: 10.1053/j.gastro.2007.03.031.

Selim, A., Ali, A.-F. and Ramadan, E. (2019) Prevalence and molecular epidemiology of Johne's disease in Egyptian cattle. *Acta Tropica* 195, 1–5. DOI: 10.1016/j.actatropica.2019.04.019.

Sergeant, E.S.G., Whittington, R.J. and More, S.J. (2002) Sensitivity and specificity of pooled faecal culture and serology as flock-screening tests for detection of ovine paratuberculosis in Australia. *Preventive Veterinary Medicine* 52(3–4), 199–211. DOI: 10.1016/S0167-5877(01)00261-6.

Shin, S. (1989) Report of the Committee on Johne's disease. *Proceedings of the 93rd Annual Meeting of the United States Animal Health Association*, Richmond, Virginia: United States Animal Health Association, pp. 380–381.

Shin, S.J., Han, J.H., Manning, E.J.B. and Collins, M.T. (2007) Rapid and reliable method for quantification of *Mycobacterium paratuberculosis* by use of the BACTEC MGIT 960 system. *Journal of Clinical Microbiology* 45(6), 1941–1948. DOI: 10.1128/JCM.02616-06.

Slana, I., Pribylova, R., Kralova, A. and Pavlik, I. (2011) Persistence of *Mycobacterium avium* subsp. *paratuberculosis* at a farm-scale biogas plant supplied with manure from paratuberculosis-affected dairy cattle. *Applied and Environmental Microbiology* 77(9), 3115–3119. DOI: 10.1128/AEM.02407-10.

Smith, H.W. (1953) Modifications of Dubos's media for the cultivation of *Mycobacterium johnei*. *The Journal of Pathology and Bacteriology* 66(2), 375–381. DOI: 10.1002/path.1700660206.

Sockett, D.C., Carr, D.J. and Collins, M.T. (1992) Evaluation of conventional and radiometric fecal culture and a commercial DNA probe for diagnosis of *Mycobacterium paratuberculosis* infection in cattle. *Canadian Journal of Veterinary Research* 56, 148–153.

Stabel, J.R. (1997) An improved method for cultivation of *Mycobacterium paratuberculosis* from bovine fecal samples and comparison to three other methods. *Journal of Veterinary Diagnostic Investigation* 9(4), 375–380. DOI: 10.1177/104063879700900406.

Stabel, J.R., Steadham, E.M. and Bolin, C.A. (1997) Heat inactivation of *Mycobacterium paratuberculosis* in raw milk: are current pasteurization conditions effective? *Applied and Environmental Microbiology* 63(12), 4975–4977. DOI: 10.1128/AEM.63.12.4975-4977.1997.

Steuer, P., de Waard, J., Collins, M.T., Troncoso, E.P., Martínez, O.A. *et al.* (2015) Identification of contaminating bacteria when attempting to isolate *Mycobacterium avium* subsp. *paratuberculosis* (MAP) from bovine faecal and tissue samples using the BACTEC MGIT 960 system. *Archivos de Medicina Veterinaria* 47(1), 97–100. DOI: 10.4067/S0301-732X2015000100016.

Steuer, P., Avilez, C., Tejeda, C., Gonzalez, N., Ramirez-Reveco, A. *et al.* (2018) In vitro inactivation of *Mycobacterium avium* subsp. *paratuberculosis* (MAP) by use of copper ions. *BMC Microbiology* 18(1), 7. DOI: 10.1186/s12866-018-1313-6.

Stevenson, K. (2015) Genetic diversity of *Mycobacterium avium* subspecies *paratuberculosis* and the influence of strain type on infection and pathogenesis: a review. *Veterinary Research* 46(1), 64. DOI: 10.1186/s13567-015-0203-2.

Stevenson, K., Hughes, V.M., de Juan, L., Inglis, N.F., Wright, F. *et al.* (2002) Molecular characterization of pigmented and nonpigmented isolates of *Mycobacterium avium* subsp. *paratuberculosis*. *Journal of Clinical Microbiology* 40(5), 1798–1804. DOI: 10.1128/JCM.40.5.1798-1804.2002.

Stich, R.W., Byrum, B., Love, B., Theus, N., Barber, L. *et al.* (2004) Evaluation of an automated system for non-radiometric detection of *Mycobacterium avium paratuberculosis* in bovine feces. *Journal of Microbiological Methods* 56(2), 267–275. DOI: 10.1016/j.mimet.2003.10.013.

Stinson, M.W. and Solotorovsky, M. (1971) Interaction of Tween 80 detergent with mycobacteria in synthetic medium. I. Effect of Tween 80 on the growth and turbidimetric response of *Mycobacterium avium* cultures. *The American Review of Respiratory Disease* 104(5), 717–727. DOI: 10.1164/arrd.1971.104.5.717.

Stringer, L.A., Wilson, P.R., Heuer, C., Hunnam, J.C., Verdugo, C. *et al.* (2013) Prevalence of *Mycobacterium avium* subsp. *paratuberculosis* in farmed red deer (*Cervus elaphus*) with grossly normal mesenteric lymph nodes. *New Zealand Veterinary Journal* 61(3), 147–152. DOI: 10.1080/00480169.2012.755888.

Sweeney, R.W., Whitlock, R.H. and Rosenberger, A.E. (1992) *Mycobacterium paratuberculosis* cultured from milk and supramammary lymph nodes of infected asymptomatic cows. *Journal of Clinical Microbiology* 30(1), 166–171. DOI: 10.1128/JCM.30.1.166-171.1992.

Sweeney, R.W., Whitlock, R.H. and McAdams, S.C. (2006) Comparison of three DNA preparation methods for real-time polymerase chain reaction confirmation of *Mycobacterium avium* subsp. *paratuberculosis* growth in an automated broth culture system. *Journal of Veterinary Diagnostic Investigation* 18(6), 587–590. DOI: 10.1177/104063870601800611.

Tamarin, R. and Landau, M. (1961) Congenital and uterine infection with *Mycobacterium johnei* in sheep. *Refuah Veterinarith* 18, 43–44.

Tavornpanich, S., Muñoz-Zanzi, C.A., Wells, S.J., Raizman, E.A., Carpenter, T.E. *et al.* (2008) Simulation model for evaluation of testing strategies for detection of paratuberculosis in midwestern US dairy herds. *Preventive Veterinary Medicine* 83(1), 65–82. DOI: 10.1016/j.prevetmed.2007.06.010.

Taylor, A.W. (1950) Observations on the isolation of *Mycobacterium johnei* in primary culture. *The Journal of Pathology and Bacteriology* 62(4), 647–650. DOI: 10.1002/path.1700620414.

Taylor, T.K., Wilks, C.R. and Mcqueen, D.S. (1981) Isolation of *Mycobacterium paratuberculosis* from the milk of a cow with Johne's disease. *Veterinary Record* 109, 532–533.

Thirumalapura, N.R., Feria, W. and Tewari, D. (2018) Comparison of three DNA extraction methods for molecular confirmation of *Mycobacterium avium* subspecies *paratuberculosis* from the VersaTrek™ liquid cultures of bovine fecal samples. *Journal of Microbiological Methods* 152, 27–30. DOI: 10.1016/j.mimet.2018.07.012.

Thorel, M.F. (1984) Review of the occurrence of mycobactin dependence among mycobacteria species. *Annales De Recherches Veterinaires* 15, 405–409.

Thorel, M.F., Krichevsky, M. and Lévy-Frébault, V.V. (1990) Numerical taxonomy of mycobactin-dependent mycobacteria, emended description of *Mycobacterium avium*, and description of *Mycobacterium avium* subsp. *avium* subsp. *nov.*, *Mycobacterium avium* subsp. *paratuberculosis* subsp. *nov.*, and *Mycobacterium avium* subsp. *silvaticum* subsp. *nov. International Journal of Systematic Bacteriology* 40(3), 254–260. DOI: 10.1099/00207713-40-3-254.

Thornton, C.G., MacLellan, K.M., Stabel, J.R., Carothers, C., Whitlock, R.H. *et al.* (2002) Application of the C(18)-carboxypropylbetaine specimen processing method to recovery of *Mycobacterium avium* subsp. *paratuberculosis* from ruminant tissue specimens. *Journal of Clinical Microbiology* 40(5), 1783–1790. DOI: 10.1128/JCM.40.5.1783-1790.2002.

Toribio, J.-A.L.M.L. and Sergeant, E.S.G. (2007) A comparison of methods to estimate the prevalence of ovine Johne's infection from pooled faecal samples. *Australian Veterinary Journal* 85(8), 317–324. DOI: 10.1111/j.1751-0813.2007.00188.x.

Trefz, P., Koehler, H., Klepik, K., Moebius, P., Reinhold, P. *et al.* (2013) Volatile emissions from *Mycobacterium avium* subsp. *paratuberculosis* mirror bacterial growth and enable distinction of different strains. *PLoS ONE* 8(10), e76868. DOI: 10.1371/journal.pone.0076868.

Twort, F.W. and Ingram, G.L.Y. (1912) A method for isolating and cultivating the *Mycobacterium enteritidis chronicae pseudotuberculosae bovis*, Johne, and some experiments on the preparation of a diagnostic vaccine for pseudo-tuberculous enteritis of bovines. *Proceedings of the Royal Society of London (Series B)* 84, 517–543.

van Boxtel, R.M., Lambrecht, R.S. and Collins, M.T. (1990) Effect of polyoxyethylene sorbate compounds (Tweens) on colonial morphology, growth, and ultrastructure of *Mycobacterium paratuberculosis*. *APMIS* 98(10), 901–908. DOI: 10.1111/j.1699-0463.1990.tb05013.x.

Van Brandt, L., Coudijzer, K., Herman, L., Michiels, C., Hendrickx, M. *et al.* (2011a) Survival of *Mycobacterium avium* ssp. *paratuberculosis* in yoghurt and in commercial fermented milk products containing probiotic cultures. *Journal of Applied Microbiology* 110(5), 1252–1261. DOI: 10.1111/j.1365-2672.2011.04979.x.

Van Brandt, L., Van der Plancken, I., De Block, J., Vlaemynck, G., Van Coillie, E. *et al.* (2011b) Adequacy of current pasteurization standards to inactivate *Mycobacterium paratuberculosis* in milk and phosphate buffer. *International Dairy Journal* 21(5), 295–304. DOI: 10.1016/j.idairyj.2010.12.014.

van Griethuysen, A.J., Jansz, A.R. and Buiting, A.G. (1996) Comparison of fluorescent BACTEC 9000 MB system, Septi-Chek AFB system, and Lowenstein-Jensen medium for detection of mycobacteria. *Journal of Clinical Microbiology* 34(10), 2391–2394. DOI: 10.1128/JCM.34.10.2391-2394.1996.

van Schaik, G., Stehman, S.M., Schukken, Y.H., Rossiter, C.R. and Shin, S.J. (2003) Pooled fecal culture sampling for *Mycobacterium avium* subsp. *paratuberculosis* at different herd sizes and prevalence. *Journal of Veterinary Diagnostic Investigation* 15(3), 233–241. DOI: 10.1177/104063870301500304.

Vazquez, P., Garrido, J.M., Molina, E., Geijo, M.V., Gomez, N. *et al.* (2014) Latent infections are the most frequent form of paratuberculosis in slaughtered Friesian cattle. *Spanish Journal of Agricultural Research* 12(4), 1049. DOI: 10.5424/sjar/2014124-5978.

Vialard, J., Lacheretz, A., Thiercy, A., Richard, Y. and Prave, M. (1993) Detection of bovine paratuberculosis by culture of pooled feces (detection de l'infection paratuberculeuse chez les bovins par une technique de coproculture de group). *Revue De Medecine Veterinaire* 144, 527–533.

Wang, J., Moolji, J., Dufort, A., Staffa, A., Domenech, P. *et al.* (2016) Iron acquisition in *Mycobacterium avium* subsp. *paratuberculosis*. *Journal of Bacteriology* 198(5), 857–866. DOI: 10.1128/JB.00922-15.

Watson, E.A. (1935) Tuberculin, Johnin and Mallein derived from non-protein media. *Canadian Public Health Journal* 26, 268–275.

Wells, S.J., Whitlock, R.H., Lindeman, C.J. and Fyock, T. (2002) Evaluation of bacteriologic culture of pooled fecal samples for detection of *Mycobacterium paratuberculosis*. *American Journal of Veterinary Research* 63(8), 1207–1211. DOI: 10.2460/ajvr.2002.63.1207.

Wells, S.J., Godden, S.M., Lindeman, C.J. and Collins, J.E. (2003) Evaluation of bacteriologic culture of individual and pooled fecal samples for detection of *Mycobacterium paratuberculosis* in dairy cattle herds. *Journal of the American Veterinary Medical Association* 223(7), 1022–1025. DOI: 10.2460/javma.2003.223.1022.

Whipple, D.L. and Merkal, R.S. (1983) Modifications in the technique for cultivation of *Mycobacterium paratuberculosis*. *Proceedings of the International Colloquium on Research in Paratuberculosis*, Ames, International Association for Paratuberculosis, pp. 82–92.

Whipple, D.L., Callihan, D.R. and Jarnagin, J.L. (1991) Cultivation of *Mycobacterium paratuberculosis* from bovine fecal specimens and a suggested standardized procedure. *Journal of Veterinary Diagnostic Investigation* 3(4), 368–373. DOI: 10.1177/104063879100300424.

Whitlock, R.H. and Rosenberger, A.E. (1990) Fecal culture protocol for *Mycobacterium paratuberculosis*. A recommended procedure. *Proceedings of the Annual Meeting of the United States Animal Health Association*, US Animal Health Association, pp. 280–285.

Whitlock, R.H., Rosenberger, A.E. and Spencer, P.A. (1989) Laboratory culture techniques for Johne's disease: a critical evaluation of contamination and incubation times. *Proceedings of the Ninety-third Annual Meeting of the United States Animal Health Association*, Richmond, Virginia, United States Animal Health Association, pp. 382–386.

Whitlock, R.H., West, S.R., Layton, B., Ellingson, J., Stabel, J. *et al.* (1999) Paratuberculosis in bison: a comparison of PCR, culture, serology and histopathology. In: Manning, E.J.B. and Collins, M.T. (eds) *Proceedings of the Sixth International Colloquium on Paratuberculosis*. International Association for Paratuberculosis, Madison, pp. 424–438.

Whittington, R.J. (2009) Factors affecting isolation and identification of *Mycobacterium avium* subsp. *paratuberculosis* from fecal and tissue samples in a liquid culture system. *Journal of Clinical Microbiology* 47(3), 614–622. DOI: 10.1128/JCM.01986-08.

Whittington, R.J. and Sergeant, E.S. (2001) Progress towards understanding the spread, detection and control of *Mycobacterium avium* subsp. *paratuberculosis* in animal populations. *Australian Veterinary Journal* 79(4), 267–278. DOI: 10.1111/j.1751-0813.2001.tb11980.x.

Whittington, R.J., Marsh, I., Turner, M.J., McAllister, S., Choy, E. *et al.* (1998) Rapid detection of *Mycobacterium paratuberculosis* in clinical samples from ruminants and in spiked environmental samples by modified BACTEC 12B radiometric culture and direct confirmation by IS900 PCR. *Journal of Clinical Microbiology* 36(3), 701–707. DOI: 10.1128/JCM.36.3.701-707.1998.

Whittington, R.J., Marsh, I., McAllister, S., Turner, M.J., Marshall, D.J. *et al.* (1999) Evaluation of modified BACTEC 12B radiometric medium and solid media for culture of *Mycobacterium avium* subsp. *paratuberculosis* from sheep. *Journal of Clinical Microbiology* 37(4), 1077–1083. DOI: 10.1128/JCM.37.4.1077-1083.1999.

Whittington, R.J., Fell, S., Walker, D., McAllister, S., Marsh, I. *et al.* (2000a) Use of pooled fecal culture for sensitive and economic detection of *Mycobacterium avium* subsp. *paratuberculosis* infection in flocks of sheep. *Journal of Clinical Microbiology* 38(7), 2550–2556. DOI: 10.1128/JCM.38.7.2550-2556.2000.

Whittington, R.J., Reddacliff, L.A., Marsh, I., McAllister, S. and Saunders, V. (2000b) Temporal patterns and quantification of excretion of *Mycobacterium avium* subsp *paratuberculosis* in sheep with Johne's disease. *Australian Veterinary Journal* 78(1), 34–37. DOI: 10.1111/j.1751-0813.2000.tb10355.x.

Whittington, R.J., Lloyd, J.B. and Reddacliff, L.A. (2001) Recovery of *Mycobacterium avium* subsp. *paratuberculosis* from nematode larvae cultured from the faeces of sheep with Johne's disease. *Veterinary Microbiology* 81(3), 273–279. DOI: 10.1016/S0378-1135(01)00345-5.

Whittington, R.J., Marsh, I.B., Taylor, P.J., Marshall, D.J., Taragel, C. *et al.* (2003a) Isolation of *Mycobacterium avium* subsp. *paratuberculosis* from environmental samples collected from farms before and after destocking sheep with paratuberculosis. *Australian Veterinary Journal* 81(9), 559–563. DOI: 10.1111/j.1751-0813.2003.tb12887.x.

Whittington, R.J., Marsh, I.B., Taylor, P.J., Marshall, D.J., Taragel, C. *et al.* (2003b) Isolation of *Mycobacterium avium* subsp. *paratuberculosis* from environmental samples collected from farms before and after destocking sheep with paratuberculosis. *Australian Veterinary Journal* 81(9), 559–563. DOI: 10.1111/j.1751-0813.2003.tb12887.x.

Whittington, R.J., Marshall, D.J., Nicholls, P.J., Marsh, I.B. and Reddacliff, L.A. (2004) Survival and dormancy of *Mycobacterium avium* subsp. *paratuberculosis* in the environment. *Applied and Environmental Microbiology* 70(5), 2989–3004. DOI: 10.1128/AEM.70.5.2989-3004.2004.

Whittington, R.J., Marsh, I.B. and Reddacliff, L.A. (2005) Survival of *Mycobacterium avium* subsp. *paratuberculosis* in dam water and sediment. *Applied and Environmental Microbiology* 71(9), 5304–5308. DOI: 10.1128/AEM.71.9.5304-5308.2005.

Whittington, R.J., Waldron, A. and Warne, D. (2010) Thermal inactivation profiles of *Mycobacterium avium* subsp. *paratuberculosis* in lamb skeletal muscle homogenate fluid. *International Journal of Food Microbiology* 137(1), 32–39. DOI: 10.1016/j.ijfoodmicro.2009.10.009.

Whittington, R.J., Marsh, I.B., Saunders, V., Grant, I.R., Juste, R. *et al.* (2011) Culture phenotypes of genomically and geographically diverse *Mycobacterium avium* subsp. *paratuberculosis* isolates from different hosts. *Journal of Clinical Microbiology* 49(5), 1822–1830. DOI: 10.1128/JCM.00210-11.

Whittington, R.J., Whittington, A.-M., Waldron, A., Begg, D.J., de Silva, K. *et al.* (2013) Development and validation of a liquid medium (M7H9C) for routine culture of *Mycobacterium avium* subsp. *paratuberculosis* to replace modified BACTEC 12B medium. *Journal of Clinical Microbiology* 51(12), 3993–4000. DOI: 10.1128/JCM.01373-13.

Whittington, R.J., Begg, D.J., de Silva, K., Purdie, A.C., Dhand, N.K. *et al.* (2017) Case definition terminology for paratuberculosis (Johne's disease). *BMC Veterinary Research* 13(1), 328. DOI: 10.1186/s12917-017-1254-6.

Wilson, G. and Miles, A. (1975) Tuberculosis. In: *Topley and Wilson's Principles of Bacteriology, Virology and Immunity*, 6th edn. Edward Arnold, London, pp. 1724–1785.

Wolf, R., Donate, K., Khol, J.L., Barkema, H.W., Kastelic, J. *et al.* (2017) Detection of *Mycobacterium avium* subspecies *paratuberculosis* infected cattle herds using environmental samples: a review. *Berliner Und Munchener Tierarztliche Wochenschrift* 130, 4–12.

19 Diagnosis of Paratuberculosis by PCR

Karren M. Plain[1]*, Ian Marsh[2] and Auriol C. Purdie[1]
[1]Sydney School of Veterinary Science, University of Sydney, Sydney, Australia;
[2]Elizabeth Macarthur Agricultural Institute, NSW Department of Primary Industries, Menangle, Australia

19.1 Introduction

Diagnosis of paratuberculosis infection can be made either by detection of the causative agent, *Mycobacterium avium* subsp. *paratuberculosis* (*MAP*), or through detection of a biological response made by an infected animal, for example using a serum antibody enzyme-linked immunosorbent assay (ELISA) to detect a humoral immune response. Culture methods for *MAP* are the gold standard diagnostic test, however the time it takes to grow the bacteria is a major issue for producers and in the control of disease spread. Nucleic acid amplification based tests (NAATs) such as the polymerase chain reaction (PCR) are an alternative, far more rapid method to detect *MAP* DNA in a sample matrix; most commonly this is faeces due to the pathobiology of paratuberculosis, but other sample types including milk, tissue, semen, blood, lymphatic fluid and culture media can also be examined.

PCR has, and will continue, to make a significant contribution to the detection of pathogens associated with infectious disease. Direct PCR-based diagnostics can have higher sensitivity than culture for the detection of pathogens and have the advantage of being rapid and unaffected by antibiotics (Obara *et al.*, 2011; Ljungström *et al.*, 2015). Conventional PCR (PCR followed by gel-based detection) was first integrated into the diagnostic regime for paratuberculosis in the early 1990s and has continued, using a range of *MAP* target sequences. Most notably, IS*900* has been used for the detection of *MAP* (Green *et al.*, 1989; Moss *et al.*, 1991) and IS*1311* for strain typing applications (Collins *et al.*, 1997; Marsh *et al.*, 1999). Following conventional PCR was the introduction of real-time or quantitative PCR (qPCR) in the late 1990s/early 2000s.

For *MAP* PCR-based diagnostic tests, though certainly much faster than culture, lack of diagnostic sensitivity has been a major issue, particularly when comparing the sensitivity of faecal PCR and faecal culture. The previous edition of this book concluded with a statement that progress was being made on this front and that 'PCR holds more promise now for the diagnosis of paratuberculosis' (Bölske and Herthnek, 2012). In the past 10 years, there have indeed been marked advancements, moving from a consensus in the literature that culture is the more sensitive detection method (Eamens *et al.*, 2007; Douarre *et al.*, 2010), to the current status where the majority of countries that have implemented a control programme for paratuberculosis employ a

*Corresponding author: karren.plain@sydney.edu.au

© CAB International 2020. *Paratuberculosis: Organism, Disease, Control, 2nd Edition* (eds M.A. Behr *et al.*)

faecal PCR diagnostic test in that programme (Yamamoto *et al.*, 2018b; Whittington *et al.*, 2019). There is support for a move towards increased PCR-based diagnostics in human tuberculosis medicine to 'rule in' cases and enable more rapid treatment, though sputum culture remains the gold standard in most low-burden settings (Babafemi *et al.*, 2017; Vinuesa *et al.*, 2018). Developments in *MAP* PCR-based diagnostics have been made possible by advancements within the paratuberculosis field, as well as those in research on related mycobacterial diseases and more distant areas of applied science such as forensics.

The ability to detect *MAP* DNA in a sample by PCR is governed by a number of factors: the disease pathogenesis, which impacts the likelihood that *MAP* will be present in a sample from an infected animal at the time that the diagnostic test is applied, the type of sample tested and the resilience of the *MAP* organism. Paratuberculosis is a chronic infection, characterized by intermittent faecal shedding and dissemination events by the *MAP* bacterium, particularly during the subclinical stages of the disease. This means that, while many antemortem diagnostic tests including those based on PCR perform well at the herd level, individual animal diagnosis is complicated by the stage of disease. Diagnosis becomes progressively easier as an animal approaches clinical stages of disease due to more consistent and high levels of faecal shedding of *MAP* and increased incidence of dissemination and presence in the milk and blood (Alonso-Hearn *et al.*, 2009; Bower *et al.*, 2011; Stabel *et al.*, 2014). The sample type is an important consideration as this can impact the ability to extract useable DNA (Stevenson and Sharp, 1997), one of the main impediments to the application of PCR directly to sample material from infected animals. The nature of *MAP* is also important to consider, in relation to the extraction approach required to lyse the tough outer cell wall of the bacterium in order to release the DNA (Odumeru *et al.*, 2001).

This chapter will examine recent developments in PCR-based diagnostic tests for paratuberculosis and the application and interpretation of these tests at the individual and herd level.

19.2 PCR Techniques

19.2.1 Conventional PCR

Early methods for *MAP* detection by PCR utilized conventional PCR (Stevenson and Sharp, 1997). In conventional PCR methods, amplification of the final DNA product only is assessed following the completion of the cycles of DNA amplification (Lorenz, 2012). This can involve gel electrophoresis or a capillary-based analytical method (Yokoyama *et al.*, 2006). Conventional PCR techniques, involving either single or nested amplification of *MAP* target regions, have been phased out in many countries in preference of qPCR assays.

The challenge with conventional PCR assays is the reduced analytical sensitivity compared with qPCR (Zemtsova *et al.*, 2015; Xia *et al.*, 2018). Nested PCR, though more sensitive, carries with it a high risk of cross-contamination due to the additional manipulation post-primary amplification. However, these methods are still applied in some settings, such as *MAP* detection in tissues from patients with Crohn's disease (Sharp *et al.*, 2018; Zarei-Kordshouli *et al.*, 2019). Strain typing assays still commonly use conventional PCR for IS*1311* and restriction enzyme digest to identify single nucleotide polymorphisms common to the differing strains (Marsh *et al.*, 1999; Park *et al.*, 2018; de Albuquerque *et al.*, 2018).

19.2.2 Real-time/Quantitative PCR

In qPCR, DNA amplification is monitored in 'real time' during the PCR cycling by the collection of fluorescent signal data associated with the amplified product. qPCR instruments measure florescence after each cycle, with multiple fluorescent molecules and detection channels available allowing for evaluation of multiple concurrent targets (multiplex). The point at which the amplification curve crosses a definable threshold (usually set in the lower section just after the amplification curve exceeds background fluorescence levels) is called the quantification cycle value (Cq, also referred to as Ct) and is related to the initial copy number of the template DNA target. A lower Cq value indicates a higher initial quantity of target DNA in the sample; with the inclusion of a standard curve, absolute quantification can be achieved. Two chemistries are available to generate fluorescence signals; base intercalating dyes (e.g. SYBR Green I) and hydrolysis probe assays (e.g. TaqMan), with each having its advantages and disadvantages. There are already excellent published reviews of qPCR assay principles (Kralik and Ricchi, 2017) and it is beyond the scope of this chapter to provide a full review of this technique.

Application of qPCR for the detection of pathogens is now a routine diagnostic approach used in clinical and veterinary testing and is commonly used in paratuberculosis diagnostic testing (Alinovi *et al.*, 2009; Park *et al.*, 2014; Plain *et al.*, 2014; Heuvelink *et al.*, 2017; Arsenault *et al.*, 2019). A broad range of assays have been developed for varying sample types (see Section 19.5) with many commercial kits now also available (see Commercial Diagnostic Kits).

One of the main differences between the different qPCR assays is the gene target and hence the primers used in the assay. Multiplex qPCR assays amplify multiple gene target-specific sequences through the inclusion and optimization of multiple primer pairs in the same reaction mix. There are a number of multiplex assays that have been developed for the detection of *MAP* sequences, often combined with an internal amplification control (Irenge *et al.*, 2009; Schönenbrücher *et al.*, 2008; Sevilla *et al.*, 2014). These offer the potential for increased specificity by targeting more than one gene from *MAP*.

19.2.3 Droplet digital PCR

While real-time PCR has improved timeliness and sensitivity of molecular diagnostic approaches to detect *MAP*, the increased sensitivity has come with its own problems. In closing the gap on detection of diseases with very low infectious doses or where subclinical forms of the disease exist, such as paucibacillary forms of mycobacterial infections, eliminating sample inhibition and ensuring sufficient target DNA makes it into the PCR reaction has become the limiting factor. Droplet Digital PCR (ddPCR) is becoming the new 'go to' technology to reduce and/or overcome PCR inhibition and respond to the most critical need, overcoming the normal distribution of target DNA more commonly associated with advanced disease, compared with a Poisson distribution in cases where very few numbers of organisms exit. Essentially, ddPCR distributes the sample into a large number of replicates following the Poisson distribution in a cost-effective manner, such that the likelihood of identifying low target copy numbers is markedly improved (Majumdar *et al.*, 2017). This process simultaneously reduces the effect of inhibitory substances.

Review papers are now appearing in the scientific literature discussing the merits of ddPCR for the detection of infectious diseases (Kuypers and Jerome, 2017; Li *et al.*, 2018) in comparison with qPCR; the advantages and disadvantages are typically described as shown in Table 19.1.

The major limitation to the introduction of ddPCR at present is the number of instrument models currently available and the cost of these instruments. This most likely explains the limited use of ddPCR in assays to diagnose paratuberculosis; however, Ricchi *et al.* (2017) have successfully demonstrated the use of ddPCR for the detection of *MAP* DNA. To date, ddPCR has most frequently being reported in the mycobacterial field for the diagnosis of tuberculosis (Pholwat *et al.*, 2013; Devonshire *et al.*, 2016; Ushio *et al.*, 2016; Yang *et al.*, 2017; Song *et al.*, 2018; Yamamoto *et al.*, 2018a; Nyaruaba *et al.*, 2019) but more recently this has extended to leprosy. Cheng *et al.* (2019) reported superior detection of paucibacillary cases of leprosy in skin biopsies using ddPCR over qPCR with sensitivities of 79.5% compared with 36.4%, respectively.

The approval by the US Food and Drug Administration of the first ddPCR-based test in 2019, monitoring the treatment of chronic myeloid in leukaemia patients (https://www.fda.gov/medical-devices/vitro-diagnostics/nucleic-acid-based-tests), suggests we can expect to see ongoing acceptance and expansion of this technology for other applications already utilizing qPCR, including those for pathogens of animals and humans. While ddPCR has yet to impact significantly in terms of new diagnostic tests, interest from researchers is sufficient to prompt a modified version of the Minimum Information for Publication of Quantitative Real-Time PCR Experiments (MIQE) guidelines specifically for ddPCR (Huggett *et al.*, 2013). As with all new technologies, the

Table 19.1. Advantages and disadvantages of Droplet Digital PCR (ddPCR) vs quantitative PCR (qPCR) (adapted from Kuypers and Jerome, 2017).

Advantage	Disadvantage
Absolute quantification, no standard curve	Limited reaction mixture volume
Improved interlaboratory commutability	Smaller dynamic range
Less affected by sample inhibitors	Molecular dropout
Less affected by poor amplification efficiency	Less accurate quantification of larger amplicons
More precise	Lower throughput
Better detection of low-copy-number variants	Limited multiplexing exacerbated if assay requires internal control
	More expensive instrumentation and reagents
	Higher risk for contamination
	More complex to perform

availability of more ddPCR platforms will increase and the cost will come down; as this happens, the use of ddPCR in the diagnosis of paratuberculosis will undoubtedly escalate.

19.2.4 Viable *MAP* detection

A growing concern with molecular diagnosis of bacterial pathogens is the question, 'Are they alive?'. PCR-based diagnostics that produce late or high cycling threshold values are often considered to be associated with low numbers of bacteria, more commonly characteristic of subclinical or paucibacillary forms of disease. However, with diseases caused by mycobacteria, this is in itself to be expected. However, if the bacteria are not alive, how can they cause disease? Traditional culture has been undertaken to confirm suspect or index cases, but this is expensive and protracted. A number of alternative approaches have been developed to address this (Emerson *et al.*, 2017). Most recently, phage-based assays have been integrated into research studies to detect viable *MAP* (Botsaris *et al.*, 2016). However, such assays introduce their own logistical issues for the laboratory, especially in a diagnostic context, partly due to the risk of the bacteriophage contaminating routine mycobacterial cultures.

Another alternative is intercalating dye-based PCR assays, where DNA from dead cells is rendered ineffective for amplification. Two dyes are commonly used: propidium monoazide (PMA) and ethidium monoazide (EMA) (Emerson *et al.*, 2017). In both cases, compromised (dead) cells allow the dye to enter the cell and intercalate into the DNA, thus blocking subsequent amplification, whereas intact cells do not and therefore the DNA is able to be amplified in PCR. As a result, positive PCR results are considered to be only from viable bacteria. PMA-based assays have been described for paratuberculosis, primarily as a research tool for *MAP* detection in various sample types (Pribylova *et al.*, 2012; Dalton and Hill, 2013; Ricchi *et al.*, 2014; Kralik *et al.*, 2010, 2014a, 2018), but have yet to be adopted more broadly. Unfortunately, these assays contribute complexity to an already complex situation. Removing non-viable bacteria from a diagnostic PCR assay, where low target numbers are anticipated, will no doubt limit detection. This might explain the paucity of these types of assays, but this issue may be overcome with the advent of ddPCR, as previously discussed. This appears to be consistent with the use of PMA-based PCR for the detection of *M. tuberculosis* (Mbelele *et al.*, 2018).

19.3 Choice of Target Genes for PCR

The gene target in any pathogen PCR detection assay needs to be specific for that pathogen. The discovery of IS*900*, a multicopy insertion sequence (17 copies/*MAP* genome), has provided a sensitive and specific target for molecular

Table 19.2. Genes from *Mycobacterium avium* subsp. *paratuberculosis* (*MAP*) used as target genes for PCR-based diagnostic tests for paratuberculosis, with examples of published assays for each gene (adapted from Bölske and Herthnek, 2012).

Gene	Copy no.[a]	Selected references regarding validation and use in PCR assays
IS*900*	17	Kawaji *et al.* (2007); Donaghy *et al.* (2011); Schwalm *et al.* (2018); Arsenault *et al.* (2019)
ISMap02	6	Stabel and Bannantine (2005); Irenge *et al.* (2009)
ISMav2	3	Strommenger *et al.* (2001); Sting *et al.* (2014)
f57	1	Ricchi *et al.* (2014); Schönenbrücher *et al.* (2008)
hspX	1	Ellingson *et al.* (1998); Alinovi *et al.* (2009)
MAP2765c	1	Imirzalioglu *et al.* (2011)
MAP0865	1	Imirzalioglu *et al.* (2011)

[a]Number of copies in the *MAP* genome.

detection of *MAP* DNA (Green *et al.*, 1989; Collins *et al.*, 1993) and continues to be used in many PCR assays. The advantage of using a multicopy gene is it provides higher analytical sensitivity of the PCR assay as compared with targeting a single-copy gene. However, a number of sequences with similarity to the IS*900* sequence have been identified, following non-specific amplification using primers thought to be specific for the *MAP* IS*900*; Strain 2333 in Sweden was the most similar, having 94% sequence homology to *MAP* IS*900* (Cousins *et al.*, 1999; Englund *et al.*, 2002; Taddei *et al.*, 2008). These bacterial species are not commonly identified in the field and are usually phenotypically distinct in terms of growth in culture (e.g. faster growth on media without mycobactin), however it is advisable that the analytical specificity of IS*900* PCR assays be confirmed (Kawaji *et al.*, 2007).

Various other *MAP*-specific genes have been identified and used as targets in PCR-based diagnostic assays (Table 19.2); these may allow for increased specificity of the PCR assays, however, there are fewer copy numbers compared with IS*900* within the *MAP* genome (Bannantine *et al.*, 2002).

19.4 DNA Extraction Procedures

Detection of *MAP* can be described as a 'triple threat' the combination of a complex, chronic disease, an extremely resilient organism and sample matrices that are some of the most difficult to work with in a molecular biology setting. Mycobacteria have a complex, robust cell wall structure that limits cell permeability, confers resistance to many antibiotics and may also relate to strain virulence (Alonso-Hearn *et al.*, 2017; Gago *et al.*, 2018). This characteristic also makes mycobacteria resistant to lysis during standard DNA extraction protocols, with specialized protocols required.

An overarching principle is that the DNA extraction step is critically important and can significantly impact the detection of *MAP* in a PCR diagnostic test (Kawaji *et al.*, 2007; Schwalm *et al.*, 2018). The DNA extraction method used must be optimized for the sample type. Various methodologies have been applied, from spin column extraction methods, magnetic bead technologies, to 'old school' organic solvent phase extraction methodologies, immunomagnetic separation (IMS) (Grant *et al.*, 2002; Khare *et al.*, 2004), peptide-mediated magnetic separation (Stratmann *et al.*, 2002) and hybridization capture techniques (Marsh *et al.*, 2000). Although these show some promise in their ability to approach the sensitivity required to align with *MAP* culture, they have largely fallen out of favour due to the risk of sample cross-contamination and the need for methods that are less time-consuming, more conducive to a diagnostic setting and cost efficient. The various extraction methods will be considered below in relation to specific sample types.

A factor that is independent of the sample type, however, is the need to achieve sufficient lysis of the *MAP* bacterium. While DNA can be extracted from many bacterial species using chemical lysis alone, *MAP* can require the inclusion of both a mechanical and a chemical lysis step to obtain complete lysis (Odumeru *et al.*, 2001; Fang *et al.*, 2002; Herthnek and Bölske, 2006; Timms *et al.*, 2015). Mechanical lysis by a bead-beating process, with silica/zirconia beads and a homogenization instrument, is able to lyse *MAP* and allow release of the DNA. The addition of this single step to extraction protocols has had a major impact on the diagnostic sensitivity of PCR-based mycobacterial detection methods (Amaro *et al.*, 2008).

19.5 Sample-Specific Techniques and Challenges

19.5.1 Faeces

Faeces, as the primary test sample used for antemortem PCR-based diagnosis of paratuberculosis, is a non-invasive and easy sample to collect but is a difficult sample type to work with. In addition, shedding can be intermittent and the levels of *MAP* present can often be low, depending on the stage of the disease. It is important in terms of disease diagnosis at the individual or herd level to be able to detect and monitor these low levels of *MAP* in the faeces. Early identification of individuals that are shedding the *MAP* bacterium prior to clinical disease onset is important for control, particularly as the majority of infected animals in any herd or flock are likely to be subclinical cases.

Ruminant faeces is an extremely challenging sample due to the presence of the faecal microbiota (O' Donnell *et al.*, 2017), as well as undigested particulate and plant matter, humic acid, bile and proteins, among other components. For ruminants, undigested residues of herbivorous matter such as cellulose, hemicelluloses and lignin constitute a major part of the faecal matrix. Many of these substances can be inhibitory in qPCR reactions, due to interactions with the components involved in amplification of the DNA, leading to potential false-negative results (Schrader *et al.*, 2012).

It may not be achievable to predict and address the complex and variable nature of ruminant faecal samples given differences in breeds, production systems, diets, gut microbiome, health status, seasonal and regional variations. Thus, scientists in this area need to apply approaches to cope with this complexity in the local setting and validate these approaches. One quote that speaks to the trade-off between DNA purity and test sensitivity is 'effective removal of faecal components (from DNA extracts) has to reach a balance between obtaining a concentration of inhibitors at a level that does not affect PCR amplification, and a concentration of target DNA sufficient for PCR amplification' (Rapp, 2010, p. 1489).

Approaches to develop molecular detection assays for faeces that have high sensitivity and minimize the impact of inhibitors have been numerous. Debris present in the faecal sample is often allowed to settle out by gravity or gentle centrifugation and only the upper aqueous supernatant of the faecal suspension is included in the test (Sting *et al.*, 2014; Mita *et al.*, 2016). Immunocapture techniques have been applied as a means of reducing the carry-over of PCR inhibitors (Chui *et al.*, 2010). Extraction kits that are designed to deal with complex samples, such as soil and faeces, have been utilized in the DNA extraction step for detection of *MAP* in faeces (Imirzalioglu *et al.*, 2011; Arsenault *et al.*, 2019).

A magnetic bead-based DNA purification method has been used during the extraction step by a number of researchers in order to achieve optimal PCR results and limit the negative effects of faecal components (Petrich *et al.*, 2006; Plain *et al.*, 2014; Sevilla *et al.*, 2014; Schwalm *et al.*, 2018). Rather than column-based methods, which require unbound sample lysate to be washed through the silica membrane, the advantage of this method is that the nucleic acids bind to the magnetic beads and are physically removed from the remaining sample lysate through successive wash steps in order to remove any contaminants. This type of extraction method has been applied in forensic testing of challenging environmental samples to improve the ability to obtain a genetic profile (Bogas *et al.*, 2011).

Table 19.3 summarizes qPCR assays to detect *MAP* in faeces, including a broad range of DNA extraction approaches, that have been published in the past 10 years. The diagnostic sensitivity of these, compared with faecal

Table 19.3. Published studies on the development of direct faecal quantitative PCR (qPCR) assays to detect *Mycobacterium avium* subsp. *paratuberculosis* (*MAP*) in faecal samples, with information regarding analytical and diagnostic performance (Jan 2008 – July 2019).

				qPCR method								
Author	Year	Species	No. samples (total)	DNA extraction method	Bead beating (Y/N)	Target gene(s)	Assay chemistry	Reference test(s)	ASe[a]	ASp	DSe (No. samples)	DSp (No. samples)
Arsenault *et al.* (2019)	2019	Ovine	77	Spin column[b]	Y	*IS900*[c]	Probe, two-step	TC, LJ FC[d]	Previous[c]	Previous[c]	0.84 (*n* = 44)	0.93 (*n* = 30)[e]
				MAP DNA test kit (Tetracore)	Y	hspX	Probe, two-step		Com	Com	0.52 (*n* = 44)	0.93 (*n* = 30)[e]
Schwalm *et al.* (2018)	2018	Bovine	157	Magnetic bead isolation[f]	Y	*IS900*[g]	Com	M7H9C liquid FC[j]	0.2 GE	Com	0.86 (*n* = 107)	1.0 (*n* = 50)
				IDvet[h]	Com	*IS900*[i]	Probe, two-step		0.2 GE	Com	0.89 (*n* = 107)	1.0 (*n* = 50)
Mita *et al.* (2016)	2016	Bovine[k]	24	Five methods assessed[l]	Y	*IS900*[i]	SYBR Green, two-step	HEYM FC	Previous[m]	Previous[m]	0.71–1.0 (*n* = 24)	Previous[m]
Sevilla *et al.* (2014)	2014	Bovine	615	Magnetic bead isolation[n]	Y	*IS900*[o] ISMAP02	Triplex (IAC), probe, two-step	HEYM and LJ FC	10–100 *MAP*/g	1.0	0.59 (*n* = 88), 0.70 (*n* = 74)[p]	0.85–0.94 (*n* = 453)[e]
Plain *et al.* (2014)	2014	Bovine	1330	Magnetic bead isolation[q]	Y	*IS900* [i]	SYBR Green, two-step	BACTEC™ RM liquid FC	10 *MAP*/g	1.0	0.61 (*n* = 870)[r]	0.996 (*n* = 460)
		Ovine	596								0.84 (*n* = 507)[r]	0.99 (*n* = 89)
Sting *et al.* (2014)	2014	Bovine	116	Eight methods assessed[s]	Y	*IS900*[t]	Probe, two-step	HEYM FC	6 GE	nd	0.59–0.85 (*n* = 116)	nd
						ISMav2[u]	Probe, two-step		28 GE	nd	0.46–0.65 (*n* = 116)	nd
Salgado *et al.* (2013)	2013	Bovine	351	PMS/ethanol precipitation	Y	*IS900*	Probe, three-step	MGIT™ FC	<50 *MAP*/g	nd	1.0 (*n* = 43)[r]	nd
Logar *et al.* (2012)	2012	Bovine	141	Spin column[w]	Y	*IS900*	Probe, three-step	HEYM FC	90 *MAP*/g	1.0	1.0 (*n* = 27)	nd
Imirzalioglu *et al.* (2011)	2011	Bovine	12[v]	Spin column[x]	N	MAP2765c, MAP0865	Probe	HEYM/acid-fast staining	70 *MAP*/ml	1.0	1.0 (*n* = 9)	1.0 (*n* = 3)
Zhang and Zhang (2011)	2011	Bovine	179[y]	Phase separation/spin column[z]	Y	*IS900*	Probe	HEYM FC	3.34 CFU/g	nd	0.98 (*n* = 130)	1.0 (*n* = 49)
Douarre *et al.* (2010)	2010	Bovine	290[y]	Phase separation	Y	*IS900*	SYBR Green, three-step	HEYM FC	nd	nd	1.0 (*n* = 23)[r]	1.0 (*n* = 123)

Continued

Table 19.3. Continued

Author	Year	Species	No. samples (total)	qPCR method: DNA extraction method	Bead beating (Y/N)	Target gene(s)	Assay chemistry	Reference test(s)	ASe[a]	ASp	DSe (No. samples)	DSp (No. samples)
Kawaji *et al.* (2011)	2011	Ovine	231	Spin column[α]	Y	IS*900*	SYBR Green, two-step	BACTEC™ RM liquid FC	Previous[m]	Previous[m]	0.99 (n = 69)	Previous[m]
	2009	Bovine	55 adults, 197 calves	Phase separation	N	IS*900*, ISMAP02, f57	Triplex, probe, two-step	HEYM FC	250 *MAP*/g	1.0	Cohen's kappa=0.63	1.0 (n = 35)
Schönenbrücher *et al.* (2008)	2008	Bovine faeces	108	Spin column[β]	Y	f57, ISMav2	Triplex (IAC), probe	HEYM FC	100 *MAP*/g	1.0	0.91 (n = 53)	nd

[a]ASe units provided in *MAP*/g faeces (if available), *MAP*/ml (pure culture) or genome equivalents (GE); [b]ZR faecal DNA extraction kit (Zymo Research); [c]IS*900* TaqMan assay from Irenge *et al.* (2009); [d] infection status of culled sheep determined by TC, FC and histopathology; [e] population included animals from *MAP*-exposed herds/flocks, [f] NukEx Mag Extreme extraction kit (Gerbion), [g] diarellaMAP vet PCR (Gerbion), [h] IDvet Preparation kit/IDvet Extraction kit (IDvet Genetics); [i] IDGene™ Paratuberculosis duplex PCR (IDvet Genetics), [j] M7H9C media from Whittington *et al.* (2013); [k] pooled faeces not individual faecal samples; [l] methods: Johne-Spin kit (Fasmac, Japan), Tetracore MAP Extraction System (Tetracore Inc.), MagMAX™ Total Nucleic Acid Isolation kit (Applied Biosystems), QIAamp stool DNA Mini kit (Qiagen), ZR faecal DNA MiniPrep (Zymo Research); [m] previously published (Kawaji *et al.*, 2007); [n] MagMAX™ total nucleic acid isolation kit (Ambion, Applied Biosystems); [o] IS*900* assay from Herthnek and Bölske (2006); [p] based on follow-up testing of animals from two separate herds; [q] modified method incorporating BioSprint 96 one-for-all vet kit (Qiagen); [r] more positives detected by qPCR than FC; [s] methods: protocol from Selim and Gaede (2012) or MagMAX™ Total Nucleic Isolation Kit (Life Technologies) ± four faecal suspension method variants; [t] IS*900* assay based on Kim *et al.* (2002); [u] ISMav2 assay based on Wells *et al.* (2006); [v] study also includes 1293 human patient samples; [w] SmartHelix™ First DNAid kit (Institute of Physical Biology,Slovenia); [x] PSP Spin Stool DNA kit (Invitek); [y] proficiency kit samples; [z] Mississippi Veterinary Research and Diagnostic Laboratory (MVRDL) method, including chemical pre-treatment, cell lysis and extraction, chelex matrix absorption and spin column; [α] Johne-Spin kit (Fasmac, Japan); [β] QIAmp Stool Mini kit (Qiagen).
MAP, *Mycobacterium avium* subsp. paratuberculosis; *GE,* genome equivalents; ASe, analytical sensitivity; ASp, analytical specificity; DSe, diagnostic sensitivity; DSP, diagnostic specificity; Com, commercial kit; FC, faecal culture; TC, tissue culture; LJ, Lowenstein–Jensen; HEYM, Herrold's Egg Yolk Medium; RM, radiometric; IAC, Internal amplification control; PMS, peptide-mediated magnetic separation; ND, not determined.

culture as the reference method, ranged from 0.46–1.0; this varied depending on the DNA extraction method, sample selection approach, target gene(s) and culture method used for comparison. Diagnostic specificity ranged from 0.85–1.0. However, true diagnostic specificity can be difficult to assess when the population studied includes *MAP*-exposed animals, as qPCR can be more sensitive than other diagnostic testing methods, often detecting additional positive samples compared with faecal culture. Further, PCR methods detect non-viable bacteria and some low positives may be due to passive shedding (Kralik *et al.*, 2014b). Hence, the analytical specificity of the qPCR assay may provide a more accurate assessment of the likelihood of obtaining false-positive results.

19.5.2 Milk and dairy products

Milk has been used as a test sample for both individual and herd-level diagnosis of paratuberculosis. A range of technologies can be applied: culture of milk or milk fractions, molecular methods (PCR, genotyping) and a rapid bacteriophage-based assay that specifically detects viable *MAP*, not just the presence of *MAP* DNA (Okura *et al.*, 2012; Botsaris *et al.*, 2016; Grant *et al.*, 2017). The majority of recent publications of PCR to detect *MAP* in dairy products have been related to the food safety aspects for human consumption, as a potential zoonotic infectious transmission route, or calf milk replacers as a source of disease spread on-farm (Fig. 19.1, Table 19.4).

The occurrence of *MAP* in milk on-farm has been previously reviewed (Okura *et al.*, 2012). *MAP* is generally present in low concentrations in milk samples thus larger volumes are tested, with most methods incorporating a centrifugation step. Immunomagnetic or peptide-mediated separation prior to DNA extraction has also been used in order to 'pull out' the *MAP* organisms from the sample prior to the test being conducted (Grant *et al.*, 2002; Metzger-Boddien *et al.*, 2006; Ricchi *et al.*, 2014; Gilardoni *et al.*, 2016). The analytical limit of detection of *MAP* in milk by PCR has been reported to range between 1 and 100 *MAP*/ml of milk, using qPCR methods with gene targets including IS*900*, f57 ISMAP02 and ISMAV2 (Gao *et al.*, 2007; Slana *et al.*, 2009; Alajmi *et al.*, 2016; Acharya *et al.*, 2017b). The fraction of milk tested is significant; the majority of *MAP* is reported to be concentrated in the casein pellet rather than the whey (Acharya *et al.*, 2017b). However, the cream layer can contain significant numbers of *MAP* and may also be important to consider in testing regimes (Gao *et al.*, 2007; Herthnek *et al.*, 2008). Molecular techniques applied to milk samples in the diagnosis of bovine tuberculosis may also be of relevance to the diagnosis of paratuberculosis (Lermo *et al.*, 2010).

PCR-based assays using milk as a sample may be best suited to the identification of highly

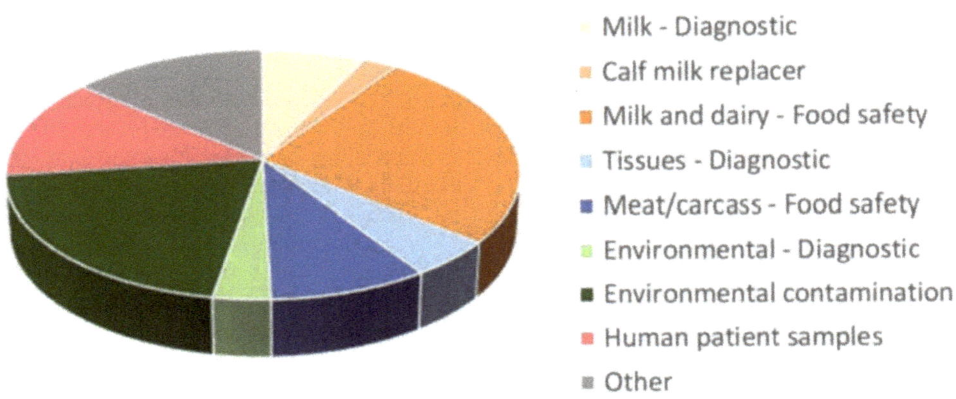

Fig. 19.1. Number of published studies by category using PCR methods to detect *Mycobacterium avium* subsp. *paratuberculosis* (*MAP*) in samples other than faeces (January 2008 – July 2019).

Table 19.4. Examples of recent published studies on PCR assays to detect *Mycobacterium avium* subsp. *paratuberculosis* (*MAP*) by sample type, for samples other than faeces.

Category	Recent published studies
Milk – diagnostic	Logar *et al.* (2012); Slana *et al.* (2012); Stabel *et al.* (2014); Laurin *et al.* (2015); Bauman *et al.* (2019)
Calf milk replacer	Grant *et al.* (2017); Khol *et al.* (2017)
Milk and dairy – food safety	
Tissues – diagnostic	Acharya *et al.* (2017a); Park *et al.* (2014)
Meat/carcass – food safety	Lorencova *et al.* (2019); Savi *et al.* (2015)
Environmental – diagnostic	Hahn *et al.* (2017); Donat *et al.* (2016)
Environmental contamination	Donat *et al.* (2019); Richardson *et al.* (2019)
Patient samples (irritable bowel disease/ Crohn's)	Sharp *et al.* (2018); Khan *et al.* (2016)
Other	Khol *et al.* (2014); Blaiotta *et al.* (2016)

infectious animals. Stabel *et al.* (2014) reported the detection of *MAP* by PCR in 13% of milk samples from subclinical cows compared with 49% of samples from clinically infected cows, with viable *MAP* mainly found in milk/colostrum collected during early lactation (≤60 days in milk). If used for herd-level surveillance, individual animal milk testing would require large numbers of animals to be tested to generate confidence in detection of low prevalence farms and is unlikely to be cost-effective. However, if the information obtained from the extracted DNA is utilized in another way, for example for epidemiological investigations (Kaur *et al.*, 2010) or multiplex testing for a range of pathogens (Parker *et al.*, 2017), then PCR-based methods may be suitable, otherwise they may be a more expensive test approach than a milk ELISA with similar diagnostic sensitivity.

19.5.3 Animal tissues and meat products

Confirmation of infection with *MAP* at the individual animal level is commonly performed by culture of the *MAP* bacterium from the tissues of the intestine and mesenteric lymph nodes, which is considered the gold standard post-mortem diagnostic test for paratuberculosis. Molecular approaches can also be applied for detection of *MAP* in these same tissues, for rapid determination of the presence of *MAP* (Table 19.4).

Challenges with detection of a pathogen in host tissues include the extraction method used and the high quantity of host DNA (Park *et al.*, 2014). This is a common issue for molecular approaches attempting to identify a proportionally minute component of sequences specific for the pathogen in a high background of host DNA (Lee *et al.*, 2017; Gu *et al.*, 2019). High input quantities of DNA can cause inhibition of qPCR reactions and also interfere with the detection of the amplified product (Restrepo *et al.*, 2006; Radomski *et al.*, 2013). There are a number of ways to deal with this, such as target enrichment, simple dilution of the DNA extract and/or optimizing the assay for input DNA quantity. Conventional PCR can provide an advantage over qPCR when applied to animal tissues as this method is less sensitive to the inhibition due to high levels of host DNA, thus researchers may choose conventional over qPCR for these applications (Zarei-Kordshouli *et al.*, 2019). Validation is key to determine the limit of detection of the assay used.

Tissue qPCR approaches may be of value in disease surveillance at slaughter or determination of farm/regional prevalence levels, as many current abattoir surveillance programmes involve examination of gross lesions and confirmation by histopathological and/or tissue culture, which is time-consuming and has relatively low sensitivity to detect infection (Münster *et al.*, 2011; Acharya *et al.*, 2017a). The majority of published studies on PCR-based detection of *MAP* in tissues relate to food safety or the detection of *MAP* DNA in human patient samples (Fig. 19.1). Methods applied in human research on Crohn's disease may utilize PCR due to anticipated low quantities of *MAP* in the intestinal tissue sample (Banche *et al.*, 2015) (see also Chapter 3, this volume). Such PCR assays face the same issues as detection of *MAP* in animal tissues; greater cross-validation of PCR used in human and veterinary research would be of value to determine the analytical performance of these assays.

19.5.4 Other sample types

Methods for detecting *MAP* by PCR have been published for samples as diverse as semen, blood, lymphatic fluids, environmental samples and culture medium (Juste *et al.*, 2008; Münster *et al.*, 2013; Khol *et al.*, 2014; Plain *et al.*, 2015; Hahn *et al.*, 2017). *MAP* has been detected in the blood of animals with late-stage paratuberculosis by PCR and bacteriophage assays (Münster *et al.*, 2013; Swift *et al.*, 2013), although *MAP* detection in the blood is not a common finding at a herd or flock level (Bower *et al.*, 2011; Bower, 2011; Swift *et al.*, 2016). Detection of *MAP* in culture medium in order to confirm specific growth once posed a significant challenge, particularly for media containing egg (Whittington *et al.*, 1998), though now commonly involves high-throughput approaches and/or commercial kits (Plain *et al.*, 2015; Bauman *et al.*, 2016; Rangel *et al.*, 2017).

Methods to detect *MAP* by PCR in slurry/liquid manure, boot swabs, dust and other types of environmental samples have been published (Aly *et al.*, 2009; Eisenberg *et al.*, 2012; Hahn *et al.*, 2017; Donat *et al.*, 2019). These are informative to the mechanisms of disease spread on-farm as well as potential avenues for herd-level diagnosis. The use of samples collected from the farm environment to identify infected herds is increasingly being investigated as an economical surveillance approach (Aly *et al.*, 2009; Donat *et al.*, 2016). There is also a substantial component of the literature regarding PCR to detect *MAP* in environmental samples that pertains to the risks and prevalence of *MAP* contamination of the environment (Fig. 19.1). A number of methods have been developed for water samples, for example to examine the effect of farm runoff on the environmental load in local waterways (Richardson *et al.*, 2019). In this study, 1 l of water was filtered (0.2 μM) and the pelleted material extracted using a method incorporating bead-beating and a magnetic bead extraction kit, followed by qPCR detection of both the IS*900* and F57 genetic elements. Water from rivers and tributaries in catchment areas within regions used for grazing by livestock (sheep and cattle) for both dairy farming and meat production exhibited up to 10^8 cell equivalents/l of *MAP*.

19.6 Quality Control of PCR-Based Diagnostic Assays

19.6.1 IAC, or no IAC, that is the question...

It is clear that, given the challenging sample types that need to be tested, the potential for PCR inhibition is a major consideration in *MAP* PCR-based diagnostic tests to avoid false negatives. This is commonly an issue with complex sample matrices such as faeces and there needs to be a balance between adding sufficient material to achieve the desired diagnostic sensitivity and carrying over too many inhibitors into the DNA extract.

The goal of an internal amplification control (IAC) is to signal a potential false-negative result. Thus, delayed (higher Cq value) or lack of amplification of the IAC is an indicator of failure of the PCR. An internal extraction control is similar to an IAC but is added to the initial step of the extraction protocol and undergoes all processing and amplification steps. However, there is an assumption implicit with such controls, which is that both the PCR amplification of the IAC and the target sequence are equally impacted by any PCR inhibitors present in the DNA extract. This assumption has been challenged with findings of differential impact on the IAC and target-specific amplification (Schrader *et al.*, 2012). This may depend on a number of factors, such as the primers used, the GC content of the target regions or competition between the reactions.

Furthermore, a repeat extraction of a sample with evidence of inhibition does not ensure that the issue will not be repeated, leading to challenges in a diagnostic setting with high sample throughout (Acharya *et al.*, 2017c).

Alternative approaches include the dilution of the DNA extract, which can both identify and rectify the inhibition (Acharya *et al.*, 2017c). This is an effective approach, however consideration must be given to possible loss of detection of low positive samples due to this effect of the dilution. This is dependent on the diagnostic test setting and sensitivity requirements. A third alternative is the detection of an alteration in the amplification efficiency of individual reactions due to the presence of inhibitors, however this is only indicated if there is amplification in the first instance (King *et al.*, 2009). In complex samples such as pooled cattle faeces, PCR inhibition can completely prevent amplification, even in samples with high quantities of specific DNA (Mita *et al.*, 2016; Ly *et al.*, 2019).

Any approach to circumvent potential inhibition and the risk of false negatives in a *MAP* PCR detection assay, be that an IAC, DNA extract dilution and/or amplification curve efficiency assessment, requires validation in the appropriate setting.

19.6.2 Validation of assays

For any molecular diagnostic assay, validation of the analytical and diagnostic performance is critical prior to the application of the test. This is described in detail in the World Organisation for Animal Health (OIE) manual (OIE, 2019). PCR assays to detect *MAP* are generally compared with *MAP* culture as the 'gold standard' test to determine the diagnostic accuracy. Though culture is considered the test for benchmarking, it is not perfect; rather it is the best option available. However, with increased analytical and diagnostic sensitivity, molecular-based tests can perform better than the gold standard, making determination of diagnostic specificity difficult in the absence of an unexposed animal population.

Considerations in relation to diagnostic test validation have been reviewed in detail (Nielsen *et al.*, 2011); briefly these include the purpose and target condition that the diagnostic test aims to identify (e.g. herd- vs individual-level diagnosis, infected vs infectious), validation sample population selection to avoid bias (e.g. including randomly selected animals rather than a targeted group at a particular disease stage) and the classification of disease/infection status in the sample population (case definition).

19.6.3 Confirming PCR identification

OIE now states it 'would be good practice to' include a subsequent sequencing step to confirm PCR positive results, especially when index cases from a previously unaffected location or low Cq positives are detected. This can be done by sequencing the PCR amplicon from a conventional PCR, which can be challenging for samples with low initial template quantities. Sequencing of *IS900* PCR products to confirm *MAP* identity has been performed in a number of studies, to confirm the sequence identity to *MAP* K10 (Lee *et al.*, 2011; Garg *et al.*, 2015).

19.6.4 International benchmarking

Standardization and harmonization of diagnostic testing (Ahmed *et al.*, 2008, 2011) is gaining momentum at many levels, from local to international, for a number of microbial pathogens (Mcgiven *et al.*, 2006; Mengoli *et al.*, 2009; White *et al.*, 2010; Nielsen *et al.*, 2011; Xing *et al.*, 2011; Makela *et al.*, 2012; He *et al.*, 2012; Johnson *et al.*, 2012; McNerney *et al.*, 2014; Henaux *et al.*, 2018). Meaningful comparisons between results, such that inter-laboratory results can be more easily and reliably compared, are becoming increasingly important to support our understanding of the epidemiology of disease, disease outbreaks and to underpin trade agreements between countries. Improved epidemiological knowledge based on accurate prevalence data, supported by consistent, standardized tests, will also benefit policy decisions that support the aims and objectives of control programmes.

An earlier review of the scientific literature on direct faecal PCR for *MAP* in cattle demonstrated very little if any standardization of conventional or qPCR (Marsh *et al.*, 2014). The majority of these assays failed to meet the minimum information for publication of quantitative real-time PCR experiments (MIQE) (Bustin *et al.*, 2009) or standards for reporting of diagnostic accuracy studies (STARD) (Bossuyt *et al.*, 2003; Bossuyt and Reitsma, 2003; Cohen *et al.*, 2016) guidelines, particularly in terms of the number of animals tested and both analytical and diagnostic sensitivity and specificity data. Furthermore, the diagnostic sensitivity values may have been inflated due to comparisons of PCR with culture techniques with differing analytical sensitivities and/or selection bias in the animal cohort tested. To date, an initiative to assist in the development and validation of standardized diagnostic PCR protocols for paratuberculosis is yet to be established internationally, but has been attempted in some countries. An inter-laboratory assessment of qPCR and culture methods was undertaken in the Netherlands to aid diagnostic test selection for the paratuberculosis control programme in that country (Heuvelink *et al.*, 2017). In Australia, standardization was successfully undertaken for the high-throughput Johne's (HT-J) qPCR test (Plain *et al.*, 2014). Following approval of the test, a national technical working group was convened that included representatives from each of the state laboratories, the test developers and representatives from the governing body for veterinary diagnostic tests in Australia, in order to develop a process to standardize the test and establish a national quality assurance programme.

A similar international initiative would directly support the level of precision and accuracy required for control programmes and as new technologies such as whole genome sequencing become a desktop reality in laboratories. This could encompass medical research, to enable cross-validation of diagnostic protocols for *MAP* in Crohn's disease patients and animals with paratuberculosis through sharing of sample panels. It is critical now that new diagnostic assays and tests meet guidelines such as OIE, MIQE and STARD. An initiative such as this has been made easier for the paratuberculosis research community through the efforts of a group of well-established paratuberculosis researchers who have published STARD-based guidelines specifically for this disease (Gardner *et al.*, 2011).

19.7 Commercial Diagnostic Kits

There are many commercial diagnostic qPCR kits for diagnostic testing that are available on the market and in use within control programmes internationally. Information on their performance is required; analytical performance information is available, however limited information is recorded in the published literature regarding diagnostic performance in differing production settings. Alinovi *et al.* (2009) evaluated the performance of the Tetracore VetAlert™ kit, which incorporates a probe-based qPCR detection assay targeting the *MAP hspX* gene, compared with HEYM and Trek ESP culture and serum antibody ELISA. The diagnostic sensitivity was estimated using a Baysian approach to be 0.6. Although the diagnostic specificity was difficult to evaluate, given that the two herds tested in this study were both infected and no unexposed animals were assessed, it was estimated to be 0.97.

A few studies have been published evaluating the performance of commercially available kits for *MAP* extraction and qPCR, determining analytical and diagnostic characteristics, comparing kits with each other, available culture methods and/or new, 'in-house' qPCR assays (Laurin *et al.*, 2015; Alajmi *et al.*, 2016; Mita *et al.*, 2016; Heuvelink *et al.*, 2017; Schwalm *et al.*, 2018; Arsenault *et al.*, 2019). Such studies can be time consuming and costly, however they are important to ultimately achieve the goals of control programmes. It would be useful to have more published data on the use of commercially available qPCR tests in the international setting, including interlaboratory comparisons, to enable informed selection of the appropriate qPCR tool to achieve the desired diagnostic outcomes.

19.8 Application and Interpretation of Findings From PCR-Based Diagnostic Tests

Interpretation of test results should consider the validation data in relation to the PCR method used; that is, the analytical and diagnostic sensitivity and specificity for the method including both the DNA extraction and PCR detection steps. Faecal PCR tests generally report higher herd- and flock-level sensitivity than other antemortem sample types such as blood, milk, semen and lymphatic fluid. A number of reports indicate current faecal PCR diagnostic methods have comparable or higher sensitivity than culture (Plain *et al.*, 2014; Meyer *et al.*, 2019). It is difficult to determine true diagnostic sensitivity and understand the nature of positive results in faecal culture with matched negative faecal PCR results (and vice versa) unless a comprehensive range of diagnostic methods is used to determine true infection status of the animal; such studies can be prohibitively expensive and still may not be conclusive unless animals are followed long term.

Thus, individual diagnostic faecal PCR test results must be considered carefully in terms of how these relate to the disease status of the animal. As qPCR provides an indication of the amount of *MAP* present in the sample, such results can be stratified into ‘heavy’ or ‘high’ shedders, compared to moderate, low or non-shedding individuals. Interpretation of individual animal results needs to consider the pathobiology of the disease and the fact that PCR can detect DNA from non-viable organisms. A negative individual faecal PCR test result, as with individual faecal culture, does not necessarily mean that the animal is not infected as this can be dependent on the stage of disease and hence the shedding level and presence of *MAP* in the faeces. In addition, the age of the animal tested is important: young calves have been demonstrated to shed *MAP* (Wolf *et al.*, 2015), however transient shedding following an exposure event may not necessarily be correlated with infection status (Begg *et al.*, 2018). Further, low DNA quantities may reflect passive shedding due to environmental contamination rather than true infection (Kralik *et al.*, 2014b). For these reasons, follow-up testing may be recommended in relation to an index case for a property or region to confirm infection.

Due to the clinical course of the disease, surveillance is often focused at the herd or flock level. Herd-level testing strategies can involve collection of a mixed ‘slurry’ from the environment, or the creation of pools from faecal samples from individual animals for evaluation by PCR (Aly *et al.*, 2009; Mita *et al.*, 2016). In an evaluation of national surveillance methods for Irish dairy herds, pooled faecal testing (pool size of 5) was one of the methods with highest sensitivity, estimated to be 76% based on previous studies, both for the identification of infected herds and assurance testing for uninfected herds (Sergeant *et al.*, 2019).

As a herd-level diagnostic sample, bulk tank milk (BTM) is cost effective and easy to obtain. PCR-based diagnostic tests have been applied to this sample type, as have *MAP*-specific milk antibody ELISAs, to examine cattle, goat and sheep dairy farms (Slana *et al.*, 2008; Okura *et al.*, 2012; Bauman *et al.*, 2019). The issue with BTM testing using PCR-based diagnostics as a method of surveillance relates to the low overall diagnostic sensitivity and specificity values in bulk milk samples in comparison with other pooled testing approaches. Bauman *et al.* (2019) evaluated a BTM PCR assay as a herd-level diagnostic surveillance test to identify paratuberculosis-affected goat and sheep dairy farms in Canada, for which the disease is a significant concern due to substantial economic and production-limiting impacts. The study compared diagnosis of flocks through testing of BTM using a PCR method, milk ELISA and faecal culture. The BTM PCR method had low diagnostic sensitivity (0–25%) when compared with individual faecal testing of 20 randomly selected >2-year-old animals by culture or PCR, with lower sensitivity than a milk ELISA on BTM (33.3–50%) based on the same comparison, but even these sensitivity estimates may be over-inflated as the true disease status of the farms was not known. One study using a PCR method for BTM on known infected dairy cattle herds and goat flocks reported a diagnostic sensitivity (DSe) of 100% (Favila-Humara *et al.*, 2010). However, BTM samples used in this study were from herds or flocks with seroprevalence of 8–24%, thus detection of *MAP* by qPCR in the BTM samples was correlated with high seroprevalence of disease. A similar correlation of BTM PCR positivity for *MAP* with strong humoral immune responses was found by Beaver *et al.* (2016).

Environmental sampling for herd-level surveillance is an option, but careful consideration needs to be given to how representative the sample type is for the purpose of the test and the production system (intensive vs extensive production, dairy vs beef). Species with pelleted faeces pose their own unique challenges and are more compatible with faecal pooling approaches. The sensitivity of herd environmental testing protocols is reported to be high, with good agreement between culture and qPCR test approaches (Berghaus *et al.*, 2006; Aly *et al.*, 2010; Lavers *et al.*, 2013).

19.9 Future Perspectives

Point-of-care (POC) diagnostic tests are increasingly being integrated into the management of human diseases, but have yet to find a significant role in veterinary diseases (Manessis *et al.*, 2019). An example is the rapid (2-h) diagnostic test for tuberculosis (Xpert MTB/RIF) with reported high sputum sensitivity and specificity compared with culture. This is an automated cartridge-based nucleic acid amplification test that simultaneously detects *M. tuberculosis* DNA through integrated purification of bacilli from sputum samples, isolation and amplification of genomic DNA combined with a test for rifampicin resistance (Dorman *et al.*, 2018).

To date, the focus of POC in veterinary diagnostics has been on lateral flow immunoassays and microfluidic-based POC technologies (Busin *et al.*, 2016). The uptake of veterinary POC applications may be suffering from two issues. First, previously developed POC tests for veterinary pathogens often rely on technologies that are invariably insensitive, and second, the cost–benefit of veterinary POCs, given their poor sensitivity, is often unfavourable and limits interest in commercial partners for their production and distribution.

However, the strengths of laboratory-based qPCR diagnostics (sensitivity, specificity and speed) have for some time now been considered to be integrated into POC, but field-based PCR technology is now a reality. For example, isothermal assays such as recombinase polymerization amplification (RPA) offer a very simple design that is capable of detecting RNA as well and DNA, thus negating the requirement for a separate cDNA production step. This technique requires amplification at a single low temperature, after which the amplified product can simply be detected on a lateral flow device. Recent publications have appeared reporting isothermal-based tests for bacterial pathogens of both terrestrial (*Pasteurella multocida* and *Mycoplasma bovis*) and aquatic animals (*Francisella noatunensis* subsp. *orientali*) (Shahin *et al.*, 2018; Zhao *et al.*, 2018a, 2019). Similarly, isothermal assays have been reported for viral pathogens of veterinary significance, including the foot-and-mouth disease virus (Liu *et al.*, 2018; Wang *et al.*, 2018). A recent publication suggests optimization of the RPA assay in detection of monkeypox virus in humans may enable detection of a target gene in both serum and whole blood samples with a very rapid turnaround (3–10 min), a specificity of 100% and sensitivity of 95% (Davi *et al.*, 2019). Were this technique optimized to the *MAP* this would be a paradigm change for diagnostics of *MAP* in the field.

Various isothermal assays have been reported for the detection of *MAP*, including loop-mediated isothermal amplification (LAMP) (Enosawa *et al.*, 2003; Trangoni *et al.*, 2015, Trangoni *et al.*, 2017; Sange *et al.*, 2019; Yashiki *et al.*, 2019) and RPA assays (Hansen *et al.*, 2016; Zhao *et al.*, 2018b). While LAMP assays offer some advantages, the complexities in the assay design can be prohibitive compared with the simpler RPA assays. Both studies using an RPA assay reported similar sensitivity to qPCR and equivalent specificity (Hansen *et al.*, 2016; Zhao *et al.*, 2018b). At present, the major drawback to the integration of isothermal POC tests is the DNA extraction step. The issues and implications of DNA extraction in conjunction with POC have been discussed (Ali *et al.*, 2017), but once these have been overcome, isothermal assays will find their place in POC diagnostics for paratuberculosis.

Alternative molecular diagnostic methodologies may be the next major development in diagnostic and POC tests for paratuberculosis. Nanoparticles are already frequently used in sample preparation, bioassays and for enrichment, as they can be readily functionalized (Hahn *et al.*, 2017; Gordillo-Marroquín *et al.*, 2018; Markwalter *et al.*, 2019). For detection of nucleic acids from a pathogen, however, signal amplification is often essential; nanoparticle-assisted signal amplification and detection methods are being developed to meet this need for diagnostic testing in tuberculosis (Teengam *et al.*, 2017). CRISPR-Cas-based molecular assays are another potential to replace qPCR in the future (Gootenberg *et al.*, 2018; Batista and

Pacheco, 2018). These involve engineered sensors that target DNA or RNA sequences specific to the pathogen and enable high sensitivity pathogen detection, with minimal processing of the clinical sample.

Detection of multiple RNA or miRNA biomarkers, rather than genetic elements of the organism itself, is an evolving field. Specifically, the identification of infection-associated gene expression biomarkers within cells found in milk or blood may be used to develop a rapid PCR-based diagnostic test capable of differentiating subclinical and clinical paratuberculosis (Purdie *et al.*, 2012, 2019). This could be combined with concurrent detection of the pathogenic organism to provide an overall picture of the biological status of an animal. Evidence has shown that in terms of biomarker-based detection assays, it is necessary to utilize multiple biomarkers to adequately evaluate disease status.

19.10 Conclusions

There is a need to accurately classify herds/flocks in order to provide producers with the required information to justify the inputs of effort and resources into implementing a disease control programme. Faecal PCR diagnostic methods for paratuberculosis are widely applied in control programmes and advances in DNA extraction techniques have meant that these methods have diagnostic sensitivity that approaches or exceeds that of faecal culture.

Evidence in the literature regarding the zoonotic potential of *MAP* has led to application of PCR test methods to investigate and monitor *MAP* levels in dairy and meat products intended for consumers. In addition, detection of *MAP* in tissue samples is a useful ante-mortem diagnostic methodology. However, such methods require validation against other diagnostic tests to determine the sensitivity, preferably tissue culture as the gold standard post-mortem diagnostic test and/or histopathology with ZN staining. This is due to challenges associated with the high levels of host DNA present in extracts from tissues, which can have a detrimental impact on the detection of *MAP* DNA.

The results of PCR diagnostics for *MAP* in milk show that this is correlated with humoral immunity, which is consistent with direct excretion of *MAP* in the milk due to bacteraemia (dissemination to other sites in the body) or infection of supramammary lymph nodes, both of which are known to occur more frequently in the later stages of the disease at which time animals are more likely to be seropositive (Sweeney *et al.*, 1992; Bower *et al.*, 2011). PCR methodologies applied to BTM do not have high diagnostic sensitivity and hence are not well suited as herd-level surveillance tools, but are of use in risk assessment regarding potential infection routes and food safety in relation to milk.

The application of PCR-based diagnostics for paratuberculosis still must contend with challenging issues such as PCR inhibition. It may be that the next advances in molecular diagnostics for *MAP* and other pathogens will come from alternative technologies other than PCR.

References

Acharya, K.R., Dhand, N.K., Whittington, R.J. and Plain, K.M. (2017a) Culture-independent identification of *Mycobacterium avium* subspecies *paratuberculosis* in ovine tissues: comparison with bacterial culture and histopathological lesions. *Frontiers in Veterinary Science* 4, 232. DOI: 10.3389/fvets.2017.00232.

Acharya, K.R., Dhand, N.K., Whittington, R.J. and Plain, K.M. (2017b) Detection of *Mycobacterium avium* subspecies *paratuberculosis* in powdered infant formula using IS*900* quantitative PCR and liquid culture media. *International Journal of Food Microbiology* 257, 1–9. DOI: 10.1016/j.ijfoodmicro.2017.06.005.

Acharya, K.R., Dhand, N.K., Whittington, R.J. and Plain, K.M. (2017c) PCR Inhibition of a quantitative PCR for detection of *Mycobacterium avium* subspecies *paratuberculosis* DNA in feces: diagnostic implications and potential solutions. *Frontiers in Microbiology* 08(381), 115. DOI: 10.3389/fmicb.2017.00115.

Ahmed, J.S., Oliva, A.G. and Seitzer, U. (2008) Animal health: harmonisation and distribution of pathogen detection and differentiation tools. *Transboundary and Emerging Diseases* 55(5–6), 187–189. DOI: 10.1111/j.1865-1682.2008.01038.x.

Ahmed, J.S., Carrondo, M.J., Gunseli Alp, H., Jongejan, F., Yin, H. *et al*. (2011) INCOME – Evaluation and improvement of integrated livestock disease control measures through distribution of molecular diagnostic tools, evaluation of disease situation, training and capacity building in Asia. *Funded under 6th FWP (Sixth Framework Programme) Research area: INCO-A.3 Food security*.

Alajmi, A., Klein, G., Grabowski, N.T., Fohler, S., Akineden, O. *et al*. (2016) Evaluation of a commercial real-time PCR kit for the detection of *Mycobacterium avium* subspecies *paratuberculosis* in milk. *Current Microbiology* 73(5), 668–675. DOI: 10.1007/s00284-016-1109-6.

Ali, N., Rampazzo, R.C.P., Costa, A.D.T. and Krieger, M.A. (2017) Current nucleic acid extraction methods and their implications to point-of-care diagnostics. *BioMed Research International* 2017(12), 9306564–13. DOI: 10.1155/2017/9306564.

Alinovi, C.A., Ward, M.P., Lin, T.L., Moore, G.E. and Wu, C.C. (2009) Real-time PCR, compared to liquid and solid culture media and ELISA, for the detection of *Mycobacterium avium* ssp. *paratuberculosis*. *Veterinary Microbiology* 136(1–2), 177–179. DOI: 10.1016/j.vetmic.2008.10.012.

Alonso-Hearn, M., Molina, E., Geijo, M., Vazquez, P., Sevilla, I. *et al*. (2009) Isolation of *Mycobacterium avium* subsp. *paratuberculosis* from muscle tissue of naturally infected cattle. *Foodborne Pathogens and Disease* 6(4), 513–518. DOI: 10.1089/fpd.2008.0226.

Alonso-Hearn, M., Abendaño, N., Ruvira, M.A., Aznar, R., Landin, M. *et al*. (2017) *Mycobacterium avium* subsp. *paratuberculosis* (MAP) fatty acids profile is strain-dependent and changes upon host macrophages infection. *Frontiers in Cellular and Infection Microbiology* 7, 89. DOI: 10.3389/fcimb.2017.00089.

Aly, S.S., Anderson, R.J., Whitlock, R.H., Fyock, T.L., McAdams, S. *et al*. (2009) Reliability of environmental sampling to quantify *Mycobacterium avium* subspecies *paratuberculosis* on California free-stall dairies. *Journal of Dairy Science* 92(8), 3634–3642. DOI: 10.3168/jds.2008-1680.

Aly, S.S., Mangold, B.L., Whitlock, R.H., Sweeney, R.W., Anderson, R.J. *et al*. (2010) Correlation between Herrold egg yolk medium culture and real-time quantitative polymerase chain reaction results for *Mycobacterium avium* subspecies *paratuberculosis* in pooled fecal and environmental samples. *Journal of Veterinary Diagnostic Investigation* 22(5), 677–683. DOI: 10.1177/104063871002200501.

Amaro, A., Duarte, E., Amado, A., Ferronha, H. and Botelho, A. (2008) Comparison of three DNA extraction methods for *Mycobacterium bovis*, *Mycobacterium tuberculosis* and *Mycobacterium avium* subsp. *avium*. *Letters in Applied Microbiology* 47(1), 8–11. DOI: 10.1111/j.1472-765X.2008.02372.x.

Arsenault, J., Singh Sohal, J., Leboeuf, A., Hélie, P., Fecteau, G. *et al*. (2019) Validation of an in-house real-time PCR fecal assay and comparison with two commercial assays for the antemortem detection of *Mycobacterium avium* subsp. *paratuberculosis* infection in culled sheep. *Journal of Veterinary Diagnostic Investigation* 31(1), 58–68. DOI: 10.1177/1040638718810744.

Babafemi, E.O., Cherian, B.P., Banting, L., Mills, G.A. and Ngianga, K. (2017) Effectiveness of real-time polymerase chain reaction assay for the detection of *Mycobacterium tuberculosis* in pathological samples: a systematic review and meta-analysis. *Systematic Reviews* 6(1), 215. DOI: 10.1186/s13643-017-0608-2.

Banche, G., Allizond, V., Sostegni, R., Lavagna, A., Bergallo, M. *et al*. (2015) Application of multiple laboratory tests for *Mycobacterium avium* ssp. *paratuberculosis* detection in Crohn's disease patient specimens. *New Microbiologica* 38, 357–367.

Bannantine, J.P., Baechler, E., Zhang, Q., Li, L. and Kapur, V. (2002) Genome scale comparison of *Mycobacterium avium* subsp. *paratuberculosis* with *Mycobacterium avium* subsp. *avium* reveals potential diagnostic sequences. *Journal of Clinical Microbiology* 40(4), 1303–1310. DOI: 10.1128/JCM.40.4.1303-1310.2002.

Batista, A.C. and Pacheco, L.G.C. (2018) Detecting pathogens with zinc-finger, tale and CRISPR-based programmable nucleic acid binding proteins. *Journal of Microbiological Methods* 152, 98–104. DOI: 10.1016/j.mimet.2018.07.024.

Bauman, C.A., Jones-Bitton, A., Jansen, J., Kelton, D. and Menzies, P. (2016) Evaluation of fecal culture and fecal RT-PCR to detect *Mycobacterium avium* ssp. *paratuberculosis* fecal shedding in dairy goats and dairy sheep using latent class Bayesian modeling. *BMC Veterinary Research* 12(1), 212. DOI: 10.1186/s12917-016-0814-5.

Bauman, C.A., Jones-Bitton, A., Jansen, J., Kelton, D. and Menzies, P. (2019) Evaluation of bulk tank milk PCR and bulk tank milk modified ELISA tests for the detection of paratuberculosis at the herd level in goat and sheep dairies in Ontario, Canada. *Journal of Dairy Science* 102(1), 511–520. DOI: 10.3168/jds.2018-15020.

Beaver, A., Cazer, C.L., Ruegg, P.L., Gröhn, Y.T. and Schukken, Y.H. (2016) Implications of PCR and ELISA results on the routes of bulk-tank contamination with *Mycobacterium avium* ssp. *paratuberculosis*. *Journal of Dairy Science* 99(2), 1391–1405. DOI: 10.3168/jds.2015-9855.

Begg, D.J., Plain, K.M., de Silva, K., Gurung, R., Gunn, A. *et al.* (2018) Immunopathological changes and apparent recovery from infection revealed in cattle in an experimental model of Johne's disease using a lyophilised culture of *Mycobacterium avium* subspecies *paratuberculosis*. *Veterinary Microbiology* 219, 53–62. DOI: 10.1016/j.vetmic.2018.03.029.

Berghaus, R.D., Farver, T.B., Anderson, R.J., Jaravata, C.C. and Gardner, I.A. (2006) Environmental sampling for detection of *Mycobacterium avium* ssp. *paratuberculosis* on large California Dairies. *Journal of Dairy Science* 89(3), 963–970. DOI: 10.3168/jds.S0022-0302(06)72161-0.

Blaiotta, G., Di Cerbo, A., Murru, N., Coppola, R. and Aponte, M. (2016) Persistence of bacterial indicators and zoonotic pathogens in contaminated cattle wastes. *BMC Microbiology* 16(1), 87. DOI: 10.1186/s12866-016-0705-8.

Bogas, V., Balsa, F., Carvalho, M., Anjos, M.J., Pinheiro, M.F. *et al.* (2011) Comparison of four DNA extraction methods for forensic application. *Forensic Science International: Genetics Supplement Series* 3(1), e194–e195. DOI: 10.1016/j.fsigss.2011.08.097.

Bölske, G. and Herthnek, D. (2012) Diagnosis of paratuberculosis by PCR. In: Behr, M.A. and Collins, D. (eds) *Paratuberculosis: Organism, Disease, Control*, 1st edn. CAB International, Wallingford, UK.

Bossuyt, P.M. and Reitsma, J.B. (2003) The STARD initiative. *The Lancet* 361(9351), 71. DOI: 10.1016/S0140-6736(03)12122-8.

Bossuyt, P.M., Reitsma, J.B., Bruns, D.E., Gatsonis, C.A., Glasziou, P.P. *et al.* (2003) The STARD statement for reporting studies of diagnostic accuracy: explanation and elaboration. *Clinical Chemistry* 49(1), 7–18. DOI: 10.1373/49.1.7.

Botsaris, G., Swift, B.M.C., Slana, I., Liapi, M., Christodoulou, M. *et al.* (2016) Detection of viable *Mycobacterium avium* subspecies *paratuberculosis* in powdered infant formula by phage-PCR and confirmed by culture. *International Journal of Food Microbiology* 216, 91–94. DOI: 10.1016/j.ijfoodmicro.2015.09.011.

Bower, K.L. (2011) Aspects of bacteraemia and dissemination in *Mycobacterium avium* subspecies *paratuberculosis* infection. PhD thesis. The University of Sydney, Sydney, Australia.

Bower, K.L., Begg, D.J. and Whittington, R.J. (2011) Culture of *Mycobacterium avium* subspecies *paratuberculosis* (MAP) from blood and extra-intestinal tissues in experimentally infected sheep. *Veterinary Microbiology* 147(1–2), 127–132. DOI: 10.1016/j.vetmic.2010.06.016.

Busin, V., Wells, B., Kersaudy-Kerhoas, M., Shu, W. and Burgess, S.T.G. (2016) Opportunities and challenges for the application of microfluidic technologies in point-of-care veterinary diagnostics. *Molecular and Cellular Probes* 30(5), 331–341. DOI: 10.1016/j.mcp.2016.07.004.

Bustin, S.A., Benes, V., Garson, J.A., Hellemans, J., Huggett, J. *et al.* (2009) The MIQE guidelines: minimum information for publication of quantitative real-time PCR experiments. *Clinical Chemistry* 55(4), 611–622. DOI: 10.1373/clinchem.2008.112797.

Cheng, X., Sun, L., Zhao, Q., Mi, Z., Yu, G. *et al.* (2019) Development and evaluation of a droplet digital PCR assay for the diagnosis of paucibacillary leprosy in skin biopsy specimens. *PLoS Neglected Tropical Diseases* 13(3), e0007284. DOI: 10.1371/journal.pntd.0007284.

Chui, L.W., King, R. and Sim, J. (2010) Development of an immunocapture-polymerase chain reaction assay using IgY to detect *Mycobacterium avium* subsp. *paratuberculosis*. *Canadian Journal of Veterinary Research* 74, 102–107.

Cohen, J.F., Korevaar, D.A., Altman, D.G., Bruns, D.E., Gatsonis, C.A. *et al.* (2016) STARD 2015 guidelines for reporting diagnostic accuracy studies: explanation and elaboration. *BMJ Open* 6(11), e012799. DOI: 10.1136/bmjopen-2016-012799.

Collins, D.M., Stephens, D.M. and de Lisle, G.W. (1993) Comparison of polymerase chain reaction tests and faecal culture for detecting *Mycobacterium paratuberculosis* in bovine faeces. *Veterinary Microbiology* 36(3-4), 289–299. DOI: 10.1016/0378-1135(93)90095-O.

Collins, D.M., Cavaignac, S. and de Lisle, G.W. (1997) Use of four DNA insertion sequences to characterize strains of the *Mycobacterium avium* complex isolated from animals. *Molecular and Cellular Probes* 11(5), 373–380. DOI: 10.1006/mcpr.1997.0131.

Cousins, D.V., Whittington, R., Marsh, I., Masters, A., Evans, R.J. *et al.* (1999) Mycobacteria distinct from *Mycobacterium avium* subsp. *paratuberculosis* isolated from the faeces of ruminants possess

IS900-like sequences detectable IS900 polymerase chain reaction: implications for diagnosis. *Molecular and Cellular Probes* 13(6), 431–442. DOI: 10.1006/mcpr.1999.0275.

Dalton, J.P. and Hill, C. (2013) Survival of *Mycobacterium avium* subsp. *paratuberculosis* in synthetic human gastric juice and acidified porcine bile. *Applied and Environmental Microbiology* 79(4), 1418–1420. DOI: 10.1128/AEM.03523-12.

Davi, S.D., Kissenkötter, J., Faye, M., Böhlken-Fascher, S., Stahl-Hennig, C. *et al.* (2019) Recombinase polymerase amplification assay for rapid detection of monkeypox virus. *Diagnostic Microbiology and Infectious Disease* 95(1), 41–45. DOI: 10.1016/j.diagmicrobio.2019.03.015.

de Albuquerque, P.P.F., de Melo, R.P.B., de Farias Brito, M., Bovino, F., de Souza, M.A. *et al.* (2018) First molecular epidemiological study of *Mycobacterium avium* subsp. *paratuberculosis* in cattle and buffalo from different regions of Brazil. *Tropical Animal Health and Production* 50(8), 1929–1935. DOI: 10.1007/s11250-018-1650-3.

Devonshire, A.S., O'Sullivan, D.M., Honeyborne, I., Jones, G., Karczmarczyk, M. *et al.* (2016) The use of digital PCR to improve the application of quantitative molecular diagnostic methods for tuberculosis. *BMC Infectious Diseases* 16(1), 366. DOI: 10.1186/s12879-016-1696-7.

Donaghy, J.A., Johnston, J. and Rowe, M.T. (2011) Detection of *Mycobacterium avium* ssp. *paratuberculosis* in cheese, milk powder and milk using IS*900* and f57-based qPCR assays. *Journal of Applied Microbiology* 110(2), 479–489. DOI: 10.1111/j.1365-2672.2010.04905.x.

Donat, K., Hahn, N., Eisenberg, T., Schlez, K., Kohler, H. *et al.* (2016) Within-herd prevalence thresholds for the detection of *Mycobacterium avium* subspecies *paratuberculosis*-positive dairy herds using boot swabs and liquid manure samples. *Epidemiology and Infection* 144(2), 413–424. DOI: 10.1017/S0950268815000977.

Donat, K., Eisenberg, S.W.F., Einax, E., Reinhold, G. and Zoche-Golob, V. (2019) Reduction of viable *Mycobacterium avium* ssp. *paratuberculosis* in slurry subjected to anaerobic digestion in biogas plants. *Journal of Dairy Science* 102(7), 6485–6494. DOI: 10.3168/jds.2018-15937.

Dorman, S.E., Schumacher, S.G., Alland, D., Nabeta, P., Armstrong, D.T. *et al.* (2018) Xpert MTB/RIF Ultra for detection of *Mycobacterium tuberculosis* and rifampicin resistance: a prospective multicentre diagnostic accuracy study. *The Lancet Infectious Diseases* 18(1), 76–84. DOI: 10.1016/S1473-3099(17)30691-6.

Douarre, P.E., Cashman, W., Buckley, J., Coffey, A. and O'Mahony, J.M. (2010) Isolation and detection of *Mycobacterium avium* subsp. *paratuberculosis* (MAP) from cattle in Ireland using both traditional culture and molecular based methods. *Gut Pathogens* 2(1), 11. DOI: 10.1186/1757-4749-2-11.

Eamens, G.J., Whittington, R.J., Turner, M.J., Austin, S.L., Fell, S.A. *et al.* (2007) Evaluation of radiometric faecal culture and direct PCR on pooled faeces for detection of *Mycobacterium avium* subsp. *paratuberculosis* in cattle. *Veterinary Microbiology* 125(1-2), 22–35. DOI: 10.1016/j.vetmic.2007.04.043.

Eisenberg, S.W.F., Nielen, M., Hoeboer, J., Rutten, V., Heederik, D. *et al.* (2012) Environmental contamination with *Mycobacterium avium* subspecies *paratuberculosis* within and around a dairy barn under experimental conditions. *Journal of Dairy Science* 95(11), 6477–6482. DOI: 10.3168/jds.2012-5548.

Ellingson, J.L.E., Bolin, C.A. and Stabel, J.R. (1998) Identification of a gene unique to *Mycobacterium avium* subspecies *paratuberculosis* and application to diagnosis of paratuberculosis. *Molecular and Cellular Probes* 12(3), 133–142. DOI: 10.1006/mcpr.1998.0167.

Emerson, J.B., Adams, R.I., Román, C.M.B., Brooks, B., Coil, D.A. *et al.* (2017) Schrödinger's microbes: tools for distinguishing the living from the dead in microbial ecosystems. *Microbiome* 5(1), 86. DOI: 10.1186/s40168-017-0285-3.

Englund, S., Bölske, G. and Johansson, K.E. (2002) An IS900-like sequence found in a *Mycobacterium* sp. other than *Mycobacterium avium* subsp. *paratuberculosis*. *FEMS Microbiology Letters* 209(2), 267–271. DOI: 10.1111/j.1574-6968.2002.tb11142.x.

Enosawa, M., Kageyama, S., Sawai, K., Watanabe, K., Notomi, T. *et al.* (2003) Use of loop-mediated isothermal amplification of the IS*900* sequence for rapid detection of cultured *Mycobacterium avium* subsp. *paratuberculosis*. *Journal of Clinical Microbiology* 41(9), 4359–4365. DOI: 10.1128/JCM.41.9.4359-4365.2003.

Fang, Y., Wu, W.-H., Pepper, J.L., Larsen, J.L., Marras, S.A.E. *et al.* (2002) Comparison of real-time, quantitative PCR with molecular beacons to nested PCR and culture methods for detection of *Mycobacterium avium* subsp. *paratuberculosis* in bovine fecal samples. *Journal of Clinical Microbiology* 40(1), 287–291. DOI: 10.1128/JCM.40.1.287-291.2002.

Favila-Humara, L.C., Chávez-Gris, G.G., Carrillo-Casas, E.M. and Hernández-Castro, R. (2010) *Mycobacterium avium* subsp. *paratuberculosis* detection in individual and bulk tank milk samples from bovine herds and caprine flocks. *Foodborne Pathogens and Disease* 7(4), 351–355. DOI: 10.1089/fpd.2009.0374.

Gago, G., Diacovich, L. and Gramajo, H. (2018) Lipid metabolism and its implication in mycobacteria–host interaction. *Current Opinion in Microbiology* 41, 36–42. DOI: 10.1016/j.mib.2017.11.020.

Gao, A., Mutharia, L., Raymond, M. and Odumeru, J. (2007) Improved template DNA preparation procedure for detection of *Mycobacterium avium* subsp. *paratuberculosis* in milk by PCR. *Journal of Microbiological Methods* 69(2), 417–420. DOI: 10.1016/j.mimet.2006.10.019.

Gardner, I.A., Nielsen, S.S., Whittington, R.J., Collins, M.T., Bakker, D. *et al.* (2011) Consensus-based reporting standards for diagnostic test accuracy studies for paratuberculosis in ruminants. *Preventive Veterinary Medicine* 101(1-2), 18–34. DOI: 10.1016/j.prevetmed.2011.04.002.

Garg, R., Patil, P.K., Singh, S.V., Sharma, S., Gandham, R.K. *et al.* (2015) Comparative evaluation of different test combinations for diagnosis of *Mycobacterium avium* subspecies *paratuberculosis* infecting dairy herds in India. *BioMed Research International* 2015(11), 983978–6. DOI: 10.1155/2015/983978.

Gilardoni, L.R., Fernández, B., Morsella, C., Mendez, L., Jar, A.M. *et al.* (2016) *Mycobacterium paratuberculosis* detection in cow's milk in Argentina by immunomagnetic separation-PCR. *Brazilian Journal of Microbiology* 47(2), 506–512. DOI: 10.1016/j.bjm.2016.01.013.

Gootenberg, J.S., Abudayyeh, O.O., Kellner, M.J., Joung, J., Collins, J.J. *et al.* (2018) Multiplexed and portable nucleic acid detection platform with Cas13, Cas12a, and Csm6. *Science* 360(6387), 439–444. DOI: 10.1126/science.aaq0179.

Gordillo-Marroquín, C., Gómez-Velasco, A., Sánchez-Pérez, H., Pryg, K., Shinners, J. *et al.* (2018) Magnetic nanoparticle-based biosensing assay quantitatively enhances acid-fast bacilli count in paucibacillary pulmonary tuberculosis. *Biosensors* 8(4), 128. DOI: 10.3390/bios8040128.

Grant, I.R., Ball, H.J. and Rowe, M.T. (2002) Incidence of *Mycobacterium paratuberculosis* in bulk raw and commercially pasteurized cows' milk from approved dairy processing establishments in the United Kingdom. *Applied and Environmental Microbiology* 68(5), 2428–2435. DOI: 10.1128/AEM.68.5.2428-2435.2002.

Grant, I.R., Foddai, A.C.G., Tarrant, J.C., Kunkel, B., Hartmann, F.A. *et al.* (2017) Viable *Mycobacterium avium* ssp. *paratuberculosis* isolated from calf milk replacer. *Journal of Dairy Science* 100(12), 9723–9735. DOI: 10.3168/jds.2017-13154.

Green, E.P., Tizard, M.L.V., Moss, M.T., Thompson, J., Winterbourne, D.J. *et al.* (1989) Sequence and characteristics or IS*900*, an insertion element identified in a human Crohn's disease isolate or *Mycobacterium paratuberculosis*. *Nucleic Acids Research* 17(22), 9063–9073. DOI: 10.1093/nar/17.22.9063.

Gu, W., Miller, S. and Chiu, C.Y. (2019) Clinical metagenomic next-generation sequencing for pathogen detection. *Annual Review of Pathology: Mechanisms of Disease* 14(1), 319–338. DOI: 10.1146/annurev-pathmechdis-012418-012751.

Hahn, N., Failing, K., Eisenberg, T., Schlez, K., Zschöck, P.-M. *et al.* (2017) Evaluation of different diagnostic methods for the detection of *Mycobacterium avium* subsp. *paratuberculosis* in boot swabs and liquid manure samples. *BMC Veterinary Research* 13(1), 259. DOI: 10.1186/s12917-017-1173-6.

Hansen, S., Schäfer, J., Fechner, K., Czerny, C.-P. and Abd El Wahed, A. (2016) Development of a recombinase polymerase amplification assay for rapid detection of the *Mycobacterium avium* subsp. *paratuberculosis*. *PLoS ONE* 11(12), e0168733. DOI: 10.1371/journal.pone.0168733.

He, Q., Barkoff, A.M., Mertsola, J., Glismann, S., Bacci, S. *et al.* (2012) High heterogeneity in methods used for the laboratory confirmation of pertussis diagnosis among European countries, 2010: integration of epidemiological and laboratory surveillance must include standardisation of methodologies and quality assurance. *Eurosurveillance* 17(32). DOI: 10.2807/ese.17.32.20239-en.

Henaux, V., Ja, Y.M., Siebeke, C., Calavas, D. and Ponsart, C. (2018) Review of bovine brucellosis surveillance in Europe in 2015. *Revue Scientifique et Technique de l'OIE* 37(3), 805–821. DOI: 10.20506/rst.37.3.2887.

Herthnek, D. and Bölske, G. (2006) New PCR systems to confirm real-time PCR detection of *Mycobacterium avium* subsp. *paratuberculosis*. *BMC Microbiology* 6(1), 87. DOI: 10.1186/1471-2180-6-87.

Herthnek, D., Nielsen, S.S., Lindberg, A. and Bölske, G. (2008) A robust method for bacterial lysis and DNA purification to be used with real-time PCR for detection of *Mycobacterium avium* subsp. *paratuberculosis* in milk. *Journal of Microbiological Methods* 75(2), 335–340. DOI: 10.1016/j.mimet.2008.07.009.

Heuvelink, A., Hassan, A.A., van Weering, H., van Engelen, E., Bülte, M. *et al.* (2017) An intralaboratory cultural and real-time PCR method comparison and evaluation for the detection of subclinical paratuberculosis in dairy herds. *Folia Microbiologica* 62(3), 197–205. DOI: 10.1007/s12223-016-0488-1.

Huggett, J.F., Foy, C.A., Benes, V., Emslie, K., Garson, J.A. *et al.* (2013) The digital MIQE guidelines: minimum information for publication of quantitative digital PCR experiments. *Clinical Chemistry* 59(6), 892–902. DOI: 10.1373/clinchem.2013.206375.

Imirzalioglu, C., Dahmen, H., Hain, T., Billion, A., Kuenne, C. *et al.* (2011) Highly specific and quick detection of *Mycobacterium avium* subsp. *paratuberculosis* in feces and gut tissue of cattle and humans by multiple real-time PCR assays. *Journal of Clinical Microbiology* 49(5), 1843–1852. DOI: 10.1128/JCM.01492-10.

Irenge, L.M., Walravens, K., Govaerts, M., Godfroid, J., Rosseels, V. *et al.* (2009) Development and validation of a triplex real-time PCR for rapid detection and specific identification of *M. avium* sub sp. *paratuberculosis* in faecal samples. *Veterinary Microbiology* 136(1–2), 166–172. DOI: 10.1016/j.vetmic.2008.09.087.

Johnson, G.L., Bibby, D.F., Wong, S., Agrawal, S.G. and Bustin, S.A. (2012) A MIQE-compliant real-time PCR assay for Aspergillus detection. *PLoS ONE* 7(7), e40022. DOI: 10.1371/journal.pone.0040022.

Juste, R.A., Elguezabal, N., Garrido, J.M., Pavon, A., Geijo, M.V. *et al.* (2008) On the prevalence of *M. avium* subspecies *paratuberculosis* DNA in the blood of healthy individuals and patients with inflammatory bowel disease. *PLoS ONE* 3(7), e2537. DOI: 10.1371/journal.pone.0002537.

Kaur, P., Filia, G., Singh, S.V., Patil, P.K. and Sandhu, K.S. (2010) Molecular detection and typing of *Mycobacterium avium* subspecies *paratuberculosis* from milk samples of dairy animals. *Tropical Animal Health and Production* 42(5), 1031–1035. DOI: 10.1007/s11250-009-9521-6.

Kawaji, S., Taylor, D.L., Mori, Y. and Whittington, R.J. (2007) Detection of *Mycobacterium avium* subsp. *paratuberculosis* in ovine faeces by direct quantitative PCR has similar or greater sensitivity compared to radiometric culture. *Veterinary Microbiology* 125(1–2), 36–48. DOI: 10.1016/j.vetmic.2007.05.002.

Kawaji, S., Begg, D.J., Plain, K.M. and Whittington, R.J. (2011) A longitudinal study to evaluate the diagnostic potential of a direct faecal quantitative PCR test for Johne's disease in sheep. *Veterinary Microbiology* 148(1), 35–44. DOI: 10.1016/j.vetmic.2010.07.022.

Khan, I.A., Pilli, S., A, S., Rampal, R., Chauhan, S.K., Tiwari, V. *et al.* (2016) Prevalence and association of *Mycobacterium avium* subspecies *paratuberculosis* with disease course in patients with ulcero-constrictive ileocolonic disease. *PLoS ONE* 11(3), e0152063. DOI: 10.1371/journal.pone.0152063.

Khare, S., Ficht, T.A., Santos, R.L., Romano, J., Ficht, A.R. *et al.* (2004) Rapid and sensitive detection of *Mycobacterium avium* subsp. *paratuberculosis* in bovine milk and feces by a combination of immunomagnetic bead separation-conventional PCR and real-time PCR. *Journal of Clinical Microbiology* 42(3), 1075–1081. DOI: 10.1128/JCM.42.3.1075-1081.2004.

Khol, J.L., Pinedo, P.J., Buergelt, C.D., Neumann, L.M. and Rae, D.O. (2014) Lymphatic fluid for the detection of *Mycobacterium avium* subsp. *paratuberculosis* in cows by PCR, compared to fecal sampling and detection of antibodies in blood and milk. *Veterinary Microbiology* 172(1–2), 301–308. DOI: 10.1016/j.vetmic.2014.05.022.

Khol, J.L., Braun, A.L., Slana, I., Kralik, P. and Wittek, T. (2017) Testing of milk replacers for *Mycobacterium avium* subsp. *paratuberculosis* by PCR and bacterial culture as a possible source for Johne's disease (paratuberculosis) in calves. *Preventive Veterinary Medicine* 144, 53–56. DOI: 10.1016/j.prevetmed.2017.05.013.

Kim, S.G., Shin, S.J., Jacobson, R.H., Miller, L.J., Harpending, P.R. *et al.* (2002) Development and application of quantitative polymerase chain reaction assay based on the ABI 7700 system (TaqMan) for detection and quantification of *Mycobacterium avium* subsp. paratuberculosis. *Journal of Veterinary Diagnostic Investigation* 14(2), 126–131. DOI: 10.1177/104063870201400206.

King, C.E., Debruyne, R., Kuch, M., Schwarz, C. and Poinar, H.N. (2009) A quantitative approach to detect and overcome PCR inhibition in ancient DNA extracts. *BioTechniques* 47(5), 941–949. DOI: 10.2144/000113244.

Kralik, P. and Ricchi, M. (2017) A basic guide to real time PCR in microbial diagnostics: definitions, parameters, and everything. *Frontiers in Microbiology* 8, 108. DOI: 10.3389/fmicb.2017.00108.

Kralik, P., Nocker, A. and Pavlik, I. (2010) Mycobacterium avium subsp. paratuberculosis viability determination using f57 quantitative PCR in combination with propidium monoazide treatment. *International Journal of Food Microbiology* 141, S80–S86. DOI: 10.1016/j.ijfoodmicro.2010.03.018.

Kralik, P., Babak, V. and Dziedzinska, R. (2014a) Repeated cycles of chemical and physical disinfection and their influence on *Mycobacterium avium* subsp. *paratuberculosis* viability measured by propidium monoazide F57 quantitative real time PCR. *The Veterinary Journal* 201(3), 359–364. DOI: 10.1016/j.tvjl.2014.05.032.

Kralik, P., Pribylova-Dziedzinska, R., Kralova, A., Kovarcik, K. and Slana, I. (2014b) Evidence of passive faecal shedding of *Mycobacterium avium* subsp. *paratuberculosis* in a Limousin cattle herd. *The Veterinary Journal* 201(1), 91–94. DOI: 10.1016/j.tvjl.2014.02.011.

Kralik, P., Babak, V. and Dziedzinska, R. (2018) The impact of the antimicrobial compounds produced by lactic acid bacteria on the growth performance of *Mycobacterium avium* subsp. *paratuberculosis*. *Frontiers in Microbiology* 9, 638. DOI: 10.3389/fmicb.2018.00638.

Kuypers, J. and Jerome, K.R. (2017) Applications of digital PCR for clinical microbiology. *Journal of Clinical Microbiology* 55(6), 1621–1628. DOI: 10.1128/JCM.00211-17.

Laurin, E., McKenna, S., Chaffer, M. and Keefe, G. (2015) Sensitivity of solid culture, broth culture, and real-time PCR assays for milk and colostrum samples from *Mycobacterium avium* ssp. *paratuberculosis*-infectious dairy cows. *Journal of Dairy Science* 98(12), 8597–8609. DOI: 10.3168/jds.2014-8758.

Lavers, C.J., Mckenna, S.L.B., Dohoo, I.R., Barkema, H.W. and Keefe, G.P. (2013) Evaluation of environmental fecal culture for *Mycobacterium avium* subspecies *paratuberculosis* detection in dairy herds and association with apparent within-herd prevalence. *Canadian Veterinary Journal* 54, 1053–1060.

Lee, A., Griffiths, T.A., Parab, R.S., King, R.K., Dubinsky, M.C. *et al.* (2011) Association of *Mycobacterium avium* subspecies paratuberculosis with Crohn disease in pediatric patients. *Journal of Pediatric Gastroenterology and Nutrition* 52(2), 170–174. DOI: 10.1097/MPG.0b013e3181ef37ba.

Lee, J.S., Mackie, R.S., Harrison, T., Shariat, B., Kind, T. *et al.* (2017) Targeted enrichment for pathogen detection and characterization in three felid species. *Journal of Clinical Microbiology* 55(6), 1658–1670. DOI: 10.1128/JCM.01463-16.

Lermo, A., Liebana, S., Campoy, S., Fabiano, S., Garcia, M. I. *et al.* (2010) A novel strategy for screening-out raw milk contaminated with *Mycobacterium bovis* on dairy farms by double-tagging PCR and electrochemical genosensing. *International Microbiology* 13, 91–97.

Li, H., Bai, R., Zhao, Z., Tao, L., Ma, M. *et al.* (2018) Application of droplet digital PCR to detect the pathogens of infectious diseases. *Bioscience Reports* 38(6), BSR20181170. DOI: 10.1042/BSR20181170.

Liu, L., Wang, J., Zhang, R., Lin, M., Shi, R. *et al.* (2018) Visual and equipment-free reverse transcription recombinase polymerase amplification method for rapid detection of foot-and-mouth disease virus. *BMC Veterinary Research* 14(1), 263. DOI: 10.1186/s12917-018-1594-x.

Ljungström, L., Enroth, H., Claesson, B.E.B., Ovemyr, I., Karlsson, J. *et al.* (2015) Clinical evaluation of commercial nucleic acid amplification tests in patients with suspected sepsis. *BMC Infectious Diseases* 15(1), 199. DOI: 10.1186/s12879-015-0938-4.

Logar, K., Kopinč, R., Bandelj, P., Starič, J., Lapanje, A. *et al.* (2012) Evaluation of combined high-efficiency DNA extraction and real-time PCR for detection of *Mycobacterium avium* subsp. *paratuberculosis* in subclinically infected dairy cattle: comparison with faecal culture, milk real-time PCR and milk ELISA. *BMC Veterinary Research* 8(1), 49. DOI: 10.1186/1746-6148-8-49.

Lorencova, A., Babak, V., Kralova, A. and Borilova, G. (2019) Survival of *Mycobacterium avium* subsp. *paratuberculosis* in raw fermented sausages during production and storage. *Meat Science* 155, 20–26. DOI: 10.1016/j.meatsci.2019.04.023.

Lorenz, T.C. (2012) Polymerase chain reaction: basic protocol plus troubleshooting and optimization strategies. *Journal of Visualized Experiments* (63), e3998. DOI: 10.3791/3998.

Ly, A., Dhand, N.K., Sergeant, E.S.G., Marsh, I. and Plain, K.M. (2019) Determining an optimal pool size for testing beef herds for Johne's disease in Australia. *PLoS ONE* 14(11), e0225524. DOI: 10.1371/journal.pone.0225524.

Majumdar, N., Banerjee, S., Pallas, M., Wessel, T. and Hegerich, P. (2017) Poisson plus quantification for digital PCR systems. *Scientific Reports* 7(1), 9617. DOI: 10.1038/s41598-017-09183-4.

Makela, P., Beloeil, P.-A., Rizzi, V., Boelaert, F. and Deluyker, H. (2012) Harmonisation of monitoring zoonoses, antimicrobial resistance and foodborne outbreaks. *EFSA Journal* 10(10), s1013. DOI: 10.2903/j.efsa.2012.s1013.

Manessis, G., Gelasakis, A. I. and Bossis, I. (2019) The challenge of introducing point of care diagnostics in farm animal health management. *Biomedical Journal of Scientific and Technical Research* 14.

Markwalter, C.F., Kantor, A.G., Moore, C.P., Richardson, K.A. and Wright, D.W. (2019) Inorganic complexes and metal-based nanomaterials for infectious disease diagnostics. *Chemical Reviews* 119(2), 1456–1518. DOI: 10.1021/acs.chemrev.8b00136.

Marsh, I., Whittington, R. and Cousins, D. (1999) PCR-restriction endonuclease analysis for identification and strain typing of *Mycobacterium avium* subsp. *paratuberculosis* and *Mycobacterium avium* subsp. *avium* based on polymorphisms in IS1311. *Molecular and Cellular Probes* 13(2), 115–126. DOI: 10.1006/mcpr.1999.0227.

Marsh, I., Whittington, R. and Millar, D. (2000) Quality control and optimized procedure of hybridization capture-PCR for the identification of *Mycobacterium avium* subsp. *paratuberculosis* in faeces. *Molecular and Cellular Probes* 14(4), 219–232. DOI: 10.1006/mcpr.2000.0309.

Marsh, I.B., Plain, K.M. and Whittington, R.J. (2014) Molecular diagnostic tests for Johne's disease; time to standardise. In: Manning, E.J.B. and Nielsen, S.S. (eds) *Twelfth International Colloquium on Paratuberculosis*. International Association for Paratuberculosis, Parma, Italy.

Mbelele, P., Mohamed, S., Sauli, E., Mpolya, E., Mfinanga, S. *et al.* (2018) Meta-narrative review of molecular methods for diagnosis and monitoring of multidrug-resistant tuberculosis treatment in adults. *International Journal of Mycobacteriology* 7(4), 299–309. DOI: 10.4103/ijmy.ijmy_135_18.

Mcgiven, J.A., Stack, J.A., Perrett, L.L., Tucker, J.D., Brew, S.D. *et al.* (2006) Harmonisation of European tests for serological diagnosis of Brucella infection in bovines. *Revue scientifique et technique* 25, 1039–1053.

McNerney, R., Sollis, K. and Peeling, R.W. (2014) Improving access to new diagnostics through harmonised regulation: priorities for action. *African Journal of Laboratory Medicine* 3(1), 123. DOI: 10.4102/ajlm.v3i1.123.

Mengoli, C., Cruciani, M., Barnes, R.A., Loeffler, J. and Donnelly, J.P. (2009) Use of PCR for diagnosis of invasive aspergillosis: systematic review and meta-analysis. *The Lancet Infectious Diseases* 9(2), 89–96. DOI: 10.1016/S1473-3099(09)70019-2.

Metzger-Boddien, C., Khaschabi, D., Schönbauer, M., Boddien, S., Schlederer, T. *et al.* (2006) Automated high-throughput immunomagnetic separation-PCR for detection of *Mycobacterium avium* subsp. *paratuberculosis* in bovine milk. *International Journal of Food Microbiology* 110(3), 201–208. DOI: 10.1016/j.ijfoodmicro.2006.01.042.

Meyer, A., McAloon, C.G., Tratalos, J.A., More, S.J., Citer, L.R. *et al.* (2019) Modeling of alternative testing strategies to demonstrate freedom from *Mycobacterium avium* ssp. *paratuberculosis* infection in test-negative dairy herds in the Republic of Ireland. *Journal of Dairy Science* 102(3), 2427–2442. DOI: 10.3168/jds.2018-14883.

Mita, A., Mori, Y., Nakagawa, T., Tasaki, T., Utiyama, K. *et al.* (2016) Comparison of fecal pooling methods and DNA extraction kits for the detection of *Mycobacterium avium* subspecies *paratuberculosis*. *MicrobiologyOpen* 5(1), 134–142. DOI: 10.1002/mbo3.318.

Moss, M.T., Green, E.P., Tizard, M.L., Malik, Z.P. and Hermon-Taylor, J. (1991) Specific detection of *Mycobacterium paratuberculosis* by DNA hybridisation with a fragment of the insertion element IS*900*. *Gut* 32(4), 395–398. DOI: 10.1136/gut.32.4.395.

Münster, P., Völkel, I., Wemheuer, W., Petschenka, J., Wemheuer, W. *et al.* (2011) Detection of *Mycobacterium avium* ssp. *paratuberculosis* in ileocaecal lymph nodes collected from elderly slaughter cows using a semi-nested IS900 polymerase chain reaction. *Veterinary Microbiology* 154(1–2), 197–201. DOI: 10.1016/j.vetmic.2011.06.033.

Münster, P., Völkel, I., Wemheuer, W., Schwarz, D., Döring, S. *et al.* (2013) A longitudinal study to characterize the distribution patterns of *Mycobacterium avium* ssp. *paratuberculosis* in semen, blood and faeces of a naturally infected bull by IS 900 semi-nested and quantitative real-time PCR. *Transboundary and Emerging Diseases* 60(2), 175–187. DOI: 10.1111/j.1865-1682.2012.01336.x.

Nielsen, S.S., Toft, N. and Gardner, I.A. (2011) Structured approach to design of diagnostic test evaluation studies for chronic progressive infections in animals. *Veterinary Microbiology* 150(1–2), 115–125. DOI: 10.1016/j.vetmic.2011.01.019.

Nyaruaba, R., Mwaliko, C., Kering, K.K. and Wei, H. (2019) Droplet digital PCR applications in the tuberculosis world. *Tuberculosis* 117, 85–92. DOI: 10.1016/j.tube.2019.07.001.

Obara, H., Aikawa, N., Hasegawa, N., Hori, S., Ikeda, Y. *et al.* (2011) The role of a real-time PCR technology for rapid detection and identification of bacterial and fungal pathogens in whole-blood samples. *Journal of Infection and Chemotherapy* 17(3), 327–333. DOI: 10.1007/s10156-010-0168-z.

O'Donnell, M.M., Harris, H.M.B., Ross, R.P. and O'Toole, P.W. (2017) Core fecal microbiota of domesticated herbivorous ruminant, hindgut fermenters, and monogastric animals. *MicrobiologyOpen* 6(5), e00509. DOI: 10.1002/mbo3.509.

Odumeru, J., Gao, A., Chen, S., Raymond, M. and Mutharia, L. (2001) Use of the bead beater for preparation of *Mycobacterium paratuberculosis* template DNA in milk. *Canadian Journal of Veterinary Research* 65, 201–205.

OIE, W.O.F. (2019) Principles and methods of validation of diagnostic assays for infectious diseases. *Manual of Diagnostic Tests for Aquatic Animals*.

Okura, H., Toft, N. and Nielsen, S.S. (2012) Occurrence of *Mycobacterium avium* subsp. *paratuberculosis* in milk at dairy cattle farms: a systematic review and meta-analysis. *Veterinary Microbiology* 157(3-4), 253–263. DOI: 10.1016/j.vetmic.2011.12.019.

Park, K.T., Allen, A.J. and Davis, W.C. (2014) Development of a novel DNA extraction method for identification and quantification of *Mycobacterium avium* subsp. *paratuberculosis* from tissue samples by real-time PCR. *Journal of Microbiological Methods* 99, 58–65. DOI: 10.1016/j.mimet.2014.02.003.

Park, H.-T., Park, H.-E., Park, W.B., Kim, S., Hur, T.Y. *et al.* (2018) Genetic diversity of bovine *Mycobacterium avium* subsp. *paratuberculosis* discriminated by IS *1311* PCR-REA, MIRU-VNTR, and MLSSR genotyping. *Journal of Veterinary Science* 19(5), 627–634. DOI: 10.4142/jvs.2018.19.5.627.

Parker, A.M., House, J.K., Hazelton, M.S., Bosward, K.L. and Sheehy, P.A. (2017) Comparison of culture and a multiplex probe PCR for identifying Mycoplasma species in bovine milk, semen and swab samples. *Plos One* 12(3), e0173422. DOI: 10.1371/journal.pone.0173422.

Petrich, A., Mahony, J., Chong, S., Broukhanski, G., Gharabaghi, F. *et al.* (2006) Multicenter comparison of nucleic acid extraction methods for detection of severe acute respiratory syndrome coronavirus RNA in stool specimens. *Journal of Clinical Microbiology* 44(8), 2681–2688. DOI: 10.1128/JCM.02460-05.

Pholwat, S., Stroup, S., Foongladda, S. and Houpt, E. (2013) Digital PCR to detect and quantify heteroresistance in drug resistant *Mycobacterium tuberculosis*. *PLoS ONE* 8(2), e57238. DOI: 10.1371/journal.pone.0057238.

Plain, K.M., Marsh, I.B., Waldron, A.M., Galea, F., Whittington, A.-M. *et al.* (2014) High-Throughput direct fecal PCR assay for detection of *Mycobacterium avium* subsp. *paratuberculosis* in sheep and cattle. *Journal of Clinical Microbiology* 52(3), 745–757. DOI: 10.1128/JCM.03233-13.

Plain, K.M., Waldron, A.M., Begg, D.J., de Silva, K., Purdie, A.C. *et al.* (2015) Efficient, validated method for detection of mycobacterial growth in liquid culture media by use of bead beating, magnetic-particle-based nucleic acid isolation, and quantitative PCR. *Journal of Clinical Microbiology* 53(4), 1121–1128. DOI: 10.1128/JCM.03521-14.

Pribylova, R., Kubickova, L., Babak, V., Pavlik, I. and Kralik, P. (2012) Effect of short- and long-term antibiotic exposure on the viability of *Mycobacterium avium* subsp. *paratuberculosis* as measured by propidium monoazide F57 real time quantitative PCR and culture. *The Veterinary Journal* 194(3), 354–360. DOI: 10.1016/j.tvjl.2012.05.002.

Purdie, A.C., Plain, K.M., Begg, D.J., de Silva, K. and Whittington, R.J. (2012) Expression of genes associated with the antigen presentation and processing pathway are consistently regulated in early *Mycobacterium avium* subsp. *paratuberculosis* infection. *Comparative Immunology, Microbiology and Infectious Diseases* 35(2), 151–162. DOI: 10.1016/j.cimid.2011.12.007.

Purdie, A.C., Plain, K.M., Begg, D.J., de Silva, K. and Whittington, R.J. (2019) Gene expression profiles during subclinical *Mycobacterium avium* subspecies *paratuberculosis* infection in sheep can predict disease outcome. *Scientific Reports* 9(1), 8245. DOI: 10.1038/s41598-019-44670-w.

Radomski, N., Kreitmann, L., McIntosh, F. and Behr, M.A. (2013) The critical role of DNA extraction for detection of mycobacteria in tissues. *PLoS ONE* 8(10), e78749. DOI: 10.1371/journal.pone.0078749.

Rangel, S., Arango-Sabogal, J.C., Labrecque, O., Paré, J., Fairbrother, J.-H. *et al.* (2017) Evaluation of a PCR assay on overgrown individual fecal samples cultured for *Mycobacterium avium* subsp. *paratuberculosis*. *Journal of Veterinary Diagnostic Investigation* 29(6), 912–915. DOI: 10.1177/1040638717724837.

Rapp, D. (2010) DNA extraction from bovine faeces: current status and future trends. *Journal of Applied Microbiology* 108(5), 1485–1493. DOI: 10.1111/j.1365-2672.2009.04606.x.

Restrepo, B.I., Gomez, D.I., Shipley, G.L., McCormick, J.B. and Fisher-Hoch, S.P. (2006) Selective enrichment and detection of mycobacterial DNA in paucibacillary specimens. *Journal of Microbiological Methods* 67(2), 220–229. DOI: 10.1016/j.mimet.2006.03.016.

Ricchi, M., De Cicco, C., Kralik, P., Babak, V., Boniotti, M.B. *et al.* (2014) Evaluation of viable *Mycobacterium avium* subsp. *paratuberculosis* in milk using peptide-mediated separation and propidium monoazide qPCR. *FEMS Microbiology Letters* 356(1), 127–133. DOI: 10.1111/1574-6968.12480.

Ricchi, M., Bertasio, C., Boniotti, M.B., Vicari, N., Russo, S. *et al.* (2017) Comparison among the quantification of bacterial pathogens by qPCR, dPCR, and cultural methods. *Frontiers in Microbiology* 8, 1174. DOI: 10.3389/fmicb.2017.01174.

Richardson, H., Rhodes, G., Henrys, P., Sedda, L., Weightman, A.J. *et al.* (2019) Presence of *Mycobacterium avium* subspecies *paratuberculosis* monitored over varying temporal and spatial

scales in river catchments: persistent routes for human exposure. *Microorganisms* 7(5), 136. DOI: 10.3390/microorganisms7050136.
Salgado, M., Steuer, P., Troncoso, E. and Collins, M.T. (2013) Evaluation of PMS–PCR technology for detection of *Mycobacterium avium* subsp. *paratuberculosis* directly from bovine fecal specimens. *Veterinary Microbiology* 167(3–4), 725–728. DOI: 10.1016/j.vetmic.2013.09.009.
Sange, M.D., Becker, A., Hassan, A.A., Bülte, M., Ganter, M. *et al.* (2019) Development and validation of a loop-mediated isothermal amplification assay—a rapid and sensitive detection tool for *Mycobacterium avium* subsp. *paratuberculosis* in small ruminants. *Journal of Applied Microbiology* 127(1), 47–58. DOI: 10.1111/jam.14284.
Savi, R., Ricchi, M., Cammi, G., Garbarino, C., Leo, S. *et al.* (2015) Survey on the presence of *Mycobacterium avium* subsp. *paratuberculosis* in ground beef from an industrial meat plant. *Veterinary Microbiology* 177(3–4), 403–408. DOI: 10.1016/j.vetmic.2015.03.013.
Schönenbrücher, H., Abdulmawjood, A., Failing, K. and Bülte, M. (2008) New triplex real-time PCR assay for detection of *Mycobacterium avium* subsp. *paratuberculosis* in bovine feces. *Applied and Environmental Microbiology* 74(9), 2751–2758. DOI: 10.1128/AEM.02534-07.
Schönenbrücher, H., Abdulmawjood, A., Failing, K. and Bülte, M. (2008) New triplex real-time PCR assay for detection of *Mycobacterium avium* subsp. *paratuberculosis* in bovine feces. *Applied and Environmental Microbiology* 74(9), 2751–2758. DOI: 10.1128/AEM.02534-07.
Schrader, C., Schielke, A., Ellerbroek, L. and Johne, R. (2012) PCR inhibitors – occurrence, properties and removal. *Journal of Applied Microbiology* 113(5), 1014–1026. DOI: 10.1111/j.1365-2672.2012.05384.x.
Schwalm, A.K., Obiegala, A., Pfeffer, M. and Sting, R. (2018) Enhanced sensitivity and fast turnaround time in laboratory diagnosis for bovine paratuberculosis in faecal samples. *Journal of Microbiological Methods* 152, 39–47. DOI: 10.1016/j.mimet.2018.07.010.
Selim, A. and Gaede, W. (2012) Optimization of DNA extraction for identification of *Mycobacterium avium* subsp. *paratuberculosis* in faeces. *Online Journal of Veterinary Research* 16, 163–171.
Sergeant, E.S.G., McAloon, C.G., Tratalos, J.A., Citer, L.R., Graham, D.A. *et al.* (2019) Evaluation of national surveillance methods for detection of Irish dairy herds infected with *Mycobacterium avium* ssp. *paratuberculosis*. *Journal of Dairy Science* 102(3), 2525–2538. DOI: 10.3168/jds.2018-15696.
Sevilla, I.A., Garrido, J.M., Molina, E., Geijo, M.V., Elguezabal, N. *et al.* (2014) Development and evaluation of a novel multicopy-element-targeting triplex PCR for detection of *Mycobacterium avium* subsp. *paratuberculosis* in feces. *Applied and Environmental Microbiology* 80(12), 3757–3768. DOI: 10.1128/AEM.01026-14.
Shahin, K., Gustavo Ramirez-Paredes, J., Harold, G., Lopez-Jimena, B., Adams, A. *et al.* (2018) Development of a recombinase polymerase amplification assay for rapid detection of *Francisella noatunensis* subsp. *orientalis*. *PLoS ONE* 13(2), e0192979. DOI: 10.1371/journal.pone.0192979.
Sharp, R.C., Naser, E.S., Alcedo, K.P., Qasem, A., Abdelli, L.S. *et al.* (2018) Development of multiplex PCR and multi-color fluorescent in situ hybridization (m-FISH) coupled protocol for detection and imaging of multi-pathogens involved in inflammatory bowel disease. *Gut Pathogens* 10(1), 51. DOI: 10.1186/s13099-018-0278-1.
Slana, I., Kralik, P., Kralova, A. and Pavlik, I. (2008) On-farm spread of *Mycobacterium avium* subsp. *paratuberculosis* in raw milk studied by IS*900* and F57 competitive real time quantitative PCR and culture examination. *International Journal of Food Microbiology* 128(2), 250–257. DOI: 10.1016/j.ijfoodmicro.2008.08.013.
Slana, I., Liapi, M., Moravkova, M., Kralova, A. and Pavlik, I. (2009) *Mycobacterium avium* subsp. *paratuberculosis* in cow bulk tank milk in Cyprus detected by culture and quantitative IS*900* and F57 real-time PCR. *Preventive Veterinary Medicine* 89(3–4), 223–226. DOI: 10.1016/j.prevetmed.2009.02.020.
Slana, I., Kralik, P., Kralova, A., Babak, V. and Pavlik, I. (2012) Short communication: Examination of milk filters by real-time PCR as a herd-level indicator of the presence of *Mycobacterium avium* subspecies *paratuberculosis* in dairy herds. *Journal of Dairy Science* 95(3), 1162–1165. DOI: 10.3168/jds.2011-4658.
Song, N., Tan, Y., Zhang, L., Luo, W., Guan, Q. *et al.* (2018) Detection of circulating *Mycobacterium tuberculosis* -specific DNA by droplet digital PCR for vaccine evaluation in challenged monkeys and TB diagnosis. *Emerging Microbes & Infections* 7(1), 1–9. DOI: 10.1038/s41426-018-0076-3.
Stabel, J.R. and Bannantine, J.P. (2005) Development of a nested PCR method targeting a unique multicopy element, ISMap02, for detection of *Mycobacterium avium* subsp. *paratuberculosis* in fecal samples. *Journal of Clinical Microbiology* 43(9), 4744–4750. DOI: 10.1128/JCM.43.9.4744-4750.2005.

Stabel, J.R., Bradner, L., Robbe-Austerman, S. and Beitz, D.C. (2014) Clinical disease and stage of lactation influence shedding of *Mycobacterium avium* subspecies *paratuberculosis* into milk and colostrum of naturally infected dairy cows. *Journal of Dairy Science* 97(10), 6296–6304. DOI: 10.3168/jds.2014-8204.

Stevenson, K. and Sharp, J.M. (1997) The contribution of molecular biology to *Mycobacterium avium* subspecies *paratuberculosis* research. *The Veterinary Journal* 153(3), 269–286. DOI: 10.1016/S1090-0233(97)80062-7.

Sting, R., Hrubenja, M., Mandl, J., Seemann, G., Salditt, A. *et al.* (2014) Detection of *Mycobacterium avium* subsp. *paratuberculosis* in faeces using different procedures of pre-treatment for real-time PCR in comparison to culture. *The Veterinary Journal* 199(1), 138–142. DOI: 10.1016/j.tvjl.2013.08.033.

Stratmann, J., Strommenger, B., Stevenson, K. and Gerlach, G.-F. (2002) Development of a peptide-mediated capture PCR for detection of *Mycobacterium avium* subsp. *paratuberculosis* in milk. *Journal of Clinical Microbiology* 40(11), 4244–4250. DOI: 10.1128/JCM.40.11.4244-4250.2002.

Strommenger, B., Stevenson, K. and Gerlach, G.-F. (2001) Isolation and diagnostic potential of IS *Mav2*, a novel insertion sequence-like element from *Mycobacterium avium* subspecies *paratuberculosis*. *FEMS Microbiology Letters* 196(1), 31–37. DOI: 10.1111/j.1574-6968.2001.tb10536.x.

Sweeney, R.W., Whitlock, R.H. and Rosenberger, A.E. (1992) *Mycobacterium paratuberculosis* cultured from milk and supramammary lymph nodes of infected asymptomatic cows. *Journal of Clinical Microbiology* 30(1), 166–171. DOI: 10.1128/JCM.30.1.166-171.1992.

Swift, B.M.C., Denton, E.J., Mahendran, S.A., Huxley, J.N. and Rees, C.E.D. (2013) Development of a rapid phage-based method for the detection of viable *Mycobacterium avium* subsp. *paratuberculosis* in blood within 48 h. *Journal of Microbiological Methods* 94(3), 175–179. DOI: 10.1016/j.mimet.2013.06.015.

Swift, B.M.C., Huxley, J.N., Plain, K.M., Begg, D.J., de Silva, K. *et al.* (2016) Evaluation of the limitations and methods to improve rapid phage-based detection of viable *Mycobacterium avium* subsp. *paratuberculosis* in the blood of experimentally infected cattle. *BMC Veterinary Research* 12(1), 115. DOI: 10.1186/s12917-016-0728-2.

Taddei, R., Barbieri, I., Pacciarini, M.L., Fallacara, F., Belletti, G.L. *et al.* (2008) *Mycobacterium porcinum* strains isolated from bovine bulk milk: implications for *Mycobacterium avium* subsp. *paratuberculosis* detection by PCR and culture. *Veterinary Microbiology* 130(3–4), 338–347. DOI: 10.1016/j.vetmic.2008.02.007.

Teengam, P., Siangproh, W., Tuantranont, A., Vilaivan, T., Chailapakul, O. *et al.* (2017) Multiplex paper-based colorimetric DNA sensor using pyrrolidinyl peptide nucleic acid-induced AgNPs aggregation for detecting MERS-CoV, MTB, and HPV oligonucleotides. *Analytical Chemistry* 89(10), 5428–5435. DOI: 10.1021/acs.analchem.7b00255.

Timms, V.J., Mitchell, H.M. and Neilan, B.A. (2015) Optimisation of DNA extraction and validation of PCR assays to detect *Mycobacterium avium* subsp. *paratuberculosis*. *Journal of Microbiological Methods* 112, 99–103. DOI: 10.1016/j.mimet.2015.03.016.

Trangoni, M.D., Gioffré, A.K., Cerón Cucchi, M.E., Caimi, K.C., Ruybal, P. *et al.* (2015) LAMP technology: rapid identification of *Brucella* and *Mycobacterium avium* subsp. *paratuberculosis*. *Brazilian Journal of Microbiology* 46(2), 619–626. DOI: 10.1590/S1517-838246220131206.

Trangoni, M.D., Gioffre, A.K. and Cravero, S.L. (2017) A closed-tube loop-mediated isothermal amplification assay for the visual endpoint detection of *Brucella* spp. and *Mycobacterium avium* subsp. *paratuberculosis*. *Methods in Molecular Biology* 1616, 221–229.

Ushio, R., Yamamoto, M., Nakashima, K., Watanabe, H., Nagai, K. *et al.* (2016) Digital PCR assay detection of circulating *Mycobacterium tuberculosis* DNA in pulmonary tuberculosis patient plasma. *Tuberculosis* 99, 47–53. DOI: 10.1016/j.tube.2016.04.004.

Vinuesa, V., Borrás, R., Briones, M.L., Clari, M.Á., Cresencio, V. *et al.* (2018) Performance of a highly sensitive *Mycobacterium tuberculosis* complex real-time PCR assay for diagnosis of pulmonary tuberculosis in a low-prevalence setting: a prospective intervention study. *Journal of Clinical Microbiology* 56(5). DOI: 10.1128/JCM.00116-18.

Wang, H., Hou, P., Zhao, G., Yu, L., Gao, Y.W. *et al.* (2018) Development and evaluation of serotype-specific recombinase polymerase amplification combined with lateral flow dipstick assays for the diagnosis of foot-and-mouth disease virus serotype A, O and Asia1. *BMC Veterinary Research* 14(1), 359. DOI: 10.1186/s12917-018-1644-4.

Wells, S.J., Collins, M.T., Faaberg, K.S., Wees, C., Tavornpanich, S. *et al.* (2006) Evaluation of a rapid fecal PCR test for detection of *Mycobacterium avium* subsp. *paratuberculosis* in dairy cattle. *Clinical and Vaccine Immunology* 13(10), 1125–1130. DOI: 10.1128/CVI.00236-06.

White, P.L., Bretagne, S., Klingspor, L., Melchers, W.J.G., McCulloch, E. *et al.* (2010) Aspergillus PCR: one step closer to standardization. *Journal of Clinical Microbiology* 48(4), 1231–1240. DOI: 10.1128/JCM.01767-09.

Whittington, R.J., Marsh, I., Turner, M.J., McAllister, S., Choy, E. *et al.* (1998) Rapid detection of *Mycobacterium paratuberculosis* in clinical samples from ruminants and in spiked environmental samples by modified BACTEC 12B radiometric culture and direct confirmation by IS900 PCR. *Journal of Clinical Microbiology* 36(3), 701–707. DOI: 10.1128/JCM.36.3.701-707.1998.

Whittington, R.J., Whittington, A.-M., Waldron, A., Begg, D.J., de Silva, K. *et al.* (2013) Development and validation of a liquid medium (M7H9C) for routine culture of *Mycobacterium avium* subsp. *paratuberculosis* to replace modified BACTEC 12B medium. *Journal of Clinical Microbiology* 51(12), 3993–4000. DOI: 10.1128/JCM.01373-13.

Whittington, R., Donat, K., Weber, M.F., Kelton, D., Nielsen, S.S. *et al.* (2019) Control of paratuberculosis: who, why and how. A review of 48 countries. *BMC Veterinary Research* 15(1), 198. DOI: 10.1186/s12917-019-1943-4.

Wolf, R., Orsel, K., De Buck, J. and Barkema, H.W. (2015) Calves shedding *Mycobacterium avium* subspecies *paratuberculosis* are common on infected dairy farms. *Veterinary Research* 46(1), 71. DOI: 10.1186/s13567-015-0192-1.

Xia, Z., Johansson, M.L., Gao, Y., Zhang, L., Haffner, G.D. *et al.* (2018) Conventional versus real-time quantitative PCR for rare species detection. *Ecology and Evolution* 8(23), 11799–11807. DOI: 10.1002/ece3.4636.

Xing, D., Markey, K., Newland, P., Rigsby, P., Hockley, J. *et al.* (2011) EUVAC.NET collaborative study: evaluation and standardisation of serology for diagnosis of pertussis. *Journal of Immunological Methods* 372(1–2), 137–145. DOI: 10.1016/j.jim.2011.07.005.

Yamamoto, M., Ushio, R., Watanabe, H., Tachibana, T., Tanaka, M. *et al.* (2018a) Detection of *Mycobacterium tuberculosis*-derived DNA in circulating cell-free DNA from a patient with disseminated infection using digital PCR. *International Journal of Infectious Diseases* 66, 80–82. DOI: 10.1016/j.ijid.2017.11.018.

Yamamoto, T., Murai, K., Hayama, Y., Kobayashi, S., Nagata, R. *et al.* (2018b) Evaluation of fecal shedding and antibody response in dairy cattle infected with paratuberculosis using national surveillance data in Japan. *Preventive Veterinary Medicine* 149, 38–46. DOI: 10.1016/j.prevetmed.2017.10.009.

Yang, J., Han, X., Liu, A., Bai, X., Xu, C. *et al.* (2017) Use of digital droplet PCR to detect *Mycobacterium tuberculosis* DNA in whole blood-derived DNA samples from patients with pulmonary and extrapulmonary tuberculosis. *Frontiers in Cellular and Infection Microbiology* 7, 369. DOI: 10.3389/fcimb.2017.00369.

Yashiki, N., Yamazaki, Y., Subangkit, M., Okabayashi, T., Yamazaki, W. *et al.* (2019) Development of a LAMP assay for rapid and sensitive detection and differentiation of *Mycobacterium avium* subsp. *avium* and subsp. *hominissuis*. *Letters in Applied Microbiology* 65, 155–160. DOI: 10.1111/lam.13188.

Yokoyama, E., Kishida, K., Uchimura, M. and Ichinohe, S. (2006) Comparison between agarose gel electrophoresis and capillary electrophoresis for variable numbers of tandem repeat typing of *Mycobacterium tuberculosis*. *Journal of Microbiological Methods* 65(3), 425–431. DOI: 10.1016/j.mimet.2005.08.014.

Zarei-Kordshouli, F., Geramizadeh, B. and Khodakaram-Tafti, A. (2019) Prevalence of *Mycobacterium avium* subspecies *paratuberculosis* IS *900* DNA in biopsy tissues from patients with Crohn's disease: histopathological and molecular comparison with Johne's disease in Fars province of Iran. *BMC Infectious Diseases* 19(1), 23. DOI: 10.1186/s12879-018-3619-2.

Zemtsova, G.E., Montgomery, M. and Levin, M.L. (2015) Relative sensitivity of conventional and real-time PCR assays for detection of SFG Rickettsia in blood and tissue samples from laboratory animals. *PLoS ONE* 10(1), e0116658. DOI: 10.1371/journal.pone.0116658.

Zhang, M.Z. and Zhang, S. (2011) An efficient DNA extraction method for polymerase chain reaction-based detection of *Mycobacterium avium* subspecies *paratuberculosis* in bovine fecal samples. *Journal of Veterinary Diagnostic Investigation* 23(1), 41–48. DOI: 10.1177/104063871102300106.

Zhao, G., Hou, P., Huan, Y., He, C., Wang, H. *et al.* (2018a) Development of a recombinase polymerase amplification combined with a lateral flow dipstick assay for rapid detection of the *Mycoplasma bovis*. *BMC Veterinary Research* 14(1), 412. DOI: 10.1186/s12917-018-1703-x.

Zhao, G., Wang, H., Hou, P., He, C. and He, H. (2018b) Rapid visual detection of *Mycobacterium avium* subsp. *paratuberculosis* by recombinase polymerase amplification combined with a lateral flow dipstick. *Journal of Veterinary Science* 19(2), 242–250. DOI: 10.4142/jvs.2018.19.2.242.
Zhao, G., He, H. and Wang, H. (2019) Use of a recombinase polymerase amplification commercial kit for rapid visual detection of *Pasteurella multocida*. *BMC Veterinary Research* 15(1), 154. DOI: 10.1186/s12917-019-1889-6.

20 Immune-Based Diagnosis of Paratuberculosis

Søren Saxmose Nielsen*
University of Copenhagen, Copenhagen, Denmark

20.1 Introduction

Immune-based diagnosis is based on the detection of processes related to the immune response against *Mycobacterium avium* subsp. *paratuberculosis* (*MAP*). This poses a challenge because there are many aspects of the immune response that we do not fully understand (see Chapter 17, this volume). However, the immune response also provides an opportunity to observe how the individual responds to infection, which is crucial if we are to understand the process. The sequence of events that may occur in an infected individual, and the temporal aspects of immune responses related to these, are inadequately characterized and understood, adversely affecting the interpretation and use of diagnostic test results. Another challenge in the 'diagnosis' of *MAP* infections is the difference in the concepts and terminology used in relation to management of the infection. This chapter is based on existing knowledge about the immune response and how this knowledge can be used for decision making in relation to the management of *MAP* infection. The utility of these tests can be optimized if one considers the context in which the diagnostic test is used. This chapter discusses immune-based diagnostics for various testing and decision making purposes and outlines the diagnostic process (Table 20.1).

20.2 Establishment of a Diagnosis

Many different decision makers rely on diagnostic test results, but often the decision makers are farmers, who wish to make inferences about individual animals or their herd/flock. Veterinary authorities can also use diagnostic tests to classify individual animals, herds/flocks or even entire regions as infected or contaminated. The reason for this classification is to enable decision making in relation to a specific goal. The goals may differ from one decision maker to another, but they are usually directed by the purpose of the efforts: why does the decision maker wish to establish a diagnosis? Ultimately, a test result should lead to an action. Actions might include: (i) doing nothing, e.g. because the tested subject is considered to be non-infected or unlikely to contribute to transmission; (ii) isolating the subject from susceptible subjects, e.g. to stop transmission; (iii) culling to remove the infected or infectious subject fully from the population; (iv) removing a subject from the food chain if the subject is considered a public health hazard; (v) treating the subject; (vi) re-testing if the classification of the subject is not clear. Other actions can also be defined, for example for research purposes, or because herds or flocks with a superior infection status may attract a financial

*saxmose@sund.ku.dk

© CAB International 2020. *Paratuberculosis: Organism, Disease, Control, 2nd Edition* (eds M.A. Behr *et al.*)

Table 20.1. Overview of aspects relevant in a diagnostic test strategy.

Aspect	Relevance
Why	Purpose of testing and efforts
Who	Decision makers Stakeholders Advisors
What	Adverse effects/outcomes (pathogenesis) Analytes (and their concentration)
How	Sample matrix Sampling Sampling frequency
Interpretation	Effect of bias
	Utility • Interpretation • Application • Uncertainty • Decision making

premium for their products. The purpose and the associated actions usually define what the test should be able to detect and how it should be interpreted. A purpose is established to reduce adverse outcomes through these actions.

MAP infections have a long and variable incubation period. During this period, a number of outcomes can occur, such as: (i) establishment of the infection; (ii) faecal shedding of *MAP* in sufficient quantities to result in transmission to susceptible herd-mates; (iii) bacteraemia, which can lead to the shedding of *MAP* in milk or the transfer of *MAP* to a fetus *in utero*; and (iv) decreased milk production, weight loss, diarrhoea or death. These outcomes are of varying interest to different decision makers (politicians, farmers, etc.) and in various settings (no reported occurrence, low prevalence or high prevalence). Consequently, the reason for using a diagnostic test will vary as well. The following are examples of purposes for testing: (i) certification, i.e. to certify a herd, animal or region free of *MAP* infection; (ii) to establish the prevalence, i.e. the proportion of *MAP*-infected animals in a target population to determine whether intervention should be instated; (iii) reducing transmission to prevent future *MAP* infections; (iv) to increase production parameters (e.g. milk and beef); (v) to increase animal welfare, e.g. to avoid animals suffering from diarrhoea, etc. These distinct purposes will serve as the background for further discussion of immune-based diagnosis and the actions that should follow the diagnosis. In addition to these different aims of testing, the target conditions can range from: (i) non-infected (for certification); (ii) *MAP*-infected (for prevalence estimation); (iii) *MAP*-infectious (for reduced transmission); and (iv) *MAP*-affected (for increased production parameters and animal welfare). The term '*MAP*-affected' is preferentially used over 'clinical disease' because of the variable definition and use of the term 'clinical' in the literature and among decision makers.

Once the purpose and target condition are established, we must identify the analytes on which the diagnostic test is based. To identify relevant analytes, we require an insight into the pathogenesis of infection and disease. The commonly detected analytes are *MAP*, *MAP* DNA or *MAP*-specific antibodies, with the latter being of primary relevance in this chapter. These are used, because we have sufficient information about their significance in the pathogenesis. Other analytes include indicators of cell-mediated immunity such as interferon-gamma (IFN-γ). However, the relation between cell-mediated immune responses and the outcomes listed above is still insufficiently described to be used in decision making. In order to detect the analytes, they must of course be present, but more importantly, we must be able to infer what their presence (or absence) means in relation to the purpose and the target condition. This often presents a challenge due to the many potential adverse outcomes, target conditions and associated purposes. Therefore, we need to be precise in the terminology used when characterizing how well the detection of the analytes relates to the purpose and target condition.

20.3 Test Characteristics and Biases

A diagnostic test can be characterized by its diagnostic sensitivity (Se) and diagnostic specificity (Sp). The standard definitions are that SE describes the probability that a test can detect an animal with a given target condition, and

Sp describes the probability that a test is negative given that the target condition is absent. However, there are two important limitations in their use in relation to the diagnosis of *MAP* infections.

First, target conditions vary with the purpose of testing, and therefore the purpose should be specified in any test validation to ensure that the test evaluation is fit for purpose (OIE, 2003). Second, the test should be evaluated in a population similar to the one in which it is used in order to avoid a substantial risk of spectrum bias (Ransohoff and Feinstein, 1978). Spectrum bias is a particular concern in diagnostic test evaluations for chronic diseases, where case-control designs have been used. A thorough description of the population in which a given test has been evaluated, along with the selection criteria for the cases and controls, are vital to the interpretation of any test results used subsequently. Lack of such a description can easily lead to scenarios where estimates presented as diagnostic sensitivities and specificities have an interpretation that is very different to the standard interpretation (Table 20.2). If the reader is not cautious, then the estimates can be misinterpreted. For example, Clark *et al.* (2008) stated that, 'direct fecal PCR was 70.2% sensitive and 85.3% specific when using culture as the gold standard', i.e. the target condition considered by Clark *et al.* (2008) may have been 'infectious with *MAP*' or 'shedding of *MAP*'. However, this was interpreted by Magombedze *et al.* (2017) as sensitivity to detect *MAP* infection, which is a very different target condition. Therefore, the conclusions of Magombedze *et al.* (2017) must be reconsidered. Even the same target condition can be presented with different case definitions and consequently different meanings within the same studies (Table 20.2). Further caution is therefore warranted when estimates are interpreted.

Se and Sp estimates are generally inconsistently reported, with highly variable target conditions. A summary of Se and Sp estimates for ante-mortem tests, stratified by target condition, was reported by Nielsen and Toft (2008). Because these estimates related to the purpose of testing and relied on the occurrence of immune responses, these aspects will be discussed further.

20.4 Immunity and Immune-Based Diagnostic Tests

20.4.1 Immunological responses

Specific immune responses to *MAP* infections can be divided into proinflammatory and anti-inflammatory reactions (see Chapter 17, this volume). The proinflammatory process is characterized by the production of IFN-γ and other cytokines involved in the cell-mediated responses, which attempt to control the infection by destroying *MAP* in activated macrophages (Coussens, 2001). Shedding of *MAP* occurs sporadically and IFN-γ stimulates the production of immunoglobulin G2 (IgG2). The proinflammatory reactions may gradually be replaced by anti-inflammatory processes, characterized by the occurrence of IgG1 antibodies. Therefore, IFN-γ and IgG2 are mainly associated with proinflammatory responses and IgG1 with anti-inflammatory responses. Both responses may occur simultaneously due to a gradual shift between the two responses. IgG2 generally occurs at lower concentrations than IgG1 (Koets *et al.*, 2001). The immune-based diagnostics are primarily based on the detection of IFN-γ, IgG1 and IgG2 antibodies using antigens derived from *MAP*.

The speed of progression between disease stages is poorly characterized. Some infected cows produce antibodies several years prior to continuous shedding of detectable amounts of *MAP*, while antibodies may not be detectable in other animals during the early stages of infection, when *MAP* shedding is minimal. This phenomenon is illustrated in Fig. 20.1. It should be noted that this example cannot be considered a definitive understanding of the antibody response, because the appearance of antibodies is yet to be fully elucidated.

20.4.2 Antigens and antibodies

The choice of antigen for an immune-based test has a major impact on the test result. Antibody reactivity is not well characterized for the majority of *MAP*-specific antigens available. Koets *et al.* (2001) characterized the ability of IgG1 and IgG2 to react with four antigens: two cytosolic antigens (heat-shock proteins), a cell wall component (lipoarabinomannan) and a

Table 20.2. Examples of reported sensitivities and specificities, where the intended target condition and the given meaning differ considerably.

Target condition	Case definition	Test under evaluation	Reported Se and Sp	Reference	Meaning of Reported Se
MAP-infected	Faecal culture positive and ELISA negative	MAP0210c antibody ELISA	Se = 33.3%[a] Sp = 93.3%[a]	Li *et al*. (2017)	Probability of testing MAP0210c positive in an animal with detectable faecal shedding of *MAP* and no antibodies according to a commercial ELISA
MAP-infected	Faecal culture and ELISA positive	MAP0210c antibody ELISA	Se = 26.7%[a] Sp = 93.3%[a]	Li *et al*. (2017)	Probability of testing MAP0210c positive in an animal with detectable faecal shedding of *MAP* and *MAP*-specific antibodies according to a commercial ELISA
Not specified; probably *MAP*-infectious	Faecal culture positive	Faecal PCR	Se = 70.2% Sp = 85.5%	Clark *et al*. (2008)	Probability of testing faecal PCR positive among animals shedding detectable amounts of *MAP*

[a]The reported diagnostic sensitivities (Se) and specificities (Sp) are for an antibody enzyme-linked immunosorbent assay (ELISA) based on the MAP0210c antigen (Li *et al*., 2017), while estimates for more antigens were presented in the paper.

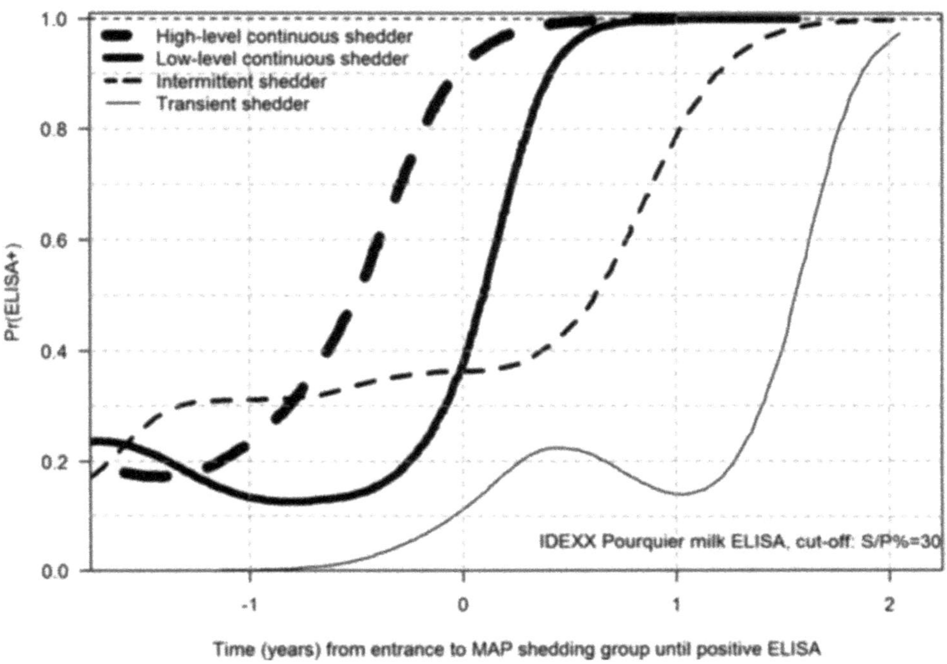

Fig. 20.1. Probability of testing positive using a commercial antibody enzyme-linked immunosorbent assay (ELISA) at various time points relative to the start of *MAP* shedding.

MAP-derived protein purified derivative (PPD-P) known as johnin, which is the *MAP* tuberculin. IgG1 reactivity to all four antigens was elevated in cows classified as *MAP* shedders when compared with non-infected animals. Only PPD-P was associated with high levels of IgG2 in 'non-shedders', '*MAP* shedders' and animals with 'clinical paratuberculosis'. PPD-P was also associated with very high levels of IgG1 in *MAP* affected animals. These results are complex and underline the importance of knowing the central components of a given immune-based test.

20.4.3 Types of tests and sample material

One of the most widely used immune-based tests is the indirect antibody enzyme-linked immunosorbent assay (ELISA), which detects antibodies in serum and milk samples. There is a consistent level of accuracy across different ELISAs, irrespective of whether they are used with milk or serum samples (Sweeney *et al.*, 1995; Nielsen *et al.*, 2002b). However, differences in the probability of testing positive during lactation exist between the two sample types. A serum ELISA is more likely to yield a positive reaction towards the end of lactation, whereas a milk ELISA is more likely to be positive at the beginning of lactation (Nielsen *et al.*, 2002a). These differences can be used to optimize sampling but will not be discussed further in this chapter.

The IFN-γ response after stimulation of a whole-blood sample with *MAP*-specific antigens (e.g. PPD-P) has also been used in an ELISA-formatted test. Detection of IFN-γ depends on the processes for successful antigen presentation and presence of viable T cells, which are stimulated by the antigen to secrete IFN-γ. Antigen presentation is reduced within 8 h of sampling. However, Jungersen *et al.* (2005) developed a method to rescue the costimulatory signals required for antigen presentation and T-cell reactivity, so sample processing can occur up to 24 h after sampling.

20.4.4 Advantages and disadvantages of immune-based tests

The main advantages of immune-based diagnostic tests include that they are relatively inexpensive, can easily be adapted to high-throughput testing and the results can be available within 1–2 weeks in routine settings. Since antibody ELISA testing is cheaper than most agent-detecting tests, farmers are more likely to test their animals frequently, if frequent testing is required (see Section 20.6.2). The advantage of sampling milk over serum is that the milk has often already been collected for other purposes in milk-recording schemes and is easy to obtain from the animal. Since immune-based tests detect host responses to a *MAP* infection, they should be able to determine that an infection has already occurred and may have persisted. A prerequisite is that the performance of the test has been characterized at the various stages of infection. Repeated testing makes determination of the stage of infection easier.

A major disadvantage of immune-based tests is that they do not provide a direct measure of a *MAP* infection, infectiousness or of being affected by a *MAP* infection (see Section 20.2). Therefore, there is a need to correlate the immune response profiles to relevant stages of infection. False-positive reactions occur because in some cases the positive test results do not relate to the purpose of testing. For example, antibodies can occur prior to shedding of *MAP*. If the testing aims to determine whether the animal is infectious prior to this shedding, the detection of antibodies will be considered false positive, whereas the animal is actually infected.

False-positive reactions may also occur due to cross-reacting antibodies or laboratory errors. It should be noted that vaccinated animals can give a positive reaction to immune-based tests for a considerable period of time (Muskens *et al.*, 2002) without necessarily having a *MAP* infection. These disadvantages make communication of test results more complex. None the less, with appropriate interpretation, immunological tests may be more reliable than microbiological tests, depending on the purpose of testing.

20.5 Cell-Mediated Tests

There are two tests available for detecting a cell-mediated immune (CMI) response to *MAP* infection: the intradermal tuberculin test (or skin test), using johnin or avian PPD, and the IFN-γ test. The skin test was used more frequently in the past, but its current use is limited (Kalis *et al.*, 2003). Since its introduction, the *in vitro* IFN-γ assay has replaced the skin test in current studies using the CMI response in the diagnosis of *MAP* infection.

Se and Sp estimates for IFN-γ tests have not been reported for the detection of *MAP*-infected animals. It has been suggested that some animals with CMI reactions may be able to control and eradicate *MAP* (Mikkelsen *et al.*, 2009). If this is the case, a test-positive reaction may be an indicator of a past infection, and the test can be expected to have a specificity of less than 100%, regardless of the antigens used. For animals deemed to be *MAP* infectious, Se estimates for IFN-γ tests have been reported between 0.13 and 0.85, with Sp estimates between 0.88 and 0.94 (Nielsen and Toft, 2008). There are limited data available to support the use of CMI diagnostics for purposes other than determining whether animals have been exposed to *MAP*. If IFN-γ tests are to be used, animals should be >15 months of age to avoid false-positive reactions caused by innate IFN-γ production by natural killer (NK) cells (Jungersen *et al.*, 2002).

20.6 Antibody ELISAs

Multiple antibody ELISAs have been evaluated, and Se and Sp estimates vary greatly within and between tests (Nielsen and Toft, 2008). It is therefore not possible to provide point estimates characterizing these tests. They should preferably be evaluated prior to their specific use in a particular population.

20.6.1 Detection of *MAP*-infected animals

Overall, the Se of antibody ELISA for detecting *MAP*-infected animals is low (~5–30%), but it increases with increasing age (Nielsen and Toft, 2006). Sp estimates are generally above 95% for commercial ELISAs.

20.6.2 Detection of *MAP*-infectious animals

The ability of antibody ELISAs to detect *MAP*-infectious animals depends on the test frequency (Fig. 20.1), the manufacturer and the cut-off chosen to deem the ordinal ELISA response 'positive' or 'negative'. For example, when assuming that only animals with high-level continuous shedding can infect a susceptible animal, one can test such animals on different schedules to reduce the risk of transmission. As shown in Fig. 20.1, using the Pourquier milk ELISA (IDEXX, Montpellier, France) to test such animals daily, one can expect to detect 91% of infectious animals (time = 0 in Fig. 20.1). However, if the test is performed once per year, and this test date is 1 year prior to the start of the high-level continuous shedding, the Se would be 20% (Fig. 20.1, time = −1 year). In a third scenario, if such infectious animals were tested four times per year, 70–91% would be detected (Fig. 20.1, time interval −0.25 to 0 years). This example illustrates the major impact of test frequency on the Se. The cumulative Se from repeated testing would be higher if at least one positive test result from the repeated test dates were considered sufficient to deem the animal positive.

Sp for the target condition 'infectious' can be interpreted as the probability that an animal will test negative given that it is not shedding *MAP* in doses deemed likely to cause infection in a susceptible animal. It should be noted that a 'non-infectious' animal can be infected with *MAP*. Therefore, animals with *MAP*-specific antibodies with negative results from a sensitive microbiological assay can be considered *MAP* infected but not *MAP*-infectious.

Some decision makers would prefer to confirm whether test-positive animals are *MAP*-infected or *MAP*-infectious, but it is not possible to exclude infection from an ELISA-positive animal by use of ante-mortem tests. For example, for cows testing positive using the Pourquier milk antibody ELISA, only 50–60% would have culture-positive faecal samples within a year following a positive antibody test (Fig. 20.2, time interval 0-1 year). Some of the cows were therefore not *MAP*-infectious at the time that the antibodies were detected. However, the majority of cows would shed detectable amounts of *MAP* within 3.5 years of testing positive using the ELISA (Fig. 20.2, time interval 0 to 3.5 years). Therefore, cows that were antibody-positive but negative in culture would be considered *MAP*-infected but non-infectious. In practice, a microbiological test for infectious animals should only be used to determine whether an animal is shedding *MAP* at the time of testing. As illustrated by this example, a microbiological test for infectious animals should not be used to rule out *MAP* infection in an antibody-positive animal.

20.6.3 Detection of *MAP*-affected animals

ELISAs tend to perform well in confirming the presence of *MAP* infection in animals with a high pre-test suspicion of disease (e.g. persistent diarrhoea). For example, Weber *et al.* (2009) reported that the Pourquier serum ELISA had a Se of 0.83–0.92, at a Sp of 0.998. However, not all cows affected by *MAP* infection will experience diarrhoea, but they may experience reduced productivity. A combination of milk production data and antibody results can therefore be used to increase the diagnostic accuracy (Wang *et al.*, 2006). Cows with fluctuating antibodies (i.e. fluctuating between test-positive and test-negative) may have increased milk production, whereas those with last-positive or repeated-positive test results may experience decreased milk production (Nielsen *et al.*, 2009). This decline may start 300 days prior to the occurrence of antibodies in some cows. To determine the *MAP* status of an animal, it may be necessary to evaluate both production data and ELISA results.

20.6.4 Use of ELISA results on an ordinal scale

Results from antibody ELISAs are measured on an ordinal scale but are often dichotomized

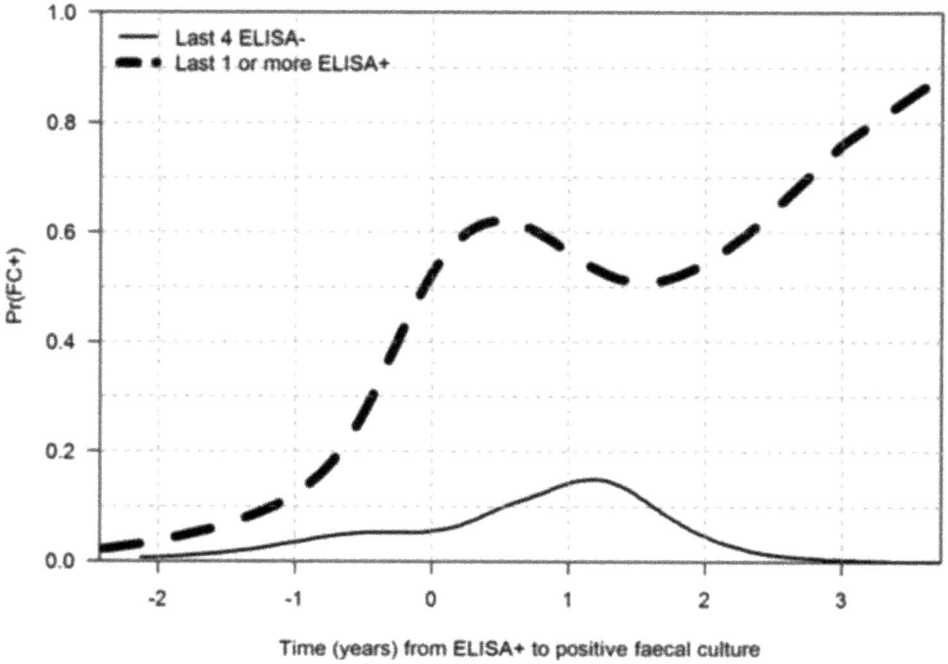

Fig. 20.2. Probability of testing positive in a faecal culture test relative to the time of testing positive in the commercial Pourquier milk enzyme-linked immunosorbent assay (ELISA) (using a cut-off sample-to-positive ratio of 30%).

to antibody-positive or antibody-negative results, making communication from the laboratory more straightforward. However, this simplification comes at the cost of a decrease in information, i.e. whether the ELISA values correlate with the probability of infection. Toft *et al.* (2005) estimated that cows with an optical density (OD) value of 0.3 had a 4% probability of shedding *MAP*, whereas if the cows had an OD value of 1.0, the probability of *MAP* shedding was 80% (Fig. 20.3). Therefore, there may be considerable advantages in using values on the ordinal scale, whether it is OD values or sample-to-positive ratios (S/P values) (Collins, 2002).

20.6.5 Predictive values of antibody ELISA

Single-antibody ELISA results may be of limited value (with the exception of prevalence estimates and herd certification) because animals with various antibody profiles are in different stages of *MAP* infection. Instead, results from frequent testing can better reveal the stage of infection as trends in results can be tracked over time. However, there is limited information on test evaluation in the context of repeated testing. Diagnostic test evaluations based on repeated tests are few, but two independent studies produced very similar age-specific test-sensitivity estimates (Nielsen *et al.*, 2013; Meyer *et al.*, 2018), with increases in the relative sensitivity from 0.3 at 2 years of age to 0.7 at 6 years of age in dairy cattle. However, the repeated test results should rather be used based on their profile, e.g. two, three or more repeated positive tests would with increasing likelihood increase the probability that the cow is infected (Meyer *et al.*, 2018). Current information suggests that the predictive properties of antibody ELISA are as shown in Table 20.3.

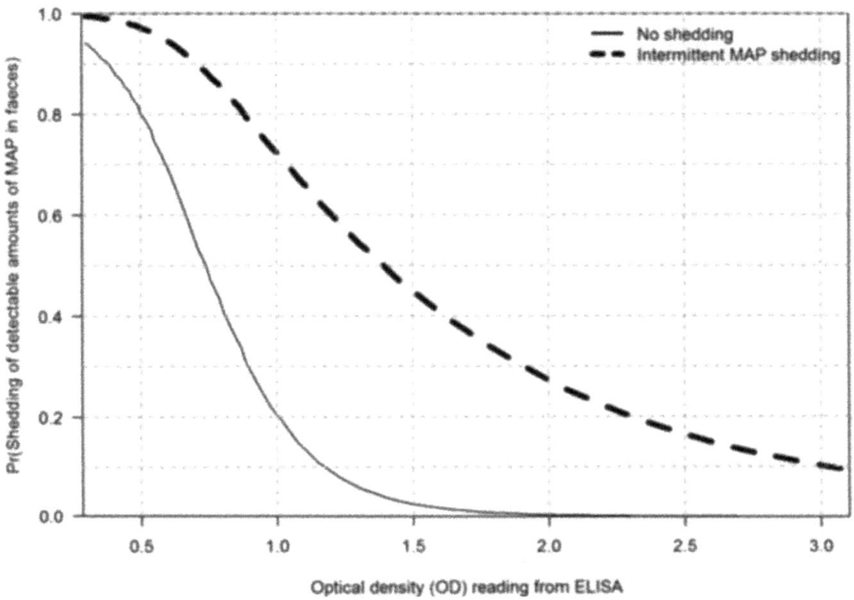

Fig. 20.3. Probability of detectable shedding of *Mycobacterium avium* subsp. *paratuberculosis* (*MAP*) in faeces. Values below the solid line show the probability of no *MAP* shedding at a given optical density (OD) value. Values between the solid and dashed line are probabilities of 'intermittent' *MAP* shedding, and values above the dashed line are probabilities of continuous *MAP* shedding. (Modified from Toft *et al*., 2005.)

Table 20.3. Prognostic properties of the antibody enzyme-linked immunosorbent assay (ELISA) for predicting *Mycobacterium avium* subsp. *paratuberculosis* (*MAP*)-affected, *MAP*-infectious, *MAP*-infected and non-infected adult cows.

	Predicted condition			
ELISA results	Non-infected	*MAP*-infected	*MAP*-infectious	*MAP*-affected
Repeated negative	Dependent on test history of herd	Possible in infected population	Rare	Rare
Fluctuating but last sample negative	Rare	Likely	Possible but likely to be low	Rare; a potential future event
Last sample positive	Rare	Very likely	Likely	Likely
Repeated positive	Rare	Very likely	Very likely	Likely; may be in near future

20.7 Decision Making

Interpretation of diagnostic test results is usually part of the decision making process related to specific purposes such as those specified in Section 20.2. It is therefore difficult to have a single recommendation for each test result. The following are suggestions related to different purposes. These suggestions are not the only decision options.

20.7.1 Certification

Animal herds cannot be certified free of *MAP* infection if insufficient numbers of animals are tested or if the animals tested are too young. The number of animals that would need to be tested depends on the Se and Sp of the specific antibody ELISA used, in addition to the herd demography. For example, Sergeant *et al.* (2008) suggested that a minimum of 80 adult animals should be tested to suggest that a herd has a low (but not zero) prevalence. An approach to correct for false-positive reactions caused by antibodies not specific to *MAP* was also suggested. Follow-up testing of ELISA-positive animals using a microbiological test cannot be recommended due to the frequency of ELISA-positive animals not shedding *MAP* (see Section 20.6.2).

The probability that a herd can be certified as free of *MAP*-infected animals (P_{Free}) depends on the number of animals tested. P_{Free} indicates that a herd has a lower prevalence than a predetermined prevalence, usually referred to as the design prevalence. Herds with a sufficient number of animals tested and with a high P_{Free} can be classified as 'potentially free of *MAP* infection'. A requirement for certifying herds 'free of *MAP* infection' is that the herd is closed, or the animals purchased originate from 'free' herds. Irrespective that a herd is classified as 'potentially free', herds and animals are difficult to verify as free of *MAP* infection. Therefore, different levels of 'freedom' can be used, e.g. using P_{Free} directly, or categorizing into different levels. The higher P_{Free} and the longer this value has been achieved, the higher is the probability that a herd is truly non-infected. Herds that are not 'free of *MAP* infection' may be classified as *MAP*-infected, and a control plan can be established.

A special case of 'certification' could be to determine whether an action plan has been effective. An action plan to reduce transmission of *MAP* is pivotal in controlling *MAP* infections in general. Testing using a cell-mediated diagnostic test such as IFN-γ in animals >15 months of age could be used to determine whether or not an action plan is working. If there are test-positive animals born after the action plan has been established, and the test is considered 100% specific, then the action plan should be revised because test-positive reactions would suggest that the animals have been exposed to *MAP*.

20.7.2 Reduction in transmission

An action plan to reduce transmission could include testing using antibody ELISA. Animals that are test-positive have a high risk of either being or becoming *MAP* infectious (Fig. 20.2). Therefore, these animals should either be culled or measures to avoid transmission of *MAP* to susceptible animals should be established. Animals that are repeatedly negative in antibody ELISA tests generally have a low probability of shedding *MAP* (Fig. 20.2). However, there is a chance that these animals might still shed *MAP*, particularly if the ELISA used has a low Se for detection of *MAP*-infectious animals. An increased test frequency can increase the overall probability of detecting infectious animals prior to the start of high bacterial shedding. According to Kudahl *et al.* (2008), a 'sufficient' proportion of the infectious animals can be detected in this way, but it is unlikely that all infectious animals will be detected. Test-negative animals should be tested repeatedly, because they might become *MAP*-infectious at some stage in their life. A more detailed description of a possible approach is described by Nielsen (2009).

20.7.3 Increase in production parameters and animal welfare

Production parameters can be evaluated on either a herd or animal level. Consequently, productivity can be related to both the individual and the herd. Decisions relating to the individual would usually involve culling prior to the animal becoming affected by *MAP* infection, whereas decisions relating to the herd could also include decisions leading to a reduction in transmission (see above). At cow level, single-antibody ELISA results may be insufficient to determine whether a cow is *MAP*-affected. The test information must be combined with production data or clinical observations of the cow. A positive ELISA result in combination with diarrhoea or a decline in milk production should lead to immediate culling to avoid further production losses. Animals with production loss or diarrhoea potentially related to *MAP* infection can be confirmed using ELISA.

Box 20.1. Summary recommendations.

An immune-based diagnostic test should be evaluated for its ability to detect *MAP*-infected and *MAP*-infectious animals in the population in which it will be used.

The purpose of testing should be clear, and the test results should be interpreted in relation to the purpose. In particular, special consideration should be given to the definition and communication of results, which could be 'false positive' for one purpose and 'true positive' for other purposes.

Cell-mediated immunodiagnostics may be used to determine whether a population has been exposed to *MAP* but cannot be used to take actions on an individual animal.

Humoral immunodiagnostics can be used to establish relative prevalence estimates, which can be used to compare estimates obtained previously with the same test.

Antibody ELISAs can only be used for certification if a sufficient number of animals are tested. A test-positive animal should not be confirmed by an agent-detecting test in a certification scheme. False-positive reactions can be excluded using mathematical formulas or retesting with other immune-based tests.

Antibody ELISA can be used for the risk-based management of *MAP*-infectious animals, since it will generally detect *MAP*-infected animals before they become *MAP*-infectious. However, frequent testing is a prerequisite if this approach is used.

20.8 Recommendations and Concluding Thoughts

Summary recommendations based on currently available test modalities are presented in Box 20.1. Priorities for future research include the characterization of immune responses in prospective studies spanning the lifetime of the animals, and factors that are involved in the variation in immune responses. In addition, the performance of diagnostic tests used in repeated testing should be characterized, and the use of CMI-detecting tests should be evaluated with respect to the purpose of testing. Current diagnostic tests have potential if they are used – and the test results interpreted – appropriately (Box 20.1). However, the use of tests that detect humoral immune responses could have greater utility with repeated testing, whereas CMI tests might be useful for early detection of *MAP*-infected or *MAP*-exposed animals.

References

Clark, D.L., Koziczkowski, J.J., Radcliff, R.P., Carlson, R.A. and Ellingson, J.L.E. (2008) Detection of *Mycobacterium avium* subspecies *paratuberculosis*: comparing fecal culture versus serum enzyme-linked immunosorbent assay and direct fecal polymerase chain reaction. *Journal of Dairy Science* 91(7), 2620–2627. DOI: 10.3168/jds.2007-0902.

Collins, M.T. (2002) Interpretation of a commercial bovine paratuberculosis enzyme-linked immunosorbent assay by using likelihood ratios. *Clinical and Diagnostic Laboratory Immunology* 9(6), 1367–1371. DOI: 10.1128/cdli.9.6.1367-1371.2002.

Coussens, P.M. (2001) *Mycobacterium paratuberculosis* and the bovine immune system. *Animal Health Research Reviews* 2(2), 141–162. DOI: 10.1079/AHRR200134.

Jungersen, G., Huda, A., Hansen, J.J. and Lind, P. (2002) Interpretation of the gamma interferon test for diagnosis of subclinical paratuberculosis in cattle. *Clinical and Diagnostic Laboratory Immunology* 9(2), 453–460. DOI: 10.1128/cdli.9.2.453-460.2002.

Jungersen, G., Grell, S.N., Clemensen, A., Roust, T. and Howard, C.J. (2005) Interleukin-12 potentiation of the interferon-gamma test rescues day-old blood samples for the diagnosis of paratuberculosis. In: Manning, E.J.B. and Nielsen, S.S. (eds) *Proceedings of the 8th International Colloquium on Paratuberculosis*. International Association for Paratuberculosis, Madison, Wisconsin, pp. 501–506.

Kalis, C.H.J., Collins, M.T., Hesselink, J.W. and Barkema, H.W. (2003) Specificity of two tests for the early diagnosis of bovine paratuberculosis based on cell-mediated immunity: the Johnin skin test and the gamma interferon assay. *Veterinary Microbiology* 97(1–2), 73–86. DOI: 10.1016/j.vetmic.2003.07.003.

Koets, A.P., Rutten, V.P., de Boer, M., Bakker, D., Valentin-Weigand, P. *et al.* (2001) Differential changes in heat shock protein-, lipoarabinomannan-, and purified protein derivative-specific immunoglobulin G1 and G2 isotype responses during bovine *Mycobacterium avium* subsp. *paratuberculosis* infection. *Infection and Immunity* 69(3), 1492–1498. DOI: 10.1128/IAI.69.3.1492-1498.2001.

Kudahl, A.B., Nielsen, S.S. and Østergaard, S. (2008) Economy, efficacy, and feasibility of a risk-based control program against paratuberculosis. *Journal of Dairy Science* 91(12), 4599–4609. DOI: 10.3168/jds.2008-1257.

Li, L., Bannantine, J.P., Campo, J.J., Randall, A., Grohn, Y.T. *et al.* (2017) Identification of sero-reactive antigens for the early diagnosis of Johne's disease in cattle. *PLoS ONE* 12(9), e0184373. DOI: 10.1371/journal.pone.0184373.

Magombedze, G., Shiri, T., Eda, S. and Stabel, J.R. (2017) Inferring biomarkers for *Mycobacterium avium* subsp. *paratuberculosis* infection and disease progression in cattle using experimental data. *Scientific Reports* 7, 44765. DOI: 10.1038/srep44765.

Meyer, A., Bond, K., Van Winden, S., Green, M. and Guitian, J. (2018) A probabilistic approach to the interpretation of milk antibody results for diagnosis of Johne's disease in dairy cattle. *Preventive Veterinary Medicine* 150, 30–37. DOI: 10.1016/j.prevetmed.2017.11.016.

Mikkelsen, H., Jungersen, G. and Nielsen, S.S. (2009) Association between milk antibody and interferon-gamma responses in cattle from *Mycobacterium avium* subsp. *paratuberculosis* infected herds. *Veterinary Immunology and Immunopathology* 127(3–4), 235–241. DOI: 10.1016/j.vetimm.2008.10.315.

Muskens, J., van Zijderveld, F., Eger, A. and Bakker, D. (2002) Evaluation of the long-term immune response in cattle after vaccination against paratuberculosis in two Dutch dairy herds. *Veterinary Microbiology* 86(3), 269–278. DOI: 10.1016/S0378-1135(02)00006-8.

Nielsen, S.S. (2009) Use of diagnostics for risk-based control of paratuberculosis in dairy herds. *In Practice* 31(4), 150–154. DOI: 10.1136/inpract.31.4.150.

Nielsen, S.S. and Toft, N. (2006) Age-specific characteristics of ELISA and fecal culture for purpose-specific testing for paratuberculosis. *Journal of Dairy Science* 89(2), 569–579. DOI: 10.3168/jds.S0022-0302(06)72120-8.

Nielsen, S.S. and Toft, N. (2008) Ante mortem diagnosis of paratuberculosis: a review of accuracies of ELISA, interferon-gamma assay and faecal culture techniques. *Veterinary Microbiology* 129(3-4), 217–235. DOI: 10.1016/j.vetmic.2007.12.011.

Nielsen, S.S., Enevoldsen, C. and Gröhn, Y.T. (2002a) The *Mycobacterium avium* subsp. *paratuberculosis* ELISA response by parity and stage of lactation. *Preventive Veterinary Medicine* 54(1), 1–10. DOI: 10.1016/S0167-5877(02)00008-9.

Nielsen, S.S., Grønbaek, C., Agger, J.F. and Houe, H. (2002b) Maximum-likelihood estimation of sensitivity and specificity of ELISAs and faecal culture for diagnosis of paratuberculosis. *Preventive Veterinary Medicine* 53(3), 191–204. DOI: 10.1016/S0167-5877(01)00280-X.

Nielsen, S.S., Krogh, M.A. and Enevoldsen, C. (2009) Time to the occurrence of a decline in milk production in cows with various paratuberculosis antibody profiles. *Journal of Dairy Science* 92(1), 149–155. DOI: 10.3168/jds.2008-1488.

Nielsen, S.S., Toft, N. and Okura, H. (2013) Dynamics of specific anti-*Mycobacterium avium* subsp. *paratuberculosis* antibody response through age. *PLoS ONE* 8(4), e63009. DOI: 10.1371/journal.pone.0063009.

OIE (2003) Validation and certification of diagnostic assays. *Resolution no. XXIX by the International Committee to the OIE. World Organisation for Animal Health (OIE), Paris, France.*

Ransohoff, D.F. and Feinstein, A.R. (1978) Problems of spectrum and bias in evaluating the efficacy of diagnostic tests. *New England Journal of Medicine* 299(17), 926–930. DOI: 10.1056/NEJM197810262991705.

Sergeant, E.S.G., Nielsen, S.S. and Toft, N. (2008) Evaluation of test-strategies for estimating probability of low prevalence of paratuberculosis in Danish dairy herds. *Preventive Veterinary Medicine* 85(1–2), 92–106. DOI: 10.1016/j.prevetmed.2008.01.005.

Sweeney, R.W., Whitlock, R.H., Buckley, C.L. and Spencer, P.A. (1995) Evaluation of a commercial enzyme-linked immunosorbent assay for the diagnosis of paratuberculosis in dairy cattle. *Journal of Veterinary Diagnostic Investigation* 7(4), 488–493. DOI: 10.1177/104063879500700411.

Toft, N., Nielsen, S.S. and Jørgensen, E. (2005) Continuous-data diagnostic tests for paratuberculosis as a multistage disease. *Journal of Dairy Science* 88(11), 3923–3931. DOI: 10.3168/jds.S0022-0302(05)73078-2.

Wang, C., Turnbull, B.W., Gröhn, Y.T. and Nielsen, S.S. (2006) Estimating receiver operating characteristic curves with covariates when there is no perfect reference test for diagnosis of Johne's disease. *Journal of Dairy Science* 89(8), 3038–3046. DOI: 10.3168/jds.S0022-0302(06)72577-2.

Weber, M.F., Verhoeff, J., van Schaik, G. and van Maanen, C. (2009) Evaluation of Ziehl–Neelsen stained faecal smear and ELISA as tools for surveillance of clinical paratuberculosis in cattle in the Netherlands. *Preventive Veterinary Medicine* 92(3), 256–266. DOI: 10.1016/j.prevetmed.2009.08.017.

21 Paratuberculosis Control Measures

Karsten Donat[1,2]*, Susanne Eisenberg[3] and Richard Whittington[4]
[1]Thuringian Animal Disease Fund, Animal Health Service, Jena, Germany; [2]Justus-Liebig-University, Gießen, Hesse, Germany; [3]Animal Disease Fund of Lower Saxony, Hanover, Germany; [4]The University of Sydney, Sydney, Australia

21.1 Background

21.1.1 History and presence of paratuberculosis control programmes

Early programmes to control paratuberculosis were set up in France in the 1920s, followed by the UK and Iceland in the 1960s, Cyprus, the USA and Japan in the 1970s, and France and Norway in the 1980s (Benedictus *et al.*, 2000). In the Netherlands, first initiatives regarding test-and-cull strategies based on different diagnostic tests have been reported since 1942 (Benedictus *et al.*, 2000). Since the 1990s, control programmes for paratuberculosis have been implemented in other high-income countries (HIC), e.g. in Australia (1996), Canada (2007) and New Zealand (2009), and middle-income countries (MIC) such as South Africa (1997). In Europe, paratuberculosis control was started in the UK (1998), Spain (2004), Denmark (2006), Belgium (2006), some regions of Germany (2003), Ireland (2013) and Italy (2014) (Bakker, 2010; Geraghty *et al.*, 2014). A mandatory control programme for elimination of clinical cases of paratuberculosis from the food chain has been established in Austria since 2006 (Khol *et al.*, 2007) and in Switzerland since 2015 (Whittington *et al.*, 2019). With respect to its very low *Mycobacterium avium* spp. *paratuberculosis* (*MAP*) prevalence, Sweden has had mandatory regulations in force since 1998 and all adult cattle submitted for necropsy are actively monitored for the presence of *MAP*. Mandatory measures including elimination of the infected herd and extensive tracing of all contact herds are applied (Sternberg *et al.*, 2007).

A review presenting an overview of paratuberculosis control in cattle and other important susceptible livestock species in 48 countries was published recently (Whittington *et al.*, 2019). In the period 2012–2018, a total of 22 of 48 countries had a control programme for paratuberculosis with more than half of the countries in Europe (Fig. 21.1). All except Ireland, Italy and Switzerland had a long-term programme that started before 2012 and will be continued after 2018. With a median starting year of 2000, these long-term programmes began as early as 1942 (the Netherlands) and 1962 (Iceland), compared with as late as 2009 (New Zealand). Control programmes changed over time in most countries. The majority of the countries without a control programme for paratuberculosis were in South and Central America, Asia and Africa. With respect to the United Nations Development Program index ranking for 2015, countries with a paratuberculosis control programme had a higher development index compared with those without a control programme (Whittington *et al.*, 2019).

*Corresponding author: kdonat@thtsk.de

© CAB International 2020. *Paratuberculosis: Organism, Disease, Control, 2nd Edition* (eds M.A. Behr *et al.*)

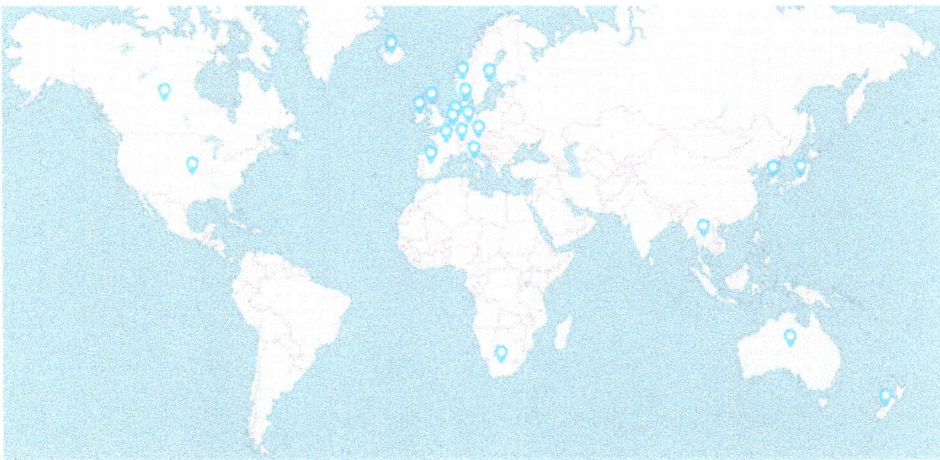

Fig. 21.1. Countries that had a control programme for paratuberculosis between 2012 and 2018. (Whittington *et al.*, 2019, reproduced with the permission of BMC Veterinary Research.)

21.1.2 Reasons for having a control programme for paratuberculosis or not

Paratuberculosis impacts animal welfare, has direct and indirect economic costs, and arouses public health concerns. Infection leads over time to a chronic granulomatous enteritis. Animals with clinical signs suffer from weight loss, diarrhoea in some species, and death. In dairy cattle, economic losses have been extensively studied and total annual economic losses were estimated to be between US$21 and €234 depending on which costs were included and the herd's origin (Whittington *et al.*, 2019).

Concerning the link between *MAP* and diseases in humans, the authors of a series of review articles concluded that human exposure exists. It was stated that 'while the zoonotic potential of *M. paratuberculosis* cannot be ignored, due to important knowledge gaps in understanding its role and importance in the development or progression of human disease, its impact on public health cannot yet be quantified or described' (Waddell *et al.*, 2015a, b, 2016). The authors concluded that steps beyond the already existing programmes for improvement of dairy and ruminant health and reduction of economic losses could not be justified by public health authorities (Waddell *et al.*, 2015b). In other words, a reliance of public health authorities on animal health authorities to reduce the exposure of humans to *MAP* by controlling paratuberculosis in livestock was identified (Whittington *et al.*, 2019).

The survey-identified reasons for having a paratuberculosis control programme or not are presented in Table 21.1 (Whittington *et al.*, 2019). Participants in the survey stated that one of the reasons for having a paratuberculosis control programme was improvement of animal health. Reduction of economic or production losses were mentioned as a driver for 90% of the respondents whereas public health reasons were cited by Austria, Belgium, Canada, Germany, Republic of Ireland, Japan, Korea and Thailand while the UK cited the precautionary principle. The most important reasons for not having a paratuberculosis control programme were lack of economic resources and that paratuberculosis was ranked as a disease with a lower priority compared with other livestock diseases.

21.1.3 Objectives of paratuberculosis control programmes

Control of paratuberculosis can have different goals ranging from a modest objective of reducing only clinical cases, through reducing the within-herd prevalence of infected or infectious animals, right up to the eradication

Table 21.1. Reasons for having a control programme or not (adopted from Whittington *et al.*, 2019, modified).

Most cited reasons for having a control programme (22 countries)	Number	Percentage
Animal health	22	100
Reducing production losses	20	90
Maintaining trade, regional or international	15	68
Animal welfare	11	50
Public health	8	36
Most cited reasons for not having a control programme (26 countries)	**Number**	**Percentage**
Economic constraints	12	46
Animal health resources are currently deployed to tackle other priority diseases	11	42
Lack of national/regional animal health capacity, infrastructure or operational resources	8	31
Paratuberculosis is not prevalent at herd or individual animal levels and is not considered to be a problem relative to other animal health issues	8	31
Lack of laboratory diagnostic services	6	23
Paratuberculosis is present but is not considered to affect the animal population	5	19

of the infectious agent from a herd or region. In addition, limiting the entry of *MAP* into the human food chain via milk or meat by eliminating clinically diseased animals from the food chain is stated as an objective of programmes in Austria, Germany, Japan, New Zealand and the UK. Overall prevalence reduction in *MAP*-infected herds has been shown to be feasible in general (Ferrouillet *et al.*, 2009; Collins *et al.*, 2010) whereas eradication of *MAP* from the herd can be achieved only in some herds (Kalis *et al.*, 2004; Donat, 2017). Stamping out can be a mandatory approach in countries where prevalence is low (Sternberg *et al.*, 2007).

Table 21.2 presents the survey-identified most common objectives for paratuberculosis control aggregated across types of livestock (Whittington *et al.*, 2019). In Germany, the Netherlands and South Africa eradication at herd level was an objective that could be chosen by a farmer. National or regional eradication of paratuberculosis was described as an objective in control programmes in Belgium, Norway and Sweden. Increasing the level of knowledge about the disease and the awareness or research on paratuberculosis were not commonly expressed as objectives.

Table 21.2. Most common objectives of paratuberculosis control programmes among 22 countries (adopted from Whittington *et al.*, 2019, modified).

Objective	Countries *n* (%)
Reduce prevalence of *MAP* (equivalent term = control)	17 (77)
Reduce incidence of clinical cases	10 (45)
Reduce *MAP* contamination in the human food chain/improve consumer safety	7 (32)
Reduce spread to new farms or regions	6 (27)
Certification of freedom or provide information on low risk herds as a source of replacement stock	6 (27)
Provide confidence or assurance to, or safeguard markets (including export)	4 (18)
Reduce production/economic losses	4 (18)
Risk management: determine risk in a herd; allow trade/marketing with an accredited or specified risk level; reduce within-herd spread	4 (18)

21.1.4 Current role of international organizations in paratuberculosis control

After decades of intensive research, paratuberculosis control is still controversially discussed among scientists, farmers, practising veterinarians and stakeholders. Although, there is sufficient evidence on which measures contribute effectively to disease control and which strategies and tools can be effectively used, there is no consensus at international level on how to deal with the disease and little progress has been made in limiting the spread of the disease between countries. Allowing member countries to impose scientifically based sanitary measures to protect human and animal health is one of the main tasks of the World Trade Organization (WTO, 2016). The World Organisation for Animal Health (OIE), whose standards are recognized by WTO, offers little guidance on paratuberculosis (OIE, 2017). Based on a discussion of scientists at the International Colloquium for Paratuberculosis, Mexico 2018 (unpublished), a major reason that the Code chapter of the OIE has not been developed was the concern about the low accuracy of diagnostic tests in individual animals. Although these tests are useful tools to identify *MAP* shedders or animals at high risk of infection within a herd, they are not adequate to certify freedom of infection at individual animal level. Since no consensus has been achieved, several WTO members require freedom of *MAP* in traded livestock even if they themselves do not carry out activities to document such freedom. Furthermore, a proportion of official movement protocols are scientifically flawed, ineffective and therefore do not enhance control of *MAP* infection. In particular, certification based on recent herd history of clinical disease or on testing of the individual animals to be moved is still common but not effective. Such protocols effectively 'reward' farmers who are not testing. This may discourage participation in organized herd classification programmes. The latter can penalize herd owners who actively test their animals to control *MAP* (Kennedy *et al*., 2017). Therefore, the International Association for Paratuberculosis provided guidelines for movement of livestock, according to the rules laid down in the Sanitary and Phytosanitary (SPS) Agreement of WTO (Kennedy *et al*., 2017). These guidelines rely on herd or population level certification as has been implemented for other diseases for which negative individual animal tests provide limited assurance, such as bovine brucellosis or bovine tuberculosis.

In 2016, in context of the new Animal Health Law (AHL), the European Union (EU) decided to list and categorize animal diseases according to specific criteria (European Union, 2016). This law will be in force in all member states of the EU in 2021 with specific rules for each disease. Previously, the European Commission asked the European Food Safety Authority (EFSA) for scientific advice regarding listing and categorizing paratuberculosis among others. EFSA recommended that paratuberculosis should be listed because the disease meets all mandatory criteria laid down in article 7 of the AHL and causes or could cause significant negative economic impact on production in the EU. Furthermore, the scientists suggested categorizing paratuberculosis as a disease that is intended for optional control programmes in the member states because it meets all mandatory criteria and has significant impact on animal welfare (More *et al*., 2017). In the light of earlier mentioned difficulties to diagnose *MAP* infection at individual level and the large effort assumed to be necessary to control the disease, this scientific opinion was ignored. In the AHL, paratuberculosis will be categorized as a disease for which surveillance is necessary but trade regulations or optional control programmes are not.

21.1.5 Legal conditions for handling paratuberculosis in different countries (notifiability)

In 2018, paratuberculosis was classified as a notifiable disease in 35 of 48 countries in at least one of the susceptible farmed species (Whittington *et al*., 2019). Presence of *MAP* was notifiable in Finland, Norway and Sweden regardless of its host. In all other 32 countries only the presence of *MAP* in farmed animals was notifiable. However, in eight countries (Argentina, Austria, Italy, Mexico, Slovenia, South Africa, the UK, Venezuela) notification was not required for some farmed species, typically camelids and deer. *MAP* in dairy cattle was notifiable in all 35 countries. In beef cattle it was notifiable only in 33, in sheep and goats in 28, in farmed deer in 15 and in camelids in 12 countries. Presence of *MAP* in buffalo was notifiable in five and in bison in two countries.

Participants of the survey responded that paratuberculosis is underreported in 26 of the 35 countries in which it was notifiable, and 29 of all 48 respondents regardless of the geographic zone. It was stated that reasons for that phenomenon were characteristics of available tests that are known to underestimate true prevalence, lack of awareness of the signs of or knowledge about the disease, reluctance to report due to farmer concerns about the consequences, farmer's fear of the stigma of being identified as a positive herd, notification of only clinical cases or lack of surveillance. Furthermore, it is common to avoid notification, by choosing tests that are not notifiable, or to cull suspicious animals to prevent detection (Whittington *et al.*, 2019).

Countries without notifiability were not clustered or within a particular geographic region. Paratuberculosis was not notifiable in any species in Belgium, the Czech Republic, Denmark, Ecuador, France, Greece, India, Iran, the Netherlands, New Zealand, Nigeria, the USA and Uruguay.

21.2 Herd-Level Prevalence and Within-Herd Prevalence in Infected Herds

According to the survey data of Whittington *et al.* (2019), respondents of very few countries reported a herd-level prevalence of less than 1%, based on objective laboratory test data (Fig. 21.2a). A significant association between herd size and herd-level prevalence of paratuberculosis for dairy cattle was present: for each one log increase in herd size the odds of a country having a higher category of prevalence increased by 9.7 ($p = 0.001$; 95% CI 1.9–48.8).

Only a few countries had an average within-herd prevalence of less than 1%, based on objective laboratory test data (Fig. 21.2b). An average within-herd prevalence estimate >10% for most species was common. Within-herd prevalence was unknown in 12 countries with dairy cattle, 18 with beef cattle, 26 with sheep and 27 with goats.

In 18 of 48 countries, *MAP* infection is present in free-living ruminants and wildlife. Those countries show no geographic clustering, however, contact with farmed livestock and wildlife has been documented. A great variety of species were affected including omnivorous, herbivorous and carnivorous mammals and even some birds (see Chapter 15 this volume for more information on infection in non-ruminant wildlife). In 26 countries, the *MAP* prevalence in wildlife was unknown.

21.3 Current Approaches of Paratuberculosis Control

21.3.1 Organization of and participation in paratuberculosis control

The data presented in this section all refer to the 22 countries with a control programme according to the recently performed survey (Whittington *et al.*, 2019).

Coordination

A single national programme was in force in 18 countries. The Netherlands and the UK had more than one national control programme for dairy and/or beef cattle. Australia, Belgium, Canada, France, Germany, Spain and the UK had several regionally coordinated programmes, whereas programmes in the other countries were coordinated nationally. Within species, the control programmes differed between regions in some countries (Australia, Belgium, France, Germany, Canada), but were similar or identical between species in other countries (Iceland, Korea, Spain, Thailand and the UK).

Leadership

Leadership of the control programmes varied, ranging from a simple structure with just one major nominated leader in 11 of the 22 countries up to seven distinct components in Canada. In four of the single-leader programmes, leadership was provided by a private organization, whereas in seven countries the government was

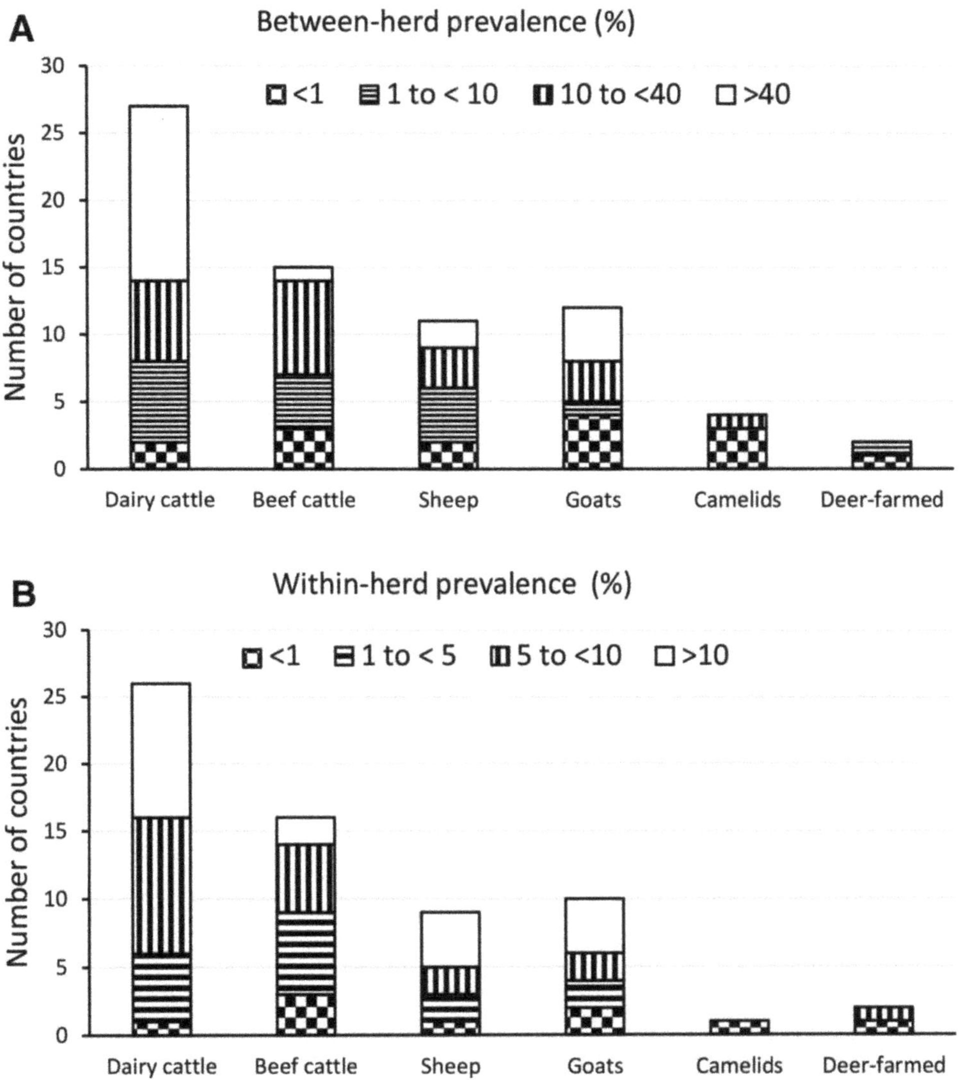

Fig. 21.2. Average between-herd prevalence (a) and within-herd prevalence (b) of paratuberculosis in countries where laboratory testing had been conducted (adopted from Whittington *et al.*, 2019).

appointed as the leader. In the remaining 11 countries, multiple organizations were involved in leadership like government, veterinary or farmers' organizations as well as industry milk or meat associations.

Funding

Sources for funding for programme control as well as operations varied considerably between countries. Main funding sources differed between programme leadership and operational costs (Whittington *et al.*, 2019). Leadership was funded by the government in most countries. In eight countries, the government was the only source, whereas in another eight the government funded part of the leadership. Farmer organizations ($n = 3$) were the next most common financial source for programme leadership.

However, programme operations were most likely to be funded partially or wholly by farmers (17 countries) compared with government alone (three countries) or by a combination of different sources.

Participation

Of the 22 countries with a control programme, participation was mandatory in nine, whereas participation was completely voluntary in the other 13 countries. Compulsory participation was legislated in Japan, Norway and a federal state of Germany. In Sweden, reporting of *MAP* suspicion in any species is mandatory, and if the presence of *MAP* is confirmed, stamping out and contact tracing is compulsory. In South Africa, animals diagnosed with *MAP* have to be isolated and slaughtered under the supervision of a state official, infected herds/flocks have to be placed under quarantine and in-contact animals have to be tested. Components of control that are compulsory vary from country to country according to the objectives of the programme. For example, in Austria and Switzerland the notification and elimination of clinical cases is compulsory, but on-farm measures to reduce the within-herd prevalence are voluntary. Active surveillance is mandatory in Japan while vaccination is mandatory in Iceland.

Manual

A manual or a legislative document describing the control programme in detail including the case definitions, rules and procedures was publicly available for 20 of 22 countries. Alternatively, some regions provided information for farmers online or upon request. In most cases, detailed descriptions of the methods of diagnosis/surveillance, control and the rules/regulations associated with control programmes as well as definitions for terms used in control programmes such as 'infected' and 'diseased' animals/herds and 'control' and 'eradication' were provided.

Incentives

Incentives for participation to enhance the enrolment of farmers in programmes were identified in 15 countries with control programmes, mostly in those with voluntary programmes.

Full or partial financial support, assistance or compensation to farmers for one or more operational aspects of the control programme was provided in 12 countries during 2012–2018. The components covered included costs of testing or conduction of risk assessments, the value of culled livestock and the cost of its culling. For example, in Austria and Switzerland, compensation is paid for clinically diseased animals that have to be culled. Sampling and testing are free of charge for cattle owners in Norway, Austria and Switzerland and partly subsidized in Belgium and Germany. Financial support may have begun or ceased due to programme review processes. For example, in Australia support was paid to beef farmers until 2015, then ceased, and in some federal states of Germany partial compensation for culled *MAP* shedders to cattle owners is paid.

Avoidance of penalties such as market access restrictions for non-participation can be an incentive as well because in some countries the consequences of diagnosis could mean exclusion from the market. In the Netherlands, participation of dairy cattle herds in the Milk Quality Assurance Program or the Intensive Paratuberculosis Program is part of the milk delivery conditions between herd owner and the dairy. However, in the Netherlands there are no incentives or restrictions for beef cattle, sheep and goats. Some milk processors in the UK impose an absolute requirement for farmers to engage in the control programme. Milk produced by Italian farms with clinical cases is not allowed to be sold when it is destined for dairy product exports.

Market Assurance Programs are available for Australian producers to mitigate the risk of between-herd transmission. In some regions of Germany there is a higher demand for breeding stock from farms with 'non-suspect for paratuberculosis' status. In the UK, herds with low herd risk level are believed to sell more (pedigree) stock. In general, higher prices for animals from low-risk herds are expected although this has not been consistently realized. Market access is used to apply pressure on dairy farmers in some European countries.

Research

Knowledge gaps constraining successful control have been reviewed recently (Barkema *et al.*, 2018). Implemented paratuberculosis control programmes were combined with research objectives regarding aspects such as disease control, vaccination, pathophysiology, microbiology and economics in 12 out of 22 countries. The most common research objective was improving diagnostic tests and/or diagnostic test validation, followed by epidemiological research including prevalence, risk factors, transmission dynamics and environmental survival of *MAP*, research on farmer attitudes, food safety and animal genetics. Denmark, Japan, the Netherlands, Germany, Ireland, the UK and the USA had research programmes on paratuberculosis that were conducted independently of control programmes, but mostly paratuberculosis control and research are closely linked. In Switzerland, research on paratuberculosis was completely independent of a control programme.

21.3.2 Communication

Communication directed towards farmers and veterinarians about the disease and its causative agent are key elements. Furthermore, the importance and the background of the programme as well as consideration of farmers' attitudes towards the implementation of control measures on their farms are essential in *MAP* control (Benedictus *et al.*, 2000; Roche *et al.*, 2015; Ritter *et al.*, 2016). An important issue that hampers the acceptance of paratuberculosis control among farmers is their perception that an 'official' control programme aims at eradication of the disease (Lorna Citer, personal communication, 8 June 2018), which is not primarily the case in most paratuberculosis control programmes. Communication, extension or education and training activities were included in 77% of the control programmes according to the results of the recent survey (Whittington *et al.*, 2019). Included activities were websites, conferences and seminars, field days and newsletters, while a consistent objective was to increase awareness of the disease. The specific target audience for this activity were veterinarians ($n = 12$) and farmers ($n = 9$), whereas two countries focused on government representatives and stakeholders in general ($n = 3$). In more than two-thirds of the countries with a control programme the respondents reported that the control programmes benefited from an active stakeholder support consisting of farmers' organizations, government, veterinary organizations and private veterinarians. In more than half of the countries, the milk industry and individual farmers supported the programmes. Industry organizations for meat, livestock trading and food processing were supportive in one-third of the countries.

21.3.3 Practices and tools

Control programmes for paratuberculosis are generally multi-component and involve many possible practices and tools. The most important factor in dairy operations is preventing calves from coming into contact with the faeces of adult cows (Doré *et al.*, 2012). The main pillars of paratuberculosis control are a combination of animal-level and population-level measures such as culling shedding animals, applying hygienic measures aimed at reducing contamination of calves with manure from cows and vaccination. Limiting paratuberculosis control to the test-and-cull approach, i.e. the identification and elimination of either clinically diseased, *MAP*-shedding or subclinical infected animals will not eradicate paratuberculosis in the long run (Dorshorst *et al.*, 2006). Actions limiting infection routes, e.g. avoiding faecal–oral transmission by improving calving area hygiene or by adequate colostrum/milk feeding management, are effective strategies to decrease *MAP* prevalence (Marcé *et al.*, 2010). Several field trials as well as simulation studies have shown that the most effective control strategy consists of a long-term combination of 'test-and-cull' and increasing on-farm biosecurity (Dorshorst *et al.*, 2006; Smith *et al.*, 2017). In a field study linked to the control programme in a German region with a test-and-cull strategy, the availability of a separate calving pen for *MAP* shedders and its hygiene were factors associated with incidence reduction (Donat *et al.*, 2016a).

Effectiveness of control measures in dairy cattle herds depends on disease prevalence among adults. In an individual-based modelling approach, the most influential measure was reduced calf exposure, followed by test frequency and the proportion of detected and culled *MAP* shedders (Camanes *et al.*, 2018). Furthermore, it was identified that combining test-and-cull options with calf management should depend on herd prevalence status with improving calf management in cases of increasing prevalence. In high-prevalence herds, moderate and high shedders should always be culled based on annual tests. For herds with a moderate prevalence, early culling of high shedders improved the effect of control measures. Culling the progeny of known infected cows was effective in low-prevalence herds (Camanes *et al.*, 2018) as *in utero* infection with *MAP* may occur (Whittington and Windsor, 2009). Considering the finite environmental survival of *MAP*, pasture and grazing management should be utilized to reduce the exposure of grazing youngstock (Whittington *et al.*, 2004; Eppleston *et al.*, 2014).

Biosecurity measures have to be in place to prevent the transmission between herds or the reintroduction of *MAP* after successful elimination and to protect *MAP*-unsuspected herds from new infection. The main risk factor for between-herd transmission is the purchase or movement of subclinical infected animals into the herd (Rangel *et al.*, 2015). This risk can be effectively mitigated by reducing movements of animals and faeces between farms. Certification of a *MAP*-herd status that reflects the risk of *MAP* transmission by animals or manure originating from that herd is essential.

As a further tool, vaccination of cattle and sheep has been shown to be effective in delaying the onset of the disease, reducing clinical incidence and faecal shedding of *MAP* and therefore reducing economic losses and the transmission rate within a herd (Reddacliff *et al.*, 2006; Bastida and Juste, 2011). In paratuberculosis control in cattle, vaccination is not widely used at present. The limited protective potency of the vaccine, its considerable side effects and the risk of interference between paratuberculosis vaccination and intradermal testing for bovine tuberculosis results in restricted use (Coad *et al.*, 2013; Serrano *et al.*, 2017). Less than one-third of the countries applied vaccination to some (never all) species; for example, only sheep were vaccinated in South Africa and only sheep and goats in the Netherlands, and in France and Germany, vaccination requires a special permit (Whittington *et al.*, 2019). In contrast, in both the Australian and the Spanish sheep industries, vaccination contributed to a significant reduction of within-herd prevalence and has been a key element in control for decades (Dhand *et al.*, 2016).

Table 21.3 compresses the most common practices and tools used in control programmes for paratuberculosis depending on

Table 21.3. Most common practices and tools used in control programmes for paratuberculosis in 22 countries (adopted from Whittington *et al.*, 2019, modified).

Tool	No. of countries	% of countries
Cull clinical cases	19	86.4
Hygienic rearing of neonates/juvenile livestock	17	77.3
Farm-level biosecurity to prevent introduction of infection	17	77.3
Test-and-cull subclinical cases	16	72.7
Environmental and pasture management	14	63.6
Communications programme	14	63.6
Herd/flock assurance certification	13	59.1
Research programme	11	50.0
Vaccination	7	31.8
Regional biosecurity to prevent introduction of infection	5	22.7
National biosecurity to prevent introduction of infection	4	18.2
Stamping out infected herds/flocks	3	13.6
Individual animal assurance certification	3	13.6

the prevalence of paratuberculosis within the country or region and on the objectives of the programme (Whittington *et al.*, 2019).

'Classical' measures of animal disease control are applied, for example, in Switzerland: any animal traffic to and from the infected farm is forbidden, clinical cases as well as their suckling calves/lambs have to be culled, and the housing has to be cleaned and disinfected before the animal traffic ban can be lifted from the farm.

21.3.4 Diagnostic tools and their purposes

Primarily due to the chronic nature of the disease, diagnostic tests used for paratuberculosis are generally imperfect. If used properly they can be useful meeting a specific purpose such as:

- confirmation of clinical disease;
- use in a test-and-cull strategy to identify and eliminate *MAP* shedders from a herd;
- identify high-risk animals to apply specific management practices that contributes to control at herd level;
- establishment of animal, herd or population freedom from infection;
- surveillance;
- prevalence estimation.

Diagnostic tools should be used in order to answer specific questions and results should lead to a decision, for example, culling of clinically diseased animals or to prevent insemination of subclinical infected cows to accelerate removal of the cow from the herd.

A wide range of tests is available as tools for the above-mentioned purposes. In general, available paratuberculosis tests offer a high specificity, but unfortunately, this is combined with a low sensitivity when used to identify individual cow status. Polymerase chain reaction (PCR) or bacterial culture are used for the direct detection of the infectious agent in faecal samples and are considered a direct measure of bacterial shedding of the sampled cow. Bacterial culture is widely accepted as the gold standard. Detection of *MAP*-specific antibodies using enzyme-linked immunosorbent assay (ELISA) (or agar gel immunodiffusion test (AGID) and complement fixation test (CFT) in the past) for serum or milk samples are indirect measures of the infection detecting the humoral immune responses in infected animals. To prove *MAP* infection in individual cows, pathological examination followed by microscopic examination and bacterial culture or PCR of target tissues is necessary. If an individual test result is used in combination with the herd status or historical results, the accuracy of the test used at animal level can be increased (More *et al.*, 2015). The status or the historical results can represent individual test results as mentioned above, or herd- or flock-level diagnostics such as culture or PCR on environmental samples (Raizman *et al.*, 2004) or boot swabs (Donat *et al.*, 2016b), or use of pooled faecal samples (Whittington *et al.*, 2000). Bulk tank milk ELISA is also used in several programmes, but its low herd-level sensitivity (Sergeant *et al.*, 2019) hampers practical application for low within-herd prevalence herds. However, the different tools for herd- or flock-level diagnostics provide a useful and economic way to identify high-prevalence herds and their results can help to convince farmers to work on risk mitigation strategies and participate within control programmes (Khol *et al.*, 2019).

The OIE (2014) provides an overview of different tests for different purposes, but a multitude of considerations are needed to establish a useful test-strategy (see e.g. Nielsen, 2014). The choice between different tests may be related to availability, logistics or costs (Whittington *et al.*, 2000).

Under different circumstances, additional tests might be considered. For example, the Ziehl–Nielsen (ZN) stain of faecal or tissue smears may serve as a fast diagnostic tool to confirm clinical diagnosis (Weber *et al.*, 2009). In the past, skin testing to detect cell-mediated immune responses was also used (Kalis *et al.*, 2003). Since its use may interfere with skin testing for *Mycobacterium bovis* and the availability of Johnin is limited, this method is no longer commonly used. *In vitro* tests for the detection of cell-mediated immune responses such as the interferon-gamma release assay may merely report exposure to *MAP*, and the clinical relevance has yet to be established. Recent research focuses on the identification of purified proteins for the improvement of the interferon-gamma release assay, the diagnostic value of volatile organic compounds (see e.g. Gierschner *et al.*, 2019) or

Table 21.4. Purposes of use of the most frequently applied types of test among control programmes for cattle in 22 countries (adopted from Whittington *et al.*, 2019, modified).

Test	Individual animal diagnosis	Individual animal certification/assurance	Herd-level screening	Herd-level certification/ assurance
Serum enzyme-linked immunosorbent assay (ELISA)	17	2	14	9
Faecal polymerase chain reaction (PCR) – individual	18	8	9	8
Faecal culture – individual	17	3	5	5
Pathology	15	1	1	2
Faecal PCR – pooled	3		11	6
Milk ELISA – individuals	10		11	6
Faecal culture – pooled	3		8	4
Milk ELISA – bulk milk			6	1
Environmental faecal test – culture or PCR	1	1	6	4

metabolomic profiling (De Buck *et al.*, 2014) for the early identification of *MAP*-infected animals or low-prevalence herds.

The answer given in the survey identified that countries with a control programme for cattle most commonly used serum ELISA, faecal PCR and pathology as diagnostic tests (Whittington *et al.*, 2019). AGID, CFT (both on blood), intradermal skin test and faecal Ziehl–Neelsen smear were least often applied (Table 21.4). Regarding the use of other tests or the testing in other species, see Whittington *et al.* (2019).

21.3.5 Goals, results and success of control programmes

Publicly available results of control programmes operating between 2012 and 2018 were the number of suspect and confirmed cases and farms (Austria, Japan), numbers of participating dairy farms (the Netherlands, Germany), incidence of paratuberculosis based on abattoir surveillance in sheep (Australia) and deer (New Zealand), weekly test prevalence data for dairy cattle (Denmark) or deer (New Zealand) farms, as well as lists of certified farms (Australia, the UK, regions of Germany) (Whittington *et al.*, 2019). Nearly half of the countries had indicators to measure the success of their control programme. The most common goal was an increase of participants in the control programme followed by reductions in the number of infected animals or clinical cases detected within each testing round and meeting targets in the number of 'low-risk', 'free' or certified herds. Norway included active surveillance targets and farm-level post-eradication checks as primary objectives. Sixteen countries reported having successful control programmes, whereas 13 commented that their programme objectives had been or were being met. Some survey participants rated specific outcomes such as market access maintained, prevented clinical cases or farmer satisfaction as their goal. Four countries with successful outcomes in general acknowledged specific problems:

- voluntary continuation of the programme among producers was much lower than hoped for;
- declining numbers of farms in market assurance programmes;
- herd-level prevalence did not diminish (Belgium);
- the winding back of the national control programme (Australia);
- funding for the programme and for research dried up (Canada, Australia);

- only a minority of herds benefited from the programme, since most of the herds were not in the programme and therefore it is unlikely that the programme has had a significant impact on the individual animal or herd-level prevalence in the region (UK);
- in six of the 22 countries, outcomes could not be assessed or it was too early to tell if success had been achieved.

21.3.6 Paratuberculosis control in low and middle income countries

Although the survey was co-authored by several authors from low and middle income countries (LMIC) only limited information about paratuberculosis control was available (Whittington *et al.*, 2019). It was observed that the size and complexities of the animal populations at risk were extraordinary in the livestock producing countries outside the major developed economies of Europe, North America and Australia. Nine LMIC provided country-specific overviews including information about the animal population, the husbandry system and the situation regarding paratuberculosis (Whittington *et al.*, 2019, additional files); five were from Central and South America (Argentina, Brazil, Colombia, Ecuador, Panama), three from Asia (Bangladesh, India, Iran) and two from Africa (Nigeria, Zambia). A total of about 440 million cattle, 202 million sheep, 232 million goats and 113 million buffalo were kept in these countries with the largest populations in India (cattle, sheep and buffalo) and Nigeria (goats). All countries had several types of farmed ruminants in multiple husbandry systems, in most countries up to six or seven types, in tens of thousands of individual farms per country.

Data showed that lack of resources of one kind or another were the most common reason for not implementing paratuberculosis control in LMIC countries. Simply put, there were other priorities for which resources were needed, e.g. controlling bovine tuberculosis. In most LMIC animal health and welfare was found to be still rudimentary because education, accessibility of rural areas and economic resources are missing. Veterinary advice to farmers in general tended to be missing in these countries, too. Lack of leadership at national and international level was identified as another relevant cause leading to only a little attention for paratuberculosis control in many LMIC (Whittington *et al.*, 2019). However, cases of paratuberculosis as well as an endemic paratuberculosis situation were reported from most countries except Nigeria (no information available). Because a surveillance system was not established in most of these LMIC countries, information regarding prevalence is lacking. In Brazil and Zambia, the disease is notifiable. Although countries with a lower socio-economic status were found to often have large animal populations and arguably from a human development perspective would benefit the most from an animal health programme, official paratuberculosis control measures were not reported from any of these countries except Thailand. However, in India and in Argentina, Colombia, Ecuador and Iran, on-farm-level control activities were reported with voluntary testing, removal of test-positive animals and separation of calves.

21.4 Issues Impeding Future Control

Paratuberculosis is a common disease with considerable impact on animal health, animal welfare and economics, and it may affect public health. Without a doubt, prevalence and incidence will increase if it is not controlled (Groenendaal *et al.*, 2002; Mitchell *et al.*, 2008). The practices and tools for control of paratuberculosis and their limitations have been known for decades. Culling of clinical cases, test-and-cull of subclinical cases, improving biosecurity to prevent new infections in young stock and environmental pasture management are the most important chosen approaches. Vaccination is seldom used, despite the potential appeal given the difficulty of long-term implementation of improved biosecurity measures and on-farm management processes. There are many challenges for controlling the disease, including the need to deal with different species, very large animal populations, disease presence in a large number of herds and the need to manage the programme over a long time-frame while participating farms stay economically stable. Although control of this disease seems to be important, relevant impediments exist.

21.4.1 Lack of an international animal health code for paratuberculosis

Due to open markets and global animal trade and the endemic and transboundary character of the disease, leadership is required to enhance the control of paratuberculosis to a global level, commencing with an agreed international code for paratuberculosis, specifying the principles and methods of control ideally adopted by the OIE. The lack of any general guidelines, let alone international regulation for paratuberculosis control, leads to heterogeneity of control programmes in different countries and even worse to fragmentation of control activities within several countries. This creates a vacuum where animal health authorities do not know what to recommend and farmers can avoid costly measures while the disease is spreading. An international dialogue to rationalize paratuberculosis control is needed to use available resources for animal disease control most efficiently and to harmonize public health assessments of *MAP* between countries.

21.4.2 Lack of paratuberculosis control in LMIC

Other priorities and missing veterinary advice in general and paratuberculosis control specifically aggravates the situation in LMIC compared with HIC since these countries will suffer the greatest impact on human wealth and human health through lost production of animal protein and potential zoonotic impacts (Whittington *et al.*, 2019). Several types of farmed ruminants in multiple, often free-ranging husbandry systems are common in LMIC in tens of thousands of individual farms per country. In this landscape, the challenges for disease control are enormous, animal diseases in general and paratuberculosis specifically can spread between farms, regions, countries and between livestock species, prior to any clinical evidence. There is a strong need to make the issues of paratuberculosis more visible worldwide, which will support the approaches for animal disease control in LMIC as well. The 'think globally and act locally' approach should be adopted. An international consensus to approve the use of vaccination could be a helpful tool to reduce economic losses and *MAP* shedding especially in LMIC when bovine tuberculosis testing of cattle is not performed. In the future, the options linking paratuberculosis and tuberculosis control activities should be evaluated because both require a similar framework including on-farm work, individual animal testing and interaction with farmers.

21.4.3 Lack of data on true prevalence

The global prevalence of paratuberculosis as reported by Whittington *et al.* (2019) is concerning. It is estimated that in about 70% of the countries more than 10% of herds and flocks were affected, with between-herd prevalences of >40% in some HIC. The prevalence of paratuberculosis within infected herds and flocks is often 'guestimated'. Respondents of many countries did not know what is going on with respect to paratuberculosis in their herds and flocks, particularly in species other than cattle. Better estimates for within-herd prevalence are needed worldwide, and therefore standardized testing schemes should be defined to establish data that are comparable between countries. This information should support countries in deciding which stages of control or eradication are appropriate. At this moment, the most common objective is prevalence reduction; however, several countries already offer an option for motivated farmers to work on elimination of the disease at farm or regional level. Sweden and Norway are even considered to be in a surveillance phase after successful eradication.

21.4.4 Lack of performance indicators

Albeit in most countries herd participation rates in control programmes are used as performance criteria, this may not be a good measure to determine success. It can be argued that reduction of the annual incidence of clinical cases should be used as a performance indicator, interpreting this as a successful reduction of transmission.

Nevertheless, this may lead to a false sense of security and premature cessation of strict control measures since improved animal management or a lower mean age of the herd may have a similar effect while subclinical infection rates may still be high. However, without short-term performance indicators, farmers, stakeholders, the industry and organizations funding the control programme tend to become less motivated over time. The negative aspects of available performance indicators should lead to exploration of other measures. Determination of within-herd prevalence over several years based on objective surveillance and testing seems a promising indication about a programme's success.

21.4.5 Lack of holistic approach

All ruminant livestock industries and wildlife must be involved to prevent the development of a reservoir of *MAP*. Some countries have control programmes only for one type of livestock (mostly for dairy cattle) or manage the control differently between species, even though the disease is present in others. For example, in Australia paratuberculosis control in beef cattle has been managed independently from that in sheep, despite the occurrence of co-grazing of pastures, or control in dairy cattle has been managed independently from beef cattle despite the fact that dairy calves enter beef production facilities. These compartmentalized approaches are ineffective and costly since there is evidence of spread between these sectors. Despite apparent host preferences of different strains of *MAP*, it is unclear whether the pathogen can persist in new niches. Evaluating spread of *MAP* from farmed livestock to wildlife populations and the role of potential reservoir hosts for *MAP* is important.

21.4.6 Cessation of funding for the programme

As is the case for several endemic diseases, e.g. bovine tuberculosis, the time frame for successful control of paratuberculosis is measured in decades. In a voluntary control environment, the availability of funding for long-term control activities has been shown to be problematic in several countries. There is a strong need for sustainability of such an effort, which arguably must be determined and justified economically. Research studies to better understand consumer decisions and forecast market demands with respect to the animal welfare and public health aspects of *MAP* infection are warranted. Mostly in HIC, several stakeholders, non-governmental organizations and even relevant political parties presently address animal welfare and health aspects politically. All those needs may be determined by a variety of objective factors, including for example the need to ameliorate animal welfare concerns, the need to meet market access specifications or the need to reduce economic losses due to the burden of disease. Clearly, most ongoing control programmes address these needs. Encouragement of the efforts of the sector by special regulations enforced by legislation and supported by public money may be required to improve *MAP* control. Without that support, control programmes are prone to be undervalued by stakeholders of the cattle industry and not sustainable, because farmer and consumer behaviour alone are unlikely to provide market changes and appropriate price signals to reward the efforts to control paratuberculosis in the long term. For example, in the USA, Canada and Australia, control programmes have expired in recent years, regardless of apparent or partial successes. In Sweden however, the compulsory activity to detect and to eradicate *MAP* was financially supported by the state, which is considered one of the reasons why prevention, control and eradication of *MAP* has been successful.

21.4.7 Lack of research

Paratuberculosis control strategies should be continuously monitored and reviewed. Ongoing research developments on improved diagnostic tests, vaccines and epidemiological insights should be implemented in existing programmes. Improvement of diagnostic tests would be desirable to assist test-and-cull strategies as well as surveillance for *MAP*. In addition, improved tests are important to enable a broader use of existing commercial vaccines for paratuberculosis. The development of a vaccine preventing infection would be advantageous for controlling *MAP* in scenarios where individual testing by use of antibody

ELISAs is not meaningful or required. New methods for studying the pathology of the disease may enhance discovery of vaccine candidates.

21.4.8 Lack of adherence to movement controls and herd status certification

The guidelines of the International Association on Paratuberculosis provide a template to improve certification in international trade and are an important first step forward (Kennedy *et al.*, 2017). They rely on the existence of areas with different *MAP* status, for example *MAP*-'free', *MAP*-eradication and *MAP*-certification areas. Presently, eradication is rarely an objective of control programmes. However, since animal trade is common, mitigating the risk of paratuberculosis spread from one herd to another is a strong need for successful control in every stage of a control programme. Guidelines for movement of livestock should relay on the status of the herd of origin in combination with individual test results rather than individual test results alone. A risk status based on either individual test results interpreted at herd level or on herd-level diagnostic approaches like repeated environmental samples or bulk milk testing may be advantageous. The results of the herd-level diagnosis can be used for monitoring and prevalence estimation over time. Ideally, this monitoring should be mandatory and funded by public money. This approach represents an economical use of public funds, particularly in the framework of the new European AHL. Figure 21.3 provides a general model how this low-cost monitoring system followed by a

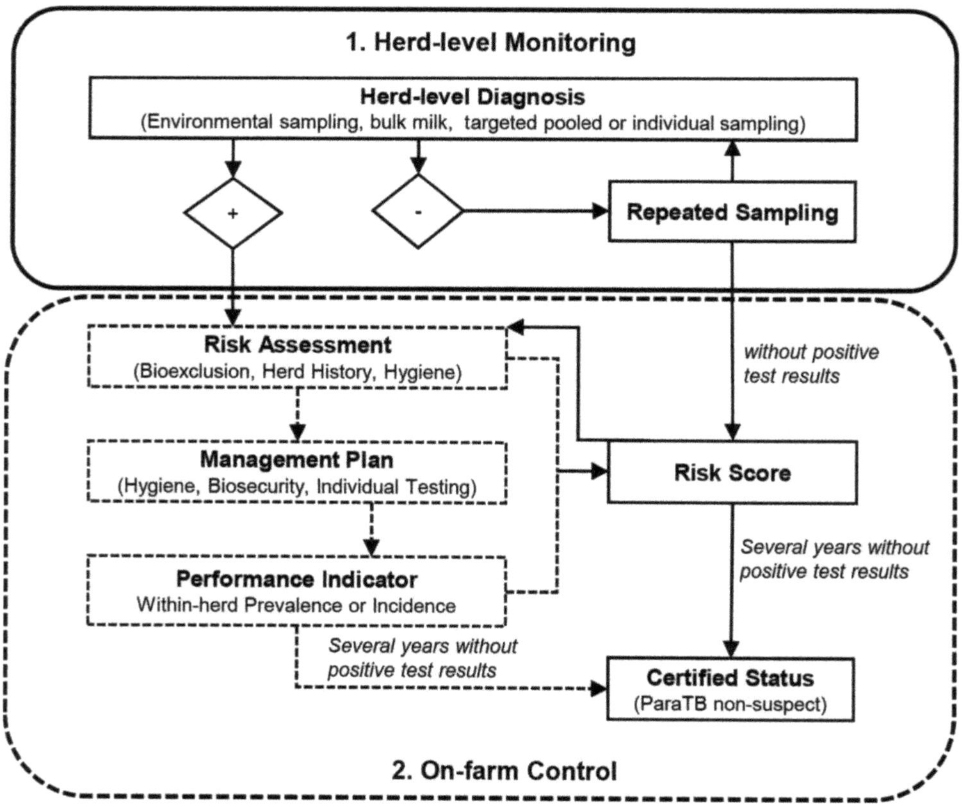

Fig. 21.3. Suggested model to integrate herd-level monitoring and on-farm control measures as a two-stage approach in paratuberculosis control (continuous lines: preferable mandatory part at regional or national level, dashed lines: preferable voluntary part at farm level).

control phase in *MAP*-positive herds asking more commitment of participating farmers can be integrated into a control programme. This approach is presently applied in some German and Austrian regions (Khol *et al.*, 2019).

21.4.9 Lack of communication

Finally, the success of paratuberculosis control is dependent on the synergistic actions of farmers, their veterinarians and breeding associations, of state veterinary authorities, diagnostic laboratories, food processors and policy makers. All involved parties should understand the advantages of and the need for paratuberculosis control as well as the risks that failure to control *MAP* may present in the future. Farmers should be formally educated about on-farm and between-farm transmission risks and at the same time, solutions to minimize risks should be offered. All Stakeholders have to take responsibility for identifying critical control points and diminish their impacts. All stakeholders must be educated about long-term goals and benefits in order to create a mutually supportive environment for successful implementation of a paratuberculosis control programme. Such an environment will be the first requirement to meet the challenges of paratuberculosis control on a national and an international level. Cooperation between countries and research groups, global initiatives and creating centres of excellence on research and diagnostic laboratory activities would improve the approaches to handle the disease and the knowledge about its spread and, as a consequence, the awareness of farmers, veterinarians, national and international authorities.

21.5 Conclusions

Paratuberculosis is prevalent globally and its impact will certainly increase if not dealt with more effectively by the international community. This must begin with leadership from peak animal health and agricultural agencies, who currently fail to provide any guidelines. Culling of clinical cases, test-and-cull based on laboratory tests, tighter biosecurity at national, regional and farm levels to hamper introduction of new infections, and environmental management are all important. Vaccination, although hampering the future use of ELISA as a diagnostic tool, must be advanced as a control tool, particularly in high-prevalence situations. Resources must be found to deal with different species and large animal populations in most countries, but in LMIC the need is acute. In the future, the options linking paratuberculosis and tuberculosis control activities might be evaluated because of their similarities. The availability of funding must be guaranteed to ensure the sustainability of control efforts. In addition, further consideration of market-driven vs government-regulated approaches are required. However, public funding of some components and supportive legislation are essential. At the same time, as control would be advanced through better tests and vaccines, the research community should be focused on these objectives. All stakeholders must be aligned to long-term goals and mutually respect the often complex socio-economic contexts of others in order for a paratuberculosis control programme to be successful.

References

Bakker, D. (2010) Paratuberculosis control measures in Europe. In: Behr, M. and Collins, D. (eds) *Paratuberculosis Organism, Disease, Control*. CAB International, Wallingford, UK, pp. 306–318.

Barkema, H.W., Orsel, K., Nielsen, S.S., Koets, A.P., Rutten, V.P.M.G. *et al.* (2018) Knowledge gaps that hamper prevention and control of *Mycobacterium avium* subspecies *paratuberculosis* infection. *Transboundary and Emerging Diseases* 65(Suppl 1), 125–148. DOI: 10.1111/tbed.12723.

Bastida, F. and Juste, R.A. (2011) Paratuberculosis control: a review with a focus on vaccination. *Journal of Immune Based Therapies and Vaccines* 9, 8. DOI: 10.1186/1476-8518-9-8.

Benedictus, G., Verhoeff, J., Schukken, Y.H. and Hesselink, J.W. (2000) Dutch paratuberculosis programme history, principles and development. *Veterinary Microbiology* 77(3–4), 399–413. DOI: 10.1016/S0378-1135(00)00325-4.

Camanes, G., Joly, A., Fourichon, C., Ben Romdhane, R. and Ezanno, P. (2018) Control measures to prevent the increase of paratuberculosis prevalence in dairy cattle herds: an individual-based modelling approach. *Veterinary Research* 49(1), 60. DOI: 10.1186/s13567-018-0557-3.

Coad, M., Clifford, D.J., Vordermeier, H.M. and Whelan, A.O. (2013) The consequences of vaccination with the Johne's disease vaccine, Gudair, on diagnosis of bovine tuberculosis. *Veterinary Record* 172(10), 266. DOI: 10.1136/vr.101201.

Collins, M.T., Eggleston, V. and Manning, E.J.B. (2010) Successful control of Johne's disease in nine dairy herds: results of a six-year field trial. *Journal of Dairy Science* 93(4), 1638–1643. DOI: 10.3168/jds.2009-2664.

De Buck, J., Shaykhutdinov, R., Barkema, H.W. and Vogel, H.J. (2014) Metabolomic profiling in cattle experimentally infected with *Mycobacterium avium* subsp. *paratuberculosis*. *PLoS ONE* 9(11), e111872. DOI: 10.1371/journal.pone.0111872.

Dhand, N.K., Eppleston, J., Whittington, R.J. and Windsor, P.A. (2016) Changes in prevalence of ovine paratuberculosis following vaccination with Gudair®: results of a longitudinal study conducted over a decade. *Vaccine* 34(42), 5107–5113. DOI: 10.1016/j.vaccine.2016.08.064.

Donat, K., Hahn, N., Eisenberg, T., Schlez, K., Köhler, H. *et al.* (2016b) Within-herd prevalence thresholds for the detection of *Mycobacterium avium* subspecies *paratuberculosis*-positive dairy herds using boot swabs and liquid manure samples. *Epidemiology and Infection* 144(2), 413–424. DOI: 10.1017/S0950268815000977.

Donat, K. (2017) The Thuringian bovine paratuberculosis control programme – results and experiences. *Berliner und Münchener Tierärztliche Wochenschrift* 130, 42–49.

Doré, E., Paré, J., Côté, G., Buczinski, S., Labrecque, O. *et al.* (2012) Risk factors associated with transmission of *Mycobacterium avium* subsp. *paratuberculosis* to calves within dairy herd: a systematic review. *Journal of Veterinary Internal Medicine* 26(1), 32–45. DOI: 10.1111/j.1939-1676.2011.00854.x.

Donat, K., Schmidt, M., Köhler, H. and Sauter-Louis, C. (2016a) Management of the calving pen is a crucial factor for paratuberculosis control in large dairy herds. *Journal of Dairy Science* 99(5), 3744–3752. DOI: 10.3168/jds.2015-10625.

Dorshorst, N.C., Collins, M.T. and Lombard, J.E. (2006) Decision analysis model for paratuberculosis control in commercial dairy herds. *Preventive Veterinary Medicine* 75(1–2), 92–122. DOI: 10.1016/j.prevetmed.2006.02.002.

Eppleston, J., Begg, D.J., Dhand, N.K., Watt, B. and Whittington, R.J. (2014) Environmental survival of *Mycobacterium avium* subsp. *paratuberculosis* in different climatic zones of eastern Australia. *Applied and Environmental Microbiology* 80(8), 2337–2342. DOI: 10.1128/AEM.03630-13.

European Union (2016) Regulation (EU) 2016/429 of the European Parliament and of the Council of 9 March 2016 on transmissible animal diseases and amending and repealing certain acts in the area of animal health ('Animal Health Law'). *Official Journal of the European Union* 59, L84.

Ferrouillet, C., Wells, S.J., Hartmann, W.L., Godden, S.M. and Carrier, J. (2009) Decrease of Johne's disease prevalence and incidence in six Minnesota, USA, dairy cattle herds on a long-term management program. *Preventive Veterinary Medicine* 88(2), 128–137. DOI: 10.1016/j.prevetmed.2008.08.001.

Geraghty, T., Graham, D.A., Mullowney, P. and More, S.J. (2014) A review of bovine Johne's disease control activities in 6 endemically infected countries. *Preventive Veterinary Medicine* 116(1–2), 1–11. DOI: 10.1016/j.prevetmed.2014.06.003.

Gierschner, P., Küntzel, A., Reinhold, P., Köhler, H., Schubert, J.K. *et al.* (2019) Crowd monitoring in dairy cattle-real-time VOC profiling by direct mass spectrometry. *Journal of Breath Research* 13(4), 046006. DOI: 10.1088/1752-7163/ab269f.

Groenendaal, H., Nielen, M., Jalvingh, A.W., Horst, S.H., Galligan, D.T. *et al.* (2002) A simulation of Johne's disease control. *Preventive Veterinary Medicine* 54(3), 225–245. DOI: 10.1016/S0167-5877(02)00027-2.

Kalis, C.H.J., Collins, M.T., Hesselink, J.W. and Barkema, H.W. (2003) Specificity of two tests for the early diagnosis of bovine paratuberculosis based on cell-mediated immunity: the Johnin skin test and the gamma interferon assay. *Veterinary Microbiology* 97(1–2), 73–86. DOI: 10.1016/j.vetmic.2003.07.003.

Kalis, C.H.J., Collins, M.T., Barkema, H.W. and Hesselink, J.W. (2004) Certification of herds as free of *Mycobacterium paratuberculosis* infection: actual pooled faecal results versus certification model predictions. *Preventive Veterinary Medicine* 65(3–4), 189–204. DOI: 10.1016/j.prevetmed.2004.07.005.

Marcé, C., Ezanno, P., Weber, M.F., Seegers, H., Pfeiffer, D.U. *et al.* (2010) Invited review: modeling within-herd transmission of *Mycobacterium avium* subspecies *paratuberculosis* in dairy cattle: a review. *Journal of Dairy Science* 93(10), 4455–4470. DOI: 10.3168/jds.2010-3139.

Kennedy, D., Benedictus, G., Nielsen, S., Lybeck, K., Schwan, E. *et al.* (2017) Guidelines for certification with respect to the movement of livestock for *Mycobacterium avium* subsp. *paratuberculosis* (MAP) infection V1.3. Paratuberculosis News. Available at: http://www.paratuberculosis.net/documents/IAP_Guidelines_for_MAP_certification_for_livestock_FINAL.pdf (accessed 6 August 2019).

Khol, J.L., Damoser, J., Dünser, M. and Baumgartner, W. (2007) Paratuberculosis, a notifiable disease in Austria – current status, compulsory measures and first experiences. *Preventive Veterinary Medicine* 82(3–4), 302–307. DOI: 10.1016/j.prevetmed.2007.06.002.

Khol, J.L., Eisenberg, S., Noll, I., Zschöck, M., Eisenberg, T. *et al.* (2019) Two-stage control of paratuberculosis: Herd-status surveillance as the basis for operational measures to reduce the prevalence. Experiences from Lower Saxony, Hesse, Thuringia and Tyrol [Article in German]. *Tierärztliche Praxis Grosstiere* 47, 171–183.

Mitchell, R.M., Whitlock, R.H., Stehman, S.M., Benedictus, A., Chapagain, P.P. *et al.* (2008) Simulation modeling to evaluate the persistence of *Mycobacterium avium* subsp. *paratuberculosis* (MAP) on commercial dairy farms in the United States. *Preventive Veterinary Medicine* 83(3–4), 360–380. DOI: 10.1016/j.prevetmed.2007.09.006.

More, S.J., Cameron, A.R., Strain, S., Cashman, W., Ezanno, P. *et al.* (2015) Evaluation of testing strategies to identify infected animals at a single round of testing within dairy herds known to be infected with *Mycobacterium avium* ssp. *paratuberculosis*. *Journal of Dairy Science* 98(8), 5194–5210. DOI: 10.3168/jds.2014-8211.

More, S., Botner, A., Butterworth, A., Calistri, P., Depner, K. *et al.* (2017) Assessment of listing and categorisation of animal diseases within the framework of the Animal Health Law (Regulation (EU) No2016/429): paratuberculosis. *EFSA Journal* 15, 7.

Nielsen, S. (2014) Developments in diagnosis and control of bovine paratuberculosis. *CAB Reviews: Perspectives in Agriculture, Veterinary Science, Nutrition and Natural Resources* 9(012), 1–12. DOI: 10.1079/PAVSNNR20149012.

OIE (2014) Chapter 2.1.11 Paratuberculosis. In: *Manual of Diagnostic Tests and Vaccines for Terrestrial Animals*. OIE, Paris, France.

OIE (2017) Chapter 8.13 Paratuberculosis. In: *Terrestrial Animal Health Code*. OIE, Paris, France.

Raizman, E.A., Wells, S.J., Godden, S.M., Bey, R.F., Oakes, M.J. *et al.* (2004) The distribution of *Mycobacterium avium* ssp. *paratuberculosis* in the environment surrounding Minnesota dairy farms. *Journal of Dairy Science* 87(9), 2959–2966. DOI: 10.3168/jds.S0022-0302(04)73427-X.

Rangel, S.J., Paré, J., Doré, E., Arango, J.C., Côté, G. *et al.* (2015) A systematic review of risk factors associated with the introduction of *Mycobacterium avium* ssp. *paratuberculosis* (MAP) into dairy herds. *Canadian Veterinary Journal* 56, 69–77.

Reddacliff, L., Eppleston, J., Windsor, P., Whittington, R. and Jones, S. (2006) Efficacy of a killed vaccine for the control of paratuberculosis in Australian sheep flocks. *Veterinary Microbiology* 115(1–3), 77–90. DOI: 10.1016/j.vetmic.2005.12.021.

Ritter, C., Jansen, J., Roth, K., Kastelic, J.P., Adams, C.L. *et al.* (2016) Dairy farmers' perceptions toward the implementation of on-farm Johne's disease prevention and control strategies. *Journal of Dairy Science* 99(11), 9114–9125. DOI: 10.3168/jds.2016-10896.

Roche, S.M., Jones-Bitton, A., Meehan, M., Von Massow, M. and Kelton, D.F. (2015) Evaluating the effect of focus farms on Ontario dairy producers' knowledge, attitudes, and behavior toward control of Johne's disease. *Journal of Dairy Science* 98(8), 5222–5240. DOI: 10.3168/jds.2014-8765.

Sergeant, E.S.G., McAloon, C.G., Tratalos, J.A., Citer, L.R., Graham, D.A. *et al.* (2019) Evaluation of national surveillance methods for detection of Irish dairy herds infected with *Mycobacterium avium* ssp. *paratuberculosis*. *Journal of Dairy Science* 102(3), 2525–2538. DOI: 10.3168/jds.2018-15696.

Serrano, M., Elguezabal, N., Sevilla, I.A., Geijo, M.V., Molina, E. *et al.* (2017) Tuberculosis detection in paratuberculosis vaccinated calves: new alternatives against interference. *PLoS ONE* 12(1), e0169735. DOI: 10.1371/journal.pone.0169735.

Smith, R.L., Al-Mamun, M.A. and Gröhn, Y.T. (2017) Economic consequences of paratuberculosis control in dairy cattle: a stochastic modeling study. *Preventive Veterinary Medicine* 138, 17–27. DOI: 10.1016/j.prevetmed.2017.01.007.

Sternberg, S., Ågren, E., Frössling, J., Bölske, G., Holmström, A. *et al.* (2007) Control of paratuberculosis in Sweden. In: Nielsen, S.S. (ed.) *Proceedings of the 9th International Colloquium on Paratuberculosis, International Association for Paratuberculosis*, pp. 319–324.

Waddell, L.A., Rajić, A., Stärk, K.D.C. and McEwen, S.A. (2015a) The potential public health impact of *Mycobacterium avium* ssp. *paratuberculosis*: global opinion survey of topic specialists. *Zoonoses and Public Health* 63(3), 212–222. DOI: 10.1111/zph.12221.

Waddell, L.A., Rajić, A., Stärk, K.D.C. and McEwen, S.A. (2015b) The zoonotic potential of *Mycobacterium avium* ssp. *paratuberculosis*: a systematic review and meta-analyses of the evidence. *Epidemiology and Infection* 143(15), 3135–3157. DOI: 10.1017/S095026881500076X.

Waddell, L., Rajić, A., Stärk, K. and McEwen, S.A. (2016) *Mycobacterium avium* ssp. *paratuberculosis* detection in animals, food, water and other sources or vehicles of human exposure: A scoping review of the existing evidence. *Preventive Veterinary Medicine* 132, 32–48. DOI: 10.1016/j.prevetmed.2016.08.003.

Weber, M.F., Verhoeff, J., van Schaik, G. and van Maanen, C. (2009) Evaluation of Ziehl-Neelsen stained faecal smear and ELISA as tools for surveillance of clinical paratuberculosis in cattle in the Netherlands. *Preventive Veterinary Medicine* 92(3), 256–266. DOI: 10.1016/j.prevetmed.2009.08.017.

Whittington, R.J. and Windsor, P.A. (2009) In utero infection of cattle with *Mycobacterium avium* subsp. *paratuberculosis*: a critical review and meta-analysis. *The Veterinary Journal* 179(1), 60–69. DOI: 10.1016/j.tvjl.2007.08.023.

Whittington, R.J., Fell, S., Walker, D., McAllister, S., Marsh, I. *et al.* (2000) Use of pooled fecal culture for sensitive and economic detection of *Mycobacterium avium* subsp. *paratuberculosis* infection in flocks of sheep. *Journal of Clinical Microbiology* 38(7), 2550–2556. DOI: 10.1128/JCM.38.7.2550-2556.2000.

Whittington, R.J., Marshall, D.J., Nicholls, P.J., Marsh, I.B. and Reddacliff, L.A. (2004) Survival and dormancy of *Mycobacterium avium* subsp. *paratuberculosis* in the environment. *Applied and Environmental Microbiology* 70(5), 2989–3004. DOI: 10.1128/AEM.70.5.2989-3004.2004.

Whittington, R.J., Donat, K., Weber, M.F., Kelton, D., Nielsen, S.S. *et al.* (2019) Control of paratuberculosis: who, why and how. A review of 48 countries. *BMC Veterinary Research* 15(1), 198. DOI: 10.1186/s12917-019-1943-4.

WTO (2016) Agreement on the application of sanitary and phytosanitary measures. Available at: https://www.wto.org/english/docs_e/legal_e/15-sps.pdf (accessed 6 August 2019).

22 Paratuberculosis Vaccines and Vaccination

Ramón A. Juste[1,2]*, Joseba M. Garrido[1], Natalia Elguezabal[1] and Iker A. Sevilla[1]
[1]NEIKER-Basque Institute for Agricultural Research and Development, Derio, Spain;
[2]SERIDA, Villaviciosa, Spain

22.1 Introduction

Paratuberculosis vaccination has a long history that starts about a decade after the isolation of the causal agent (Twort and Ingram, 1911; Vallée and Rinjard, 1926). Interference with testing for tuberculosis, however, has restricted its use in cattle and collaterally in other susceptible species, in spite of a clear positive protection balance in both experimental and field trial reports (Bastida and Juste, 2011). Even so, vaccination is still widely used throughout the world and is the only single paratuberculosis control strategy that has been in continuous use for nearly 100 years. In this chapter, we review paratuberculosis vaccination in different perspectives that will cover the challenges and advances in the search for an efficient tool to control paratuberculosis in the livestock industry.

22.2 General Considerations and Paratuberculosis Vaccination History

Vaccination has been the backbone of programmes for the only two infectious diseases that have been eradicated from the world: smallpox (WHO, 2018) and rinderpest (OIE, World Organisation for Animal Health, 2013). Unfortunately, immunity conferred by mycobacterial vaccines is not 100% effective and lifelong as it is for these two diseases. This is either because individuals are already infected when first dosed, or because immunity is not strong enough to prevent infection. The result is that even for populations 100% covered, there is a small number of vaccinated individuals that become carriers, even though none of them progresses to clinical disease. On the positive side however, paratuberculosis is a slow infection (Sigurdsson, 1954) with low rates of both transmission and clinical disease progression. This means that slight increases in individual resistance and small reductions in bacterial shedding can contribute to reducing economic losses in the short term and in the longer term, clearing herds from the infectious agent.

Paratuberculosis was recognized as a clinical infectious disease right at the time when microbes were first being discovered and demonstrated as a cause of disease (Johne and Frothingham, 1895). With the quick success of Pasteur vaccination against rabies (Berche, 2012) and the perspectives opened by Bacillus Calmette–Guérin (BCG) tuberculosis vaccination (Nieuwenhuizen and Kaufmann, 2018) it was natural that the first choice for a control strategy against a disease similar to tuberculosis was the development of a

*Corresponding author: rjuste@neiker.eus

© CAB International 2020. *Paratuberculosis: Organism, Disease, Control, 2nd Edition* (eds M.A. Behr *et al.*)

vaccine. According to contemporary knowledge, French scientists (Vallée and Rinjard, 1926) developed a product for use in cattle that had an effect owing to both the mycobacterial cell components and the strong adjuvants (pumice powder and oil emulsion). This vaccine was made available to veterinarians in France in the 1920s and was continued at least until the 1940s (Vallée *et al.*, 1941). A similar vaccination strategy was soon adopted in the USA, where a vaccine was assayed with good results regarding mortality and pathology, but not when measuring *Mycobacterium avium* subsp. *paratuberculosis* (*MAP*) isolation (Hagan, 1935). No further studies on vaccination in cattle were reported until almost 25 years later, when a series of studies carried out in the UK (Doyle, 1960; Stuart, 1962, 1965) provided results considered highly satisfactory to the farmers (Doyle, 1964). By that time, studies on vaccination were going on in the National Animal Disease Center (NADC) (Larsen, 1950; Larsen *et al.*, 1964; Merkal *et al.*, 1965; Larsen *et al.*, 1969, 1974, 1978) and later in the field (Hurley, 1983; Hurley and Ewing, 1983). Then there were reports from Denmark (Jorgensen, 1983, 1988), France (Hillion and Argente, 1987; Argenté, 1992; Saint-Marc, 1992) and the Netherlands (Benedictus *et al.*, 1988; Kalis *et al.*, 1992, 1995, 2001, 2002; van Schaik *et al.*, 1996; Kalis *et al.*, 1999). By the late 1980s, even though test-and-cull programmes based on faecal culture were being abandoned, scientific reports on vaccination were becoming scarce.

The cattle experience set the ground for the use of vaccination when paratuberculosis appeared in an epidemic form in sheep in Iceland in the 1930s after the introduction of a few *MAP*-infected Karakul rams that were also carriers of the then unknown maedi-visna. While the new viral disease was eradicated in a few years by a culling strategy, paratuberculosis could not be brought under control until a vaccine was produced and vaccination was made compulsory (Sigurdsson and Tryggvadóttir, 1949; Sigurdsson, 1960; Gunnarsson *et al.*, 1984), as it remains until now. The Icelandic reports drew the attention of the Moredun Research Institute, which initiated a research programme on sheep paratuberculosis that generated important knowledge on pathogenesis and vaccination (Brotherston and Gilmour, 1961; Gilmour and Brotherston, 1962; Nisbet *et al.*, 1962; Gilmour and Angus, 1973). Vaccination was also adopted as a solution for paratuberculosis in goats in Norway (Saxegaard, 1984; Saxegaard and Fodstad, 1985). In Spain, paratuberculosis vaccination was implemented in the late 1970s as an official measure with both a French and a Spanish live vaccine freely issued to farmers until the late 1990s, where it was left at farmers' cost. During that period about 200,000 doses/year were used according to the Ministry of Agriculture (Tejedor, 1993).

The 1990s and the first decade of the 21st century saw a few experiments in the search for new vaccines taking advantage of the new molecular technologies available. Then, Australia's change of control policy against paratuberculosis in sheep – from eradication by stamping out infected flocks to control by vaccination – started generating a large amount of literature on paratuberculosis in general and on its control by vaccination in particular (Windsor *et al.*, 2002; Bush *et al.*, 2006).

This information was reviewed in a meta-analysis in 2011, where it was shown that vaccination induced beneficial results in all species in almost any setting (Bastida and Juste, 2011), despite results that fail to reach 100% protection. Still, although the idea of using vaccination in small ruminants, deer and camelids seems to be well established, the cattle industry is reluctant to use it for two reasons: interference with bovine tuberculosis diagnosis and fears of delay in eliminating risks for food safety due to the presence of *MAP* in milk and other products. Even so, a recent review on paratuberculosis control throughout the world showed that vaccination was in place in about 32% of 48 countries (Whittington *et al.*, 2019). Vaccine production has always been a reduced circle, with fabrication limited to a handful of laboratories, mostly from governments. In addition to the French vaccine, originally produced by Rhône-Merieux, Keldur central veterinary laboratory in Iceland and Weybridge central laboratory in the UK produced vaccine for national industry, as well as the national laboratories in Denmark and Norway. In the USA, the vaccine was produced by Solvay, later absorbed by Boehringer-Ingelheim, while in Spain, there was Ovejero in the 1980s, who produced a live vaccine that was later replaced by a killed one manufactured by CZVeterinaria (now CZVaccines). Currently, however, after discontinuation of commercial

Table 22.1. Paratuberculosis vaccine sales in the world in 2018.

Country	Doses
Australia	5,618,000
Spain	179,250
Republic of South Africa	159,000
The Netherlands[a]	105,510
Greece	90,090
New Zealand	70,000
France	50,550
Iceland	45,000
United Arab Emirates[b]	35,850
UK	30,000
Cyprus	24,000
Germany	22,250
USA[c]	>22,400
Denmark[d]	14,400
India[e]	10,000
Portugal	9000
Oman[b]	6000
Turkey[f]	2100
World	6,493,000

Source: CZVaccines Gudair sales except otherwise indicated.
[a] Mainly used in goats.
[b] Some used in camels.
[c] Used in cattle. The doses figure was provided by Dr Elisabeth Patton. The vaccine manufacturer, Boehringer Ingelheim, informed of discontinued vaccine production.
[d] Used in the Faroe islands.
[e] Vaccine licensed, but currently only in non-commercial production according to Professor Shoor Vir Singh.
[f] Used in mouflon. Local company (Vetal) advertise a paratuberculosis vaccine, but did not answer enquiries.

production of most of these vaccines, it seems that all the international paratuberculosis vaccine market is supplied by CZVaccines (Table 22.1), with the exception of some production in Turkey (Vetal) and in India.

22.3 Vaccine Types

Initially, the two main types of vaccine were live vaccines comprising attenuated strains of *MAP* and inactivated (killed) whole-cell vaccines. Later as vaccination technology advanced, fraction vaccines, subunit vaccines and vector DNA vaccines have been introduced in an attempt to address some of the problems that were identified in practice. Although each particular vaccine has used a different formulation and different strain, the most commonly used strain has been strain 316F, which was laboratory adapted and attenuated by the original French researchers (Table 22.2). Live, inactivated or mixed *MAP* strains have always been used with a strong adjuvant for parenteral inoculation that nearly exclusively has been an oil or wax (olive oil, mineral oil, paraffin, etc.). Although in the early times attenuated vaccines were the standard, storing and handling issues and concerns with the spread of bacteria in the environment led to substitution with inactivated cells (Shin *et al.*, 2008; Bull *et al.*, 2013). It is generally considered that inactivated vaccines do not trigger long-lasting immune responses. However, parenterally administered water-in-oil formulations of inactivated *MAP* produce an inflammation nodule from where the antigens are slowly released permanently stimulating the host immune system in such a way that a single dose at an early age could be sufficient for the whole productive life of vaccinated animals (Bastida and Juste, 2011). The main commercial vaccines developed to date are water-in-oil emulsions with killed (Gudair®, Silirum® and Mycopar®) (Bastida and Juste, 2011) or live-attenuated *MAP* (Neoparasec®, LioJohne®) (Tejedor, 1993; Molina *et al.*, 1996) (Table 22.2). Vaccination with these types of vaccines has been shown to be the most cost-effective measure, reducing paratuberculosis pathology, *MAP* excretion and environmental contamination, while also improving production figures (Bastida and Juste, 2011). However, fears of problems of interference with the immune tests used in bovine tuberculosis eradication programmes (Park and Yoo, 2016; Barkema *et al.*, 2018) as well as loss of differentiation between infected and vaccinated animals (DIVA) have heavily weighted against vaccination. Other issues include a limited efficacy in preventing infection and, to a lesser extent, injection-site tissue damage and accidental self-inoculation. While generation of tuberculosis false positives among paratuberculosis-vaccinated individuals can be readily eliminated by using the official (OIE) comparative tuberculin test with both PPD-B and PPD-A or including *M. tuberculosis* complex-specific reagents (Pérez de Val *et al.*, 2012; Garrido *et al.*, 2013; Park and Yoo, 2016;

Table 22.2. Vaccine types (Bastida and Juste, 2011); [a](Park and Yoo, 2016; Shippy *et al*., 2017); [b](Rosseels *et al*., 2006); [c](Bull *et al*., 2013).

	Route	Inactivation	Strain or antigen	Adjuvant/carrier	Natural host studied	Months at vaccination	Country
Live-non-attenuated	P		High-passage strains	oil, none	cattle	1	USA
Live-attenuated	P		316F	POP, oil paraffin, oil, saline, lipid	Cattle, sheep, goat	0.25, 0.5–1, 1, 1.5, 2, 4, adult	UK, France, Australia, Greece, Spain, New Zealand, Denmark, Germany
	P		316F+2E	POP	Goat	1	Norway
	P		avirulent strain	POP	Cattle	0.5	USA
	O		K10relA, K10pknG	Saline	Goat	8	USA
	P		989WAg906, 989WAg913, K10WAg915	Saline	None (mouse)		
	P		K10leuD, K10mpt64, K10secA2	Saline	None (mouse)		
	P[a]		K10sigL, K10sigH, K10lipN	Saponin-based (QuilA), none	Goat (sigL only in mouse)	1	USA
Inactivated (killed)	P	Heat, –	*Mycobacterium avium avium* Strain18	Oil, oil + hIL-12	Cattle	0–1, 0.25, 1, 2–5.5	USA
	P	Heat	–	Oil, none, CFA, oil paraffin	Cattle, sheep, goat	1, 1–24, 3, adult	USA, Netherlands, Spain, Iceland, Greece
	P	Heat	316F	Oil, lipid	Cattle, sheep, goat	0.5, 0.75, 1, 1–4, 2, 2–3, 3, 4, 4–6, 8, 24, adult	Australia, New Zealand, Spain, USA, UK, India
	P	Heat	5889 Bergey	Oil	Cattle	1	Hungary
	P, O	Heat	Field strains	Oil, –, alum	Sheep, goat	0.25, 0.5, 0.75, 1, 1.75, 2, 3, 4–6	UK, India

Continued

Table 22.2. Continued

	Route	Inactivation	Strain or antigen	Adjuvant/carrier	Natural host studied	Months at vaccination	Country
	P	Heat	Field strain CWD, CWC	Saponin-based (QS21), alum	Goat	0.25–1	USA
	P[b]	Irradiation	ATCC19698	Oil	None (mouse)		
Subunit	P		rec. hsp70	DDA	Cattle	1 + 11 (boost)	Netherlands
	P		rec. 85A, 85B, 85C, SOD	RAS	Cattle	0.3 + 1 (boost)	USA
	P		rec. 85A, 85B, 85C, SOD	RAS +bIL-12	Cattle	0.3 + 1 (boost)	USA
	P		rec. 85A, 85B, SOD, 74F	DDA, none	Goats	0.3 + 1 (boost)	USA
DNA	P		pVAX1 MEV encoding *MAP* genes	Gold microbeads	None (mouse)		
	P		pEGFP-N1 MEV encoding p85A-Mav, p85A-BCG or hsp65	Saline	Sheep	5 + 5.7 (boost) +6 (boost)	Italy
	P		pV1J.ns-tPA-his encoding MAP0586c or MAP4308c + protein boost	IFA (boost only)	None (mouse)		
	P[c]		Ad5.HAV priming +MVA.HAV boosting (encoding MAP genes)	–	Cattle	2 + 3.5 (boost)	UK

–, not indicated; P, parenteral; O, oral; POP, paraffin + olive oil + pumice; DDA, dimethyl dioctadecyl ammonium bromide; SOD, superoxide dismutase; hIL, human interleukin; bIL, bovine interleukin; RAS, Ribi's adjuvant system, consisting of bacterial monophosphoryl lipid A, trehalose dicorynomycolate and mycobacterial cell wall skeleton; CWD, cell wall deficient; CWC, cell wall competent; MEV, mammalian expression vector; CFA, complete Freund adjuvant; IFA, incomplete Freund adjuvant; HAV, priming with non-replicative human Adenovirus 5 followed by boosting with Modified Vaccinia virus Ankara delivery vectors expressing a fusion of *MAP* antigens.

Serrano *et al.*, 2017; Roy *et al.*, 2018), resolving the DIVA problem remains one of the main goals in paratuberculosis vaccine research.

Subunit vaccines or vectored DNA vaccine formulations based on immunogenic antigens have been developed with the goal of overcoming the above-reported potential diagnostic interference (Park and Yoo, 2016). A major role has been attributed to cell-mediated immunity in terms of protection from mycobacterial infection, and thus researchers have focused chiefly on antigens able to enhance this kind of immune response. The first reported paratuberculosis subunit vaccine evaluated in cattle consisted of recombinant heat shock protein 70 (Hsp70) adjuvanted with dimethyl dioctadecyl ammonium bromide for subcutaneous administration (Koets *et al.*, 2006). Other antigens with potential for being included in a subunit vaccine include lipoproteins (LprG and MAP0261c), PPE family proteins (MAP1518 and MAP3184), superoxide dismutase and alkyl hydroperoxide reductases (AhpC, AhpD), however these have not been evaluated as vaccine candidates in target animals (Park and Yoo, 2016). Initial investigations on DNA vaccination using expression vectors or viruses to deliver immunogenic *MAP* antigens showed promising results (Park and Yoo, 2016). According to interferon-gamma expression levels and other microbiological and pathological parameters, vaccination with plasmids encoding for 85A-BCG and Hsp65 seemed to elicit strong protective immune responses against *MAP* in sheep (Sechi *et al.*, 2006). A prime-boost vaccination strategy using non-replicative human Adenovirus 5 (prime) and Modified Vaccinia virus Ankara recombinant for *MAP*-specific antigens (HAV) showed a degree of protection against challenge in a calf model as assessed by different immunological markers and bacterial faecal shedding, while no cross-reactivity with tuberculin was observed and differentiation of vaccinated animals was enabled by a DIVA test (Bull *et al.*, 2014). Despite promise, the results indicate that these subunit or DNA vaccines have been less effective than was expected in terms of protection. This research line remains still active since more subunit vaccines have been proposed as having potential to prevent infection (Barkema *et al.*, 2018; Chapter 23, this volume).

Recent research on human anti-tuberculosis vaccines seemed to indicate that live genetically modified attenuated vaccines could better stimulate both cellular and humoral immune responses providing better protection than other formulations. Therefore, the search for suitable attenuated strains as vaccine candidates regained researchers' interest (Rathnaiah *et al.*, 2017). Many live-attenuated vaccine candidates have been produced by phage-mediated, transposon and allelic exchange mutagenesis (Park and Yoo, 2016). Whole-genome expression studies led to the identification of exploitable *MAP* genes for the construction of targeted mutant *MAP* strains as live-attenuated vaccine candidates (Wu *et al.*, 2007). Some of these vaccines induced strong protection in the mouse model, protection that was not replicated in other studies using the goat model (Barkema *et al.*, 2018). Very recently, superior protection has been reported for a live-attenuated lipN mutant as it was able to eliminate faecal shedding from experimentally challenged goats (Shippy *et al.*, 2017). However, the efficacy of such knockout vaccine candidates needs to be confirmed in cattle, under natural conditions, on a large scale and in long-term evaluation studies. In spite of their potential, disadvantages of live-attenuated vaccines need to be assessed: shedding of living and mutated microorganisms to the environment, virulence recovery, possible severe reactions, activation of immune modulation pathways, stability issues, producing-storing-handling difficulties, diagnostic interference and the lack of DIVA tests.

22.4 Vaccine Models

The chronic nature of paratuberculosis, as well as the difficulties in reproducing the clinical disease, together with the need to assay different approaches to better know the disease, its diagnosis, and the ways to prevent and eradicate it have required the development of experimental models (Hines *et al.*, 2007). In particular, the development of new vaccines or the design of novel vaccination strategies against paratuberculosis must be followed by a thorough evaluation involving experimental trials in the target species, which can be technically challenging, time consuming and very expensive. Accordingly, in addition to experimental models in the susceptible species cattle, sheep and goats, more convenient models have been developed. *In vitro* and *ex vivo* models can be useful for vaccine prototype testing in initial phases of vaccine development and experimental challenge models are essential for screening potential candidates that can be subsequently assessed in field trials with natural challenge. Finally, simulation studies based on computerized models can also be useful for vaccination programme designing.

22.4.1 *In vivo* models

Classical methods in the target species have been the standard since the first trials (Vallée *et al.*, 1941; Sigurdsson and Tryggvadóttir, 1949; Sigurdsson, 1960). However, the more recent research has been increasingly carried out on the mouse and rabbit model because of new animal experimentation rules and the need of higher numbers of individuals for testing more alternatives and to more strongly support conclusions (see Chapter 23 this volume). Small laboratory animal models also offer many other advantages that include thorough characterization of the immune response in immunologically defined animals, faster development of infection and disease, ease of handling and reduced cost. The mouse model has been used for vaccine evaluation of live-attenuated vaccines (Scandurra *et al.*, 2010; Chen *et al.*, 2012) against *MAP* and, although lesions do not develop in the intestine, histological and immunological changes similar to those observed in ruminants occur (Shin *et al.*, 2006; Scandurra *et al.*, 2010). The low cost and availability of reagents of this species make this model useful for preliminary analysis of vaccine candidates (Chiodini and Buergelt, 1993). On the other hand, rabbits are naturally susceptible to infection with *MAP* (Fuentes and Cebrian, 1988; Angus, 1990) and they develop granulomatous lesions in gut-associated lymphoid tissue as early as 20 weeks after oral challenge (Arrazuria *et al.*, 2015) providing an attractive animal to study aspects that cannot be studied in the mouse model. The rabbit model has been used to evaluate vaccination sequence (Arrazuria *et al.*, 2016a) and vaccination routes Ladero *et al.* (2018).

After initial screening of vaccines in small animal models, final testing should be done in ruminants. Ruminant challenge models include bovine, caprine, ovine and cervine (see Chapter 16, this volume). Subunit vaccines have been tested in cattle and sheep (Sechi *et al.*, 2006; Kathaperumal *et al.*, 2008). Nowadays, the goat model is the most widely used ruminant model for vaccination evaluation as it provides a direct homologue of the disease in cattle. It has been used to study live-attenuated vaccines (Scandurra *et al.*, 2010; Park *et al.*, 2011; Faisal *et al.*, 2013; Hines *et al.*, 2014; Shippy *et al.*, 2017), to evaluate protection and interference with bovine tuberculosis (Pérez de Val *et al.*, 2012) and also to evaluate vaccination sequence (Arrazuria *et al.*, 2016b).

22.4.2 *In vitro* and *ex vivo* models

It has been suggested that *in vitro* screening results do not translate to *in vivo* efficacy. *Ex vivo* assays combine both *in vivo* and *in vitro* aspects, since they begin with the vaccination of animals so that the immune response is mounted *in vivo* and then immune cells are isolated from these animals and an *in vitro MAP* infection assay is run. Monocytes are the main cell type used in these assays (Park *et al.*, 2011), although cocultures of monocytes with autologous lymphocytes have shown to be promising predictors of vaccine non-response (Pooley *et al.*, 2018).

22.4.3 Mathematical modelling and simulation studies

The availability of powerful computing capabilities has led to attempts to reproduce the main epidemiological characteristics of infectious diseases through mathematical modelling. Paratuberculosis is a good candidate for modelling studies aimed at designing improved control approaches, given the low sensitivity of conventional diagnostics and the slow nature of infection and disease. Thus, the first published model was based on a relatively simple matrix calculation method that allowed estimation of prevalence evolution and costs of two control strategies vs inaction. It clearly demonstrated that, although eradication could be reached by the two approaches, vaccination was by far more cost-effective than testing and culling (Juste and

Casal, 1993). Mathematical models have shown that they can aid in designing paratuberculosis control programmes bringing out critical issues and indicating the ideal way to integrate vaccination in control plans (Groenendaal *et al.*, 2002; Lu *et al.*, 2013a). Simulation studies with hypothetical vaccines have been used to increase knowledge on *MAP* transmission dynamics in the herds and to evaluate whether vaccination is able to prevent *MAP* invasion (Lu *et al.*, 2013b). Other models have confirmed this conclusion with more sophisticated methods (Groenendaal *et al.*, 2015). The drawback is that these predictive models need to be fuelled by data in order to be developed. However, they are useful and complementary to experimental challenge and field trial studies.

22.5 Particularities of Field Paratuberculosis Vaccination Evaluation

Paratuberculosis is an infection that has been unchecked for at least a century throughout most of the world and therefore is widespread nowadays. Since it is also chronic and subclinical in the majority of cases, defining protection criteria for vaccine evaluation is not an easy task. Therefore, as mentioned above, vaccine evaluation can be either summarized (Table 22.3) or focused on different variables according to the specific objective. In order to reduce direct paratuberculosis losses, studies can evaluate productive losses, specific and overall mortality, clinical incidence or pathological lesions. For transmission prevention, microbiological variables such as tissue bacterial load or faecal shedding are the most appropriate. Even more important for paratuberculosis vaccine evaluation are two issues: (i) interference with widespread official tuberculosis eradication programmes, which is the main reason for its limited use in cattle as mentioned before; and (ii) the risk of self-inoculation by operators that could cause local inflammation due to oil adjuvants.

Table 22.3. Average protection conferred by paratuberculosis vaccines (Bastida and Juste, 2011).

		Reports	Animals	Protection
Cattle	Killed	26	81,808	82.4
	Live	17	247,093	95.7
	All	**43**	328,901	**92.4**
Sheep	Killed	24	38,827	89.9
	Live	22	13,629	55.0
	All	**46**	52,456	**80.9**
Goats	Killed	15	1847	76.2
	Live	5	15,539	63.7
	All	**20**	17,386	**65.0**
Overall		**109**	398,743	**89.7**

22.5.1 Microbiological variables

Excretion of *MAP* in faeces is an objective parameter of vaccine effect. In this sense there are two related goals: (i) reduction of the number of shedding individual animals, and (ii) reduction of the amount of bacteria that accumulates in their tissues and can be spread in the environment. To detect *MAP*, isolation in pure culture has been the gold standard, although it is being increasingly replaced by the faster, cheaper and even more sensitive specific DNA amplification by polymerase chain reaction. It has been seen that, although a reduction in the number of shedders does not immediately occur, a decrease in the amount of bacteria in faeces is quickly and consistently observed (Argenté, 1992; Reddacliff *et al.*, 2006; Juste *et al.*, 2009; Sweeney *et al.*, 2009; Bannantine *et al.*, 2014; Hines *et al.*, 2014; Dhand *et al.*, 2016). This bacterial load can also be determined in post-mortem studies through the processing of tissue samples using microbiological techniques. This load is closely related to the type of lesion observed (Alonso-Hearn *et al.*, 2012; Hines *et al.*, 2014).

22.5.2 Pathological variables

Paratuberculosis vaccine effect is containment of the evolution of the lesion at resistance focal forms with low bacterial load. Therefore the analysis of the lesion type of the vaccinated animals in comparison with the others is another way to evaluate its effect (Juste *et al.*, 1994; Juste and Perez, 2011).

22.5.3 Clinical and productive indices

Reduction of clinical cases and of negative effects on milk or meat production are the target variables for industrial use of a paratuberculosis vaccine. However, for the reasons above mentioned, this might be difficult to determine and therefore several indicators have been used. In some studies changes in body condition scores or in the number of animals with diarrhoea have been assessed (Hagan, 1935; Doyle, 1959; Kalis *et al.*, 2001; Stringer *et al.*, 2013). In dairy cattle, field studies have compared milk production before and after establishing a paratuberculosis control programme through vaccination. Although factors other than vaccination could affect the time evolution of the infection in a herd, some studies conclude that there are increases in milk production (Juste *et al.*, 2009), as well as an improvement in the general state of health of the farm. As a consequence of this, some authors value the productive lifespan of vaccinated animals compared with unvaccinated animals (Alonso-Hearn *et al.*, 2012; Pérez de Val *et al.*, 2012).

22.5.4 Vaccination versus testing and culling

Although vaccine protection is not 100%, a reasonable estimate based on the literature reviewed above can provide an estimated protective efficacy of 89%. Using a simulation model (Juste and Casal, 1993) on a herd of 1000 individuals with a 10% infection and 25% replacement rate, this protection will drive infection to disappearance in 7 years. The simulation allows us to estimate that protection rate is not so critical above 60% and that infection will not disappear faster with a 100% protection rate, nor much slower with any protection above 60%. This must be contrasted with the alternative strategy of testing and culling that, with the highest sensitivity of any diagnostic test being about 50%, can only eliminate infection in 19 years. Only with a sensitivity of 100% is it possible to reduce this time to eradication to a period shorter than with vaccination. The difference is that at a cost of €10 per test per animal per year and €15 per vaccination per animal once in its lifetime, the accumulated cost–benefit ratio at 20 years is 2.72 and 9.3, respectively (Fig. 22.1). Vaccination thus appears to be a sustainable approach to paratuberculosis control that can be much easier to accept both by farmers (Doyle, 1964) and the general population since it preserves life instead of destroying it through culling. The model also indicates that any protection above 50% will eliminate infection faster than testing and culling with a test with a sensitivity below 50%. These results even seem to be improved in real field conditions as has been recently shown in a long-term vaccine field trial where vaccination

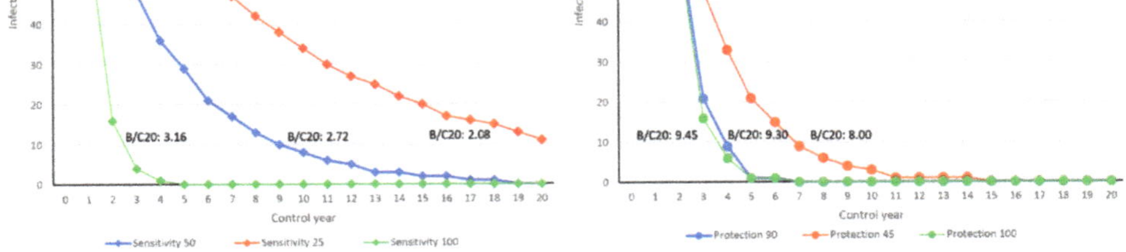

Fig. 22.1. Comparison of number of infected animals evolution according to test and vaccine performance (Juste and Casal, 1993). Notice that changes in sensitivity strongly affect time to eradication, while changes in protection have small effect.

eliminated shedding after 6 years while testing and culling took 8 years (Garrido *et al.*, 2018).

22.5.5 Interference with bovine tuberculosis diagnosis

One of the reasons why vaccination against paratuberculosis in cattle has been strongly limited is the interference with the immune techniques used in bovine tuberculosis eradication programmes, which have the highest priority because of the zoonotic nature of tuberculosis. However the problem can be easily fixed using either better interpretation algorithms or more specific antigens (Stabel *et al.*, 2011; Pérez de Val *et al.*, 2012; Garrido *et al.*, 2013; Hines *et al.*, 2014; Roy *et al.*, 2017; Serrano *et al.*, 2017).

22.5.6 Non-specific effects (NSE)

There is controversy regarding the NSE of vaccines, and in particular of BCG in humans (Wenner Moye, 2019). There are observations on field results that indicate that early administration of vaccines reduces overall child mortality (Timmermann *et al.*, 2015; Aaby *et al.*, 2017; Jenum *et al.*, 2018; Welaga *et al.*, 2018; Wenner Moye, 2019). This effect appears to be related to the so-called trained innate immunity (Kleinnijenhuis *et al.*, 2015; Gyssens and Netea, 2019), which could respond to primitive mechanisms such as an increase in macrophage lytic capacity (Juste *et al.*, 2016b; Pooley *et al.*, 2018). That seems to be confirmed in field conditions as there is about 36% of overall mortality reduction in vaccinated herds vs testing and culling that corresponds to animals younger than 2 years (Juste *et al.*, 2016a); that is, before the accepted paratuberculosis clinical onset age.

22.6 Conclusions

Vaccination is a useful tool that is widely and continually used in different species and settings throughout the world with highly satisfactory results for veterinarians and farmers. In spite of this, interference with bovine tuberculosis has hampered extensive use, an issue that cleaner DIVA vaccines could address. The potential for current killed vaccines to allow eradication of infection from susceptible domestic animals (detailed in 22.5.4) presents as an opportunity where farmer investment could secure uptake at a small additional production cost. Spread of the infection in the environment, however, poses a challenge to complete eradication of the infection, thus making necessary vaccination of replacement animals to keep them safe during their productive life. In summary, vaccination is a tool that deserves further scientific evaluation and research to strengthen its advantages and reduce its drawbacks, in order to maximize its potential to globally reduce the impact of paratuberculosis.

References

Aaby, P., Andersen, A., Ravn, H. and Zaman, K. (2017) Co-administration of BCG and diphtheria-tetanus-pertussis (DTP) vaccinations may reduce infant mortality more than the WHO-schedule of BCG first and then DTP. A re-analysis of demographic surveillance data from rural Bangladesh. *EBioMedicine* 22, 173–180. DOI: 10.1016/j.ebiom.2017.07.012.

Alonso-Hearn, M., Molina, E., Geijo, M., Vazquez, P., Sevilla, I.A. *et al.* (2012) Immunization of adult dairy cattle with a new heat-killed vaccine is associated with longer productive life prior to cows being sent to slaughter with suspected paratuberculosis. *Journal of Dairy Science* 95(2), 618–629. DOI: 10.3168/jds.2009-2860.

Angus, K.W. (1990) Intestinal lesions resembling paratuberculosis in a wild rabbit (*Oryctolagus cuniculus*). *Journal of Comparative Pathology* 103(1), 101–105. DOI: 10.1016/S0021-9975(08)80140-5.

Argenté, G. (1992) Efficiency of vaccination and other control measures estimated by fecal culturing in a regional program. In: Chiodini, R.J. and Kreeger, J.M. (eds) *Proceedings of the 3rd International Colloquium on Paratuberculosis*. IAP (Proc Int Coll PTBC), Providence, Rhode Island, USA, pp. 495–503.

Arrazuria, R., Molina, E., Mateo-Abad, M., Arostegui, I., Garrido, J.M. *et al.* (2015) Effect of various dietary regimens on oral challenge with *Mycobacterium avium* subsp. *paratuberculosis* in a rabbit model. *Research in Veterinary Science* 101, 80–83. DOI: 10.1016/j.rvsc.2015.06.006.

Arrazuria, R., Molina, E., Garrido, J.M., Pérez, V., Juste, R.A. *et al.* (2016a) Vaccination sequence effects on immunological response and tissue bacterial burden in paratuberculosis infection in a rabbit model. *Veterinary Research* 47(1). DOI: 10.1186/s13567-016-0360-y.

Arrazuria, R., Molina, E., Fernández, M. and Royo, M. (2016b) Effects of vaccination before or after *Mycobacterium avium* subsp. *paratuberculosis* experimental infection in goats. In: *13th International Colloquim on Paratuberculosis*. International Association for Paratuberculosis, Nantes, France.

Bannantine, J.P., Hines, M.E., Bermudez, L.E., Talaat, A.M., Sreevatsan, S. *et al.* (2014) A rational framework for evaluating the next generation of vaccines against *Mycobacterium avium* subspecies *paratuberculosis*. *Frontiers in Cellular and Infection Microbiology* 4(88). DOI: 10.3389/fcimb.2014.00126.

Barkema, H.W., Orsel, K., Nielsen, S.S., Koets, A.P., Rutten, V.P.M.G. *et al.* (2018) Knowledge gaps that hamper prevention and control of *Mycobacterium avium* subspecies *paratuberculosis* infection. *Transboundary and Emerging Diseases* 65, 125–148. DOI: 10.1111/tbed.12723.

Bastida, F. and Juste, R.A. (2011) Paratuberculosis control: a review with a focus on vaccination. *Journal of Immune Based Therapies and Vaccines* 9(1), 8. DOI: 10.1186/1476-8518-9-8.

Benedictus, G., Dinkla, E.T.B. and Wentink, G.H. (1988) Preliminary results of vaccination against paratuberculosis in adult dairy cattle. In: Thorel, M.F. and Merkal, R.S. (eds) *Proceedings of the Second International Colloquium on Paratuberculosis*. Laboratoire Central de Recherches Veterinaires (Proc Int Coll PTBC), Maisons-Alfort, France, pp. 136–140.

Berche, P. (2012) Louis Pasteur, from crystals of life to vaccination. *Clinical Microbiology and Infection* 18(Suppl 5), 1–6. DOI: 10.1111/j.1469-0691.2012.03945.x.

Brotherston, J.G. and Gilmour, N.J.L. (1961) Quantitative studies of *Mycobacterium johnei* in the tissues of sheep. I Routes of infection and assay of viable *M. johnei*. *Journal of Comparative Pathology* 71, 286–299.

Bull, T.J., Schock, A., Sharp, J.M., Greene, M., McKendrick, I.J. *et al.* (2013) Genomic variations associated with attenuation in *Mycobacterium avium* subsp. *paratuberculosis* vaccine strains. *BMC Microbiology* 13(1), 11. DOI: 10.1186/1471-2180-13-11.

Bull, T.J., Vrettou, C., Linedale, R., McGuinnes, C., Strain, S. *et al.* (2014) Immunity, safety and protection of an Adenovirus 5 prime--Modified Vaccinia virus Ankara boost subunit vaccine against *Mycobacterium avium* subspecies *paratuberculosis* infection in calves. *Veterinary Research* 45(1), 112. DOI: 10.1186/s13567-014-0112-9.

Bush, R.D., Windsor, P.A. and Toribio, J.A. (2006) Losses of adult sheep due to ovine Johne's disease in 12 infected flocks over a 3-year period. *Australian Veterinary Journal* 2570(84), 246–253.

Chen, J.-W., Faisal, S.M., Chandra, S., McDonough, S.P., Moreira, M.A.S. *et al.* (2012) Immunogenicity and protective efficacy of the *Mycobacterium avium* subsp. *paratuberculosis* attenuated mutants against challenge in a mouse model. *Vaccine* 30(19), 3015–3025. DOI: 10.1016/j.vaccine.2011.11.029.

Chiodini, R.J. and Buergelt, C.D. (1993) Susceptibility of BALB/c, C57/B6 and C57/B10 mice to infection with *Mycobacterium paratuberculosis*. *Journal of Comparative Pathology* 109(4), 309–319. DOI: 10.1016/S0021-9975(08)80295-2.

Dhand, N.K., Eppleston, J., Whittington, R.J. and Windsor, P.A. (2016) Changes in prevalence of ovine paratuberculosis following vaccination with Gudair®: results of a longitudinal study conducted over a decade. *Vaccine* 34(42), 5107–5113. DOI: 10.1016/j.vaccine.2016.08.064.

Doyle, T.M. (1959) Johne's disease. In: Stableforth, A.W. and Galloway, I.A. (eds). Butterworths Scientific Publications, pp. 319–345.

Doyle, T.M. (1960) Vaccination against Johne's disease. *British Veterinary Journal* 116(9), 294–301. DOI: 10.1016/S0007-1935(17)44051-6.

Doyle, T.M. (1964) Vaccination against Johne's disease. *The Veterinary Record* 76(3), 73–76.

Faisal, S.M., Chen, J.-W., Yan, F., Chen, T.-T., Useh, N.M. *et al.* (2013) Evaluation of a *Mycobacterium avium* subsp. *paratuberculosis leuD* mutant as a vaccine candidate against challenge in a caprine model. *Clinical and Vaccine Immunology* 20(4), 572–581. DOI: 10.1128/CVI.00653-12.

Fuentes, O. and Cebrian, J. (1988) El conejo de monte ¿reservorio de la paratuberculosis ovina y caprina? *Zaragoza: AIDA*, 32.

Garrido, J.M., Vazquez, P., Molina, E., Plazaola, J.M., Sevilla, I.A. *et al.* (2013) Paratuberculosis vaccination causes only limited cross-reactivity in the skin test for diagnosis of bovine tuberculosis. *PLoS ONE* 8(11), e80985. DOI: 10.1371/journal.pone.0080985.

Garrido, J.M., Geijo, M.V., Elguezabal, N., Juste, R.A. and Serrano, M. (2018) Long term results of an experimental vaccination trial in dairy cattle. *Proceedings of the 14th International Colloquim on Paratuberculosis*. Available at: http://www.paratuberculosis.net/proceedings/proc14. pdf (accessed 18 June 2020).

Gilmour, N.J.L. and Angus, K.W. (1973) Effect of revaccination on *Mycobacterium johnei* infection in sheep. *Journal of Comparative Pathology* 83(3), 437–445. DOI: 10.1016/0021-9975(73)90067-4.

Gilmour, N.J. and Brotherston, J.G. (1962) Quantitative studies of *Mycobacterium johnei* in the tissues of sheep. IV. The distribution of *M. johnei* shortly after oral dosing. *Journal of Comparative Pathology* 72, 165–169. DOI: 10.1016/s0368-1742(62)80020-4.

Groenendaal, H., Nielen, M., Jalvingh, A.W., Horst, S.H., Galligan, D.T. *et al.* (2002) A simulation of Johne's disease control. *Preventive Veterinary Medicine* 54(3), 225–245. DOI: 10.1016/S0167-5877(02)00027-2.

Groenendaal, H., Zagmutt, F.J., Patton, E.A. and Wells, S.J. *et al.* (2015) Cost-benefit analysis of vaccination against Mycobacterium avium ssp. paratuberculosis in dairy cattle, given its cross-reactivity with tuberculosis tests. *Journal of Dairy Science* 98(9), 6070–6084. DOI: 10.3168/jds.2014-8914.

Gunnarsson, E., Palsson, P.A. and Sigurdarson, S. (1984) *The control of paratuberculosis in sheep in Iceland by vaccination*. CEC (Proc Workshop PTBC), Copenhagen, pp. 123–130.

Gyssens, I.C. and Netea, M.G. (2019) Heterologous effects of vaccination and trained immunity. *Clinical Microbiology and Infection* 25(12), 1457–1458. DOI: 10.1016/j.cmi.2019.05.024.

Hagan, W.A. (1935) Vaccination against Johne's disease. *The Cornell Veterinarian* 25, 344–353.

Hillion, E. and Argente, G. (1987) Plan de lutte contre la paratuberculose dans les Cotes-du-Nord: aspects sanitaires et économiques. *Le Point Vétérinaire* 19(104), 123–130.

Hines, M.E., Stabel, J.R., Sweeney, R.W., Griffin, F., Talaat, A.M. *et al.* (2007) Experimental challenge models for Johne's disease: a review and proposed international guidelines. *Veterinary Microbiology* 122(3-4), 197–222.

Hines, M.E., Turnquist, S.E., Ilha, M.R.S., Rajeev, S., Jones, A.L. *et al.* (2014) Evaluation of novel oral vaccine candidates and validation of a caprine model of Johne's disease. *Frontiers in Cellular and Infection Microbiology* 4, 26. DOI: 10.3389/fcimb.2014.00026.

Hurley, S.S. (1983) Merkal, R.S. (ed.) *Efficacy of bacterins*. NADC, USDA (Proc Int Coll PTBC), Ames, Iowa, pp. 289–290.

Hurley, S.S. and Ewing, E. (1983) Merkal, R.S. (ed.) *Results of a field evaluation of a whole cell bacterin*, Edited by R. S. Merkal. NADC, USDA (Proc Int Coll PTBC), Ames, Iowa, pp. 244–248.

Jenum, S., Sumithra, S., Nelson, J., Christian, R., Anneke C, H. *et al.* (2018) Incidence of tuberculosis and the influence of surveillance strategy on tuberculosis case-finding and all-cause mortality: a cluster randomised trial in Indian neonates vaccinated with BCG. *BMJ Open Respiratory Research* 5(1), e000304. DOI: 10.1136/bmjresp-2018-000304.

Johne, H.A. and Frothingham, L. (1895) Ein eigenthümlicher fall von Tuberculose beim Rind. *Deutsche Zeitschrift fur Tiermedicin und Vergleichende Pathologie* 21, 438–454.

Jorgensen, J.B. (1983) Merkal, R.S. (ed.) *The effect of vaccination on the excretion of Mycobacterium paratuberculosis*. NADC, USDA (Proc Int Coll PTBC), Ames, Iowa, pp. 249–254.

Jorgensen, J.B. (1988) *Experiences with Vaccination Against Paratuberculosis in Danish Cattle*. CEC (CEC Workshop), Crete, Greece.

Juste, R., Vazquez, P., Geijo, M.V., Serrano, M., Elguezabal, N. *et al.* (2016a) Paratuberculosis vaccinated cattle lifespan and unspecific protection. *Proceedings of the 13th International Colloquim on Paratuberculosis*. Nantes, France: IAP. Available at: http://www.paratuberculosis.net/proceedings/proc13.pdf

Juste, R.A., Alonso-Hearn, M., Garrido, J.M., Abendaño, N., Sevilla, I.A. *et al.* (2016b) Increased lytic efficiency of bovine macrophages trained with killed mycobacteria. *Plos One* 11(11), e0165607. DOI: 10.1371/journal.pone.0165607.

Juste, R.A. and Casal, J. (1993) An economic and epidemiologic simulation of different control strategies for ovine paratuberculosis. *Preventive Veterinary Medicine* 15(2–3), 101–115. DOI: 10.1016/0167-5877(93)90106-4.

Juste, R.A. and Perez, V. (2011) Control of paratuberculosis in sheep and goats. *The Veterinary Clinics of North America. Food Animal Practice* 27(1), 127–138. DOI: 10.1016/j.cvfa.2010.10.020.

Kalis, C.H.J., Benedictus, G. and van Weering, H.J. (1992) Experiences with the use of an experimental vaccine in the control of paratuberculosis in the Netherlands. *Proceedings of the Third International Colloquium on Paratuberculosis*484–494.

Juste, R.A., García Marín, J.F., Peris, B., Sáez de Ocáriz, C., Badiola, J.J. *et al.* (1994) Experimental infection of vaccinated and non-vaccinated lambs with *Mycobacterium paratuberculosis*. *Journal of Comparative Pathology* 110(2), 185–194. DOI: 10.1016/S0021-9975(08)80189-2.

Juste, R.A., Alonso-Hearn, M., Molina, E., Geijo, M., Vazquez, P. *et al.* (2009) Significant reduction in bacterial shedding and improvement in milk production in dairy farms after the use of a new inactivated paratuberculosis vaccine in a field trial. *BMC Research Notes* 2(1), 233. DOI: 10.1186/1756-0500-2-233.

Kalis, C.H.J., Van Schaik, G., Dijkhuizen, A.A. and Benedictus, G. *et al.* (1995) *Economic significance of vaccination against paratuberculosis*, Edited by R. J. Chiodini, M. T. Collins, and E. O. E. Bassey. IAP (Proc Int Coll PTBC), Rehoboth, MA, USA, pp. 136–139.

Kalis, C.H.J., Hesselink, J.W. and Barkema, H.W. (1999) Long term use of killed vaccine does not prevent faecal shedding of *Mycobacterium avium* subsp. *paratuberculosis*. In: Manning, E.J.B. and Collins, M.T. (eds). IAP (Proc Int Coll PTBC), Madison, Wisconsin, pp. 153–156.

Kalis, C.H., Hesselink, J.W., Barkema, H.W. and Collins, M.T. (2001) Use of long-term vaccination with a killed vaccine to prevent fecal shedding of *Mycobacterium avium* subsp *paratuberculosis* in dairy herds. *American Journal of Veterinary Research* 62(2), 270–274. DOI: 10.2460/ajvr.2001.62.270.

Kalis, C.H.J., Barkema, H.W., Hesselink, J.W., van Maanen, C., Collins, M.T. *et al.* (2002) Evaluation of two absorbed enzyme-linked immunosorbent assays and a complement fixation test as replacements for fecal culture in the detection of cows shedding *Mycobacterium avium* subspecies paratuberculosis. *Journal of Veterinary Diagnostic Investigation* 14(3), 219–224. DOI: 10.1177/104063870201400305.

Kathaperumal, K., Park, S.-U., McDonough, S., Stehman, S., Akey, B. *et al.* (2008) Vaccination with recombinant *Mycobacterium avium* subsp. *paratuberculosis* proteins induces differential immune responses and protects calves against infection by oral challenge. *Vaccine* 26(13), 1652–1663. DOI: 10.1016/j.vaccine.2008.01.015.

Kleinnijenhuis, J., van Crevel, R. and Netea, M.G. (2015) Trained immunity: consequences for the heterologous effects of BCG vaccination. *Transactions of the Royal Society of Tropical Medicine and Hygiene* 109(1), 29–35. DOI: 10.1093/trstmh/tru168.

Koets, A., Hoek, A., Langelaar, M., Overdijk, M., Santema, W. *et al.* (2006) Mycobacterial 70 kD heat-shock protein is an effective subunit vaccine against bovine paratuberculosis. *Vaccine* 24(14), 2550–2559. DOI: 10.1016/j.vaccine.2005.12.019.

Ladero, I., Arrazuria, R., Molina, E. and Fernández, M. (2018) Effects of vaccination route on paratuberculosis infection in a rabbit model. In: *14 Int. Colloq. on Paratuberculosis*. 38. International Association for Paratuberculosis, Riviera Maya, Mexico.

Larsen, A.B. (1950) Vaccination against Johne's disease. *Journal of the American Veterinary Medical Association*, 201–206.

Larsen, A.B., Hawkins, W.W. and Merkal, R.S. (1964) Experimental vaccination of sheep against Johne's disease. *American Journal of Veterinary Research* 25(107), 974–976.

Larsen, A.B., Merkal, R.S., Kopecky, K.E. and Boothe, A.D. (1969) Hypersensitivity and serologic responses in cattle vaccinated with disrupted *Mycobacterium paratuberculosis* cells and subsequently infected with *Mycobacterium bovis*. *American Journal of Veterinary Research* 30(12), 2167–2172.

Larsen, A.B., Merkal, R.S. and Moon, H.W. (1974) Evaluation of a paratuberculosis vaccine given to calves before infection. *American Journal of Veterinary Research* 35(3), 367–369.

Larsen, A.B., Moyle, A.I. and Himes, E.M. (1978) Experimental vaccination of cattle against paratuberculosis (Johne's disease) with killed bacterial vaccines: a controlled field study. *American Journal of Veterinary Research* 39(1), 65–69.

Lu, Z., Schukken, Y.H., Smith, R.L. and Gröhn, Y.T. (2013a) Using vaccination to prevent the invasion of *Mycobacterium avium* subsp. *paratuberculosis* in dairy herds: a stochastic simulation study. *Preventive Veterinary Medicine* 110(3–4), 335–345. DOI: 10.1016/j.prevetmed.2013.01.006.

Lu, Z., Schukken, Y.H., Smith, R.L., Mitchell, R.M., Gröhn, Y.T. *et al.* (2013b) Impact of imperfect *Mycobacterium avium* subsp. *paratuberculosis* vaccines in dairy herds: a mathematical modeling approach. *Preventive Veterinary Medicine* 108(2–3), 148–158. DOI: 10.1016/j.prevetmed.2012.08.001.

Merkal, R.S., Larsen, A.B., Velicer, L.F. and Thurston, J.R. (1965) Serologic and allergenic effects of 3 *Mycobacterium paratuberculosis* antigens. *American Journal of Veterinary Research* 26(115), 1267–1270.

Molina, J.M., Anguiano, A. and Ferrer, O. (1996) Study on immune response of goats vaccinated with a live strain of *Mycobacterium paratuberculosis*. *Comparative Immunology, Microbiology and Infectious Diseases* 19(1), 9–15. DOI: 10.1016/0147-9571(95)00022-4.

Nieuwenhuizen, N.E. and Kaufmann, S.H.E. (2018) Next-generation vaccines based on Bacille Calmette–Guérin. *Frontiers in Immunology* 9, 121. DOI: 10.3389/fimmu.2018.00121.

Nisbet, D. I., Gilmour, N.J. and Brotherston, J.G. (1962) Quantitative studies of *Mycobacterium johnei* in tissues of sheep. III. intestinal histopathology. *Journal of Comparative Pathology* 72, 80–91. DOI: 10.1016/s0368-1742(62)80009-5.

OIE, World Organisation for Animal Health (2013) OIE. Available at: http://www.oie.int/en/for-the-media/rinderpest/ (accessed 17 June 2019).

Park, H.-T. and Yoo, H.S. (2016) Development of vaccines to *Mycobacterium avium* subsp. *paratuberculosis* infection. *Clinical and Experimental Vaccine Research* 5(2), 108–116. DOI: 10.7774/cevr.2016.5.2.108.

Park, K.T., Allen, A.J., Bannantine, J.P., Seo, K.S., Hamilton, M.J. *et al.* (2011) Evaluation of two mutants of *Mycobacterium avium* subsp. *paratuberculosis* as candidates for a live attenuated vaccine for Johne's disease. *Vaccine* 29(29–30), 4709–4719. DOI: 10.1016/j.vaccine.2011.04.090.

Pérez de Val, B., Nofrarías, M., López-Soria, S., Garrido, J.M., Vordermeier, H. *et al.* (2012) Effects of vaccination against paratuberculosis on tuberculosis in goats: diagnostic interferences and cross-protection. *BMC Veterinary Research* 8(1), 191. DOI: 10.1186/1746-6148-8-191.

Pooley, H.B., Plain, K.M., Purdie, A.C., Begg, D.J., Whittington, R.J. *et al.* (2018) Integrated vaccine screening system: using cellular functional capacity in vitro to assess genuine vaccine protectiveness in ruminants. *Pathogens and Disease* 76(3). DOI: 10.1093/femspd/fty029.

Rathnaiah, G., Zinniel, D.K., Bannantine, J.P., Stabel, J.R., Gröhn, Y.T. *et al.* (2017) Pathogenesis, molecular genetics, and genomics of *Mycobacterium avium* subsp. *paratuberculosis*, the etiologic agent of Johne's disease. *Frontiers in Veterinary Science* 4, 187. DOI: 10.3389/fvets.2017.00187.

Reddacliff, L., Eppleston, J., Windsor, P., Whittington, R., Jones, S. *et al.* (2006) Efficacy of a killed vaccine for the control of paratuberculosis in Australian sheep flocks. *Veterinary Microbiology* 115(1–3), 77–90. DOI: 10.1016/j.vetmic.2005.12.021.

Rosseels, V., Marché, S., Roupie, V., Govaerts, M., Godfroid, J. *et al.* (2006) Members of the 30- to 32-kilodalton mycolyl transferase family (Ag85) from culture filtrate of *Mycobacterium avium* subsp. *paratuberculosis* are immunodominant Th1-type antigens recognized early upon infection in mice and cattle. *Infection and Immunity* 74(1), 202–212. DOI: 10.1128/IAI.74.1.202-212.2006.

Roy, A., Risalde, M.A., Casal, C., Romero, B., de Juan, L. *et al.* (2017) Oral vaccination with heat-inactivated *Mycobacterium bovis* does not interfere with the antemortem diagnostic techniques for tuberculosis in goats. *Frontiers in Veterinary Science* 4, 124. DOI: 10.3389/fvets.2017.00124.

Roy, Á., Infantes-Lorenzo, J.A., Blázquez, J.C., Venteo, Ángel., Mayoral, F.J. *et al.* (2018) Temporal analysis of the interference caused by paratuberculosis vaccination on the tuberculosis diagnostic tests in goats. *Preventive Veterinary Medicine* 156, 68–75. DOI: 10.1016/j.prevetmed.2018.05.010.

Saint-Marc, B., Argenté, G., Guillemin, F., Milward, F., Lacoste, F. *et al.* (1992) Efficiency of a live 316 F strain vaccine against paratuberculosis in ruminants. *Anonymous*, 96–101.

Saxegaard, F. (1984) Controle de la paratuberculose (Maladie de Johne) des chévres POR vaccination en Noruége. In: Yvore, P. and Perrin, G.G. (eds). INRA, Niort, France, pp. 541–550.

Saxegaard, F. and Fodstad, F. (1985) Control of paratuberculosis (Johne's disease) in goats by vaccination. *Veterinary Record* 116(16), 439–441. DOI: 10.1136/vr.116.16.439.

Scandurra, G.M., de Lisle, G.W., Cavaignac, S.M., Young, M., Kawakami, R.P. *et al.* (2010) Assessment of live candidate vaccines for paratuberculosis in animal models and macrophages. *Infection and Immunity* 78(3), 1383–1389. DOI: 10.1128/IAI.01020-09.

Sechi, L.A., Mara, L., Cappai, P., Frothingam, R., Ortu, S. *et al.* (2006) Immunization with DNA vaccines encoding different mycobacterial antigens elicits a Th1 type immune response in lambs and protects against *Mycobacterium avium* subspecies *paratuberculosis* infection. *Vaccine* 24(3), 229–235. DOI: 10.1016/j.vaccine.2005.08.086.

Serrano, M., Elguezabal, N., Sevilla, I.A., Geijo, M.V., Molina, E. *et al.* (2017) Tuberculosis detection in paratuberculosis vaccinated calves: new alternatives against interference. *PLoS ONE* 12(1), e0169735. DOI: 10.1371/journal.pone.0169735.

Shin, S.J., Wu, C.-W., Steinberg, H. and Talaat, A.M. (2006) Identification of novel virulence determinants in *Mycobacterium paratuberculosis* by screening a library of insertional mutants. *Infection and Immunity* 74(7), 3825–3833. DOI: 10.1128/IAI.01742-05.

Shin, S.J., Cho, D. and Collins, M.T. (2008) Diagnosis of bovine paratuberculosis by a novel enzyme-linked immunosorbent assay based on early secreted antigens of *Mycobacterium avium* subsp. *paratuberculosis*. *Clinical and Vaccine Immunology* 15(8), 1277–1281. DOI: 10.1128/CVI.00105-08.

Shippy, D.C., Lemke, J.J., Berry, A., Nelson, K., Hines, M.E. *et al.* (2017) Superior protection from live-attenuated vaccines directed against Johne's disease. *Clinical and Vaccine Immunology* 24(1). DOI: 10.1128/CVI.00478-16.

Sigurdsson, B. (1954) RIDA, a chronic encephalitis of sheep: with general remarks on infections which develop slowly and some of their special characteristics. *The British Veterinary Journal* 110(9), 341–354.

Sigurdsson, B. (1960) A killed vaccine against paratuberculosis (Johne's disease) in sheep. *American Journal of Veterinary Research* 21, 54–67.

Sigurdsson, B. and Tryggvadóttir, A.G. (1949) Immunization with heat-killed *Mycobacterium paratuberculosis* in mineral oil. *Journal of Bacteriology* 58(3), 271–278. DOI: 10.1128/JB.58.3.271-278.1949.

Stabel, J.R., Waters, W.R., Bannantine, J.P. and Lyashchenko, K. (2011) Mediation of host immune responses after immunization of neonatal calves with a heat-killed *Mycobacterium avium* subsp. *paratuberculosis* vaccine. *Clinical and Vaccine Immunology* 18(12), 2079–2089. DOI: 10.1128/CVI.05421-11.

Stringer, L.A., Wilson, P.R., Heuer, C. and Mackintosh, C.G. (2013) A randomised controlled trial of Silirum vaccine for control of paratuberculosis in farmed red deer. *Veterinary Record* 173(22), 551. DOI: 10.1136/vr.101799.

Stuart, P. (1962) The diagnosis of Johne's disease in cattle and the effect of vaccination on tuberculin and johnin tests. *Bulletin de l'Office International de Epizooties* 58, 33–50.

Stuart, P. (1965) Vaccination against Johne's disease in cattle exposed to experiment infection. *British Veterinary Journal* 121(7), 289–318. DOI: 10.1016/S0007-1935(17)41102-X.

Sweeney, R.W., Whitlock, R.H., Bowersock, T.L., Cleary, D.L., Meinert, T.R. *et al.* (2009) Effect of subcutaneous administration of a killed *Mycobacterium avium* subsp *paratuberculosis* vaccine on colonization of tissues following oral exposure to the organism in calves. *American Journal of Veterinary Research* 70(4), 493–497. DOI: 10.2460/ajvr.70.4.493.

Tejedor, F.J. (1993) *Estudio epidemiológico de la paratuberculosis ovina en la provincia de Segovia.* Universidad Complutense de Madrid.

Timmermann, C.A.G., Biering-Sørensen, S., Aaby, P., Fisker, A.B., Monteiro, I. *et al.* (2015) Tuberculin reaction and BCG scar: association with infant mortality. *Tropical Medicine & International Health* 20(12), 1733–1744. DOI: 10.1111/tmi.12614.

Twort, F.W. and Ingram, G.L.Y. (1911) A method for isolating and cultivating the *Mycobacterium enteriditis chronicae pseudotuberculosae bovis*, Johne, and some experiments on the preparation of a diagnostic vaccine for pseudo-tuberculous enteritis of bovines. *Royal Society Proceedings on Biology* 84, 517–542.

Vallée, H. and Rinjard, P. (1926) Etudes sur l'entérite paratuberculeuse des bovidés. *Revue Générale de Médicine Vétérinaire* 409, 1–9.

Vallée, H., Rinjard, P. and Vallée, M. (1941) Sur la prémunisation de l'entérite paratuberculeuse due au bacille de Johne. *Bulletin de l'Academie de médecine*, 195–198.

van Schaik, G., Kalis, C.H., Benedictus, G., Dijkhuizen, A.A., Huirne, R.B. *et al.* (1996) Cost-benefit analysis of vaccination against paratuberculosis in dairy cattle. *The Veterinary Record* 139(25), 624–627.

Welaga, P., Debpuur, C., Aaby, P., Hodgson, A., Azongo, D.K. *et al.* (2018) Is the decline in neonatal mortality in northern Ghana, 1996–2012, associated with the decline in the age of BCG vaccination? An ecological study. *BMJ Open* 8(12), e023752. DOI: 10.1136/bmjopen-2018-023752.

Wenner Moye, M. (2019) Could a single live vaccine protect against a multitude of diseases?. *Scientific American*. Available at: https://www.scientificamerican.com/article/could-a-single-live-vaccine-protect-against-a-multitude-of-diseases/?redirect=1 (accessed 19 June 2019).

Whittington, R., Donat, K., Weber, M.F., Kelton, D., Nielsen, S.S. *et al.* (2019) Control of paratuberculosis: who, why and how. A review of 48 countries. *BMC Veterinary Research* 15(1), 198. DOI: 10.1186/s12917-019-1943-4.

WHO (2018) Smallpox. World Health Organization. Available at: https://www.who.int/csr/disease/smallpox/en/ (accessed 17 June 2019).

Windsor, P.A.*et al.* (2002) Efficacy of a killed Mycobacterium avium subsp. paratuberculosis vaccine for the control of OJD in Australian sheep flocks. *Proceedings of the International Colloquium on Paratuberculosis*, pp. 420–423.

Wu, C.-W., Schmoller, S.K., Shin, S.J. and Talaat, A.M. (2007) Defining the stressome of *Mycobacterium avium* subsp. *paratuberculosis* in vitro and in naturally infected cows. *Journal of Bacteriology* 189(21), 7877–7886. DOI: 10.1128/JB.00780-07.

23 Development of New Paratuberculosis Vaccines

Tim Bull*
St George's University, London, UK

23.1 Introduction

Infection of ruminants, particularly cattle, by *Mycobacterium avium* subsp. *paratuberculosis* (*MAP*) is common in most countries and leads to sufficient clinical paratuberculosis to result in significant economic losses (Garcia and Shalloo, 2015; Kirkeby *et al.*, 2016). Most national control programmes for paratuberculosis place priority on the reduction of clinical disease, because eradication of infection remains unrealistic. This is because the insensitivity of diagnostic testing enables a proportion of subclinical animals to remain as a continuing source of herd infection in test-and-cull programmes (Nielsen, 2008; Dernivoix *et al.*, 2017). Against this background, the implementation of large-scale vaccination programmes appears an attractive alternative. However, the provision of an effective *MAP* vaccine has been frustratingly elusive. Despite 80 years of vaccine development, no formulation has yet been found that can protect all animals against disease, nor provide total protection from infection. Current whole-cell vaccines often cause large granulomas at the inoculation site (Windsor *et al.*, 2005; Musk *et al.*, 2019) and induce cross-reactivity to tuberculin screening tests (Roy *et al.*, 2018; Roupie *et al.*, 2018) providing significant disincentive towards their usage, particularly in cattle (Rosseels and Huygen, 2008).

The principal reasons behind these failures remain far from clear but stem from the exquisite ability of *MAP* to subvert immune responses allowing it to establish long-term intracellular persistence (Bastida and Juste, 2011; Frie *et al.*, 2017). It is probable that factors inherent within the choice of strains, method of attenuation or vaccine delivery provide suboptimal induction of early innate and adaptive cellular mechanisms essential to clear *MAP* effectively during primary infection, especially in particularly susceptible animals such as neonates (Mortier *et al.*, 2013). Chronic *MAP* persistence in the host is not an inactive state as intracellular *MAP* infection has the capacity to induce suppressive immune responses that promote the development of dysregulated immune responses predisposing to clinical disease. Suppression can also interfere with expansion of vaccine-primed, long-term protective mechanisms, which, although sufficient to contain clinical manifestations in some animals, remain ineffective against intestinal infection and shedding (Coussens, 2004). Contributing factors also include: the use of non-*MAP* strains for whole-cell vaccines (Uzonna *et al.*, 2003); the reliance on non-immunodominant *MAP*-specific antigens; insufficient attenuation of live vaccine strains resulting in mycobacterial mechanisms that subvert immune responses to remain active (Barkema *et al.*, 2018); the use of vaccines that

*tim.bull@sgul.ac.uk

© CAB International 2020. *Paratuberculosis: Organism, Disease, Control, 2nd Edition* (eds M.A. Behr *et al.*)

skew towards major humoral responses that are not fully protective (Santema *et al.*, 2011); and the use of modes of antigen delivery that fail to promote correct Th1 imprinting and immune memory (Griffin *et al.*, 2009).

The history of vaccine use (covered more fully in Chapter 22, this volume) includes the use of a variety of approaches including classically derived, but poorly characterized, live attenuated or killed vaccine strains, *MAP* protein subunits and even *MAP* glycolipid extractions (Jolly *et al.*, 2013). More recent studies reverted back to using live attenuated strains (LAV) but with improved definitions, derived by directed knockout of *MAP* genes involved in various aspects of virulence. Early vaccine strains from France (Vallée *et al.*, 1934) were probably not markedly attenuated (Doyle, 1964). They were used live (Stuart, 1965) and relied on subcutaneous compartmentalization in an oil and pumice-based adjuvant to optimize antigen presentation, inhibit growth and prevent dissemination (Wilesmith, 1982). Later, a combinational preparation of live UK strains (316F, strain II and 2e) administered in an oil-based emulsion to sheep (Sigurdsson, 1960) and goats (Saxegaard and Fodstad, 1985) achieved some success but stocks were poorly maintained and are no longer available. Genomic analysis showed strains 2e and II, but not 316F, to have significant deletions in *MAP*-specific genomic regions (Bull *et al.*, 2013) and were attenuated to some extent in mice or when delivered orally to cattle (Watkins *et al.*, 2010).

A single culture of 316F from the 1907s was used as an original seed stock for various live vaccine formulations tested in France and Hungary in the 1980s (Körmendy, 1994), in New Zealand up until 2002 (Begg and Griffin, 2005) and in a more contemporary study showed significant reduction in mortalities in Greek sheep over a 4-year period (Dimareli-Malli *et al.*, 2013). Preliminary genomic studies of 316F and 316 v, a similar Weybridge strain used for *MAP* enzyme-linked immunosorbent assay (ELISA) testing in Australia from 1986 (Milner *et al.*, 1987) suggested them to have a similar gene complement to the virulent reference strain K-10. This was confirmed by microarray analysis, which also revealed that some subculture stocks of 316F had gained large genomic deletions generated during long-term exposure in suboptimal minimal culture media used for maintenance (Bull *et al.*, 2013).

Concerns about the use of live strains on the grounds of health and safety, short shelf life and potential spread to the environment led to the introduction of heat-killed, whole-cell vaccines (Emery and Whittington, 2004). The only *MAP* vaccine licensed for use in the USA uses a killed strain of *Mycobacterium avium* subspecies *avium* (Rathnaiah *et al.*, 2017). An indigenous killed *MAP* strain isolated from bison has been used in some Indian studies (Singh *et al.*, 2013); however, all other major formulations derive from a series of 316F strain subcultures grown in various types of liquid cultures and administered at different concentrations (Emery and Whittington, 2004) or as *MAP* cell wall-deficient preparations (Hines *et al.*, 2007b). One commercial killed 316F vaccine (Gudair) has been trialled extensively and registered in a national control programme (Reddacliff *et al.*, 2006). Yet, even with improvements in adjuvant and delivery (Silirum), 316F killed whole-cell vaccines are unable to eliminate shedding (Hines *et al.*, 2014) and *MAP* infection from herds (Juste *et al.*, 2009). Further details can be found in Chapter 22, this volume.

A shift in direction beginning after 2005 towards development of subunit vaccines incorporating immunodominant *MAP* proteins was initiated primarily to avoid the induction of suppressive immune responses by whole-cell vaccine components and circumvent problems of cross-reactions with tuberculin in cattle. The availability of genomic and proteomic (Leroy *et al.*, 2007; Hughes *et al.*, 2008) arrays provided tools identifying a range of potential immunodominant targets. The breadth of immune stimulation, longevity and type of immunological memory provided by these vaccines is still poorly understood and, so far, despite testing with a variety of delivery systems and immune-enhancing substrates, very few subunit vaccines have shown significantly better reduction of faecal shedding than whole-cell killed adjuvanted vaccines and none has been taken sufficiently far to show protection from disease. The observations that live Bacillus Calmette–Guérin (BCG) can prime or generate 'trained immunity' (Kleinnijenhuis *et al.*, 2012) that can be enhanced through subsequent directed subunit boosting and protect against subsequent mycobacterial challenge

Table 23.1. Summary of live attenuated strain (LAV) candidates tested in animals.

Gene knockout target and parent strain	Other name(s)	Function	Selection tag	Attenuation	Tested species	Challenge strain	Dose and route of infection	Comments
				Scandurra *et al.*, 2010; Kabara and Coussens, 2012				
MAP1566 (MAP989)	TM58 WAg906 MAP_RS07970 JDIP322	Unknown hypothetical	Kanamycin	Positive trend in MDM survival after 7 days. Two-log decrease in bovine macrophages at 14 days. Reduced MDM apoptosis vs wild type. 10^8 CFU ip. in BALB/c mice gave four-log decrease in spleen and liver at 12 weeks. 2×10^9 CFU iv. in goats not detected in tissue after 6–10 months	BALB/c mice	MAPK10	2×10^8 CFU ip. 7 weeks post-immunization	Immunized 2×10^6 CFU in PBS sc. No protection in spleen and liver at 12 weeks
MAP0053c (MAP989)	WAg913 MAP_RS00280	GntR family transcriptional regulator	Kanamycin	Two-log decrease in bovine macrophages at 14 days. Reduced MDM apopotsis vs wild type. No decrease in mice spleen and liver at 12 weeks. 90% clearance in goats challenged 10^9 iv. after 6–10 months				
MAP0011 (MAPK10)	WAg915 ppiA MAP_RS00065 JDIP323	Peptidyl-prolyl cis-trans isomerase	Hygromycin	One-log decrease in bovine MDM at 14 days. Reduced MDM apoptosis vs wild type. No decrease in mice spleen and liver at 12 weeks. 80% clearance in goats challenged 10^9 iv after 6 months				
				Shin *et al.*, 2006; Settles *et al.*, 2014; Bannantine *et al.*, 2014a				
MAP3963 (ATCC19698)	umaA1 MAP_RS20325 JDIP326 pgs3965	SAM-dependent methyltransferase	Kanamycin	Negative trend in bovine MDM survival after 7 days. 1×10^7 CFU innoculated ip. C57BL/6 mice gave one-log decrease in liver and intestinal tissue at 12 weeks	C57BL/6 mice	MAPK10	1×10^8 CFU ip. 5 weeks after last immunization	Immunized 1×10^6 CFU sc. boosted at week 2 gave one log decrease in spleen, liver and gut mucosa relative to PBS only at 18 weeks
MAP2408c (ATCC19698)	fabG2_2 MAP_RS12280 JDIP329 pgs1360	3-oxoacyl-[acyl-carrier-protein] reductase	Kanamycin	Positive trend in MDM survival after 7 days. Increased MDM apoptosis vs wild type ATCC19698. 1×10^7 CFU ip. C57BL/6 mice gave three-log decrease in liver and intestinal tissue at 12 weeks. 1 × 105 CFU ip. in C57BL/6 gave one-log decrease in spleen and liver at 18 weeks				
				Chen *et al.*, 2012; Faisal *et al.*, 2013a				
MAP3025c (MAPK10)	leuD MAP_RS15490	3-isopropylmalate dehydratase	Hygromycin	5×10^8 CFU innoculated ip. Into C57BL/6 mice. Negative AFB stain in liver but positive in spleen at 12 weeks	Goats	MAP 6615-98 clinical isolate from cow	5×10^8 CFU oral 3 weeks after last immunization	Immunized 5 × 108 CFU sc. boosted at week 3. Two out of five vaccinated animals negative gut tissue culture at 6 months. All animals shedding MAP in faeces throughout but decreased in vaccinated group
MAP3291c (MAPK10)	mpt64 MAP_RS16915	Membrane protein homologue involved in apoptosis of multinucleated giant cells in *M. tuberculosis*	Hygromycin	5×10^8 CFU inoculated ip. into C57BL/6 mice. Positive AFB present in spleen and liver at 12 weeks	C57BL/6 mice			
MAP1534 (MAPK10)	secA2 MAP_RS07815	Accessory Sec system translocase	Hygromycin					

Continued

Table 23.1. Continued

Gene knockout target and parent strain	Other name(s)	Function	Selection tag	Attenuation	Tested species	Challenge strain	Dose and route of infection	Comments
		Kabara and Coussens, 2012; Lamont *et al.*, 2014; Bannantine *et al.*, 2014a, b; Hines *et al.*, 2014; Rathnaiah *et al.*, 2014						
MAP1566 (MAPK10)	30H9 MAP_RS07970 JDIP319	Unknown protein	Kanamycin	Negative trend in bovine MDM survival after 7 days. Increased MDM apoptosis vs wild type ATCC19698. 1×10^5 CFU ip. in C57BL/6 not found in spleen and liver at 18 weeks	C57BL/6 mice	MAPK10	6.5×10^6 CFU ip. 6 weeks after last immunization	Immunized 3×10^3–4×10^5 CFU ip. No significant protection at 12 weeks post- challenge relative to PBS sham vaccination control
MAP1566 (MAPK10)	STM68 MAP_RS07970 JDIP315	Unknown protein	Kanamycin	Negative trend in bovine MDM survival after 7 days. 1 × 105 CFU ip. in C57BL/6 not found in spleen and liver at 18 weeks				
MAP3694c–MAP3695 (MAPK10)	2E11 MAP_RS18945–MAP_ RS18950 JDIP316	Intragenic	Kanamycin					
MAP0282c–MAP0283c (MAPK10)	40A9 unannotated sequences JDIP318	Intragenic	Kanamycin	Negative trend in bovine MDM survival after 7 days. 1 × 105 CFU ip. in C57BL/6 gave one-log decrease in spleen and liver at 18 weeks				
MAP1150c–MAP1151c (MAPK10)	4H2 MAP_RS05855–MAP_ RS22585 JDIP321	Intragenic	Kanamycin	Positive trend in MDM survival after 7 days. No reduction in MDM apoptosis. 1×10^5 CFU ip. in C57Bl/6 not found in spleen and liver at 18 weeks				
MAP0460 (MAPK10)	22F4 MAP_RS02360 JDIP317	Lsr2 family protein	Kanamycin	Positive trend in MDM survival after 7 days. Increased MDM apoptosis vs wild type ATCC19698. 1×10^7 CFU ip. C57BL/6 mice gave three-log decrease in liver and intestinal tissue at 12 weeks. 1 × 105 CFU ip. in C57BL/6 gave one-log decrease in spleen and liver at 18 weeks				
MAP2408c (ATCC19698)	fabG2_2 MAP_RS12280 JDIP329 pgs1360	3-oxoacyl-[acyl-carrier-protein] reductase	Kanamycin	Positive trend in MDM survival after 7 days. 1 × 105 CFU ip. in C57BL/6 gave one-log decrease in spleen and liver at 18 weeks				
MAP2296c–MAP2297c (MAPK10)	3H4 MAP_RS11700–MAP_ RS11705 JDIP320	Intergenic	Kanamycin	Negative trend in bovine MDM survival after 7 days. 1 × 105 CFU ip. in C57BL/6 gave one-log decrease in spleen and liver at 18 weeks				

Continued

Table 23.1. Continued

Gene knockout target and parent strain	Other name(s)	Function	Selection tag	Attenuation	Tested species	Challenge strain	Dose and route of infection	Comments
Ghosh *et al.*, 2013, 2015; Shippy *et al.*, 2017								
MAP3323c (MAPK10)	sigH MAP_RS17080	Mycothiol system anti-sigma-R factor	Hygromycin	Decreased survival vs MAPK10 in MDM at 7 days. 2 × 10^8 CFU ip. in BALB/c mice gave one-log decrease in gut mucosa and spleen at 12 weeks	C57BL/6 mice	MAP JTC-1285 clinical isolate from goat	5 × 10^8 CFU ip. 6 weeks after last immunization	Immunized 2 × 10^6 CFU sc. in PBS or QuilA adjuvent boosted at week 2. One log decrease in spleen and two log in liver and gut mucosa at 9 weeks relative to Mycopar vaccination
MAP3006c (MAPK10)	lipN MAP_RS15385 JDIP325	Lipase	Hygromycin	Positive trend in MDM survival after 7 days. 1 × 10^7 CFU ip. in BALB/c mice gave three- to four-log decrease in liver at 12 weeks	244 week-old goat kids		3 × 10^9 CFU oral 8 weeks after last immunization	Single immunization 1 × 10^9 in QuilA adjuvant sc. gave one to two log decrease in LN and gut tissue samples at 12 months. No faecal shedding in lipN vaccine group
Kabara and Coussens, 2012; Lamont *et al.*, 2014; Bannantine *et al.*, 2014a, b; Hines *et al.*, 2014; Rathnaiah *et al.*, 2014								
MAP1566 (MAPK10)	30H9 MAP_RS07970 JDIP319	Unknown protein	Kanamycin	Negative trend in bovine MDM survival after 7 days. 1 × 105 CFU ip. in C57BL/6 not found in spleen and liver at 18 weeks	***2-month-old goat kids	MAP-700535 cow isolate	Two doses oral of 1 × 10^9 CFU 3 weeks after last immunization	Immunized 1 × 10^8 CFU oral boosted at 2 weeks. No significant protection at 13 months post-challenge. Faecal shedding greater than Silirum control vaccination
MAP1566 (MAPK10)	STM68 MAP_RS07970 JDIP315	Unknown protein	Kanamycin					
MAP3694c–MAP3695 (MAPK10)	2E11 MAP_RS18945–MAP_RS18950 JDIP316	Intragenic	Kanamycin					
MAP0282c–MAP0283c (MAPK10)	40A9 unannotated sequences JDIP318	Intragenic	Kanamycin	Positive trend in MDM survival after 7 days. Increased MDM apoptosis vs wild type ATCC19698. 1 × 10^7 CFU ip. C57BL/6 mice gave three-log decrease in liver and intestinal tissue at 12 weeks. 1 × 105 CFU ip. in C57BL/6 gave one-log decrease in spleen and liver at 18 weeks				

MDM, monocyte-derived macrophage;CFU, colony-forming units;ip., intraperitoneal;iv., intravenous;sc, subcutaneous;PBS, phosphate buffered saline;LN, lymph node.

(Hawkridge *et al.*, 2008; Kaveh *et al.*, 2016) suggests live attenuated prime, subunit boost strategies may also be applicable for paratuberculosis; however, this approach has not as yet been explored.

Stimulating the complex balance of immunological responses required to achieve effective immunity from any *MAP* vaccine over long periods is particularly challenging. Humoral (Pooley *et al.*, 2019) and mucosal (Weiss *et al.*, 2006) immunity may have some impact, but the priming and maintenance of appropriate Th1 responses is crucial (Coussens, 2004; Vordermeier and Hewinson, 2006). Vaccination approaches need optimizing against contributory factors including the possible requirement for boosting, type or breadth of antigens presented, differential responses to vaccination between neonates and adults (Koets and Gröhn, 2015), and sensitization from encounters with other mycobacteria (Orme and Collins, 1985; Hope *et al.*, 2005; Zimmermann *et al.*, 2018). The mode of vaccine delivery is also likely to provide a significant variable. Recent studies on *Mycobacterium tuberculosis* vaccination in non-human primates show that mucosal (Kaveh *et al.*, 2016) or intravenous vaccination (iv.) with whole-cell mycobacterial vaccines provides better protection than parenteral subcutaneous (sc.), intradermal (id.) or intraperitoneal (ip.) vaccination, possibly elicited through improved induction of more T helper Th17 cells, resident memory T cells and effector T cells at these sites (Voss *et al.*, 2018). As described in this chapter, the means of delivery can be crucial and can affect efficacy. The optimal method however may be related to the type of vaccine being administered and this as yet remains an unknown.

The size of *MAP* reservoirs also suggests that exposure to *MAP* is almost inevitable and that even eradication programmes that include vaccination may not be successful unless the vaccine used also has a therapeutic effect. Vaccines are aimed principally at prophylactic protection; however, there is evidence that selected delivery systems could be used therapeutically on infected animals (Bull *et al.*, 2007; Santema *et al.*, 2013). The possibility that chronic *MAP* infection in humans may be involved in the development of Crohn's disease raises the possibility that a vaccine could provide an alternative direction for treatment. A vaccine adapted for humans would need to act therapeutically and pass rigorous Phase I and Phase II safety trials, and one human vaccine has already begun development, with some success (Folegatti *et al.*, 2019). The current goal in vaccine development however is firmly focused on providing better protection of ruminants against disease and elimination of shedding from subclinically infected animals. Many challenges lie ahead. Devising effective ways of designing, screening and testing vaccine candidates needs more work. Despite attempts to better standardize animal models (Hines *et al.*, 2007a) there is little agreement on how vaccine testing in one species will predict protection in another (Bannantine *et al.*, 2014a; Hines *et al.*, 2014; Park and Yoo, 2016; Rathnaiah *et al.*, 2017; Barkema *et al.*, 2018).

Techniques for monitoring and predicting disease progression clearly remain suboptimal. This includes little knowledge of the relationship between faecal shedding and subclinical cattle disease (Corbett *et al.*, 2018), low specificity with serum testing using *MAP* antigens and disease status (Bannantine *et al.*, 2017), and the absence of a gold standard vaccine able to provide sterilizing protection in any model. The lack of agreed correlates of protection able to screen for candidate vaccines prior to full challenge trials also remains a major roadblock to progress (Ganusov *et al.*, 2015). Nevertheless, against this background of unknowns a few promising avenues of research have emerged, including two different prophylactic vaccines (Bull *et al.*, 2014; Shippy *et al.*, 2017) that have managed to abrogate *MAP* faecal shedding in cattle after high dose challenge, within a test environment.

23.2 Live Attenuated Whole-Cell Vaccines

23.2.1 Overview

The current trend towards testing novel live attenuated *MAP* strains (LAV) produced by molecular genetic techniques as vaccine candidates follows the growing evidence that specific attenuation of *M. tuberculosis* to provide a stable and safe strain is capable of improving on BCG, an attenuated *Mycobacterium bovis* strain, to control *M. tuberculosis* infection (Tameris *et al.*,

2019). Tailoring this approach to *MAP* assumes that the wild-type genome encodes mechanisms that can interfere with immune processing of the multiplicity of antigens present in the *MAP* cell wall or secreted during intracellular persistence and that these critical pathways will be disabled in LAV strains. This assumption then predicts that by freeing these road-blocks a range of protective responses will be enacted, able to clear the vaccine strain, reducing its dissemination into the environment and any subsequent wild-type challenge. As with all live strains, an additional confounder is the high probability that any LAV inoculation will induce reactivities that interfere with existing bovine tuberculosis skin testing, thus a further assumption, predicting that a more specific test differentiating infected from vaccinated animals (DIVA) is likely to emerge in the near future, is inherent (Serrano *et al.*, 2017).

Initial attempts to identify candidate targets determining *MAP* virulence have relied in the most part, on *in vitro* assays to provide a preliminary screen before entering animal studies. Techniques employed have included the onset of auxotrophy (Cavaignac *et al.*, 2000) and, more widely, differential survival of candidate LAV vs wild-type strain controls in short-lived monocyte-derived macrophage (MDM) assays involving either uptake and entry, short-time survival or ability to induce apoptosis (Scandurra *et al.*, 2009; Park *et al.*, 2011). It has been emphatically demonstrated however that this rationale, while intuitive, does not predict generation of protective immunity *in vivo* (Bannantine *et al.*, 2014a; Lamont *et al.*, 2014). This revelation severely limits screening options for LAV candidates and underlines the current lack of knowledge about *in vivo MAP* pathogenic mechanisms and the genes critical to enact them. Most LAV entered into expensive animal trials have thus included *MAP* gene homologues of targets already validated as virulence factors in *M. tuberculosis* (Forrellad *et al.*, 2013). In all cases these have involved single-gene knockouts (Table 23.1), which have, as yet, not had their individual virulence factor status verified by gene complementation. It is likely, however, that as in *M. tuberculosis* vaccine development, any novel LAV with commercial promise will require double gene knockouts, excision of undesired antibiotic selection markers inserted during construction and definitive environmental survival studies, to pass the strict regulatory safety requirements for authorized genetically modified organism release (Walker *et al.*, 2010). As double knockout LAV of *MAP* have not yet been generated or tested, the influence this preferred alteration may have on protective outcomes is a current unknown and will undoubtedly require investigation in future studies.

Genomic manipulation approaches used to generate novel LAV candidates have included random transposon mutagenesis (Harris *et al.*, 1999; Shin *et al.*, 2006) and directed allelic exchange through specialized shuttle plasmidphage transduction, first developed in *M. tuberculosis* (Bardarov *et al.*, 2002; Park *et al.*, 2008). Strain libraries generated by random insertion mutagenesis using a kanamycin-marked transposon (TN5367) were pre-screened for potential attenuation using auxotrophic media (Cavaignac *et al.*, 2000) or ip. challenge in mice 12 weeks/goats 6–12 months (Shin *et al.*, 2006; Scandurra *et al.*, 2010). This identified a shortlist of LAV candidates (Table 23.1) with defined *MAP* gene knockouts (MAP1566, MAP0053c, MAP0011, MAP3963, MAP2408c) that were chosen for downstream challenge studies in animals (Scandurra *et al.*, 2010; Settles *et al.*, 2014). In a consortium approach (Bannantine *et al.*, 2014b) aimed at deriving a rational framework for trialling *MAP* vaccines, this list was extended further to include more TN5367-derived LAV candidates (Rathnaiah *et al.*, 2014) targeting both genes (MAP0460) and intragenic regions (MAP3695-3696, MAP0282c-0283c, MAP1150c-1151c). Additionally, two allelic exchange-derived knockout strains (MAP1047, MAP3893c) were included that had shown decreased *in vitro* survival characteristics in MDM and evidence of attenuation in an ileal cannulation challenge (12 weeks) in young calves (Park *et al.*, 2011). The candidate panel (22 strains) when tested for survival in bovine MDM over 7 days indicated that none of the strains was significantly attenuated for survival relative to a wild-type (MAPK10) strain (Lamont *et al.*, 2014). Some strains were flagged as having a lower capacity for intracellular persistence measured by a sharper slope of decrease in colony-forming unit (CFU) load between 2 and 7 days (Lamont *et al.*, 2014) and others a reduced capacity to induce MDM apoptosis (Kabara and Coussens,

2012) or entry into MDM cultures (Rathnaiah *et al.*, 2014). Using these screening assays, eight candidates were promoted to mouse studies and five candidates for goat challenge model studies to be compared against a heat-killed vaccine standard (Silirum) using previously defined animal models (Hines *et al.*, 2007a).

Despite these high likelihood 'indicator' parameters, each of eight LAV chosen when tested in mice (CL57BL/6) using a low-dose immunization given ip. followed by low-dose ip. challenge after 6 weeks, failed to show any significant protection 12 weeks post-challenge relative to a phosphate buffered saline (PBS) sham vaccinated group (Bannantine *et al.*, 2014a). Parallel study of two further strains LAV MAP3963 and MAP2408c using a higher dose immunization but administered sc., boosted after 2 weeks, followed by low-dose ip. challenge given 5 weeks later showed only minor improvement in protection, with only one log decrease in mucosal tissue loads relative to PBS sham vaccination 18 weeks post-challenge (Settles *et al.*, 2014). The LAV panel tested in goats using a higher immunization dose plus 2-week boost given orally, followed by high-dose challenge also given orally 3 weeks after immunization, also failed to show any significant protection in tissue and shedding 13 weeks post-challenge with all LAV strains showing greater faecal shedding than animals protected with the sc. administered commercial control vaccine (Hines *et al.*, 2014). These authors concluded that the use of a non-ruminant animal model such as mice for screening was suboptimal due to unpredictable infection rates, the lack of inducible chronic intestinal granulomatous infection and the inability to monitor faecal excretion. They also concluded that determining strain attenuation *in vitro* using cell culture models was not a good predictor for survival of LAV *in vivo* (in this case goats) or of protective potential.

These observations are somewhat in line with studies on other LAV candidates. LAV strains MAP3025c, MAP3291c and MAP1534 screened for attenuated persistence in C57BL/6 mice using a high ip. dose, showed significant attenuation (but no clearance) after 12 weeks (Chen *et al.*, 2012). High ip. dose immunization of C57BL/6 mice with these candidates followed by high oral dose challenge, 3 weeks post-immunization, provided a decrease in visible acid-fast organisms by microscopy 12 weeks post-challenge relative to PBS sham vaccination. Further study of LAV MAP3025c in goats using high sc. dose immunization, boosted at 3 weeks, followed by high oral dose challenge, 3 weeks post-immunization, provided a decrease in tissue loads 6 months post-challenge and some evidence of decreases in faecal shedding relative to PBS sham but no reduction superior to commercial (Mycopar) vaccination (Faisal *et al.*, 2013a). Of note here is the choice to deliver LAV immunizations ip., a route previously shown to be ineffective for live 316F-based vaccines (Griffin *et al.*, 2009).

A separate study using LAV MAP1566, MAP0053c and MAP0011 showed differential induction of apoptosis in MDM and reduced persistence after 12-day infections relative to wild-type strains (Kabara and Coussens, 2012). Goats challenged with a high LAV dose delivered unusually iv., showed a significant attenuation in MAP1566 after 6–10 months (Scandurra *et al.*, 2010). Disappointingly, single low sc. dose immunization of BALB/c mice, followed by medium ip. dose challenge, 7 weeks post-immunization with these strains, showed no significant decrease in overall challenge load in spleen and liver 12 weeks post-challenge compared with sham PBS vaccination. The authors suggested that failure in protection could have been due to excessive attenuation of the LAVs being tested. This situation applies in the case of tuberculosis vaccination, where *M. bovis* BCG is known to replicate for some time in a host and engenders a good protective response against tuberculosis (Andersen and Doherty, 2005), whereas strains of *M. bovis* that are too attenuated do not (Collins *et al.*, 2002). In contrasting studies however, LAV MAP3893c and LAV MAP1047 were shown to have low and high attenuation, respectively, in a cattle cannulation model using 12 weeks post-challenge (Park *et al.*, 2011). The highly attenuated strain given as a single high oral dose immunization in goats, followed by high oral dose challenge, 8 weeks post-immunization, showed an increased degree of protection and LAV strain clearance after 8 weeks over the lower attenuated example. When repeated in cattle with a single high oral dose immunization, followed by high oral dose challenge, 4 weeks post-immunization a small relative increase in the degree of protection from wild-type MAPK10 challenge was again apparent in the LAV MAP1047 strain plus apparent clearance of the LAV strain 16 weeks post-immunization (Park *et al.*, 2014).

Study of single-gene allelic exchange mutants MAP4201 or MAP3323c, both global signalling factors, showed LAV attenuation *in vitro* using MDM at 7 days and some attenuation when given ip. at high dose into immunodeficient BALB/c mice (Ghosh *et al.*, 2013, 2014, 2015). Testing in C57BL/6 mice using a low sc. dose immunization with a 2-week boost followed by high ip. dose challenge, 4 weeks post-immunization, provided two logs of protection 9 weeks post-challenge relative to a PBS sham or a commercial heat-killed (Mycopar) vaccination. Challenge studies continued in goats with LAV MAP3323c and another strain LAV MAP3006c (encoding a lipase), a strain that had been rejected in previous screening (Bannantine *et al.*, 2014b) due to positive growth characteristics in MDM. Animals given a single high s.c. dose immunization followed by high oral dose challenge, 8 weeks post-immunization, provided two logs of protection in tissue 12 months post-challenge and encouragingly, in the LAV MAP3006c strain only, no detection of faecal shedding (Shippy *et al.*, 2017). This is the only LAV reported to include possible abrogation of faecal shedding in a ruminant animal group and, if transferable to cattle without compromising tuberculin testing, it could represent a significant step forward in this field.

Notably, despite some attempts at standardizing methodology and encouraging the use of comparable controls, none of the studies described above followed the same protocol for any species of animal and mouse models using a variety of strains. In some studies power calculations on group sizes prior to testing were absent or not reported and test groups in larger animals were often very low. Other parameters seemed to have been arbitrarily chosen. Immunization dosing ranged from $1 \times 10^4 - 2 \times 10^9$ CFU/animal and multiple variations of ip., iv., sc., id., im. and oral were employed to immunize and/or challenge. Some LAV were boosted and some, particularly those given orally, were only single dose. Several different challenge strains were used among the published studies. Times between the last immunizing dose and challenge ranged from 3–8 weeks, longitudinal monitoring of LAV and challenge strain excretion was often omitted or deemed impossible to perform and the important final time point chosen to evaluate the degree of protection post-challenge were even greater in range. The use of Silirum heat-killed vaccine in mice was reported to be ineffective (Bannantine *et al.*, 2014a) thus in many cases there was no valid control group allowing comparison with one of the current commercial killed whole-cell vaccine 'gold standards' and protections relative to sham vaccination with PBS were used. Surprisingly, LAV were delivered in most cases without adjuvant, a major factor in subunit-based vaccines. In the only case where one was used, comparison was made on one LAV only (Ghosh *et al.*, 2015).

The reasons for not considering adjuvanted LAV are unclear. Early *MAP* vaccines using classically attenuated strains were all delivered sc., live in aggressive adjuvants such as mineral oil or pumice and, as a result, stimulated a wide range of immunological responses, in a similar way to the dead whole vaccines that have replaced them. One attempt to induce protection using a non-adjuvanted live strain of *MAP316F* in sheep however was unsuccessful (Begg and Griffin, 2005). This is contrary to the situation pertaining to the use of BCG in protecting against tuberculosis, where live BCG without an adjuvant gives very much better protection than dead BCG administered in a mineral oil adjuvant (Griffin *et al.*, 1999). Using an appropriate adjuvant however could be critical. A study comparing adjuvanted vaccination in sheep (Griffin *et al.*, 2009) suggested standard mineral oil-based vaccine sc. delivery of a killed 316F strain invoked vigorous cell-mediated and humoral reactivity while live 316F delivered in lipid-based adjuvant formulations invoked lower cell-mediated responses and no humoral response.

The development of novel live *MAP* strains to use as vaccines has several important regulatory, bio-safety and diagnostic issues that will need to be addressed, similar to some of those pertaining to live tuberculosis vaccines (Walker *et al.*, 2010), before commercialization is feasible. Nevertheless, the use of well-characterized stable attenuated *MAP* mutants as vaccines remains attractive as they would be cheaper than subunit-based vaccines to manufacture and at this moment appear to offer the opportunity of delivering superior protection possibly without the need for a logistically difficult requirement for administering a boost dose.

23.3 Subunit Protein-Based Vaccines

23.3.1 Overview

The identification of immunodominant protein antigens inducing strong Th1-type immune responses during the first asymptomatic stage of the disease and the demonstration of their protective potential in experimental infection models (mouse and target species) is central to the development of subunit-based vaccines. If effective immunization of animals with recombinant proteins in adjuvant or with delivery vectors encoding immunogenic antigens or combinations of both can be achieved, this has the potential to overcome bovine tuberculosis skin test interference issues linked to whole-cell-based vaccines (Santema *et al.*, 2009) and provide an opportunity to engineer-in DIVA testing. Whole genome sequencing of *MAP* strains from a range of animals has provided an invaluable tool for the identification of *MAP* antigens with potential for more effective immunoprophylaxis (Li *et al.*, 2005; Bannantine *et al.*, 2014c; Stevenson, 2015; Möbius *et al.*, 2017).

23.3.2 Purified protein subunit vaccines

MAP comprises a large number of proteins, not all of which are recognized immunologically (Bannantine *et al.*, 2008, Bannantine *et al.*, 2017). This has prompted studies to target soluble or secreted *MAP* antigens, preferably with immunologically active homologues, involved in protective responses among other pathogenic mycobacteria (Thakur *et al.*, 2013). Those few that have been trialled in animals (Table 23.2) and used as expressed purified protein subunit targets, each have varying identity to homologues in *M. tuberculosis* and include: MAP2121c (superoxide dismutase, SOD) a soluble exported protein associated with resistance to killing by intracellular host mechanisms and anti-apoptotic properties that is highly immunogenic in mice (Mullerad *et al.*, 2002a), stimulates γδ T cells, and is thought to be important in the early stages of infection and in granuloma formation in cattle (Shin *et al.*, 2005); three secreted mycolyl-transferases of the Ag85 complex, Ag85A (MAP1609c), Ag85B (MAP0126) and Ag85C (MAP3531c), immunodominant in experimentally infected cattle and mice (Mullerad *et al.*, 2002b; Rosseels *et al.*, 2006a), eliciting strong T-cell responses (proliferation, IL-2 and IFN-γ) in low- and medium-shedder animals but not in culture-negative cows (Shin *et al.*, 2005); MAP Ag74F, a fusion protein of MAP1519, a member of the PPE family, able to elicit significant IFN-γ levels in macrophages of experimentally infected calves (Nagata *et al.*, 2005); and MAP3527, a serine protease whose homologue in *M. tuberculosis* is capable of generating strong Th1 responses (Skeiky *et al.*, 2004). Further targets have included MAP1087 (a cell wall permease), MAP1204 and MAP1272c (putative invasion proteins), and MAP2077c (an anti-sigma factor agonist), shown to generate robust antigen-specific IFN-γ and antibody responses in infected cattle (Stabel *et al.*, 2012).

Immunization of C57BL/6 mice sc. with 50 μg/animal Ag74F in a monophosphoryl lipid A adjuvant (MPL), boosted at 3 weeks and challenged with a virulent *MAP* cattle strain after 3 weeks showed one-log reductions in bacterial loads in spleen and liver at 16 weeks with apparent *MAP* absence in lymph nodes relative to adjuvant alone (Chen *et al.*, 2008). However, immunization of BALB/c mice sc. with Ag74F alone without adjuvant, boosted at 3 weeks and challenged with a virulent *MAP* cattle strain after 3 weeks fared worse, showing no significant protection in spleen and lymph node tissue and 0.5-log decreases in liver and ileal tissue 12 weeks post-challenge (Stabel *et al.*, 2012). In the same study when Ag74F was used similarly in various combination with MAP1087, MAP1204, MAP1272c and MAP2077c also without adjuvant, results were not significantly improved. Several parameters and different animal breeds were used in these studies; thus, conclusions concerning adjuvant usage as a requirement or individual contributions to protection are tentative. The same Ag74F had been used in combination with SOD, and Ag85 complex units A and B in goats, immunizing sc. with 100 μg/animal using another adjuvant (a

Table 23.2. Summary of protein vaccine candidates tested in animals.

Vaccine target and delivery type	Other name(s)	Size (kDa)	Function	Tested species	Challenge strain	Dose and route of infection	Comments
Stabel *et al.*, 2012							
MAP1087		15.4	ABC transporter permease	Male BALB/c mice	*MAP* strain 167 (from an infected cow)	10^8 CFU intraperitoneally 3 weeks after the last immunization	Immunized subcutaneously with 100 µg total protein of cocktail variations (three in each) and 50 µg 74F in PBS. Control received PBS alone. No reduction in bacterial load at 3 months in spleen or MLN, some reduction in liver and one log in ileum tissue
MAP1204		25.4	Putative invasion protein; NipC/P60 family; cell wall-associated hydrolase				
MAP1272c		33.4	Putative invasion protein; NipC/P60 family; cell wall-associated hydrolase				
MAP2077c		11.1	Sulfate transporter antagonist of anti-sigma factor				
MAP1519	74F	74	MAP3527 (PepA) trypsin-like serine protease				
MAP3527			MAP1519 PPE protein family				
Kathaperumal *et al.*, 2008							
MAP0216	Ag85A	32	Mycolyl-transferases involved in cell wall synthesis	24 calves	*MAP* strain 66115-98	1×10^7 CFU orally for 7 consecutive days, 4 weeks after the last immunization	Immunized subcutaneously with mix of 100 µg of each protein in MPL or intramuscularly with MPL + 100 µg bovine IL-12 DNA. Some protection induced but no significant differences between any vaccinated groups
MAP1609c	Ag85B	30					
MAP3531c	Ag85C	32					
MAP0187c	SOD	23	Superoxide dismutase				
Kathaperumal *et al.*, 2009							
MAP0216	Ag85A	32	Mycolyl-transferases involved in cell wall synthesis	25 goat kids	*MAP* strain 66115-98	1×10^7 CFU orally for 7 consecutive days, 3 weeks after the last immunization	Immunized subcutaneously with mix of 100 µg of each protein with and without cationic surfactant dimethyl dioctadecyl ammonium bromide (DDA). Boosted at 3 weeks. More than two-log protection with Ag mix + DDA group in seven out of eight animals' mucosal tissue culture at 38 weeks
MAP1609c	Ag85B	30					
Fusion protein: 17.6 kDa C-terminal fragment of MAP3527 & 14.6 kDa N-terminal fragment of MAP1519	MAP 74F	74	MAP3527 (PepA) trypsin-like serine protease MAP1519 PPE protein family				
MAP0187c	SOD	23	Superoxide dismutase				

Continued

Table 23.2. Continued

Vaccine target and delivery type	Other name(s)	Size (kDa)	Function	Tested species	Challenge strain	Dose and route of infection	Comments
				Chen *et al*., 2008			
Fusion protein: 17.6 kDa C-terminal fragment of MAP3527 & 14.6 kDa N-terminal fragment of MAP1519	MAP 74F	74	MAP3527 (PepA) trypsin-like serine protease MAP1519 PPE protein family	C57BL/6 mice	*MAP* strain 66115-98 (from an infected cow)	10^9 CFU intraperitoneally 3 weeks after the last immunization	Immunized subcutaneously with 50 µg/animal of fusion protein in MPL. Boosted at 3 weeks. Control group received MPL alone. One-log reduction in in spleen and liver and no detection in lymph nodes 16 weeks post-challenge
				Koets *et al*., 2006; Santema *et al*., 2012			
MAP3840	Hsp70, dnaK	70	Heat shock chaperonin	40 female calves	*MAP* from infected cow	At least 2×10^4 CFU, orally; nine gavages over 21-day period	Immunized with 200 µg recombinant Hsp70 in DDA boosted at week 44. Reduced pattern of shedding of *MAP* in the faeces during 2 years in vaccine group
				56 male calves	*MAP* from infected cow	At least 2×10^4 CFU, orally; nine gavages over 21-day period	Immunized with 200 µg recombinant Hsp70 in DDA boosted at week 12, 24 and 52. No reduction in faecel shedding over 2 years and no protection against establishment of chronic infection
				455-year-old female cows	Naturally acquired chronic *MAP* infection		Therapeutic vaccination subcutaneously with 200 µg recombinant Hsp70 in DDA 0, 4, 16, 28 weeks. Significant reduction in faecal shedding, no overall protection

ABC, ATP-binding cassette; MLN, mesenteric lymph node; MAP, *Mycobacterium avium* subsp. *paratuberculosis*; CFU, colony-forming units; PBS, phosphate buffered saline; MPL, monophosphoryl lipid A.

cationic surfactant: dimethyl dioctadecyl ammonium bromide; DDA), boosted at 3 weeks and challenged with a moderate oral dose of a virulent *MAP* cattle strain after 3 weeks. In this instance protection greater than two logs mucosal tissue culture from seven out of eight animals at 38 weeks (Kathaperumal *et al.*, 2009) relative to adjuvant alone was seen. Faecal shedding was not monitored. In a previous study (Kathaperumal *et al.*, 2008) the Ag85 complex had been used in combination with SOD administered as 100 μg/animal sc. in MPL adjuvant using a similar regimen in calves. This combination again showed some protection relative to adjuvant alone in intestinal tissue, although this was only after 18 weeks. Faecal shedding was not reduced and no commercial heat-killed vaccine group was included as a control, making it difficult to draw accurate conclusions concerning the individual contribution towards protection; nor did it establish the likelihood that these vaccines could improve over commercial killed whole-cell combinations, or compare the results with those of similar subunit combinations delivered as DNA vaccine in a previous study (Chen *et al.*, 2008).

A series of studies has been made examining the potential of a single-target subunit vaccination using MAP3840, a soluble heat shock protein (Hsp70, DnaK) initially chosen for its ability to induce specific immune responses in *MAP*-infected and *MAP*-vaccinated cattle and change in detection parameters with disease progression (Koets *et al.*, 2006). Initial trials in cattle immunized with 200 μg sc. adjuvanted with DDA and boosted at week 44 and challenged with a series of low oral doses administered over 3 weeks was initially shown to provide a reduced pattern of faecal shedding relative to unvaccinated group over the 2-year study. Subsequent retesting in cattle, however, using a similar regimen but with extra boosts at weeks 12, 24 and 52 showed a more similar pattern of faecal shedding between vaccinated and sham vaccinated groups over a similar 2-year period (Santema *et al.*, 2012). A third cattle experiment given therapeutically with boosting at 4, 16 and 32 weeks to cattle with naturally acquired chronic *MAP* infection was able to induce a modest 10–20% reduction in shedding (Santema *et al.*, 2013). The authors attribute the differences observed to unusually high humoral component and suboptimal Th1 cell-mediated responses elicited by the subunit deployed in this vaccine, as measured by a lack of relative increases in INF-γ positive cells and INF-γ release response to *Mycobacterium avium*-purified protein derivative antigen stimulation in vaccinated animals. This underlines the need to better characterize the subset of immune responses that are actively protective against *MAP* persistence and dissemination, long-term, in ruminants, and suggests that using a single subunit target in vaccines may be a disadvantageous approach. It highlights the diverse and currently rather unpredictable ways in which some *MAP* proteins are recognized and handled by the host and intriguingly supports lines of thought in tuberculosis vaccine design that advocate the need for a mix of cell-mediated and humoral responses in an effective mycobacterial vaccine (Kawahara *et al.*, 2019).

Overall, attempts to use purified *MAP* protein preparations as vaccines have not been as successful as hoped. This is not to say however that this approach should be discarded. Only a very few subunit targets and types of adjuvant have been investigated to date. As the pathways of *MAP* pathogenesis and the immunoreactivite subunits that, when targeted, protect the vast majority of naturally infected subclinical animals are better described, this method of vaccination could still provide insights and solutions.

23.3.3 DNA delivery subunit vaccines

This mode of vaccination involves direct introduction of cloned or constructed DNA regions encoding target subunit *MAP* proteins engineered to be expressed from functional eukaryotic promoters into host dendritic cells leading to subunit expression, delivery and immune processing. They are relatively stable *ex vivo*, easily generated from known sequences and cheap to produce. They induce a variety of humoral and cellular immune responses and have been able to offer various degrees of protection against intracellular pathogens with some approved by the US Food and Drug Administration (FDA) and the United States Department of Agriculture (USDA) for veterinary use (Huygen, 2006; Hobernik and Bros, 2018). The protective potential of this approach was first demonstrated by immunization of BALB/c mice sc. with 2 μg/animal plasmid mixes encoding clone pools of 75 antigens in each attached to gold gene gun delivery beads, boosted at 3 weeks then challenged with high load wild-type *MAP* ip. after 2 weeks, which conferred up to a two-log increase in protection relative to a sham plasmid control (Huntley *et al.*, 2005). Twenty-six genes in the protective mix encoding transport/binding, membrane and virulence proteins and mycobactin/polyketide synthases were suggested as important; however, individual contributions of respective antigens were not examined. Another combined plasmid approach immunized C57BL/6 mice using a mix of five plasmids encoding the Ag85 ABC complex (MAP0216, MAP1609c, MAP3531c) and SOD (MAP0187c) and MAP2121c (membrane protein) given sc. in a total 250 μg/animal, boosted three times over 3 weeks and challenged iv. with a high dose of a clinical *MAP* isolate 3 weeks after the last immunization, which gave significant reductions in spleen and liver relative to adjuvant alone (Chen *et al.*, 2008). Further testing comparing C57BL/6 and BALB/c mice using a plasmid expressing MAP0586c (a transglycolase) given im. in a total 100 μg/animal, boosted four times at 3-week intervals and challenged iv. with a medium dose of a *MAP* ATCC19698 reference strain 6 weeks after the last immunization, also gave significant reductions in load in spleen and liver at 8 weeks post-challenge (Roupie *et al.*, 2008). A more extensive study (Roupie *et al.*, 2012) using histidine-tagged constructs was made using eight candidate *MAP* genes (MAP1963c, MAP2677c, MAP1637c, MAP0388, MAP3743, MAP3198-9, MAP2151-52c, MAP0863-5) delivered using the pV1.ns-tPA DNA vector in BALB/c and C57BL/6 and boosted with 50 μg of purified recombinant protein. This showed some antigen specific interferon gamma responses but no protection in any of the constructs when challenged 6 weeks post-immunization with a medium dose of a luminescent mutant of *MAP* strain S23 (Table 23.3). None of the above was tested subsequently in larger animals. A single study using DNA delivery alone has been reported in sheep (Sechi *et al.*, 2006). In this instance animals were immunized im. 1 mg/animal with plasmids expressing MAP3966 or BCG or *Mycobacterium avium* gene constructs, boosted three times at 20-day intervals then challenged 3 months later with a high oral dose of a human isolate of *MAP*. Culture and shedding were not performed but histopathological acid-fast staining of small intestinal tissue sections was suggestive of increased protection in the vaccine group relative to heat-killed (Gudair) vaccine control after 1 year.

23.3.4 Attenuated live vector subunit delivery vaccines

The delivery of pathogen-specific subunit antigens by recombinant live heterologous vectors including attenuated bacteria (Ding *et al.*, 2018) and replication-deficient viruses (Lauer *et al.*, 2017) is an alternative vaccination strategy already trialled in other animal and human diseases. The rationale for this approach relies on the properties of infectious vectors that are not an aetiological agent of disease, allowing transient intracellular infection without disease and naturally adjuvant antigen processing. Optimal live vector delivery regimens can require multiple vaccinations usually as a prime-boost, but complete cycles of intracellular replication are not required for successful antigen delivery and presentation. Repeated exposure to the same vector is potentially an option (Gabitzsch *et al.*, 2009); however, heterologous vectors for prime and boost are preferable. Vector combinations have included naked DNA and bacterial or viral priming, followed by a viral vector boost (Wu *et al.*, 2018). BCG used as a bacterial prime with the recombinant modified viral boost provides some protection against *M. tuberculosis*/*M. bovis* infections in both humans and cattle (Dean *et al.*, 2014; Kaveh *et al.*, 2016; Voss *et al.*, 2018) suggesting that this type of approach may also be suitable for *MAP* vaccination. BCG itself appears to give some protection in mice against infection with *MAP* (Roupie *et al.*, 2008) and this protection increased when BCG expressed a group of genes from an operon in a putative pathogenicity island in *MAP* (Heinzmann *et al.*, 2008). Using recombinant BCG for the delivery of *MAP* antigens would produce problems with cross-reactivity to conventional skin tuberculin testing in cattle (Roy *et al.*, 2018). Greater specificity in tuberculin testing could overcome this problem (Srinivasan *et al.*, 2019); however,

Table 23.3. Summary of DNA vaccine candidates tested in animals.

Vaccine target and delivery type	Other name(s)	Size (kDa)	Function	Tested species	Challenge strain	Dose and route of infection	Comments
Roupie *et al.*, 2012							
Plasmid DNA expressing MAP1963c	MAP_RS09975	18.3	Peptidyl-prolyl cis-trans isomerase	BALB/c and C57BL/6 mice	Luminescent MAP-23 containing hygromycin cassette	2 × 10^6 CFU intravenously 6 weeks after the last immunization	Immunized intramuscularly with 100 µg DNA four times at 3-week intervals. Low immunogenicity and no significant decrease in *MAP* load in spleen or liver with any *MAP* protein at 8 weeks post-challenge
Plasmid DNA expressing MAP2677c	MAP_RS13640	14.5	Glyoxalase family				
Plasmid DNA expressing MAP1637c	MAP_RS08320	52	UbiD family decarboxylase				
Plasmid DNA expressing MAP0388	MAP_RS01975	43.8	Hypothetical protein				
Plasmid DNA expressing MAP3743	MAP_RS19185	36.5	NAD(P)H-binding protein				
Plasmid DNA expressing MAP3198–MAP3199	Ag3 MAP_RS16415-MAP_RS16420	26.1	Intragenic – not annotated				
Plasmid DNA expressing MAP2151–MAP2152c	Ag5 MAP_RS10940–MAP_RS22710	25	Intragenic – not annotated				
Plasmid DNA expressing MAP0863 – MAP0865	Ag6 MAP_RS04380– MAP_RS22535	15	Intragenic – not annotated				
Sechi *et al.*, 2006							
Plasmid DNA expressing MAP3936	Hsp65	65	GroEL-like type I chaperonin	5-month-old lambs (25)	*MAP* from a patient with Crohn's disease	2 × 10^9 CFU orally 3 months after the last vaccination	Immunized intramuscularly 1 mg three times at 20-day intervals. Histopathology of post-mortem tissue sections revealed absence of lesions or bacteria in the groups vaccinated with the three DNA vaccine constructs

Continued

Table 23.3. Continued

Vaccine target and delivery type	Other name(s)	Size (kDa)	Function	Tested species	Challenge strain	Dose and route of infection	Comments
Plasmid DNA expressing BCG3866c	Ag85A	32					
Plasmid DNA expressing MAV0214	Ag85A	32					

CFU, colony-forming units; MAP, Mycobacterium avium subsp. paratuberculosis.

alternative priming vectors, such as attenuated *Salmonella* (Faisal *et al.*, 2013b), *Lactobaccillus* (Johnston *et al.*, 2014), recombinant human adenovirus (Bull *et al.*, 2014), simian adenovirus (Folegatti *et al.*, 2019) and attenuated vaccinia virus (Bull *et al.*, 2014), have the potential for bovine vaccination against *MAP* without inducing interference (Table 23.4).

A double-gene knockout attenuated strain of *Salmonella enterica* transformed with a gene cassette expressing *MAP* antigens including MAP0216(Ag85A), MAP1609c(Ag85B), MAP0187c(SOD) and MAP3527–MAP1519 (fusion Ag74F) was used to immunize C57BL/6 mice sc. with 5 × 10^8 CFU/animal with a single boost at 3 weeks followed 6 weeks later by challenge with a high-dose clinical *MAP* strain given ip., generated a significant three-log reduction in spleen and liver loads after 16 weeks (Chandra *et al.*, 2012). However, when a similar construct (removing Ag74F) was immunized into goats with a similar inoculation sc. plus a single boost at 3 weeks followed 3 weeks later by oral challenge with a high oral dose of a wild-type MAP66115-98 (Faisal *et al.*, 2013b), there was no significant decrease in tissue loads and no decrease in faecal shedding relative to a 316F control vaccine after 16 weeks. The authors speculate that lack of protection in this study could have been due to suboptimal antigen processing, low vector expression and low secretion of the subunit mycobacterial proteins able to sufficiently stimulate in a small animal model but not a large one.

To try and address the problem of mycobacterial subunit toxicity towards the delivery vector, reduction of secretion through insolubility and low subunit expression as a result of rare codon usage, common in mycobacterial transcription, studies have attempted to alter codon usage at source. Instead of cloning genes directly from the *MAP* genome, DNA constructed from synthesized oligonucleotides has been used allowing engineered substitution of rare codons commonly used by mycobacteria, addition of efficient terminal secretion sequences or eukaryotic processing tags and removal of insoluble transmembrane regions. An approach of this type constructed a eukaryotic codon optimized cassette encoding a fusion protein of soluble regions from four *MAP* antigens (MAP1589c, MAP1234, MAP2444c, MAP1235) with a 5′ ubiquitin expression terminus added to optimize intracellular processing (AgHAV). This construct was then cloned into both a prime viral vector, a replication-deficient human adenovirus serotype 5 (Ad5.HAV) and a boost viral vector, a replication-deficient vaccinia Ankara (MVA.HAV). Prophylactic immunization of C57BL/6 with 10^8 plaque-forming units (PFU)/animal sc. Ad5.HAV, boosted with 10^8 PFU/animal sc. MVA.HAV at 2 weeks followed by ip. challenge with low- or high-dose MAPK103 weeks later showed a one-log decrease in spleen and liver after 6 weeks (Bull *et al.*, 2007). Therapeutic immunization using the same dosing but employing a MAPK10 challenge 3 weeks prior to prime/boost regimen also gave similar protection. When trialled in cattle, animals were immunized with 10^9 PFU/animal id. prime Ad5. HAV, boosted with 10^9 PFU/animal id. MVA. HAV at 2 weeks followed 5 weeks later by oral challenge with high-dose clinical isolate *MAP* R0808. Vaccinated animals showed a significant two-log decrease in gut mucosal tissue load relative to sham viral vaccinated animals after 38 weeks. No faecal shedding was observed in the vaccinated group (Bull *et al.*, 2014) and this represents the only vaccine thus far to have abrogated *MAP* shedding in a cattle model. Similar constructs have successfully completed Phase I safety trials and are entering Phase II trials in humans (Folegatti *et al.*, 2019).

23.4 Assessing Vaccine Efficacy in Different Animal Models

Testing efficacy of any new vaccine requires a disease model and gold standard controls to act as quantitative markers. The slow progression of *MAP* pathogenesis, the lack of reliable correlates of protection and the inability of current vaccines to effectively stop transmission make delivering these a logistical challenge. Ruminant models are certainly required for the final testing of vaccines but are very expensive. As discussed above, cheaper mouse models have been used for initial screening but have shown poor predictability for protection in larger animals. The problem of how to screen new LAV candidates quickly and effectively has been compounded

Table 23.4. Summary of live vector subunit and other vaccine candidates tested in animals.

Vaccine target and delivery type	Other name(s)	Size (kDa)	Function	Tested species	Challenge strain	Dose and route of infection	Comments
Bull *et al*., 2007, 2014							
Fusion construct of non-transmembrane regions from four *MAP* genes MAP1589c, MAP1234, MAP2444c, MAP1235 inserted as an expression construct into modified Adenovirus 5 and Vaccinia Ankara delivery vectors	AphC MAP_ RS08085	95	Alkyl hydroperoxidase	C57BL/6 mice	Low-passage isolate of *MAP* (K37) from cow with JD	1 × 10^9 CFU ip. 2 weeks after (prophylactic) or 3 weeks before (therapeutic) vaccination	Immunized sc. with Ad5 prime. Boost after 2 weeks. One-log decrease in spleen and liver in both therapeutic and prophylactic administration 6 weeks (prophylactic) and 12 weeks (therapeutic) post-challenge
	gsd MAP_ RS06265		GPL fucosyl transferase				
	p12 MAP_ RS12460		Antisense protein to IS*900* transferase	3-week-old calves	Low-passage isolate of *MAP* (R0808) from cow with JD	1 × 10^9 CFU total orally in two doses 5 weeks after MVA boost vaccination	Immunized id. with adenoviral prime (11 weeks prior to challenge) and vaccinia boost (5 weeks prior to challenge). Significant two log decrease in gut mucosal load. No faecal shedding in vaccinated group at 36 weeks post-challenge
	mpa MAP_ RS22605		GPL acetyl transferase				
Chandra *et al*., 2012							
Fusion constructs of partial proteins from MAP0216, MAP1609c, MAP3527, MAP1519 and MAP0187c expressing in attenuated *Salmonella* species. Also purified fusion partial protein mix	Ag85A (202-347)	34.5	Secretable fragments of mycolyl-transferases involved in cell wall synthesis	180 C57BL/6 mice	*MAP* strain 66115-98	1 × 10^9 CFU in PBS ip. 6 weeks after last immunization	Immunized ip. with 5 × 10^8 CFU in PBS or 10 µg of each purified antigen in MPL adjuvant. Boosted at 3 weeks. Both purified Ag and *Salmonella* delivery induced three-log reduction in spleen and liver CFU at 16 weeks post-challenge. Some decrease in histopathological and granuloma scores
	Ag85B (173-330)	33.3					
	MAP74F (1-148 & 669-786)	40.7	Secretable serine protease & PPE gene fragments				
	SOD (1–72)	19.5	Secretable fragments of superoxide dismutase				
Faisal *et al*., 2013b							
MAP0216	Ag85A	32	Secretable fragments of mycolyl-transferases involved in cell wall synthesis	Goats	*MAP* strain 66115-98	Seven doses 5 × 10^8 oral 3 weeks after last immunization	Immunized sc. 5 × 108 CFU/ animal. Boosted at 3 weeks
MAP1609c	Ag85B	30					

Continued

Table 23.4. Continued

Vaccine target and delivery type	Other name(s)	Size (kDa)	Function	Tested species	Challenge strain	Dose and route of infection	Comments
MAP0187c	SOD	23	superoxide dismutase				
				Jolly *et al.*, 2013			
Lipoarabinomannan (LAM) produced by phenol/ methanol extraction of *MAP* culture	LAM	20	Glycolipid	106-month-old Aberdeen Angus calves	Local *MAP* strain from JD cow	1 × 10^9 CFU oral 10 days after last immunization	Immunized and boosted at 6, 12 and 24 weeks with 2 mg of LAM extract dissolved in 1 ml of PBS and emulsified in 1 ml of Freund's Incomplete Adjuvant. No faecal shedding in two out of two cattle in vaccine group

CFU, colony-forming units; MAP, *Mycobacterium avium* subsp. *paratuberculosis*; sc., subcutaneous; ip., intraperitoneal; id., intradermal; PBS, phosphate buffered saline; MPL, monophosphoryl lipid A; JD, Johne's disease.

by the apparent failure of *in vitro* macrophage assays.

Protective efficacy of a prophylaxic vaccine is generally demonstrated by comparing challenge loads present in naïve vs vaccinated animals after a designated immune clearance period. It is unlikely that any vaccine directed against pathogenic mycobacteria however will have sterilizing capacity; thus, comparisons against current 'gold standard' vaccine efficacy are also essential. Replication in liver and spleen is usually monitored by CFU plating and when testing LAV commonly includes a selection marker to monitor immunogen persistence. Examples include kanamycin or hygromycin but could also include luminometry using constructs expressing luciferase or GFP (Rosseels *et al.*, 2006b; Park *et al.*, 2011). Standardized protocols for certain animal models have been proposed (Hines *et al.*, 2007a) but not validated as authoritative. If new vaccines are to be brought to the market place, a significant improvement in the manner in which trials are performed is required. New tests are needed to detect the direct presence of *MAP* and the host's specific responses to challenge by live *MAP* and many of its individual components; models require rigorous standardizing and a realistic achievable output measure of efficacy related to faecal shedding is needed.

23.5 Vaccine Regulatory and Production Issues

To successfully introduce a new *MAP* vaccine into the market will require provision of clear and repeatable results from experiments demonstrating significant protection plus a validated manufacturing process capable of regulating a consistently safe and effective product. National regulation authorities (NRAs) are responsible for control of these processes. Consultation with authorities to obtain recommendations that will guide how best to obtain marketing authorization is now required from very early design stages. NRAs, initially instigated to standardize methods in the light of tragic vaccine contamination incidents such as smallpox in 1901 (Lilienfeld, 2008), BCG in 1931 (Calmette, 1931) and polio in 1955 (Nathanson and Langmuir, 1995), exist in all developed countries, and some who are licensing vaccine products as a pre-qualification for sale to United Nations agencies (Saidu *et al.*, 2013). The degree and complexity of regulations will differ depending on the type of vaccine to be used. All LAV or live attenuated vector subunit vaccines will be classified as a genetically modified organism (GMO), while most of the European Union (EU) Member States do not regulate DNA vaccines as a GMO as this is not considered an organism. Likewise, human cells transfected with plasmids are not GMOs, provided the plasmid is not replicative and is unlikely to integrate into the cell genome. Legislation in these areas is not globally uniform and international cooperation is hindered by significant complexities in the way regulations are interpreted and administered between countries due to diverse divisions in the responsible departments. In the EU, all GMO vaccine authorizations are granted through the European Commission, with applications scientifically evaluated by the European Medicines Agency (EMA). In the US GMO legislation is currently less defined and animal vaccines reside under the FDA and Center for Veterinary Medicine mantle; similarly in China, with the China Institute for Veterinary Drug Control. While some improvements are being targeted (Crager, 2014) and fast-track mechanisms for getting vaccines where they are most needed are active for human diseases, animal vaccines are less well supported.

The current indication that live attenuated vaccine strains or replication-deficient delivery vectors may provide modes for improved vaccine efficacy has presented a new wave of applications for trials of GMOs in animals and humans. Concerns for the safe release of GMOs has prompted the need for NRAs to provide precise definitions of what constitutes a GMO while still allowing dynamic review. However, there is no universally accepted definition of a GMO. The EU defines a GMO as 'an organism, with the exception of human beings, in which the genetic material has been altered in a way that does not occur naturally by mating and/or natural recombination' (EU Directive, 2001).

Controlled storage of master seed stocks of a stringently defined vaccine strain is mandatory in all countries to ensure consistent quality. Standard testing for strain purity and contamination that conforms to a country's good manufacturing practice (GMP) regulations is also required (Schmeer *et al.*, 2017). In the EU, for example, GMP is required for the 'active' antigen alone and for the finished vaccine product containing the 'active' antigen in combination with delivery organisms and/or adjuvant preparations (Heldens *et al.*, 2008).

Regular testing of manufactured vaccine batches is needed throughout the designated shelf life to determine the degree of safety, potency and efficacy to the target species from single, repeated and up to ten-fold overdoses. This includes a recommendation for an *in vivo* test to predict clinical efficacy and remove the need to regularly test vaccine batches on animals (EMeA, 2016). As discussed in this chapter the relationship between vaccine efficacy and *in vitro* attenuation testing parameters is still highly uncertain; thus, demonstrations of immunological activity as a measure of potency may be more practical. Exceptionally, for some recombinant proteins and polysaccharide vaccines, physico-chemical testing that correlates well with biological effectiveness is an acceptable and preferred replacement under regulatory pathways for licensing vaccines aimed for developing countries (Milstien and Belgharbi, 2004). This may also be applicable to certain *MAP* subunit vaccines.

Many countries require vaccines that comprise or derive from live GMOs to have additional assessments, particularly with reference to virulence and biosafety. In the EU, for example, important factors include: zero transfer of genetic material from the vaccine into the host chromosome (EMeA, 2004); a minimum requirement to show genetic stability and an absence of reversion to virulence in a GMO during at least five consecutive passages within a designated test animal species (EMeA, 2007); no significant increase in capacity for survival or shedding over the parent wild-type organisms in either the environment or susceptible hosts; use of a registered adjuvant (EMeA, 2018); and an optimum route of administration, particularly relevant to orally administered vaccines.

The requirement for complete regulatory compliance has been escalated by the expectation of zero risk demanded from the public, and this has led to introduction of larger clinical trials and widening of the safety margins for GMO release. Many older registered vaccines would not be acceptable in the current regulatory climate. GMOs that contain antibiotic markers are now unacceptable (Leunda and Pauwels, 2019). This affects vaccines using plasmids expressing immunologically active agents delivered via bacterial vectors and requires that the majority of novel DNA constructs referred to in the works discussed above would need some heavy restructuring, and restabilizing. Resistance to heavy metals may be acceptable, provided that this phenotype does not confer a fitness advantage on the GMO (Favre and Viret, 2006). Of additional interest, however, and a point that is still not clearly defined in the law of some countries, is the status of the recipient of a GMO or DNA vaccine and whether the actual act of vaccination demands that the recipient be reclassified as a GMO in itself. If agreed, this would certainly impact on animal importation regulations. As an additional control on GMO release, some countries have also introduced the concept of a 'new organism'. New Zealand regards all organisms that were not present in the country immediately before 1998 or subsequently eradicated as 'new' and thus subject to special conditional release application (SCNZ, 2003).

While none of the new generation of *MAP* vaccines has yet shown sufficient promise to warrant commercial development, these regulatory hurdles suggest that a long and involved process awaits.

References

Andersen, P. and Doherty, T.M. (2005) Learning from BCG: designing a better tuberculosis vaccine. *Discovery Medicine* 5, 383–387.

Bannantine, J.P., Bayles, D.O., Waters, W.R., Palmer, M.V., Stabel, J.R. *et al.* (2008) Early antibody response against *Mycobacterium avium* subspecies *paratuberculosis* antigens in subclinical cattle. *Proteome Science* 6(1), 5. DOI: 10.1186/1477-5956-6-5.

Bannantine, J.P., Everman, J.L., Rose, S.J., Babrak, L., Katani, R. *et al.* (2014a) Evaluation of eight live attenuated vaccine candidates for protection against challenge with virulent *Mycobacterium avium* subspecies *paratuberculosis* in mice. *Frontiers in Cellular and Infection Microbiology* 4(11), 88. DOI: 10.3389/fcimb.2014.00088.

Bannantine, J.P., Hines, M.E., Bermudez, L.E., Talaat, A.M., Sreevatsan, S. *et al.* (2014b) A rational framework for evaluating the next generation of vaccines against *Mycobacterium avium* subspecies *paratuberculosis*. *Frontiers in Cellular and Infection Microbiology* 4(88), 126. DOI: 10.3389/fcimb.2014.00126.

Bannantine, J.P., Li, L., Mwangi, M., Cote, R., Raygoza Garay, J.A. *et al.* (2014c) Complete genome sequence of *Mycobacterium avium* subsp. *paratuberculosis*, isolated from human breast milk. *Genome Announcements* 2(1). DOI: 10.1128/genomeA.01252-13.

Bannantine, J.P., Campo, J.J., Li, L., Randall, A., Pablo, J. *et al.* (2017) Identification of novel Seroreactive antigens in Johne's disease cattle by using the *Mycobacterium tuberculosis* protein array. *Clinical and Vaccine Immunology* 24(7). DOI: 10.1128/CVI.00081-17.

Bardarov, S., Bardarov, S., Pavelka, M.S., Sambandamurthy, V., Larsen, M. *et al.* (2002) Specialized transduction: an efficient method for generating marked and unmarked targeted gene disruptions in *Mycobacterium tuberculosis*, *M. bovis* BCG and *M. smegmatis*. *Microbiology* 148(10), 3007–3017. DOI: 10.1099/00221287-148-10-3007.

Barkema, H.W., Orsel, K., Nielsen, S.S., Koets, A.P., Rutten, V. *et al.* (2018) Knowledge gaps that hamper prevention and control of *Mycobacterium avium* subspecies *paratuberculosis* infection. *Transboundary and Emerging Diseases* 65(Suppl 1), 125–148. DOI: 10.1111/tbed.12723.

Bastida, F. and Juste, R.A. (2011) Paratuberculosis control: a review with a focus on vaccination. *Journal of Immune Based Therapies and Vaccines* 9(1), 8. DOI: 10.1186/1476-8518-9-8.

Begg, D.J. and Griffin, J.F.T. (2005) Vaccination of sheep against *M. paratuberculosis*: immune parameters and protective efficacy. *Vaccine* 23(42), 4999–5008. DOI: 10.1016/j.vaccine.2005.05.031.

Bull, T.J., Gilbert, S.C., Sridhar, S., Linedale, R., Dierkes, N. *et al.* (2007) A novel multi-antigen virally vectored vaccine against *Mycobacterium avium* subspecies *paratuberculosis*. *PLoS ONE* 2(11), e1229. DOI: 10.1371/journal.pone.0001229.

Bull, T.J., Schock, A., Sharp, J.M., Greene, M., McKendrick, I.J. *et al.* (2013) Genomic variations associated with attenuation in *Mycobacterium avium* subsp. *paratuberculosis* vaccine strains. *BMC Microbiology* 13(1), 11. DOI: 10.1186/1471-2180-13-11.

Bull, T.J., Vrettou, C., Linedale, R., McGuinnes, C., Strain, S. *et al.* (2014) Immunity, safety and protection of an Adenovirus 5 prime – Modified Vaccinia virus Ankara boost subunit vaccine against *Mycobacterium avium* subspecies *paratuberculosis* infection in calves. *Veterinary Research* 45(1), 112. DOI: 10.1186/s13567-014-0112-9.

Calmette, A. (1931) Preventive vaccination against tuberculosis with BCG. *Proceedings of the Royal Society of Medicine* 24(11), 1481–1490. DOI: 10.1177/003591573102401109.

Chen, L.-H., Kathaperumal, K., Huang, C.-J., McDonough, S.P., Stehman, S. *et al.* (2008) Immune responses in mice to *Mycobacterium avium* subsp. *paratuberculosis* following vaccination with a novel 74F recombinant polyprotein. *Vaccine* 26(9), 1253–1262. DOI: 10.1016/j.vaccine.2007.12.014.

Cavaignac, S.M., White, S.J., de Lisle, G.W. and Collins, D.M. (2000) Construction and screening of *Mycobacterium paratuberculosis* insertional mutant libraries. *Archives of Microbiology* 173(3), 229–231. DOI: 10.1007/s002039900132.

Chandra, S., Faisal, S.M., Chen, J.-W., Chen, T.-T., McDonough, S.P. *et al.* (2012) Immune response and protective efficacy of live attenuated Salmonella vaccine expressing antigens of *Mycobacterium avium* subsp. *paratuberculosis* against challenge in mice. *Vaccine* 31(1), 242–251. DOI: 10.1016/j.vaccine.2012.09.024.

Chen, J.-W., Faisal, S.M., Chandra, S., McDonough, S.P., Moreira, M.A.S. *et al.* (2012) Immunogenicity and protective efficacy of the *Mycobacterium avium* subsp. *paratuberculosis* attenuated mutants against challenge in a mouse model. *Vaccine* 30(19), 3015–3025. DOI: 10.1016/j.vaccine.2011.11.029.

Collins, D.M., Wilson, T., Campbell, S., Buddle, B.M., Wards, B.J. *et al.* (2002) Production of avirulent mutants of *Mycobacterium bovis* with vaccine properties by the use of illegitimate recombination and screening of stationary-phase cultures. *Microbiology* 148(10), 3019–3027. DOI: 10.1099/00221287-148-10-3019.

Corbett, C.S., Barkema, H.W. and De Buck, J. (2018) Quantifying fecal shedding of *Mycobacterium avium* ssp. *paratuberculosis* from calves after experimental infection and exposure. *Journal of Dairy Science* 101(2), 1478–1487. DOI: 10.3168/jds.2017-13544.

Coussens, P.M. (2004) Model for immune responses to *Mycobacterium avium* subspecies *paratuberculosis* in cattle. *Infection and Immunity* 72(6), 3089–3096. DOI: 10.1128/IAI.72.6.3089-3096.2004.

Crager, S.E. (2014) Improving global access to new vaccines: intellectual property, technology transfer, and regulatory pathways. *American Journal of Public Health* 104(11), e85–e91. DOI: 10.2105/AJPH.2014.302236.

Dean, G., Clifford, D., Gilbert, S., McShane, H., Hewinson, R.G. *et al.* (2014) Effect of dose and route of immunisation on the immune response induced in cattle by heterologous Bacille Calmette–Guerin priming and recombinant adenoviral vector boosting. *Veterinary Immunology and Immunopathology* 158(3–4), 208–213. DOI: 10.1016/j.vetimm.2014.01.010.

Dernivoix, K., Roupie, V., Welby, S., Roelandt, S., Viart, S. *et al.* (2017) Field performance of six *Mycobacterium avium* subsp. *paratuberculosis* antigens in a 20 H interferon gamma release assay in Belgium. *Veterinary Immunology and Immunopathology* 189, 17–27. DOI: 10.1016/j.vetimm.2017.05.008.

Dimareli-Malli, Z., Mazaraki, K., Stevenson, K., Tsakos, P., Zdragas, A. *et al.* (2013) Culture phenotypes and molecular characterization of *Mycobacterium avium* subsp. *paratuberculosis* isolates from small ruminants. *Research in Veterinary Science* 95(1), 49–53. DOI: 10.1016/j.rvsc.2013.03.010.

Ding, C., Ma, J., Dong, Q. and Liu, Q. (2018) Live bacterial vaccine vector and delivery strategies of heterologous antigen: a review. *Immunology Letters* 197, 70–77. DOI: 10.1016/j.imlet.2018.03.006.

Doyle, T.M. (1964) Strains of *Mycobacterium johnei* used for the preparation of vaccine. *State Veterinary Journal* 19–20, 154–155.

EMeA (2004) Committee for medicinal products for veterinary use: EMEA/CVMP/004/04-FINAL: guideline on live recombinant vector vaccines for veterinary use. European Medicines Agency. Available at: https://www.ema.europa.eu/en/documents/scientific-guideline/guideline-live-recombinant-vector-vaccines-veterinary-use_en.pdf (accessed 13 June 2020).

EMeA (2007) VICH Topic VL41. Guideline on target animal safety: examination of live veterinary vaccines in target animals for absence of reversion to virulence.. European Medicines Agency. Available at: https://www.ema.europa.eu/en/documents/scientific-guideline/vich-gl41-target-animal-safety-examination-live-veterinary-vaccines-target-animals-absence-reversion_en (accessed 13 June 2020).

EMeA (2016) VICH GL55 on harmonisation of criteria to waive target animal batch safety testing for live vaccines for veterinary use. European Medicines Agency. Available at: https://www.ema.europa.eu/en/documents/scientific-guideline/vich-gl55-harmonization-criteria-waive-target-animal-batch-safety-testing-live-vaccines-veterinary_en.pdf (accessed 18 June 2020).

EMeA (2018) Guideline on the use of adjuvanted veterinary vaccines. European Medicines Agency. Available at: https://www.ema.europa.eu/en/documents/scientific-guideline/draft-guideline-use-adjuvanted-veterinary-vaccines_en.pdf (accessed 13 June 2020).

Emery, D.L. and Whittington, R.J. (2004) An evaluation of mycophage therapy, chemotherapy and vaccination for control of *Mycobacterium avium* subsp. *paratuberculosis* infection. *Veterinary Microbiology* 104, 143–155.

EU Directive (2001) EU Directive 2001/18/EC of the European Parliament and of the Council of 12 March 2001 on the Deliberate Release into the Environment of Genetically Modified Organisms and Repealing Council Directive 90/220/EEC, art. 2(2), 2001 O.J. (L 106) 1, amended by Directive 2008/27/EC, 2008 O.J. (L 81) 45/EC. Available at: https://ec.europa.eu/health/sites/health/files/files/eudralex/vol-1/dir_2001_18/dir_2001_18_en.pdf (accessed accessed 13 June 2020).

Faisal, S.M., Chen, J.-W., Yan, F., Chen, T.-T., Useh, N.M. *et al.* (2013a) Evaluation of a *Mycobacterium avium* subsp. *paratuberculosis leuD* mutant as a vaccine candidate against challenge in a caprine model. *Clinical and Vaccine Immunology* 20(4), 572–581. DOI: 10.1128/CVI.00653-12.

Faisal, S.M., Yan, F., Chen, T.-T., Useh, N.M., Guo, S. *et al.* (2013b) Evaluation of a Salmonella vectored vaccine expressing *Mycobacterium avium* subsp. *paratuberculosis* antigens against challenge in a goat model. *PLoS ONE* 8(8), e70171. DOI: 10.1371/journal.pone.0070171.

Favre, D. and Viret, J.-F. (2006) Biosafety evaluation of recombinant live oral bacterial vaccines in the context of European regulation. *Vaccine* 24(18), 3856–3864. DOI: 10.1016/j.vaccine.2005.07.018.

Folegatti, P.M., Bellamy, D., Roberts, R., Powlson, J., Edwards, N.J. *et al.* (2019) Safety and immunogenicity of a novel recombinant simian adenovirus ChAdOx2 as a vectored vaccine. *Vaccines* 7(2), 40. DOI: 10.3390/vaccines7020040.

Forrellad, M.A., Klepp, L.I., Gioffré, A., Sabio y García, J., Morbidoni, H.R. *et al.* (2013) Virulence factors of the *Mycobacterium tuberculosis* complex. *Virulence* 4(1), 3–66. DOI: 10.4161/viru.22329.

Frie, M.C., Sporer, K.R.B., Kirkpatrick, B.W. and Coussens, P.M. (2017) T and B cell activation profiles from cows with and without Johne's disease in response to in vitro stimulation with *Mycobacterium avium* subspecies *paratuberculosis*. *Veterinary Immunology and Immunopathology* 193–194, 50–56. DOI: 10.1016/j.vetimm.2017.10.005.

Gabitzsch, E.S., Xu, Y., Yoshida, L.H., Balint, J., Amalfitano, A. *et al.* (2009) Novel adenovirus type 5 vaccine platform induces cellular immunity against HIV-1 Gag, Pol, Nef despite the presence of Ad5 immunity. *Vaccine* 27(46), 6394–6398. DOI: 10.1016/j.vaccine.2009.06.028.

Ganusov, V.V., Klinkenberg, D., Bakker, D. and Koets, A.P. (2015) Evaluating contribution of the cellular and humoral immune responses to the control of shedding of *Mycobacterium avium* spp. *paratuberculosis* in cattle. *Veterinary Research* 46(1), 62. DOI: 10.1186/s13567-015-0204-1.

Garcia, A.B. and Shalloo, L. (2015) Invited review: the economic impact and control of paratuberculosis in cattle. *Journal of Dairy Science* 98(8), 5019–5039. DOI: 10.3168/jds.2014-9241.

Ghosh, P., Wu, C.W. and Talaat, A.M. (2013) Key role for the alternative sigma factor, SigH, in the intracellular life of *Mycobacterium avium* subsp. *paratuberculosis* during macrophage stress. *Infection and Immunity* 81(6), 2242–2257. DOI: 10.1128/IAI.01273-12.

Ghosh, P., Steinberg, H. and Talaat, A.M. (2014) Virulence and immunity orchestrated by the global gene regulator *sigL* in *Mycobacterium avium* subsp. *paratuberculosis*. *Infection and Immunity* 82(7), 3066–3075. DOI: 10.1128/IAI.00001-14.

Ghosh, P., Shippy, D.C. and Talaat, A.M. (2015) Superior protection elicited by live-attenuated vaccines in the murine model of paratuberculosis. *Vaccine* 33(51), 7262–7270. DOI: 10.1016/j.vaccine.2015.10.116.

Griffin, J.F.T., Mackintosh, C.G., Slobbe, L., Thomson, A.J. and Buchan, G.S. (1999) Vaccine protocols to optimise the protective efficacy of BCG. *Tubercle and Lung Disease* 79(3), 135–143. DOI: 10.1054/tuld.1998.0202.

Griffin, J.F.T., Hughes, A.D., Liggett, S., Farquhar, P.A., Mackintosh, C.G. *et al.* (2009) Efficacy of novel lipid-formulated whole bacterial cell vaccines against *Mycobacterium avium* subsp. *paratuberculosis* in sheep. *Vaccine* 27(6), 911–918. DOI: 10.1016/j.vaccine.2008.11.053.

Harris, N.B., Feng, Z., Liu, X., Cirillo, S.L.G., Cirillo, J.D. *et al.* (1999) Development of a transposon mutagenesis system for *Mycobacterium avium* subsp. *paratuberculosis*. *FEMS Microbiology Letters* 175(1), 21–26. DOI: 10.1111/j.1574-6968.1999.tb13597.x.

Hawkridge, T., Scriba, T. J., Gelderbloem, S., Smit, E., Tameris, M. *et al.* (2008) Safety and immunogenicity of a new tuberculosis vaccine, MVA85A, in healthy adults in South Africa. *The Journal of Infectious Diseases* 198(4), 544–552. DOI: 10.1086/590185.

Heinzmann, J., Wilkens, M., Dohmann, K. and Gerlach, G.-F. (2008) *Mycobacterium avium* subsp. *paratuberculosis*-specific MPT operon expressed in M. bovis BCG as vaccine candidate. *Veterinary Microbiology* 130(3–4), 330–337. DOI: 10.1016/j.vetmic.2008.01.014.

Heldens, J.G.M., Patel, J.R., Chanter, N., ten Thij, G.J., Gravendijck, M. *et al.* (2008) Veterinary vaccine development from an industrial perspective. *The Veterinary Journal* 178(1), 7–20. DOI: 10.1016/j.tvjl.2007.11.009.

Hines, M.E., Stabel, J.R., Sweeney, R.W., Griffin, F., Talaat, A.M. *et al.* (2007a) Experimental challenge models for Johne's disease: a review and proposed international guidelines. *Veterinary Microbiology* 122(3–4), 197–222. DOI: 10.1016/j.vetmic.2007.03.009.

Hines, M.E., Stiver, S., Giri, D., Whittington, L., Watson, C. *et al.* (2007b) Efficacy of spheroplastic and cell-wall competent vaccines for *Mycobacterium avium* subsp. *paratuberculosis* in experimentally-challenged baby goats. *Veterinary Microbiology* 120(3–4), 261–283. DOI: 10.1016/j.vetmic.2006.10.030.

Hines, M.E., Turnquist, S.E., Ilha, M.R.S., Rajeev, S., Jones, A.L. *et al.* (2014) Evaluation of novel oral vaccine candidates and validation of a caprine model of Johne's disease. *Frontiers in Cellular and Infection Microbiology* 4, 26. DOI: 10.3389/fcimb.2014.00026.

Hobernik, D. and Bros, M. (2018) DNA Vaccines—How far from clinical use? *International Journal of Molecular Sciences* 19(11), 3605. DOI: 10.3390/ijms19113605.

Hope, J.C., Thom, M.L., Villarreal-Ramos, B., Vordermeier, H.M., Hewinson, R.G. *et al.* (2005) Exposure to *Mycobacterium avium* induces low-level protection from *Mycobacterium bovis* infection but compromises diagnosis of disease in cattle. *Clinical and Experimental Immunology* 141(3), 432–439. DOI: 10.1111/j.1365-2249.2005.02882.x.

Hughes, V., Bannantine, J.P., Denham, S., Smith, S., Garcia-Sanchez, A. *et al.* (2008) Immunogenicity of proteome-determined *Mycobacterium avium* subsp. *paratuberculosis*-specific proteins in sheep with paratuberculosis. *Clinical and Vaccine Immunology* 15(12), 1824–1833. DOI: 10.1128/CVI.00099-08.

Huntley, J.F., Stabel, J.R., Paustian, M.L., Reinhardt, T.A. and Bannantine, J.P. (2005) Expression library immunization confers protection against *Mycobacterium avium* subsp. *paratuberculosis* infection. *Infection and Immunity* 73(10), 6877–6884. DOI: 10.1128/IAI.73.10.6877-6884.2005.

Huygen, K. (2006) DNA vaccines against mycobacterial diseases. *Future Microbiology* 1(1), 63–73. DOI: 10.2217/17460913.1.1.63.

Johnston, C.D., Bannantine, J.P., Govender, R., Endersen, L., Pletzer, D. (2014) Enhanced expression of codon optimized *Mycobacterium avium* subsp. *paratuberculosis* antigens in *Lactobacillus salivarius*. *Frontiers in Cellular and Infection Microbiology* 4, 120. DOI: 10.3389/fcimb.2014.00120.

Jolly, A., Morsella, C., Bass, L., Fiorentino, M.A., Paolicchi, F.A. *et al.* (2013) Bovine response to lipoarabinomannan vaccination and challenge with *Mycobacterium paratuberculosis*. *Brazilian Journal of Microbiology* 44(2), 511–514. DOI: 10.1590/S1517-83822013000200029.

Juste, R.A., Alonso-Hearn, M., Molina, E., Geijo, M., Vazquez, P. *et al.* (2009) Significant reduction in bacterial shedding and improvement in milk production in dairy farms after the use of a new inactivated paratuberculosis vaccine in a field trial. *BMC Research Notes* 2(1), 233. DOI: 10.1186/1756-0500-2-233.

Kabara, E. and Coussens, P.M. (2012) Infection of primary bovine macrophages with *Mycobacterium avium* subspecies *paratuberculosis* suppresses host cell apoptosis. *Frontiers in Microbiology* 3, 215. DOI: 10.3389/fmicb.2012.00215.

Kathaperumal, K., Park, S.-U., McDonough, S., Stehman, S., Akey, B. *et al.* (2008) Vaccination with recombinant *Mycobacterium avium* subsp. *paratuberculosis* proteins induces differential immune responses and protects calves against infection by oral challenge. *Vaccine* 26(13), 1652–1663. DOI: 10.1016/j.vaccine.2008.01.015.

Kathaperumal, K., Kumanan, V., McDonough, S., Chen, L.-H., Park, S.-U. *et al.* (2009) Evaluation of immune responses and protective efficacy in a goat model following immunization with a cocktail of recombinant antigens and a polyprotein of *Mycobacterium avium* subsp. *paratuberculosis*. *Vaccine* 27(1), 123–135. DOI: 10.1016/j.vaccine.2008.10.019.

Kaveh, D.A., Garcia-Pelayo, M.C., Webb, P.R., Wooff, E.E., Bachy, V.S. *et al.* (2016) Parenteral adenoviral boost enhances BCG induced protection, but not long term survival in a murine model of bovine TB. *Vaccine* 34(34), 4003–4011. DOI: 10.1016/j.vaccine.2016.06.032.

Kawahara, J.Y., Irvine, E.B. and Alter, G. (2019) A case for antibodies as mechanistic correlates of immunity in tuberculosis. *Frontiers in Immunology* 10, 996. DOI: 10.3389/fimmu.2019.00996.

Kirkeby, C., Græsbøll, K., Nielsen, S.S., Christiansen, L.E., Toft, N. *et al.* (2016) Simulating the epidemiological and economic impact of paratuberculosis control actions in dairy cattle. *Frontiers in Veterinary Science* 3, 90. DOI: 10.3389/fvets.2016.00090.

Kleinnijenhuis, J., Quintin, J., Preijers, F., Joosten, L.A.B., Ifrim, D.C. *et al.* (2012) Bacille Calmette-Guerin induces Nod2-dependent nonspecific protection from reinfection via epigenetic reprogramming of monocytes. *Proceedings of the National Academy of Sciences* 109(43), 17537–17542. DOI: 10.1073/pnas.1202870109.

Koets, A., Hoek, A., Langelaar, M., Overdijk, M., Santema, W. *et al.* (2006) Mycobacterial 70kD heat-shock protein is an effective subunit vaccine against bovine paratuberculosis. *Vaccine* 24(14), 2550–2559. DOI: 10.1016/j.vaccine.2005.12.019.

Koets, A.P. and Gröhn, Y.T. (2015) Within- and between-host mathematical modeling of *Mycobacterium avium* subspecies *paratuberculosis* (MAP) infections as a tool to study the dynamics of host-pathogen interactions in bovine paratuberculosis. *Veterinary Research* 46(1), 60. DOI: 10.1186/s13567-015-0205-0.

Körmendy, B. (1994) The effect of vaccination on the prevalence of paratuberculosis in large dairy herds. *Veterinary Microbiology* 41(1–2), 117–125. DOI: 10.1016/0378-1135(94)90141-4.

Lamont, E.A., Talaat, A.M., Coussens, P.M., Bannantine, J.P., Grohn, Y.T. *et al.* (2014) Screening of *Mycobacterium avium* subsp. *paratuberculosis* mutants for attenuation in a bovine monocyte-derived macrophage model. *Frontiers in Cellular and Infection Microbiology* 4(88), 87. DOI: 10.3389/fcimb.2014.00087.

Lauer, K.B., Borrow, R. and Blanchard, T.J. (2017) Multivalent and multipathogen viral vector vaccines. *Clinical and Vaccine Immunology* 24(1). DOI: 10.1128/CVI.00298-16.

Leroy, B., Roupie, V., Noël-Georis, I., Rosseels, V., Walravens, K. *et al.* (2007) Antigen discovery: a post-genomic approach to paratuberculosis diagnosis. *Proteomics* 7(7), 1164–1176. DOI: 10.1002/pmic.200600988.

Leunda, A. and Pauwels, K. (2019) Gmo regulatory aspects of novel investigational vaccine candidates. *IntechOpen Vaccines*. DOI: 10.5772/intechopen.85341.

Li, L., Bannantine, J.P., Zhang, Q., Amonsin, A., May, B.J. *et al.* (2005) The complete genome sequence of *Mycobacterium avium* subspecies *paratuberculosis*. *Proceedings of the National Academy of Sciences* 102(35), 12344–12349. DOI: 10.1073/pnas.0505662102.

Lilienfeld, D.E. (2008) The first pharmacoepidemiologic investigations: national drug safety policy in the United States, 1901–1902. *Perspectives in Biology and Medicine* 51(2), 188–198. DOI: 10.1353/pbm.0.0010.

Milner, A.R., Lepper, A.W.D., Symonds, W.N. and Gruner, E. (1987) Analysis by ELISA and Western blotting of antibody reactivities in cattle infected with *Mycobacterium paratuberculosis* after absorption of serum with M phlei. *Research in Veterinary Science* 42(2), 140–144. DOI: 10.1016/S0034-5288(18)30675-1.

Milstien, J. and Belgharbi, L. (2004) Regulatory pathways for vaccines for developing countries. *Bulletin of the World Health Organization* 82, 128–133.

Möbius, P., Nordsiek, G., Hölzer, M., Jarek, M., Marz, M. *et al.* (2017) Complete genome sequence of JII-1961, a bovine *Mycobacterium avium* subsp. *paratuberculosis* field isolate from Germany. *Genome Announcements* 5(34). DOI: 10.1128/genomeA.00870-17.

Mortier, R.A.R., Barkema, H.W., Bystrom, J.M., Illanes, O., Orsel, K. *et al.* (2013) Evaluation of age-dependent susceptibility in calves infected with two doses of *Mycobacterium avium* subspecies *paratuberculosis* using pathology and tissue culture. *Veterinary Research* 44(1), 94. DOI: 10.1186/1297-9716-44-94.

Mullerad, J., Hovav, A.-H., Fishman, Y., Barletta, R.G. and Bercovier, H. (2002a) Antigenicity of *Mycobacterium paratuberculosis* superoxide dismutase in mice. *FEMS Immunology & Medical Microbiology* 34(1), 81–88. DOI: 10.1111/j.1574-695X.2002.tb00606.x.

Mullerad, J., Michal, I., Fishman, Y., Hovav, A.-H., Barletta, R.G. *et al.* (2002b) The immunogenicity of *Mycobacterium paratuberculosis* 85B antigen. *Medical Microbiology and Immunology* 190(4), 179–187. DOI: 10.1007/s00430-001-0104-z.

Musk, G.C., Kershaw, H., Tano, K., Niklasson, A., von Unge, M. *et al.* (2019) Reactions to Gudair® vaccination identified in sheep used for biomedical research. *Australian Veterinary Journal* 97(3), 56–60. DOI: 10.1111/avj.12788.

Nagata, R., Muneta, Y., Yoshihara, K., Yokomizo, Y. and Mori, Y. (2005) Expression cloning of gamma interferon-inducing antigens of *Mycobacterium avium* subsp. *paratuberculosis*. *Infection and Immunity* 73(6), 3778–3782. DOI: 10.1128/IAI.73.6.3778-3782.2005.

Nathanson, N. and Langmuir, A.D. (1995) The cutter incident. poliomyelitis following formaldehyde-inactivated poliovirus vaccination in the United States during the spring of 1955. II. Relationship of poliomyelitis to cutter vaccine. 1963. *American Journal of Epidemiology* 142(2), 109–140. DOI: 10.1093/oxfordjournals.aje.a117611.

Nielsen, S.S. (2008) Transitions in diagnostic tests used for detection of *Mycobacterium avium* subsp. *paratuberculosis* infections in cattle. *Veterinary Microbiology* 132(3–4), 274–282. DOI: 10.1016/j.vetmic.2008.05.018.

Orme, I.M. and Collins, F.M. (1985) Prophylactic effect in mice of BCG vaccination against non-tuberculous mycobacterial infections. *Tubercle* 66(2), 117–120. DOI: 10.1016/0041-3879(85)90076-5.

Park, H.-T. and Yoo, H.S. (2016) Development of vaccines to *Mycobacterium avium* subsp. *paratuberculosis* infection. *Clinical and Experimental Vaccine Research* 5(2), 108–116. DOI: 10.7774/cevr.2016.5.2.108.

Park, K.T., Dahl, J.L., Bannantine, J.P., Barletta, R.G., Ahn, J. *et al.* (2008) Demonstration of allelic exchange in the slow-growing bacterium *Mycobacterium avium* subsp. *paratuberculosis*, and generation of mutants with deletions at the pknG, relA, and LSR2 loci. *Applied and Environmental Microbiology* 74(6), 1687–1695. DOI: 10.1128/AEM.01208-07.

Park, K.T., Allen, A.J., Bannantine, J.P., Seo, K.S., Hamilton, M.J. *et al.* (2011) Evaluation of two mutants of *Mycobacterium avium* subsp. *paratuberculosis* as candidates for a live attenuated vaccine for Johne's disease. *Vaccine* 29(29–30), 4709–4719. DOI: 10.1016/j.vaccine.2011.04.090.

Park, K.T., Allen, A.J., Barrington, G.M. and Davis, W.C. (2014) Deletion of relA abrogates the capacity of *Mycobacterium avium paratuberculosis* to establish an infection in calves. *Frontiers in Cellular and Infection Microbiology* 4, 64. DOI: 10.3389/fcimb.2014.00064.

Pooley, H.B., Begg, D.J., Plain, K.M., Whittington, R.J., Purdie, A.C. *et al.* (2019) The humoral immune response is essential for successful vaccine protection against paratuberculosis in sheep. *BMC Veterinary Research* 15(1), 223. DOI: 10.1186/s12917-019-1972-z.

Rathnaiah, G., Lamont, E.A., Harris, N.B., Fenton, R.J., Zinniel, D.K. *et al.* (2014) Generation and screening of a comprehensive *Mycobacterium avium* subsp. *paratuberculosis* transposon mutant bank. *Frontiers in Cellular and Infection Microbiology* 4(88), 144. DOI: 10.3389/fcimb.2014.00144.

Rathnaiah, G., Zinniel, D.K., Bannantine, J.P., Stabel, J.R., Gröhn, Y.T. *et al.* (2017) Pathogenesis, molecular genetics, and genomics of *Mycobacterium avium* subsp. *paratuberculosis*, the etiologic agent of Johne's disease. *Frontiers in Veterinary Science* 4, 187. DOI: 10.3389/fvets.2017.00187.

Reddacliff, L., Eppleston, J., Windsor, P., Whittington, R. and Jones, S. (2006) Efficacy of a killed vaccine for the control of paratuberculosis in Australian sheep flocks. *Veterinary Microbiology* 115(1-3), 77–90. DOI: 10.1016/j.vetmic.2005.12.021.

Rosseels, V., Marche, S., Roupie, V., Govaerts, M., Godfroid, J. *et al.* (2006a) Members of the 30- to 32-kilodalton mycolyl transferase family (Ag85) from culture filtrate of *Mycobacterium avium* subsp. *paratuberculosis* are immunodominant Th1-type antigens recognized early upon infection in mice and cattle. *Infection and Immunity* 74(1), 202–212. DOI: 10.1128/IAI.74.1.202-212.2006.

Rosseels, V., Roupie, V., Zinniel, D., Barletta, R.G. and Huygen, K. (2006b) Development of luminescent *Mycobacterium avium* subsp. *paratuberculosis* for rapid screening of vaccine candidates in mice. *Infection and Immunity* 74(6), 3684–3686. DOI: 10.1128/IAI.01521-05.

Rosseels, V. and Huygen, K. (2008) Vaccination against paratuberculosis. *Expert Review of Vaccines* 7(6), 817–832. DOI: 10.1586/14760584.7.6.817.

Roupie, V., Leroy, B., Rosseels, V., Piersoel, V., Noël-Georis, I. *et al.* (2008) Immunogenicity and protective efficacy of DNA vaccines encoding MAP0586c and MAP4308c of *Mycobacterium avium* subsp. *paratuberculosis* secretome. *Vaccine* 26(37), 4783–4794. DOI: 10.1016/j.vaccine.2008.07.009.

Roupie, V., Viart, S., Leroy, B., Romano, M., Trinchero, N. *et al.* (2012) Immunogenicity of eight *Mycobacterium avium* subsp. *paratuberculosis* specific antigens in DNA vaccinated and MAP infected mice. *Veterinary Immunology and Immunopathology* 145(1-2), 74–85. DOI: 10.1016/j.vetimm.2011.10.012.

Roupie, V., Alonso-Velasco, E., Van Der Heyden, S., Holbert, S., Duytschaever, L. *et al.* (2018) Evaluation of mycobacteria-specific gamma interferon and antibody responses before and after a single intradermal skin test in cattle naturally exposed to *M. avium* subsp. *paratuberculosis* and experimentally infected with *M. bovis*. *Veterinary Immunology and Immunopathology* 196, 35–47. DOI: 10.1016/j.vetimm.2017.12.007.

Roy, Á., Infantes-Lorenzo, J.A., Blázquez, J.C., Venteo, Á., Mayoral, F.J. *et al.* (2018) Temporal analysis of the interference caused by paratuberculosis vaccination on the tuberculosis diagnostic tests in goats. *Preventive Veterinary Medicine* 156, 68–75. DOI: 10.1016/j.prevetmed.2018.05.010.

Saidu, Y., De Angelis, D., Aiolli, S., Stefano, G. and Georges, A.M. (2013) Product registration in developing countries: a proposal for an integrated regional licensing system among countries in regional economic Blocs. *Therapeutic innovation & regulatory science* 47(3), 327–335. DOI: 10.1177/2168479013478952.

Santema, W., Overdijk, M., Barends, J., Krijgsveld, J., Rutten, V. *et al.* (2009) Searching for proteins of *Mycobacterium avium* subspecies *paratuberculosis* with diagnostic potential by comparative qualitative proteomic analysis of mycobacterial tuberculins. *Veterinary Microbiology* 138(1–2), 191–196. DOI: 10.1016/j.vetmic.2009.03.021.

Santema, W., van Kooten, P., Hoek, A., Leeflang, M., Overdijk, M. *et al.* (2011) Hsp70 vaccination-induced antibodies recognize B cell epitopes in the cell wall of *Mycobacterium avium* subspecies *paratuberculosis*. *Vaccine* 29(7), 1364–1373. DOI: 10.1016/j.vaccine.2010.12.071.

Santema, W.J., Poot, J., Segers, R.P.A.M., Van den Hoff, D.J.P., Rutten, V.P.M.G. *et al.* (2012) Early infection dynamics after experimental challenge with *Mycobacterium avium* subspecies *paratuberculosis* in calves reveal limited calf-to-calf transmission and no impact of Hsp70 vaccination. *Vaccine* 30(49), 7032–7039. DOI: 10.1016/j.vaccine.2012.09.065.

Santema, W., Rutten, V., Segers, R., Poot, J., Hensen, S. *et al.* (2013) Postexposure subunit vaccination against chronic enteric mycobacterial infection in a natural host. *Infection and Immunity* 81(6), 1990–1995. DOI: 10.1128/IAI.01121-12.

Saxegaard, F. and Fodstad, F.H. (1985) Control of paratuberculosis (Johne's disease) in goats by vaccination. *Veterinary Record* 116(16), 439–441. DOI: 10.1136/vr.116.16.439.

Scandurra, G.M., Young, M., de Lisle, G.W. and Collins, D.M. (2009) A bovine macrophage screening system for identifying attenuated transposon mutants of *Mycobacterium avium* subsp. *paratuberculosis* with vaccine potential. *Journal of Microbiological Methods* 77(1), 58–62. DOI: 10.1016/j.mimet.2009.01.005.

Scandurra, G.M., de Lisle, G.W., Cavaignac, S.M., Young, M., Kawakami, R.P. *et al.* (2010) Assessment of live candidate vaccines for paratuberculosis in animal models and macrophages. *Infection and Immunity* 78(3), 1383–1389. DOI: 10.1128/IAI.01020-09.

Schmeer, M., Buchholz, T. and Schleef, M. (2017) Plasmid DNA manufacturing for indirect and direct clinical applications. *Human Gene Therapy* 28(10), 856–861. DOI: 10.1089/hum.2017.159.

SCNZ (2003) Hazardous substance and new organisms act 2003. Sustainability Council of New Zealand. Available at: http://www.legislation.govt.nz/act/public/1996/0030/latest/DLM381222.html (accessed 13 June 2020).

Sechi, L.A., Mara, L., Cappai, P., Frothingam, R., Ortu, S. *et al.* (2006) Immunization with DNA vaccines encoding different mycobacterial antigens elicits a Th1 type immune response in lambs and protects against *Mycobacterium avium* subspecies *paratuberculosis* infection. *Vaccine* 24(3), 229–235. DOI: 10.1016/j.vaccine.2005.08.086.

Serrano, M., Elguezabal, N., Sevilla, I.A., Geijo, M.V., Molina, E. *et al.* (2017) Tuberculosis detection in paratuberculosis vaccinated calves: new alternatives against interference. *PLoS ONE* 12(1), e0169735. DOI: 10.1371/journal.pone.0169735.

Settles, E.W., Kink, J.A. and Talaat, A. (2014) Attenuated strains of *Mycobacterium avium* subspecies *paratuberculosis* as vaccine candidates against Johne's disease. *Vaccine* 32(18), 2062–2069. DOI: 10.1016/j.vaccine.2014.02.010.

Shin, S.J., Chang, C.-F., Chang, C.-D., McDonough, S.P., Thompson, B. *et al.* (2005) In vitro cellular immune responses to recombinant antigens of *Mycobacterium avium* subsp. *paratuberculosis*. *Infection and Immunity* 73(8), 5074–5085. DOI: 10.1128/IAI.73.8.5074-5085.2005.

Shin, S.J., Wu, C.-W., Steinberg, H. and Talaat, A.M. (2006) Identification of novel virulence determinants in *Mycobacterium paratuberculosis* by screening a library of insertional mutants. *Infection and Immunity* 74(7), 3825–3833. DOI: 10.1128/IAI.01742-05.

Shippy, D.C., Lemke, J.J., Berry, A., Nelson, K., Hines, M.E. *et al.* (2017) Superior protection from live-attenuated vaccines directed against Johne's disease. *Clinical and Vaccine Immunology* 24(1). DOI: 10.1128/CVI.00478-16.

Sigurdsson, B. (1960) A killed vaccine against paratuberculosis (Johne's disease) in sheep. *American Journal of Veterinary Research* 21, 54–67.

Singh, K., Chandel, B.S., Chauhan, H.C., Dadawala, A., Singh, S.V. *et al.* (2013) Efficacy of 'indigenous vaccine' using native 'Indian bison type' genotype of *Mycobacterium avium* subspecies *paratuberculosis* for the control of clinical Johne's disease in an organized goat herd. *Veterinary Research Communications* 37(2), 109–114. DOI: 10.1007/s11259-013-9551-4.

Skeiky, Y.A.W., Alderson, M.R., Ovendale, P.J., Guderian, J.A., Brandt, L. *et al.* (2004) Differential immune responses and protective efficacy induced by components of a tuberculosis polyprotein vaccine, Mtb72F, delivered as naked DNA or recombinant protein. *The Journal of Immunology* 172(12), 7618–7628. DOI: 10.4049/jimmunol.172.12.7618.

Srinivasan, S., Jones, G., Veerasami, M., Steinbach, S., Holder, T. *et al.* (2019) A defined antigen skin test for the diagnosis of bovine tuberculosis. *Science Advances* 5(7), eaax4899. DOI: 10.1126/sciadv.aax4899.

Stabel, J.R., Barnhill, A., Bannantine, J.P., Chang, Y.F. and Osman, M.A. (2012) Evaluation of protection in a mouse model after vaccination with *Mycobacterium avium* subsp. *paratuberculois* protein cocktails. *Vaccine* 31(1), 127–134. DOI: 10.1016/j.vaccine.2012.10.090.

Stevenson, K. (2015) Genetic diversity of *Mycobacterium avium* subspecies *paratuberculosis* and the influence of strain type on infection and pathogenesis: a review. *Veterinary Research* 46, 64. DOI: 10.1186/s13567-015-0203-2.

Stuart, P. (1965) Vaccination against Johne's disease in cattle exposed to experimental infection. *British Veterinary Journal* 121, 289–318. DOI: 10.1016/S0007-1935(17)41102-X.

Tameris, M., Mearns, H., Penn-Nicholson, A., Gregg, Y., Bilek, N. *et al.* (2019) Live-Attenuated *Mycobacterium tuberculosis* vaccine MTBVAC versus BCG in adults and neonates: a randomised controlled, double-blind dose-escalation trial. *The Lancet Respiratory Medicine* 7(9), 757–770. DOI: 10.1016/S2213-2600(19)30251-6.

Thakur, A., Aagaard, C., Stockmarr, A., Andersen, P. and Jungersen, G. (2013) Cell-mediated and humoral immune responses after immunization of calves with a recombinant multiantigenic *Mycobacterium avium* subsp. *paratuberculosis* subunit vaccine at different ages. *Clinical and Vaccine Immunology* 20(4), 551–558. DOI: 10.1128/CVI.05574-11.

Uzonna, J.E., Chilton, P., Whitlock, R.H., Habecker, P.L., Scott, P. *et al.* (2003) Efficacy of commercial and field-strain *Mycobacterium paratuberculosis* vaccinations with recombinant IL-12 in a bovine experimental infection model. *Vaccine* 21(23), 3101–3109. DOI: 10.1016/S0264-410X(03)00261-5.

Vallée, H., Rinjard, P. and Vallée, M. (1934) Sur la préimmunisation de l'entérite paratuberculeuse des bovidés. *Revue Générale de Médécine Vétérinaire* 43, 777–779.

Vordermeier, M. and Hewinson, R.G. (2006) Development of cattle TB vaccines in the UK. *Veterinary Immunology and Immunopathology* 112(1–2), 38–48. DOI: 10.1016/j.vetimm.2006.03.010.

Voss, G., Casimiro, D., Neyrolles, O., Williams, A., Kaufmann, S.H.E. *et al.* (2018) Progress and challenges in TB vaccine development. *F1000Research* 7, 199. DOI: 10.12688/f1000research.13588.1.

Walker, K.B., Brennan, M.J., Ho, M.M., Eskola, J., Thiry, G. *et al.* (2010) The second Geneva consensus: recommendations for novel live TB vaccines. *Vaccine* 28(11), 2259–2270. DOI: 10.1016/j.vaccine.2009.12.083.

Watkins, C., Schock, A., May, L., Denham, S., Sales, J. *et al.* (2010) Assessing virulence of vaccine strains of *Mycobacterium avium* subspecies *paratuberculosis* in a calf model. *Veterinary Microbiology* 146(1–2), 63–69. DOI: 10.1016/j.vetmic.2010.04.017.

Weiss, D.J., Evanson, O.A. and Souza, C.D. (2006) Mucosal immune response in cattle with subclinical Johne's disease. *Veterinary Pathology* 43(2), 127–135. DOI: 10.1354/vp.43-2-127.

Wilesmith, J.W. (1982) Johne's disease: a retrospective study of vaccinated herds in Great Britain. *British Veterinary Journal* 138(4), 321–331. DOI: 10.1016/S0007-1935(17)31037-0.

Windsor, P.A., Bush, R., Links, I. and Eppleston, J. (2005) Injury caused by self-inoculation with a vaccine of a Freund's complete adjuvant nature (Gudair) used for control of ovine paratuberculosis. *Australian Veterinary Journal* 83(4), 216–220. DOI: 10.1111/j.1751-0813.2005.tb11654.x.

Wu, Y., Cai, M., Ma, J., Teng, X., Tian, M. *et al.* (2018) Heterologous boost following *Mycobacterium bovis BCG* reduces the late persistent, rather than the early stage of intranasal tuberculosis challenge infection. *Frontiers in Immunology* 9, 2439. DOI: 10.3389/fimmu.2018.02439.

Zimmermann, P., Finn, A. and Curtis, N. (2018) Does BCG vaccination protect against nontuberculous mycobacterial infection? A systematic review and meta-analysis. *The Journal of Infectious Diseases* 218(5), 679–687. DOI: 10.1093/infdis/jiy207.

Index

Note: Page numbers in **bold** type refer to **figures**
Page numbers in *italic* type refer to *tables*

www.ingramcontent.com/pod-product-compliance
Lightning Source LLC
LaVergne TN
LVHW071335210926
848486LV00003B/152